CLASSICAL AND QUANTUM GRAVITY RESEARCH

Classical and Quantum Gravity Research

Mikkel N. Christiansen
and
Tobias K. Rasmussen
Editors

Nova Science Publishers, Inc.
New York

For permission to use material from this book please contact us:
Telephone 631-231-7269; Fax 631-231-8175
Web Site: http://www.novapublishers.com

Library of Congress Cataloging-in-Publication Data

Classical and quantum gravity research / Mikkel N. Christiansen and Tobias K. Rasmussen (editor).
p. cm.
ISBN 978-1-60456-366-5 (hardcover)
1. Quantum gravity. 2. Gravitation. 3. Space and time. I. Christiansen, Mikkel N. II. Rasmussen, Tobias K.
QC178.C455 2008
531'.14--dc22 2008003387

Published by Nova Science Publishers, Inc. ✦ New York

CONTENTS

PREFACE

This new book presents recent research from around the globe in gravitational physics and the theory of spacetime.

The General Theory of Relativity has been an extremely successful theory, with a well established experimental footing, at least for weak gravitational fields. It's predictions range from the existence of black holes, gravitational radiation to the cosmological models, predicting a primordial beginning, namely the big-bang. All these solutions have been obtained by first considering plausible distributions of matter, i.e., a plausible stress-energy tensor, and through the Einstein field equation, the spacetime metric of the geometry is determined. However, one may solve the Einstein field equation in the reverse direction, namely, one first considers an interesting and exotic spacetime metric, then finds the matter source responsible for the respective geometry. In this manner, it was found that some of these solutions possess a peculiar property, namely "exotic matter," involving a stress-energy tensor that violates the null energy condition. These geometries also allow closed timelike curves, with the respective causality violations. Another interesting feature of these spacetimes is that they allow "effective" superluminal travel, although, locally, the speed of light is not surpassed. These solutions are primarily useful as "gedanken-experiments" and as a theoretician's probe of the foundations of general relativity, and include traversable wormholes and superluminal "warp drive" spacetimes. Thus, one may be tempted to denote these geometries as "exotic" solutions of the Einstein field equation, as they violate the energy conditions and generate closed timelike curves. In Chapter 1, in addition to extensively exploring interesting features, in particular, the physical properties and characteristics of these "exotic spacetimes," the authors also analyze other non-trivial general relativistic geometries which generate closed timelike curves.

In Chapter 2 the authors review the field-theoretical approach. In this framework perturbations in general relativity as well as in an arbitrary D-dimensional metric theory are described and studied. A background, on which the perturbations propagate, is a solution (arbitrary) of the theory. Lagrangian for perturbations is defined, and field equations for perturbations are derived from the variational principle. These equations are exact, equivalent to the equations in the standard formulation and have a form that permits an easy and natural expansion to an arbitrary order. Being covariant, the field-theoretical description is also invariant under gauge (inner) transformations, which can be presented both in exact and approximate forms. Following the usual field-theoretical prescriptions, conserved quantities for perturbations are constructed. Conserved currents are expressed through divergences of superpotentials — antisymmetric tensor densities. This form allows to relate a necessity to consider local properties of perturbations with a theoretical representation of the quasi-local nature of conserved quantities in metric theories. Properties of the conserved quantities under gauge transformations are established and analyzed, this allows to describe the well known non-localization problem in explicit mathematical expressions and operate with them. Applications of the formalism in general relativity for studying 1) the falloff at spatial infinity in asymptotically flat spacetimes, 2) linear perturbations on Friedmann-Robertson-Walker backgrounds, 3) a closed Friedmann world and 4) black holes presented as gravitationally-field configurations in a Minkowski space, are reviewed. Possible applications of the formalism in cosmology and astrophysics are also discussed. Generalized formulae for an arbitrary metric D-dimensional theory are tested to calculate the mass of a Schwarzschild-anti-de Sitter black hole in the Einstein-Gauss-Bonnet gravity.

Chapter 3 discusses the progress in the path-integral approach to quantum gravity, particularly in the context of the divergence of the Euclidean classical action which is unbounded from below. The authors show that the effective action in the path-integral can be made positive definite by isolating the trace part from the measure.

Chapter 4 studies Open Quantum Relativity, which is a theory based on two fundamental features: the assumption of a General Conservation Principle which states that the conservation laws can never be violated and the achievement that both General Relativity and Quantum Mechanics can be described under the standard of a covariant symplectic formalism. These facts lead to some important consequences. First of all the existence of a dynamical unification scheme of fundamental interactions achieved by assuming a 5D space which allows that the conservation laws are always and absolutely valid as a natural necessity. Then what we usually describe as violations of conservation laws can be described by a process of topology change, embedding and dimensional reduction, which gives rise to an induced-gravity-matter theory in the 4D space-time by which the usual masses, spins and charges of particles, naturally spring out. As results the theory leads to a dynamical explanation of several problems of modern physics (e.g. entanglement of quantum states, quantum teleportation, gamma ray bursts origin, black hole singularities, cosmic primary antimatter absence). Moreover the theory provides a selfconsistent picture of the observed accelerated cosmological behavior with the correct reproduction of experimental cosmological parameters and new predictions as gravitationally induced neutrino oscillations and further scalar modes in gravitational waves. A fundamental role in this approach is the link between the geodesic structure and the field equations of the theory before and after the dimensional reduction process. The emergence of an extra force term in the reduction process and the possibility to recover the masses of particles, allow to reinterpret the Equivalence Principle as a dynamical consequence which naturally "selects" geodesics from metric structure and vice-versa the metric structure from the geodesics. It is worth noting that, in the Einstein General Relativity, geodesic structure is "imposed" by choosing a Levi-Civita connection and this fact can be criticized considering a more general completely "affine" approach like in the Palatini formalism (in agreement with our covariant symplectic approach). As the authors will show, the dimensional reduction process gives rise to the generation of the masses of particles which emerge both from the field equations and the embedded geodesics. Due to this result, the coincidence of chronological and geodesic structures is derived from the embedding and a new dynamical formulation of the Equivalence Principle is the direct consequence of dimensional reduction. The dynamically derived structure becomes more general since two time arrows and closed time-like paths naturally emerge, so opening the doors to even more fundamental consequences, first of all a reinterpretation of the standard notion of causality which can be, in this way, always recovered, even in the case in which it is questioned (like in entanglement phenomena and quantum teleportation), since it is generalized to a "forward" and a "backward" causation.

Several authors have studied the generation of gravitational fields by condensed-matter systems in non-extreme density conditions (i.e., conditions not like those of collapsed stars, but such to be possibly obtained in a laboratory). General Relativity and lowest-order perturbative Quantum Gravity predict in this case an extremely small emission rate, so these phenomena can become relevant only if some strong quantum effect occurs. Quantum aspects of gravity are still poorly understood. It is believed that they could play a role in systems which exhibit macroscopic quantum coherence, like superconductors and superfluids, leading to an "anomalous" coupling between matter and field. The authors mention here recent work in this field by Woods, Chiao, Becker, Agop et al., Ummarino, Kiefer and Weber. Many of these theoretical works were stimulated by the experimental claims of Podkletnov. His results have not yet been confirmed, but the published replication attempts have admittedly been incomplete. Recently, Tajmar claimed to have detected a gravitomagnetic field generated by a spinning superconductor. Chiao also made some attempts at the construction of a gravity/e.m. transducer based on quantum effects. In our previous theoretical work, the authors sought an interpretation of the anomalous emission reported by Podkletnov as a consequence of the local modification of the vacuum energy density in the superconductor. The authors hypothesed that the vacuum energy density term could interfere with a set of strong gravitational fluctuations called "dipolar fluctuations". In Chapter 5 the authors improve our earlier model and also present new results concerning anomalous stimulated gravitational emission in a layered superconductor like YBCO. The authors model the superconductor as an array of intrinsic Josephson junctions. The superconducting parameters are defined by our preliminary measurements with melt-textured samples. Coherent e.m. emission by synchronized Josephson junctions arrays was first reported by Barbara et al. in 1999. The authors write explicitly and solve numerically the Josephson equations which give the normal and super components of the total current in the superconductor, and derive from this the total available power $P=IV$. Then, assuming that the coefficients A and B for spontaneous and stimulated gravitational emission are known, the authors apply to this case the Frantz-Nodvik equation for a laser amplifier. The equation is suitably modified in order to allow for a "continuous pumping" given by an oscillating transport current. The conclusions are relevant for the evaluation of gravitational emission from superconductors. The authors find that even if the A and B coefficients are anomalously large (possibly because of the Quantum Gravity effects mentioned above), the conditions for stimulated emission are quite strict and the emission rate strongly limited by the IV value, for reasons intrinsic to the nature of the superconductor.

In Chapter 6 the authors construct a class of cosmological models with the adequately defined topological field configurations (Blau-Thompson-Horowitz-Baez BF-systems) on a specific collection of plumbed V-cobordisms (pV-cobordisms) and corresponding plumbed Vmanifolds (pV-manifolds), modeling a four-dimensional space-time in the Euclidean regime. The explicit expressions for transition amplitudes (partition functions) are written in these BF-models and it is shown that the basic topological invariants of the pV-cobordisms and pV-manifolds (intersection matrices) play the role of coupling constants between the formal analogues of electric and magnetic fluxes quantized `a la Dirac, using Poicar′e–Lefschetz duality. The Hartle-Hawking quantum amplitudes (in semi-classical approximation) for the BF-systems on pV-manifolds are calculated, they turn out to be the space-time topological invariants and correspond to the topdown approach to cosmology. The diagonal elements and eigenvalues of the intersection matrix for a definite pV-cobordism and corresponding pV-manifold reproduce the hierarchy of dimensionless low-energy coupling constants of the fundamental interactions acting in the real universe at the present time.

In Chapter 7 the authors derive semiclassical physics of black holes using coherent states in loop quantum gravity (LQG). The authors find an explanation for the origin of horizon entropy in this framework by tracing over the wavefunction within the horizon. The coherent state within the horizon is shown to be correlated with the coherent state outside the horizon, and the physics for the outside observer is described by a reduced density matrix which yields the entropy. They then examine the next order quantum fluctuations as measured in the coherent state, and show that information emerges from behind the horizon. This appears to be the origin of Hawking radiation.

According to the black hole thermodynamics, a black hole itself is regarded as a self-gravitating system being in thermal equilibrium state of Hawking temperature. If the outside environment around black hole is cooler than the Hawking temperature, then the black hole loses its mass energy. This is the black hole evaporation. When a black hole evaporates, there arises a net energy flow from black hole into its outside environment due to the Hawking radiation and the energy accretion onto black hole. The existence of energy flow means that the black hole evaporation is a nonequilibrium process: Look at each moment during the evaporation process of a black hole whose horizon scale is larger than the Planck scale. Then it is recognized that, although the black hole itself is regarded as in an (quasi-)equilibrium state, the outside environment is in a nonequilibrium state whose nonequilibrium nature arises by the net energy flow. Therefore, to study the detail of evaporation process, nonequilibrium effects of the net energy flow in outside environment should be taken into account. The nonequilibrium nature of black hole evaporation is a challenging topic which includes not only black hole physics but also nonequilibrium physics.

In Chapter 8 the authors simplify the situation so that the Hawking radiation consists of non-self-interacting massless matter fields and also the energy accretion onto the black hole consists of the same fields. Then the nonequilibrium nature of black hole evaporation is described by a nonequilibrium state of that field. Hence the authors formulate nonequilibrium thermodynamics of non-self-interacting massless fields. Then, by applying it to black hole evaporation, followings are shown: (1) Nonequilibrium effects of the energy flow tends to accelerate the black hole evaporation, and, consequently, a specific nonequilibrium phenomenon of semi-classical black hole evaporation is suggested. Furthermore a suggestion about the end state of quantum size black hole evaporation is proposed in the context of information loss paradox. (2) Negative heat capacity of black hole is the physical essence of the generalized second law of black hole thermodynamics, and self-entropy production inside the matter around black hole is not necessary to ensure the generalized second law. Furthermore a lower bound for total entropy at the end of black hole evaporation is given.

The possible existence of black holes has fascinated scientists at least sinceMichell and Laplace's proposal that a gravitating object could exist from which light could not escape. In the 20th century, in light of the general theory of relativity, it became apparent that, were such objects to exist, their structure would be far richer than originally imagined. Today, astronomical observations strongly suggest that either black holes, or objects with similar properties, not only exist but may well be abundant in our universe. In light of this, black hole research is now not only motivated by the fascinating theoretical properties such objects must possess but also as an attempt to better understand the universe around us. The authors review here some selected developments in black hole research, from a review of its early history to current topics in black hole physics research. Black holes have been studied at all levels; classically, semi-classically, and more recently, as an arena to test predictions of candidate theories of quantum gravity. They will review in Chapter 9 progress and current research at all these levels as well as discuss some proposed alternatives to black holes.

The aim of the study of anisotropic universe models is to explain the apparent isotropy the Universe currently has. In Chapter 10 the authors will review what we know about the evolution of the anisotropic Bianchi universes. In this regard the investigation of everexpanding universes of Bianchi type with a tilted perfect fluid as a source have recently been completed using the dynamical systems approach. The dynamical systems approach has proven to be extremely useful for this analysis and, therefore, some detail will be given in determining the evolution equations for the various Bianchi models. The authors will also review some of the different behaviours found for these models and the authors will discuss what they have learnt regarding the evolution of anisotropies from this analysis. Furthermore, the authors will discuss the issue of isotropy in detail and point out which of these models isotropise in the future (and in what sense). From a more mathematical point of view, the analysis has also shown a wealth of different phenomena, like centre manifolds, attracting closed curves and even attracting tori. The authors will also discuss some aspects of alternative theories of gravity and point out certain different behaviours that might appear for these theories.

The possible extensions of GR for description of fermions on a curved space, for supergravity and for loop quantum gravity require a richer set of 16 independent variables. These variables can be assembled in a coframe field, i.e., a local set of four linearly independent 1-forms. In Chapter 11 the authors study the gravity field models based on a coframe variable alone. The authors give a short review of the coframe gravity. This model has the viable Schwarzschild solutions even being an alternative to the standard GR. Moreover, the coframe model treating of the gravity energy may be preferable to the ordinary GR where the gravity energy cannot be defined at all. A principal problem that the coframe gravity does not have any connection to a specific geometry even being constructed from the geometrical meaningful objects. A geometrization of the coframe gravity is an aim of this chapter. The authors construct a complete class of the coframe connections which are linear in the first order derivatives of the coframe field on an n dimensional manifolds with and without a metric. The subclasses of the torsion-free, metric-compatible and flat connections are derived. They also study the behavior of the geometrical structures under local transformations of the coframe. The remarkable fact is an existence of a subclass of connections which are invariant when the infinitesimal transformations satisfy the Maxwell-like system of equations. In the framework of the coframe geometry construction, the authors propose a geometrical action for the coframe gravity. It is similar to the Einstein-Hilbert action of GR, but the scalar curvature is constructed from the general coframe connection. The authors show that this geometric Lagrangian is equivalent to the coframe Lagrangian up to a total derivative term. Moreover there is a family of coframe connections in which

Lagrangian does not include the higher order terms at all. In this case, the equivalence is complete.

Chapter 12 concerns relational first principles from which the Dirac procedure exhaustively picks out the geometrodynamics corresponding to general relativity as one of a handful of consistent theories. This was accompanied by a number of results and conjectures about matter theories and general features of physics – such as gauge theory, the universal light cone principle of special relativity and the equivalence principle – being likewise picked out. The authors have previously shown that many of these matter results and conjectures are contingent on further unrelational simplicity assumptions. In this paper, they point out 1) that the exhaustive procedure in these cases with matter fields is slower than it was previously held to be. 2) While the example of equivalence principle violating matter theory that they previously showed how to accommodate on relational premises has a number of pathological features, in this paper the authors point out that there is another closely related equivalence principle violating theory that also follows from those premises and is less pathological. This example being known as an 'Einstein–aether theory', it also serves for 3) illustrating limitations on the conjectured emergence of the universal light cone special relativity principle.

In: Classical and Quantum Gravity Research
Editors: M.N. Christiansen et al, pp. 1-78
ISBN 978-1-60456-366-5

Chapter 1

EXOTIC SOLUTIONS IN GENERAL RELATIVITY: TRAVERSABLE WORMHOLES AND "WARP DRIVE" SPACETIMES

Francisco S.N. Lobo*
Centro de Astronomia e Astrofísica da Universidade de Lisboa,
Campo Grande, Ed. C8 1749-016 Lisboa, Portugal
Institute of Gravitation & Cosmology, University of Portsmouth,
Portsmouth PO1 2EG, UK

Abstract

The General Theory of Relativity has been an extremely successful theory, with a well established experimental footing, at least for weak gravitational fields. It's predictions range from the existence of black holes, gravitational radiation to the cosmological models, predicting a primordial beginning, namely the big-bang. All these solutions have been obtained by first considering plausible distributions of matter, i.e., a plausible stress-energy tensor, and through the Einstein field equation, the spacetime metric of the geometry is determined. However, one may solve the Einstein field equation in the reverse direction, namely, one first considers an interesting and exotic spacetime metric, then finds the matter source responsible for the respective geometry. In this manner, it was found that some of these solutions possess a peculiar property, namely "exotic matter," involving a stress-energy tensor that violates the null energy condition. These geometries also allow closed timelike curves, with the respective causality violations. Another interesting feature of these spacetimes is that they allow "effective" superluminal travel, although, locally, the speed of light is not surpassed. These solutions are primarily useful as "gedanken-experiments" and as a theoretician's probe of the foundations of general relativity, and include traversable wormholes and superluminal "warp drive" spacetimes. Thus, one may be tempted to denote these geometries as "exotic" solutions of the Einstein field equation, as they violate the energy conditions and generate closed timelike curves. In this article, in addition to extensively exploring interesting features, in particular, the physical properties and characteristics of these "exotic spacetimes," we also analyze other non-trivial general relativistic geometries which generate closed timelike curves.

*E-mail address: flobo@cosmo.fis.fc.ul.pt

1. Introduction

1.1. Review of Wormhole Physics

Wormholes act as tunnels from one region of spacetime to another, possibly through which observers may freely traverse. Interest in traversable wormholes, as hypothetical shortcuts in spacetime, has been rekindled by the classical paper by Morris and Thorne [1]. It was first introduced as a tool for teaching general relativity, as well as an allurement to attract young students into the field, but it also served to stimulate research in several branches. These developments culminated with the publication of the book *Lorentzian Wormholes: From Einstein to Hawking* by Visser [2], where a review on the subject up to 1995, as well as new ideas are developed and hinted at. However, it seems that wormhole physics can originally be traced back to Flamm in 1916 [3], when he analyzed the then recently discovered Schwarzschild solution. Paging through history one finds next that wormhole-type solutions were considered, in 1935, by Einstein and Rosen [4], where they constructed an elementary particle model represented by a "bridge" connecting two identical sheets. This mathematical representation of physical space being connected by a wormhole-type solution was denoted an "Einstein-Rosen bridge". The field laid dormant, until Wheeler revived the subject in the 1950s. Wheeler considered wormholes, such as Reissner-Nordström or Kerr wormholes, as objects of the quantum foam connecting different regions of spacetime and operating at the Planck scale [5, 6], which were transformed later into Euclidean wormholes by Hawking [7] and others. However, these Wheeler wormholes were not traversable, and furthermore would, in principle, develop some type of singularity [8]. These objects were obtained from the coupled equations of electromagnetism and general relativity and were denoted "geons", i.e., gravitational-electromagnetic entities.

Geons possess curious properties such as: firstly, the gravitational mass originates solely from the energy stored in the electromagnetic field, and in particular, there are no material masses present (this gave rise to the term "mass without mass"); and secondly, no charges are present ("charge without charge"). These entities were further explored by several authors in different contexts [9], but due to the extremely ambitious program and the lack of experimental evidence soon died out. Nevertheless, it is interesting to note that Misner inspired in Wheeler's geon representation, found wormhole solutions to the source-free Einstein equations in 1960 [10]. With the introduction of multi-connected topologies in physics, the question of causality inevitably arose, as a light signal travelling through the short-cut, i.e., the wormhole, could outpace another light signal. Thus, Wheeler and Fuller examined this situation in the Schwarzschild solution and found that causality is preserved [11], as the Schwarzschild throat pinches off in a finite time, preventing the traversal of a signal from one region to another through the wormhole. However, Graves and Brill [12], considering the Reissner-Nordström metric also found wormhole-type solutions between two asymptotically flat spaces, but with an electric flux flowing through the wormhole. They found that the region of minimum radius, the "throat", contracted, reaching a minimum and re-expanded after a finite proper time, rather than pinching off as in the Schwarzschild case. The throat, "cushioned" by the pressure of the electric field through the throat, pulsated periodically in time. The modern renaissance of wormhole physics was mainly brought about by the classic Morris-Thorne paper [1]. Thorne together with his student Morris [1], understanding that the energy conditions lay on shaky ground [13, 14], considered that wormholes, with two mouths and a throat, might be objects of nature, as stars and black holes are.

Wormhole physics is a specific example of adopting the reverse philosophy of solving the Einstein field equation, by first constructing the spacetime metric, then deducing the stress-energy tensor components. Thus, it was found that these traversable wormholes possess a stress-energy tensor that violates the null energy condition [1, 2]. In fact, they violate all the known pointwise energy conditions and averaged energy conditions, which are fundamental to the singularity theorems and theorems of classical black hole thermodynamics. The weak energy condition (WEC) assumes that the local energy density is non-negative and states that $T_{\mu\nu}U^{\mu}U^{\nu} \geq 0$, for all timelike vectors U^{μ}, where $T_{\mu\nu}$ is the stress energy tensor (in the frame of the matter this amounts to $\rho \geq 0$ and $\rho + p \geq 0$). By continuity, the WEC implies the null energy condition (NEC), $T_{\mu\nu}k^{\mu}k^{\nu} \geq 0$, where k^{μ} is a null vector. The null energy condition is the weakest of the energy conditions, and its violation signals that the other energy conditions are also violated. Although classical forms of matter are believed to obey these energy conditions [15], it is a well-known fact that they are violated by certain quantum fields, amongst which we may refer to the Casimir effect and Hawking evaporation (see [16] for a short review). It was further found that for quantum systems in classical gravitational backgrounds the weak or null energy conditions could only be violated in small amounts, and a violation at a given time through the appearance of a negative energy state, would be overcompensated by the appearance of a positive energy state soon after. Thus, violations of the pointwise energy conditions led to the averaging of the energy conditions over timelike or null geodesics [17, 18]. For instance, the averaged weak energy condition (AWEC) states that the integral of the energy density measured by a geodesic observer is non-negative, i.e., $\int T_{\mu\nu}U^{\mu}U^{\nu}\, d\tau \geq 0$, where τ is the observer's proper time. Thus, the averaged energy conditions permit localized violations of the energy conditions, as long as they hold when averaged along a null or timelike geodesic [17].

Pioneering work by Ford in the late 1970's on a new set of energy constraints [19], led to constraints on negative energy fluxes in 1991 [20]. These eventually culminated in the form of the Quantum Inequality (QI) applied to energy densities, which was introduced by Ford and Roman in 1995 [21]. The QI was proven directly from Quantum Field Theory, in four-dimensional Minkowski spacetime, for free quantized, massless scalar fields. The inequality limits the magnitude of the negative energy violations and the time for which they are allowed to exist, yielding information on the distribution of the negative energy density in a finite neighborhood [21, 22, 23, 24]. The basic applications to curved spacetimes is that these appear flat if restricted to a sufficiently small region. The application of the QI to wormhole geometries is of particular interest [22, 25]. A small spacetime volume around the throat of the wormhole was considered, so that all the dimensions of this volume are much smaller than the minimum proper radius of curvature in the region. Thus, the spacetime can be considered approximately flat in this region, so that the QI constraint may be applied. The results of the analysis is that either the wormhole possesses a throat size which is only slightly larger than the Planck length, or there are large discrepancies in the length scales which characterize the geometry of the wormhole. The analysis imply that generically the exotic matter is confined to an extremely thin band, and/or that large red-shifts are involved, which present severe difficulties for traversability, such as large tidal forces [22]. Due to these results, Ford and Roman concluded that the existence of macroscopic traversable wormholes is very improbable (see [26, 27] for a review). It was also shown that, by using the QI, enormous amounts of exotic matter are needed to support the Alcubierre warp drive and the superluminal Krasnikov tube [28, 29, 30].

Relative to the energy conditions, the situation has changed drastically, as it has been now shown that even classical systems, such as those built from scalar fields non-minimally coupled to gravity, violate all the energy conditions [31]. It is interesting to note that recent observations in cosmology strongly suggest that the cosmological fluid violates the strong energy condition (SEC), and provides tantalizing hints that the NEC *might* possibly be violated in a classical regime [32, 33, 34]. Thus, gradually the weak and null energy conditions, and with it the other energy conditions, are losing their status of a kind of law [35]. Surely, this has had implications on the construction of wormholes. In the original paper [1], Morris and Thorne first provided a spherically symmetric spacetime metric, then deduced that it needed exotic matter to sustain the wormhole geometry. The engineering work was left to an absurdly advanced civilization, which could manufacture such matter and construct these wormholes. Then, once it was understood that quantum effects should enter in the stress-energy tensor, a self-consistent wormhole solution of semiclassical gravity was found [36], presumably obeying the quantum inequalities. Thus, it seems that these exotic spacetimes arise naturally in the quantum regime, as a large number of quantum systems have been shown to violate the energy conditions, such as the Casimir effect. Indeed, various wormhole solutions in semi-classical gravity have been considered in the literature. For instance, semi-classical wormholes were found in the framework of the Frolov-Zelnikov approximation for $\langle T_{\mu\nu} \rangle$ [37]. Analytical approximations of the stress-energy tensor of quantized fields in static and spherically symmetric wormhole spacetimes were also explored in Refs. [38]. However, the first self-consistent wormhole solution coupled to a quantum scalar field was obtained in Ref. [36]. The ground state of a massive scalar field with a non-conformal coupling on a short-throat flat-space wormhole background was computed in Ref. [39], by using a zeta renormalization approach. The latter wormhole model, which was further used in the context of the Casimir effect [40], was constructed by excising spherical regions from two identical copies of Minkowski spacetime, and finally surgically grafting the boundaries (A more realistic geometry was considered in Ref. [41]). Recently, semi-classical wormholes have also been obtained using a one-loop graviton contribution approach [42, 43]. Finally with the realization that nonminimal scalar fields violate the weak energy condition, a set of self-consistent classical wormholes was found [44].

It is fair to say that, though outside this mainstream, classical wormholes were found by Homer Ellis back in 1973 [45] and further explored in [46], and related self-consistent solutions were found by Kirill Bronnikov in 1973 [47], Takeshi Kodama in 1978 [48], and Gérard Clément in 1981 [49]. These papers written much before the wormhole boom originated from Morris and Thorne's work [1] (see [50] for a short account of these previous solutions). A self-consistent Ellis wormhole was found again by Harris [51] by solving, through an exotic scalar field, an exercise for students posed in [1]. Visser [52] motivated by the aim of minimizing the violation of the energy conditions and the possibility of a traveller not encountering regions of exotic matter in a traversal through a wormhole, constructed polyhedral wormholes and, in particular, cubic wormholes. These contained exotic matter concentrated only at the edges and the corners of the geometrical structure, and a traveller could pass through the flat faces without encountering matter, exotic or otherwise. He further generalized a suggestion of Roman for a configuration with two wormholes [1] into a Roman ring [53], and in 1999 analyzed generic dynamical traversable wormhole throats [54]. Furthermore, in 1999, together with Barcelo, he found classically consistent solutions with scalar fields [31], and in collaboration with Dadhich, Kar and Mukherjee has also found self-dual solutions [55]. Other authors have also made interesting studies and significant contributions.

Before Visser's book we can quote an interesting application to wormhole physics by Frolov and Novikov, where they relate wormhole and black hole physics [56]. An interesting application of wormholes to cosmology was treated by Hochberg and Kephart relatively to the horizon problem [57]. These authors speculated that wormholes allow the two-way passage of signals between spatially separated regions of spacetime, and could permit the thermalization of these respective regions. After the Matt Visser book, González-Díaz generalized the static spherically symmetric traversable wormhole solution to that of a torus-like topology [58]. This geometrical construction was denoted as a ringhole. González-Díaz went on to analyze the causal structure of the solution, i.e., the presence of closed timelike curves, and has recently studied the ringhole evolution due to the accelerating expansion of the universe, in the presence of dark energy [59]. Wormhole solutions inside cosmic strings were found by Clément [60], and Aros and Zamorano [61]; wormholes supported by strings by Schein and Aichelburg [62, 63]; the maintenance of a wormhole with a scalar field was considered by Vollick [64], solutions with minimal and non-minimal scalar fields were explored by Kim and Kim [65] and exact solutions of charged wormholes considering the back reaction to the traversable Lorentzian wormhole spacetime by a scalar field or electric charge were found by Kim and Lee [66]; rotating wormholes solutions were analyzed by Teo [67], and further generalized by Kuhfittig [68]; a solution was found in which the exotic matter is controlled by an external magnetic field by Parisio [69]; wormholes with stress-energy tensor of massless neutrinos and other massless fields were considered by Krasnikov [70]; wormholes made of a crossflow of dust null streams were discussed by Hayward [71] and Gergely [72]; self consistent charged solutions were found by Bronnikov and Grinyok [73]; and the possible existence of wormhole geometries in the context of nonlinear electrodynamics was also explored [74]. It is interesting to note that building on [71], Hayward and Koyama, using a model of pure phantom radiation, i.e., pure radiation with negative energy density, and the idealization of impulsive radiation, considered analytic solutions describing the theoretical construction of a traversable wormhole from a Schwarzschild black hole [75], and the respective enlargement of the wormhole [76]. More recently, exact solutions of traversable wormholes were found under the assumption of spherical symmetry and the existence of a *non-static* conformal symmetry, which presents a more systematic approach in searching for exact wormhole solutions [77].

One of the main areas in wormhole research is to try to avoid as much as possible the violation of the null energy condition. For static wormholes the null energy condition is violated [1, 2], and thus, several attempts have been made to overcome somehow this problem. In the original article [1], Morris and Thorne had already tried to minimize the violating region by constructing specific examples of wormhole geometries. As mentioned before, Visser [52] found solutions where observers can pass the throat without interacting with the exotic matter, which was pushed to the corners, and Kuhfittig [78] has found that the region made of exotic matter can be made arbitrarily small. For dynamic wormholes, the null energy condition, more precisely the averaged null energy condition can be avoided in certain regions [54, 79, 80, 81, 82, 83]. More recently, Visser *et al* [84, 85], noting the fact that the energy conditions do not actually quantify the "total amount" of energy condition violating matter, developed a suitable measure for quantifying this notion by introducing a "volume integral quantifier". This notion amounts to calculating the definite integrals $\int T_{\mu\nu}U^{\mu}U^{\nu}\,dV$ and $\int T_{\mu\nu}k^{\mu}k^{\nu}\,dV$, and the amount of violation is defined as the extent to which these integrals become negative. Although the null energy and averaged null energy conditions are always violated for wormhole spacetimes, Visser *et al* considered specific examples of spacetime geometries containing wormholes that are supported by arbitrarily small quantities of averaged null energy condition violating matter.

Some papers have added a cosmological constant to the wormhole construction analysis. Thin-shell wormhole solutions with Λ, in the spirit of Visser [2, 86] were analyzed in [87, 88]. Roman [89] found a wormhole solution inflating in time to test whether one could evade the violation of the energy conditions, and Delgaty and Mann [90] looked for new wormhole solutions with Λ. Construction of wormhole solutions by matching an interior wormhole spacetime to an exterior vacuum solution, at a junction surface, were also recently analyzed extensively [91, 92, 93]. In particular, a thin shell around a traversable wormhole, with a zero surface energy density was analyzed in [91], and with generic surface stresses in [92]. A similar analysis for the plane symmetric case, with a negative cosmological constant, is done in [94]. A general class of wormhole geometries with a cosmological constant and junction conditions was analyzed by DeBenedictis and Das [95], and further explored in higher dimensions [96]. To know the stability of an object against several types of perturbation is always an important issue. In particular, the stability of thin-shell wormholes, constructed using the cut-and-paste technique, by considering specific equations of state [86, 87, 90, 97, 98, 99], or by applying a linearized radial perturbation around a stable solution [100, 101, 102, 88, 103], were analyzed. For the Ellis' drainhole [45, 46], Armendáriz-Picón [104] finds that it is stable against linear perturbations, whereas Shinkai and Hayward [105] find this same class unstable to nonlinear perturbations. Bronnikov and Grinyok [73, 106] found that the consistent wormholes of Barceló and Visser [44] are unstable.

In alternative theories to general relativity wormhole solutions have been worked out. In higher dimensions, solutions have been found by Chodos and Detweiler [107], Clément [108], and DeBenedictis and Das [96]; in the nonsymmetric gravitational theory traversable wormholes were found by Moffat and Svoboda [109]; in Brans-Dicke theory by Agnese and Camera [110], Anchordoqui *at al* [111], Nandi and collaborators [112, 113, 114, 115], and He and Kim [116]; in Kaluza-Klein theory by Shen and collaborators [117]; in Einstein-Gauss-Bonnet by Bhawal and Kar [118]; and Koyama, Hayward and Kim [119] examined wormholes in a two-dimensional dilatonic theory. Anchordoqui and Bergliaffa found a wormhole solution in a brane world scenario [120], further examined by Barceló and Visser [121], and Bronnikov and Kim considered possible traversable wormhole solutions in a brane world, by imposing the condition $R = 0$, where R is the four-dimensional scalar curvature [122]; the latter solution was generalized in Ref. [123]. La Camera using the simplest form of the Randall-Sundrum model, considered the metric generated by a static, spherically symmetric distribution of matter on the physical brane, and found that the solution to the five-dimensional Einstein equations, obtained numerically, describes a wormhole geometry [124]. Recently, wormhole throats were also analyzed in a higher derivative gravity model governed by the Einstein-Hilbert Lagrangian, supplemented with $1/R$ and R^2 curvature scalar terms [125]. Using the resulting equations of motion, it was found that the weak energy condition may be respected in the throat vicinity, with conditions compatible with those required for stability [126] and an acceptable Newtonian limit [127]. The R^2 theory was meticulously studied by Ghoroku and Soma [128], where it was concluded that, under the assumption that an asymptotically flat global solution exists, no weak energy condition respecting wormhole solution may exist in such a theory.

If it is true that wormholes act as shortcuts between two regions of spacetime, then it is interesting to note that shortcuts also exist in the context of brane cosmology [129]. The latter model stipulates that our Universe is a three-brane embedded in a five-dimensional anti-se Sitter spacetime, in which matter is confined to the brane and gravity exists throughout the bulk. This implies the causal propagation of light and gravitational signals is in general different [130]. A gravitational signal travelling between two points on the brane may propagate through the bulk, taking a shortcut, and appearing quicker than a photon which propagates on the brane between the two respective points [131, 132]. It is then expected that these shortcuts would play an important role in solving the horizon problem [132, 133].

An important side effect of wormholes is that they can theoretically generate closed timelike curves with relative ease, by performing a sufficient delay to the time of one mouth in relation to the other (see [134] for a review on closed timelike curves). This can be done either by the special relativistic twin paradox method [135] or by the general relativistic redshift way [136]. The importance of wormholes in the study of time machines is that they provide a non-eternal time machine, where closed timelike curves appear to the future of some hypersurface, the chronology horizon (a special case of a Cauchy horizon which is the onset of the nonchronal region containing closed timelike curves) which is generated in a compact region in this case [137, 138]. Since time travel to the past is in general unwelcome, it is possible to test whether classical or semiclassical effects will destroy the time machine. It is found that classically it can be easily stabilized [135, 2]. Semiclassiclaly, there are calculations that favor the destruction [139, 140, 141], leading to chronology protection [140], others that maintain the time machine [53, 142, 143]. Other simpler systems that simulate a wormhole, such as Misner spacetime which is a species of two-dimensional wormhole, have been studied more thoroughly, with no conclusive answer. For Misner spacetime the debate still goes on, favoring chronology protection [144], disfavoring it [145], and back in favoring [146]. The upshot is that semiclassical calculations will not settle the issue of chronology protection [147], one needs a quantum gravity, as has been foreseen sometime before by Thorne [148].

In a cosmological context, it is extraordinary that recent observations have confirmed that the Universe is undergoing a phase of accelerated expansion. Evidence of this cosmological expansion, coming from measurements of supernovae of type Ia (SNe Ia) [149, 150] and independently from the cosmic microwave background radiation [151, 152], shows that the Universe additionally consists of some sort of negative pressure "dark energy". The Wilkinson Microwave Anisotropy Probe (WMAP), designed to measure the CMB anisotropy with great precision and accuracy, has recently confirmed that the Universe is composed of approximately 70 percent of dark energy [151]. Several candidates representing dark energy have been proposed in the literature, namely, a positive cosmological constant, the quintessence fields, generalizations of the Chaplygin gas and so-called tachyon models. A simple way to parameterize the dark energy is by an equation of state of the form $\omega \equiv p/\rho$, where p is the spatially homogeneous pressure and ρ the energy density of the dark energy [153]. A value of $\omega < -1/3$ is required for cosmic expansion, and $\omega = -1$ corresponds to a cosmological constant [154]. A possibility that has been widely explored, is that of quintessence, where the parameter range is $-1 < \omega < -1/3$. However, a note on the choice of the imposition $\omega > -1$ is in order. This is considered to ensure that the null energy condition, $T_{\mu\nu}\, k^\mu\, k^\nu \geq 0$, is satisfied. If $\omega < -1$ [155, 156, 157], a case certainly not excluded by observation, then the null energy condition is violated, $\rho + p < 0$, and consequently all of the other energy conditions. Matter with the property $\omega < -1$ has been denoted "phantom energy" [158]. As the possibility of phantom energy implies the violation of the null energy condition, this leads us back to wormhole physics. This possibility has been explored with wormholes being supported by phantom energy [159, 160, 161], the generalized Chaplygin gas [162], and the van der Waals equation of state [163].

Stars are common for everyone to see, black holes also inhabit the universe in billions, so one may tentatively assume that wormholes, formed or constructed from one way or another, can also appear in large amounts. If they inhabit the cosmological space, they will produce microlensing effects on point sources at non-cosmological distances [164], as well as at cosmological distances, in this case gamma-ray burts could be the objects microlensed [165, 166]. If peculiarly large, then wormholes will produce macrolensing effects [167]. There is now a growing consensus that wormholes are in the same chain of stars and black holes. For instance, González-Días [58] understood that an enormous pressure on the center ultimately meant a negative energy density to open up the tunnel; DeBenedectis and Das [95] mention that the stress-energy supporting the structure consists of an anisotropic brown dwarf 'star'; and the wormhole joining one Friedmann-Robertson-Walker universe with Minkowski spacetime or joining two Friedmann-Robertson-Walker universes [54] could be interpreted, after further matchings, as a wormhole joining a collapsing (or expanding) star to Minkowski spacetime or a wormhole joining two dynamical stars, respectively. It has also been recognized, and emphasized by Hayward [168], that wormholes and black holes can be treated in a unified way, the black hole being described by a null outer trapped surface, and the wormhole by a timelike outer trapped surface, this surface being the throat where incoming null rays start to diverge [80, 168]. Thus, it seems there is a continuum of objects from stars to wormholes passing through black holes, where stars are made of normal matter, black holes of vacuum, and wormholes of exotic matter. Although not so appealing perhaps, wormholes could be called "exotic stars".

1.2. "Warp drive" Spacetimes and Superluminal Travel

Much interest has been revived in superluminal travel in the last few years. Despite the use of the term superluminal, it is not "really" possible to travel faster than light, in any *local* sense. The point to note is that one can make a round trip, between two points separated by a distance D, in an arbitrarily short time as measured by an observer that remained at rest at the starting point, by varying one's speed or by changing the distance one is to cover. Providing a general *global* definition of superluminal travel is no trivial matter [169, 170], but it is clear that the spacetimes that allow "effective" superluminal travel generically suffer from the severe drawback that they also involve significant negative energy densities. More precisely, superluminal effects are associated with the presence of *exotic* matter, that is, matter that violates the null energy condition (see [28] for a review). In fact, superluminal spacetimes violate all the known energy conditions, and Ken Olum demonstrated that negative energy densities and superluminal travel are intimately related [171].

Apart from wormholes [1, 2], two spacetimes which allow superluminal travel are the Alcubierre warp drive [172] and the solution known as the Krasnikov tube [173, 30]. Alcubierre demonstrated that it is theoretically possible, within the framework of general relativity, to attain arbitrarily large velocities [172]. A warp bubble is driven by a local expansion behind the bubble, and an opposite contraction ahead of it. However, by introducing a slightly more complicated metric, José Natário [174] dispensed with the need for expansion. The Natário version of the warp drive can be thought of as a bubble sliding through space.

It is interesting to note that Krasnikov [173] discovered a fascinating aspect of the warp drive, in which an observer on a spaceship cannot create nor control on demand an Alcubierre bubble, with $v > c$, around the ship [173], as points on the outside front edge of the bubble are always spacelike separated from the centre of the bubble. However, causality considerations do not prevent the crew of a spaceship from arranging, by their own actions, to complete a *round trip* from the Earth to a distant star and back in an arbitrarily short time, as measured by clocks on the Earth, by altering the metric along the path of their outbound trip. Thus, Krasnikov introduced a two-dimensional metric with an interesting property that although the time for a one-way trip to a distant destination cannot be shortened, the time for a round trip, as measured by clocks at the starting point (e.g. Earth), can be made arbitrarily short. Soon after, Everett and Roman generalized the Krasnikov two-dimensional analysis to four dimensions, denoting the solution as the *Krasnikov tube* [30], where they analyzed the superluminal features, the energy condition violations, the appearance of closed timelike curves and applied the Quantum Inequality.

Recently, linearized gravity was applied to warp drive spacetimes, testing the energy conditions at first and second order of the non-relativistic warp-bubble velocity [175], $v \ll 1$. Thus, attention was not focussed on the "superluminal" aspects of the warp bubble [176], such as the appearance of horizons [177, 178, 179] and of closed timelike curves [180], but rather on a secondary unremarked effect: The warp drive (*if it can be realised in nature*) appears to be an example of a "reaction-less drive" wherein the warp bubble moves by interacting with the geometry of spacetime instead of expending reaction mass. A particularly interesting aspect of this construction is that one may place a finite mass spaceship at the origin and consequently analyze how the warp field compares with the mass-energy of the spaceship. This is not possible in the usual finite-strength warp field, since by definition the point in the center of the warp bubble moves along a geodesic and is "massless". That is, in the usual formalism the spaceship is always treated as a test particle, while in the linearized theory one can treat the spaceship as a finite mass object.

For warp drive spacetimes, by using the "quantum inequality" deduced by Ford and Roman [21], it was soon verified that enormous amounts of energy are needed to sustain superluminal warp drive spacetimes [22, 29]. To reduce the enormous amounts of exotic matter needed in the superluminal warp drive, van den Broeck proposed a slight modification of the Alcubierre metric which considerably ameliorates the conditions of the solution [181]. It is also interesting to note that, by using the "quantum inequality", enormous quantities of negative energy densities are needed to support the superluminal Krasnikov tube [30]. Gravel and Plante [182, 183] in a way similar in spirit to the van den Broeck analysis, showed that it is theoretically possible to lower significantly the mass of the Krasnikov tube. However, in the linearized analysis, no *a priori* assumptions as to the ultimate source of the energy condition violations were made, so that the quantum inequalities were not used nor needed. This means that the restrictions derived on warp drive spacetimes are more generic than those derived using the quantum inequalities – the restrictions derived in [175] hold regardless of whether the warp drive is assumed to be classical or quantum in its operation. It was not meant to suggest that such a "reaction-less drive" is achievable with current technology, as indeed extremely stringent conditions on the warp bubble were obtained, in the weak-field limit. These conditions are so stringent that it appears unlikely that the "warp drive" will ever prove technologically useful.

1.3. Closed Timelike Curves

As time is incorporated into the proper structure of the fabric of spacetime, it is interesting to note that general relativity is contaminated with non-trivial geometries which generate *closed timelike curves* [2, 26, 28, 134, 91, 184, 185]. A closed timelike curve (CTC) allows time travel, in the sense that an observer which travels on a trajectory in spacetime along this curve, returns to an event which coincides with the departure. The arrow of time leads forward, as measured locally by the observer, but globally he/she may return to an event in the past. This fact apparently violates causality, opening Pandora's box and producing time travel paradoxes [186], throwing a veil over our understanding of the fundamental nature of Time. The notion of causality is fundamental in the construction of physical theories, therefore time travel and it's associated paradoxes have to be treated with great caution. The paradoxes fall into two broad groups, namely the *consistency paradoxes* and the *causal loops*.

The consistency paradoxes include the classical grandfather paradox. Imagine travelling into the past and meeting one's grandfather. Nurturing homicidal tendencies, the time traveller murders his grandfather, impeding the birth of his father, therefore making his own birth impossible. In fact, there are many versions of the grandfather paradox, limited only by one's imagination. The consistency paradoxes occur whenever possibilities of changing events in the past arise.

The paradoxes associated to causal loops are related to self-existing information or objects, trapped in spacetime. Imagine a time traveller going back to his past, handing his younger self a manual for the construction of a time machine. The younger version then constructs the time machine over the years, and eventually goes back to the past to give the manual to his younger self. The time machine exists in the future because it was constructed in the past by the younger version of the time traveller. The construction of the time machine was possible because the manual was received from the future. Both parts considered by themselves are consistent, and the paradox appears when considered as a whole. One is liable to ask, what is the origin of the manual, for it apparently surges out of nowhere. There is a manual never created, nevertheless existing in spacetime, although there are no causality violations. An interesting variety of these causal loops was explored by Gott and Li [187], where they analyzed the idea of whether there is anything in the laws of physics that would prevent the Universe from creating itself. Thus, tracing backwards in time through the original inflationary state a region of CTCs may be encountered, giving *no* first-cause.

A great variety of solutions to the Einstein Field Equations (EFEs) containing CTCs exist, but, two particularly notorious features seem to stand out. Solutions with a tipping over of the light cones due to a rotation about a cylindrically symmetric axis; and solutions that violate the Energy Conditions of GTR, which are fundamental in the singularity theorems and theorems of classical black hole thermodynamics [15]. A great deal of attention has also been paid to the quantum aspects of closed timelike curves [188, 189, 190].

Thus, as shall be shown in Section 4., it is possible to find solutions to the EFEs, with certain ease, which generate CTCs, which This implies that if we consider general relativity valid, we need to include the *possibility* of time travel in the form of CTCs. A typical reaction is to exclude time travel due to the associated paradoxes. But the paradoxes do not prove that time travel is mathematically or physically impossible. Consistent mathematical solutions to the EFEs have been found, based on plausible physical processes. What they do seem to indicate is that local information in spacetimes containing CTCs are restricted in unfamiliar ways. The grandfather paradox, without doubt, does indicate some strange aspects of spacetimes that contain CTCs. It is logically inconsistent that the time traveller murders his grandfather. But, one can ask, what exactly impeded him from accomplishing his murderous act if he had ample opportunities and the free-will to do so. It seems that certain conditions in local events are to be fulfilled, for the solution to be globally self-consistent. These conditions are denominated *consistency constraints* [191]. To eliminate the problem of free-will, mechanical systems were developed as not to convey the associated philosophical speculations on free-will [192, 193]. Much has been written on two possible remedies to the paradoxes, namely the Principle of Self-Consistency [137, 193, 194, 195] and the Chronology Protection Conjecture [140, 147, 196].

One current of thought, led by Igor Novikov, is the Principle of Self-Consistency, which stipulates that events on a CTC are self-consistent, i.e., events influence one another along the curve in a cyclic and self-consistent way. In the presence of CTCs the distinction between past and future events are ambiguous, and the definitions considered in the causal structure of well-behaved spacetimes break down. What is important to note is that events in the future can influence, but cannot change, events in the past. The Principle of Self-Consistency permits one to construct local solutions of the laws of physics, only if these can be prolonged to a unique global solution, defined throughout non-singular regions of spacetime. Therefore, according to this principle, the only solutions of the laws of physics that are allowed locally, reinforced by the consistency constraints, are those which are globally self-consistent.

Hawking's Chronology Protection Conjecture [140] is a more conservative way of dealing with the paradoxes. Hawking notes the strong experimental evidence in favour of the conjecture from the fact that "we have not been invaded by hordes of tourists from the future". An analysis reveals that the value of the renormalized expectation quantum stress-energy tensor diverges in the imminence of the formation of CTCs. This conjecture permits the existence of traversable wormoles, but prohibits the appearance of CTCs. The transformation of a wormhole into a time machine results in enormous effects of the vacuum polarization, which destroys it's internal structure before attaining the Planck scale. Nevertheless, Li has shown given an example of a spacetime containing a time machine that might be stable against vacuum fluctuations of matter fields [197], implying that Hawking's suggestion that the vacuum fluctuations of quantum fields acting as a chronology protection might break down. There is no convincing demonstration of the Chronology Protection Conjecture, but the hope exists that a future theory of quantum gravity may prohibit CTCs.

Visser still considers the possibility of two other conjectures [2]. The first is the radical reformulation of physics conjecture, in which one abandons the causal structure of the laws of physics and allows, without restriction, time travel, reformulating physics from the ground up. The second is the boring physics conjecture, in which one simply ceases to consider the solutions to the EFEs generating CTCs. Perhaps an eventual quantum gravity theory will provide us with the answers. But, as stated by Thorne [148], it is by extending the theory to it's extreme predictions that one can get important insights to it's limitations, and probably ways to overcome them. Therefore, time travel in the form of CTCs, is more than a justification for theoretical speculation, it is a conceptual tool and an epistemological instrument to probe the deepest levels of GTR and extract clarifying views.

2. Traversable Lorentzian Wormholes

One adopt the reverse philosophy in solving the Einstein field equation, namely, one first considers an interesting and exotic spacetime metric, then finds the matter source responsible for the respective geometry. In this manner, it was found that some of these solutions possess a peculiar property, namely "exotic matter", involving a stress-energy tensor that violates the null energy condition, $T_{\mu\nu}k^\mu k^\nu \geq 0$, where k^μ is a null vector. These geometries also allow closed timelike curves, with the respective causality violations. Another interesting feature of these spacetimes is that they allow "effective" superluminal travel, although, locally, the speed of light is not surpassed. These solutions are primarily useful as "gedanken-experiments" and as a theoretician's probe of the foundations of general relativity, and include traversable wormholes, which shall be extensively reviewed in this Section.

2.1. Spacetime Metric

Consider the following spherically symmetric and static wormhole solution

$$ds^2 = -e^{2\Phi(r)}\,dt^2 + \frac{dr^2}{1 - b(r)/r} + r^2\,(d\theta^2 + \sin^2\theta\,d\phi^2)\,, \tag{1}$$

where $\Phi(r)$ and $b(r)$ are arbitrary functions of the radial coordinate r. $\Phi(r)$ is denoted the redshift function, for it is related to the gravitational redshift, and $b(r)$ is denoted the shape function, because as can be shown by embedding diagrams, it determines the shape of the wormhole [1]. The coordinate r is non-monotonic in that it decreases from $+\infty$ to a minimum value r_0, representing the location of the throat of the wormhole, where $b(r_0) = r_0$, and then it increases from r_0 to $+\infty$. The proper circumference of a circle of fixed r is given by $2\pi r$. Although the metric coefficient g_{rr} becomes divergent at the throat, which is signalled by the coordinate singularity, the proper radial distance

$$l(r) = \pm \int_{r_0}^{r} \frac{dr}{(1 - b(r)/r)^{1/2}}, \tag{2}$$

is required to be finite everywhere. Note that as $0 \leq 1 - b(r)/r \leq 1$, the proper distance is greater than or equal to the coordinate distance, i.e., $|l(r)| \geq r - r_0$. The metric (1) may be written in terms of the proper radial distance as

$$ds^2 = -e^{2\Phi(l)}dt^2 + dl^2 + r^2(l)(d\theta^2 + \sin^2\theta \, d\phi^2) \, . \tag{3}$$

The proper distance decreases from $l = +\infty$, in the upper universe, to $l = 0$ at the throat, and then from zero to $-\infty$ in the lower universe. For the wormhole to be traversable it must have no horizons, which implies that $g_{tt} = -e^{2\Phi(r)} \neq 0$, so that $\Phi(r)$ must be finite everywhere.

The four-velocity of a static observer is $U^\mu = dx^\mu/d\tau = (U^t, 0, 0, 0) = (e^{-\Phi(r)}, 0, 0, 0)$. The observer's four-acceleration is $a^\mu = U^\mu{}_{;\nu} U^\nu$, so that taking into account eq. (1) we have

$$\begin{aligned} a^t &= 0 \, , \\ a^r &= \Gamma^r_{tt} \left(\frac{dt}{d\tau} \right)^2 = \Phi' \, (1 - b/r) \, , \end{aligned} \tag{4}$$

where the prime denotes a derivative with respect to the radial coordinate r. From the geodesic equation, a radially moving test particle which starts from rest initially has the equation of motion

$$\frac{d^2 r}{d\tau^2} = -\Gamma^r_{tt} \left(\frac{dt}{d\tau} \right)^2 = -a^r \, . \tag{5}$$

Therefore, a^r is the radial component of proper acceleration that an observer must maintain in order to remain at rest at constant r, θ, ϕ. Note that from eq. (4), a static observer at the throat for generic $\Phi(r)$ is a geodesic observer. In particular, for a constant redshift function, $\Phi'(r) = 0$, static observers are also geodesic. It is interesting to note that a wormhole is "attractive" if $a^r > 0$, i.e., observers must maintain an outward-directed radial acceleration to keep from being pulled into the wormhole; and "repulsive" if $a^r < 0$, i.e., observers must maintain an inward-directed radial acceleration to avoid being pushed away from the wormhole. This distinction depends on the sign of Φ', as is transparent from eq. (4).

2.2. The Mathematics of Embedding and Generic Static Throat

We can use embedding diagrams to represent a wormhole and extract some useful information for the choice of the shape function, $b(r)$. Due to the spherically symmetric nature of the problem, one may consider an equatorial slice, $\theta = \pi/2$, without loss of generality. The respective line element, considering a fixed moment of time, $t = \text{const}$, is given by

$$ds^2 = \frac{dr^2}{1 - b(r)/r} + r^2 \, d\phi^2 \, . \tag{6}$$

To visualize this slice, one embeds this metric into three-dimensional Euclidean space, in which the metric can be written in cylindrical coordinates, (r, ϕ, z), as

$$ds^2 = dz^2 + dr^2 + r^2\, d\phi^2 \,. \tag{7}$$

Now, in the three-dimensional Euclidean space the embedded surface has equation $z = z(r)$, and thus the metric of the surface can be written as,

$$ds^2 = \left[1 + \left(\frac{dz}{dr}\right)^2\right] dr^2 + r^2\, d\phi^2 \,. \tag{8}$$

Comparing eq. (8) with (6), we have the equation for the embedding surface, given by

$$\frac{dz}{dr} = \pm \left(\frac{r}{b(r)} - 1\right)^{-1/2} \,. \tag{9}$$

To be a solution of a wormhole, the geometry has a minimum radius, $r = b(r) = r_0$, denoted as the throat, at which the embedded surface is vertical, i.e., $dz/dr \to \infty$, see Figure 1. Far from the throat consider that space is asymptotically flat, $dz/dr \to 0$ as $r \to \infty$.

To be a solution of a wormhole, one needs to impose that the throat flares out, as in Figure 1. Mathematically, this flaring-out condition entails that the inverse of the embedding function $r(z)$, must satisfy $d^2r/dz^2 > 0$ at or near the throat r_0. Differentiating $dr/dz = \pm(r/b(r) - 1)^{1/2}$ with respect to z, we have

$$\frac{d^2r}{dz^2} = \frac{b - b'r}{2b^2} > 0 \,. \tag{10}$$

At the throat we verify that the form function satisfies the condition $b'(r_0) < 1$. We will see below that this condition plays a fundamental role in the analysis of the violation of the energy conditions.

This treatment has the drawback of being highly coordinate dependent. However, for a covariant treatment we follow the analysis by Hochberg and Visser [79, 198]. Consider a generic static spacetime given by the following metric

$$ds^2 = g_{\mu\nu}\, dx^\mu\, dx^\nu = -e^{2\Phi(r)}\, dt^2 + g_{ij}\, dx^i\, dx^j \,. \tag{11}$$

Consider that Greek indices run from 0 to 3; latin indices $(i, j, k, ...)$ run from 1 to 3; and $(a, b, c, ...)$ run from 1 to 2 and refer to the wormhole throat and direction parallel to the throat.

A wormhole throat Σ is defined to be a two-dimensional hypersurface of minimal area taken in one of the constant-time spatial slices, and is given by

$$A(\Sigma) = \int \sqrt{^{(2)}g}\;\, d^2x \,. \tag{12}$$

The two-surface is embedded in a three-dimensional space, so that the definition of the extrinsic curvature is well defined. Consider Gaussian coordinates $x^i = (x^a, n)$, so that the hypersurface Σ lies at $n = 0$, the three-dimensional spatial metric is given by

$${}^{(3)}g_{ij}\, dx^i\, dx^j = {}^{(2)}g_{ab}\, dx^a\, dx^b + dn^2 \,. \tag{13}$$

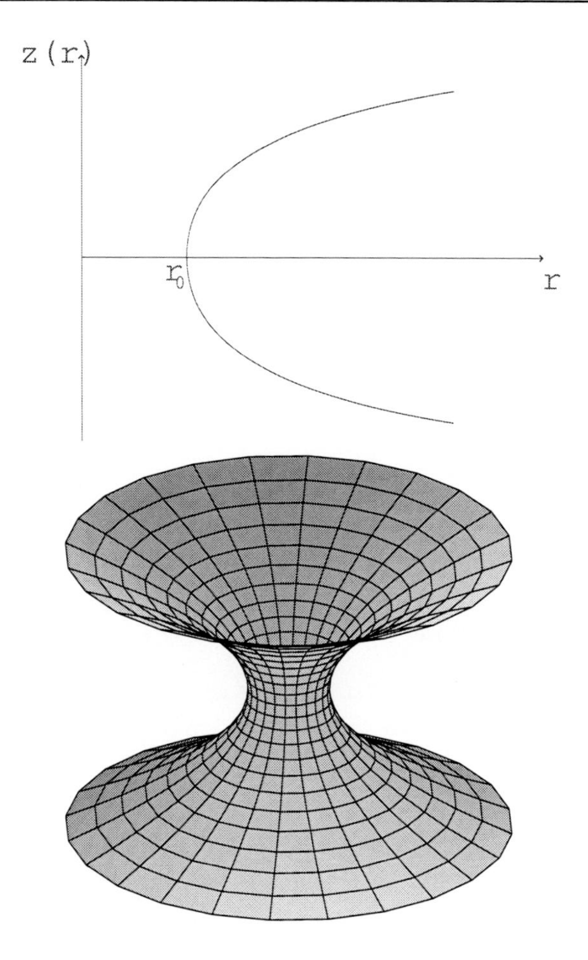

Figure 1. The embedding diagram of a two-dimensional section along the equatorial plane ($t = \text{const}, \theta = \pi/2$) of a traversable wormhole. For a full visualization of the surface sweep through a 2π rotation around the $z-$axis, as can be seen from the graphic on the right.

The variation in surface area is given by

$$\delta A(\Sigma) = \int \frac{\partial \sqrt{^{(2)}g}}{\partial n}\, \delta n(x)\, d^2x = \int \sqrt{^{(2)}g}\, \frac{1}{2}\, g^{ab}\, \frac{\partial g_{ab}}{\partial n}\, \delta n(x)\, d^2x\,. \tag{14}$$

Using Gaussian normal coordinates the extrinsic curvature is defined by [199]

$$K_{ab} = -\frac{1}{2}\,\frac{\partial g_{ab}}{\partial n}\,, \tag{15}$$

which substituted into eq. (14), provides

$$\delta A(\Sigma) = -\int \sqrt{^{(2)}g}\ \text{tr}(K)\, \delta n(x)\, d^2x\,, \tag{16}$$

where the definition $\text{tr}(K) = g^{ab}K_{ab}$ is used. The condition that the area be extremal, for arbitrary $\delta n(x)$, is then simply $\text{tr}(K) = 0$.

For the area to be minimal, we have the additional requirement that $\delta^2 A(\Sigma) > 0$. We have then

$$\delta^2 A(\Sigma) = -\int \sqrt{^{(2)}g}\,\left(\frac{\partial\,\mathrm{tr}(K)}{\partial n} - \mathrm{tr}(K)^2\right)\,\delta n(x)\,\delta n(x)\,d^2x\,. \tag{17}$$

Taking into account the extremal condition, $\mathrm{tr}(K) = 0$, reduces the minimality constraint to

$$\delta^2 A(\Sigma) = -\int \sqrt{^{(2)}g}\,\left(\frac{\partial\,\mathrm{tr}(K)}{\partial n}\right)\,\delta n(x)\,\delta n(x)\,d^2x\,. \tag{18}$$

Considering an arbitrary $\delta n(x)$, at the throat we have the following important condition

$$\frac{\partial\,\mathrm{tr}(K)}{\partial n} < 0\,, \tag{19}$$

which is the covariant generalization of the Morris-Thorne flaring-out condition to arbitrary static wormhole throats.

2.3. Einstein Field Equation

The mathematical analysis and the physical interpretation will be simplified using a set of orthonormal basis vectors. These may be interpreted as the proper reference frame of a set of observers who remain at rest in the coordinate system (t, r, θ, ϕ), with (r, θ, ϕ) fixed. Denote the basis vectors in the coordinate system as $(\mathbf{e}_t, \mathbf{e}_r, \mathbf{e}_\theta, \mathbf{e}_\phi)$. Thus, the orthonormal basis vectors are given by

$$\begin{cases} \mathbf{e}_{\hat{t}} = e^{-\Phi}\,\mathbf{e}_t \\ \mathbf{e}_{\hat{r}} = (1 - b/r)^{1/2}\,\mathbf{e}_r \\ \mathbf{e}_{\hat{\theta}} = r^{-1}\,\mathbf{e}_\theta \\ \mathbf{e}_{\hat{\phi}} = (r\sin\theta)^{-1}\,\mathbf{e}_\phi\,. \end{cases} \tag{20}$$

The Einstein tensor, given in the orthonormal reference frame by $G_{\hat{\mu}\hat{\nu}} = R_{\hat{\mu}\hat{\nu}} - \frac{1}{2}R\,g_{\hat{\mu}\hat{\nu}}$, yields for the metric (1), the following non-zero components

$$G_{\hat{t}\hat{t}} = \frac{b'}{r^2}\,, \tag{21}$$

$$G_{\hat{r}\hat{r}} = -\frac{b}{r^3} + 2\left(1 - \frac{b}{r}\right)\frac{\Phi'}{r}\,, \tag{22}$$

$$G_{\hat{\theta}\hat{\theta}} = \left(1 - \frac{b}{r}\right)\left[\Phi'' + (\Phi')^2 - \frac{b'r - b}{2r(r-b)}\Phi' - \frac{b'r - b}{2r^2(r-b)} + \frac{\Phi'}{r}\right]\,, \tag{23}$$

$$G_{\hat{\phi}\hat{\phi}} = G_{\hat{\theta}\hat{\theta}}\,. \tag{24}$$

The Einstein field equation, $G_{\hat{\mu}\hat{\nu}} = 8\pi T_{\hat{\mu}\hat{\nu}}$, stipulates that the stress energy tensor $T_{\hat{\mu}\hat{\nu}}$ should be proportional to the Einstein tensor. Thus $T_{\hat{\mu}\hat{\nu}}$ has the same algebraic structure as $G_{\hat{\mu}\hat{\nu}}$, eqs. (21)-(24), and the only nonzero components are precisely the diagonal terms $T_{\hat{t}\hat{t}}$, $T_{\hat{r}\hat{r}}$, $T_{\hat{\theta}\hat{\theta}}$ and $T_{\hat{\phi}\hat{\phi}}$. Using the orthonormal basis, these components carry a simple physical interpretation, i.e.,

$$T_{\hat{t}\hat{t}} = \rho(r)\,, \qquad T_{\hat{r}\hat{r}} = -\tau(r)\,, \qquad T_{\hat{\theta}\hat{\theta}} = T_{\hat{\phi}\hat{\phi}} = p(r)\,, \tag{25}$$

in which $\rho(r)$ is the energy density, $\tau(r)$ is the radial tension, with $\tau(r) = -p_r(r)$, i.e., it is the negative of the radial pressure, $p(r)$ is the pressure measured in the tangential directions, orthogonal to the radial direction.

Using the Einstein field equation, $G_{\hat{\mu}\hat{\nu}} = 8\pi T_{\hat{\mu}\hat{\nu}}$, we verify the following stress-energy scenario

$$\rho(r) = \frac{1}{8\pi}\frac{b'}{r^2}, \tag{26}$$

$$\tau(r) = \frac{1}{8\pi}\left[\frac{b}{r^3} - 2\left(1-\frac{b}{r}\right)\frac{\Phi'}{r}\right], \tag{27}$$

$$p(r) = \frac{1}{8\pi}\left(1-\frac{b}{r}\right)\left[\Phi'' + (\Phi')^2 - \frac{b'r-b}{2r^2(1-b/r)}\Phi' - \frac{b'r-b}{2r^3(1-b/r)} + \frac{\Phi'}{r}\right]. \tag{28}$$

Evaluated at the throat they assume the following simplified form

$$\rho(r_0) = \frac{1}{8\pi}\frac{b'(r_0)}{r_0^2}, \tag{29}$$

$$\tau(r_0) = \frac{1}{8\pi r_0^2}, \tag{30}$$

$$p(r_0) = \frac{1}{8\pi}\frac{1-b'(r_0)}{2r_0^2}\left(1 + r_0\Phi'(r_0)\right). \tag{31}$$

Integrating eq. (26), we have

$$b(r) = b(r_0) + \int_{r_0}^{r} 8\pi\,\rho(r')\,r'^2\,dr' = 2m(r). \tag{32}$$

This can be expressed in the following manner

$$m(r) = \frac{r_0}{2} + \int_{r_0}^{r} 4\pi\,\rho(r')\,r'^2\,dr', \tag{33}$$

which is the effective mass contained in the interior of a sphere of radius r. Therefore, the form function has an interpretation which depends on the mass distribution of the wormhole. Moving out to spatial infinity, we have

$$\lim_{r\to\infty} m(r) = \frac{r_0}{2} + \int_{r_0}^{\infty} 4\pi\,\rho(r')\,r'^2\,dr' = M. \tag{34}$$

By taking the derivative with respect to the radial coordinate r, of eq. (27), and eliminating b' and Φ'', given in eq. (26) and eq. (28), respectively, we obtain the following equation

$$\tau' = (\rho - \tau)\Phi' - \frac{2}{r}(p + \tau). \tag{35}$$

Equation (35) is the relativistic Euler equation, or the hydrostatic equation for equilibrium for the material threading the wormhole, and can also be obtained using the conservation of the stress-energy tensor, $T^{\hat{\mu}\hat{\nu}}{}_{;\hat{\nu}} = 0$, inserting $\hat{\mu} = r$. The conservation of the stress-energy tensor, in turn can be deduced from the Bianchi identities,

$$R^{\hat{\alpha}}{}_{\hat{\beta}[\hat{\lambda}\hat{\mu};\hat{\nu}]} = 0, \tag{36}$$

which are equivalent to $G^{\hat{\mu}\hat{\nu}}{}_{;\hat{\nu}} = 0$, also called the contracted Bianchi identities.

2.4. Exotic Matter

To gain some insight into the matter threading the wormhole, Morris and Thorne defined the dimensionless function $\xi = (\tau - \rho)/|\rho|$ [1]. Using equations (26)-(27) one finds

$$\xi = \frac{\tau - \rho}{|\rho|} = \frac{b/r - b' - 2r(1 - b/r)\Phi'}{|b'|} . \tag{37}$$

Combining eq. (37) with the flaring-out condition, eq. (10), the exoticity function takes the form

$$\xi = \frac{2b^2}{r|b'|} \frac{d^2r}{dz^2} - 2r\left(1 - \frac{b}{r}\right)\frac{\Phi'}{|b'|} . \tag{38}$$

Considering the finite character of ρ, and therefore of b', and the fact that $(1 - b/r)\Phi' \to 0$ at the throat, we have the following relationship

$$\xi(r_0) = \frac{\tau_0 - \rho_0}{|\rho_0|} > 0 . \tag{39}$$

The restriction $\tau_0 > \rho_0$ is an extremely troublesome condition, as it states that the radial tension at the throat should exceed the energy density. Thus, Morris and Thorne coined matter restricted by this condition "exotic matter"[1]. We shall verify below that this is matter that violates the null energy condition (in fact, it violates all the energy conditions)[1, 2, 54].

For instance, consider a specific class of particularly simple solutions corresponding to the choice of $b = b(r)$ and $\Phi(r) = 0$ [1]. Equations (26)-(28) reduce to

$$\rho(r) = \frac{b'(r)}{8\pi r^2} , \qquad \tau(r) = \frac{b(r)}{8\pi r^3} , \qquad p(r) = \frac{b(r) - b'r}{16\pi r^3} . \tag{40}$$

Note that the sign of the energy density depends on the sign of $b'(r)$. In particular, consider the form function given by $b(r) = r_0^2/r$. This corresponds to an embedding function $z(r)$ given by $z(r) = r_0 \cosh^{-1}(r/r_0)$, which has the shape of a catenary, i.e.,

$$\frac{dz}{dr} = \frac{r_0}{\sqrt{r^2 - {r_0}^2}} . \tag{41}$$

The wormhole material is everywhere exotic, i.e., $\xi > 0$ everywhere, extending outward from the throat, with ρ, τ, and p tending to zero as $r \to +\infty$.

Exotic matter is particularly troublesome for measurements made by observers traversing through the throat with a radial velocity close to the speed of light. Consider a Lorentz transformation, $x^{\hat{\mu}'} = \Lambda^{\hat{\mu}'}{}_{\hat{\nu}}\, x^{\hat{\nu}}$, with $\Lambda^{\hat{\mu}}{}_{\hat{\alpha}'}\, \Lambda^{\hat{\alpha}'}{}_{\hat{\nu}} = \delta^{\hat{\mu}}{}_{\hat{\nu}}$ and $\Lambda^{\hat{\mu}}{}_{\hat{\nu}'}$ defined as

$$(\Lambda^{\hat{\mu}}{}_{\hat{\nu}'}) = \begin{bmatrix} \gamma & 0 & 0 & \gamma v \\ 0 & 1 & 0 & 0 \\ 0 & 0 & 1 & 0 \\ \gamma v & 0 & 0 & \gamma \end{bmatrix} . \tag{42}$$

The energy density measured by these observers is given by $T_{\hat{0}'\hat{0}'} = \Lambda^{\hat{\mu}}{}_{\hat{0}'}\, \Lambda^{\hat{\nu}}{}_{\hat{0}'}\, T_{\hat{\mu}\hat{\nu}}$, i.e.,

$$T_{\hat{0}'\hat{0}'} = \gamma^2 \left(\rho_0 - v^2 \tau_0\right) , \tag{43}$$

with $\gamma = (1 - v^2)^{-1/2}$. For sufficiently high velocities, $v \to 1$, the observer will measure a negative energy density, $T_{\hat{0}'\hat{0}'} < 0$.

This feature also holds for any traversable, nonspherical and nonstatic wormhole. To see this, one verifies that a bundle of null geodesics that enters the wormhole at one mouth and emerges from the other must have a cross-sectional area that initially increases, and then decreases. This conversion of decreasing to increasing is due to the gravitational repulsion of matter, requiring a negative energy density, through which the bundle of null geodesics traverses.

2.5. Traversability Conditions

We will be interested in specific solutions for traversable wormholes and assume that a traveller of an absurdly advanced civilization, with human traits, begins the trip in a space station in the lower universe, at proper distance $l = -l_1$, and ends up in the upper universe, at $l = l_2$. Assume that the traveller has a radial velocity $v(r)$, as measured by a static observer positioned at r. One may relate the proper distance travelled dl, radius travelled dr, coordinate time lapse dt, and proper time lapse as measured by the observer $d\tau$, by the following relationships

$$v = e^{-\Phi}\,\frac{dl}{dt} = \mp\, e^{-\Phi}\left(1-\frac{b}{r}\right)^{-1/2}\frac{dr}{dt}\,, \tag{44}$$

$$v\,\gamma = \frac{dl}{d\tau} = \mp\left(1-\frac{b}{r}\right)^{-1/2}\frac{dr}{d\tau}\,. \tag{45}$$

It is also important to impose certain conditions at the space stations [1]. Firstly, consider that space is asymptotically flat at the stations, i.e., $b/r \ll 1$. Secondly, the gravitational redshift of signals sent from the stations to infinity should be small, i.e., $\Delta\lambda/\lambda = e^{-\Phi} - 1 \approx -\Phi$, so that $|\Phi| \ll 1$. The condition $|\Phi| \ll 1$ imposes that the proper time at the station equals the coordinate time. Thirdly, the gravitational acceleration measured at the stations, given by $g = -(1 - b/r)^{-1/2}\,\Phi' \simeq -\Phi'$, should be less than or equal to the Earth's gravitational acceleration, $g \leq g_{\oplus}$, so that the condition $|\Phi'| \leq g_{\oplus}$ is met.

For a convenient trip through the wormhole, certain conditions should also be imposed [1]. Firstly, the entire journey should be done in a relatively short time as measured both by the traveller and by observers who remain at rest at the stations. Secondly, the acceleration felt by the traveller should not exceed the Earth's gravitational acceleration, $g_{\oplus}$. Finally, the tidal accelerations between different parts of the traveller's body, should not exceed, once again, Earth's gravity.

Total time in a traversal. The trip should take a relatively short time, for instance Morris and Thorne considered one year, as measured by the traveler and for observers that stay at rest at the space stations, $l = -l_1$ and $l = l_2$, i.e.,

$$\Delta\tau_{\rm traveler} = \int_{-l_1}^{+l_2}\frac{dl}{v\gamma} \leq 1 \text{ year}, \tag{46}$$

$$\Delta t_{\rm space station} = \int_{-l_1}^{+l_2}\frac{dl}{v e^{\Phi}} \leq 1 \text{ year}, \tag{47}$$

respectively.

Acceleration felt by a traveler. An important traversability condition required is that the acceleration felt by the traveller should not exceed Earth's gravity [1]. Consider an orthonormal basis of the traveller's proper reference frame, $(\mathbf{e}_{\hat{0}'}, \mathbf{e}_{\hat{1}'}, \mathbf{e}_{\hat{2}'}, \mathbf{e}_{\hat{3}'})$, given in terms of the orthonormal basis vectors of eqs. (20) of the static observers, by a Lorentz transformation, i.e.,

$$\mathbf{e}_{\hat{0}'} = \gamma\, \mathbf{e}_{\hat{t}} \mp \gamma\, v\, \mathbf{e}_{\hat{r}}\,, \qquad \mathbf{e}_{\hat{1}'} = \mp \gamma\, \mathbf{e}_{\hat{r}} + \gamma\, v\, \mathbf{e}_{\hat{t}}\,, \qquad \mathbf{e}_{\hat{2}'} = \mathbf{e}_{\hat{\theta}}\,, \qquad \mathbf{e}_{\hat{3}'} = \mathbf{e}_{\hat{\phi}}\,, \tag{48}$$

where $\gamma = (1-v^2)^{-1/2}$, and $v(r)$ being the velocity of the traveller as he passes r, as measured by a static observer positioned there. Thus, the traveller's four-acceleration expressed in his proper reference frame, $a^{\hat{\mu}'} = U^{\hat{\nu}'} U^{\hat{\mu}'}{}_{;\hat{\nu}'}$, yields the following restriction

$$|\vec{a}| = \left| \left(1 - \frac{b}{r}\right)^{1/2} e^{-\Phi} \left(\gamma\, e^{\Phi}\right)' \right| \leq g_{\oplus}\,. \tag{49}$$

For the particular case of $\Phi' = 0$, this restriction reduces to

$$|\vec{a}| = \left| \left(1 - \frac{b}{r}\right)^{1/2} \gamma' c^2 \right| \leq g_{\oplus}\,. \tag{50}$$

For observers traversing the wormhole with a constant velocity, $v = \text{const}$, one has $|\vec{a}| = 0$, of course!

Tidal acceleration felt by a traveler. Its important that an observer traversing through the wormhole should not be ripped apart by enormous tidal forces. Thus, another of the traversability conditions required is that the tidal accelerations felt by the traveller should not exceed, for instance, the Earth's gravitational acceleration [1]. The tidal acceleration felt by the traveller is given by $\Delta a^{\hat{\mu}'} = -R^{\hat{\mu}'}{}_{\hat{\nu}'\hat{\alpha}'\hat{\beta}'}\, U^{\hat{\nu}'} \eta^{\hat{\alpha}'} U^{\hat{\beta}'}$, where $U^{\hat{\mu}'} = \delta^{\hat{\mu}'}{}_{\hat{0}'}$ is the traveller's four velocity and $\eta^{\hat{\alpha}'}$ is the separation between two arbitrary parts of his body. Note that $\eta^{\hat{\alpha}'}$ is purely spatial in the traveller's reference frame, as $U^{\hat{\mu}'} \eta_{\hat{\mu}'} = 0$, so that $\eta^{\hat{0}'} = 0$. For simplicity, assume that $|\eta^{\hat{i}'}| \approx 2\,\mathrm{m}$ along any spatial direction in the traveller's reference frame. Taking into account the antisymmetric nature of $R^{\hat{\mu}'}{}_{\hat{\nu}'\hat{\alpha}'\hat{\beta}'}$ in its first two indices, we verify that $\Delta a^{\hat{\mu}'}$ is purely spatial with the components

$$\Delta a^{\hat{i}'} = -R^{\hat{i}'}{}_{\hat{0}'\hat{j}'\hat{0}'}\, \eta^{\hat{j}'} = -R_{\hat{i}'\hat{0}'\hat{j}'\hat{0}'}\, \eta^{\hat{j}'}\,. \tag{51}$$

By using a Lorentz transformation of the Riemann tensor components in the static observer's frame, $(\mathbf{e}_{\hat{t}}, \mathbf{e}_{\hat{r}}, \mathbf{e}_{\hat{\theta}}, \mathbf{e}_{\hat{\phi}})$, to the traveller's frame, $(\mathbf{e}_{\hat{0}'}, \mathbf{e}_{\hat{1}'}, \mathbf{e}_{\hat{2}'}, \mathbf{e}_{\hat{3}'})$, the nonzero components of $R_{\hat{i}'\hat{0}'\hat{j}'\hat{0}'}$ are given by

$$\begin{aligned} R_{\hat{1}'\hat{0}'\hat{1}'\hat{0}'} &= R_{\hat{r}\hat{t}\hat{r}\hat{t}} \\ &= -\left(1 - \frac{b}{r}\right)\left[-\Phi'' - (\Phi')^2 + \frac{b'r - b}{2r(r-b)}\Phi'\right], \end{aligned} \tag{52}$$

$$\begin{aligned} R_{\hat{2}'\hat{0}'\hat{2}'\hat{0}'} &= R_{\hat{3}'\hat{0}'\hat{3}'\hat{0}'} = \gamma^2 R_{\hat{\theta}\hat{t}\hat{\theta}\hat{t}} + \gamma^2 v^2 R_{\hat{\theta}\hat{r}\hat{\theta}\hat{r}} \\ &= \frac{\gamma^2}{2r^2}\left[v^2\left(b' - \frac{b}{r}\right) + 2(r-b)\Phi'\right]. \end{aligned} \tag{53}$$

Thus, eq. (51) takes the form

$$\Delta a^{\hat{1}'} = -R_{\hat{1}'\hat{0}'\hat{1}'\hat{0}'}\,\eta^{\hat{1}'}, \qquad \Delta a^{\hat{2}'} = -R_{\hat{2}'\hat{0}'\hat{2}'\hat{0}'}\,\eta^{\hat{2}'}, \qquad \Delta a^{\hat{3}'} = -R_{\hat{3}'\hat{0}'\hat{3}'\hat{0}'}\,\eta^{\hat{3}'}. \tag{54}$$

The constraint $|\Delta a^{\hat{\mu}'}| \leq g_{\oplus}$ provides the tidal acceleration restrictions as measured by a traveller moving radially through the wormhole, given by the following inequalities

$$\left|\left(1-\frac{b}{r}\right)\left[\Phi'' + (\Phi')^2 - \frac{b'r-b}{2r(r-b)}\Phi'\right]\right| \, |\eta^{\hat{1}'}| \leq g_{\oplus}, \tag{55}$$

$$\left|\frac{\gamma^2}{2r^2}\left[v^2\left(b'-\frac{b}{r}\right) + 2(r-b)\Phi'\right]\right| \, |\eta^{\hat{2}'}| \leq g_{\oplus}. \tag{56}$$

The radial tidal constraint, eq. (55), constrains the redshift function, and the lateral tidal constraint, eq. (56), constrains the velocity with which observers traverse the wormhole. These inequalities are particularly simple at the throat, r_0,

$$|\Phi'(r_0)| \leq \frac{2g_{\oplus}\, r_0}{(1-b')\,|\eta^{\hat{1}'}|}, \tag{57}$$

$$\gamma^2 v^2 \leq \frac{2g_{\oplus}\, r_0^2}{(1-b')\,|\eta^{\hat{2}'}|}, \tag{58}$$

For the particular case of a constant redshift function, $\Phi' = 0$, the radial tidal acceleration is zero, and eq. (56) reduces to

$$\frac{\gamma^2 v^2}{2r^2}\left|\left(b'-\frac{b}{r}\right)\right| \, |\eta^{\hat{2}'}| \leq g_{\oplus}, \tag{59}$$

For this specific case one verifies that stationary observers with $v = 0$ measure null tidal forces.

It is interesting to note that if the tidal forces are velocity independent, then the wormhole is not traversable. For instance, consider the lateral tidal constraint, eq. (56), which is the only component of the tidal acceleration that is velocity dependent. This velocity dependence cancels out if and only if $R_{\hat{\theta}\hat{t}\hat{\theta}\hat{t}} = -R_{\hat{\theta}\hat{r}\hat{\theta}\hat{r}}$ (see eq. (53)), or

$$b' - \frac{b}{r} = -2r\left(1-\frac{b}{r}\right)\Phi'. \tag{60}$$

Integrating this restriction yields

$$e^{2\Phi(r)} = e^{2\Phi(\infty)}\left(1-\frac{b}{r}\right), \tag{61}$$

which indicates that a horizon is present at $r = r_0$, and that the wormhole is not traversable.

2.6. Energy Conditions

2.6.1. Pointwise Energy Conditions

Given the fact that wormhole spacetimes are supported by exotic matter, we shall specify the energy conditions for the specific case in which the stress-energy tensor is diagonal [15], i.e.,

$$T^{\mu\nu} = \mathrm{diag}(\rho, p_1, p_2, p_3)\,, \tag{62}$$

where ρ is the mass density and the p_j are the three principal pressures. In the case that $p_1 = p_2 = p_3$ this reduces to the perfect fluid stress-energy tensor. Although classical forms of matter are believed to obey these energy conditions, it is a well-known fact that they are violated by certain quantum fields, amongst which we may refer to the Casimir effect.

Null energy condition (NEC). The NEC asserts that for any null vector k^μ

$$T_{\mu\nu}k^\mu k^\nu \geq 0\,. \tag{63}$$

In the case of a stress-energy tensor of the form (62), we have

$$\forall i\,, \quad \rho + p_i \geq 0\,. \tag{64}$$

Weak energy condition (WEC). The WEC states that for any timelike vector U^μ

$$T_{\mu\nu}U^\mu U^\nu \geq 0\,. \tag{65}$$

One can physically interpret $T_{\mu\nu}U^\mu U^\nu$ as the energy density measured by any timelike observer with four-velocity U^μ. Thus, the WEC requires that this quantity to be positive. In terms of the principal pressures this gives

$$\rho \geq 0 \quad \text{and} \quad \forall i\,, \quad \rho + p_i \geq 0\,. \tag{66}$$

By continuity, the WEC implies the NEC.

Strong energy condition (SEC). The SEC asserts that for any timelike vector U^μ the following inequality holds

$$\left(T_{\mu\nu} - \frac{T}{2}\, g_{\mu\nu}\right) U^\mu U^\nu \geq 0\,, \tag{67}$$

where T is the trace of the stress energy tensor.

In terms of the diagonal stress energy tensor (62) the SEC reads

$$\forall i\,, \quad \rho + p_i \geq 0 \quad \text{and} \quad \rho + \sum_i p_i \geq 0\,. \tag{68}$$

The SEC implies the NEC but not necessarily the WEC.

Dominant energy condition (DEC). The DEC states that for any timelike vector U^μ

$$T_{\mu\nu}U^\mu U^\nu \geq 0 \quad \text{and} \quad T_{\mu\nu}U^\nu \text{ is not spacelike} \tag{69}$$

These conditions imply that the locally observed energy density be positive and that the energy flux should be timelike or null. The DEC implies the WEC, and therefore the NEC, but not necessarily the SEC. In the case of a stress-energy tensor of the form (62), we have

$$\rho \geq 0 \quad \text{and} \quad \forall i\,, \quad p_i \in [-\rho, +\rho]\,. \tag{70}$$

One may readily verify that wormhole spacetimes violate all the pointwise energy conditions. Taking into account eqs. (26)-(27), we have

$$\rho(r) - \tau(r) = \frac{1}{8\pi}\left[\frac{b'r - b}{r^3} + 2\left(1 - \frac{b}{r}\right)\frac{\Phi'}{r}\right]. \tag{71}$$

Due to the flaring out condition of the throat deduced from the mathematics of embedding, eq. (10), i.e., $(b - b'r)/b^2 > 0$ [1, 91, 2], we verify that at the throat $b(r_0) = r = r_0$, and due to the finiteness of $\Phi(r)$, from eq. (71) we have $\rho(r) - \tau(r) < 0$. From this we verify that all the energy conditions are violated. However, eq. (71) is precisely the definition of the NEC, i.e., $T_{\hat{\mu}\hat{\nu}}k^{\hat{\mu}}k^{\hat{\nu}} = \rho(r) - \tau(r)$, with $k^{\hat{\mu}} = (1, 1, 0, 0)$. Matter that violates the NEC is denoted as "exotic matter".

2.6.2. Averaged Energy Conditions

Violations of the pointwise energy conditions led to the averaging of the energy conditions over timelike or null geodesics [17]. The averaged energy conditions are somewhat weaker than the pointwise energy conditions, as they permit localized violations of the energy conditions, as long on average the energy conditions hold when integrated along timelike or null geodesics.

Averaged null energy condition (ANEC). The ANEC is satisfied along a null curve, Γ, if the following holds

$$\int_\Gamma T_{\mu\nu}k^\mu k^\nu \, d\lambda \geq 0\,, \tag{72}$$

where λ is the generalized affine parameter, and k^μ is a null vector. If the curve Γ is a null geodesic, then λ is reduced to the ordinary affine parameter.

Averaged weak energy condition (AWEC). The AWEC is satisfied along a timelike curve, Γ, if

$$\int_\Gamma T_{\mu\nu}U^\mu U^\nu \, ds \geq 0\,, \tag{73}$$

where s is a parameterization, the proper time of the timelike curve, and U^μ is the respective tangent vector.

It can be shown, under general conditions, that traversable wormholes violate the ANEC in the region of the throat using the Raychaudhuri equation for null geodesics [2, 15, 200].

2.6.3. Volume Integral Quantifier

Unfortunately the ANEC involves a line integral, with dimensions (mass)/(area), not a volume integral, and therefore gives no useful information regarding the "total amount" of energy-condition violating matter. Therefore, this prompted Visser *et al* [84, 85] to propose a "volume integral quantifier" which amounts to calculating the following definite integrals

$$\int T_{\mu\nu}\, U^{\mu}\, U^{\nu}\, dV \quad \text{and} \quad \int T_{\mu\nu}\, k^{\mu}\, k^{\nu}\, dV. \tag{74}$$

The amount of energy condition violations is then the extent that these integrals become negative. A more precise measure was analyzed in Ref. [201] by considering the proper volume in the integral.

To develop the key volume-integral result note that $T_{\hat{\mu}\hat{\nu}}\, k^{\hat{\mu}}\, k^{\hat{\nu}}$, with the null vector given by $k^{\hat{\mu}} = (1, 1, 00)$, can be written in the following manner

$$\rho - \tau = \frac{1}{8\pi r}\left(1 - \frac{b}{r}\right)\left[\ln\left(\frac{e^{2\Phi}}{1 - b/r}\right)\right]' . \tag{75}$$

Thus, integrating by parts, we have

$$\begin{aligned} I_V = \int (\rho - \tau)\, dV &= \left[(r - b)\ln\left(\frac{e^{2\Phi}}{1 - b/r}\right)\right]_{r_0}^{\infty} \\ &\quad - \int_{r_0}^{\infty} (1 - b')\left[\ln\left(\frac{e^{2\Phi}}{1 - b/r}\right)\right] dr. \end{aligned} \tag{76}$$

The boundary term at r_0 vanishes by the construction of the wormhole, and the boundary term at infinity also vanishes because of the assumed asymptotic behaviour. Thus, eq. (76) reduces to

$$I_V = \int (\rho - \tau)\, dV = -\int_{r_0}^{\infty} (1 - b')\left[\ln\left(\frac{e^{2\Phi}}{1 - b/r}\right)\right] dr. \tag{77}$$

This volume-integral theorem provides information about the "total amount" of ANEC violating matter in the spacetime, and one may now consider specific cases, by choosing the form function and the redshift function.

Consider the solution obtained by setting the following choices for the redshift function and form function: $\Phi = 0$ and $b = {r_0}^2/r$, respectively. In fact, this is a solution obtained by Homer Ellis [45] in 1973. The properties are commented in [1] and [50]. Harris showed that it is a solution of the EFE with a stress-energy tensor of a peculiar massless scalar field [51]. In terms of the proper radial distance $l(r)$, the metric takes the form

$$ds^2 = -dt^2 + dl^2 + ({r_0}^2 + l^2)\,(d\theta^2 + \sin^2\theta\, d\phi^2)\,, \tag{78}$$

where $l = \pm(r^2 - {r_0}^2)^{1/2}$. The stress-tensor components are given by

$$\rho = -\tau = -p = -\frac{{r_0}^2}{8\pi r^4} = -\frac{{r_0}^2}{8\pi({r_0}^2 + l^2)^2} \,. \tag{79}$$

Suppose now that the wormhole extends from the throat, r_0, to a radius situated at a. Evaluating the volume integral, eq. (77), one deduces

$$I_V = \int (\rho - \tau)\, dV = \frac{1}{a}\left[\left(a^2 - r_0^2\right)\, \ln\left(1 - \frac{r_0^2}{a^2}\right) + 2r_0(r_0 - a)\right]. \tag{80}$$

Taking the limit as $a \to r_0^+$, one verifies that $\int(\rho - \tau)\, dV \to 0$. Thus, as in the examples presented in [84, 85], one may construct a traversable wormhole with arbitrarily small quantities of ANEC violating matter. The exotic matter threading the wormhole extends from the throat at r_0 to the junction boundary situated at a, where the interior solution is matched to an exterior vacuum spacetime.

2.6.4. Quantum Inequality

Pioneering work by Ford in the late 1970's on a new set of energy constraints [19], led to constraints on negative energy fluxes in 1991 [20]. These eventually culminated in the form of the Quantum Inequality (QI) applied to energy densities, which was introduced by Ford and Roman in 1995 [21]. The QI was proven directly from Quantum Field Theory, in four-dimensional Minkowski spacetime, for free quantized, massless scalar fields and takes the following form

$$\frac{\tau_0}{\pi}\int_{-\infty}^{+\infty} \frac{\langle T_{\mu\nu}U^\mu U^\nu\rangle}{\tau^2 + \tau_0^2}d\tau \geq -\frac{3}{32\pi^2\tau_0^4}\,, \tag{81}$$

in which, U^μ is the tangent to a geodesic observer's wordline; τ is the observer's proper time and τ_0 is a sampling time. The expectation value $\langle\rangle$ is taken with respect to an arbitrary state $|\Psi\rangle$. Contrary to the averaged energy conditions, one does not average over the entire wordline of the observer, but weights the integral with a sampling function of characteristic width, τ_0. The inequality limits the magnitude of the negative energy violations and the time for which they are allowed to exist. The physical interpretation of eq. (81) is that the more negative the energy density is in an interval, the shorter must be the duration of the interval.

The basic applications to curved spacetimes is that these appear flat if restricted to a sufficiently small region. The application of the QI to wormhole geometries is of particular interest [22]. A small spacetime volume around the throat of the wormhole is considered, so that all the dimensions of this volume are much smaller than the minimum proper radius of curvature in the region. Thus, the spacetime can be approximately flat in this region, so that the QI constraints may be applied. The sampling time τ_0 is restricted to be small compared to the local proper radii of curvature and the proper distance to any boundaries in the spacetime.

The Riemann tensor components will play a fundamental role in the analysis that follows. At the throat, the components of the Riemann tensor reduce to

$$R_{\hat{t}\hat{r}\hat{t}\hat{r}}|_{r_0} = \frac{\Phi_0'}{2r_0}\,(1 - b_0')\,, \tag{82}$$

$$R_{\hat{t}\hat{\theta}\hat{t}\hat{\theta}}|_{r_0} = R_{\hat{t}\hat{\phi}\hat{t}\hat{\phi}}|_{r_0} = 0\,, \tag{83}$$

$$R_{\hat{r}\hat{\theta}\hat{r}\hat{\theta}}|_{r_0} = R_{\hat{r}\hat{\phi}\hat{r}\hat{\phi}}|_{r_0} = -\frac{1}{2{r_0}^2}\,(1 - b_0')\,, \tag{84}$$

$$R_{\hat{\theta}\hat{\phi}\hat{\theta}\hat{\phi}}|_{r_0} = \frac{1}{{r_0}^2} \,. \tag{85}$$

All the other components vanish, except for those related to the above by symmetry.

Let the magnitude of the maximum curvature component be $R_{\max}$, and the smallest proper radius of curvature be given by $r_c \approx 1/\sqrt{R_{\max}}$. The QI-bound is applied to a small spacetime volume around the throat of the wormhole such that all dimensions of this volume are much smaller than r_c, the smallest proper radius of curvature anywhere in the region, so that in the absence of boundaries, spacetime can be considered to be approximately Minkowskian in the respective region [22].

As specific example, consider QI-bound applied to the Ellis drainhole geometry. Consider the Ellis drainhole, given by $\Phi = 0$ and $b = {r_0}^2/r$. The metric is given by eq. (78), and the Riemann curvature components are

$$R_{\hat{\theta}\hat{\phi}\hat{\theta}\hat{\phi}} = -R_{\hat{l}\hat{\theta}\hat{l}\hat{\theta}} = -R_{\hat{l}\hat{\phi}\hat{l}\hat{\phi}} = \frac{{r_0}^2}{({r_0}^2 + l^2)^2} \,. \tag{86}$$

Note that all the curvature components are equal in magnitude, and have their maximum magnitude at the throat, i.e., $1/r_0^2$. The same holds true for the stress-tensor components given by eq. (79).

Applying the QI-bound to a static observer at $r = r_0$, and as the energy density seen by this static observer is constant, we have

$$\frac{\tau_0}{\pi}\int_{-\infty}^{\infty}\frac{\langle T_{\mu\nu}u^\mu u^\nu\rangle\, d\tau}{\tau^2 + {\tau_0}^2} = \rho_0 \geq -\frac{3}{32\pi^2\,{\tau_0}^4} \,, \tag{87}$$

where τ is the observer's proper time, and τ_0 is the sampling time. If the sampling time is chosen to be $\tau_0 = f r_m = f r_0 \ll r_c$, with $f \ll 1$ (recall that the QI is applicable if τ_0 is smaller than the local proper radius of curvature), using $\rho_0 = -1/(8\pi r_0^2)$ and from eq. (87), one finally deduces

$$r_0 \leq \frac{l_p}{2f^2} \,, \tag{88}$$

where l_p is the Planck length. For any reasonable choice of f gives a value of r_0 which is not much larger than l_p. For example, for $f \approx 0.01$ one has $r_0 \leq 10^4\, l_p = 10^{-31}$ m. Note from eqs. (79) and (86) that if the spacetime region is such that $l \ll r_0$, then the curvature and stress-tensor components do not change very much [22].

Ford and Roman considered more specific examples by choosing appropriate definitions of length scales. They also found general bounds on the relative size scales of arbitrary static and spherically symmetric Morris-Thorne wormholes, i.e., for generic $\Phi(r)$ and $b(r)$. The results of the analysis is that either the wormhole possesses a throat size which is only slightly larger than the Planck length, or there are large discrepancies in the length scales which characterize the geometry of the wormhole. The analysis imply that generically the exotic matter is confined to an extremely thin band, and/or that large red-shifts are involved, which present severe difficulties for traversability, such as large tidal forces [22].

Due to these results, Ford and Roman argued that the existence of macroscopic traversable wormholes is very improbable. But, there are a series of considerations that can be applied to the QI [26]. Firstly, the QI is only of interest if one is relying on quantum field theory to provide the exotic matter to support the wormhole throat. But there are classical systems (non-minimally coupled scalar fields) that violate the null and the weak energy conditions [31], whilst presenting plausible results when applying the QI. Secondly, even if one relies on quantum field theory to provide exotic matter, the QI does not rule out the existence of wormholes, although they do place serious constraints on the geometry. Thirdly, it may be possible to reformulate the QI in a more transparent covariant notation, and to prove it for arbitrary background geometries.

2.7. Rotating Wormholes

Now, consider the stationary and axially symmetric $(3+1)-$dimensional spacetime, it possesses a time-like Killing vector field, which generates invariant time translations, and a spacelike Killing vector field, which generates invariant rotations with respect to the angular coordinate ϕ. We have the following metric

$$ds^2 = -N^2dt^2 + e^{\mu}\, dr^2 + r^2K^2[d\theta^2 + \sin^2\theta(d\phi - \omega\, dt)^2] \tag{89}$$

where N, K, ω and μ are functions of r and θ [67]. $\omega(r,\theta)$ may be interpreted as the angular velocity $d\phi/dt$ of a particle that falls freely from infinity to the point (r,θ). For simplicity, we shall consider the definition [67]

$$e^{-\mu(r,\theta)} = 1 - \frac{b(r,\theta)}{r}\,, \tag{90}$$

which is well suited to describe a traversable wormhole. Assume that $K(r,\theta)$ is a positive, nondecreasing function of r that determines the proper radial distance R, i.e., $R \equiv rK$ and $R_r > 0$ [67], as for the $(2+1)-$dimensional case. We shall adopt the notation that the subscripts $_r$ and $_\theta$ denote the derivatives in order of r and θ, respectively [67].

Note that an event horizon appears whenever $N = 0$ [67]. The regularity of the functions N, b and K are imposed, which implies that their θ derivatives vanish on the rotation axis, $\theta = 0,\ \pi$, to ensure a non-singular behavior of the metric on the rotation axis. The metric (89) reduces to the Morris-Thorne spacetime metric (1) in the limit of zero rotation and spherical symmetry

$$N(r,\theta) \to e^{\Phi(r)}, \quad b(r,\theta) \to b(r)\,, \quad K(r,\theta) \to 1\,, \quad \omega(r,\theta) \to 0\,. \tag{91}$$

In analogy with the Morris-Thorne case, $b(r_0) = r_0$ is identified as the wormhole throat, and the factors N, K and ω are assumed to be well-behaved at the throat.

The scalar curvature of the space-time (89) is extremely messy, but at the throat $r = r_0$ simplifies to

$$\begin{aligned} R \;=\; & -\frac{1}{r^2K^2}\left(\mu_{\theta\theta} + \frac{1}{2}\mu_\theta^2\right) - \frac{\mu_\theta}{Nr^2K^2}\,\frac{(N\sin\theta)_\theta}{\sin\theta} \\ & - \frac{2}{Nr^2K^2}\,\frac{(N_\theta\sin\theta)_\theta}{\sin\theta} - \frac{2}{r^2K^3}\,\frac{(K_\theta\sin\theta)_\theta}{\sin\theta} \\ & + e^{-\mu}\,\mu_r\left[\ln(Nr^2K^2)\right]_r + \frac{\sin^2\theta\,\omega_\theta^2}{2N^2} + \frac{2}{r^2K^4}\,(K^2 + K_\theta^2)\,. \end{aligned} \tag{92}$$

The only troublesome terms are the ones involving the terms with μ_θ and $\mu_{\theta\theta}$, i.e.,

$$\mu_\theta = \frac{b_\theta}{(r-b)}\,, \qquad \mu_{\theta\theta} + \frac{1}{2}\mu_\theta^2 = \frac{b_{\theta\theta}}{r-b} + \frac{3}{2}\frac{{b_\theta}^2}{(r-b)^2}\,. \tag{93}$$

Note that one needs to impose that $b_\theta = 0$ and $b_{\theta\theta} = 0$ at the throat to avoid curvature singularities. This condition shows that the throat is located at a constant value of r.

Thus, one may conclude that the metric (89) describes a rotating wormhole geometry, with an angular velocity ω. The factor K determines the proper radial distance. N is the analog of the redshift function in the Morris-Thorne wormhole and is finite and nonzero to ensure that there are no event horizons or curvature singularities. b is the shape function which satisfies $b \leq r$; it is independent of θ at the throat, i.e., $b_\theta = 0$; and obeys the flaring out condition $b_r < 1$.

The analysis is simplified using an orthonormal reference frame, with the following orthonormal basis vectors

$$\mathbf{e}_{\hat{t}} = \frac{1}{N}\,\mathbf{e}_t + \frac{\omega}{N}\,\mathbf{e}_\phi\,, \quad \mathbf{e}_{\hat{r}} = \left(1-\frac{b}{r}\right)^{1/2}\mathbf{e}_r\,, \quad \mathbf{e}_{\hat{\theta}} = \frac{1}{rK}\,\mathbf{e}_\theta\,, \quad \mathbf{e}_{\hat{\phi}} = \frac{1}{rK\sin\theta}\,\mathbf{e}_\phi\,. \tag{94}$$

Now the stress-energy tensor components are extremely messy, but assume a more simplified form using the orthonormal reference frame and evaluated at the throat. They have the following non-zero components

$$\begin{aligned}
8\pi T_{\hat{t}\hat{t}} &= -\frac{(K_\theta\sin\theta)_\theta}{r^2K^3\sin\theta} - \frac{\omega_\theta^2\sin^2\theta}{4N^2} + e^{-\mu}\,\mu_r\,\frac{(rK)_r}{rK} + \frac{K^2+K_\theta^2}{r^2K^4}\,, &(95)\\
8\pi T_{\hat{r}\hat{r}} &= \frac{(K_\theta\sin\theta)_\theta}{r^2K^3\sin\theta} - \frac{\omega_\theta^2\sin^2\theta}{4N^2} + \frac{(N_\theta\sin\theta)_\theta}{Nr^2K^2\sin\theta} - \frac{K^2+K_\theta^2}{r^2K^4}\,, &(96)\\
8\pi T_{\hat{\theta}\hat{\theta}} &= \frac{N_\theta(K\sin\theta)_\theta}{Nr^2K^3\sin\theta} + \frac{\omega_\theta^2\sin^2\theta}{4N^2} - \frac{\mu_r\,e^{-\mu}(NrK)_r}{2NrK}\,, &(97)\\
8\pi T_{\hat{\phi}\hat{\phi}} &= -\frac{\mu_r\,e^{-\mu}\,(NKr)_r}{2NKr} - \frac{3\sin^2\theta\,\omega_\theta^2}{4N^2} + \frac{N_{\theta\theta}}{Nr^2K^2} - \frac{N_\theta K_\theta}{Nr^2K^3}\,, &(98)\\
8\pi T_{\hat{t}\hat{\phi}} &= \frac{1}{4N^2K^2r}\Big(6NK\,\omega_\theta\,\cos\theta + 2NK\,\sin\theta\,\omega_{\theta\theta} \\
&\quad -\mu_r e^{-\mu} r^2 NK^3\,\sin\theta\,\omega_r + 4N\,\omega_\theta\,\sin\theta\,K_\theta - 2K\,\sin\theta\,N_\theta\,\omega_\theta\Big)\,. &(99)
\end{aligned}$$

The components $T_{\hat{t}\hat{t}}$ and $T_{\hat{i}\hat{j}}$ have the usual physical interpretations, and in particular, $T_{\hat{t}\hat{\phi}}$ characterizes the rotation of the matter distribution. Taking into account the Einstein tensor components above, the NEC at the throat is given by

$$8\pi\,T_{\hat{\mu}\hat{\nu}}k^{\hat{\mu}}k^{\hat{\nu}} = \mathrm{e}^{-\mu}\mu_r\,\frac{(rK)_r}{rK} - \frac{{\omega_\theta}^2\sin^2\theta}{2N^2} + \frac{(N_\theta\sin\theta)_\theta}{(rK)^2N\sin\theta}\,. \tag{100}$$

Rather than reproduce the analysis here, we refer the reader to Ref. [67], where it was shown that the NEC is violated in certain regions, and is satisfied in others. Thus, it is possible for an infalling observer to move around the throat, and avoid the exotic matter supporting the wormhole. However, it is important to emphasize that one cannot avoid the use of exotic matter altogether.

2.8. Evolving Wormholes in a Cosmological Background

Consider the metric element of a wormhole in a cosmological background given by

$$ds^2 = \Omega^2(t)\left[-e^{2\Phi(r)}\,dt^2 + \frac{dr^2}{1-kr^2-\frac{b(r)}{r}} + r^2\left(d\theta^2+\sin^2\theta\,d\phi^2\right)\right] \tag{101}$$

where $\Omega^2(t)$ is the conformal factor, which is finite and positive definite throughout the domain of t. It is also possible to write the metric (101) using "physical time" instead of "conformal time", by replacing t by $\tau = \int \Omega(t)dt$ and therefore $\Omega(t)$ by $R(\tau)$, where the latter is the functional form of the metric in the τ coordinate [81, 82]. When the form function and the redshift function vanish, $b(r) \to 0$ and $\Phi(r) \to 0$, respectively, the metric (101) becomes the FRW metric. As $\Omega(t) \to$ const and $k \to 0$, it approaches the static wormhole metric, eq. (1).

The Einstein field equation will be written $G_{\hat{\mu}\hat{\nu}} = R_{\hat{\mu}\hat{\nu}} - \frac{1}{2}g_{\hat{\mu}\hat{\nu}}R = 8\pi T_{\hat{\mu}\hat{\nu}}$,, in an orthonormal reference frame, so that any cosmological constant terms will be incorporated as part of the stress-energy tensor $T_{\hat{\mu}\hat{\nu}}$. The components of the stress-energy tensor $T_{\hat{\mu}\hat{\nu}}$ are given by

$$T_{\hat{t}\hat{t}} = \rho(r,t)\,, \qquad T_{\hat{r}\hat{r}} = -\tau(r,t)\,, \qquad T_{\hat{t}\hat{r}} = -f(r,t)\,, \qquad T_{\hat{\phi}\hat{\phi}} = T_{\hat{\theta}\hat{\theta}} = p(r,t)\,, \tag{102}$$

with

$$\rho(r,t) = \frac{1}{8\pi}\frac{1}{\Omega^2}\left[3e^{-2\Phi}\left(\frac{\dot{\Omega}}{\Omega}\right)^2 + \left(3k+\frac{b'}{r^2}\right)\right], \tag{103}$$

$$\tau(r,t) = -\frac{1}{8\pi} \times \frac{1}{\Omega^2}\left\{e^{-2\Phi(r)}\left[\left(\frac{\dot{\Omega}}{\Omega}\right)^2 - 2\frac{\ddot{\Omega}}{\Omega}\right] - \left[k+\frac{b}{r^3} - 2\frac{\Phi'}{r}\left(1-kr^2-\frac{b}{r}\right)\right]\right\}, \tag{104}$$

$$f(r,t) = -\frac{1}{8\pi}\left[2\frac{\dot{\Omega}}{\Omega^3}\,e^{-\Phi}\,\Phi'\left(1-kr^2-\frac{b}{r}\right)^{1/2}\right], \tag{105}$$

$$p(r,t) = \frac{1}{8\pi}\frac{1}{\Omega^2}\left\{e^{-2\Phi(r)}\left[\left(\frac{\dot{\Omega}}{\Omega}\right)^2 - 2\frac{\ddot{\Omega}}{\Omega}\right] + \left(1-kr^2-\frac{b}{r}\right)\times \left[\Phi'' + (\Phi')^2 - \frac{2kr^3+b'r-b}{2r(r-kr^3-b)}\Phi' - \frac{2kr^3+b'r-b}{2r^2(r-kr^3-b)} + \frac{\Phi'}{r}\right]\right\}. \tag{106}$$

The overdot denotes a derivative with respect to t, and the prime a derivative with respect to r. The physical interpretation of $\rho(r,t)$, $\tau(r,t)$, $f(r,t)$, and $p(r,t)$ are the following: the energy density, the radial tension per unit area, energy flux in the (outward) radial direction, and lateral pressures as measured by observers stationed at constant r, θ, ϕ, respectively. The stress-energy tensor has a non-diagonal component due to the time dependence of $\Omega(t)$ and/or the dependence of the redshift function on the radial coordinate. The stress-energy tensor of an imperfect fluid was analyzed in [202].

A particularly interesting case of the metric (101) is that of a wormhole in a time-dependent inflationary background, considered by Thomas Roman [89]. The primary goal in the Roman analysis was to use inflation to enlarge an initially small [89], possibly sub-microscopic, wormhole. $\Phi(r)$ and $b(r)$ are chosen to give a reasonable wormhole at $t = 0$, which is assumed to be the onset of inflation. Roman [89] went on to explore interesting properties of the inflating wormholes, in particular, by analyzing constraints placed on the initial size of the wormhole, if the mouths were to remain in causal contact throughout the inflationary period; and the maintenance of the wormhole during and after the decay of the false vacuum. It is also possible that the wormhole will continue to be enlarged by the subsequent FRW phase of expansion. One could perform a similar analysis to ours by replacing the deSitter scale factor by an FRW scale factor $a(t)$ [81, 82, 83]. In particular, in Refs.[81, 82] specific examples for evolving wormholes that exist only for a finite time were considered, and a special class of scale factors which exhibit 'flashes' of the WEC violation were also analyzed.

2.9. Thin Shells

Consider two distinct spacetime manifolds, $\mathcal{M}_+$ and $\mathcal{M}_-$, with metrics given by $g^+_{\mu\nu}(x^\mu_+)$ and $g^-_{\mu\nu}(x^\mu_-)$, in terms of independently defined coordinate systems x^μ_+ and x^μ_-. The manifolds are bounded by hypersurfaces Σ_+ and Σ_-, respectively, with induced metrics g^+_{ij} and g^-_{ij}. The hypersurfaces are isometric, i.e., $g^+_{ij}(\xi) = g^-_{ij}(\xi) = g_{ij}(\xi)$, in terms of the intrinsic coordinates, invariant under the isometry. A single manifold $\mathcal{M}$ is obtained by gluing together $\mathcal{M}_+$ and $\mathcal{M}_-$ at their boundaries, i.e., $\mathcal{M} = \mathcal{M}_+ \cup \mathcal{M}_-$, with the natural identification of the boundaries $\Sigma = \Sigma_+ = \Sigma_-$. In particular, assuming the continuity of the four-dimensional coordinates $x^\mu_\pm$ across Σ, then $g^-_{\mu\nu} = g^+_{\mu\nu}$ is required, which together with the continuous derivatives of the metric components $\partial g_{\mu\nu}/\partial x^\alpha|_- = \partial g_{\mu\nu}/\partial x^\alpha|_+$, provide the Lichnerowicz conditions [203].

The three holonomic basis vectors $\mathbf{e}_{(i)} = \partial/\partial\xi^i$ tangent to Σ have the following components $e^\mu_{(i)}|_\pm = \partial x^\mu_\pm/\partial\xi^i$, which provide the induced metric on the junction surface by the following scalar product

$$g_{ij} = \mathbf{e}_{(i)} \cdot \mathbf{e}_{(j)} = g_{\mu\nu} e^\mu_{(i)} e^\nu_{(j)}|_\pm . \tag{107}$$

We shall consider a timelike junction surface Σ, defined by the parametric equation of the form $f(x^\mu(\xi^i)) = 0$. The unit normal 4–vector, n^μ, to Σ is defined as

$$n_\mu = \pm \left| g^{\alpha\beta} \frac{\partial f}{\partial x^\alpha} \frac{\partial f}{\partial x^\beta} \right|^{-1/2} \frac{\partial f}{\partial x^\mu} , \tag{108}$$

with $n_\mu \, n^\mu = +1$ and $n_\mu e^\mu_{(i)} = 0$. The Israel formalism requires that the normals point from $\mathcal{M}_-$ to $\mathcal{M}_+$ [204].

The extrinsic curvature, or the second fundamental form, is defined as $K_{ij} = n_{\mu;\nu} e^\mu_{(i)} e^\nu_{(j)}$, or

$$K^\pm_{ij} = -n_\mu \left(\frac{\partial^2 x^\mu}{\partial\xi^i \, \partial\xi^j} + \Gamma^{\mu\pm}{}_{\alpha\beta} \frac{\partial x^\alpha}{\partial\xi^i} \frac{\partial x^\beta}{\partial\xi^j} \right) . \tag{109}$$

Note that for the case of a thin shell K_{ij} is not continuous across Σ, so that for notational convenience, the discontinuity in the second fundamental form is defined as $\kappa_{ij} = K^{+}_{ij} - K^{-}_{ij}$. In particular, the condition that $g^{-}_{ij} = g^{+}_{ij}$, together with the continuity of the extrinsic curvatures across Σ, $K^{-}_{ij} = K^{+}_{ij}$, provide the Darmois conditions [205].

Now, the Lanczos equations follow from the Einstein equations for the hypersurface, and are given by

$$S^{i}{}_{j} = -\frac{1}{8\pi}\left(\kappa^{i}{}_{j} - \delta^{i}{}_{j}\kappa^{k}{}_{k}\right), \tag{110}$$

where $S^{i}{}_{j}$ is the surface stress-energy tensor on Σ.

The first contracted Gauss-Kodazzi equation or the "Hamiltonian" constraint

$$G_{\mu\nu}n^{\mu}n^{\nu} = \frac{1}{2}\left(K^2 - K_{ij}K^{ij} - {}^{3}R\right), \tag{111}$$

with the Einstein equations provide the evolution identity

$$S^{ij}\overline{K}_{ij} = -\left[T_{\mu\nu}n^{\mu}n^{\nu} - \Lambda/8\pi\right]^{+}_{-}. \tag{112}$$

The convention $[X]^{+}_{-} \equiv X^{+}|_{\Sigma} - X^{-}|_{\Sigma}$ and $\overline{X} \equiv (X^{+}|_{\Sigma} + X^{-}|_{\Sigma})/2$ is used.

The second contracted Gauss-Kodazzi equation or the "ADM" constraint

$$G_{\mu\nu}e^{\mu}_{(i)}n^{\nu} = K^{j}_{i|j} - K_{,i}\,, \tag{113}$$

with the Lanczos equations gives the conservation identity

$$S^{i}_{j|i} = \left[T_{\mu\nu}e^{\mu}_{(j)}n^{\nu}\right]^{+}_{-}. \tag{114}$$

The momentum flux term in the right hand side corresponds to the net discontinuity in the momentum which impinges on the shell.

In particular, considering spherical symmetry considerable simplifications occur, namely $\kappa^{i}{}_{j} = \text{diag}\left(\kappa^{\tau}{}_{\tau}, \kappa^{\theta}{}_{\theta}, \kappa^{\theta}{}_{\theta}\right)$. The surface stress-energy tensor may be written in terms of the surface energy density, σ, and the surface pressure, $\mathcal{P}$, as $S^{i}{}_{j} = \text{diag}(-\sigma, \mathcal{P}, \mathcal{P})$. The Lanczos equations then reduce to

$$\sigma = -\frac{1}{4\pi}\kappa^{\theta}{}_{\theta}\,, \tag{115}$$

$$\mathcal{P} = \frac{1}{8\pi}\left(\kappa^{\tau}{}_{\tau} + \kappa^{\theta}{}_{\theta}\right). \tag{116}$$

Taking into account the wormhole spacetime metric (1) and the Schwarzschild solution, the non-trivial components of the extrinsic curvature are given by

$$K^{\tau\,+}_{\ \tau} = \frac{\frac{M}{a^2} + \ddot{a}}{\sqrt{1 - \frac{2M}{a} + \dot{a}^2}}\,, \tag{117}$$

$$K^{\tau\,-}_{\ \tau} = \frac{\Phi'\left(1 - \frac{b}{a} + \dot{a}^2\right) + \ddot{a} - \frac{\dot{a}^2(b - b'a)}{2a(a-b)}}{\sqrt{1 - \frac{b(a)}{a} + \dot{a}^2}}\,, \tag{118}$$

and

$$K^{\theta\ +}_{\ \theta} = \frac{1}{a}\sqrt{1-\frac{2M}{a}+\dot{a}^2}\,, \tag{119}$$

$$K^{\theta\ -}_{\ \theta} = \frac{1}{a}\sqrt{1-\frac{b(a)}{a}+\dot{a}^2}\,. \tag{120}$$

The Lanczos equation, Eq. (110), then provide us with the following expressions for the surface stresses

$$\sigma = -\frac{1}{4\pi a}\left(\sqrt{1-\frac{2M}{a}+\dot{a}^2}-\sqrt{1-\frac{b(a)}{a}+\dot{a}^2}\right), \tag{121}$$

$$\mathcal{P} = \frac{1}{8\pi a}\left[\frac{1-\frac{M}{a}+\dot{a}^2+a\ddot{a}}{\sqrt{1-\frac{2M}{a}+\dot{a}^2}}-\frac{(1+a\Phi')\left(1-\frac{b}{a}+\dot{a}^2\right)+a\ddot{a}-\frac{\dot{a}^2(b-b'a)}{2(a-b)}}{\sqrt{1-\frac{b(a)}{a}+\dot{a}^2}}\right] \tag{122}$$

where σ and $\mathcal{P}$ are the surface energy density and the tangential surface pressure, respectively.

Using $S^{i}_{\tau|i} = -\left[\dot{\sigma}+2\dot{a}(\sigma+\mathcal{P})/a\right]$, Eq. (114) provides us with

$$\sigma' = -\frac{2}{a}\left(\sigma+\mathcal{P}\right)+\Xi\,, \tag{123}$$

where Ξ, defined for notational convenience, is given by

$$\Xi = -\frac{1}{4\pi a^2}\left[\frac{b'a-b}{2a\left(1-\frac{b}{a}\right)}+a\Phi'\right]\sqrt{1-\frac{b}{a}+\dot{a}^2}\,. \tag{124}$$

For self-completeness, we shall also include the $\sigma+\mathcal{P}$ term, which is given by

$$\sigma+\mathcal{P} = \frac{1}{8\pi a}\left[\frac{(1-a\Phi')\left(1-\frac{b}{a}+\dot{a}^2\right)-a\ddot{a}+\frac{\dot{a}^2(b-b'a)}{2(a-b)}}{\sqrt{1-\frac{b(a)}{a}+\dot{a}^2}}-\frac{1-\frac{3M}{a}+\dot{a}^2-a\ddot{a}}{\sqrt{1-\frac{2M}{a}+\dot{a}^2}}\right]. \tag{125}$$

Thus, taking into account Eq. (125), and the definition of Ξ, we verify that Eq. (123) finally takes the form

$$\sigma' = \frac{1}{4\pi a^2}\left(\frac{1-\frac{3M}{a}+\dot{a}^2-a\ddot{a}}{\sqrt{1-\frac{2M}{a}+\dot{a}^2}}-\frac{1-\frac{3b}{2a}+\frac{b'}{2}+\dot{a}^2-a\ddot{a}}{\sqrt{1-\frac{b}{a}+\dot{a}^2}}\right), \tag{126}$$

which, evaluated at a static solution a_0, shall play a fundamental role in determining the stability regions. Note that Eq. (126) can also be deduced by taking the radial derivative of the surface energy density, Eq. (121).

The construction of dynamic shells in wormholes have been extensively analyzed in Ref. [206], where the stability of generic spherically symmetric thin shells to linearized perturbations around static solutions were considered, and applying the analysis to traversable wormhole geometries, by considering specific choices for the form function, the stability regions were deduced. It was found that the latter may be significantly increased by considering appropriate choices for the redshift function (The linearized stability analysis was also applied to dark energy stars [207]).

2.10. Late-Time Cosmic Accelerated Expansion and Traversable Wormholes

In this section, we shall explore the possibility that traversable wormholes be supported by specific equations of state responsible for the late time accelerated expansion of the Universe, namely, phantom energy, the generalized Chaplygin gas, and the van der Waals quintessence equation of state. Firstly, phantom energy possesses an equation of state of the form $\omega \equiv p/\rho < -1$, consequently violating the null energy condition (NEC), which is a fundamental ingredient necessary to sustain traversable wormholes. Thus, this cosmic fluid presents us with a natural scenario for the existence of wormhole geometries [159, 160, 161]. Secondly, the generalized Chaplygin gas (GCG) is a candidate for the unification of dark energy and dark matter, and is parametrized by an exotic equation of state given by $p_{ch} = -A/\rho_{ch}^{\alpha}$, where A is a positive constant and $0 < \alpha \leq 1$. Within the framework of a flat Friedmann-Robertson-Walker cosmology the energy conservation equation yields the following evolution of the energy density $\rho_{ch} = \left[A + Ba^{-3(1+\alpha)}\right]^{1/(1+\alpha)}$, where a is the scale factor, and B is normally considered to be a positive integration constant to ensure the dominant energy condition (DEC). However, it is also possible to consider $B < 0$, consequently violating the DEC, and the energy density is an increasing function of the scale function [208]. It is in the latter context that we shall explore exact solutions of traversable wormholes supported by the GCG [162]. Thirdly, the van der Waals quintessence equation of state, $p = \gamma\rho/(1-\beta\rho) - \alpha\rho^2$, is an interesting scenario for describing the late universe, and seems to provide a solution to the puzzle of dark energy, without the presence of exotic fluids or modifications of the Friedmann equations. Note that $\alpha, \beta \to 0$ and $\gamma < -1/3$ reduces to the dark energy equation of state. The existence of traversable wormholes supported by the VDW equation of state shall also be explored [163]. Despite of the fact that, in a cosmological context, these cosmic fluids are considered homogeneous, inhomogeneities may arise through gravitational instabilities, resulting in a nucleation of the cosmic fluid due to the respective density perturbations. Thus, the wormhole solutions considered in this work may possibly originate from density fluctuations in the cosmological background.

The strategy we shall adopt is to impose an equation of state, $p_r = p_r(\rho)$, which provides four equations, together with the Einstein field equations. However, we have five unknown functions of r, i.e., $\rho(r)$, $p_r(r)$, $p_t(r)$, $b(r)$ and $\Phi(r)$. Therefore, to fully determine the system we impose restricted choices for $b(r)$ or $\Phi(r)$ [159, 162, 163]. It is also possible to consider plausible stress-energy components, and through the field equations determine the metric fields[161].

Now, using the equation of state representing phantom energy, $p_r = \omega\rho$ with $\omega < -1$, and taking into account Eqs. (26), we have the following condition

$$\Phi'(r) = \frac{b + \omega r b'}{2r^2\left(1 - b/r\right)}. \tag{127}$$

For instance, consider a constant $\Phi(r)$, so that Eq. (127) provides $b(r) = r_0(r/r_0)^{-1/\omega}$, which corresponds to an asymptotically flat wormhole geometry. It was shown that this solution can be constructed, in principle, with arbitrarily small quantities of averaged null energy condition violating phantom energy, and the traversability conditions were explored[159]. The dynamic stability of these phantom wormholes were also analyzed[160], and we refer the reader to [159, 161] for further examples.

Relative to the GCG gas equation of state, $p_r = -A/\rho^\alpha$, using the field equations, we have the following condition

$$2r\left(1-\frac{b}{r}\right)\Phi'(r) = -Ab'\left(\frac{8\pi r^2}{b'}\right)^{1+\alpha} + \frac{b}{r} \,. \tag{128}$$

Solutions of the metric (1), satisfying Eq. (128) are denoted "Chaplygin wormholes". To be a generic solution of a wormhole, the GCG equation of state imposes the following restriction $A < (8\pi r_0^2)^{-(1+\alpha)}$, consequently violating the NEC. However, for the GCG cosmological models it is generally assumed that the NEC is satisfied, which implies $\rho \geq A^{1/(1+\alpha)}$. The NEC violation is a fundamental ingredient in wormhole physics, and it is in this context that the construction of traversable wormholes, i.e., for $\rho < A^{1/(1+\alpha)}$, are explored. Note that as emphasized in [208], considering $B < 0$ in the evolution of the energy density, one also deduces that $\rho_{ch} < A^{1/(1+\alpha)}$, which violates the DEC. We refer the reader to [162] for specific examples of Chaplygin wormholes, where the physical properties and characteristics of these geometries were analyzed in detail. The solutions found are not asymptotically flat, and the spatial distribution of the exotic GCG is restricted to the throat vicinity, so that the dimensions of these Chaplygin wormholes are not arbitrarily large.

Finally, consider the VDW equation of state for an inhomogeneous spherically symmetric spacetime, given by $p_r = \gamma\rho/(1-\beta\rho) - \alpha\rho^2$. The Einstein field equations provide the following relationship

$$2r\left(1-\frac{b}{r}\right)\Phi' = \frac{b}{r} + \frac{\gamma b'}{1-\frac{\beta b'}{8\pi r^2}} - \frac{\alpha b'^2}{8\pi r^2} \,. \tag{129}$$

It was shown that traversable wormhole solutions may be constructed using the VDW equation of state, which are either asymptotically flat or possess finite dimensions, where the exotic matter is confined to the throat neighborhood [163]. The latter solutions are constructed by matching an interior wormhole geometry to an exterior vacuum Schwarzschild vacuum, and we refer the reader to [163] for further details.

In concluding, it is noteworthy the relative ease with which one may theoretically construct traversable wormholes with the exotic fluid equations of state used in cosmology to explain the present accelerated expansion of the Universe. These traversable wormhole variations have far-reaching physical and cosmological implications, namely, apart from being used for interstellar shortcuts, an absurdly advanced civilization may convert them into time-machines, probably implying the violation of causality.

3. "Warp Drive" Spacetimes and Superluminal Travel

Much interest has been revived in superluminal travel in the last few years. Despite the use of the term superluminal, it is not "really" possible to travel faster than light, in any *local* sense. The point to note is that one can make a round trip, between two points separated by a distance D, in an arbitrarily short time as measured by an observer that remained at rest at the starting point, by varying one's speed or by changing the distance one is to cover.

Providing a general *global* definition of superluminal travel is no trivial matter [169, 170], but it is clear that the spacetimes that allow "effective" superluminal travel generically suffer from the severe drawback that they also involve significant negative energy densities. More precisely, superluminal effects are associated with the presence of *exotic* matter, that is, matter that violates the null energy condition [NEC].

In fact, superluminal spacetimes violate all the known energy conditions, and Ken Olum demonstrated that negative energy densities and superluminal travel are intimately related [171]. Although most classical forms of matter are thought to obey the energy conditions, they are certainly violated by certain quantum fields [35]. Additionally, certain classical systems (such as non-minimally coupled scalar fields) have been found that violate the null and the weak energy conditions [31, 44]. It is also interesting to note that recent observations in cosmology strongly suggest that the cosmological fluid violates the strong energy condition [SEC], and provides tantalizing hints that the NEC *might* possibly be violated in a classical regime [32, 33, 34].

Apart from wormholes [1, 2], two spacetimes which allow superluminal travel are the Alcubierre warp drive [172] and the solution known as the Krasnikov tube [173, 30]. Alcubierre demonstrated that it is theoretically possible, within the framework of general relativity, to attain arbitrarily large velocities [172]. A warp bubble is driven by a local expansion behind the bubble, and an opposite contraction ahead of it. However, by introducing a slightly more complicated metric, José Natário [174] dispensed with the need for expansion. Thus, the Natário version of the warp drive can be thought of as a bubble sliding through space.

It is interesting to note that Krasnikov [173] discovered a fascinating aspect of the warp drive, in which an observer on a spaceship cannot create nor control on demand an Alcubierre bubble, with $v > c$, around the ship [173], as points on the outside front edge of the bubble are always spacelike separated from the centre of the bubble. However, causality considerations do not prevent the crew of a spaceship from arranging, by their own actions, to complete a *round trip* from the Earth to a distant star and back in an arbitrarily short time, as measured by clocks on the Earth, by altering the metric along the path of their outbound trip. Thus, Krasnikov introduced a two-dimensional metric with an interesting property that although the time for a one-way trip to a distant destination cannot be shortened, the time for a round trip, as measured by clocks at the starting point (e.g. Earth), can be made arbitrarily short. Soon after, Everett and Roman generalized the Krasnikov two-dimensional analysis to four dimensions, denoting the solution as the *Krasnikov tube* [30], where they analyzed the superluminal features, the energy condition violations, the appearance of closed timelike curves and applied the Quantum Inequality.

3.1. "Warp Drive" Spacetime Metric

Within the framework of general relativity, Alcubierre demonstrated that it is in principle possible to warp spacetime in a small *bubble-like* region, in such a way that the bubble may attain arbitrarily large velocities. Inspired by the inflationary phase of the early universe, the enormous speed of separation arises from the expansion of spacetime itself. The simplest model for hyper-fast travel is to create a local distortion of spacetime, producing an expansion behind the bubble, and an opposite contraction ahead of it. Natário's version of the warp drive dispensed with the need for expansion at the cost of introducing a slightly more complicated metric.

The warp drive spacetime metric, in cartesian coordinates, is given by (with $G = c = 1$)

$$ds^2 = -dt^2 + \left[d\vec{x} - \vec{\beta}(x, y, z - z_0(t))\, dt\right] \cdot \left[d\vec{x} - \vec{\beta}(x, y, z - z_0(t))\, dt\right]. \tag{130}$$

In terms of the well-known ADM formalism this corresponds to a spacetime wherein *space* is flat, while the "lapse function" is identically unity, and the only non-trivial structure lies in the "shift vector" $\beta(t, \vec{x})$. Thus warp drive spacetimes can also be viewed as specific examples of "shift-only" spacetimes. The Alcubierre warp drive corresponds to taking the shift vector to always lie in the direction of motion

$$\vec{\beta}(x, y, z - z_0(t)) = v(t)\ \hat{z}\ f(x, y, z - z_0(t)), \tag{131}$$

in which $v(t) = dz_0(t)/dt$ is the velocity of the warp bubble, moving along the positive z-axis, whereas in the Natário warp drive the shift vector is constrained by being divergence-free

$$\nabla \cdot \vec{\beta}(x, y, z) = 0. \tag{132}$$

3.2. Alcubierre Warp Drive

In the Alcubierre warp drive the spacetime metric is

$$ds^2 = -dt^2 + dx^2 + dy^2 + \left[dz - v(t)\ f(x, y, z - z_0(t))\ dt\right]^2. \tag{133}$$

The form function $f(x, y, z)$ possesses the general features of having the value $f = 0$ in the exterior and $f = 1$ in the interior of the bubble. The general class of form functions, $f(x, y, z)$, chosen by Alcubierre was spherically symmetric: $f(r)$ with $r = \sqrt{x^2 + y^2 + z^2}$. Then

$$f(x, y, z - z_0(t)) = f(r(t)) \qquad \text{with} \qquad r(t) = \left\{[(z - z_0(t)]^2 + x^2 + y^2\right\}^{1/2}. \tag{134}$$

Whenever a more specific example is required we adopt

$$f(r) = \frac{\tanh\left[\sigma(r + R)\right] - \tanh\left[\sigma(r - R)\right]}{2\tanh(\sigma R)}, \tag{135}$$

in which $R > 0$ and $\sigma > 0$ are two arbitrary parameters. R is the "radius" of the warp-bubble, and σ can be interpreted as being inversely proportional to the bubble wall thickness. If σ is large, the form function rapidly approaches a *top hat* function, i.e.,

$$\lim_{\sigma \to \infty} f(r) = \begin{cases} 1, & \text{if } r \in [0, R], \\ 0, & \text{if } r \in (R, \infty). \end{cases} \tag{136}$$

It can be shown that observers with the four velocity

$$U^\mu = (1, 0, 0, vf), \qquad U_\mu = (-1, 0, 0, 0). \tag{137}$$

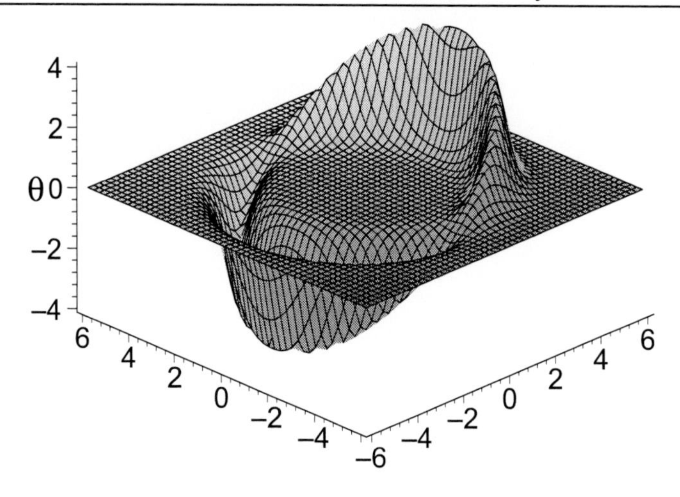

Figure 2. The expansion of the volume elements. These are expanding behind the spaceship, and contracting in front of it.

move along geodesics, as their 4-acceleration is zero, *i.e.*, $a^\mu = U^\nu \, U^\mu{}_{;\nu} = 0$. They were denoted Eulerian observers by Alcubierre. The spaceship, which in the original formulation is treated as a test particle which moves along the curve $z = z_0(t)$, can easily be seen to always move along a timelike curve, regardless of the value of $v(t)$. One can also verify that the proper time along this curve equals the coordinate time, by simply substituting $z = z_0(t)$ in eq. (133). This reduces to $d\tau = dt$, taking into account $dx = dy = 0$ and $f(0) = 1$.

Consider a spaceship placed within the Alcubierre warp bubble. The expansion of the volume elements, $\theta = U^\mu{}_{;\mu}$, is given by $\theta = v \; (\partial f/\partial z)$. Taking into account eq. (135), we have (for Alcubierre's version of the warp bubble)

$$\theta = v \, \frac{z - z_0}{r} \, \frac{df(r)}{dr} . \tag{138}$$

The center of the perturbation corresponds to the spaceship's position $z_0(t)$. The volume elements are expanding behind the spaceship, and contracting in front of it, as shown in Figure 2.

3.3. The Violation of the Energy Conditions

If we attempt to treat the spaceship as more than a test particle, we must confront the fact that by construction we have forced $f = 0$ outside the warp bubble. [Consider, for instance, the explicit form function of eq. (135) in the limit $r \to \infty$.] This implies that the spacetime geometry is asymptotically Minkowski space, and in particular the ADM mass (defined by taking the limit as one moves to spacelike infinity i^0) is zero. That is, the ADM mass of the spaceship and the warp field generators must be exactly compensated by the ADM mass due to the stress-energy of the warp-field itself. Viewed in this light it is now patently obvious that there must be massive violations of the classical energy conditions (at least in the original version of the warp-drive spacetime), and the interesting question becomes "Where are these energy condition violations localized?".

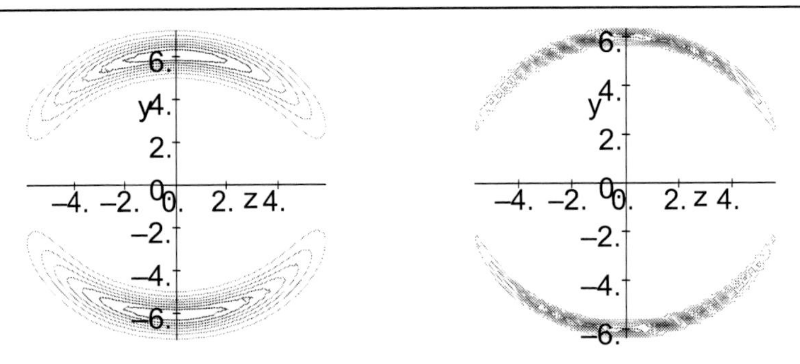

Figure 3. The energy density is distributed in a toroidal region perpendicular to the direction of travel of the spaceship, which is situated at $z_0(t)$. We have considered the following values, $v = 2$ and $R = 6$, with $\sigma = 1$ and $\sigma = 4$ in (a) and (b), respectively.

One of our tasks in the current and next Chapter will be to see if we can first avoid this exact cancellation of the ADM mass, and second, to see if we can make qualitative and quantitative statements concerning the localization and "total amount" of energy condition violations. (A similar attempt at quantification of the "total amount" of energy condition violation in traversable wormholes was recently presented in [84, 85]).

By using the Einstein field equation, $G_{\mu\nu} = 8\pi\ T_{\mu\nu}$, we can make rather general statements regarding the nature of the stress energy required to support a warp bubble.

3.3.1. The Violation of the WEC

The WEC states $T_{\mu\nu}\, U^\mu\, U^\nu \geq 0$, in which U^μ is a timelike vector and $T_{\mu\nu}$ is the stress-energy tensor. Its physical interpretation is that the local energy density is positive. By continuity it implies the NEC. We verify that for the warp drive metric, the WEC is violated, *i.e.*,

$$T_{\mu\nu}\, U^\mu\, U^\nu = -\frac{v^2}{32\pi}\left[\left(\frac{\partial f}{\partial x}\right)^2 + \left(\frac{\partial f}{\partial y}\right)^2\right] < 0\,, \tag{139}$$

or by taking into account the Alcubierre form function (135), we have

$$T_{\mu\nu}\, U^\mu\, U^\nu = -\frac{1}{32\pi}\frac{v^2(x^2+y^2)}{r^2}\left(\frac{df}{dr}\right)^2 < 0\,. \tag{140}$$

By considering the Einstein tensor component, $G_{\hat{t}\hat{t}}$, in an orthonormal basis, and taking into account the Einstein field equation, we verify that the energy density of the warp drive spacetime is given by $T_{\hat{t}\hat{t}} = T_{\hat{\mu}\hat{\nu}}\, U^{\hat{\mu}}\, U^{\hat{\nu}}$, that is, eq. (140). It is easy to verify that the energy density is distributed in a toroidal region around the z-axis, in the direction of travel of the warp bubble [29], as may be verified from Figure 3. It is perhaps instructive to point out that the energy density for this class of spacetimes is nowhere positive. That the total ADM mass can nevertheless be zero is due to the intrinsic nonlinearity of the Einstein equations.

It is interesting to note that the inclusion of a generic lapse function $\alpha(x, y, z, t)$, decreases the negative energy density, which is given by

$$T_{\hat{t}\hat{t}} = -\frac{v^2}{32\pi\,\alpha^2}\left[\left(\frac{\partial f}{\partial x}\right)^2 + \left(\frac{\partial f}{\partial y}\right)^2\right]\,. \tag{141}$$

Now, α may be taken as unity in the exterior and interior of the warp bubble, so proper time equals coordinate time. In order to significantly decrease the negative energy density in the bubble walls, one may impose an extremely large value for the lapse function. However, the inclusion of the lapse function suffers from an extremely severe drawback, as proper time as measured in the bubble walls becomes absurdly large, $d\tau = \alpha\, dt$, for $\alpha \gg 1$.

We can (in analogy with the definitions in [84, 85]) quantify the "total amount" of energy condition violating matter in the warp bubble by defining

$$\begin{aligned} M_{\text{warp}} &= \int \rho_{\text{warp}}\, d^3x = \int T_{\mu\nu}\, U^\mu\, U^\nu\, d^3x \\ &= -\frac{v^2}{32\pi} \int \frac{x^2+y^2}{r^2} \left(\frac{df}{dr}\right)^2 r^2\, dr\, d^2\Omega = -\frac{v^2}{12} \int \left(\frac{df}{dr}\right)^2 r^2\, dr. \end{aligned} \quad (142)$$

This is emphatically not the total mass of the spacetime, but it characterizes how much (negative) energy one needs to localize in the walls of the warp bubble. For the specific shape function (135) we can estimate

$$M_{\text{warp}} \approx -v^2\, R^2\, \sigma. \quad (143)$$

(The integral can be done exactly, but the exact result in terms of polylog functions is unhelpful.) Note that the energy requirements for the warp bubble scale quadratically with bubble velocity, quadratically with bubble size, and inversely as the thickness of the bubble wall.

3.3.2. The Violation of the NEC

The NEC states that $T_{\mu\nu}\, k^\mu\, k^\nu \geq 0$, where k^μ is *any* arbitrary null vector and $T_{\mu\nu}$ is the stress-energy tensor. The NEC for a null vector oriented along the $\pm\hat{z}$ directions takes the following form

$$T_{\mu\nu}\, k^\mu\, k^\nu = -\frac{v^2}{8\pi}\left[\left(\frac{\partial f}{\partial x}\right)^2 + \left(\frac{\partial f}{\partial y}\right)^2\right] \pm \frac{v}{8\pi}\left(\frac{\partial^2 f}{\partial x^2} + \frac{\partial^2 f}{\partial y^2}\right). \quad (144)$$

In particular if we average over the $\pm\hat{z}$ directions we have

$$\frac{1}{2}\left\{T_{\mu\nu}\, k^\mu_{+\hat{z}}\, k^\nu_{+\hat{z}} + T_{\mu\nu}\, k^\mu_{-\hat{z}}\, k^\nu_{-\hat{z}}\right\} = -\frac{v^2}{8\pi}\left[\left(\frac{\partial f}{\partial x}\right)^2 + \left(\frac{\partial f}{\partial y}\right)^2\right], \quad (145)$$

which is manifestly negative, and so the NEC is violated for all v. Furthermore, note that even if we do not average, the coefficient of the term linear in v must be nonzero *somewhere* in the spacetime. Then at low velocities this term will dominate and at low velocities the un-averaged NEC will be violated in either the $+\hat{z}$ or $-\hat{z}$ directions.

To be a little more specific about how and where the NEC is violated consider the Alcubierre form function. We have

$$T_{\mu\nu}\, k^\mu_{\pm\hat{z}}\, k^\nu_{\pm\hat{z}} = -\frac{1}{8\pi}\frac{v^2(x^2+y^2)}{r^2}\left(\frac{df}{dr}\right)^2$$

$$\pm \frac{v}{8\pi}\left[\frac{x^2+y^2+2(z-z_0(t))^2}{r^3}\,\frac{df}{dr}+\frac{x^2+y^2}{r^2}\,\frac{d^2f}{dr^2}\right]. \tag{146}$$

The first term is manifestly negative everywhere throughout the space. As f decreases monotonically from the center of the warp bubble, where it takes the value of $f=1$, to the exterior of the bubble, with $f \approx 0$, we verify that df/dr is negative in this domain. The term d^2f/dr^2 is also negative in this region, as f attains its maximum in the interior of the bubble wall. Thus, the term in square brackets unavoidably assumes a negative value in this range, resulting in the violation of the NEC.

Equation (146) is plotted in Figures 4 and 5, for various values of the parameters. Figure 4 represents the null vector oriented along the $+\hat{z}$ direction, and Figure 5 along the $-\hat{z}$ direction. We have considered the following values of $v=2$, $\sigma=2$ and $R=6$, for the parameters.

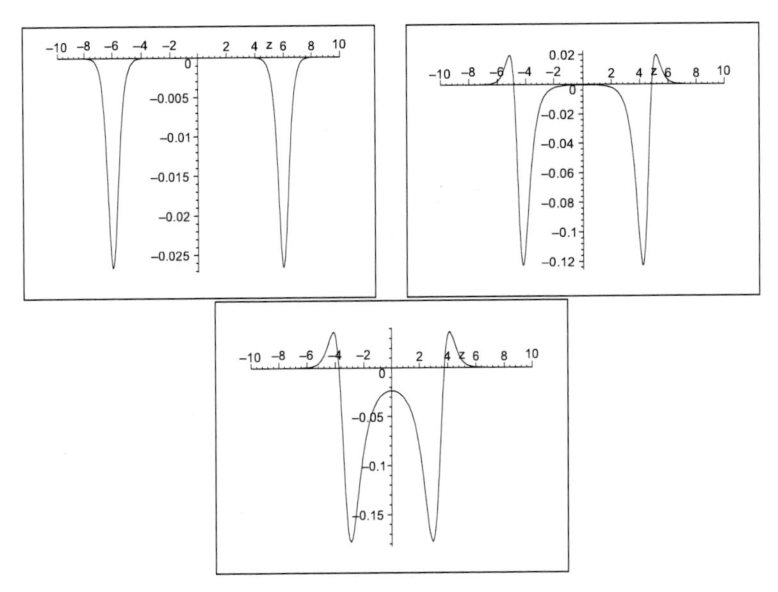

Figure 4. The NEC for a null vector oriented along the $+\hat{z}$ direction. Taking into account the Alcubierre form function, we have considered the parameters $v=2$, $\sigma=2$ and $R=6$. Considering the definition $\rho=\sqrt{x^2+y^2}$, the plots have the respective values of $\rho=0$, $\rho=4$ and $\rho=5$.

For a null vector oriented perpendicular to the direction of motion (for definiteness take $\hat{k}=\pm\hat{x}$) the NEC takes the following form

$$T_{\mu\nu}\,k^{\mu}_{\pm\hat{x}}\,k^{\nu}_{\pm\hat{x}}=-\frac{v^2}{8\pi}\left[\frac{1}{2}\left(\frac{\partial f}{\partial y}\right)^2+\left(\frac{\partial f}{\partial z}\right)^2-(1-f)\frac{\partial^2 f}{\partial z^2}\right]\mp\frac{v}{8\pi}\left(\frac{\partial^2 f}{\partial x\partial z}\right). \tag{147}$$

Again, note that the coefficient of the term linear in v must be nonzero *somewhere* in the spacetime. Then at low velocities this term will dominate, and at low velocities the NEC will be violated in one or other of the transverse directions. Upon considering the specific form of the spherically symmetric Alcubierre form function, we have

$$T_{\mu\nu}\,k^{\mu}_{\pm\hat{x}}\,k^{\nu}_{\pm\hat{x}}=-\frac{v^2}{8\pi}\left[\frac{y^2+2(z-z_0(t))^2}{2r^2}\left(\frac{df}{dr}\right)^2\right.$$

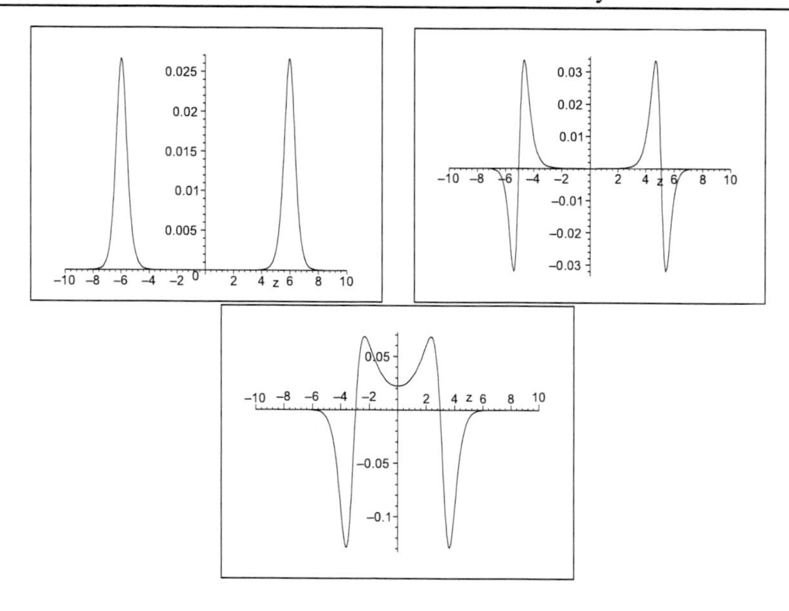

Figure 5. The NEC for a null vector oriented along the $-\hat{z}$ direction. Taking into account the Alcubierre form function, we have considered the parameters $v = 2, \sigma = 2$ and $R = 6$. Considering the definition $\rho = \sqrt{x^2 + y^2}$, the plots have the respective values of $\rho = 0$, $\rho = 3$ and $\rho = 5$.

$$
\begin{aligned}
&-(1-f)\left(\frac{x^2+y^2}{r^3}\,\frac{df}{dr} + \frac{(z-z_0(t))^2}{r^2}\,\frac{d^2 f}{dr^2}\right)\Bigg] \\
&\mp \frac{v}{8\pi}\,\frac{x\,(z-z_0(t))}{r^2}\left(\frac{d^2 f}{dr^2} - \frac{1}{r}\frac{df}{dr}\right) .
\end{aligned} \tag{148}
$$

Again, the message to take from this is that localized NEC violations are ubiquitous and persist to arbitrarily low warp bubble velocities.

Using the "volume integral quantifier" (as defined in [84, 85]), we may estimate the "total amount" of averaged null energy condition violating matter in this spacetime, given by

$$
\int T_{\mu\nu}\, k^\mu_{\pm\hat{z}}\, k^\nu_{\pm\hat{z}}\, d^3x \approx \int T_{\mu\nu}\, k^\mu_{\pm\hat{x}}\, k^\nu_{\pm\hat{x}}\, d^3x \approx -v^2\, R^2\, \sigma \approx M_{\rm warp} \,. \tag{149}
$$

The key things to note here are that the net volume integral of the $O(v)$ term is zero, and that the net volume average of the NEC violations is approximately the same as the net volume average of the WEC violations, which are $O(v^2)$.

3.4. The Quantum Inequality Applied to the "Warp Drive"

It is of a particular interest to apply the Quantum Inequality (QI), outlined in Section 2.6.4., to "warp drive" spacetimes [29]. Inserting the energy density, eq. (140), into the QI, eq. (81), one gets

$$
t_0 \int_{-\infty}^{+\infty} \frac{v(t)^2}{r^2}\left(\frac{df}{dr}\right)^2 \frac{dt}{t^2+t_0^2} \leq \frac{3}{\rho^2\, t_0^4} \,, \tag{150}
$$

where $\rho = \left(x^2 + y^2\right)^{1/2}$ is defined for notational convenience.

The warp bubble's velocity can be considered roughly constant, $v_s(t) \approx v_b$, if the time scale of the sampling is sufficiently small compared to the time scale over which the bubble's velocity is changing. Taking into account the small sampling time, the $(t^2 + t_0^2)^{-1}$ term becomes strongly peaked, so that only a small portion of the geodesic is sampled by the QI integral. Consider that the observer is at the equator of the warp bubble at $t = 0$ [29], so that the geodesic is approximated by

$$x(t) \approx f(\rho) v_b t \,, \tag{151}$$

so that we have $r(t) = \left[(v_b t)^2 (f(\rho) - 1)^2 + \rho^2\right]^{1/2}$.

Without a significant loss of generality, one may consider a piece-wise continuous form of the shape function given by

$$f_{p.c.}(r) = \begin{cases} 1 & r < R - \frac{\Delta}{2} \\ -\frac{1}{\Delta}(r - R - \frac{\Delta}{2}) & R - \frac{\Delta}{2} < r < R + \frac{\Delta}{2} \\ 0 & r > R + \frac{\Delta}{2} \end{cases} \tag{152}$$

where R is the radius of the bubble, and Δ the bubble wall thickness [29]. Δ is related to the Alcubierre parameter σ by setting the slopes of the functions $f(r)$ and $f_{p.c.}(r)$ to be equal at $r = R$, which provides the following relationship

$$\Delta = \frac{\left[1 + \tanh^2(\sigma R)\right]^2}{2\,\sigma\,\tanh(\sigma R)} \,, \tag{153}$$

Note that in the limit of large σR one obtains the approximation $\Delta \simeq 2/\sigma$. The QI-bound then becomes

$$t_0 \int_{-\infty}^{+\infty} \frac{dt}{(t^2 + \beta^2)(t^2 + t_0^2)} \leq \frac{3\Delta^2}{v_b^2\, t_0^4\, \beta^2} \tag{154}$$

where

$$\beta = \frac{\rho}{v_b\left[1 - f(\rho)\right]} \,, \tag{155}$$

and yields the following inequality

$$\frac{\pi}{3} \leq \frac{\Delta^2}{v_b^2\, t_0^4} \left[\frac{v_b t_0}{\rho}(1 - f(\rho)) + 1\right] . \tag{156}$$

It is important to emphasize that the above inequality is only valid for sampling times on which the spacetime may be considered approximately flat. Considering the Riemann tensor components in an orthonormal frame [29], the largest component is given by

$$|R_{\hat{t}\hat{y}\hat{t}\hat{y}}| = \frac{3 v_b^2\, y^2}{4\,\rho^2} \left[\frac{df(\rho)}{d\rho}\right]^2 \,, \tag{157}$$

which yields $r_{\min} \equiv 1/\sqrt{|R_{\hat{t}\hat{y}\hat{t}\hat{y}}|} \sim \frac{2\Delta}{\sqrt{3}\,v_b}$, when $y = \rho$ and the piece-wise continuous form of the shape function is used. The sampling time must be smaller than this length scale, so that one my define

$$t_0 = \alpha \frac{2\Delta}{\sqrt{3}\,v_b} \,, \qquad 0 < \alpha \ll 1 \,. \tag{158}$$

Considering $\Delta/\rho \sim v_b\, t_0/\rho \ll 1$, the term involving $1-f(\rho)$ in eq. (156) may be neglected, which provides

$$\Delta \leq \frac{3}{4}\sqrt{\frac{3}{\pi}}\,\frac{v_b}{\alpha^2}\,. \tag{159}$$

Now, for instance considering $\alpha = 1/10$, one obtains

$$\Delta \leq 10^2\, v_b\, L_{\text{Planck}}\,, \tag{160}$$

where L_{Planck} is the Planck length. Thus, unless v_b is extremely large, the wall thickness cannot be much above the Planck scale.

It is also interesting to find an estimate of the total amount of negative energy that is necessary to maintain a warp metric. It was found that the energy required for a warp bubble is on the order of

$$E \leq -3 \times 10^{20}\, M_{\text{galaxy}}\, v_b\,, \tag{161}$$

which is an absurdly enormous amount of negative energy, roughly ten orders of magnitude greater than the total mass of the entire visible universe [29].

3.5. Linearized Warp Drive

To ever bring a "warp drive" into a strong-field regime, any highly-advanced civilization would first have to take it through the "weak-field" regime. The central point of this Section is to demonstrate that there are significant problems that already arise even in the weak-field regime, and long before strong field effects come into play. In the weak-field regime, applying linearized theory, the physics is much simpler than in the strong-field regime and this allows us to ask and answer questions that are difficult to even formulate in the strong field regime.

Our goal now is to try to build a more realistic model of a warp drive spacetime where the warp bubble is interacting with a finite mass spaceship. To do so we first consider the linearized theory applied to warp drive spacetimes, for non-relativistic velocities, $v \ll 1$.

3.5.1. The WEC Violation to First Order of v

It is interesting to consider the specific case of an observer which moves with an arbitrary velocity, $\tilde{\beta}$, along the positive z axis measure a negative energy density [at $O(v)$]. That is, $T_{\hat{0}\hat{0}} < 0$. The $\tilde{\beta}$ occurring here is completely independent of the shift vector $\beta(x, y, z - z_0(t))$, and is also completely independent of the warp bubble velocity v.

We have

$$T_{\hat{0}\hat{0}} = \frac{\gamma^2\tilde{\beta}v}{8\pi}\left[\left(\frac{x^2+y^2}{r^2}\right)\frac{d^2f}{dr^2} + \left(\frac{x^2+y^2+2(z-z_0(t))^2}{r^3}\right)\frac{df}{dr}\right] + O(v^2)\,. \tag{162}$$

A number of general features can be extracted from the terms in square brackets, without specifying an explicit form of f. In particular, f decreases monotonically from its value at $r = 0$, $f = 1$, to $f \approx 0$ at $r \geq R$, so that df/dr is negative in this domain. The form function attains its maximum in the interior of the bubble wall, so that d^2f/dr^2 is also negative in this region. Therefore there is a range of r in the immediate interior neighbourhood of the bubble wall that necessarily provides negative energy density, as seen by the observers considered above. Again we find that WEC violations persist to arbitrarily low warp bubble velocities. The negative character of the energy density can be seen from Figure 6.

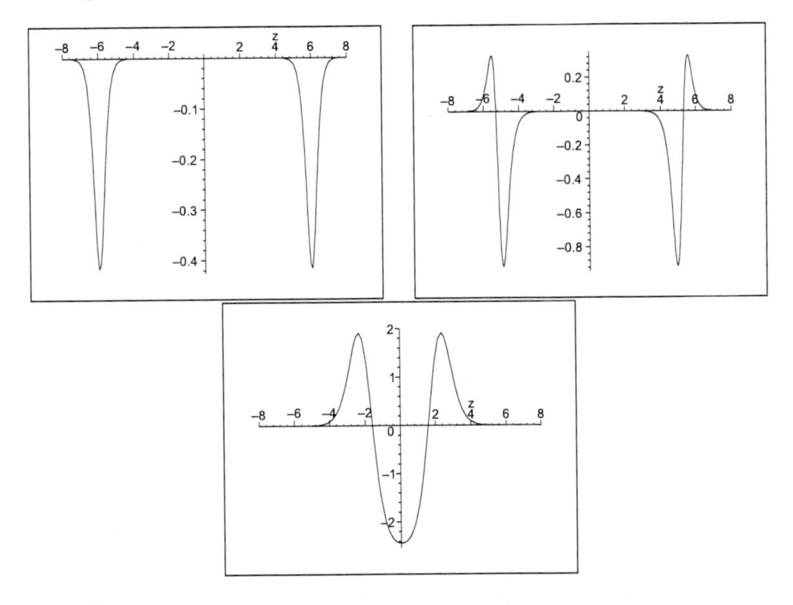

Figure 6. The term in square brackets of eq. (162) is plotted as a function of the z coordinate. Taking into account the Alcubierre form function, we have considered the following values for the parameters $\sigma = 2.5$ and $R = 6$. Considering the definition $\rho = \sqrt{x^2 + y^2}$, the plots have the respective values of $\rho = 0$, $\rho = 3$ and $\rho = 5.8$.

3.5.2. Spaceship within the Warp Bubble

Consider now a spaceship in the interior of an Alcubierre warp bubble, which is moving along the positive z axis with a non-relativistic constant velocity. That is, $v \ll 1$. The metric is given by

$$\begin{aligned} ds^2 &= -dt^2 + dx^2 + dy^2 + [dz - v\; f(x,y,z-vt)\,dt]^2 \\ &\quad -2\Phi(x,y,z-vt)\,\left[dt^2 + dx^2 + dy^2 + (dz - v\; f(x,y,z-vt)\,dt)^2\right] . \end{aligned} \tag{163}$$

If $\Phi = 0$, the metric (163) reduces to the warp drive spacetime of eq. (133). If $v = 0$, we have the metric representing the gravitational field of a static source.

Consider the approximation in which we keep the exact v dependence but linearize in the gravitational field of the spaceship Φ.

The WEC is given by

$$T_{\hat{\mu}\hat{\nu}}\, U^{\hat{\mu}}\, U^{\hat{\nu}} = \rho - \frac{v^2}{32\pi}\left[\left(\frac{\partial f}{\partial x}\right)^2 + \left(\frac{\partial f}{\partial y}\right)^2\right] + O(\Phi^2)\,, \tag{164}$$

or by taking into account the Alcubierre form function, we have

$$T_{\mu\nu}\, U^{\mu}\, U^{\nu} = \rho - \frac{1}{32\pi}\frac{v^2(x^2+y^2)}{r^2}\left(\frac{df}{dr}\right)^2 + O(\Phi^2)\,. \tag{165}$$

Once again, using the "volume integral quantifier", we find the following estimate

$$\int T_{\hat{\mu}\hat{\nu}}\, U^{\hat{\mu}}\, U^{\hat{\nu}}\, d^3x = M_{\text{ship}} - v^2\, R^2\, \sigma + \int O(\Phi^2)\, d^3x\,, \tag{166}$$

which we can recast as

$$M_{\text{ADM}} = M_{\text{ship}} + M_{\text{warp}} + \int O(\Phi^2)\, d^3x\,. \tag{167}$$

Now suppose we demand that the volume integral of the WEC at least be positive, then

$$v^2\, R^2\, \sigma \leq M_{\text{ship}}. \tag{168}$$

This equation is effectively the quite reasonable condition that the net total energy stored in the warp field be less than the total mass-energy of the spaceship itself, which places a powerful constraint on the velocity of the warp bubble. Re-writing this in terms of the size of the spaceship R_{ship} and the thickness of the warp bubble walls $\Delta = 1/\sigma$, we have

$$v^2 \leq \frac{M_{\text{ship}}}{R_{\text{ship}}}\, \frac{R_{\text{ship}}\, \Delta}{R^2}. \tag{169}$$

For any reasonable spaceship this gives extremely low bounds on the warp bubble velocity.

In a similar manner, the NEC, with $k^\mu = (1, 0, 0, \pm 1)$, is given by

$$T_{\hat{\mu}\hat{\nu}}\, k^{\hat{\mu}}\, k^{\hat{\nu}} = \rho \pm \frac{v}{8\pi}\left(\frac{\partial^2 f}{\partial x^2} + \frac{\partial^2 f}{\partial y^2}\right) - \frac{v^2}{8\pi}\left[\left(\frac{\partial f}{\partial x}\right)^2 + \left(\frac{\partial f}{\partial y}\right)^2\right] + O(\Phi^2)\,. \tag{170}$$

Considering the "volume integral quantifier", we verify that, as before, the exact solution in terms of polylogarithmic functions is unhelpful, although we may estimate that

$$\int T_{\hat{\mu}\hat{\nu}}\, k^{\hat{\mu}}\, k^{\hat{\nu}}\, d^3x = M_{\text{ship}} - v^2\, R^2\, \sigma + \int O(\Phi^2)\, d^3x\,, \tag{171}$$

which is [to order $O(\Phi^2)$] the same integral we encountered when dealing with the WEC. This volume integrated NEC is now positive if

$$v^2\, R^2\, \sigma \leq M_{\text{ship}}. \tag{172}$$

Finally, considering a null vector oriented perpendicularly to the direction of motion (for definiteness take $\hat{k} = \pm\hat{x}$), the NEC takes the following form

$$T_{\hat{\mu}\hat{\nu}}\, k^{\hat{\mu}}\, k^{\hat{\nu}} = \rho - \frac{v^2}{8\pi}\left[\frac{1}{2}\left(\frac{\partial f}{\partial y}\right)^2 + \left(\frac{\partial f}{\partial z}\right)^2 - (1-f)\frac{\partial^2 f}{\partial z^2}\right] \mp \frac{v}{8\pi}\left(\frac{\partial^2 f}{\partial x \partial z}\right) + O(\Phi^2)\,. \tag{173}$$

Once again, evaluating the "volume integral quantifier", we have

$$\int T_{\hat{\mu}\hat{\nu}}\, k^{\hat{\mu}}\, k^{\hat{\nu}}\, d^3x = M_{\text{ship}} - \frac{v^2}{4}\int \left(\frac{df}{dr}\right)^2\, r^2\, dr$$

$$+\frac{v^2}{6}\int(1-f)\left(2r\frac{df}{dr}+r^2\frac{d^2f}{dr^2}\right)\,dr+\int O(\Phi^2)d^3x\,, \tag{174}$$

which, as before, may be estimated as

$$\int T_{\hat{\mu}\hat{\nu}}\,k^{\hat{\mu}}\,k^{\hat{\nu}}\,d^3x \approx M_{\rm ship} - v^2R^2\,\sigma + \int O(\Phi^2)\,d^3x\,. \tag{175}$$

If we do not want the total NEC violations in the warp field to exceed the mass of the spaceship itself we must again demand

$$v^2\,R^2\,\sigma \leq M_{\rm ship}. \tag{176}$$

This places an extremely stringent condition on the warp drive spacetime, namely, that for all conceivably interesting situations the bubble velocity should be absurdly low, and it therefore appears unlikely that, by using this analysis, the warp drive will ever prove to be technologically useful. Finally, we point out that any attempt at building up a "strong-field" warp drive starting from an approximately Minkowski spacetime will inevitably have to pass through a weak-field regime. Since the weak-field warp drives are already so tightly constrained, the analysis above implies additional difficulties for developing a "strong field" warp drive.

3.6. Interesting Aspects of the Alcubierre Spacetime

3.6.1. Superluminal Travel in the Warp Drive

To demonstrate that it is possible to travel to a distant point and back in an arbitrary short time interval, let us consider two distant stars, A and B, separated by a distance D in flat spacetime. Suppose that, at the instant t_0, a spaceship initiates it's movement using the engines, moving away from A with a velocity $v<1$. It comes to rest at a distance d from A. For simplicity, assume that $R \ll d \ll D$. It is at this instant that the perturbation of spacetime appears, centered around the spaceship's position. The perturbation pushes the spaceship away from A, rapidly attaining a constant acceleration, a. Half-way between A and B, the perturbation is modified, so that the acceleration rapidly varies from a to $-a$. The spaceship finally comes to rest at a distance, d, from B, in which the perturbation disappears. It then moves to B at a constant velocity in flat spacetime. The return trip to A is analogous.

If the variations of the acceleration are extremely rapid, the total coordinate time, T, in a one-way trip will be

$$T = 2\left(\frac{d}{v}+\sqrt{\frac{D-2d}{a}}\right)\,. \tag{177}$$

The proper time of the stars are equal to the coordinate time, because both are immersed in flat spacetime. The proper time measured by observers within the spaceship is given by:

$$\tau = 2\left(\frac{d}{\gamma v}+\sqrt{\frac{D-2d}{a}}\right)\,, \tag{178}$$

with $\gamma = (1 - v^2)^{-1/2}$. The time dilation only appears in the absence of the perturbation, in which the spaceship is moving with a velocity v, using only it's engines in flat spacetime.

Using $R \ll d \ll D$, we can then obtain the following approximation

$$\tau \approx T \approx 2\sqrt{\frac{D}{a}}\,. \tag{179}$$

Note that T can be made arbitrarily short, by increasing the value of a. The spaceship may travel faster than the speed of light. However, it moves along a spacetime temporal trajectory, contained within it's light cone, as light suffers the same distortion of spacetime [172].

3.6.2. The Krasnikov Analysis

Krasnikov discovered a fascinating aspect of the warp drive, in which an observer on a spaceship cannot create nor control on demand an Alcubierre bubble, with $v > c$, around the ship [173]. It is easy to understand this, as an observer at the origin (with $t = 0$), cannot alter events outside of his future light cone, $|r| \leq t$, with $r = (x^2 + y^2 + z^2)^{1/2}$. Applied to the warp drive, points on the outside front edge of the bubble are always spacelike separated from the centre of the bubble.

The analysis is simplified in the proper reference frame of an observer at the centre of the bubble. Using the transformation $z' = z - z_0(t)$, the metric is given by

$$ds^2 = -dt^2 + dx^2 + dy^2 + \left[dz' + (1 - f)v dt\right]^2\,. \tag{180}$$

Consider a photon emitted along the $+Oz$ axis (with $ds^2 = dx = dy = 0$):

$$\frac{dz'}{dt} = 1 - (1 - f)v\,. \tag{181}$$

If the spaceship is at rest at the center of the bubble, then initially the photon has $dz/dt = v + 1$ or $dz'/dt = 1$ (because $f = 1$ in the interior of the bubble). However, at some point $z' = z'_c$, with $f = 1 - 1/v$, we have $dz'/dt = 0$ [30]. Once photons reach z'_c, they remain at rest relative to the bubble and are simply carried along with it. Photons emitted in the forward direction by the spaceship never reach the outside edge of the bubble wall, which therefore lies outside the forward light cone of the spaceship. The bubble thus cannot be created (or controlled) by any action of the spaceship crew. This behaviour is reminiscent of an *event horizon*. This does not mean that Alcubierre bubbles, if it were possible to create them, could not be used as a means of superluminal travel. It only means that the actions required to change the metric and create the bubble must be taken beforehand by some observer whose forward light cone contains the entire trajectory of the bubble.

3.6.3. Reminiscence of an Event Horizon

The appearance of an event horizon becomes evident in the 2-dimensional model of the Alcubierre space-time [177, 178, 179]. The axis of symmetry coincides with the line element of the spaceship. The metric, eq. (133), reduces to

$$ds^2 = -(1 - v^2 f^2)dt^2 - 2vf dz dt + dz^2\,. \tag{182}$$

For simplicity, we consider the velocity of the bubble constant, $v(t) = v_b$, and we have $r = [(z - v_b t)^2]^{1/2}$. If $z > v_b t$, we consider the transformation $r = (z - v_b t)$. Note that the metric components of eq. (182) only depend on r, which may be adopted as a coordinate.

Using the transformation, $dz = dr + v_b\, dt$, the metric, eq. (182) is given by

$$ds^2 = -A(r)\left[dt - \frac{v_b(1 - f(r))}{A(r)}\, dr\right]^2 + \frac{dr^2}{A(r)}\,. \tag{183}$$

The function $A(r)$, denoted by the Hiscock function, is given by

$$A(r) = 1 - v_b^2\, [1 - f(r)]^2\,. \tag{184}$$

Its possible to represent the metric, eq. (183), in a diagonal form, using a new time coordinate

$$d\tau = dt - \frac{v_b\, [1 - f(r)]}{A(r)}\, dr\,, \tag{185}$$

with which eq. (183) reduces to

$$ds^2 = -A(r)\, d\tau^2 + \frac{dr^2}{A(r)}\,. \tag{186}$$

This form of the metric is manifestly static. The τ coordinate has an immediate interpretation in terms of an observer on board of a spaceship: τ is the proper time of the observer, because $A(r) \to 1$ in the limit $r \to 0$. We verify that the coordinate system is valid for any value of r, if $v_b < 1$. If $v_b > 1$, we have a coordinate singularity and an event horizon at the point r_0 in which $f(r_0) = 1 - 1/v_b$ and $A(r_0) = 0$.

3.7. Superluminal Subway: The Krasnikov tube

It was pointed out in Section 3.6., that Krasnikov discovered an interesting aspect of the warp drive, in which an observer on a spaceship cannot create nor control on demand an Alcubierre bubble, i.e., points on the outside front edge of the bubble are always spacelike separated from the centre of the bubble. However, causality considerations do not prevent the crew of a spaceship from arranging, by their own actions, to complete a *round trip* from the Earth to a distant star and back in an arbitrarily short time, as measured by clocks on the Earth, by altering the metric along the path of their outbound trip. Thus, Krasnikov introduced a metric with an interesting property that although the time for a one-way trip to a distant destination cannot be shortened, the time for a round trip, as measured by clocks at the starting point (e.g. Earth), can be made arbitrarily short, as will be demonstrated below.

3.7.1. The 2-dimensional Krasnikov Solution

The 2-dimensional metric is given by

$$\begin{aligned} ds^2 &= -(dt - dx)(dt + k(t,x)dx) \\ &= -dt^2 + [1 - k(x,t)]\; dx\, dt + k(x,t)\, dx^2\,. \end{aligned} \tag{187}$$

The form function $k(x,t)$ is defined by

$$k(t,x) = 1 - (2-\delta)\theta_\varepsilon(t-x)\left[\theta_\varepsilon(x) - \theta_\varepsilon(x+\varepsilon-D)\right] , \tag{188}$$

where δ and ε are arbitrarily small positive parameters. θ_ε denotes a smooth monotone function

$$\theta_\varepsilon(\xi) = \begin{cases} 1, & \text{if } \xi > \varepsilon , \\ 0, & \text{if } \xi < 0 , \end{cases}$$

which is depicted in Figure 7.

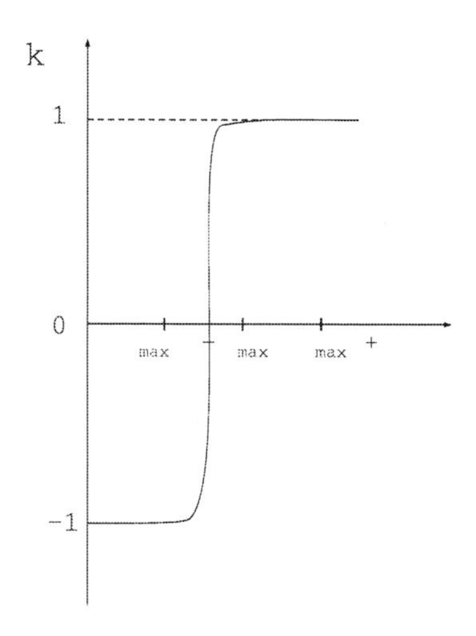

Figure 7. Graph of the Krasnikov form function k vs ρ at constant x and t, considering only the region of $\epsilon < x < D - \epsilon$.

There are three distinct regions in the Krasnikov two-dimensional spacetime, which we shall summarize in the following manner.

The outer region. The outer region is given by the following set

$$\{x < 0\} \cup \{x > D\} \cup \{x > t\} . \tag{189}$$

The two time-independent θ_ϵ-functions between the brackets in eq. (188) vanish for $x < 0$ and cancel for $x > D$, ensuring $k = 1$ for all t except between $x = 0$ and $x = D$. When this behavior is combined with the effect of the factor $\theta_\epsilon(t-x)$, one sees that the metric (187) is flat, $k = 1$, and reduces to the Minkowski spacetime everywhere for $t < 0$ and at all times outside the range $0 < x < D$. Future light cones are generated by the vectors:

$$\begin{cases} r_O = \partial_t + \partial_x , \\ l_O = \partial_t - \partial_x . \end{cases}$$

The inner region. The inner region is given by the following set

$$\{x < t - \varepsilon\} \cap \{\varepsilon < x < D - \varepsilon\} , \tag{190}$$

so that the first two θ_ϵ-functions in eq. (188) both equal 1, while $\theta_\epsilon(x+\epsilon-D)=0$, giving $k=\delta-1$ everywhere within this region. This region is also flat, but the light cones are *more open*, being generated by the following vectors

$$\begin{cases} r_I = \partial_t + \partial_x \\ l_I = -(1-\delta)\partial_t - \partial_x \,. \end{cases}$$

The transition region. The transition region is a narrow curved strip in spacetime, with width $\sim \varepsilon$. Two spatial boundaries exist between the inner and outer regions. The first lies between $x=0$ and $x=\varepsilon$, for $t>0$. The second lies between $x=D-\varepsilon$ and $x=D$, for $t>D$. It is possible to view this metric as being produced by the crew of a spaceship, departing from point A ($x=0$), at $t=0$, travelling along the x-axis to point B ($x=D$) at a speed, for simplicity, infinitesimally close to the speed of light, therefore arriving at B with $t\approx D$.

The metric is modified by changing k from 1 to $\delta-1$ along the x-axis, in between $x=0$ and $x=D$, leaving a transition region of width $\sim\varepsilon$ at each end for continuity. But, as the boundary of the forward light cone of the spaceship at $t=0$ is $|x|=t$, it is not possible for the crew to modify the metric at an arbitrary point x before $t=x$. This fact accounts for the factor $\theta_\varepsilon(t-x)$ in the metric, ensuring a transition region in time between the inner and outer region, with a duration of $\sim\varepsilon$, lying along the wordline of the spaceship, $x\approx t$. The geometry is shown in the (x,t) plane in Figure 8.

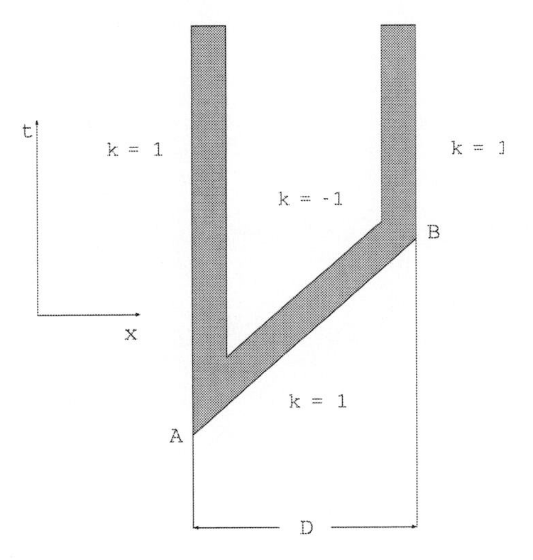

Figure 8. The Krasnikov spacetime in the (x,t) plane. The vertical lines A and B are the world lines of the stars A and B, respectively. The world line of the spaceship is approximately represented by the line segment AB.

3.7.2. Superluminal Travel within the Krasnikov Tube

The properties of the modified metric with $\delta-1\leq k\leq 1$ can be easily seen from the factored form of $ds^2=0$. The two branches of the forward light cone in the (t,x) plane are given by $dx/dt = 1$ and $dx/dt = -k$. As k becomes smaller and then negative, the slope of the left-hand branch of the light cone becomes less negative and then changes sign, i. e., the light cone along the negative x-axis "opens out". See Figure 9.

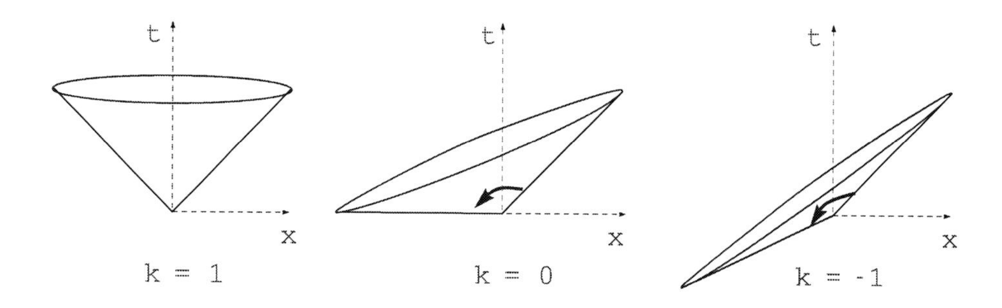

Figure 9. Forward light cones in the 2-dimensional Krasnikov spacetime for $k = 1$, $k = 0$ and $k = \delta - 1$.

The inner region, with $k = \delta - 1$, is flat because the metric, eq. (187), may be cast into the Minkowski form, applying the following coordinate transformations

$$dt' = dt + \left(\frac{\delta}{2} - 1\right) dx\,, \qquad dx' = \left(\frac{\delta}{2}\right) dx\,. \tag{191}$$

The transformation is singular at $\delta = 0$, i.e., $k = -1$. Note that the left branch of the region is given by $dx'/dt' = -1$.

From the above equations, one may easily deduce the following expression

$$\frac{dt}{dt'} = 1 + \left(\frac{2-\delta}{\delta}\right)\frac{dx'}{dt'}\,. \tag{192}$$

For an observer moving along the positive x' and x directions, with $dx'/dt' < 1$, we have $dt' > 0$ and consequently $dt > 0$, if $0 < \delta \leq 2$. However, if the observer is moving sufficiently close to the left branch of the light cone, given by $dx'/dt' = -1$, eq. (192) provides us with $dt/dt' < 0$, for $\delta < 1$. Therefore $dt < 0$, the observer traverses backward in time, as measured by observers in the outer region, with $k = 1$.

The superluminal travel analysis is as follows. Imagine a spaceship leaving star A and arriving at star B, at the instant $t \approx D$. The crew of the spaceship modify the metric, so that $k \approx -1$, for simplicity, along the trajectory. Now suppose the spaceship returns to star A, travelling with a velocity arbitrarily close to the speed of light, i.e., $\frac{dx'}{dt'} \approx -1$. Therefore, from eq. (191), one obtains the following relation

$$v_{\text{return}} = \frac{dx}{dt} \approx -\frac{1}{k} = \frac{1}{1-\delta} \approx 1 \tag{193}$$

and $dt < 0$, for $dx < 0$. The return trip from star B to A is done in an interval of $\Delta t_{\text{return}} = -D/v_{\text{return}} = D/(\delta - 1)$. The total interval of time, measured at A, is given by $T_A = D + \Delta t_{\text{return}} = D\delta$. For simplicity, consider ε negligible. Superluminal travel is implicit, because $|\Delta t_{\text{return}}| < D$, if $\delta > 0$, i.e., we have a spatial spacetime interval between A and B. Note that T_A is always positive, but may attain a value arbitrarily close to zero, for an appropriate choice of δ.

Note that for the case $\delta < 1$, it is always possible to choose an allowed value of dx'/dt' for which $dt/dt' = 0$, meaning that the return trip is instantaneous as seen by observers in the external region. This follows easily from eq. (192), which implies that $dt/dt' = 0$ when dx'/dt' satisfies

$$\frac{dx'}{dt'} = -\frac{\delta}{(2-\delta)}\,, \tag{194}$$

which lies between 0 and -1 for $0 < \delta < 1$.

3.7.3. The 4-dimensional Generalization

Soon after the Krasnikov two-dimensional solution, Everett and Roman [30] generalized the analysis to four dimensions, denoting the solution as the *Krasnikov tube*. Consider that the 4-dimensional modification of the metric begins along the path of the spaceship, which is moving along the x-axis, occurring at position x at time $t \approx x$, the time of passage of the spaceship. Also assume that the disturbance in the metric propagates radially outward from the x-axis, so that causality guarantees that at time t the region in which the metric has been modified cannot extend beyond $\rho = t - x$, where $\rho = \left(y^2 + z^2\right)^{1/2}$. The modification in the metric should also not extend beyond some maximum radial distance $\rho_{max} \ll D$ from the x-axis. Thus, the metric in the 4-dimensional spacetime, written in cylindrical coordinates, is given by [30]

$$ds^2 = -dt^2 + (1 - k(t,x,\rho))dxdt + k(t,x,\rho)dx^2 + d\rho^2 + \rho^2 d\phi^2\,, \tag{195}$$

with

$$k(t,x,\rho) = 1 - (2-\delta)\theta_\varepsilon(\rho_{max} - \rho)\theta_\varepsilon(t - x - \rho)\left[\theta_\varepsilon(x) - \theta_\varepsilon(x + \varepsilon - D)\right]. \tag{196}$$

For $t \gg D + \rho_{max}$ one has a tube of radius ρ_{max} centered on the x-axis, within which the metric has been modified. This structure is denoted by the *Krasnikov tube*. In contrast with the Alcubierre spacetime metric, the metric of the Krasnikov tube is static once it has been created.

The stress-energy tensor element T_{tt} given by

$$T_{tt} = \frac{1}{32\pi(1+k)^2}\left[-\frac{4(1+k)}{\rho}\frac{\partial k}{\partial \rho} + 3\left(\frac{\partial k}{\partial \rho}\right)^2 - 4(1+k)\frac{\partial^2 k}{\partial \rho^2}\right], \tag{197}$$

can be shown to be the energy density measured by a static observer [30], and violates the WEC in a certain range of ρ, i.e., $T_{\mu\nu}U^\mu U^\nu < 0$.

To verify the violation of the WEC, consider the energy density in the middle of the tube and at a time long after it's formation, i.e., $x = D/2$ and $t \gg x + \rho + \varepsilon$, respectively. In this region we have $\theta_\varepsilon(x) = 1$, $\theta_\varepsilon(x + \varepsilon - D) = 0$ and $\theta_\varepsilon(t - x - \rho) = 1$. With this simplification the form function, eq. (196), reduces to

$$k(t,x,\rho) = 1 - (2-\delta)\theta_\varepsilon(\rho_{max} - \rho)\,. \tag{198}$$

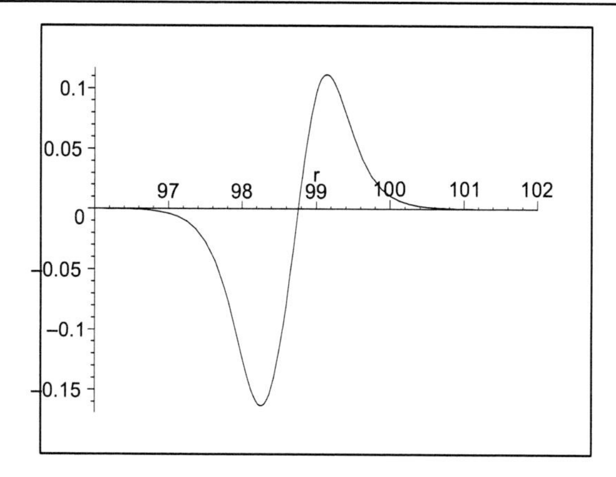

Figure 10. Graph of the energy density, T_{tt}, as a function of ρ at the middle of the Krasnikov tube, $x = D/2$, and long after it's formation, $t \gg x + \rho + \varepsilon$. We consider the following values for the parameters: $\delta = 0.1$, $\varepsilon = 1$ and $\rho_{max} = 100\varepsilon = 100$.

Consider the following specific form for $\theta_\varepsilon(\xi)$ [30] given by

$$\theta_\varepsilon(\xi) = \frac{1}{2}\left\{\tanh\left[2\left(\frac{2\xi}{\varepsilon} - 1\right)\right] + 1\right\}, \tag{199}$$

so that the form function of eq. (198) is provided by

$$k = 1 - \left(1 - \frac{\delta}{2}\right)\left\{\tanh\left[2\left(\frac{2\xi}{\varepsilon} - 1\right)\right] + 1\right\}. \tag{200}$$

Choosing the following values for the parameters: $\delta = 0.1$, $\varepsilon = 1$ and $\rho_{max} = 100\varepsilon = 100$, the negative character of the energy density is manifest in the immediate inner vicinity of the tube wall, as shown in Figure 10.

4. Closed Timelike Curves and Causality Violation

As time is incorporated into the proper structure of the fabric of spacetime, it is interesting to note that GTR is contaminated with non-trivial geometries which generate *closed timelike curves* [2, 26, 28, 134, 91, 184]. A closed timelike curve (CTC) allows time travel, in the sense that an observer which travels on a trajectory in spacetime along this curve, returns to an event which coincides with the departure. The arrow of time leads forward, as measured locally by the observer, but globally he/she may return to an event in the past. This fact apparently violates causality, opening Pandora's box and producing time travel paradoxes [186]. The notion of causality is fundamental in the construction of physical theories, therefore time travel and it's associated paradoxes have to be treated with great caution. A great variety of solutions to the Einstein Field Equations (EFEs) containing CTCs exist, but, two particularly notorious features seem to stand out. Solutions with a tipping over of the light cones due to a rotation about a cylindrically symmetric axis; and solutions that violate the Energy Conditions of GTR, which are fundamental in the singularity theorems and theorems of classical black hole thermodynamics [15].

4.1. Stationary and Axisymmetric Solutions Generating CTCs

The tipping over of light cones seem to be a generic feature of some solutions with a rotating cylindrical symmetry. The general metric for a stationary, axisymmetric solution with rotation is given by [2, 200]

$$ds^2 = -F(r)\,dt^2 + H(r)\,dr^2 + L(r)\,d\phi^2 + 2\,M(r)\,d\phi\,dt + H(r)\,dz^2\,, \tag{201}$$

in which z is the distance along the axis of rotation; r is the radial distance from the axis; ϕ is the angular coordinate; and t is the temporal coordinate. The metric components are functions of r alone. It is clear that the determinant, $g = \det(g_{\mu\nu}) = -(FL + M^2)H^2$ is Lorentzian, provided that $(FL + H^2) > 0$.

Due to the periodic nature of the angular coordinate, ϕ, an azimuthal curve with $\gamma = \{t = \text{const}, r = \text{const}, z = \text{const}\}$ is a closed curve of invariant length $s_\gamma^2 \equiv L(r)(2\pi)^2$. If $L(r)$ is negative then the integral curve with (t, r, z) fixed is a CTC. If $L(r) = 0$, then the azimuthal curve is a closed null curve, CNC. Alternatively, consider a null azimuthal curve in the (ϕ, t) plane with (r, z) fixed. It is not necessarily a geodesic, nor will it be a closed curve. The null condition, $ds^2 = 0$, implies

$$0 = -F + 2M\dot{\phi} + L\dot{\phi}^2\,, \tag{202}$$

with $\dot{\phi} = d\phi/dt$. Solving the quadratic, we have

$$\frac{d\phi}{dt} = \dot{\phi} = \frac{-M \pm \sqrt{M^2 + FL}}{L}\,. \tag{203}$$

In virtue of the Lorentzian signature constraint, $FL + H^2 > 0$, the roots are real. If $L(r) < 0$ then the light cones are tipped over sufficiently far to permit a trip to the past. By going once around the azimuthal direction, the total backward time-jump for a null curve is

$$\Delta T = \frac{2\pi|L|}{-M + \sqrt{M^2 - F|L|}}\,. \tag{204}$$

Roughly, light cones which are tilted over are generic features of spacetimes which contain CTCs.

If $L(r) < 0$ for even a single value of r, then there is a closed causal curve passing through every point of the spacetime. To visualize this consider a null curve beginning at an arbitrary x, reaching r such that $L(r) < 0$, Then follow the null curve that wraps around the azimuth a total of N times. The total backward time-jump is then $N\Delta T$. Finally, follow an ordinary null curve back to the starting point x. So, if $L(r) < 0$ for even a single value of r, the chronology-violation region covers the entire spacetime.

The present Chapter is far from making an exhaustive search of all the EFE solutions generating CTCs with these features, but the best known spacetimes will be briefly analyzed, namely, the van Stockum spacetime, the Gödel universe, the spinning cosmic strings and the Gott two-string time machine, which is a variation on the theme of the spinning cosmic string.

4.1.1. Van Stockum Spacetime

The earliest solution to the EFEs containing CTCs, is probably that of the van Stockum spacetime. It is a stationary, cylindrically symmetric solution describing a rapidly rotating infinite cylinder of dust, surrounded by vacuum. The centrifugal forces of the dust are balanced by the gravitational attraction. The metric, assuming the respective symmetries, takes the form of eq. (201). Consider a frame in which the matter is at rest, and it can be shown that the source is simply positive density dust, implying that all of the energy condition are satisfied [2].

The metric for the interior solution $r < R$, where the surface of the cylinder is located at $r = R$, is

$$ds^2 = -dt^2 + 2\omega r^2 d\phi dt + r^2(1 - \omega^2 r^2)d\phi^2 + \exp(-\omega^2 r^2)(dr^2 + dz^2) \qquad (205)$$

where ω is the angular velocity of the cylinder. It is readily verified that CTCs arise if $\omega r > 1$, i.e., for $r > 1/\omega$ the azimuthal curves with (t, r, z) fixed are CTCs. The condition $M^2 + FL = \omega^2 r^4 + r^2(1 - \omega^2 r^2) = r^2 > 0$ is imposed.

The causality violation region could be eliminated by requiring that the boundary of the cylinder to be at $r = R < 1/a$. The interior solution would then be joined to an exterior solution, which would be causally well-behaved. The resulting upper bound to the "velocity" ωR would be 1, although the orbits of the particles creating the field are timelike for all r.

Van Stockum also developed a procedure which generates an exterior solution for all $\omega R > 0$ [209]. It can be shown that the causality violation is avoided for $\omega R \leq 1/2$, but in the region $\omega R > 1/2$, CTCs appear. Consider the exterior metric components for this range, $\omega R > 1/2$:

$$H(r) = \exp(-\omega^2 r^2)\,(r/R)^{-2\omega^2 r^2}, \qquad L(r) = \frac{Rr\sin(3\beta + \gamma)}{2\sin(2\beta)\cos(\beta)},$$

$$M(r) = \frac{r\sin(\beta + \gamma)}{\sin(2\beta)}, \qquad F(r) = \frac{r\sin(\beta - \gamma)}{R\sin(\beta)},$$

with

$$\gamma = \gamma(r) = (4\omega^2 R^2 - 1)^{1/2}\ln(r/R) \qquad \text{and} \qquad \beta = \beta(r) = \arctan(4\omega^2 R^2 - 1)^{1/2}.$$

As is the interior solution, one may verify that $FL + M^2 = r^2$, so that the metric signature is Lorentzian for $R \leq r < \infty$.

The causality violations arise from the sinusoidal factors of the metric components. Thus, causality violation occur in the matter-free space surrounding a rapidly rotating infinite cylinder, as shown in Figure 11. However, it is not clear that the properties of such a cylinder also hold for realistic cylinders.

The van Stockum spacetime is not asymptotically flat. But, the gravitational potential of the cylinder's Newtonian analog also diverges at radial infinity. Shrinking the cylinder down to a "ring" singularity, one ends up with the Kerr solution, which also has CTCs (The causal structure of the Kerr spacetime has been extensively analyzed by de Felice and collaborators [210, 211, 212, 213, 214]).

In summary, the van Stockum solution contains CTC provided $\omega R > 1/2$. The causality-violating region covers the entire spacetime. Reactions to the van Stockum solution is that it is unphysical, as it applies to an infinitely long cylinder and it is not asymptotically flat.

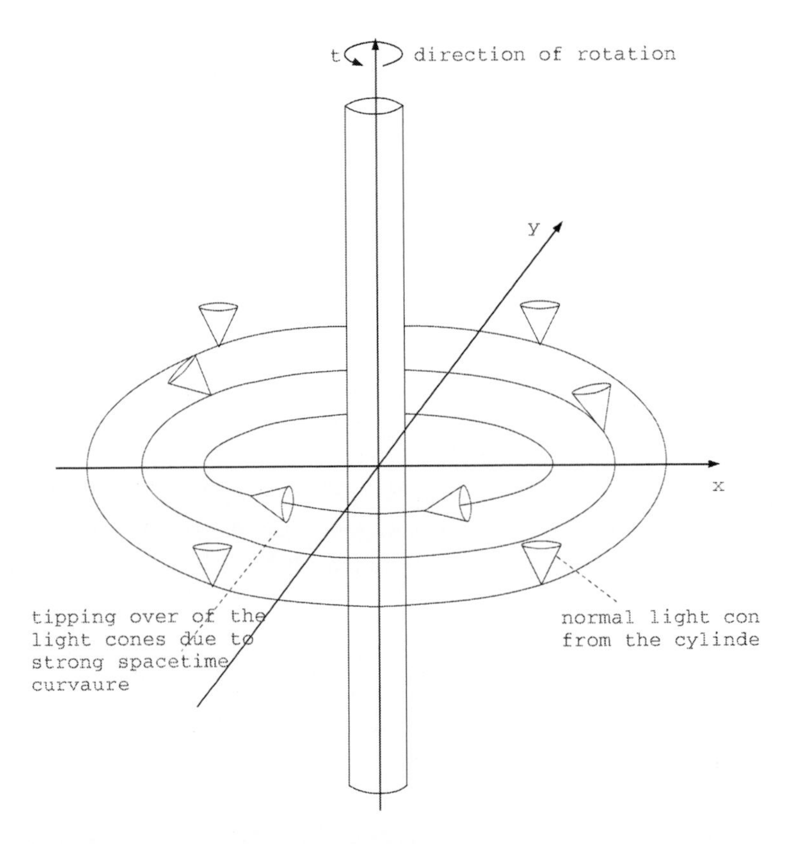

Figure 11. Van Stockum spacetime showing the tipping over of light cones close to the cylinder, due to the strong curvature of spacetime, which induce closed timelike curves.

4.1.2. The Gödel Universe

Kurt Gödel in 1949 discovered an exact solution to the EFEs of a uniformly rotating universe containing dust and a nonzero cosmological constant [215]. It can be shown that the null, weak and dominant energy conditions are satisfied. However, the dominant energy condition is in the imminence of being violated.

Consider a set of alternative coordinates, which explicitly manifest the rotational symmetry of the solution, around the axis $r = 0$, and suppressing the irrelevant z coordinate [15, 215], the metric of the Gödel solution is provided by

$$ds^2 = 2w^{-2}(-dt'^2 + dr^2 - (\sinh^4 r - \sinh^2 r)\, d\phi^2 + 2(\sqrt{2}) \sinh^2 r\, d\phi\, dt) \,. \qquad (206)$$

Moving away from the axis, the light cones open out and tilt in the ϕ-direction. The azimuthal curves with $\gamma = \{t = \text{const}, r = \text{const}, z = \text{const}\}$ are CTCs if the condition $r > \ln(1 + \sqrt{2})$ is satisfied [2].

It is interesting to note that in the Gödel spacetime, closed timelike curves are not geodesics. However, Novello and Rebouças [216] discovered a new generalized solution of the Gödel metric, of a shear-free nonexpanding rotating fluid, in which successive concentric causal and noncausal regions exist, with closed timelike curves which are geodesics. A complete study of geodesic motion in Gödel's universe, using the method of the effective potential was further explored by Novello *et al* [217]. Much interest has been aroused in time travel in the Gödel spacetime, from which we may mention the analysis of the geodesical and non-geodesical motions considered by Pfarr [218] and Malament [219, 220].

4.1.3. Spinning Cosmic String

Consider an infinitely long straight string that lies and spins around the z-axis. The symmetries are analogous to the van Stockum spacetime, but the asymptotic behavior is different [2, 221]. We restrict the analysis to an infinitely long straight string, with a delta-function source confined to the z-axis. It is characterized by a mass per unit length, μ; a tension, τ, and an angular momentum per unit length, J. For cosmic strings, the mass per unit length is equal to the tension, $\mu = \tau$.

In cylindrical coordinates the metric takes the following form

$$ds^2 = -\left[d(t + 4GJ\varphi)\right]^2 + dr^2 + (1 - 4G\mu)^2\, r^2\, d\varphi^2 + dz^2\,, \tag{207}$$

with the following coordinate range

$$-\infty < t < +\infty, \qquad 0 < r < \infty, \qquad 0 \leq \varphi \leq 2\pi, \qquad -\infty < z < +\infty\,. \tag{208}$$

Consider an azimuthal curve, i.e., an integral curve of φ. Closed timelike curves appear whenever

$$r < \frac{4GJ}{1 - 4G\mu}\,. \tag{209}$$

These CTCs can be deformed to cover the entire spacetime, consequently, the chronology-violating region covers the entire manifold.

4.1.4. Gott Cosmic String Time Machine

An extremely elegant model of a time-machine was constructed by Gott [222]. The Gott time-machine is an exact solution of the EFE for the general case of two moving straight cosmic strings that do not intersect [222]. This solution produces CTCs even though they do not violate the WEC, have no singularities and event horizons, and are not topologically multiply-connected as the wormhole solution. The appearance of CTCs relies solely on the gravitational lens effect and the relativity of simultaneity.

It is also interesting to verify whether the CTCs in the Gott solution appear at some particular moment, i.e., when the strings approach each other's neighborhood, or if they already pre-exist, i.e., they intersect any spacelike hypersurface. These questions are particularly important in view of Hawking's Chronology Protection Conjecture [140]. This conjecture states that the laws of physics prevent the creation of CTCs. If correct, then the solutions of the EFE which admit CTCs are either unrealistic or are solutions in which the CTCs are pre-existing, so that the time -machine is not created by dynamical processes.

Amos Ori proved that in Gott's spacetime, CTCs intersect every $t = \text{const}$ hypersurface [223], so that it is not a counter-example to the Chronology Protection Conjecture.

The global structure of the Gott spacetime was further explored by Cutler [224], and it was shown that the closed timelike curves are confined to a certain region of the spacetime, and that the spacetime contains complete spacelike and achronal hypersurfaces from which the causality violating regions evolve. Grant also examined the global structure of the two-string spacetime and found that away from the strings, the space is identical to a generalized Misner space [225]. The vacuum expectation value of the energy-momentum tensor for a conformally coupled scalar field was then calculated on the respective generalized Misner space, which was found to diverge weakly on the chronology horizon, but diverge strongly on the polarized hypersurfaces. Thus, the back reaction due to the divergent behaviour around the polarized hypersurfaces are expected to radically alter the structure of spacetime, before quantum gravitational effects become important, suggesting that Hawking's chronology protection conjecture holds for spaces with a noncompactly generated chronology horizon. Soon after, Laurence [226] showed that the region containing CTCs in Gott's two-string spacetime is identical to the regions of the generalized Misner space found by Grant, and constructed a family of isometries between both Gott's and Grant's regions. This result was used to argue that the slowly diverging vacuum polarization at the chronology horizon of the Grant space carries over without change to the Gott space. Furthermore, it was shown that the Gott time machine is unphysical in nature, for such an acausal behaviour cannot be realized by physical and timelike sources [227, 228, 229, 230, 231].

4.2. Solutions Violating the Energy Conditions

The traditional manner of solving the EFEs, $G_{\mu\nu} = 8\pi G T_{\mu\nu}$, consists in considering a plausible stress-energy tensor, $T_{\mu\nu}$, and finding the geometrical structure, $G_{\mu\nu}$. But one can run the EFE in the reverse direction by imposing an exotic metric $g_{\mu\nu}$, and eventually finding the matter source for the respective geometry. In this fashion, solutions violating the energy conditions have been obtained. Adopting the reverse philosophy, solutions such as traversable wormholes, the warp drive, the Krasnikov tube and the Ori-Soen spacetime have been obtained. These solutions violate the energy conditions and with simple manipulations generate CTCs.

4.2.1. Conversion of Traversable Wormholes into Time Machines

Much interest has been aroused in traversable wormholes since the classical article by Morris and Thorne [1]. A wormhole is a hypothetical tunnel which connects different regions in spacetime. These solutions are multiply-connected and probably involve a topology change, which by itself is a problematic issue. One of the most fascinating aspects of wormholes is their apparent ease in generating CTCs [135]. There are several ways to generate a time machine using multiple wormholes [2], but a manipulation of a single wormhole seems to be the simplest way [135, 232]. The basic idea is to create a time shift between both mouths. This is done invoking the time dilation effects in special relativity or in general relativity, i.e., one may consider the analogue of the twin paradox, in which the mouths are moving one with respect to the other, or simply the case in which one of the mouths is placed in a strong gravitational field.

To create a time shift using the twin paradox analogue, consider that the mouths of the wormhole may be moving one with respect to the other in external space, without significant changes of the internal geometry of the handle. For simplicity, consider that one of the mouths A is at rest in an inertial frame, whilst the other mouth B, initially at rest practically close by to A, starts to move out with a high velocity, then returns to its starting point. Due to the Lorentz time contraction, the time interval between these two events, ΔT_B, measured by a clock comoving with B can be made to be significantly shorter than the time interval between the same two events, ΔT_A, as measured by a clock resting at A. Thus, the clock that has moved has been slowed by $\Delta T_A - \Delta T_B$ relative to the standard inertial clock. Note that the tunnel (handle), between A and B remains practically unchanged, so that an observer comparing the time of the clocks through the handle will measure an identical time, as the mouths are at rest with respect to one another. However, by comparing the time of the clocks in external space, he will verify that their time shift is precisely $\Delta T_A - \Delta T_B$, as both mouths are in different reference frames, frames that moved with high velocities with respect to one another. Now, consider an observer starting off from A at an instant T_0, measured by the clock stationed at A. He makes his way to B in external space and enters the tunnel from B. Consider, for simplicity, that the trip through the wormhole tunnel is instantaneous. He then exits from the wormhole mouth A into external space at the instant $T_0 - (\Delta T_A - \Delta T_B)$ as measured by a clock positioned at A. His arrival at A precedes his departure, and the wormhole has been converted into a time machine. See Figure 12.

For concreteness, following the Morris *et al* analysis [135], consider the metric of the accelerating wormhole given by

$$ds^2 = -(1 + glF(l)\cos\theta)^2\, e^{2\Phi(l)}\, dt^2 + dl^2 + r^2(l)\,(d\theta^2 + \sin^2\theta\, d\phi^2)\,, \tag{210}$$

where the proper radial distance, $dl = (1 - b/r)^{-1/2}\, dr$, is used. $F(l)$ is a form function that vanishes at the wormhole mouth A, at $l \leq 0$, rising smoothly from 0 to 1, as one moves to mouth B; $g = g(t)$ is the acceleration of mouth B as measured in its own asymptotic rest frame. Consider that the external metric to the respective wormhole mouths is $ds^2 \cong -dT^2 + dX^2 + dY^2 + dZ^2$. Thus, the transformation from the wormhole mouth coordinates to the external Lorentz coordinates is given by

$$T = t\,, \qquad Z = Z_A + l\,\cos\theta\,, \qquad X = l\,\sin\theta\,\cos\phi\,, \qquad X = l\,\sin\theta\,\sin\phi\,, \tag{211}$$

for mouth A, where Z_A is the time-independent Z location of the wormhole mouth A, and

$$T = T_B + v\gamma\, l\,\cos\theta\,,\; Z = Z_B + \gamma\, l\,\cos\theta\,,\; X = l\,\sin\theta\,\cos\phi\,,\; X = l\,\sin\theta\,\sin\phi\,, \tag{212}$$

for the accelerating wormhole mouth B. The world line of the center of mouth B is given by $Z = Z_B(t)$ and $T = T_B(t)$ with $ds^2 = dT_B^2 - dZ_B^2$; $v(t) \equiv dZ_B/dT_B$ is the velocity of mouth B and $\gamma = (1 - v^2)^{-1/2}$ the respective Lorentz factor; the acceleration appearing in the wormhole metric is given $g(t) = \gamma^2\, dv/dt$ [199].

Novikov considered other variants of inducing a time shift through the time dilation effects in special relativity, by using a modified form of the metric (210), and by considering a circular motion of one of the mouths with respect to the other [233]. Another interesting manner to induce a time shift between both mouths is simply to place one of the mouths in a strong external gravitational field, so that times slows down in the respective mouth. The time shift will be given by $T = \int_i^f \left(\sqrt{g_{tt}(x_A)} - \sqrt{g_{tt}(x_A)}\,\right) dt$ [2, 136].

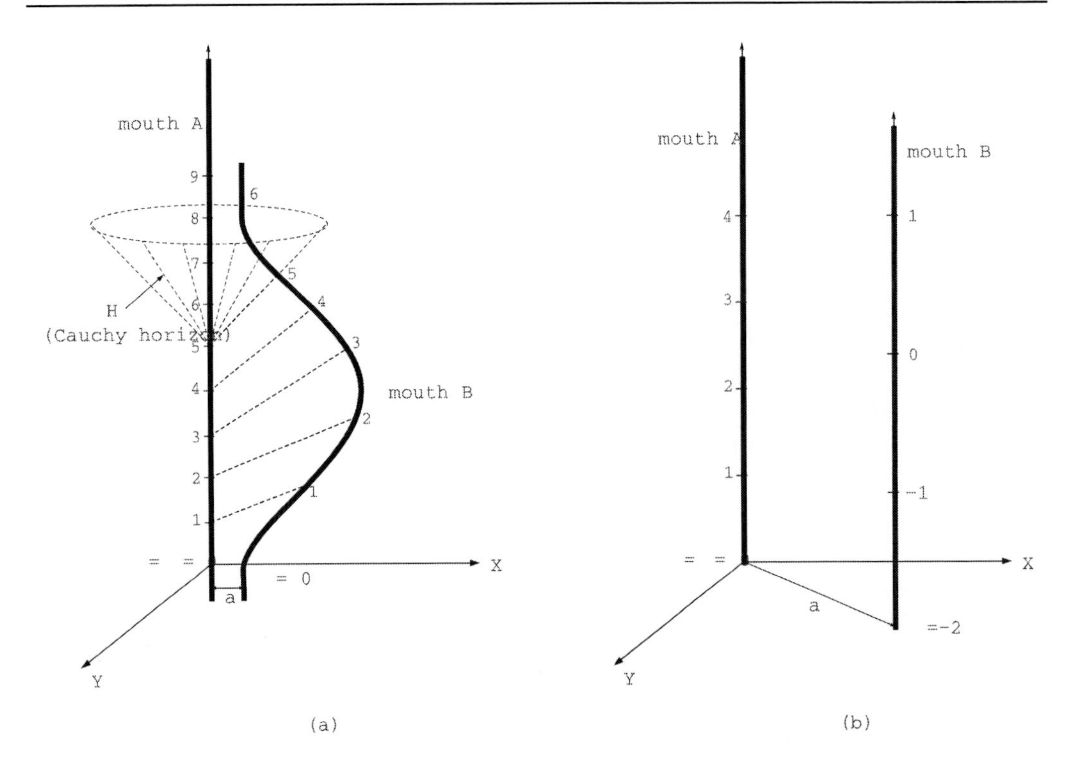

Figure 12. Depicted are two examples of wormhole spacetimes with closed timelike curves. The wormholes tunnels are arbitrarily short, and its two mouths move along two world tubes depicted as thick lines in the figure. Proper time τ at the wormhole throat is marked off, and note that identical values are the same event as seen through the wormhole handle. In Figure (a), mouth A remains at rest, while mouth B accelerates from A at a high velocity, then returns to its starting point at rest. A time shift is induced between both mouths, due to the time dilation effects of special relativity. The light cone-like hypersurface H shown is a Cauchy horizon. Through every event to the future of H there exist CTCs, and on the other hand there are no CTCs to the past of H. In Figure (b), a time shift between both mouths is induced by placing mouth B in strong gravitational field. See text for details.

4.2.2. The Ori-Soen Time Machine

A time-machine model was also proposed by Amos Ori and Yoav Soen which significantly ameliorates the conditions of the EFE's solutions which generate CTCs [234, 235, 236, 237]. The Ori-Soen model presents some notable features. It was verified that CTCs evolve from a well-defined initial slice, a partial Cauchy surface, which does not display causality violation. The partial Cauchy surface and spacetime are asymptotically flat, contrary to the Gott spacetime, and topologically trivial, contrary to the wormhole solutions. The causality violation region is constrained within a bounded region of space, and not in infinity as in the Gott solution. The WEC is satisfied up until and beyond a time slice $t = 1/a$, on which the CTCs appear.

4.2.3. Warp Drive and Closed Timelike Curves

Within the framework of general relativity, it is possible to warp spacetime in a small *bubblelike* region [172], in such a way that the bubble may attain arbitrarily large velocities, $v(t)$. Inspired in the inflationary phase of the early Universe, the enormous speed of separation arises from the expansion of spacetime itself. The model for hyperfast travel is to create a local distortion of spacetime, producing an expansion behind the bubble, and an opposite contraction ahead of it. See Section 3.2. for details.

One may consider a hypothetical spaceship immersed within the bubble, moving along a timelike curve, regardless of the value of $v(t)$. Due to the arbitrary value of the warp bubble velocity, the metric of the warp drive permits superluminal travel, which raises the possibility of the existence of CTCs. Although the solution deduced by Alcubierre by itself does not possess CTCs, Everett demonstrated that these are created by a simple modification of the Alcubierre metric [30], by applying a similar analysis as in tachyons.

4.2.4. The Krasnikov Tube and Closed Timelike Curves

Krasnikov discovered an interesting feature of the warp drive, in which an observer in the center of the bubble is causally separated from the front edge of the bubble. Therefore he/she cannot control the Alcubierre bubble on demand. Krasnikov proposed a two-dimensional metric [173], which was later extended to a four-dimensional model [180], as outlined in Section 3.7.. One Krasnikov tube in two dimensions does not generate CTCs. But the situation is quite different in the 4-dimensional generalization. Using two such tubes it is a simple matter, in principle, to generate CTCs. The analysis is similar to that of the warp drive, so that it will be treated in summary.

Imagine a spaceship travelling along the x-axis, departing from a star, S_1, at $t = 0$, and arriving at a distant star, S_2, at $t = D$. An observer on board of the spaceship constructs a Krasnikov tube along the trajectory. It is possible for the observer to return to S_1, travelling along a parallel line to the x-axis, situated at a distance ρ_0, so that $D \gg \rho_0 \gg 2\rho_{max}$, in the exterior of the first tube. On the return trip, the observer constructs a second tube, analogous to the first, but in the opposite direction, i.e., the metric of the second tube is obtained substituting x and t, for $X = D - x$ and $T = t - D$, respectively in eq. (195). The fundamental point to note is that in three spatial dimensions it is possible to construct a system of two non-overlapping tube separated by a distance ρ_0.

After the construction of the system, an observer may initiate a journey, departing from S_1, at $x = 0$ and $t = 2D$. One is only interested in the appearance of CTCs in principle, therefore the following simplifications are imposed: δ and ε are infinitesimal, and the time to travel between the tubes is negligible. For simplicity, consider the velocity of propagation close to that of light speed. Using the second tube, arriving at S_2 at $x = D$ and $t = D$, then travelling through the first tube, the observer arrives at S_1 at $t = 0$. The spaceship has completed a CTC, arriving at S_1 before it's departure.

5. Conclusion

In this Chapter we have considered two particular spacetimes which violate the energy conditions of general relativity, namely, traversable wormholes and "warp drive" spacetimes. It is important to emphasize that these solutions are primarily useful as "gedanken-experiments" and as a theoretician's probe of the foundations of general relativity. They have also been important to stimulate research in the issues of the energy condition violations, closed timelike curves and the associated causality violations and "effective" superluminal travel.

In the Introduction, we have outlined a review of wormhole physics dating from the "Einstein-Rosen" bridge, the revival of the issue by Wheeler with the introduction of the "geon" concept in the 1960s, the full renaissance of the subject by Thorne and collaborators in the late 1980s, culminating in the monograph by Visser, and detailed the issues that branched therefrom to the present date. In Section 2., we have presented a mathematical overview of the Morris-Thorne wormhole, paying close attention to the pointwise and averaged energy condition violations, the concept of the Quantum Inequality and the respective constraints on wormhole geometries, and the introduction of the "Volume Integral Quantifier" which in a certain measure quantifies the amount of energy condition violating matter needed to sustain wormhole spacetimes. We then, treated rotating wormholes and evolving wormholes in a cosmological background, focussing mainly on the energy condition violations, and in particular, on the inflating wormhole and the respective evolution in a flat FRW universe.

In Section 3., we have considered the superluminal "warp drive" spacetimes, which also violate the energy conditions and generate closed timelike curves with slight modifications to the spacetime metric. In particular, we have analyzed the Alcubierre and Natário spacetimes, paying close attention to the energy condition violations, reviewing the application of the Quantum Inequality and the superluminal features of these spacetimes, in particular, the appearance of horizons. The discovery that the outer frontal regions of the warp bubble is not in causal contact with a hypothetical observer placed in the center on the bubble, and thus cannot be controlled on demand, inspired the solution known as the Krasnikov spacetime. This spacetime in two and four dimensions were also briefly analyzed. Finally, an important theorem deduced by Olum regarding the intimate relationship between superluminal travel and negative energy densities is also presented.

Using linearized theory, we have also considered a more realistic model of the warp drive spacetime where the warp bubble interacts with a finite mass spaceship. We have tested and quantified the energy conditions to first and second order of the warp bubble velocity. By doing so we have been able to safely ignore the causality problems associated with "superluminal" motion, and so have focussed attention on a previously unremarked feature of the "warp drive" spacetime. If it is possible to realize even a weak-field warp drive in nature, such a spacetime appears to be an example of a "reaction-less drive". That is, the warp bubble moves by interacting with the geometry of spacetime instead of expending reaction mass, and the spaceship (which in linearized theory can be treated as a finite mass object placed within the warp bubble), is simply carried along with it. We have verified that in this case, the "total amount" of energy condition violating matter (the "net" negative energy of the warp field) must be an appreciable fraction of the positive mass of the spaceship carried along by the warp bubble. This places an extremely stringent condition on the warp drive spacetime, namely, that for all conceivably interesting situations the bubble velocity should be absurdly low, and it therefore appears unlikely that, by using this analysis, the warp drive will ever prove to be technologically useful. Finally, we point out that any attempt at building up a "strong-field" warp drive starting from an approximately Minkowski spacetime will inevitably have to pass through a weak-field regime. Since the weak-field warp drives are already so tightly constrained, the analysis of this work implies additional difficulties for developing a "strong field" warp drive.

Finally, in Section 4., we have analyzed some solutions to the Einstein field equation that generate closed timelike curves. Far from attempting at an exhaustive search of spacetimes that possess closed timelike curves, we have focussed on two particularly notorious properties of these geometries, namely, solutions with a tipping over of the light cones due to a rotation about a cylindrically symmetric axis and solutions that violate the energy conditions of general relativity. Closed timelike curves are troublesome as they apparently violate causality, and it is not clear that even an eventual theory of quantum gravity will provide us with an answer. However, as stated by Kip Thorne, time travel in the form of closed timelike curves, is more than a justification for theoretical speculation, it is a conceptual tool and an epistemological instrument to probe the fundamental foundations of general relativity and to extract some eventual clarifying views.

Acknowledgments

The author was funded by Fundação para a Ciência e a Tecnologia (FCT)–Portugal through the grant SFRH/BPD/26269/2006.

References

[1] M. S. Morris and K. S. Thorne, "Wormholes in spacetime and their use for interstellar travel: A tool for teaching General Relativity," *Am. J. Phys.* 56, 395 (1988).

[2] M. Visser, *Lorentzian Wormholes: From Einstein to Hawking* (American Institute of Physics, New York, 1995).

[3] L. Flamm, "Beitrage zur Einsteinschen Gravitationstheorie," *Phys. Z.* 17, 448 (1916).

[4] A. Einstein and N. Rosen, "The particle problem in the General Theory of Relativity," *Phys. Rev.* 48, 73-77 (1935).

[5] J. A. Wheeler, "Geons," *Phys. Rev.* 97, 511-536 (1955).

[6] J. A. Wheeler, *Geometrodynamics*, (Academic Press, New York, 1962).

[7] S. W. Hawking, "Wormholes in spacetime," *Phys. Rev.* D 37, 904 (1988).

[8] R. P. Geroch, "Topology in General Relativity," *J. Math. Phys.* 8, 782 (1967).

[9] F. J. Ernst, Jr., "Variational calculations in geon theory," *Phys. Rev.* 105, 1662-1664 (1957);

F. J. Ernst, Jr., "Linear and toroidal geons," *Phys. Rev.* 105, 1665-1670 (1957);

D. R. Brill and J. B. Hartle, "Method of the self-consistent field in General Relativity and its application to the gravitational geon," *Phys. Rev.* 135, B271-B278 (1964);

D. R. Brill and J. B. Hartle, "Method of the self-consistent field in General Relativity and its application to the gravitational geon," *Phys. Rev.* 135, B271-B278 (1964);

A. Komar, "Bootstrap gravitational geons," *Phys. Rev.* 137, B462-B466 (1965);

C. H. Brans, "Singularities in bootstrap gravitational geons," *Phys. Rev.* 140, B1174-B1176 (1965);

D. J. Kaup, "Klein-Gordon geon," *Phys. Rev.* 172, 1331-1342 (1968);

M. Lunetta, I. Wolk, and A. F. d. F. Teixeira, "Pure massless scalar geon," *Phys. Rev.* D 21, 3281-3283 (1980);

P. R. Anderson and D. R. Brill, "Gravitational geons revisited," *Phys. Rev.* D 56, 4824-4833 (1997);

T. Diemer and M. J. Hadley, "Charge and topology of spactime," *Class. Quant. Grav.* 16, 3567-3577 (1999).

[10] C. W. Misner, "Wormhole initial conditions," *Phys. Rev.* 118, 1110-1111 (1960).

[11] R. W. Fuller and J. A. Wheeler, "Causality and multiply connected space-Time," *Phys. Rev.* 128, 919-929 (1962).

[12] J. C. Graves and D. R. Brill, "Oscillatory character of Reissner-Nordstrm metric for an ideal charged wormhole," *Phys. Rev.* 120, 1507-1513 (1960).

[13] B. K. Harrison, K. S. Thorne, M. Wakano and J. A. Wheeler, *Gravitational Theory and Gravitational Collapse*, (University of Chicago Press, Chicago, 1965).

[14] Ya. B. Zel'dovich and I. D. Novikov, *Relativistic Astrophysics, Vol. I: Stars and Relativity*, (University of Chicago Press, Chicago, 1971).

[15] S. W. Hawking and G.F.R. Ellis, *The Large Scale Structure of Spacetime*, (Cambridge University Press, Cambridge 1973).

[16] G. Klinkhammer, "Averaged energy conditions for free scalar fields in flat spacetime," *Phys. Rev.* D 43, 2542 (1991).

[17] F. J. Tipler, "Energy conditions and spacetime singularities," *Phys. Rev.* D 17, 2521 (1978).

[18] T. A. Roman, "Quantum stress-energy tensors and the weak energy condition," *Phys. Rev.* D 33, 3526 (1986).

[19] L. H. Ford, "Quantum coherence effects and the second law of thermodynamics," *Proc. Roy. Soc. Lond.* A 364, 227 (1978).

[20] L. H. Ford, "Constraints on negative-energy fluxes," *Phys. Rev.* D 43, 3972 (1991).

[21] L. H. Ford and T. A. Roman, "Averaged energy conditions and quantum inequalities," Phys. Rev. D 51, 4277 (1995) [arXiv:gr-qc/9410043].

[22] L. H. Ford and T. A. Roman, "Quantum field theory constrains traversable wormhole geometries," *Phys. Rev.* D 53, 5496 (1996) [arXiv:gr-qc/9510071].

[23] L. H. Ford and T. A. Roman, "The quantum interest conjecture," *Phys. Rev.* D 60, 104018 (1999) [arXiv:gr-qc/9901074].

[24] M.J. Pfenning, L.H. Ford, "Quantum inequalities on the energy density in static Robertson-Walker spacetimes," *Phys. Rev.* D 55, 4813 (1997) [arXiv:gr-qc/9608005].

[25] P. K. F. Kuhfittig, "Static and dynamic traversable wormhole geometries satisfying the Ford-Roman constraints," *Phys. Rev.* D 66, 024015 (2002) [arXiv:gr-qc/0401023].

[26] F. Lobo, P. Crawford: *"Constraints on wormhole geometries,"* The Ninth Marcel Grossmann Meeting, Proceedings, Rome, Italy, 2000, edited by Vahe G. Gurzadyan, Robert Jantzen and Remo Ruffini (World Scientific, Singapore, 2002), 855-856.

[27] T. A. Roman, *"Some Thoughts on Energy Conditions and Wormholes,"* [arXiv:gr-qc/0409090].

[28] F. Lobo and P. Crawford, "Weak energy condition violation and superluminal travel," Current Trends in Relativistic Astrophysics, Theoretical, Numerical, Observational, *Lecture Notes in Physics* 617, Springer-Verlag Publishers, L. Fernández et al. eds, pp. 277–291 (2003) [arXiv:gr-qc/0204038].

[29] M. J. Pfenning and L. H. Ford, "The unphysical nature of warp drive," *Class. Quant. Grav.* 14, 1743, (1997) [arXiv:gr-qc/9702026].

[30] A. E. Everett and T. A. Roman, "A Superluminal Subway: The Krasnikov Tube," *Phys. Rev. D* 56, 2100 (1997) [arXiv:gr-qc/9702049].

[31] C. Barcelo and M. Visser, "Scalar fields, energy conditions, and traversable wormholes," *Class. Quant. Grav.* 17, 3843 (2000) [arXiv:gr-qc/0003025].

[32] A. G. Riess *et al.*, "Type Ia Supernova Discoveries at $z > 1$ From the Hubble Space Telescope: Evidence for Past Deceleration and Constraints on Dark Energy Evolution," *Astrophys. J.* 607 (2004) 665-687 [arXiv:astro-ph/0402512].

[33] M. Visser, "Jerk, snap, and the cosmological equation of state," *Class. Quant. Grav.* 21, 2603 (2004) [arXiv:gr-qc/0309109].

[34] R. R. Caldwell, M. Kamionkowski and N. N. Weinberg, "Phantom Energy and Cosmic Doomsday," *Phys. Rev. Lett.* 91 071301 (2003) [arXiv:astro-ph/0302506].

[35] C. Barcelo and M. Visser, "Twilight for the energy conditions?," *Int. J. Mod. Phys. D* 11, 1553 (2002) [arXiv:gr-qc/0205066].

[36] D. Hochberg, A. Popov and S. Sushkov, "Self-consistent wormhole solutions of semiclassical gravity," *Phys. Rev. Lett.* 78, 2050 (1997) [arXiv:gr-qc/9701064].

[37] S. V. Sushkov, "A selfconsistent semiclassical solution with a throat in the theory of gravity," *Phys. Lett.* **A164**, 33-37 (1992).

[38] A. A. Popov and S. V. Sushkov, "Vacuum polarization of a scalar field in wormhole space-times," *Phys. Rev. D* **63**, 044017 (2001) [arXiv:gr-qc/0009028];

A. A. Popov, "Stress energy of a quantized scalar field in static wormhole space-times," *Phys. Rev. D* **64**, 104005 (2001) [arXiv:hep-th/0109166].

[39] N. R. Khusnutdinov and S. V. Sushkov, "Ground state energy in a wormhole space-time," *Phys. Rev. D* **65**, 084028 (2002) [arXiv:hep-th/0202068].

[40] A. R. Khabibullin, N. R. Khusnutdinov and S. V. Sushkov, "Casimir effect in a wormhole spacetime," *Class. Quant. Grav.* **23** 627-634 (2006) [arXiv:hep-th/0510232].

[41] N. R. Khusnutdinov, "Semiclassical wormholes," *Phys. Rev. D* **67**, 124020 (2003) [arXiv:hep-th/0304176].

[42] R. Garattini, "Self Sustained Traversable Wormholes?" *Class. Quant. Grav.* **22** 1105-1118 (2005) [arXiv:gr-qc/0501105];

[43] R. Garattini and F. S. N. Lobo, "Self sustained phantom wormholes in semi-classical gravity," [arXiv:gr-qc/0701020].

[44] C. Barcelo and M. Visser, "Traversable wormholes from massless conformally coupled scalar fields," *Phys. Lett. B* 466 (1999) 127 [arXiv:gr-qc/9908029].

[45] H. G. Ellis, "Ether flow through a drainhole: A particle model in general relativity," *J. Math. Phys.* 14, 104 (1973).

[46] H. G. Ellis, "The evolving, flowless drain hole: a nongravitating particle model in general relativity theory," *Gen. Rel. Grav.* 10, 105-123 (1979).

[47] K. A. Bronnikov, "Scalar-tensor theory and scalar charge," *Acta Phys. Pol. B* 4, 251 (1973).

[48] T. Kodama, "General-relativistic nonlinear field: A kink solution in a generalized geometry," *Phys. Rev. D* 18, 3529 (1978).

[49] G. Clément, "Einstein-Yang-Mills-Higgs solitons," *Gen. Rel. Grav.* 13, 763 (1981).

[50] G. Clément, "The Ellis geometry," *Am. J. Phys.* 57, 967 (1989).

[51] E. G. Harris, "Wormhole connecting two Reissner-Nordstrom universes," *Am. J. Phys.* 61, 1140 (1993).

[52] M. Visser, "Traversable wormholes: Some simple examples," *Phys. Rev. D* 39, 3182 (1989).

[53] M. Visser, "Traversable wormholes: The Roman ring," *Phys. Rev. D* 55, 5212 (1997) [arXiv:gr-qc/9702043].

[54] M. Visser and D. Hochberg, "Generic wormhole throats," in The Internal Structure of Black Holes and Spacetime Singularities, *Institute of Physics*, Bristol, edited by L.M. Burko and A. Ori, pp.249-295 (1997) [arXiv:gr-qc/9710001].

[55] N. Dadhich, S. Kar, S. Mukherjee and M. Visser, "$R = 0$ spacetimes and self-dual Lorentzian wormholes," *Phys. Rev. D* 65, 064004 (2002) [arXiv:gr-qc/0109069].

[56] V. P. Frolov and I. D. Novikov, "Wormhole as a device for studying a black hole interior," *Phys. Rev. D* 48, 1607 (1993).

[57] D. Hochberg and T. W. Kephart, "Wormhole Cosmology and the Horizon Problem," *Phys. Rev. Lett.* 70, 2665 (1993).

[58] P. F. González-Díaz, "Ringholes and closed timelike curves," *Phys. Rev. D* 54, 6122 (1996) [arXiv:gr-qc/9608059].

[59] P. F. González-Díaz, "Wormholes and ringholes in a dark-energy universe," *Phys. Rev. D* 68, 084016 (2003) [arXiv:astro-ph/0308382].

[60] G. Clément, "Flat wormholes from cosmic strings," *J. Math. Phys.* 38, 5807 (1997) [arXiv:gr-qc/9701060].

[61] R. O. Aros and N. Zamorano, "Wormhole at the core of an infinite cosmic string," *Phys. Rev. D* 56, 6607 (1997) [arXiv:gr-qc/9711044].

[62] F. Schein, P. C. Aichelburg and W. Israel, "String-supported wormhole spacetimes containing closed timelike curves," *Phys. Rev. D* 54, 3800 (1996) [arXiv:gr-qc/9602053].

[63] F. Schein and P. C. Aichelburg, "Traversable wormholes in geometries of charged shells," *Phys. Rev. Lett.* 77, 4130 (1996) [arXiv:gr-qc/9606069].

[64] D. N. Vollick, "Maintaining a wormhole with a scalar field," *Phys. Rev. D* 56, 4724-4728 (1997) [arXiv:gr-qc/9806071].

[65] S.-W. Kim and S. P. Kim, "Traversable wormhole with classical scalar fields," *Phys. Rev. D* 58, 087703 (1998).

[66] S.-W. Kim and H. Lee, "Exact solutions of a charged wormhole," *Phys. Rev. D* 63, 064014 (2001).

[67] E. Teo, "Rotating traversable wormholes," *Phys. Rev. D* 58, 024014 (1998) [arXiv:gr-qc/9803098].

[68] P. K. F. Kuhfittig, "Axially symmetric rotating traversable wormhole," *Phys. Rev. D* 67, 064015 (2003) [arXiv:gr-qc/0401028].

[69] F. Parisio, "Wormholes: Controlling exotic matter with a magnetic field," *Phys. Rev. D* 63, 087502 (2001).

[70] S. Krasnikov, "Traversable wormhole," *Phys. Rev. D* 62, 084028 (2002) [arXiv:gr-qc/9909016].

[71] S. A. Hayward, "Wormholes supported by pure ghost radiation," *Phys. Rev. D* 65, 124016 (2002) [arXiv:gr-qc/0202059].

[72] L. Á. Gergely, "Wormholes, naked singularities, and universes of ghost radiation," *Phys. Rev. D* 65, 127502 (2002) [arXiv:gr-qc/0204016].

[73] K. Bronnikov and S. Grinyok, "Charged wormholes with non-minimally coupled scalar fields. Existence and stability," [arXiv:gr-qc/0205131].

[74] K. A. Bronnikov, "Regular magnetic black holes and monopoles from nonlinear electrodynamics," *Phys. Rev.* D63, 044005 (2001) [arXiv:gr-qc/0006014];

A. V. B. Arellano and F. S. N. Lobo, "Evolving wormhole geometries within nonlinear electrodynamics," *Class. Quant. Grav.* 23, 5811 (2006) [arXiv:gr-qc/0608003];

A. V. B. Arellano and F. S. N. Lobo, "Non-existence of static, spherically symmetric and stationary, axisymmetric traversable wormholes coupled to nonlinear electrodynamics," *Class. Quant. Grav.* 23, 7229 (2006) [arXiv:gr-qc/0604095].

[75] S. A. Hayward and H. Koyama, "How to make a traversable wormhole from a Schwarzschild black hole," *Phys. Rev. D* 70 101502 (2004) [arXiv:gr-qc/0406080].

[76] H. Koyama and S. A. Hayward, "Construction and enlargement of traversable wormholes from Schwarzschild black holes," *Phys. Rev. D* 70 084001 (2004) [arXiv:gr-qc/0406113].

[77] C. G. Boehmer, T. Harko and F. S. N. Lobo, "Conformally symmetric traversable wormholes," *Phys. Rev. D* 76 084014 (2007), arXiv:0708.1537 [gr-qc].

[78] P. K. Kuhfittig, "A wormhole with a special shape function," *Am. J. Phys.* 67, 125 (1999);

P. K. F. Kuhfittig, "Can a wormhole supported by only small amounts of exotic matter really be traversable?," *Phys. Rev. D* 68, 067502 (2003) [arXiv:gr-qc/0401048].

[79] D. Hochberg and M. Visser, "Null energy condition in dynamic wormholes," *Phys. Rev. Lett.* 81, 746 (1998) [arXiv:gr-qc/9802048].

[80] D. Hochberg and M. Visser, "Dynamic wormholes, antitrapped surfaces, and energy conditions," *Phys. Rev. D* 58, 044021 (1998) [arXiv:gr-qc/9802046].

[81] S. Kar, "Evolving wormholes and the energy conditions," *Phys. Rev. D* 49, 862 (1994).

[82] S. Kar and D. Sahdev, "Evolving Lorentzian wormholes," *Phys. Rev. D* 53, 722 (1996) [arXiv:gr-qc/9506094].

[83] S. W. Kim, "Cosmological model with a traversable wormhole," *Phys. Rev. D* 53, 6889 (1996).

[84] M. Visser, S. Kar and N. Dadhich, "Traversable wormholes with arbitrarily small energy condition violations," *Phys. Rev. Lett.* 90, 201102 (2003) [arXiv:gr-qc/0301003].

[85] S. Kar, N. Dadhich and M. Visser, "Quantifying energy condition violations in traversable wormholes," *Pramana* 63, 859-864 (2004) [arXiv:gr-qc/0405103].

[86] M. Visser, "Quantum wormholes," *Phys. Rev. D* 43, 402-409 (1991).

[87] S. Kim, "Schwarzschild-de Sitter type wormhole," *Phys. Lett. A* 166, 13 (1992).

[88] F. S. N. Lobo and P. Crawford, "Linearized stability analysis of thin-shell wormholes with a cosmological constant," *Class. Quant. Grav.* 21, 391 (2004) [arXiv:gr-qc/0311002].

[89] T. A. Roman, "Inflating Lorentzian wormholes," *Phys. Rev. D* 47, 1370 (1993) [arXiv:gr-qc/9211012].

[90] M. S. R. Delgaty and R. B. Mann, "Traversable wormholes in (2+1) and (3+1) dimensions with a cosmological constant," *Int. J. Mod. Phys.* D 4, 231 (1995) [arXiv:gr-qc/9404046].

[91] J. P. S. Lemos, F. S. N. Lobo and S. Q. de Oliveira, "Morris-Thorne wormholes with a cosmological constant," *Phys. Rev. D* 68, 064004 (2003) [arXiv:gr-qc/0302049].

[92] F. S. N. Lobo, "Surface stresses on a thin shell surrounding a traversable wormhole," *Class. Quant. Grav.* 21 4811 (2004) [arXiv:gr-qc/0409018].

[93] F. S. N. Lobo, "Energy conditions, traversable wormholes and dust shells," *Gen. Rel. Grav.* 37, 2023 (2005) [arXiv:gr-qc/0410087].

[94] J. P. S. Lemos and F. S. N. Lobo, "Plane symmetric traversable wormholes in an anti-de Sitter background," *Phys. Rev. D* 69 (2004) 104007 [arXiv:gr-qc/0402099].

[95] A. DeBenedictis and A. Das, "On a general class of wormholes," *Class. Quant. Grav.* 18, 1187 (2001) [arXiv:gr-qc/0009072].

[96] A. DeBenedictis and A. Das, "Higher dimensional wormhole geometries with compact dimensions," *Nucl. Phys.* B653 279 (2003) [arXiv:gr-qc/0207077].

[97] M. Visser, "Quantum mechanical stabilization of Minkowski signature wormholes," *Phys. Lett.* B 242, 24 (1990).

[98] S. W. Kim, H. Lee, S. K. Kim and J. Yang, "$(2+1)-$dimensional Schwarschild-de Sitter wormhole," *Phys. Lett. A* 183, 359 (1993).

[99] G. P. Perry and R. B. Mann, “Traversible wormholes in $(2+1)-$dimensons,” *Gen. Rel. Grav.* 24, 305 (1992).

[100] M. Visser, “Traversable Wormholes From Surgically Modified Schwarzschild Space-Times,” *Nucl. Phys. B* 328 (1989) 203.

[101] E. Poisson and M. Visser, “Thin-shell wormholes: Linearization stability,” *Phys. Rev. D* 52, 7318 (1995) [arXiv:gr-qc/9506083].

[102] E. F. Eiroa and G. E. Romero, “Linearized stability of charged thin-shell wormoles,” *Gen. Rel. Grav.* 36, 651 (2004), [arXiv:gr-qc/0303093].

[103] M. Ishak and K. Lake, “Stability of transparent spherically symmetric thin shells and wormholes,” *Phys. Rev. D* 65, 044011 (2002) [arXiv:gr-qc/0108058].

[104] C. Armendáriz-Picón, “On a class of stable, traversable Lorentzian wormholes in classical general relativity”. *Phys. Rev. D* 65, 104010 (2002) [arXiv:gr-qc/0201027].

[105] H. Shinkai and S. A. Hayward, “Fate of the first traversable wormhole: Black hole collapse or inflationary expansion,” *Phys. Rev. D* 66, 044005 (2002) [arXiv:gr-qc/0205041].

[106] K. A. Bronnikov and S. Grinyok, “Instability of wormholes with a non-minimally coupled scalar field,” *Grav. Cosmol.* 7, 297 (2001) [arXiv:gr-qc/0201083].

[107] A. Chodos and S. Detweiler, “Spherical symmetric solutions in five-dimensional general relativity,” *Gen. Rel. Grav.* 14, 879 (1982).

[108] G. Clément, “A class of wormhole solutions to higher-dimensional general relativity,” *Gen. Rel. Grav.* 16, 131 (1984).

[109] J. W. Moffat and T. Svoboda, “Traversible wormholes and the negative-stress-energy problem in the nonsymmetric gravitational theory,” *Phys. Rev. D* 44, 429-432 (1991).

[110] A. G. Agnese and M. La Camera, “Wormholes in the Brans-Dicke theory of gravitation,” *Phys. Rev. D* 51, 2011 (1995).

[111] L. A. Anchordoqui, S. Perez Bergliaffa, and D. F. Torres, “Brans-Dicke wormholes in nonvacuum spacetime,” *Phys. Rev. D* 55, 5226 (1997).

[112] K. K. Nandi, B. Bhattacharjee, S. M. K. Alam and J. Evans, “Brans-Dicke wormholes in the Jordan and Einstein frames,” *Phys. Rev. D* 57, 823 (1998).

[113] P. E. Bloomfield, “Comment On Brans-Dicke Wormholes In The Jordan And Einstein Frames,” *Phys. Rev. D* 59, 088501 (1999).

[114] K. K. Nandi, “Reply To Comment On ’Brans-Dicke Wormholes In The Jordan And Einstein Frames’,” *Phys. Rev. D* 59, 088502 (1999).

[115] K. K. Nandi, A. Islam and J. Evans, “Brans Wormholes,” *Phys. Rev. D* 55, 2497-2500 (1997).

[116] F. He and S.-W. Kim, "New Brans-Dicke wormholes," *Phys. Rev. D* 65, 084022 (2002) .

[117] S. You-Gen, G. Han-Ying, T. Zhen-Qiang, and D. Hao-Gang, "Wormholes in Kaluza-Klein theory," *Phys. Rev. D* 44, 1330 (1991).

[118] B. Bhawal and S. Kar, "Lorentzian wormholes in Einstein-Gauss-Bonnet theory," *Phys. Rev. D* **46**, 2464 (1992).

[119] H. Koyama, S. A. Hayward and S. Kim, "Construction and enlargement of dilatonic wormholes by impulsive radiation," *Phys. Rev. D* 67, 084008 (2003) [arXiv:gr-qc/0212106].

[120] L. A. Anchordoqui and S. E. Perez Bergliaffa, "Wormhole surgery and cosmology on the brane: The world is not enough," *Phys. Rev.* D 62, 076502 (2000) [arXiv:gr-qc/0001019].

[121] C. Barceló and M. Visser, "Brane surgery: energy conditions, traversable wormholes, and voids," *Nucl. Phys.* B584, 415 (2000) [arXiv:gr-qc/0004022].

[122] K. A. Bronnikov and S.-W. Kim, "Possible wormholes in a brane world," *Phys. Rev. D* 67, 064027 (2003) [arXiv:gr-qc/0212112].

[123] F. S. N. Lobo, "General class of braneworld wormholes," *Phys. Rev.* D75, 064027 (2007) [arXiv:gr-qc/0701133].

[124] M. La Camera, "Wormhole solutions in the Randall-Sundrum scenario," *Phys. Lett.* B 573, 27-32 (2003) [arXiv:gr-qc/0306017].

[125] N. Furey and A. DeBenedictis, "Wormhole throats in R^m gravity," *Class. Quant. Grav.* 22, 313 (2005) [arXiv:gr-qc/0410088].

[126] S. Nojiri and S. D. Odintsov, "Modified gravity with negative and positive powers of the curvature: unification of the inflation and of the cosmic acceleration," *Phys. Rev. D* 68, 123512 (2003) [arXiv:hep-th/0307288].

[127] R. Dick, "On the Newtonian limit in gravity models with inverse powers of R," *Gen. Rel. Grav.* 36, 217 (2004) [arXiv:gr-qc/0307052].

[128] K. Ghoroku and T. Soma, "Lorentzian wormholes in higher-derivative gravity and the weak energy condition," *Phys. Rev. D* 46, 1507 (1992).

[129] R. R. Caldwell and D. Langlois, "Shortcuts in the fifth dimension," *Phys. Lett. B* 511, 129 (2001) [arXiv:gr-qc/0103070].

[130] G. Kaelbermann, "Communication through an extra dimension," *Int. J. Mod. Phys. A* 15, 3197 (2000) [arXiv:gr-qc/9910063].

[131] E. Abdalla, B. Cuadros-Melgar, S. Feng and B. Wang, "Shortest cut in brane cosmology," *Phys. Rev. D* 65, 083512 (2002).

[132] E. Abdalla and B. Cuadros-Melgar, "Shortcuts in domain walls and the horizon problem," *Phys. Rev. D* 67, 084012 (2003).

[133] D. Chung and K. Freese, "Can geodesics in extra dimensions solve the cosmological horizon problem?," *Phys. Rev. D* 62, 063513 (2000).

[134] F. Lobo and P. Crawford, "Time, Closed Timelike Curves and Causality," in The Nature of Time: *Geometry, Physics and Perception*, NATO Science Series II. *Mathematics, Physics and Chemistry* - Vol. 95, Kluwer Academic Publishers, R. Buccheri et al. eds, pp.289-296 (2003) [arXiv:gr-qc/0206078].

[135] M. S. Morris, K. S. Thorne and U. Yurtsever, "Wormholes, Time Machines and the Weak Energy Condition," *Phys. Rev. Lett.* 61, 1446 (1988).

[136] V. P. Frolov and I. D. Novikov, "Physical effects in wormholes and time machines," *Phys. Rev. D* 42, 1057 (1990).

[137] J. L. Friedman, M. S. Morris, I. D. Novikov, F. Echeverria, G. Klinkhammer, K. S. Thorne, and U. Yurtsever, "Cauchy problem in spacetimes with closed timelike curves," *Phys. Rev. D* 42, 1915-1930 (1990).

[138] J. L. Friedman and M. S. Morris, "The Cauchy problem for the scalar wave equation is well defined on a class of spacetimes with closed timelike curves," *Phys. Rev. Lett.* 66, 401-404 (1991)

[139] S. Kim and K. S. Thorne, "Do vacuum fluctuations prevent the creation of closed timelike curves?," *Phys. Rev. D* 43, 3929 (1991).

[140] S. W. Hawking, "Chronology protection conjecture," *Phys. Rev. D* 46, 603 (1992).

[141] G. Klinkhammer, "Vacuum polarization of scalar and spinor fields near closed null geodesics," *Phys. Rev. D* 46, 3388-3394 (1992).

[142] S. W. Kim, "Particle creation for time travel through a wormhole," *Phys. Rev. D* 46, 2428 (1992).

[143] M. Lyutikov, "Vacuum polarization at the chronology horizon of the Roman spacetime," *Phys. Rev. D* 49, 4041 (1994).

[144] W. A. Hiscock and D. A. Konlowski, "Quantum vacuum energy in Taub-NUT-type cosmologies," *Phys. Rev. D* 26, 1225 (1982).

[145] L.-X. Li and J. R. Gott, "Self-consistent vacuum for Misner space and the chronology protection conjecture," *Phys. Rev. Lett.* 80, 2980 (1998) [arXiv:gr-qc/9711074].

[146] W. A. Hiscock, *"Quantized fields and chronology protection,"* [arXiv:gr-qc/0009061].

[147] M. Visser, *"The quantum physics of chronology protection,"* in The future of theoretical physics and cosmology, Cambridge University Press, edited by G. Gibbons *et al*, pp.161-176 (2003) [arXiv:gr-qc/0204022].

[148] K. S. Thorne, "Closed timelike curves," in *General Relativity and Gravitation*, Proceedings of the 13th Conference on General Relativity and Gravitation, edited by R. J. Gleiser et al (Institute of Physics Publishing, Bristol, 1993), p. 295.

[149] A. Grant *et al*, "The Farthest known supernova: Support for an accelerating Universe and a glimpse of the epoch of deceleration," *Astrophys. J.* 560 49-71 (2001) [arXiv:astro-ph/0104455].

[150] S. Perlmutter, M. S. Turner and M. White, "Constraining dark energy with SNe Ia and large-scale strucutre," *Phys. Rev. Lett.* 83 670-673 (1999) [arXiv:astro-ph/9901052].

[151] C. L. Bennett *et al*, "First year *Wilkinson Microwave Anisotropy Probe* (WMAP) observations: Preliminary maps and basic results," *Astrophys. J. Suppl.* 148 1 (2003) [arXiv:astro-ph/0302207].

[152] G. Hinshaw *et al*, "First year *Wilkinson Microwave Anisotropy Probe* (WMAP) observations: The angular power spectrum," [arXiv:astro-ph/0302217].

[153] R. Cai and A. Wang , *"Cosmology with Interaction between Phantom Dark Energy and Dark Matter and the Coincidence Problem,"* [arXiv:hep-th/0411025].

[154] M. Carmelli, *"Accelerating universe, cosmological constant and dark energy,"* [arXiv:astro-ph/0111259].

[155] A. Melchiorri, L. Mersini, C. J. Ödman and M. Trodden, "The state of the dark energy equation of state," *Phys. Rev. D* 68 043509 (2003) [arXiv:astro-ph/0211522].

[156] J. S. Alcaniz, "Testing dark energy beyond the cosmological constant barrier," *Phys. Rev. D* 69 083521 (2004) [arXiv:astro-ph/0312424].

[157] S. M. Carroll, M. Hoffman and M. Trodden, "Can the dark energy equation-of-state parameter w be less than -1?," *Phys. Rev. D* 68 023509 (2003) [arXiv:astro-ph/0301273].

[158] R. R. Caldwell, "A phantom menace: Cosmological consequences of a dark energy component with super-negative equation of state," *Phys. Lett.* B545 23-29 (2002) [arXiv:astro-ph/9908168]; P. F. González-Díaz, "You need not be afraid of phantom energy," *Phys. Rev. D* 68 021303(R) (2003) [arXiv:astro-ph/0305559]; P. F. González-Díaz, "Achronal cosmic future," *Phys. Rev. Lett.* 93 071301 (2004) [arXiv:astro-ph/0404045]; P. F. González-Díaz and J. A. Jimenez-Madrid, "Phantom inflation and the 'Big Trip'," *Phys. Lett.* B596 16-25 (2004) [arXiv:hep-th/0406261]; P. F. González-Díaz, "Stable accelerating universe with no hair," *Phys. Rev. D* 685 104035 (2002) [arXiv:astro-ph/0305559]; R. R. Caldwell, M. Kamionkowski and N. N. Weinberg, "Phantom Energy and Cosmic Doomsday," *Phys. Rev. Lett.* 91 071301 (2003) [arXiv:astro-ph/0302506].

[159] F. S. N. Lobo, "Phantom energy traversable wormholes," *Phys. Rev.* D71, 084011 (2005) [arXiv:gr-qc/0502099];

[160] F. S. N. Lobo, "Stability of phantom wormholes," *Phys. Rev.* D71, 124022 (2005) [arXiv:gr-qc/0506001].

[161] S. Sushkov, "Wormholes supported by a phantom energy," *Phys. Rev.* D71, 043520 (2005) [arXiv:gr-qc/0502084];

O. B. Zaslavskii, "Exactly solvable model of wormhole supported by phantom energy," *Phys. Rev.* D72, 061303 (2005) [arXiv:gr-qc/0508057].

P. K. F. Kuhfittig, *Class. Quant. Grav.* 23, 5853 (2006) [arXiv:gr-qc/0608055].

[162] F. S. N. Lobo, "Chaplygin traversable wormholes," *Phys. Rev.* D73, 064028 (2006) [arXiv:gr-qc/0511003];

[163] F. S. N. Lobo, "Van der Waals quintessence stars," *Phys. Rev.* D75, 024023 (2007) [arXiv:gr-qc/0610118];

[164] J. G. Cramer, R. L. Forward, M. S. Morris, M. Visser, G. Benford and G. A. Landis, "Natural wormholes as gravitational lenses," *Phys. Rev. D* 51, 3117 (1995) [arXiv:astro-ph/9409051].

[165] D. F. Torres, G. E. Romero and L. A. Anchordoqui, "Wormholes, gamma ray bursts and the amount of negative mass in the Universe," *Mod. Phys. Lett. A* 13, 1575 (1998) [arXiv:gr-qc/9805075].

[166] M. Safonova, D. F. Torres and G. E. Romero, "Microlensing by natural wormholes: theory and simulations," *Phys. Rev. D* 65, 023001 (2002) [arXiv:gr-qc/0105070].

[167] M. Safonova, D. F. Torres, G. E. Romero, "Macrolensing signatures of large-scale violations of the weak energy condition," *Mod. Phys. Lett. A* 16, 153 (2001) [arXiv:astro-ph/0104075].

[168] S. A. Hayward, *"Black holes and traversible wormholes: a synthesis,"* [arXiv:gr-qc/0203051].

[169] M. Visser, B. Bassett and S. Liberati, *"Perturbative superluminal censorship and the null energy condition,"* Proceedings of the Eighth Canadian Conference on General Relativity and Relativistic Astrophysics (AIP Press) (1999) [arXiv:gr-qc/9908023].

[170] M. Visser, B. Bassett and S. Liberati, "Superluminal censorship," *Nucl. Phys. Proc. Suppl.* 88, 267-270 (2000) [arXiv:gr-qc/9810026].

[171] K. Olum, "Superluminal travel requires negative energy density," *Phys. Rev. Lett.*, 81, 3567-3570 (1998) [arXiv:gr-qc/9805003].

[172] M. Alcubierre, "The warp drive: hyper-fast travel within general relativity," *Class. Quant. Grav.* 11, L73-L77 (1994) [arXiv:gr-qc/0009013].

[173] S. V. Krasnikov, "Hyper-fast Interstellar Travel in General Relativity," *Phys. Rev. D* 57, 4760 (1998) [arXiv:gr-qc/9511068].

[174] J. Natário, “Warp drive with zero expansion,” *Class. Quant. Grav.* 19, 1157, (2002) [arXiv:gr-qc/0110086].

[175] F. S. N. Lobo and M. Visser, “Fundamental limitations on ‘warp drive’ spacetimes,” *Class. Quant. Grav.* 21, 5871 (2004). [arXiv:gr-qc/0406083].

[176] D. H. Coule, “No warp drive,” *Class. Quant. Grav.* 15, 2523-2527 (1998).

[177] W. A. Hiscock, “Quantum effects in the Alcubierre warp drive spacetime,” *Class. Quant. Grav.* 14, L183 (1997) [arXiv:gr-qc/9707024].

[178] C. Clark, W. A. Hiscock and S. L. Larson, “Null geodesics in the Alcubierre warp drive spacetime: the view from the bridge,” *Class. Quant. Grav.* 16, 3965 (1999) [arXiv:gr-qc/9907019].

[179] P. F. González-Díaz, “On the warp drive space-time,” *Phys. Rev. D* 62, 044005 (2000) [arXiv:gr-qc/9907026].

[180] A. E. Everett, “Warp Drive and Causality,” *Phys. Rev. D* 53 7365 (1996).

[181] C. van den Broeck, “A warp drive with more reasonable total energy requirements,” *Class. Quant. Grav.* 16, 3973, (1999) [arXiv:gr-qc/9905084].

[182] P. Gravel and J. Plante, “Simple and double walled Krasnikov tubes: I. Tubes with low masses,” *Class. Quant. Grav.* 21, L7, (2004).

[183] P. Gravel, “Simple and double walled Krasnikov tubes: II. Primordial microtubes and homogenization,” *Class. Quant. Grav.* 21, 767, (2004).

[184] F. J. Tipler, “Causality violation in asymptotically flat space-times,” *Phys. Rev. Lett* 37, 879-882 (1976).

[185] V. P. Frolov and I. D. Novikov, “Physical effects in wormholes and time machines,” *Phys. Rev. D* 42, 1057 (1990).

[186] P. J. Nahin, *Time Machines: Time Travel in Physics, Metaphysics and Science Fiction*, Springer-Verlag and AIP Press, New York (1999).

[187] J. R. Gott and L.-X. Li, “Can the Universe create itself?,” *Phys. Rev. D* 58, 023501 (1998).

[188] S. W. Kim and K. S. Thorne, “Do vacuum fluctuations prevent the creation of closed timelike curves?,” *Phys. Rev. D* 43 3929 (1991).

[189] D. Deutsch, “Quantum mechanics near closed timelike lines,” *Phys. Rev. D* 44 3197 (1991).

[190] S. V. Krasnikov, “On the quantum stability of the time machine,” *Phys. Rev. D* 54 7322 (1996) [arXiv:gr-qc/9508038].

[191] J. Earman, *Bangs, Crunches, Whimpers, and Shrieks: Singularities and Acausalities in Relativistic Spacetimes*, Oxford University Press (1995).

[192] F. G. Echeverria, G. Klinkhammer, K. S. and Thorne, "Billiard Balls in Wormhole Spacetimes with Closed Timelike Curves: Classical Theory," *Phys. Rev. D* 44 1077 (1991).

[193] I. D. Novikov, "Time machine and self-consistent evolution in problems with self-interaction," *Phys. Rev. D* 45 1989 (1992).

[194] A. Carlini, V.P. Frolov, M. B. Mensky, I. D. Novikov and H. H. Soleng, "Time machines: The principle of self-consistency as a consequence of the principle of minimal action," *Int. J. Mod. Phys. D* 4 557 (1995); erratum-ibid D 5 99 (1996), [arXiv:gr-qc/9506087].

[195] A. Carlini and I. D. Novikov, "Time machines and the principle of self-consistency as a consequence of the principle of stationary action. ii: The Cauchy problem for a self-interacting relativistic particle," *Int. J. Mod. Phys. D* 5 445 (1996) [arXiv:gr-qc/9607063].

[196] J. D. E. Grant, "Cosmic strings and chronology protection," *Phys. Rev. D* 47 2388 (1993).

[197] L.-X. Li, "New light on time machines: Against the chronology protection conjecture," *Phys. Rev. D* 50, R6037 (1994).

[198] D. Hochberg and M. Visser, "Geometric structure of the generic static traversable wormhole throat," *Phys. Rev. D* 56, 4745 (1997) [arXiv:gr-qc/9704082].

[199] C. W. Misner, K. S. Thorne and J. A. Wheeler, *Gravitation* (W. H. Freeman and Company, San Francisco, 1973).

[200] R. M. Wald, *General Relativity*, (University of Chicago Press, Chicago, 1984).

[201] K. K. Nandi, Y. Z. Zhang and K. B. Vijaya Kumar, *Phys. Rev.* D 70, 127503 (2004) [arXiv:gr-qc/0407079].

[202] L. A. Anchordoqui, D. F. Torres and M. L. Trobo, "Evolving wormhole geometries," *Phys. Rev. D* 57, 829 (1998) [arXiv:gr-qc/9710026].

[203] A. Lichnerowicz, *"Théories Relativistes de la Gravitation et de l'Electromagnetisme,"* Masson, Paris (1955).

[204] W. Israel, "Singular hypersurfaces and thin shells in general relativity," *Nuovo Cimento* 44B, 1 (1966); and corrections in *ibid.* 48B, 463 (1966).

[205] G. Darmois, *"Mémorial des sciences mathématiques XXV,"* Fasticule XXV ch V (Gauthier-Villars, Paris, France, 1927).

[206] F. S. N. Lobo and P. Crawford, "Stability analysis of dynamic thin shells," *Class. Quant. Grav.* 22, 4869 (2005), [arXiv:gr-qc/0507063].

[207] F. S. N. Lobo, "Stable dark energy stars," *Class. Quant. Grav.* 23, 1525 (2006) [arXiv:gr-qc/0508115].

[208] M. Bouhmadi-Lopez and J. A. Jimenez Madrid, *JCAP* 0505, 005 (2005) [arXiv:astro-ph/0404540].

[209] F. J. Tipler, "Rotating Cylinders and the Possibility of Global Causality Violation," *Phys. Rev. D* 9, 2203 (1974).

[210] F. de Felice and M. Calvani, "Causality Violation In The Kerr Metric," *Gen. Rel. Grav.* 10, 335 (1979).

[211] M. Calvani, F. de Felice, B. Muchotrzeb and F. Salmistraro, "Time Machine And Geodesic Motion In Kerr Metric," *Gen. Rel. Grav.* 9, 155 (1978).

[212] C. J. S. Clarke and F. de Felice, "Globally Non-Causal Space-Times," *J. Phys.* A 15, 2415 (1982).

[213] F. de Felice, "Timelike nongeodesic trajectories which violate causality: A rigorous derivation," *Nuovo Cimento* 65B 224-232 (1981).

[214] C. J. S. Clarke and F. de Felice, "Globally non-causal spacetimes II: Naked singularities and curvature conditions," *Gen. Rel. Grav.* 16 139-148 (1984).

[215] K. Gödel, "An Example of a New Type of Cosmological Solution of Einstein's Field Equations of Gravitation," *Rev. Mod. Phys.* 21, 447 (1949).

[216] M. Novello and M. J. Rebouças, "Rotating universe with successive causal and non-causal regions," *Phys. Rev. D* 19, 2850 (1979).

[217] M. Novello, I. Damião Soares and J. Tiomno , "Geodesic motion and confinement in Gödel's universe," *Phys. Rev. D* 27, 779 (1983).

[218] J. Pfarr, "Time travel in Gödel's space," *Gen. Rel. Grav.* 13 1073 (1981).

[219] D. B. Malament, "Minimal acceleration requirements for "time travel" in Gödel's space-time," *J. Math. Phys.* 26 774 (1985).

[220] D. B. Malament, "A note about closed timelike curves in Gödel's space-time," *J. Math. Phys.* 28 2427 (1987).

[221] B. P. Jensen and H. H. Soleng, "General relativistic model of a spinning cosmic string," *Phys. Rev. D* 39 1130 (1989).

[222] J. R. Gott, "Closed Timelike Curves Produced by Pairs of Moving Cosmic Strings: Exact Solutions," *Phys. Rev. Lett.* 66 1126 (1991).

[223] A. Ori, "Rapidly moving cosmic strings and chronology protection," *Phys. Rev. D* 44, R2214 (1991).

[224] C. Cutler, "Global structure of Gott's two-string spacetime," *Phys. Rev. D* 45, 487 (1992).

[225] J. D. E. Grant, "Cosmic strings and chronology protection," *Phys. Rev. D* 47, 2388 (1993).

[226] D. Laurence, "Isometries between Gott's two-string spacetime and Grant's generalization of Misner space," *Phys. Rev. D* 50, 4957 (1994).

[227] S. Deser, R. Jackiw and G. t'Hooft, "Physical cosmic strings do not generate closed timelike curves," *Phys. Rev. Lett.* 68 267 (1992).

[228] S. Deser and R. Jackiw, "Time travel?," *Comments Nucl. Part. Phys.* 20 337 (1992).

[229] S. Deser, "Physical obstacles to time-travel," *Class. Quant. Grav.* 10 S67 (1993).

[230] S. M. Carroll, E. Farhi and A. H. Guth , "An obstacle to building a time machine," *Phys. Rev. Lett.* 68, 263-266 (1992).

[231] S. M. Carroll, E. Farhi, A. H. Guth and K. D. Olum , "Energy-momentum restrictions on the creation of Gott time machines," *Phys. Rev. D* 50, 6190-6206 (1994).

[232] M. Visser, "Wormholes, baby universes, and causality," *Phys. Rev. D* 41 1116 (1990).

[233] I. D. Novikov, "An analysis of the operation of a time machine," *Sov. Phys. JETP* 68 3 (1989).

[234] A. Ori, "Must Time-Machine Construction Violate the Weak Energy Condition?," *Phys. Rev. Lett.* 71 2517 (1993).

[235] Y. Soen and A. Ori, "Causality Violation and the Weak Energy Condition," *Phys. Rev. D* 49 3990 (1994).

[236] Y. Soen and A. Ori, "Improved time-machine model," *Phys. Rev. D* 54 4858 (1996).

[237] K. D. Olum, "The Ori-Soen time machine," *Phys. Rev. D* 61 124022 (2002) [arXiv:gr-qc/9907007].

In: Classical and Quantum Gravity Research
Editors: M.N. Christiansen et al, pp. 79-159
ISBN 978-1-60456-366-5

Chapter 2

NONLINEAR PERTURBATIONS AND CONSERVATION LAWS ON CURVED BACKGROUNDS IN GR AND OTHER METRIC THEORIES

A.N. Petrov*
Relativistic Astrophysics group, Sternberg Astronomical institute,
Universitetskii pr., 13, Moscow, 119992, RUSSIA

Abstract

In this paper we review the field-theoretical approach. In this framework perturbations in general relativity as well as in an arbitrary D-dimensional metric theory are described and studied. A background, on which the perturbations propagate, is a solution (arbitrary) of the theory. Lagrangian for perturbations is defined, and field equations for perturbations are derived from the variational principle. These equations are exact, equivalent to the equations in the standard formulation and have a form that permits an easy and natural expansion to an arbitrary order. Being covariant, the field-theoretical description is also invariant under gauge (inner) transformations, which can be presented both in exact and approximate forms. Following the usual field-theoretical prescriptions, conserved quantities for perturbations are constructed. Conserved currents are expressed through divergences of superpotentials — antisymmetric tensor densities. This form allows to relate a necessity to consider local properties of perturbations with a theoretical representation of the quasi-local nature of conserved quantities in metric theories. Properties of the conserved quantities under gauge transformations are established and analyzed, this allows to describe the well known non-localization problem in explicit mathematical expressions and operate with them. Applications of the formalism in general relativity for studying 1) the falloff at spatial infinity in asymptotically flat spacetimes, 2) linear perturbations on Friedmann-Robertson-Walker backgrounds, 3) a closed Friedmann world and 4) black holes presented as gravitationally-field configurations in a Minkowski space, are reviewed. Possible applications of the formalism in cosmology and astrophysics are also discussed. Generalized formulae for an arbitrary metric D-dimensional theory are tested to calculate the mass of a Schwarzschild-anti-de Sitter black hole in the Einstein-Gauss-Bonnet gravity.

PACS 04.20.Cv, 04.25.Nx, 04.50.+h, 11.30.-j.

*E-mail address: anpetrov@rol.ru; Telephone number: +7 (495) 7315222.

1. Introduction and Preliminaries

1.1. Perturbations in Gravitational Theories, Cosmology and Relativistic Astrophysics

Much of research in general relativity (GR) is frequently carried out under the assumption that perturbations of different kinds propagate in a given (fixed) background spacetime (exact solution to the Einstein equations) [1] - [4]. A majority of cosmological and astrophysical problems are also studied in the framework of a perturbation approach. It is quite impossible to give a more or less full bibliography on this topic. Nevetherless, to stress the importance of such studies, we shall outline shortly some of the related directions.

Cosmological perturbations on Freidmann-Robertson-Walker (FRW) backgrounds have been considered beginning from the famous work by Lifshitz [5]. The Lifshitz principles were developed by many authors in a variety of approaches (see, e.g., the popular review [6], and a recent review [7]). As well known examples, one can note the following. The effect of amplification of gravitational waves in an isotropic world was discovered by Grishchuk [8]; the gauge invariant theory of perturbations was formulated by Lukash [9] and Bardeen [10]. Lately, non-trivial perturbations in FRW worlds are being considered more frequently. For example, note the recent papers [11, 12] where, using the quasi-isotropic expansions, the authors describe non-decreasing modes of adiabatic and isocurvature scalar perturbations, and gravitational waves close to cosmological singularity. Deviations of a space-time metric from the homogeneous isotropic background become large; while locally measurable quantities, like Riemann tensor components, are still close to their FRW values. An approach, where integrals of FRW models are presented in a time independent form both for "vacuum" and for "usual" matter, has been developed [13]. In the two last decades, cosmological perturbations are considered not only in the linear approximation, but including the second order also (see the recent review [14]).

The evolution of quantized fields, including gravitational and electromagnetic fields, is intensively studied on curved backgrounds of various general classes, such as globally hyperbolic, static with symmetry groups, *etc.* In this framework, exact solutions, like FRW, anti-de Sitter (AdS) and Bianchi of different types, are also exploited as backgrounds. As an example, note the theory which is being developed by Grishchuk (see the reviews [15, 16] and references there in). Relic gravitational waves and primordial density perturbations are generated by strong variable gravitational field of the early FRW universe. The generating mechanism is the parametric amplification of the zero-point quantum oscillations. These generated fields have specific statistical properties of squeezed vacuum quantum states. Cosmological perturbations at the early inflation stage [17] also continue to be studied (see, e.g., recent papers [18] - [20] and, e.g., reviews [21] - [23] and references there in). Separately, the great interest to AdS spaces has been initiated by the discovery of the present (not at the earlier inflation stage) accelerated cosmological expansion (see review [24]). To explain the acceleration new cosmological solutions are searched [25]. On backgrounds of such solutions perturbations are also to be considered.

Several black hole solutions [26], which could represent the neighborhoods of relativistic astrophysics objects, also play a role of backgrounds for evolution of different kinds of perturbations. Amplification and dispersion of metric perturbations (including gravitational waves), electromagnetic field and massless and mass scalar fields are studied on these backgrounds (see, e.g., resent works [27] - [33]).

The rapid development of the detecting technique stimulates a development of the gravitational-wave physics (see the recent detailed review [34] and references there in). Thus the theoretical study of propagation and interaction of gravitational waves becomes especially important. In the works by Alekseev (see [35, 36] and references there in), the monodromy transform approach was developed for constructing exact solutions of the Einstein equations with spacetime isometries. This method has been applied for exact solving the characteristic initial value problem of the collision and subsequent nonlinear interaction of plane gravitational or gravitational and electromagnetic waves with distinct wave fronts in a Minkowski space [37] - [39].

Currently, following an extraordinary interest with brane models, D-dimensional metric theories of gravitation have been examined more and more intensively. Fundamental works on brane worlds have appeared two decades ago (as we know the works by Rubakov and Shaposhnikov [40] and by Akama [41] are the first in this direction). Later, especially after the works by Randall and Sundrum [42, 43], the interest to these models has risen significantly (see, e.g., a review [44]). Perturbations, including gravitational waves, in the framework of D-dimensional metric theories and brane models are also studied intensively (see, e.g., [45] - [51] and references there in).

The list of directions, where perturbations on curved or flat backgrounds are studied, could be continued. Impressing results were obtained in this direction. However, we would like to note the problems associated with methods of such investigations, rather than the results. As a rule, these methods are restricted since they are constructed for examination of particular tasks only. Thus:

- Although the modern cosmic experimental and observable data require more detailed theoretical results, frequently studies are carried out in the linear approximation only, without taking into account the "back reaction".
- Applying particular methods, one generally uses many additional assumptions. Thus. it is not clear: what results are more general and what results change under a change of these assumptions.
- It is difficult to understand: could an approach developed for a one concrete background be applied to other backgrounds. Frequently only simplified backgrounds are used; *etc.*
- In each particular case, considering perturbations, one needs to fix gauge freedoms. A concrete fixation is connected with a concrete mapping of a perturbed spacetime onto a background spacetime. It turns out, that it is not so simple to understand what gauge is "better", or how to find a connection between different gauges, *etc.*

1.2. Conservation Laws and Their Properties

Very important characteristics for studying perturbations are such quantities as energy-momentum, angular momentum, their densities, fluxes, *etc.* However, as is well known, the definition of energy and other conserved quantities in GR has principal problems, which are well described in many textbooks (see, e.g., [3]). We repeat some related issues to stress importance of these notions in a theoretical development of gravitational theory also.

It is useful to reconsider works by Einstein who paid special attention to conservation laws for "energy components" $t_\mu{}^\nu$ of the gravitational field. From the early stage, when the theory was not yet presented in a satisfactory form (its equations were $R_{\mu\nu} = \kappa T_{\mu\nu}$), he examined the conservation law $\partial_\nu(t_\mu{}^\nu + T_\mu{}^\nu) = 0$ [52, 53]. Then, the final form $R_{\mu\nu} = \kappa\left(T_{\mu\nu} - \frac{1}{2}g_{\mu\nu}T_\alpha{}^\alpha\right)$ of the GR equations was given in the work [54]. Einstein himself explains this change [54] by saying that only the additional term $-\frac{1}{2}g_{\mu\nu}T_\alpha{}^\alpha$ leads to a situation, when energy complexes both for the gravitational field, $t_\mu{}^\nu$, and for matter, $T_\mu{}^\nu$, enter the field equations in the *same* manner. Thus, historically the analysis of conserved quantities and conservation laws was *crucial* for constructing GR.

In the work [55], Einstein finally suggested the canonical energy-momentum complex for the gravitational field $t_\mu{}^\nu$. Later it was called as the Einstein pseudotensor. As an application, Einstein used $t_\mu{}^\nu$ to outline gravitational waves [56] - [58]. From the beginning he stressed that $t_\mu{}^\nu$ is a tensor *only* under linear transformations. Under the general coordinate transformations, $t_\mu{}^\nu$ can change and even be equal to zero. Einstein interpreted this as a "non-localization" of gravitational energy, which is a special property of the gravitational field, and not a defect of the theory. This was a reason for numerous criticism and discussions. Protecting his theory, Einstein himself was the first who gave physically reasonable arguments (see, e.g., [59]).

The criticism far from killing the theory, was a reason for the intensive study of properties of gravitational field in GR and its further development. Following Einstein, in next decades a great number of methods were suggested for defining conserved quantities in GR. As a result of these efforts, important theoretical tests, which restrict an ambiguity in the definition of conserved quantities, were elaborated. Thus, mathematical expressions have to give acceptable quantities for black hole masses, for angular momentum in the Kerr solution, for fluxes of energy and momentum in the Bondi solution; also, a positive energy density for weak gravitational waves. However, due to nontrivial peculiar properties of conserved quantities in GR, up to now:

- Frequently, it is very difficult to find a connection between different definitions; sometimes definitions even contradict one another; sometimes definitions (especially earlier) are not-covariant; *etc.*

- Sometimes definitions are not connected with perturbations.

Now, let us present the modern point of view on the non-localization problem. Considering the physical foundation of the theory it is clear that the non-localization is directly connected with the equivalence principle (see, e.g., [3]). On the other hand, the situation can be also explained by the fact that GR is a geometric theory where spacetime, in which all physical fields propagate, itself is a dynamical object. (Of course, the non-localization problem is related to all the metric theories of gravity, not only to GR.) Due to these objective reasons, sometimes the problem of conserved quantities in metric theories is presented as ill-defined. However, only the fact of non-localization cannot imply that these notions are meaningless. Without a doubt gravitational interaction contributes to the *total* energy-momentum of gravitating system [3]. Indeed, describing a binary star system one needs to consider gravitational energy as a binding energy; considering gravitational waves in an empty domain of space one finds a positive energy of this domain as a whole; *etc.* All of these are related to non-local characteristics. This conclusion is supported by the mathematical content of GR. Szabados [60] clearly expresses it as follows. "... the Christoffel symbols are not tensorial, but they do have geometric, and hence physical content, namely the linear connection. Indeed, the connection is a *non-local* geometric object, connecting the fibers of vector bundle over *different* points of the base manifold. Hence any expression of the connection coefficients, in particular the gravitational energy-momentum or angular momentum, must also be non-local. In fact, although the connection coefficients at a given point can be taken zero by appropriate coordinate/gauge transformation, they cannot be transformed to zero *on an open domain* unless the connection is flat."

Thus the non-localization is natural and inevitable. However, up to now there is no simple and clear description of it. Therefore:

- It is important to give mathematical expressions, which constructively present the non-localization of conserved quantities in metric theories.

The non-localization has to be connected with gauge properties in the description of perturbations. Thus, a solution of this problem has to give a possibility to find reasonable assumptions for a gauge fixation, which allow to describe certain quantities as localized ones.

It is natural that due to the aforementioned peculiar properties of gravitational field much attention has been paid just to non-local characteristics. Thus the *total* energy-momentum and angular momentum of a gravitating system in *whole* spacetime were studied intensively. Such quantities are frequently called as global ones. In this context asymptotically flat spacetimes are considered in details (see, for example, earlier reviews [61, 62], also recent papers and reviews [63] - [67] and numerous references there in). One of the great achievements was the proof of the positivity of the total energy for an isolated system [70] - [73] (see also the review [74]). The global conserved quantities for asymptotically curved backgrounds (like AdS space and some others) are also studied intensively (see, e.g., [68] - [78]).

The aforementioned development has initiated a more intensive examination of the energy problem in GR. Conserved quantities became to be associated with finite spacetime domains. Such quantities are called as *quasilocal* ones and can give a more detailed information than the global quantities. In the last two-three decades the quasilocal approach has became very popular. It is not our goal to present it here, moreover, recently a nice review by Szabados [60] has appeared. Nevetherless, below we shortly outline some of important quasilocal methods.

The Brown and York approach [79] is based on the generalized Hamilton-Jacobi analysis. It considers a spatially restricted gravitating system on 3-dimensional spacelike section Σ. A history of the boundary is a 3-dimensional timelike surface S (cylinder). It is assumed that a 3-metric γ_{ij} on S is fixed and plays a role of a time interval in the usual non-relativistic mechanics, which defines initial and final configurations. They define the energy-momentum tensor τ^{ij} on S, as a functional derivative of an action with respect to γ_{ij}. An intersection of Σ with S is a 2-sphere B which is just a spatial boundary of the system. Normal and tangential projections of τ^{ij} onto B give surface densities of energy, momentum and space tensions on B which are quasi-local expressions. It is crucial to determine a reference flat space, which is *uniquely* defined by the isometric embedding B (with a positive inner curvature) into a flat space. The Brown-York method received a significant development in the works by Brown, Lau and York [80] - [85] and in works of other authors (see review [60]). The recent work [86] could be considered as a mathematical textbook on this approach.

At the first stages of constructing the Hamiltonian dynamics of GR by Arnowitt, Deser and Misner [87] (ADM), surface integrals were neglected *a priori* and reappeared only after disregarding non-physical degrees of freedom. One of the way of developing the standard Hamiltonian description is the symplectic approach by Kijowski and Tulczyiew [88], where it is noted that surface integrals are not less important than the volume ones. Jezierski and Kijowski developed this approach in GR [89] - [91]. They use an "affine formulation", where the connection coefficients $\Gamma^{\lambda}_{\mu\nu}$ are used rather, than the metric ones. The gravitational field is considered inside a closed tube, at a boundary of which some conditions are fixed to construct a closed Hamiltonian system. The Hamiltonian describes the full energy inside the boundary and has a quasilocal sense [90]. In the linear gravity, the requirement of positiveness of the Hamiltonian [89] leads to a "localization" of gravitational energy with *unique* boundary conditions. As an application, gravitational waves on the background of the Schwarzschild geometry were studied [91].

Based on the symplectic method Nester with co-authors [92] - [98] developed a so-called 4-covariant Hamiltonian formulation both for GR and for generalized geometrical gravitational theories. The Hamiltonian on-shell is a surface integral, which defines a quasilocal conserving quantity inside a closed volume. For this approach a displacement 4-vector constructed from the lapse and shift, and a flat space defined at the boundary of the volume are necessary. For the recent development and achievements of this fruitful approach see the review paper [97].

Returning to the discussion of the previous subsection, we recall again on a necessity to operate with local quantities in cosmological and astrophysical applications. Therefore, to conclude the subsection let us formulate also the next task:

- It is important to connect local conserved characteristics with non-local quantities, which appear in the theoretical considerations.

1.3. Goals of the Review and Plan of the Presentation

Analyzing the problems accented in the previous subsections 1.1. and 1.2. a necessity in a generalized and universal approach for describing perturbations, both in GR and in generalized metric gravitational theories, becomes evident. A description, where perturbations in a geometrical theory are considered on a curved or flat background, in fact, converts this theory into the rank of a field theory, like electrodynamics in a fixed spacetime. A set of all the perturbations acquires the sense of the dynamic field configuration. Then, it is desirable to represent the perturbed gravitational metric theory in the *field-theoretical* form (or simply, *field* form) with all the properties of a field theory. We formulate these properties as the following requirements:

(a) The field-theoretical formulation has to be covariant.

(b) One can use an arbitrary curved background spacetimes (solutions to GR or another metric theory).

(c) The perturbed system has to be represented as a dynamic field configuration, which is associated with Lagrangian and corresponding action.

(d) The field equations (perturbation equations) have to be derivable from the action principle.

(e) The conserved quantities and conservation laws also have to be derivable using the variational principle.

(f) The field-theoretical formulation has to have gauge freedoms. Gauge transformations and their properties have to be connected with the action.

(g) In order not to have restrictions in the use of orders of perturbations, it is required to have an exact formulation for perturbed equations, conservation laws and gauge transformations.

(h) Lastly, it is desirable to have a simple and explicit form convenient for applications.

Only such a derivation will permit the required universality and give a full description of perturbations. Of course, the field-theoretical formulation has to be equivalent to the geometrical one, without changing the physical content of the theory.

Thus, the goal of the paper is to suggest an approach, which gives a possibility to present a perturbed metric (geometrical) gravitational theory in a field-theoretical form, which satisfies the above requirements (a) - (h). In last two decades, all the necessary elements of such an approach were developed in works by the author together with his co-authors, and in other works all of which will be cited later. Therefore, from one point of view, the present paper is a review of these works. On the other hand, the paper just unites these works into a generalized and universal approach. Our task is to give an outline of mathematical development of the approach with necessary mathematical expressions, to demonstrate possibilities of the approach and its advantages, and to outline some of its applications. Therefore we hope that the present work could be interesting both to the experts in gravitational physics, e.g., in conservation laws in GR, and to cosmologists and astrophysicists studying the evolution of perturbations on curved backgrounds. This paper is not a review of all the numerous perturbation approaches and methods developed during the history of GR. Therefore we apologize to the authors whose works are not referred here.

Let us discuss some of important points. First, a possibility to use an arbitrary curved background means that any solution of the initial metric theory can be considered as a background. Thus, the background can be flat, curved vacuum, or even curved including background matter, i.e. it can be arbitrary. Second, as was accented, the field and geometrical formulations of the theory have to be equivalent. This means that a solution of the field formulation united together with a background solution have to be transformable into a solution of the geometric formulation (initial metric theory). Symbolically this situation is explained on the figure 1. Let the slightly sloping curve be related to a background and let the oscillating curve mean a solution in the geometrical form, then the difference between them symbolizes the solution in the field-theoretical form. Third, usually it is assumed that

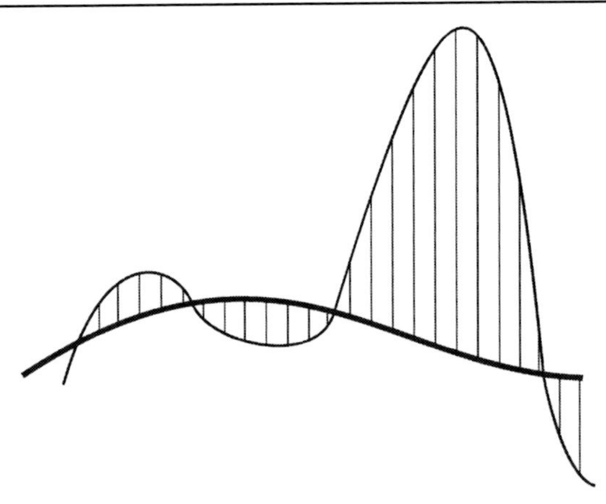

Figure 1. A symbolical connection between solutions of the same metric theory in the geometrical and field-theoretical forms.

a perturbation of a quantity is less than a quantity itself. We do not impose this restriction here. A realization of the requirements (a) - (h) means that the field-theoretical formulation can be thought of as an independent exact field theory. Then, of course, the "amplitude" of an exact solution of the exact theory can be more than "amplitude" of the background (see, e.g., the right side of the figure 1).

At the earlier stage, Einstein was trying to construct a gravitational theory as a field theory in Minkowski space, in the framework of special relativity. However, step by step he had concluded that one needs to operate with curved spacetime only, Minkowski background space had disappeared from the consideration all together. The construction of the field-theoretical formulation in the framework of the geometrical theory is, in a definite sense, the revival of the special relativity view. Moreover, as we remarked above, the field-theoretical formulation really is an independent field theory and can be constructed by independent ways (see below the discussion in subsections 2.1. and 2.2.). However, there is no contradiction here. The geometrical and field-theoretical formulations are two different formulations of the same theory with the same physical content. The background spacetime turns out non-observable and has a sense of only an auxiliary structure.

Up to now the Einstein theory retains its position as the most popular theory of gravity, leading in all the applications. Thus, in this paper we pay a significant attention to constructing the field-theoretical formulation for GR which is presented in detail. On the other hand, due to the rising precision of experiments and observables in cosmos and due to the great interest to the brane models, other metric theories generalizing GR become more and more popular and necessary. One of the main properties of the approach presented here is its universality. We use this advantage and develop the field theoretical approach applied to an arbitrary metric theory.

The paper is organized as follows. The next section 2. is devoted to the detail description of the field-theoretical formulation of GR on an arbitrary curved background, which have all the properties of a self-dependent field theory. The essential attention is paid to an invariance with respect to exact gauge transformations. The last do not effect both coordinates and background quantities in the field-theoretical formulation and are connected with the general covariance of GR.

In section 3., we construct conservation laws for perturbations in GR in the framework of the field-theoretical approach. Conserved currents and corresponding superpotentials are presented. As an important instruments we use the canonical Nœther method and the Belinfante symmetrization prescription. The conserved quantities and conservation laws are used for examination of asymptotically flat spacetime at spatial infinity, the closed Friedmann model, the Schwarzschild solution and linear perturbations on FRW backgrounds.

In section 4., we develop the field-theoretical approach to describe perturbations in an arbitrary D-dimensional metric theory of gravity on a fixed background. We construct *generalized* conserved currents and corresponding superpotentials with again the essential use of the Nœther and Belinfante methods. The conserved quantities are tested in the Einstein-Gauss-Bonnet gravity for calculating the mass of the Schwarzschild-anti-de Sitter black hole.

1.4. Notations

Here, we present notations, which will appear more frequently and which are more important. In the text, together with these notations, numerous other notations will be used, they will be outlined currently.

- Greek indexes numerate 4-dimensional spacetime coordinates as well as D-dimensional spacetime ones. Usually x^0 means a time coordinate, whereas small Latin indexes from the middle of alphabet $i,\ j,\ k,\ \ldots$ mean 3-dimensional space coordinates or $(D-1)$-dimensional hyperspace coordinates;

- Large Latin indexes $A,\ B,\ C,\ \ldots$ are used as generalized ones for an arbitrary set of tensor densities, for example, $Q^A = \{\sqrt{-g}g^{\mu\nu},\ \phi,\ T_{\alpha\beta}\}$;

- A dynamic metric of a metric theory is $g_{\mu\nu}$ $(g = \det g_{\mu\nu})$;

- Bar means that a quantity "$\overline{Q}^A$" is a background one;

- Thus, $\overline{g}_{\mu\nu}$ $(\overline{g} = \det \overline{g}_{\mu\nu})$ is a background metric. Indexes of all the quantities of a *perturbed* system are raised and lowered by the background metric;

- Many expressions are presented as densities of the weight $+1$. The reasons are as follows. First, all the Nœther identities, which are explored intensively, are such densities in the initial derivation. Second, conservation laws have to be covariant, however partial derivatives are crucial for application of the Gauss theorem. Divergences of both vector densities (conserved currents) and antisymmetric tensor densities (superpotentials) have this duality property. To accent these expressions we use "hats", as more economical notations in this situation. Thus, a quantity "$\hat{Q}^A$" is such a density, it could be a tensor multiplied by $\sqrt{-g}$ or $\sqrt{-\overline{g}}$ (for example, $\hat{Q}^{\alpha\beta}{}_\gamma = \sqrt{-\overline{g}}Q^{\alpha\beta}{}_\gamma$), or $\hat{Q}^A$ could be be independent on these determinants, a situation will be clarified from the context;

- $\eta_{\mu\nu}$ is a Minkowskian metric in the Lorentzian coordinates; sometimes $\sqrt{-\eta}$ is used explicitly instead of 1 to stress that a quantity, say $\hat{Q}^{\alpha\beta}{}_{\gamma} = \sqrt{-\eta}Q^{\alpha\beta}{}_{\gamma}$, is a density of the weight $+1$;
- $\hat{l}^{\mu\nu} = \hat{g}^{\mu\nu} - \hat{\overline{g}}^{\mu\nu} = \sqrt{-g}g^{\mu\nu} - \sqrt{-\overline{g}}\,\overline{g}^{\mu\nu}$ is a more important form of the metric perturbations;
- Partial derivatives are denoted by $(_{,i})$, $(_{,\alpha})$, or ∂_i, ∂_α,;
- D_α and $\overline{D}_\alpha$ are covariant derivatives with respect to $g_{\mu\nu}$ and $\overline{g}_{\mu\nu}$ with the Chistoffel symbols $\Gamma^\alpha_{\beta\gamma}$ and $\overline{\Gamma}^\alpha_{\beta\gamma}$, respectively;
- $\Delta^\alpha_{\beta\gamma} = \Gamma^\alpha_{\beta\gamma} - \overline{\Gamma}^\alpha_{\beta\gamma}$ is the tensor actively used in the paper;
- ξ^α is an arbitrary displacement vector, whereas λ^α is a Killing vector of a background;
- The Lie derivative is defined as

$$\pounds_\xi Q^A = -\xi^\alpha \overline{D}_\alpha Q^A + Q^A\Big|^\alpha_\beta \overline{D}_\alpha \xi^\beta\,,$$

  note the opposite sign to the usual one, $Q^A\Big|^\alpha_\beta$ is defined by the transformation properties of Q^A;
- $\zeta_{\mu\nu} = -\frac{1}{2}\pounds_\xi \overline{g}_{\mu\nu}$ is the tensor actively used in the paper;
- The Lagrangian derivative is defined as usual:

$$\frac{\delta Q^A(q^B,\, q^B_{,\alpha},\, q^B_{,\alpha\beta},\, \ldots)}{\delta q^C} = \frac{\partial Q^A}{\partial q^C} - \partial_\alpha\left(\frac{\partial Q^A}{\partial q^C_{,\alpha}}\right) + \partial_{\alpha\beta}\left(\frac{\partial Q^A}{\partial q^C_{,\alpha\beta}}\right) - \ldots$$

- $R^\alpha{}_{\mu\beta\nu}$, $R_{\mu\nu}$, $G_{\mu\nu}$, $T_{\mu\nu}$, R and $\overline{R}^\alpha{}_{\mu\beta\nu}$, $\overline{R}_{\mu\nu}$, $\overline{G}_{\mu\nu}$, $\overline{T}_{\mu\nu}$,$\overline{R}$ are the Riemannian, Ricci, Einstein, matter energy-momentum tensors and the curvature scalar for the physical and background spacetimes.
- Usually index "L" means a linearization, for example, $G^L_{\mu\nu}$ and $\Phi^L_{\mu\nu}$ mean linearized pure gravitational and matter parts of the gravitational field equations;
- The conserved currents are defined in the framework of different approaches as follows. In the case of GR: $\hat{J}^\mu_{(c)}$ is defined with the use of the canonical Nœther procedure; $\hat{J}^\mu_{(s)}$ is defined with the use of the field-theoretical prescription, based on the symmetrical energy-momentum tensor; $\hat{J}^\mu_{(B)}$ is the canonical current corrected with the use of the Belinfante method. The correspondent superpotentials in GR are $\hat{J}^{\mu\nu}_{(c)}$, $\hat{J}^{\mu\nu}_{(s)}$ and $\hat{J}^{\mu\nu}_{(B)}$;
- Analogous currents in superpotentials in an arbitrary D-dimensional metric theory are respectively $\hat{\mathcal{I}}^\mu_{(c)}$, $\hat{\mathcal{I}}^\mu_{(s)}$, $\hat{\mathcal{I}}^\mu_{(B)}$ and $\hat{\mathcal{I}}^{\mu\nu}_{(c)}$, $\hat{\mathcal{I}}^{\mu\nu}_{(s)}$, $\hat{\mathcal{I}}^{\mu\nu}_{(B)}$;
- $(\alpha\beta)$, (ik) and $[\alpha\beta]$, [ik] mean symmetrization and antisymmetrization;
- κ — the "Einstein" constant both in GR and in an arbitrary metric theory.

2. The Exact Field-Theoretical Formulation of GR

2.1. Development of the Field Approach

The study of perturbations in GR, in fact, was begun by Einstein himself. However, as a separate field, the history of the field-theoretical approach in GR began in 40's — 50's of XX century. Perturbed Einstein equations are rewritten as follows. Define the metric perturbations on a flat background in the Lorentzian coordinates as $\hat{l}^{\mu\nu} = \hat{g}^{\mu\nu} - \sqrt{-\eta}\eta^{\mu\nu}$; $l^{\mu\nu} = (\sqrt{-\eta})^{-1}\hat{l}^{\mu\nu}$. The terms linear in metric perturbations are placed on the left hand side of the Einstein equations, whereas all the nonlinear terms are transported to the right hand side, and together with a matter energy-momentum tensor are treated as a total (effective) energy-momentum tensor $t^{(tot)}_{\mu\nu}$. Then Einstein equations are rewritten in the equivalent perturbed form as

$$G^L_{\mu\nu} = \kappa t^{(tot)}_{\mu\nu} \tag{2..1}$$

where, raising the indexes by $\eta^{\alpha\beta}$, one has the left hand side in the form:

$$G^{\mu\nu}_L \equiv \tfrac{1}{2}(l^{\mu\nu,\alpha}{}_{,\alpha} + \eta^{\mu\nu}l^{\alpha\beta}{}_{,\alpha\beta} - l^{\alpha\mu,\nu}{}_{,\alpha} - l^{\alpha\nu,\mu}{}_{,\alpha})\,. \tag{2..2}$$

Its divergence identically is equal to zero: $\partial_\nu G^{\mu\nu}_L \equiv 0$. Then, one obtains directly the differential conservation law

$$\partial_\nu t^{\mu\nu}_{(tot)} = 0\,. \tag{2..3}$$

This picture was developed in a form of a Lagrangian based field theory with self-interaction in a fixed background spacetime, where $t^{(tot)}_{\mu\nu}$ is obtained by variation of an action with respect to a background metric. Following the introduction in the Deser work [99], below, we shall present the main steps in this derivation, the corresponding bibliography can also be found in [99]. Assume that a field theory of gravity in a Minkowski space is constructed. By known observable tests (see, e.g., textbook [3]), the most preferable type of the gravitational field is the tensor field, say $l^{\mu\nu}$. The linear (approximate) equations have to have the form $G^L_{\mu\nu} = 0$ and are defined by the quadratic Lagrangian $\hat{\mathcal{L}}^g_{(2)}$. Keeping a symmetrical energy-momentum tensor of matter fields ϕ^A as a source of $G^L_{\mu\nu}$ one obtains

$$G^L_{\mu\nu}(l) = \kappa T_{\mu\nu}(\phi, \eta)\,. \tag{2..4}$$

Identically $\partial_\nu G^{L\nu}_\mu \equiv 0$, therefore $\partial_\nu T^\nu_\mu = 0$. However, there is a contradiction between the conservation law $\partial_\nu T^\nu_\mu = 0$ and equations of motion for *interacting* fields ϕ^A. How does one avoid this? The right hand side of Eq. (2..4) is to be obtained by variation (conventionally) both with respect to $\eta^{\mu\nu}$ and $l^{\mu\nu}$. Therefore one needs to make an exchange $\left\{\phi^A, \eta^{\mu\nu}\right\} \to \left\{\phi^A, \eta^{\mu\nu} + l^{\mu\nu}\right\}$ both in the matter Lagrangian and in the matter energy-momentum tensor. This just means the universality of gravitational interaction. Next, one has to include the gravitational self-interaction. Therefore one adds the symmetrical energy-momentum tensor of the gravitational field $t^{(2)g}_{\mu\nu}(l)$, corresponding to $\hat{\mathcal{L}}^g_{(2)}$, to the right hand side of (2..4) together with $T_{\mu\nu}(\phi, \eta + l)$. But the equations that include $t^{(2)g}_{\mu\nu}(l)$ can be obtained if a cubic Lagrangian is added, $\hat{\mathcal{L}}^g_{(2)} + \hat{\mathcal{L}}^g_{(3)}$. After this one needs to consider the next level, and so on. In the result, one obtains the final variant of the gravitational equations:

$$G^L_{\mu\nu}(l) = \kappa\left[T_{\mu\nu}(\phi, l+\eta) + \sum_{n=2}^{\infty} t^{(n)g}_{\mu\nu}(l)\right] . \tag{2..5}$$

It turns out that the equations (2..5) are exactly the Einstein equations, i.e. equations (2..1). After the identification $\sqrt{-\eta}\eta^{\mu\nu} + \hat{l}^{\mu\nu} \equiv \sqrt{-g}g^{\mu\nu}$ one has only the dynamical metric $g^{\mu\nu}$, whereas the background metric $\eta^{\mu\nu}$ and the field $l^{\mu\nu}$ completely disappear from the consideration.

Deser himself [99], unlike (2..5), has suggested the field formulation of GR without expansions. As dynamical variables he used the two independent tensor fields $l^{\mu\nu}$ and $\Delta^\alpha_{\mu\nu}$ of the 1-st order formalism. After variation of the corresponding action he derives the equations in the form:

$$G^L_{\mu\nu}(l) = \kappa\left[t^g_{\mu\nu}(l, \Delta) + t^m_{\mu\nu}\right] , \tag{2..6}$$

instead of (2..5). After identifications $\sqrt{-\eta}\eta^{\mu\nu} + \hat{l}^{\mu\nu} \equiv \sqrt{-g}g^{\mu\nu}$ and $\Delta^\alpha_{\mu\nu} \equiv \Gamma^\alpha_{\mu\nu}$ the equivalence with the Einstein equations is confirmed, but only in the Palatini form, where $g^{\mu\nu}$ and $\Gamma^\alpha_{\mu\nu}$ are used as independent variables.

In the work [100] we have generalized the Deser approach [99]. Instead of the background Minkowski space with the Lorenzian coordinates we consider an arbitrary curved background spacetime with a given metric $\overline{g}_{\mu\nu}$ and given matter fields $\overline{\Phi}^A$ satisfying the background Einstein gravitational and matter equations. We also use the 1-st order formalism. The gravitational equations get the generalized form:

$$\hat{G}^L_{\mu\nu}(l) + \hat{\Phi}^L_{\mu\nu}(l, \phi) = \kappa\left[\hat{t}^g_{\mu\nu}(l, \Delta) + \hat{t}^m_{\mu\nu}\right] \tag{2..7}$$

where the left hand side is linear in $\hat{l}^{\mu\nu}$ and ϕ^A and defined later in Eqs. (2..25) - (2..27). The term $\hat{\Phi}^L_{\mu\nu}$ appears due to $\overline{\Phi}^A$. Equivalence with GR in the ordinary derivation is stated after identifications: $\overline{\hat{g}}^{\mu\nu} + \hat{l}^{\mu\nu} \equiv \hat{g}^{\mu\nu}$, $\overline{\Gamma}^\alpha_{\mu\nu} + \Delta^\alpha_{\mu\nu} \equiv \Gamma^\alpha_{\mu\nu}$ and $\overline{\Phi}^A + \phi^A \equiv \Phi^A$. In the following years we have developed the principles of the work [100] and have used this approach in many applications. Our results are presented in the papers [101] - [121], on the basis of which, in a more part, the present review was written.

Elements of the field approach in gravity are also actively developing nowadays in other approaches. Thus, in [122], a requirement only of the first derivatives of metrical perturbations in the total symmetrical energy-momentum tensor has led to a *new* field formulation of GR in Minkowski spacetime, which is different from the formulation in [100]. The new total energy-momentum tensor is the source for the non-linear left hand side. On the basis of this new field formulation an interesting variant of the gravitational theory with non-zero masses of gravitons was developed [123]. A comparison and a connection of the works [100, 122, 123] are discussed in [124] in detail. In [125], the work [100] was developed to construct the total energies and angular momenta for $d+1$-dimensional asymptotically anti-de Sitter spacetime. The properties of the field approach [100] appear independently in many concrete problems. For example, in [126, 127] a consideration of linear perturbations on FRW backgrounds leads to the linear approximation of the exact field formulation of GR [100]. In [128], as a development of the field approach, a class of so-called "slightly bimetric" gravitation theories was constructed. In [129, 130], a behaviour of light cones in Minkowski space and effectively curved spacetimes was examined. Then, based on the causality principle, a special criterium was stated. In [131] this criterium was used to show that if spatially flat FRW big bang model is considered as a configuration on a flat background, then the cosmological singularity is banished to past infinity in Minkowski space. The references to earlier works and the theoretical foundation for the field approach can be found in the works [132] - [134]. To the best of our knowledge up to date bibliography related to the field approach in gravity can be found in [128] - [131].

2.2. Various Directions in the Construction

There are various possibilities to approach the field-theoretical formulation of GR, which are based on different foundations. Here, we shall discuss the well known and important ones. The principle used by Deser [99] could be formulated as follows:

- The source of the linear massless field of spin two (of gravitational field) in Minkowski space is to be the total symmetrical (metric) energy-momentum tensor of all the dynamical fields, including the gravitational field itself.

Using this principle Deser has constructed a corresponding Lagrangian and field equations and energy-momentum tensor following from it. We have generalized the Deser approach on arbitrary curved backgrounds [100]. His principle has been reformulated in a way that the linear left hand side of perturbed gravitational equations has to be of the form in Eq. (2..7), i.e. together with $\hat{G}^L_{\mu\nu}$ one has to include the, linear in matter perturbations, part $\hat{\Phi}^L_{\mu\nu}$. The analogous principle was suggested for perturbed matter equations.

The next known method was most clearly presented by Grishchuk [134] and shortly can be formulated as:

- A transformation from gravistatic (Newton law) to gravidynamics, i.e. to a relativistic theory of gravitational field (general relativity), equations of which (Einstein equations) describe gravitational waves.

Following this direction, one has to transform the Newton law $\Delta\phi = -4\pi G\rho$ into a special relativity description. To satisfy the relativistic requirement a) the mass density ρ has to be generalized to 10 components of the matter stress-energy tensor $T_{\mu\nu}$; b) the single component ϕ should also be replaced by 10 gravitational potentials $l^{\mu\nu}$; c) the Laplace operator should be replaced by the d'Alembert operator; d) the gravitational field has to be nonlinear and, thus, has to be a source for itself. Following this reformations one obtains the generalized equations: $\frac{1}{2}l_{\mu\nu}{}^{;\alpha}{}_{,\alpha} = \kappa(t^g_{\mu\nu} + T_{\mu\nu})$. They imply that the gauge condition $l^{\mu\nu}{}_{,\nu} = 0$ is already chosen. c) To reconstruct the gauge invariance properties it is necessary to add to the left hand side the terms $\frac{1}{2}\left(\eta_{\mu\nu}l^{\alpha\beta}{}_{,\alpha,\beta} - l^{\alpha}{}_{\mu,\nu,\alpha} - l^{\alpha}{}_{\nu,\mu,\alpha}\right)$. As a result, one obtains Eq. (2..1), which is just the Einstein equations.

The field formulation of GR can be also constructed based on the gauge properties (see our work [107]). This direction is analogous to the method in the gauge theories of the Yang-Mills type, which are constructed by localizing parameters of a gauge group. However, unlike usual, we postulate a non-standard way of localization, namely:

- A "localization" of Killing vectors of the background spacetime.

This assumes the existence of a fixed background spacetime with symmetries presented by a Killing vector field λ^α, in which initial dynamic fields ϕ^A are propagated. It is noted that an action for the initial fields is invariant, up to a surface term, under the transformation $\phi^A \to \pounds_\lambda \phi^A$. Next, the Killing vector is changed for an arbitrary vector ξ^α, that is localized. Then the invariance is destroyed. To restore it the compensating (gauge) field has to be included. In doing so the coordinates and the background metric do not change. The requirements to have the gauge field as an universal field and to have the simplest sought-for action for the free gauge field lead just to the field formulation of GR, developed in [100].

As a rule, a fixed background spacetime, in which perturbations are studied, is determined by the problem under consideration (see subsections 1.1. and 1.2. in Introduction). Thus the background could be assumed as a known solution to the Einstein equations. Then one has to study the perturbed (with respect to this background) Einstein equations. As already was noted above, this picture can be developed as a Lagrangian based field theory, where the first step is:

- The decomposition of dynamical variables of GR into background variables and dynamic perturbations.

This method is evident in itself and has the explicit connection with the ordinary geometrical formulation of GR. Namely this method is easily adopted for constructing the field-theoretical formulation in the framework of an arbitrary metric theory. In the next subsections, basing on the works [100, 103, 104], we present it in detail. However, although the construction was developed for both in the 1-st and in the 2-nd order formalisms, here, we use the 2-nd order formalism only since it is more convenient and suitable. To the best of our knowledge, Barnebey [135] was the first who suggested to use the 2-nd order formalism for an exact (without expansions) description of perturbations in GR.

2.3. A Dynamical Lagrangian

Let us consider the Einstein theory and write out its action

$$S = \frac{1}{c}\int d^4x \hat{\mathcal{L}}^E \equiv -\frac{1}{2\kappa c}\int d^4x \hat{R}(g_{\mu\nu}) + \frac{1}{c}\int d^4x \hat{\mathcal{L}}^M(\Phi^A,\, g_{\mu\nu})\,, \tag{2..8}$$

$\hat{\mathcal{L}}^M$ depends on Φ^A and their derivatives up to a finite order. Then, the gravitational and matter equations corresponding to the Lagrangain (2..8) are

$$\frac{\delta\hat{\mathcal{L}}^E}{\delta\hat{g}^{\mu\nu}} = -\frac{1}{2\kappa}\frac{\delta\hat{R}}{\delta\hat{g}^{\mu\nu}} + \frac{\delta\hat{\mathcal{L}}^M}{\delta\hat{g}^{\mu\nu}} = 0\,, \tag{2..9}$$

$$\frac{\delta\hat{\mathcal{L}}^E}{\delta\Phi^A} = \frac{\delta\hat{\mathcal{L}}^M}{\delta\Phi^A} = 0\,. \tag{2..10}$$

The form of Eq. (2..9) corresponds to the form of the Einstein equations

$$R_{\mu\nu} - \kappa\left(T_{\mu\nu} - \tfrac{1}{2}g_{\mu\nu}T^\alpha_\alpha\right) = 0\,, \tag{2..11}$$

whereas a more customary form is

$$\sqrt{-g}\,(G_{\mu\nu} - \kappa T_{\mu\nu}) \equiv \hat{G}_{\mu\nu} - \kappa \hat{T}_{\mu\nu} = 0\,, \tag{2..12}$$

which is obtained by variation with respect to $g^{\mu\nu}$.

Next, let us define the metric and matter perturbations with the use of the decompositions:

$$\hat{g}^{\mu\nu} \equiv \overline{\hat{g}}^{\mu\nu} + \hat{l}^{\mu\nu}\,, \tag{2..13}$$

$$\Phi^A \equiv \overline{\Phi}^A + \phi^A\,. \tag{2..14}$$

It is assumed that the background quantities $\overline{\hat{g}}^{\mu\nu}$ and $\overline{\Phi}^A$ are known and satisfy the background (given) system, which is defined as follows. Its action is

$$\overline{S} = \frac{1}{c}\int d^4x \overline{\hat{\mathcal{L}}^E} \equiv -\frac{1}{2\kappa c}\int d^4x \overline{\hat{R}} + \frac{1}{c}\int d^4x \overline{\hat{\mathcal{L}}^M}\,. \tag{2..15}$$

The corresponding to the Lagrangian in (2..15) background gravitational and matter equations have the form of the barred equations (2..9) and (2..10):

$$-\frac{1}{2\kappa}\frac{\delta \overline{\hat{R}}}{\delta \overline{\hat{g}}^{\mu\nu}} + \frac{\delta \overline{\hat{\mathcal{L}}^M}}{\delta \overline{\hat{g}}^{\mu\nu}} = 0\,, \tag{2..16}$$

$$\frac{\delta \overline{\hat{\mathcal{L}}^M}}{\delta \overline{\Phi}^A} = 0\,. \tag{2..17}$$

Frequently we use a Ricci-flat background with the background equations

$$\overline{\hat{G}}_{\mu\nu} = \overline{\hat{R}}_{\mu\nu} = 0. \tag{2..18}$$

Now let us classify the perturbations $\hat{l}^{\mu\nu}$ and ϕ^A as *independent dynamic* variables, which present a field configuration on the background of the system (2..16) and (2..17). To describe this dynamical configuration we construct a corresponding Lagrangian called as a *dynamical* one [104]. After substituting the decompositions (2..13) and (2..14) into the Lagrangian $\hat{\mathcal{L}}^E$ of the action (2..8) and subtracting zero's and linear in $\hat{l}^{\mu\nu}$ and ϕ^A terms of the functional expansion of the Lagrangian $\hat{\mathcal{L}}^E$ one has

$$\hat{\mathcal{L}}^{dyn} = \hat{\mathcal{L}}^E(\overline{g}+l,\, \overline{\Phi}+\phi) - \hat{l}^{\mu\nu}\frac{\delta\overline{\hat{\mathcal{L}}^E}}{\delta\overline{\hat{g}}^{\mu\nu}} - \phi^A\frac{\delta\overline{\hat{\mathcal{L}}^E}}{\delta\overline{\Phi}^A} - \overline{\hat{\mathcal{L}}^E} - \frac{1}{2\kappa}\partial_\alpha \hat{k}^\alpha = -\frac{1}{2\kappa}\hat{\mathcal{L}}^g + \hat{\mathcal{L}}^m\,. \tag{2..19}$$

As is seen, zero's order term is the background Lagrangian, whereas the linear term is proportional to the left hand sides of the background equations (2..16) and (2..17).

In Eq. (2..19) a vector density $\hat{k}^\alpha$ is not concreted. However, consider $\hat{k}^\alpha$ defined as

$$\hat{k}^\alpha \equiv \hat{g}^{\alpha\nu}\Delta^\mu_{\mu\nu} - \hat{g}^{\mu\nu}\Delta^\alpha_{\mu\nu} \tag{2..20}$$

with $\Delta^\rho_{\mu\nu}$ presented as the perturbations of the Cristoffel symbols

$$\Delta^\alpha_{\mu\nu} \equiv \Gamma^\alpha_{\mu\nu} - \overline{\Gamma}^\alpha_{\mu\nu} = \tfrac{1}{2}g^{\alpha\rho}\left(\overline{D}_\mu g_{\rho\nu} + \overline{D}_\nu g_{\rho\mu} - \overline{D}_\rho g_{\mu\nu}\right)\,, \tag{2..21}$$

and depending on $\hat{l}^{\mu\nu}$ through the decomposition (2..13). Then a pure gravitational part in the Lagrangian (2..19) is presented in the form:

$$\begin{aligned}\hat{\mathcal{L}}^g &= \hat{R}(\bar{\hat{g}}^{\mu\nu} + \hat{l}^{\mu\nu}) - \hat{l}^{\mu\nu}\overline{R}_{\mu\nu} - \bar{\hat{g}}^{\mu\nu}\overline{R}_{\mu\nu} + \partial_\mu \hat{k}^\mu \\ &= -(\Delta^\rho_{\mu\nu} - \Delta^\sigma_{\mu\sigma}\delta^\rho_\nu)\overline{D}_\rho \hat{l}^{\mu\nu} + (\bar{\hat{g}}^{\mu\nu} + \hat{l}^{\mu\nu})\left(\Delta^\rho_{\mu\nu}\Delta^\sigma_{\rho\sigma} - \Delta^\rho_{\mu\sigma}\Delta^\sigma_{\rho\nu}\right). \end{aligned} \quad (2..22)$$

It depends only on the first derivatives of the gravitational variables $\hat{l}^{\mu\nu}$. In the case of a flat background the Lagrangian (2..22) transfers to the covariant Lagrangian suggested by Rosen [136], which has been rediscovered in [137] and [101]. The matter part of (2..19) is

$$\hat{\mathcal{L}}^m = \hat{\mathcal{L}}^M\left(\bar{g} + l,\, \overline{\Phi} + \phi\right) - \hat{l}^{\mu\nu}\frac{\delta\overline{\hat{\mathcal{L}}^M}}{\delta\bar{\hat{g}}^{\mu\nu}} - \phi^A\frac{\delta\overline{\hat{\mathcal{L}}^M}}{\delta\overline{\Phi}^A} - \overline{\hat{\mathcal{L}}^M}. \quad (2..23)$$

2.4. The Einstein Equations in the Field Formulation

The variation of the action with the Lagrangian (2..19) with respect to $\hat{l}^{\alpha\beta}$ and the contraction with $\sqrt{-\bar{g}}(\delta^\alpha_\mu\delta^\beta_\nu - \frac{1}{2}\bar{g}_{\mu\nu}\bar{g}^{\alpha\beta})$ give the field equations in the form:

$$\hat{G}^L_{\mu\nu} + \hat{\Phi}^L_{\mu\nu} = \kappa\left(\hat{t}^g_{\mu\nu} + \hat{t}^m_{\mu\nu}\right) \equiv \kappa\hat{t}^{(tot)}_{\mu\nu}. \quad (2..24)$$

They coincide with the form (2..7), only now in the 2-nd order formalism. The left hand side of Eq. (2..24) is linear in $\hat{l}^{\mu\nu}$ and ϕ^A. It consists of the pure gravitational part

$$\hat{G}^L_{\mu\nu}(\hat{l}) \equiv \frac{\delta}{\delta\bar{g}^{\mu\nu}}\hat{l}^{\rho\sigma}\frac{\delta\hat{\overline{R}}}{\delta\bar{\hat{g}}^{\rho\sigma}} \quad (2..25)$$

$$\equiv \tfrac{1}{2}\left(\overline{D}_\rho\overline{D}^\rho\hat{l}_{\mu\nu} + \bar{g}_{\mu\nu}\overline{D}_\rho\overline{D}_\sigma\hat{l}^{\rho\sigma} - \overline{D}_\rho\overline{D}_\nu\hat{l}^\rho_\mu - \overline{D}_\rho\overline{D}_\mu\hat{l}^\rho_\nu\right), \quad (2..26)$$

which is the covariantized expression (2..2), and of the matter part

$$\hat{\Phi}^L_{\mu\nu}(\hat{l},\phi) \equiv -2\kappa\frac{\delta}{\delta\bar{g}^{\mu\nu}}\left(\hat{l}^{\rho\sigma}\frac{\delta\overline{\hat{\mathcal{L}}^M}}{\delta\bar{\hat{g}}^{\rho\sigma}} + \phi^A\frac{\delta\overline{\hat{\mathcal{L}}^M}}{\delta\overline{\Phi}^A}\right). \quad (2..27)$$

It disappears for Ricci-flat backgrounds (2..18), and then Eq. (2..24) acquires the form of Eq. (2..1). The right hand side of Eq. (2..24) is the symmetrical energy-momentum tensor density

$$\hat{t}^{(tot)}_{\mu\nu} \equiv 2\frac{\delta\hat{\mathcal{L}}^{dyn}}{\delta\bar{g}^{\mu\nu}} \equiv 2\frac{\delta}{\delta\bar{g}^{\mu\nu}}\left(-\frac{1}{2\kappa}\hat{\mathcal{L}}^g + \hat{\mathcal{L}}^m\right) \equiv \hat{t}^g_{\mu\nu} + \hat{t}^m_{\mu\nu}. \quad (2..28)$$

The explicit form of the gravitational part is

$$\hat{t}^g_{\mu\nu} = \frac{1}{\kappa}\left[\sqrt{-\bar{g}}\left(-\delta^\rho_\mu\delta^\sigma_\nu + \tfrac{1}{2}\bar{g}_{\mu\nu}\bar{g}^{\rho\sigma}\right)\left(\Delta^\alpha_{\rho\sigma}\Delta^\beta_{\alpha\beta} - \Delta^\alpha_{\rho\beta}\Delta^\beta_{\alpha\sigma}\right) + \overline{D}_\tau\hat{Q}^\tau_{\mu\nu}\right]; \quad (2..29)$$

$$\begin{aligned}2\hat{Q}^\tau_{\mu\nu} \equiv\; & -\bar{g}_{\mu\nu}\hat{l}^{\alpha\beta}\Delta^\tau_{\alpha\beta} + \hat{l}_{\mu\nu}\Delta^\tau_{\alpha\beta}\bar{g}^{\alpha\beta} - \hat{l}^\tau_\mu\Delta^\alpha_{\nu\alpha} - \hat{l}^\tau_\nu\Delta^\alpha_{\mu\alpha} + \hat{l}^{\beta\tau}\left(\Delta^\alpha_{\mu\beta}\bar{g}_{\alpha\nu} + \Delta^\alpha_{\nu\beta}\bar{g}_{\alpha\mu}\right) \\ & + \hat{l}^\beta_\mu\left(\Delta^\tau_{\nu\beta} - \Delta^\alpha_{\beta\rho}\bar{g}^{\rho\tau}\bar{g}_{\alpha\nu}\right) + \hat{l}^\beta_\nu\left(\Delta^\tau_{\mu\beta} - \Delta^\alpha_{\beta\rho}\bar{g}^{\rho\tau}\bar{g}_{\alpha\mu}\right). \end{aligned} \quad (2..30)$$

The matter part is expressed through the usual matter energy-momentum tensor density $\hat{T}_{\mu\nu}$ of the Einstein theory as

$$\hat{t}^{m}_{\mu\nu} = \hat{T}_{\mu\nu} - \tfrac{1}{2}g_{\mu\nu}\hat{T}_{\alpha\beta}g^{\alpha\beta} - \tfrac{1}{2}\overline{g}_{\mu\nu}\overline{g}^{\alpha\beta}\left(\hat{T}_{\alpha\beta} - \tfrac{1}{2}g_{\alpha\beta}\hat{T}_{\pi\rho}g^{\pi\rho}\right) - 2\frac{\delta}{\delta\overline{g}^{\mu\nu}}\left(\hat{l}^{\rho\sigma}\frac{\delta\overline{\hat{\mathcal{L}}^{M}}}{\delta\overline{g}^{\rho\sigma}} + \phi^{A}\frac{\delta\overline{\hat{\mathcal{L}}^{M}}}{\delta\overline{\Phi}^{A}}\right) - \overline{\hat{T}}_{\mu\nu}\,. \tag{2..31}$$

Taking into account the definitions (2..27), (2..28) and (2..31) in the field equations (2..24) one can rewrite them in the form:

$$\hat{G}^{L}_{\mu\nu} = \kappa\left(\hat{t}^{g}_{\mu\nu} + \delta\hat{t}^{M}_{\mu\nu}\right) = \kappa\hat{t}^{(eff)}_{\mu\nu}\,; \tag{2..32}$$

$$\delta\hat{t}^{M}_{\mu\nu} \equiv \hat{t}^{M}_{\mu\nu} - \overline{\hat{t}^{M}}_{\mu\nu} = \hat{T}_{\mu\nu} - \tfrac{1}{2}g_{\mu\nu}g^{\rho\sigma}\hat{T}_{\rho\sigma} - \tfrac{1}{2}\overline{g}_{\mu\nu}\overline{g}^{\rho\sigma}\left(\hat{T}_{\rho\sigma} - \tfrac{1}{2}g_{\rho\sigma}g^{\lambda\tau}\hat{T}_{\lambda\tau}\right) - \overline{\hat{T}}_{\mu\nu}\,. \tag{2..33}$$

The equation (2..32) has the form of Eq. (2..1) even on *arbitrary* backgrounds. The price is that the effective source $\hat{t}^{(eff)}_{\mu\nu}$, including the matter part as $\delta\hat{t}^{M}_{\mu\nu}$, does not follow from the Lagrangian (2..19) directly. However, this matter part could be classified as a perturbation of

$$\hat{t}^{M}_{\mu\nu} \equiv \frac{\delta\hat{\mathcal{L}}^{M}\left(\overline{\Phi}^{A} + \phi^{A};\, \overline{g}^{\mu\nu} + \hat{l}^{\mu\nu}\right)}{\delta\overline{g}^{\mu\nu}}\,. \tag{2..34}$$

Return to the equations (2..24) in the whole. Transfer the energy-momentum tensor density to the left hand side and use the definitions (2..25), (2..27) and (2..28) with (2..19):

$$\begin{aligned}
&\hat{G}^{L}_{\mu\nu} + \hat{\Phi}^{L}_{\mu\nu} - \kappa\hat{t}^{(tot)}_{\mu\nu} \equiv \\
&-2\kappa\frac{\partial\overline{g}^{\rho\sigma}}{\partial\overline{g}^{\mu\nu}}\frac{\delta}{\delta\hat{l}^{\rho\sigma}}\left[-\frac{1}{2\kappa}\hat{R}\left(\overline{g}^{\alpha\beta} + \hat{l}^{\alpha\beta}\right) + \hat{\mathcal{L}}^{M}\left(\overline{\Phi}^{A} + \phi^{A};\, \overline{g}^{\mu\nu} + \hat{l}^{\mu\nu}\right)\right] \\
&+ 2\kappa\frac{\delta}{\delta\overline{g}^{\mu\nu}}\left(-\frac{1}{2\kappa}\overline{\hat{R}} + \overline{\hat{\mathcal{L}}}^{M}\right) = 0\,.
\end{aligned} \tag{2..35}$$

Because the second line contains the operator of the background Einstein equations in (2..16) one can assert that Eq. (2..24) is equivalent to the Einstein equations (2..9).

By the same way the perturbed matter equations are constructed. They have the form:

$$\Phi^{L}_{A} = t^{m}_{A} \tag{2..36}$$

where

$$\Phi^{L}_{A}(\hat{l},\phi) \equiv -\frac{\delta}{\delta\overline{\Phi}^{A}}\left(\hat{l}^{\rho\sigma}\frac{\delta\overline{\hat{\mathcal{L}}^{M}}}{\delta\overline{g}^{\rho\sigma}} + \phi^{B}\frac{\delta\overline{\hat{\mathcal{L}}^{M}}}{\delta\overline{\Phi}^{B}}\right), \tag{2..37}$$

$$t^{m}_{A} \equiv \frac{\delta\hat{\mathcal{L}}^{m}}{\delta\overline{\Phi}^{A}}\,. \tag{2..38}$$

2.5. Expansions

The methods of the exact field-theoretical formulation give a possibility to construct an approximate scheme easily and in clear expressions up to an arbitrary order in perturbations. Let us show this. Assuming enough smooth functions, expand the Lagrangian $\hat{\mathcal{L}}^E(\overline{g}+l,\ \overline{\Phi}+\phi)$ as

$$\begin{aligned}\hat{\mathcal{L}}^E &= \overline{\hat{\mathcal{L}}^E} + \hat{l}^{\rho\sigma}\frac{\delta\overline{\hat{\mathcal{L}}^E}}{\delta\overline{g}^{\rho\sigma}} + \phi^B\frac{\delta\overline{\hat{\mathcal{L}}^E}}{\delta\overline{\Phi}^B} + \\ &+ \frac{1}{2!}\hat{l}^{\alpha\beta}\frac{\delta}{\delta\overline{\hat{g}}^{\alpha\beta}}\hat{l}^{\rho\sigma}\frac{\delta\overline{\hat{\mathcal{L}}^E}}{\delta\overline{\hat{g}}^{\rho\sigma}} + \hat{l}^{\rho\sigma}\frac{\delta}{\delta\overline{\hat{g}}^{\rho\sigma}}\phi^A\frac{\delta\overline{\hat{\mathcal{L}}^E}}{\delta\overline{\Phi}^A} + \frac{1}{2!}\phi^B\frac{\delta}{\delta\overline{\Phi}^B}\phi^A\frac{\delta\overline{\hat{\mathcal{L}}^E}}{\delta\overline{\Phi}^A} + \\ &+ \frac{1}{3!}\hat{l}^{\mu\nu}\frac{\delta}{\delta\overline{\hat{g}}^{\mu\nu}}\hat{l}^{\alpha\beta}\frac{\delta}{\delta\overline{\hat{g}}^{\alpha\beta}}\hat{l}^{\rho\sigma}\frac{\delta\overline{\hat{\mathcal{L}}^E}}{\delta\overline{\hat{g}}^{\rho\sigma}} + \ldots + div\,. \end{aligned} \tag{2..39}$$

Then the dynamical Lagrangian (2..19) acquires the form:

$$\hat{\mathcal{L}}^{dyn} = \frac{1}{2!}\hat{l}^{\alpha\beta}\frac{\delta}{\delta\overline{\hat{g}}^{\alpha\beta}}\hat{l}^{\rho\sigma}\frac{\delta\overline{\hat{\mathcal{L}}^E}}{\delta\overline{\hat{g}}^{\rho\sigma}} + \hat{l}^{\rho\sigma}\frac{\delta}{\delta\overline{\hat{g}}^{\rho\sigma}}\phi^A\frac{\delta\overline{\hat{\mathcal{L}}^E}}{\delta\overline{\Phi}^A} + \frac{1}{2!}\phi^B\frac{\delta}{\delta\overline{\Phi}^B}\phi^A\frac{\delta\overline{\hat{\mathcal{L}}^E}}{\delta\overline{\Phi}^A} + \ldots + div\,. \tag{2..40}$$

First, note that a divergence in (2..40) is not so important because divergences vanish under the Lagrange derivative. Second, now we can explain why in the linear terms of the Lagrangian (2..19) the background equations are not taken into account before variation. Indeed, the Lagrangian $\hat{\mathcal{L}}^E(\overline{g}+l,\ \overline{\Phi}+\phi)$ in (2..19) contains the same linear terms with only the opposite sign not explicitly (the formula (2..39) shows this). Therefore, in fact, these linear terms are compensated, and the real lowest order in the expansion of $\hat{\mathcal{L}}^{dyn}$ is a quadratic one (2..40). In this relation, recall that in the usual geometrical description of GR the attempt to define the energy-momentum tensor through $\delta\hat{\mathcal{L}}^E/\delta g^{\mu\nu}$ leads to zero on the solutions of the Einstein equations. In contrast, $\hat{t}^{(tot)}_{\mu\nu}$ in (2..28) does not vanish on the field equations. The reason is just in the presence in the Lagrangian (2..19) of these linear terms.

Third, under necessary assumptions the series (2..40) can be interrupted at a corresponding order. Thus the approximate Lagrangian for the perturbed system can be obtained. Its variation gives both approximate field equations and energy-momentum tensor. For example, the quadratic approximation of (2..40) gives a possibility a) to construct the linear equations

$$-\frac{1}{2\kappa}\left(\hat{G}^L_{\mu\nu}(\hat{l},\phi) + \hat{\Phi}^L_{\mu\nu}(\hat{l},\phi)\right) \equiv \frac{\delta}{\delta\overline{g}^{\mu\nu}}\left(\hat{l}^{\rho\sigma}\frac{\delta\overline{\hat{\mathcal{L}}^E}}{\delta\overline{\hat{g}}^{\rho\sigma}} + \phi^B\frac{\delta\overline{\hat{\mathcal{L}}^E}}{\delta\overline{\Phi}^B}\right) = 0\,, \tag{2..41}$$

$$-\Phi^L_A(\hat{l},\phi) \equiv \frac{\delta}{\delta\overline{\Phi}^A}\left(\hat{l}^{\rho\sigma}\frac{\delta\overline{\hat{\mathcal{L}}^E}}{\delta\overline{\hat{g}}^{\rho\sigma}} + \phi^B\frac{\delta\overline{\hat{\mathcal{L}}^E}}{\delta\overline{\Phi}^B}\right) = 0\,, \tag{2..42}$$

b) to construct the quadratic energy-momentum tensor:

$$t^{(tot)}_{\mu\nu} = \frac{2}{\sqrt{\overline{g}}}\frac{\delta}{\delta\overline{g}^{\mu\nu}}\left(\frac{1}{2!}\hat{l}^{\alpha\beta}\frac{\delta}{\delta\overline{\hat{g}}^{\alpha\beta}}\hat{l}^{\rho\sigma}\frac{\delta\overline{\hat{\mathcal{L}}^E}}{\delta\overline{\hat{g}}^{\rho\sigma}} + \hat{l}^{\rho\sigma}\frac{\delta}{\delta\overline{\hat{g}}^{\rho\sigma}}\phi^A\frac{\delta\overline{\hat{\mathcal{L}}^E}}{\delta\overline{\Phi}^A} + \frac{1}{2!}\phi^B\frac{\delta}{\delta\overline{\Phi}^B}\phi^A\frac{\delta\overline{\hat{\mathcal{L}}^E}}{\delta\overline{\Phi}^A}\right). \tag{2..43}$$

The cubic approximation of (2..40) gives a possibility a) to construct the field equations including quadratic terms (which are related to the energy-momentum tensor), and b) to construct the energy-momentum tensor, including quadratic and cubic parts, *etc.*

Fourth, the expansions, like (2..39) - (2..43), are used in quantum field theories [4] and called as functional expansions. As is seen, in the framework of the classic theory the functional expansions (2..40) - (2..42) give, in fact, the algorithm for constructing the approximate systems, thus an each order can be obtained automatically.

2.6. Gauge Transformations and Their Properties

The important property of the field formulation is the gauge invariance. Usually gauge transformations in GR and other metric theories are connected with mapping a spacetime onto itself that is connected with differentiable coordinate transformations

$$x'^{\alpha} = f^{\alpha}(x^{\beta}) . \tag{2..44}$$

These transformations can be connected with the smooth vector field ξ^{α}:

$$x'^{\alpha} = x^{\alpha} + \xi^{\alpha}(x) + \frac{1}{2!}\,\xi^{\beta}\xi^{\alpha}{}_{,\beta} + \frac{1}{3!}\,\xi^{\pi}(\xi^{\beta}\xi^{\alpha}{}_{,\beta})_{,\pi} + \ldots . \tag{2..45}$$

To map the spacetime onto itself one has to follow the standard way [138] - [140]. After the coordinate transformations (2..44) (or (2..45)) the metric density, for example, is transformed as $\hat{g}^{\mu\nu}(x) \to \hat{g}'^{\mu\nu}(x')$. Then return to the points with quantities x within a new frame $\{x'\}$. After that one has to compare geometrical objects of the initial spacetime and of the mapped spacetime in the points with quantities x. The comparison can be carried out both without ξ^{α} in the terms of (2..44) and with ξ^{α} included in (2..45):

$$\hat{g}'^{\mu\nu}(x) = \hat{g}^{\mu\nu}(x) + \delta_f \hat{g}^{\mu\nu} = \hat{g}^{\mu\nu}(x) + \sum_{k=1}^{\infty} \frac{1}{k!}\,\pounds^k_{\xi}\hat{g}^{\mu\nu}(x) , \tag{2..46}$$

$$\Phi'^{A}(x) = \Phi^{A}(x) + \delta_f \Phi^{A}(x) = \Phi^{A}(x) + \sum_{k=1}^{\infty} \frac{1}{k!}\,\pounds^k_{\xi}\Phi^{A}(x) . \tag{2..47}$$

The next property is very useful. Assume that geometrical objects Ψ^{B} are differentiable functions of other geometrical objects ψ^{A} and their derivatives, but are not explicit functions of coordinates. Then it is clear that a simple substitution gives $\Psi^{B}(\psi'^{A}) = \Psi'^{B}$, and one has

$$\Psi^{B}(\psi'^{A}(x)) = \Psi^{B}(\psi^{A}(x)) + \delta_f \Psi^{B} = \Psi^{B}(\psi^{A}(x)) + \sum_{k=1}^{\infty} \frac{1}{k!} \pounds_{\xi}{}^{k}\Psi^{B} . \tag{2..48}$$

Now let us define the gauge transformations for the dynamical variables in the framework of the field formulation of GR:

$$\hat{l}'^{\mu\nu} = \hat{l}^{\mu\nu} + \delta_f \left(\overline{\hat{g}}^{\mu\nu} + \hat{l}^{\mu\nu}\right) = \hat{l}^{\mu\nu} + \sum_{k=1}^{\infty} \frac{1}{k!}\,\pounds^k_{\xi}\left(\overline{\hat{g}}^{\mu\nu} + \hat{l}^{\mu\nu}\right), \tag{2..49}$$

$$\phi'^{A} = \phi^{A} + \delta_f \left(\overline{\Phi}^{A} + \phi^{A}\right) = \phi^{A} + \sum_{k=1}^{\infty} \frac{1}{k!}\,\pounds^k_{\xi}\left(\overline{\Phi}^{A} + \phi^{A}\right). \tag{2..50}$$

The assumptions above, in fact, state that the operators δ_f and $\sum_{k=1}^{\infty}\frac{1}{k!}\pounds_\xi^k$ are equivalent. However, the operator δ_f could present a wider class of transformations, which cannot be expressed through infinite series. We conserve a possibility to consider such transformations, however more frequently we will use the sums, keeping in mind that they can be changed by δ_f also.

Now, let us substitute (2..49) and (2..50) into the dynamical Lagrangian (2..19). One finds that this substitution into $\hat{\mathcal{L}}^E(\overline{g}+\hat{l},\overline{\Phi}+\phi)$ just permits to use the property (2..48). Then finally one obtains

$$\begin{aligned}\hat{\mathcal{L}}'^{dyn} &= \hat{\mathcal{L}}^{dyn} + \sum_{k=1}^{\infty}\frac{1}{k!}\pounds_\xi{}^k\hat{\mathcal{L}}^E(\overline{g}+\hat{l},\overline{\Phi}+\phi) - \frac{1}{2\kappa}\partial_\alpha\left(\hat{k}'^\alpha - \hat{k}^\alpha\right)\\ &- \left(\hat{l}'^{\mu\nu}-\hat{l}^{\mu\nu}\right)\frac{\delta\overline{\hat{\mathcal{L}}^E}}{\delta\overline{\hat{g}}^{\mu\nu}} - \left(\phi'^A-\phi^A\right)\frac{\delta\overline{\hat{\mathcal{L}}^M}}{\delta\overline{\Phi}^A}\,. \end{aligned}\qquad(2..51)$$

Because $\hat{\mathcal{L}}^E$ is the scalar density of the weight $+1$ all the terms under the sum in (2..51) are divergences. Thus $\hat{\mathcal{L}}^{dyn}$ is gauge invariant up to a divergence if the background equations (2..16) and (2..17) hold.

Considering the gauge invariance properties of the field equations we use their form (2..35). The substitution of (2..49) and (2..50) with the use of the property (2..48) gives

$$\begin{aligned}&\left[\hat{G}^L_{\mu\nu}+\hat{\Phi}^L_{\mu\nu}-\kappa\hat{t}^{(tot)}_{\mu\nu}\right]' = \left[\hat{G}^L_{\mu\nu}+\hat{\Phi}^L_{\mu\nu}-\kappa\hat{t}^{(tot)}_{\mu\nu}\right]\\ &+ \frac{\partial\overline{\hat{g}}^{\rho\sigma}}{\partial\overline{g}^{\mu\nu}}\sum_{k=1}^{\infty}\frac{1}{k!}\pounds_\xi{}^k\left[\frac{\partial\overline{g}^{\delta\pi}}{\partial\overline{\hat{g}}^{\rho\sigma}}\left(\hat{G}^L_{\delta\pi}+\hat{\Phi}^L_{\delta\pi}-\kappa\hat{t}^{(tot)}_{\delta\pi}\right)-2\kappa\frac{\delta\overline{\hat{\mathcal{L}}^E}}{\delta\overline{\hat{g}}^{\rho\sigma}}\right]\,. \end{aligned}\qquad(2..52)$$

Thus the field equations are gauge invariant on solutions of themselves and with using the background equations (2..16) and (2..17). Analogous transformations could be presented for the matter equations (2..36).

Concerning the energy-momentum tensor, it is not gauge invariant. Even on the dynamical equations, as it follows from (2..52), under the transformations (2..49) and (2..50) one has

$$\kappa\hat{t}'^{(tot)}_{\mu\nu} = \kappa\hat{t}^{(tot)}_{\mu\nu} + \hat{G}^L_{\mu\nu}(l'-l) + \hat{\Phi}^L_{\mu\nu}(l'-l;\,\phi'-\phi)\,. \qquad(2..53)$$

The mathematical reason is that the background equations in the Lagrangian (2..51) cannot be taken into account before variation. In the case of the Ricci-flat backgrounds (2..18) one has $\hat{\Phi}^L_{\mu\nu}=0$, therefore the energy-momentum $\hat{t}^{(tot)}_{\mu\nu}$ is gauge invariant up to $\hat{G}^L_{\mu\nu}$ — covariant divergence (see (2..26)). Let us turn also to the equivalent form (2..32) of the field equations with the effective source $\hat{t}^{(eff)}_{\mu\nu}$. For the operator of the field equations $\hat{G}^L_{\mu\nu}-\kappa\hat{t}^{(eff)}_{\mu\nu}$ the form of the transformations (2..52) can be used without changing. Then on the field equations one has

$$\kappa\hat{t}'^{(eff)}_{\mu\nu} = \kappa\hat{t}^{(eff)}_{\mu\nu} + \hat{G}^L_{\mu\nu}(l'-l)\,, \qquad(2..54)$$

i.e. for all the kinds of backgrounds $\hat{t}^{(eff)}_{\mu\nu}$ is gauge invariant up to a covariant divergence. It is not surprising that both the energy-momentum tensors are not gauge invariant. It is a

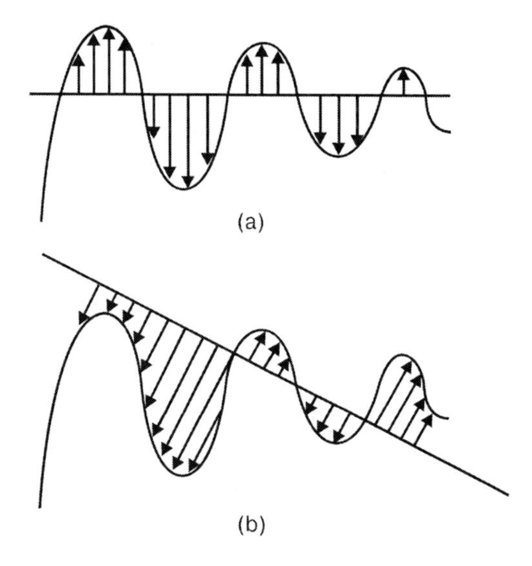

Figure 2. The perturbations (a) and (b) connected by the gauge transformations.

waited result. Indeed, this reflects the fact that energy and other conserved quantities in GR are not localized. Moreover, the formulae (2..53) and (2..54) are very useful because they give a *quantitative* and *constructive* description of the non-localization. See the discussion in subsection 1.2..

The transformations (2..49) and (2..50) are directly connected with a way of mapping a perturbed spacetime onto a given background spacetime (a perturbed solution onto a given solution). In the other words, they are connected with a definition of perturbations with respect to a given background. Let us consider a solution in the geometrical form $\hat{g}^{\mu\nu}(x)$. Next let us map a spacetime onto itself following the prescription at the beginning this subsection. Then we obtain $\hat{g}'^{\mu\nu}(x)$. After that we decompose both of the solutions into dynamic and background parts:

$$\hat{g}^{\mu\nu}(x) = \overline{g}^{\mu\nu}(x) + \hat{l}^{\mu\nu}(x)\,, \tag{2..55}$$

$$(\hat{g}'^{\mu\nu})(x) = \overline{g}^{\mu\nu}(x) + \hat{l}'^{\mu\nu}(x)\,. \tag{2..56}$$

The main property of (2..55) and (2..56) is that the background metric is the same. Then it turns out that $\hat{l}^{\mu\nu}(x)$ and $\hat{l}'^{\mu\nu}(x)$ are connected by the transformations in (2..49). This situation is interpreted as follows. The same background in (2..55) and (2..56) is chosen by different ways that symbolically is illustrated on the figure 2. The curves in both the cases (a) and (b) symbolize the same solution of GR in the geometrical form, whereas the straight lines present a background, say, a Minkowski space. The perturbations in the cases (a) and (b) are different, but they are solutions to the equations of the field formulation connected by the gauge transformations, and, thus, they are the same solution in different forms. In spite of that the gauge transformations in the field formulation have the evident geometrical origin, they can be interpreted as inner gauge transformations, like in standard gauge theories. Indeed, they act only onto the dynamical variables (perturbations), whereas the backgrounds variables and coordinates do not change.

As discussed in Introduction, in many of applications it is important to consider equations and gauge transformations in linear and quadratic approximations. Assume that perturbations and their derivatives are small: $\hat{l}^{\mu\nu} \ll \overline{\hat{g}}^{\mu\nu}$, $\phi^A \ll \overline{\Phi}^A$, $\hat{l}^{\mu\nu} \approx \partial_\alpha \hat{l}^{\mu\nu} \ldots$ and $\phi^A \approx \partial_\alpha \phi^A \approx \ldots$. Assuming that the background equations (2..16) give a connection $\overline{\hat{g}}^{\mu\nu} \approx h(\kappa)\overline{\Phi}^A$ with a coefficient $h(\kappa)$ depending on the Einstein constant one has to set $\hat{l}^{\mu\nu} \approx h(\kappa)\phi^A$, *etc.* Now, rewrite the equations (2..24) up to the second order in perturbations:

$$\hat{G}^L_{\mu\nu}(\hat{l}) + \hat{\Phi}^L_{\mu\nu}(\hat{l},\, \phi) - \kappa \hat{t}^{(tot2)}_{\mu\nu}(\hat{l}\hat{l},\, \hat{l}\phi,\, \phi\phi) = 0\,. \qquad (2..57)$$

Assuming iterations the perturbations can be expanded as $\hat{l}^{\mu\nu} = \hat{l}^{\mu\nu}_1 + \hat{l}^{\mu\nu}_2 + \ldots$, and $\phi^A = \phi^A_1 + \phi^A_2 + \ldots$. Then the linear equations acquire the form:

$$\hat{G}^L_{\alpha\beta}(\hat{l}_1) + \hat{\Phi}^L_{\alpha\beta}(\hat{l}_1,\, \phi_1) = 0\,, \qquad (2..58)$$

whereas the quadratic equations have the form:

$$\hat{G}^L_{\alpha\beta}(\hat{l}_2) + \hat{\Phi}^L_{\alpha\beta}(\hat{l}_2,\, \phi_2) - \kappa\left(\hat{t}^{(gr2)}_{\alpha\beta}(\hat{l}_1\hat{l}_1) + \hat{t}^{(m2)}_{\alpha\beta}(\hat{l}_1\hat{l}_1,\, \hat{l}_1\phi_1,\, \phi_1\phi_1)\right) = 0\,. \qquad (2..59)$$

Besides, assuming $\xi^\mu = \xi^\mu_1 + \xi^\mu_2 + \ldots$ with $\xi^\mu_1 \approx \partial_\alpha \xi^\mu_1 \approx \ldots \approx \hat{l}^{\mu\nu}_1 \approx h(\kappa)\phi^A_1$ and $\xi^\mu_2 \approx \partial_\alpha \xi^\mu_2 \approx \ldots \approx \hat{l}^{\mu\nu}_2 \approx h(\kappa)\phi^A_2$, one has a linear version of (2..49) in the simple form:

$$\begin{aligned} \hat{l}'^{\mu\nu}_1 &= \hat{l}^{\mu\nu}_1 + \pounds_{\xi_1}\overline{\hat{g}}^{\mu\nu} = \hat{l}^{\mu\nu}_1 + \overline{D}^\mu \hat{\xi}^\nu_1 + \overline{D}^\nu \hat{\xi}^\mu_1 - \overline{\hat{g}}^{\mu\nu}\, \overline{D}_\rho \hat{\xi}^\rho_1\,, & (2..60)\\ \phi'^A_1 &= \phi^A_1 + \pounds_{\xi_1}\overline{\Phi}^A. & (2..61) \end{aligned}$$

Under these transformations the equations (2..58) are transformed as

$$\begin{aligned} &\left[\hat{G}^L_{\mu\nu} + \hat{\Phi}^L_{\mu\nu}\right]' = \left[\hat{G}^L_{\mu\nu} + \hat{\Phi}^L_{\mu\nu}\right]\\ &+ \left(\sqrt{-\overline{g}}\delta^\rho_\mu\delta^\sigma_\mu - \tfrac{1}{2}\overline{\hat{g}}_{\mu\nu}\overline{\hat{g}}^{\rho\sigma}\right)\pounds_{\xi_1}\left[\overline{R}_{\rho\sigma} - \kappa\left(\overline{T}_{\rho\sigma} - \tfrac{1}{2}\overline{g}_{\rho\sigma}\overline{T}\right)\right] \qquad (2..62) \end{aligned}$$

and are, thus, gauge invariant on the background equations. In the simple case of the Ricci-flat background (2..18) the linear transformations have only the form (2..60), without (2..61). Then the formula (2..62) transfers to the formula $\hat{G}'^L_{\mu\nu} = \hat{G}^L_{\mu\nu}$, which expresses the gauge invariance of the linear field of spin 2. It is the well known gauge transformations in the linear gravity [1, 3].

The quadratic order of the gauge transformations has a form;

$$\begin{aligned} \hat{l}'^{\mu\nu}_2 &= \hat{l}^{\mu\nu}_2 + \pounds_{\xi_2}\overline{\hat{g}}^{\mu\nu} + \frac{1}{2!}\pounds^2_{\xi_1}\overline{\hat{g}}^{\mu\nu} + \pounds_{\xi_1}\hat{l}^{\mu\nu}_1\\ \phi'^A_2 &= \phi^A_2 + \pounds_{\xi_2}\overline{\Phi}^A + \frac{1}{2!}\pounds^2_{\xi_1}\overline{\Phi}^{\mu\nu} + \pounds_{\xi_1}\phi^A_1\,. \qquad (2..63) \end{aligned}$$

Substitution of Eqs. (2..60) and (2..63) into (2..59) give

$$\begin{aligned} &\left[\hat{G}^L_{\mu\nu}(\hat{l}_2) + \hat{\Phi}^L_{\mu\nu}(\hat{l}_2,\, \phi_2) - \kappa\hat{t}^{(tot2)}_{\mu\nu}\right]' = \left[\hat{G}^L_{\mu\nu}(\hat{l}_2) + \hat{\Phi}^L_{\mu\nu}(\hat{l}_2,\, \phi_2) - \kappa\hat{t}^{(tot2)}_{\mu\nu}\right]\\ &+ \frac{\partial\overline{g}^{\rho\sigma}}{\partial\overline{g}^{\mu\nu}}\left(\pounds_{\xi_2} + \frac{1}{2!}\pounds^2_{\xi_1}\right)\left[\overline{R}_{\rho\sigma} - \kappa\left(\overline{T}_{\rho\sigma} - \tfrac{1}{2}\overline{g}_{\rho\sigma}\overline{T}\right)\right] +\\ &+ \frac{\partial\overline{g}^{\rho\sigma}}{\partial\overline{g}^{\mu\nu}}\pounds_{\xi_1}\left[\frac{\partial\overline{g}^{\delta\pi}}{\partial\overline{g}^{\rho\sigma}}\left[\hat{G}^L_{\delta\pi}(\hat{l}_1) + \hat{\Phi}^L_{\delta\pi}(\hat{l}_1,\, \phi_1)\right]\right]. \qquad (2..64) \end{aligned}$$

Thus, equations (2..59) are gauge invariant on the background equations (2..16) and on the linear equations (2..58). Of course, the procedure can be continued in the next orders.

In this subsection, we were based on the *exact* theory of gauge transformations developed in our works [100, 104]. Together with this, for the presentation we have used some of details given in [101, 112, 113]. An arising interest to cosmological perturbations stimulates their more detailed study including a second order approximation [14]. Thus it turns out necessary to examine the gauge transformations up to a second order also. Such studies (independently on ours, but repeating them in main properties) were carried out, for examples, in [141, 142].

2.7. Differential Conservation Laws on Special Backgrounds

The energy-momentum tensor is the one of important objects of a field theory in Minkowski space. Its differential conservation together with symmetries of the Minkowski space permit to construct integral conserved quantities (see subsection 3.2.). The field formulation of GR has also the conserved energy-momentum in Minkowski space with the same properties (2..3). In this short subsection we demonstrate that the conservation law, like (2..3), also has a place on curved backgrounds, which are important in many applications. Although conservation laws and conserved quantities are constructed and studied in the next section 3. in detail, by the logic of the presentation we include this subsection here.

Firstly, consider Ricci-flat (including flat) backgrounds (2..18), which have an independent meaning. This means that $\overline{\Phi}^A \equiv 0$ and $\overline{\hat{\mathcal{L}}^M} \equiv 0$. Then the Lagrangian (2..19) is simplified to

$$\hat{\mathcal{L}}^{dyn} = -\frac{1}{2\kappa}\hat{\mathcal{L}}^g + \hat{\mathcal{L}}^m = -\frac{1}{2\kappa}\hat{\mathcal{L}}^g + \hat{\mathcal{L}}^M\left(\phi^A; \,\overline{\mathfrak{g}}^{\mu\nu} + \hat{l}^{\mu\nu}\right). \tag{2..65}$$

The field equations (2..24) transform into the form

$$\hat{G}^L_{\mu\nu} = \kappa\left(\hat{t}^g_{\mu\nu} + \hat{t}^m_{\mu\nu}\right) \equiv \kappa\hat{t}^{(tot)}_{\mu\nu}. \tag{2..66}$$

For (2..18) one has identically $\overline{D}_\nu \hat{G}^{L\nu}_\mu \equiv 0$ and, thus, a divergence of Eq. (2..66) leads to the differential conservation law:

$$\overline{D}_\nu \hat{t}^{(tot)\nu}_\mu = 0\,. \tag{2..67}$$

Now consider backgrounds presented by Einstein spaces in Petrov's definition [143], then the background equations are

$$\overline{R}_{\mu\nu} = \Lambda\overline{g}_{\mu\nu} \tag{2..68}$$

where Λ is a constant. Ricci-flat and AdS backgrounds are particular cases of Einstein's spaces. The Lagrangian of the background system has the form:

$$\overline{\hat{\mathcal{L}}^E} = -\frac{1}{2\kappa}\overline{\hat{R}} + \overline{\hat{\mathcal{L}}^M} = -\frac{1}{2\kappa}\left(\overline{\hat{R}} - 2\Lambda\sqrt{-\overline{g}}\right). \tag{2..69}$$

Here, the constant Λ is rather interpreted as 'degenerated' matter. Then, the dynamical Lagrangian (2..19) transforms into

$$\hat{\mathcal{L}}^{dyn} = -\frac{1}{2\kappa}\hat{\mathcal{L}}^g + \hat{\mathcal{L}}^m = -\frac{1}{2\kappa}\hat{\mathcal{L}}^g + \left[\hat{\mathcal{L}}^M\left(\phi^A;\, \overline{\hat{g}}^{\mu\nu} + \hat{l}^{\mu\nu}\right) - \hat{l}^{\mu\nu}\frac{\delta\overline{\hat{\mathcal{L}}^M}}{\delta\overline{\hat{g}}^{\mu\nu}} - \overline{\hat{\mathcal{L}}^M}\right] \tag{2..70}$$

and leads to the field equations (2..24) in the simple form:

$$\hat{G}^L_{\mu\nu} + \Lambda\hat{l}^{\mu\nu} = \kappa\left(\hat{t}^g_{\mu\nu} + \hat{t}^m_{\mu\nu}\right) \equiv \kappa\hat{t}^{(tot)}_{\mu\nu}\,. \tag{2..71}$$

Taking into account the background equations (2..68) one has identically

$$\overline{D}_\nu\left(\hat{G}^{L\nu}_\mu + \Lambda\hat{l}^\nu_\mu\right) \equiv 0. \tag{2..72}$$

Thus, the differentiation of Eq. (2..71) gives the same conservation law (2..67). In heuristic form the differential conservation law on AdS and de Sitter (dS) backgrounds was used in [68]; in the Lagrangian description it was shortly noted in [100]; and, in the paper [144], it was studied in more detail.

2.8. Different Definitions of Metric Perturbations

In GR, the metric perturbations can be defined by different decompositions:

$$\begin{aligned} g_{\mu\nu} &= \overline{g}_{\mu\nu} + h_{\mu\nu}, \\ \hat{g}^{\mu\nu} &= \overline{\hat{g}}^{\mu\nu} + \hat{l}^{\mu\nu}, \\ g^{\mu\nu} &= \overline{g}^{\mu\nu} + r^{\mu\nu}, \\ \ldots &= \ldots + \ldots, \end{aligned} \tag{2..73}$$

not only by (2..13). Denoting components of metrical densities in the united way

$$g^a = \{g^{\mu\nu},\, g_{\mu\nu},\, \sqrt{-g}g^{\mu\nu},\, \sqrt{-g}g_{\mu\nu},\, (-g)g^{\mu\nu},\, \ldots\} \tag{2..74}$$

we rewrite the action of GR as

$$S = \frac{1}{c}\int d^4x\hat{\mathcal{L}}^{E(a)} \equiv -\frac{1}{2\kappa c}\int d^4x\hat{R}(g^a) + \frac{1}{c}\int d^4x\hat{\mathcal{L}}^M(\Phi^A,\, g^a)\,. \tag{2..75}$$

After its variation the Einstein equations take the generalized form:

$$-\frac{1}{2\kappa}\frac{\delta\hat{R}}{\delta g^a} + \frac{\delta\hat{\mathcal{L}}^M}{\delta g^a} = 0\,. \tag{2..76}$$

The background action and equations have the corresponding to (2..75) and (2..76) barred form. After that we present the united form for decompositions (2..73)

$$g^a = \overline{g}^a + h^a\,, \tag{2..77}$$

and, following the rules of (2..19), construct the generalized dynamical Lagrangian:

$$\hat{\mathcal{L}}^{dyn}_{(a)} = -\frac{1}{2\kappa}\hat{R}\left(\overline{g}^a + h^a\right) + \hat{\mathcal{L}}^M\left(\overline{\Phi}^A + \phi^A;\, \overline{g}^a + h^a\right)$$
$$- \; h^a\left(-\frac{1}{2\kappa}\frac{\delta\overline{\hat{R}}}{\delta\overline{g}^a} + \frac{\delta\overline{\hat{\mathcal{L}}^M}}{\delta\overline{g}^a}\right) - \phi^A\frac{\delta\overline{\hat{\mathcal{L}}^M}}{\delta\overline{\Phi}^A} - \left(-\frac{1}{2\kappa}\overline{\hat{R}} + \overline{\hat{\mathcal{L}}^M}\right) - \frac{1}{2\kappa}\partial_\nu\hat{k}^\nu. \quad (2..78)$$

The total symmetrical energy-momentum tensor density is defined as usual:

$$\hat{t}^{(tot\ a)}_{\mu\nu} \equiv 2\frac{\delta\hat{\mathcal{L}}^{dyn}_{(a)}}{\delta\overline{g}^{\mu\nu}}\,. \quad (2..79)$$

After substituting the expression (2..78) into this definition (identity) and taking into account Eq. (2..76) and the barred Eq. (2..76) we obtain the Einstein equations in the form (2..24):

$$\hat{G}^{L(a)}_{\mu\nu} + \hat{\Phi}^{L(a)}_{\mu\nu} = \kappa\hat{t}^{(tot\ a)}_{\mu\nu}\,. \quad (2..80)$$

Here, the linear left hand side is defined by the same operators (2..25) - (2..27), only now with independent variables

$$\hat{l}^{\mu\nu}_{(a)} \equiv h^a\frac{\partial\overline{\hat{g}}^{\mu\nu}}{\partial\overline{g}^a}\,. \quad (2..81)$$

However, a choice of two different arbitrary decompositions as $g^a_1 = \overline{g}^a_1 + h^a_1$ and $g^a_2 = \overline{g}^a_2 + h^a_2$, gives the difference

$$\hat{l}^{\mu\nu}_{(a2)} - \hat{l}^{\mu\nu}_{(a1)} = \hat{\beta}^{\mu\nu}_{(a)12}\,, \quad (2..82)$$

which is not less than quadratic in perturbations. Because differences enter the linear expressions of equations (2..80) the energy-momentum tensor densities $\hat{t}^{(tot\ 1a)}_{\mu\nu}$ and $\hat{t}^{(tot\ 2a)}_{\mu\nu}$ have the same differences too. For the case of flat backgrounds this ambiguity was noted by Boulware and Deser [145]. Later we [104] have examined it for arbitrary curved backgrounds and arbitrary metric theories. However, only in our works [116, 118] this ambiguity has been resolved, and we present this solution in subsection 3.5..

3. Conservation Laws in GR

3.1. Classical Pseudotensors and Superpotentials

During many decades after constructing GR pseudotensors and superpotentials were main objects in constructing conservation laws and conserved quantities. In the framework of the field-theoretical approach these notions and quantities, in a definite sense, are developed and generalized. Therefore in this subsection we give a short review describing classical pseudotensors and superpotentials, only which are necessary in our own presentation. On their examples we outline the general properties of such objects, their problems and some of modern applications.

Let us present the general way for constructing pseudotensors and superpotentials. Using the metric and its derivations construct an *arbitrary* quantity $\hat{\mathcal{U}}^{\mu\alpha}_\nu$, satisfying the condition

$$\partial_{\mu\alpha}\hat{\mathcal{U}}^{\mu\alpha}_\nu \equiv 0\,. \quad (3..1)$$

Next, define the complex

$$\hat{\theta}^{\mu}_{\nu} \equiv \partial_{\alpha}\hat{\mathcal{U}}^{\mu\alpha}_{\nu} - \frac{1}{\kappa}\hat{G}^{\mu}_{\nu}\,, \tag{3..2}$$

which usually is called as an energy-momentum pseudotensor of gravitational field. Using the Einstein equations one obtains

$$\hat{T}^{\mu}_{\nu} + \hat{\theta}^{\mu}_{\nu} = \partial_{\alpha}\hat{\mathcal{U}}^{\mu\alpha}_{\nu} \tag{3..3}$$

where $\hat{\mathcal{U}}^{\mu\alpha}_{\nu}$ plays the role of a superpotential. Thus Eq. (3..3) is another form of the Einstein equations. Due to (3..1) one has a differential conservation law

$$\partial_{\mu}\left(\hat{T}^{\mu}_{\nu} + \hat{\theta}^{\mu}_{\nu}\right) = 0\,. \tag{3..4}$$

Concerning the physical sense, it is the 4-dimensional continuity equation, which is the differential conservation law derived directly from the field equations. As is seen, the above construction has problems. First, it is an ambiguity in a definition of $\hat{\theta}^{\mu}_{\nu}$ and $\hat{\mathcal{U}}^{\mu\alpha}_{\nu}$, second, these expressions are not covariant in general.

The presented picture, as a generalization, has been formulated after prolonged history of constructing pseudotensors and superpotentials in GR (see a review [146]). Below we recall constructing some of well known expressions. In the work [55], following the standard rules Einstein had suggested his famous pseudotensor. Firstly, a non-covariant Lagrangian

$$\hat{\mathcal{L}}^{cut} = -\frac{1}{2\kappa}\hat{g}^{\mu\nu}\left(\Gamma^{\sigma}_{\mu\rho}\Gamma^{\rho}_{\sigma\nu} - \Gamma^{\rho}_{\mu\nu}\Gamma^{\sigma}_{\rho\sigma}\right) \tag{3..5}$$

had been suggested. It differs from the covariant Hilbert Lagrangian

$$\hat{\mathcal{L}}^{H} = -\frac{1}{2\kappa}\hat{R} \tag{3..6}$$

by a divergence and leads to the same field equations (2..12). Next, the corresponding to (3..5) canonical complex had been constructed:

$$\hat{t}^{E\mu}_{\nu} = \frac{\partial\hat{\mathcal{L}}^{cut}}{\partial(\partial_{\mu}g_{\alpha\beta})}\partial_{\nu}g_{\alpha\beta} - \delta^{\mu}_{\nu}\hat{\mathcal{L}}^{cut}\,, \tag{3..7}$$

which is just the Einstein pseudotensor. Both the Lagrangian (3..5) and the pseudotensor (3..7) depend on the metric $g_{\mu\nu}$ and only its first derivatives. Combining (3..7) and the field equations Einstein had obtained the conservation law

$$\partial_{\mu}\left(\hat{T}^{\mu}_{\nu} + \hat{t}^{E\mu}_{\nu}\right) = 0\,, \tag{3..8}$$

which is a variant of (3..4).

Later Tolman [147] had found the quantity $\hat{\mathcal{T}}^{\mu\alpha}_{\nu}$ with a property (3..1), and for which, as a particular case of (3..3), one has $\hat{T}^{\mu}_{\nu} + \hat{t}^{E\mu}_{\nu} \equiv \partial_{\alpha}\hat{\mathcal{T}}^{\mu\alpha}_{\nu}$. From here the conservation law (3..8) follows directly. However, $\hat{\mathcal{T}}^{\mu\alpha}_{\nu}$ is not antisymmetrical in α and β, whereas the use of antisymmetrical superpotentials $\mathcal{U}^{\mu\alpha}_{\nu} = -\mathcal{U}^{\alpha\mu}_{\nu}$ is more reasonable because it makes explicit the identity (3..1) and presents less difficulties under covariantization, *etc*. Assuming an antisymmetry Freud [148] had found out the superpotential

$$\hat{\mathcal{F}}^{\mu\alpha}_{\nu} \equiv \frac{1}{2\kappa}\frac{g_{\nu\rho}}{\sqrt{-g}}\partial_\lambda\left[(-g)\left(g^{\mu\rho}g^{\alpha\lambda} - g^{\mu\lambda}g^{\alpha\rho}\right)\right], \tag{3..9}$$

such that

$$\hat{T}^{\mu}_{\nu} + \hat{t}^{E\mu}_{\nu} = \partial_\alpha \hat{\mathcal{F}}^{\mu\alpha}_{\nu}, \tag{3..10}$$

and connected with the Tolman one by $\hat{\mathcal{F}}^{\mu\alpha}_{\nu} \equiv \hat{\mathcal{T}}^{\mu\alpha}_{\nu} + \partial_\beta \hat{\Phi}^{\mu\alpha\beta}_{\nu}$; $\partial_{\alpha\beta}\hat{\Phi}^{\mu\alpha\beta}_{\nu} \equiv 0$.

Many authors (see, e.g., more known publications [149] - [151] and the review [146]) considered the problem of an ambiguity in definition of conserved quantities described in (3..1) - (3..4). It has turned out that some of pseudotensors and superpotentials are directly connected with a Lagrangian through the Nœther procedure. This restricts some ambiguities in their definitions. Naturally, the Eienstein pseudotensor is defined by the Lagrangian (3..5) which, being generally non-covariant, is covariant with respect to linear coordinate transformations. Then for translations $\lambda^{\mu}_{(\alpha)} = \delta^{\mu}_{\alpha}$ one can write out the Nœther identity:

$$\pounds_\lambda \hat{\mathcal{L}}^{cut} + \partial_\mu\left(\lambda^{\mu}_{(\alpha)}\hat{\mathcal{L}}^{cut}\right) \equiv 0\,. \tag{3..11}$$

Direct calculations lead to $\partial_\mu\left(\kappa^{-1}\hat{G}^{\mu}_{\nu} + \hat{t}^{E\mu}_{\nu}\right) \equiv 0$ with the definition (3..7) that after using Eq. (2..12) gives the conservation law (3..8). The general conclusion is formulated as follows.

- In the sense of the Nœther procedure the pseudotensor (3..7) is uniquely defined by the Lagrangian (3..5); other Lagrangians $\hat{\mathcal{L}} = \hat{\mathcal{L}}^{cut} + div$ lead to other pseudotensors.

Thus the identity (3..11) with the Hilbert Lagrangian (3..6) leads to

$$\hat{T}^{\mu}_{\nu} + \hat{\mathcal{M}}^{\mu}_{\nu} = \partial_\alpha \hat{\chi}^{\mu\alpha}_{\nu}\,, \tag{3..12}$$

instead of (3..10), where the Møller [151] superpotential

$$\hat{\chi}^{\mu\alpha}_{\nu} \equiv \frac{1}{2\kappa}\sqrt{-g}g^{\mu\beta}g^{\alpha\rho}\left(\partial_\beta g_{\nu\rho} - \partial_\rho g_{\nu\beta}\right) \tag{3..13}$$

is used. The components $\hat{\mathcal{M}}^{0}_{\nu}(x^\alpha)$ of the pseudotensor in Eq. (3..12) are transformed as a 4-dimensional vector density under transformations $x'^k = f^k(x^l)$, $x'^0 = x^0 + g(x^l)$ [151]. In contrary, the corresponding components of the Einstein pseudotensor (3..7) do not. This property of $\hat{\mathcal{M}}^{0}_{\nu}(x^\alpha)$, in a part, improves the general non-covariant picture.

The other problem is in a definition of angular momentum of a system. Both Einsten's and Mœller's pseudotensors themselves, as canonical Nœther complexes, cannot define it. One needs to modify these expressions or to construct new ones (see the review [146]). One of methods to construct angular momentum, using *unique* energy-momentum complex, is to construct symmetrical pseudotensors. Thus, Landau and Lifshitz have suggested a such expression [1]. Their famous symmetrical pseudotensor $t^{\mu\nu}_{LL}$, like $\hat{t}^{E\nu}_{\mu}$, has only the first derivatives, and is expressed through Einstein's (3..7) and the Freud's (3..9) expressions: $t^{\mu\nu}_{LL} = \hat{g}^{\mu\sigma}_{\ \ ,\rho}\hat{\mathcal{F}}^{[\nu\rho]}_{\sigma} + \hat{g}^{\mu\rho}\hat{t}^{E\nu}_{\rho}$. However, as is seen, it has an anomalous weight $+2$, unlike expressions in (3..10) and (3..12) with the weight $+1$. Goldberg [150] generalized the Landau-Lifshitz approach, suggesting a symmetrical pseudotensor of an arbitrary weight.

Concerning symmetrical expressions, return also to the the equation (2..1) and the expression (2..2). The last can be rewritten as

$$\hat{G}^{\mu\nu}_L \equiv \kappa\partial_\beta\hat{\mathcal{P}}^{\mu[\nu\beta]}\,, \qquad (3..14)$$

$$\hat{\mathcal{P}}^{\mu[\nu\beta]} \equiv \frac{1}{\kappa}\partial_\alpha\left(\hat{l}^{\nu[\mu}\eta^{\alpha]\beta} - \hat{l}^{\beta[\mu}\eta^{\alpha]\nu}\right). \qquad (3..15)$$

One can find that $\hat{\mathcal{P}}^{\mu[\nu\beta]}$ is the well known superpotential by Papapetrou [152]; in the linear approximation it is Weinberg's superpotential [153]; and, up to factor $\sqrt{-\eta}$, it coincides also with the linear version of the Landau-Lifshitz superpotential [1]. Then the equation (2..1) can be rewritten as

$$\hat{t}^{\mu\nu}_{(tot)} = \partial_\beta\hat{\mathcal{P}}^{\mu[\nu\beta]}\,. \qquad (3..16)$$

This means that the conservation law (2..3) can be interpreted by the same way as the conservation law (3..4), i.e. in the framework of the superpotential approach.

It was a problem to relate *modified* superpotentials to Lagrangians, like it was done for Freud's (3..9) and Møller's (3..13) ones. Only recently with the use of a covariant Hamiltonian formalism [94] (see subsection 1.2.) Chang, Nester and Chen [95] closed this gap. They describe GR by Hamiltonians with different surface terms corresponding to different boundary conditions under variation. For each of known (covariantized) superpotentials one can find its own Hamiltonian. Thus, the Dirichlet conditions correspond to the Freud superpotential (3..9), Neuman's ones — to Møller's superpotential (3..13). All the other superpotentials require more complicated boundary conditions. Recently Babak and Grishchuk [122] suggested a special Lagrangian including additional terms with Lagrangian multipliers, which leads to a covariantized Landau-Lifshitz pseudotensor of normal weight $+1$.

Of course, a non-covariance of pseudotensors and superpotentials lead to the evident problems described in all the textbooks. In this relation, return to the Hilbert Lagrangian (3..6) recalling that it is generally covariant. Thus, for constructing conservation laws one can use arbitrary diplacement vectors ξ^μ, not just the linear translations $\lambda^\mu_{(\alpha)}$, like in (3..11). Hence there has to exist a generally covariant energy-momentum complex and superpotential. Such quantities were found by Komar [154], his superpotential is

$$\hat{\mathcal{K}}^{[\mu\alpha]} = \frac{\sqrt{-g}}{2\kappa}\left(D^\mu\xi^\alpha - D^\alpha\xi^\mu\right) = \frac{\sqrt{-g}}{\kappa}D^{[\mu}\xi^{\alpha]} = \frac{1}{\kappa}D^{[\mu}\hat{\xi}^{\alpha]}\,. \qquad (3..17)$$

With $\xi^\mu = \mathrm{const}$ it goes to the Møller superpotential (3..13). However, in spite of the advantage with covariance, there are problems. The superpotential (3..17) does not make a correct ratio of angular momentum to mass in the Kerr solution (see, e.g., [137, 140]).

The mentioned above problems could be neglected if resonable assumptions are made. Then many classical pseudotensors and superpotentials could be useful for a description of differerent physical systems. In this sense it is useful to return to Einstein's arguments. First, in spite of that the local conservation laws (3..8) are not covariant they have the same form in every coordinate system. Being the continuity equations they give a *balance* between a loss and a gain of a density of *particularly* defined physical quantities, and hence they state the *equilibrium* of a system. The second remarkable Einstein's example [59] is related to masses connected by a rigid rod. Tensions in the road (which exist because $\hat{T}^{\mu}_{\nu}$ is present in (3..8)) can be interpreted *only* as a compensation to tensions of gravitational field included in $t_{\mu}{}^{E\nu}$. Third, Einstein had suggested also a simple method for the "localization" [58], when an isolated system is considered as placed into "Galileian space", and "Galileian coordinates" are used for description. Finally, he had used his energy-momentum complex to describe weak gravitational waves as metric perturbations with respect to a Mikowski metric [56] - [58].

At reasonable assumptions, pseudotensors and superpotentials are left useful nowadays too. Thus, usually quantities of global energy-momentum and angular momentum for asymptotically flat spacetimes are calculated. For example, Bergmann-Tomson's and Landau-Lifshitz's complexes give the same expressions for global angular momentum in an isolated system [155]. In [156, 157], with the use of the Einstein, Tolman and Landau-Lifshitz complexes reasonable energy and momentum densities of cylindrical gravitational waves were obtained. Recently, in the work [158] with a careful set of assumptions it was shown that a tidal work in GR calculated for different non-covariant energy-momentum complexes is the same and unambiguous. Authors frequently use pseudotensors and superpotentials to describe many popular exact solutions of GR (see [159] - [166] and references there in). Particulary, it was shown that in many cases *different* pseudotensors give *the same* energy distribution for the same solution.

However, in spite of some successes in applying pseudotensors and superpotentials, it is clear that more universal and unambiguous quantities are more preferable. The next subsection is devoted to possible ways in improving properties of the classical complexes.

3.2. Requirements for Constructing Conservation Laws on Curved Backgrounds

From the beginning we recall how conservation laws are constructed in a field theory in a Minkowski space. These principles are useful both to understand how classical pseudotensors and superpotentials could be modified to avoid their problems and to better describe and explain conservation laws in the field-theoretical formulation.

Consider an arbitrary theory of fields ψ^A in the Minkowski space in curved coordinates with the Lagrangian

$$\hat{\mathcal{L}}_{(\psi)} = \hat{\mathcal{L}}_{(\psi)}\left(\psi^A, \overline{D}_\alpha\psi^A, \overline{g}_{\mu\nu}\right). \qquad (3..18)$$

For the sake of simplicity assume that it contains the first derivatives only. The symmetrical (metric) energy-momentum tensor density is defined as usual:

$$\hat{t}_{(s)\alpha\beta} \equiv 2\frac{\delta\hat{\mathcal{L}}_{(\psi)}}{\delta\overline{g}^{\alpha\beta}}. \qquad (3..19)$$

The density of the canonical energy-momentum tensor has the form:

$$\hat{t}^{\alpha}_{(c)\sigma} \equiv \frac{\partial\hat{\mathcal{L}}_{(\psi)}}{\partial\left(\overline{D}_\alpha\psi^A\right)}\overline{D}_\sigma\psi^A - \hat{\mathcal{L}}_{(\psi)}\delta^{\alpha}_{\sigma}. \qquad (3..20)$$

Both of them are differentially conserved

$$\overline{D}_\alpha \hat{t}^\alpha_{(s)\sigma} = 0\,, \tag{3..21}$$

$$\overline{D}_\alpha \hat{t}^\alpha_{(c)\sigma} = 0\,, \tag{3..22}$$

on the field equations. To construct integral conserved quantities it is necessary to use Killing vectors of the Minkowski space. Contracting $\hat{t}^\alpha_{(s)\sigma}$ with an each of the 10 Killing vectors λ^ν one obtains the currents

$$\hat{I}^\mu_{(s)}(\lambda) \equiv \hat{t}^\mu_{(s)\nu}\lambda^\nu\,, \tag{3..23}$$

which are conserved differentially

$$\overline{D}_\alpha \hat{I}^\alpha_{(s)}(\lambda) \equiv \partial_\alpha \hat{I}^\alpha_{(s)}(\lambda) = 0\,. \tag{3..24}$$

The other situation with the canonical energy-momentum $\hat{t}^\alpha_{(c)\sigma}$. If one tries to contract it with the *4-rotation* Killing vectors, then the correspondent currents, like (3..23), are not conserved. The reason is that $\hat{t}^{\alpha\sigma}_{(c)}$ is not symmetrical in general. A one of the ways to construct differentially conserved currents is

$$\hat{I}^\alpha_{(c)}(\lambda) \equiv \hat{t}^\alpha_{(c)\sigma}\lambda^\sigma + \hat{\sigma}^{\alpha\beta}_{(\psi)\sigma}\overline{D}_\beta\lambda^\sigma \tag{3..25}$$

where

$$\hat{\sigma}^{\alpha\beta}_{(\psi)\sigma} \equiv -\frac{\partial\hat{\mathcal{L}}_{(\psi)}}{\partial\left(\overline{D}_\alpha\psi^A\right)}\,\psi^A\Big|^\beta_\sigma \tag{3..26}$$

is a spin density. Another way follows the Belinfante symmetrization [167], who has introduced the antisymmetric in α and β expression

$$\hat{S}^{\alpha\beta\gamma}_{(\psi)} = \hat{\sigma}^{\gamma[\alpha\beta]}_{(\psi)} + \hat{\sigma}^{\alpha[\gamma\beta]}_{(\psi)} - \hat{\sigma}^{\beta[\gamma\alpha]}_{(\psi)} \tag{3..27}$$

and has defined the new (symmetrized) energy-momentum tensor density as follows

$$\hat{t}^\alpha_{(B)\sigma} = \hat{t}^\alpha_{(c)\sigma} + \overline{D}_\beta\hat{S}^{\alpha\beta}_{(\psi)\sigma}. \tag{3..28}$$

It is also conserved, $\overline{D}_\alpha \hat{t}^\alpha_{(B)\sigma} = 0$, on the equations of motion. Besides, it is symmetrical on the equations of motion. Then the symmetrized current

$$\hat{I}^\alpha_{(B)}(\lambda) \equiv \hat{t}^\alpha_{(B)\sigma}\lambda^\sigma \tag{3..29}$$

is also conserved for all the Killing vectors, i.e. one can use the *unique* object $\hat{t}^\alpha_{(B)\sigma}$ without additional quantities, like the spin tensor in (3..25).

For the Minkowski background it is easy to show

$$\hat{t}^\alpha_{(s)\sigma} = \hat{t}^\alpha_{(c)\sigma} + \overline{D}_\beta\hat{S}^{\alpha\beta}_{(\psi)\sigma} = \hat{t}^\alpha_{(B)\sigma} \tag{3..30}$$

on the field equations. Thus the *metric* energy-momentum (3..19) is equivalent to the Belinfante *symmetrized* quantity (3..28) (in this relation see also [168]), i.e. $\hat{I}^\mu_{(s)} = \hat{I}^\mu_{(B)}$. Note

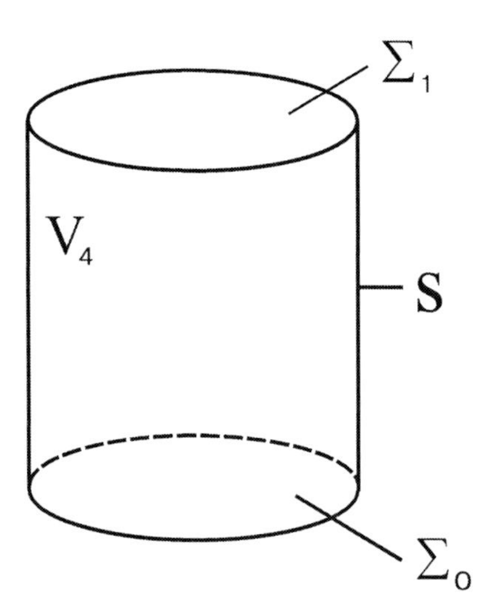

Figure 3. 4-dimensional volume V_4 restricted by a truncated cylinder.

also that $\hat{I}^{\mu}_{(s)}$ differs from $\hat{I}^{\mu}_{(c)}$ by a divergence. One can see that the Belinfante procedure suppresses the spin term in the current (3..25). Classical electrodynamics has just the Lagrangian of the type (3..18), and is a good illustration of the above.

The first example where the Belinfante method was used in GR is the paper by Papapertrou [152]. He has symmetrized the Einstein pseudotensor and obtained his remarkable superpotential. By the original definition [167] it is necessary to use a background metric. Thus, Papapetrou used an auxiliary Minkowski metric. Later Berezin [169], in the framework of the field approach in GR on a flat background, has shown that effective energy-momentum tensor can be constructed by the Belinfante method. Applications of the Belinfante method in GR to the pseudotensors without using an auxiliary background metric [168, 170] lead uniquely to the Einstein tensor. It is important and interesting result. However, this cannot be useful for description of energy characteristics, e.g., in vacuum case. Recently Borokhov [171] generalized the Belinfante method for an arbitrary field theory on arbitrary backgrounds, with arbitrary Killing vectors and with non-minimal coupling. With simplification to minimal coupling his model goes to the standard description [60, 168, 170].

Denoting $\hat{I}^{\mu}_{(\psi)} = \left\{\hat{I}^{\mu}_{(s)}, \hat{I}^{\mu}_{(c)}\right\}$ consider 4-dimensional volume V_4 in the Mikowski space, the boundary of which consists of timelike "wall" S (cylinder) and two spacelike sections: $\Sigma_0 := t_0 = \mathrm{const}$ and $\Sigma_1 := t_1 = \mathrm{const}$ (see figure 3). Because the conservation laws, like (3..24), are presented by the scalar densities, they can be integrated through the 4-volume: $\int_{V_4} \partial_\mu \hat{I}^{\mu}_{(\psi)}(\lambda) d^4x = 0$. The generalized Gauss theorem gives

$$\int_{\Sigma_1} \hat{I}^{0}_{(\psi)}(\lambda) d^3x - \int_{\Sigma_0} \hat{I}^{0}_{(\psi)}(\lambda) d^3x + \oint_S \hat{I}^{\mu}_{(\psi)}(\lambda) dS_\mu = 0 \qquad (3..31)$$

where dS_μ is the element of integration on S. If in Eq. (3..31)

$$\oint_S \hat{I}^\mu_{(\psi)}(\lambda)dS_\mu = 0\,, \tag{3..32}$$

then the quantity

$$\mathcal{P}_{(\psi)}(\lambda) = \int_\Sigma \hat{I}^0_{(\psi)}(\lambda)d^3x \tag{3..33}$$

is conserved on Σ restricted by $\partial\Sigma$ — intersection of Σ with S. If the equality (3..32) does not hold, then Eq. (3..31) describes changing the quantity (3..33), i.e. its flux trough $\partial\Sigma$. Due to the difference between $\hat{I}^\mu_{(s)}$ and $\hat{I}^\mu_{(c)}$ the quantities $\mathcal{P}_{(s)}(\lambda)$ and $\mathcal{P}_{(c)}(\lambda)$ differ one from another by surface integrals. However, as a rule, boundary conditions suppress these surface terms, and then $\mathcal{P}_{(s)}(\lambda) = \mathcal{P}_{(c)}(\lambda)$.

Now let us return to pseudotenors and superpotentials. To avoid some problems they have to be covariantized. Usually a covariantization is carried out by including an auxiliary metric of a Minkowski space. After that partial derivatives are related to Lorentzian coordinates of this flat spacetime. In arbitrary coordinates, partial derivatives naturally transform into covariant ones. Thus, one can keep in mind that in usual expressions, like (3..3) in general form, already there exist the Minkowski metric $\eta_{\mu\nu}$ and its determinant $\eta \equiv \det\eta_{\mu\nu} = -1$. For constructing integral conservation laws based on covariantized pseudotensors it is also natural to use the Killing vectors of the Minkowski space.

Let us present the program of the covariantization in detail. Consider the generalized conservation law (3..4) and rewrite it in the equivalent form:

$$\partial_\mu \left[\sqrt{-\eta}\left((\hat{\theta}^\mu_\rho + \hat{T}^\mu_\rho)/\sqrt{-\eta}\right)\lambda^\rho_{(\nu)}\right] = 0\,. \tag{3..34}$$

Here, the left hand side is a scalar density. The translation Killing vector $\lambda^\rho_{(\nu)}$ of the background Minkowski space is expressed in the Lorentzian coordinates, therefore one has $\lambda^\rho_{(\nu)} = \delta^\rho_\nu$ (the lower index is not coordinate one). Thus, in (3..34) the differentially conserved vector density (current)

$$\hat{\mathcal{J}}^\mu(\lambda) \equiv \sqrt{-\eta}\left((\hat{\theta}^\mu_\rho + \hat{T}^\mu_\rho)/\sqrt{-\eta}\right)\lambda^\rho_{(\nu)}\,, \tag{3..35}$$

$$\overline{D}_\mu\hat{\mathcal{J}}^\mu(\lambda) \equiv \partial_\mu\hat{\mathcal{J}}^\mu(\lambda) = 0 \tag{3..36}$$

is presented. Here, one has to accent that if $\hat{\theta}^{\mu\nu}$ is symmetrical, then the conservation law (3..36) could be extended for 4-rotation Killing vectors. However, if $\hat{\theta}^{\mu\nu}$ is not symmetrical, then the simple and direct covariantization is restricted to using only translation Killing vectors.

Next, as is seen, the programm of constructing integral quantities (3..31) - (3..33) can be repeated for the covariantized current (3..35). The superpotential in Eq. (3..3) can also be rewritten in the covariantized form:

$$\hat{\mathcal{J}}^{\mu\alpha}(\lambda) = \sqrt{-\eta}\left(\hat{\mathcal{U}}^{\mu\alpha}_\rho/\sqrt{-\eta}\right)\lambda^\rho_{(\nu)}\,. \tag{3..37}$$

Thus the expression (3..3) can be rewritten in a fully covariant form:

$$\hat{\mathcal{J}}^\mu(\lambda) = \partial_\alpha\hat{\mathcal{J}}^{\mu\alpha}(\lambda) \equiv \overline{D}_\alpha\hat{\mathcal{J}}^{\mu\alpha}(\lambda) \tag{3..38}$$

that has a sense of the conservation law because $\hat{\mathcal{J}}^{\mu\alpha}(\lambda)$ is an antisymmetric tensor density. Due to antisymmetry of the superpotential (3..37), the 4-momentum presented in the volume form, like (3..33), is expressed as a surface integral

$$\mathcal{P}(\lambda) = \oint_{\partial\Sigma} \hat{\mathcal{J}}^{0k}(\lambda) ds_k \tag{3..39}$$

where ds_k is the element of integration on $\partial\Sigma$.

Up to now in constructing conservation laws we considered Killing vectors of a background only. However, other displacement vectors can be used also. Let us give two examples. First, for study of perturbations on FRW backgrounds a so-called notion of an *integral constraint*, which connects a volume integral of only matter perturbations with a surface integral of only metric perturbations, frequently is very important [172]. With the use of such constraints, e.g., measurable effects of the cosmic background radiation were analyzed [173]. In the definition of integral constraints *integral constraint vectors*, not necessarily Killing vectors, play a crucial role [172]. Second, in [174], a new conserved energy-momentum pseudotensor was found and used in an effort to integrate Einstein's equations with scalar perturbations and topological defects on FRW backgrounds for a sign of spatial curvature $k = 0$. In [175] it was realized that these conservation laws might be associated with the *conformal* Killing vector of time translation, and can be also generalized for $k = \pm 1$.

Taking into account the above we formulate the generalized requirements for constructing conservation laws for perturbations on curved backgrounds as follows.

- Expressions have to be covariant on a chosen background and valid for arbitrary curved backgrounds.

- One has to construct differentially conserved currents (vector densities) $\hat{\mathcal{J}}^{\mu}(\xi)$, such that $\partial_\mu \hat{\mathcal{J}}^{\mu}(\xi) = 0$, and which are defined by the canonical Nœther procedure applied to a Lagrangian of a perturbed system.

- The currents have to be expressed through corresponding superpotentials $\hat{\mathcal{J}}^{\mu\nu}$ (antisymmetric tensor densities), such that $\partial_{\mu\nu} \hat{\mathcal{J}}^{\mu\nu}(\xi) \equiv 0$, by the way $\hat{\mathcal{J}}^{\mu}(\xi) = \partial_\nu \hat{\mathcal{J}}^{\mu\nu}(\xi)$.

- There has to be a possibility to use arbitrary displacement vectors ξ^μ, not only Killing vectors of the background.

- Applications of suggested conserved quantities and conservation laws have to satisfy the known tests (see discussion in subsection 1.2.).

3.3. The Katz, Bičák and Lynden-Bell Conservation Laws

Katz, Bičák and Lynden-Bell [176] (later we call as KBL) have satisfied the requirements at the end of previous subsection. They describe a perturbed system by the Lagrangian:

$$\hat{\mathcal{L}}_{KBL} = \hat{\mathcal{L}}^E - \overline{\hat{\mathcal{L}}^E} - \frac{1}{2\kappa}\partial_\alpha \hat{k}^\alpha , \tag{3..40}$$

see notations in (2..8) and (2..15). In fact they use bimetric formalism, where $g_{\mu\nu}$ and $\overline{g}_{\mu\nu}$ are thought as independent metric coefficients. The gravitational part of (3..40) is

$$\hat{\mathcal{L}}_G = -\frac{1}{2\kappa}\left(\hat{R} - \hat{\overline{R}} + \partial_\alpha \hat{k}^\alpha\right). \tag{3..41}$$

Here, for the physical scalar curvature density $\hat{R}$ an external background metric $\overline{g}_{\mu\nu}$ is introduced with the use of the presentation for the curvature tensor:

$$R^\lambda{}_{\tau\rho\sigma} = \overline{D}_\rho \Delta^\lambda_{\tau\sigma} - \overline{D}_\sigma \Delta^\lambda_{\tau\rho} + \Delta^\lambda_{\rho\eta}\Delta^\eta_{\tau\sigma} - \Delta^\lambda_{\sigma\eta}\Delta^\eta_{\tau\rho} + \overline{R}^\lambda{}_{\tau\rho\sigma}\,. \tag{3..42}$$

It is also very important to note that $\hat{k}^\alpha$ is defined in (2..20).

For the Lagrangian (3..41), as for a scalar density, one has the identity:

$$\pounds_\xi \hat{\mathcal{L}}_G + \partial_\mu\left(\xi^\mu \hat{\mathcal{L}}_G\right) \equiv 0\,. \tag{3..43}$$

It is a generalization of Eq. (3..11) for arbitrary displacement vectors ξ^μ. Consequent identical transformations give:

$$\partial_\mu \hat{\jmath}^\mu \equiv 0\,, \tag{3..44}$$

$$\begin{aligned}\hat{\jmath}^\mu &\equiv \left[\frac{\partial \hat{\mathcal{L}}_G}{\partial\left(\overline{D}_\mu g_{\rho\sigma}\right)}\overline{D}_\nu g_{\rho\sigma} - \hat{\mathcal{L}}_G \delta^\mu_\nu\right]\xi^\nu - \frac{\partial \hat{\mathcal{L}}_G}{\partial\left(\overline{D}_\mu g_{\rho\sigma}\right)}\, g_{\rho\sigma}|^\rho_\lambda\, \overline{g}^{\lambda\sigma}\overline{D}_\rho \xi_\sigma \\ &- \frac{\delta \hat{\mathcal{L}}_G}{\delta g_{\rho\sigma}}\, g_{\rho\sigma}|^\mu_\nu\, \xi^\nu - \frac{\delta \hat{\mathcal{L}}_G}{\delta \overline{g}_{\rho\sigma}}\, \overline{g}_{\rho\sigma}\Big|^\mu_\nu\, \xi^\nu + \hat{Z}^\mu_{(c)} \equiv 0\,.\end{aligned} \tag{3..45}$$

Note that here

$$-\frac{\delta \hat{\mathcal{L}}_G}{\delta g_{\rho\sigma}}\, g_{\rho\sigma}|^\mu_\nu\, \xi^\nu \equiv \frac{1}{\kappa}\hat{G}^\mu_\nu \xi^\nu\,, \qquad -\frac{\delta \hat{\mathcal{L}}_G}{\delta \overline{g}_{\rho\sigma}}\, \overline{g}_{\rho\sigma}\Big|^\mu_\nu\, \xi^\nu \equiv -\frac{1}{\kappa}\hat{\overline{G}}{}^\mu_\nu \xi^\nu\,. \tag{3..46}$$

Then substituting the Einstein equations (2..12) and their barred version into Eq. (3..44) one obtains the differential conservation law

$$\partial_\mu \hat{J}^\mu_{(c)}(\xi) = 0\,, \tag{3..47}$$

$$\hat{J}^\mu_{(c)}(\xi) \equiv \hat{\Theta}^\mu_{(c)\nu}\xi^\nu + \hat{\sigma}^{\mu\rho\sigma}\overline{D}_\rho \xi_\sigma + \hat{Z}^\mu_{(c)}(\xi)\,. \tag{3..48}$$

The first term in (3..48), the generalized total energy-momentum tensor density, is

$$\hat{\Theta}^\mu_{(c)\nu} = \hat{t}^\mu_\nu + \left(\hat{T}^\mu_\nu - \hat{\overline{T}}{}^\mu_\nu\right) + \frac{1}{2\kappa}\hat{l}^{\rho\sigma}\overline{R}_{\rho\sigma}\delta^\mu_\nu\,, \tag{3..49}$$

with $\hat{l}^{\rho\sigma} = \hat{g}^{\rho\sigma} - \hat{\overline{g}}{}^{\rho\sigma}$. The first term in Eq. (3..49) is the canonical energy-momentum tensor density of the gravitational field:

$$\begin{aligned}2\kappa \hat{t}^\mu_\nu &= \hat{g}^{\rho\sigma}\left(\Delta^\lambda_{\rho\lambda}\Delta^\mu_{\sigma\nu} + \Delta^\mu_{\rho\sigma}\Delta^\lambda_{\lambda\nu} - 2\Delta^\mu_{\rho\lambda}\Delta^\lambda_{\sigma\nu}\right) \\ &- \hat{g}^{\rho\sigma}\left(\Delta^\eta_{\rho\sigma}\Delta^\lambda_{\eta\lambda} - \Delta^\eta_{\rho\lambda}\Delta^\lambda_{\eta\sigma}\right)\delta^\mu_\nu + \hat{g}^{\mu\lambda}\left(\Delta^\sigma_{\rho\sigma}\Delta^\rho_{\lambda\nu} - \Delta^\sigma_{\lambda\sigma}\Delta^\rho_{\rho\nu}\right),\end{aligned} \tag{3..50}$$

the second term in Eq. (3..49) is a perturbation of the matter energy-momentum, the third term describes an interaction with the background. The second term in (3..48) is the spin tensor density:

$$
\begin{aligned}
2\kappa\hat{\sigma}^{\mu\rho\sigma} \equiv & - 2\kappa\frac{\partial\hat{\mathcal{L}}_G}{\partial\left(\overline{D}_\mu g_{\rho\sigma}\right)} g_{\rho\sigma}|^{\rho}_{\lambda}\,\bar{g}^{\lambda\sigma} = (\hat{g}^{\mu\rho}\overline{g}^{\sigma\nu} + \overline{g}^{\mu\sigma}\hat{g}^{\rho\nu} - \hat{g}^{\mu\nu}\overline{g}^{\rho\sigma})\Delta^{\lambda}_{\nu\lambda} \\
& - (\hat{g}^{\nu\rho}\overline{g}^{\sigma\lambda} + \overline{g}^{\nu\sigma}\hat{g}^{\rho\lambda} - \hat{g}^{\nu\lambda}\overline{g}^{\rho\sigma})\Delta^{\mu}_{\nu\lambda}\,. \qquad (3..51)
\end{aligned}
$$

If one simplifies to a Minkowski background then it is the quantity presented by Papapetrou [152] to construct angular momentum with the use of the Einstein pseudotensor. The last term in (3..48) has the form:

$$
\hat{Z}^{\mu}_{(c)}(\xi) = \frac{1}{2\kappa}\left[\hat{\imath}^{\mu\lambda}\partial_\lambda\zeta^{\rho}_{\rho} + \hat{\imath}^{\rho\sigma}\left(\overline{D}^{\mu}\zeta_{\rho\sigma} - 2\overline{D}_{\rho}\zeta^{\mu}_{\sigma}\right)\right], \qquad (3..52)
$$

with $2\zeta_{\rho\sigma} = -\pounds_\xi\overline{g}_{\rho\sigma} = 2\overline{D}_{(\rho}\xi_{\sigma)}$. It, thus, disappears if $\xi^\mu = \lambda^\mu$ is a Killing vector of the background spacetime. Besides, in this case the part of the second term in (3..48): $\hat{\sigma}^{\mu\rho\sigma}\overline{D}_{(\rho}\lambda_{\sigma)}$ disappears also.

Because Eq. (3..44) is the identity then the quantity $\hat{\jmath}^{\mu}(\xi)$ must be presented through a superpotential $\hat{J}^{\mu\nu}_{(c)}(\xi)$, thus indeed

$$
\begin{aligned}
\hat{\jmath}^{\mu}(\xi) &\equiv \partial_\nu\hat{J}^{\mu\nu}_{(c)}(\xi)\,, \qquad (3..53)\\
\hat{J}^{\mu\nu}_{(c)}(\xi) &\equiv \frac{1}{\kappa}(D^{[\mu}\hat{\xi}^{\nu]} - \overline{D^{[\mu}\hat{\xi}^{\nu]}}) + \frac{1}{\kappa}\hat{\xi}^{[\mu}k^{\nu]}\,. \qquad (3..54)
\end{aligned}
$$

This superpotential has been found earlier independently by Katz for flat backgrounds [137] and by Chruściel for Ricci-flat backgrounds [177]. However, as it turns out, it has the same form (3..54) for generally curved backgrounds. Note that the expression (3..54) modifies the Komar superpotential (3..17). With the use of the Einstein equations, the identity (3..53) transforms into

$$
\hat{J}^{\mu}_{(c)}(\xi) = \partial_\nu\hat{J}^{\mu\nu}_{(c)}(\xi)\,, \qquad (3..55)
$$

that is another form of the conservation law (3..47).

Now, let us compare the KBL expressions with the Einstein prescription. For the simplification $\bar{g}_{\mu\nu} \rightarrow \eta_{\mu\nu}$ the KBL Lagrangian $\hat{\mathcal{L}}_G$ transforms into the Einstein Lagrangian (3..5), and the KBL gravitational energy-momentum $\hat{t}^{\mu}_{\nu}$ transforms into the Einstein pseudotensor (3..7). Next, if $\overline{g}_{\mu\nu} \rightarrow \eta_{\mu\nu}$ and $\xi^\nu \rightarrow \lambda^\nu = \delta^{\nu}_{(\mu)}$, then the KBL current $\hat{J}^{\mu}_{(c)}(\xi)$ goes to $\hat{t}^{E\mu}_{\nu} + \hat{T}^{\mu}_{\nu}$ in (3..8), the KBL superpotential $\hat{J}^{\mu\nu}_{(c)}(\xi)$ transforms into the Freud superpotential (3..9). That is generally Eq. (3..55) transforms into Eq. (3..10). However, both the current and superpotential in Eq. (3..55) *are not a simple direct* covariantization of the quantities (3..7) and (3..9). Indeed, first, the KBL expressions hold on arbitrary curved backgrounds, not only on flat backgrounds in curved coordinates. Second, $\hat{J}^{\mu}_{(c)}$ includes the spin term (3..51) analogous to the quantity (3..26). It plays its crucial role because then rotation Killing vectors can be used giving a reasonable definition of angular momentum. Thus the KBL current gives the right ratio of mass to angular momentum for rotating black hole solutions.

The KBL approach received a significant development in applications. Thus, in [178, 179] the problem of localization was considered. In [180], the conserved quantities and their fluxes at null infinity for an isolated sysytem were studied in detail. In the works [175, 181] perturbations on FRW backgrounds and conservation laws for them were examined. At last, recently [182] the approach was developed for constructing the conserved charges in Gauss-Bonnet 5-dimensional cosmology.

The KBL quantities were also checked from the point of view of the problem of uniqueness. Thus, Julia and Silva [183, 184], and independently Chen and Nester [94], stated that the KBL quantities are uniquely defined as associated with the Dirichlet boundary conditions, under which the action with the Lagrangian (3..41) is variated. This gives evident advantages. Both the Lagrangian (3..41) and the energy-momentum (3..50) have only the first derivatives. Thus, the Cauchy problem is stated simply.

Keeping in mind successes presented by the KBL approach it is necessary to note the next generic problem related to the canonical derivation. If one chooses different boundary conditions, adding different divergences to the Lagrangian, different expressions both for currents and for superpotentials appear by the Nœther procedure. On the one hand, such a situation could be useful and interesting in gravitational theory, when analogies, e.g., with thermodynamics are carried out [95]. On the other hand, it is evident that *unique* expressions independent on boundary consditions are necessary. As examples, the symmetric energy-momentum tensor in classical electrodynamics is such a quantity. The expressions in the field-theoretical formulation of GR are also independent on boundary conditions (see section 2.). In the next subsection we just construct conserved currents and superpotentials in the framework of the field approach.

3.4. Conserved Quantities in the Field-Theoretical Formulation

Already the form of Eq. (3..16) indicates that there is a possible to construct currents and superpotentials in the framework of the field approach. Recall that for backgrounds presented by an Einstein space the conservation law (2..67) has a place. Then if the Einstein space has Killing vectors λ^α one can construct the conserved currents by the same way as in the usual field theory (3..23). At the same time the left hand sides both of Eq. (2..66) and of Eq. (2..71) contracted with λ^μ are presented as divergences of superpotentials [68, 100, 144]. However, in the general case of a curved background the conservation law (2..67) has no a place. The reason is that there is no an identical conservation both in Eq. (2..24): $\overline{D}_\nu\left(\hat{G}^{L\nu}_\mu + \hat{\Phi}^{L\nu}_\mu\right) \neq 0$ and in Eq. (2..32): $\overline{D}_\nu \hat{G}^{L\nu}_\mu \neq 0$. The physical reason is that the system (2..19) interacts with a complicated background geometry defined by the background matter fields $\overline{\Phi}^A$.

To construct conserved quantities in the general case we [118] combine the technique of previous subsection with the results [100]. Therefore, in the framework of the field approach, we are going to the KBL independent variables of the bimetrical approach by changing $\hat{l}^{\mu\nu} \to \hat{g}^{\mu\nu} - \hat{\overline{g}}^{\mu\nu}$. Then the dynamical Lagrangian $\hat{\mathcal{L}}^g$ (2..22) transforms into $2\kappa\hat{\mathcal{L}}_{G2}$. After that we connect it with the KBL Lagrangian $\hat{\mathcal{L}}_G$ (3..41):

$$\hat{\mathcal{L}}_{G2} \equiv \hat{\mathcal{L}}_G - \hat{\mathcal{L}}_{G1} \equiv \frac{1}{2\kappa}\hat{g}^{\mu\nu}\left(\Delta^\rho_{\mu\nu}\Delta^\sigma_{\rho\sigma} - \Delta^\rho_{\mu\sigma}\Delta^\sigma_{\rho\nu}\right) \tag{3..56}$$

where $\hat{\mathcal{L}}_{G1} = -(2\kappa)^{-1}(\hat{g}^{\mu\nu} - \overline{\hat{g}}^{\mu\nu})\overline{R}_{\mu\nu}$. Returning to the comparison with the Rosen Lagrangian $\hat{\mathcal{L}}_R$ [136] we note that $\hat{\mathcal{L}}_{G2}$ is a direct generalization of $\hat{\mathcal{L}}_R$ to arbitrary backgrounds, whereas $\hat{\mathcal{L}}_G$ is reduced to $\hat{\mathcal{L}}_R$ for the Ricci-flat backgrounds.

Now, consider the identity

$$\pounds_\xi \hat{\mathcal{L}}_{G2} + \partial_\mu(\xi^\mu \hat{\mathcal{L}}_{G2}) \equiv 0 \tag{3..57}$$

for the Lagrangian (3..56) and transform it identically to

$$\left[\frac{\partial \hat{\mathcal{L}}_{G2}}{\partial\left(\bar{D}_\mu g_{\rho\sigma}\right)}\overline{D}_\nu g_{\rho\sigma} - \hat{\mathcal{L}}_{G2}\delta^\mu_\nu\right]\xi^\nu - {}^2\hat{S}_\lambda{}^{\mu\rho}\overline{g}^{\sigma\lambda}\overline{D}_\rho\xi_\sigma$$
$$-\frac{\delta\hat{\mathcal{L}}_{G2}}{\delta g_{\rho\sigma}}\, g_{\rho\sigma}\big|^\mu_\nu\,\xi^\nu - \frac{\delta\hat{\mathcal{L}}_{G2}}{\delta\overline{g}_{\rho\sigma}}\,\overline{g}_{\rho\sigma}\Big|^\mu_\nu\,\xi^\nu \equiv \overline{D}_\nu\left[-{}^2\hat{S}_\lambda{}^{\mu\nu}\xi^\lambda\right]. \tag{3..58}$$

The quantity

$${}^2\hat{S}_\lambda{}^{\mu\nu} \equiv \frac{\partial\hat{\mathcal{L}}_{G2}}{\partial\left(\overline{D}_\mu g_{\rho\sigma}\right)}\, g_{\rho\sigma}\big|^\nu_\lambda + \frac{\partial\hat{\mathcal{L}}_{G2}}{\partial\left(\partial_\mu\overline{g}_{\rho\sigma}\right)}\,\overline{g}_{\rho\sigma}\Big|^\nu_\lambda \tag{3..59}$$

is antisymmetric in μ and ν. Thus on the right hand side of (3..58) the quantity $[-{}^2\hat{S}_\lambda{}^{\mu\nu}\xi^\lambda]$ plays the role of a superpotential. Besides, the expression (3..59) is connected with the spin term (3..51) as

$${}^2\hat{S}_\lambda{}^{\mu\nu}\overline{g}^{\lambda\rho} = \hat{\sigma}^{\rho[\mu\nu]} + \hat{\sigma}^{\mu[\rho\nu]} - \hat{\sigma}^{\nu[\rho\mu]}. \tag{3..60}$$

One has also to note that in (3..58)

$$\frac{\delta\hat{\mathcal{L}}_{G2}}{\delta\overline{g}_{\rho\sigma}}\,\overline{g}_{\rho\sigma}\Big|^\mu_\nu \equiv \frac{1}{\kappa}\hat{G}^{L\mu}_\nu, \tag{3..61}$$

it is exactly the expression defined in (2..25) and (2..26).

Now, substitute $\hat{\mathcal{L}}_{G2} \equiv \hat{\mathcal{L}}_G - \hat{\mathcal{L}}_{G1}$ into Eq. (3..58) and subtract it from the identity (3..53) where $\hat{\jmath}^\mu$ is defined in (3..45). Returning to the variables $\hat{l}^{\mu\nu}$: $\hat{g}^{\mu\nu} - \overline{\hat{g}}^{\mu\nu} \to \hat{l}^{\mu\nu}$, one obtains the new identity:

$$\frac{1}{\kappa}\hat{G}^{L\mu}_\nu\xi^\nu + \frac{1}{\kappa}\hat{l}^{\mu\lambda}\overline{R}_{\lambda\nu}\xi^\nu + \hat{Z}^\mu_{(s)} \equiv \overline{D}_\nu\hat{J}^{\mu\nu}_{(s)} \tag{3..62}$$

with the new superpotential connected with the KBL superpotential $\hat{J}^{\mu\nu}_{(c)}$ given in (3..54) as

$$\begin{aligned}
\hat{J}^{\mu\nu}_{(s)} &\equiv \hat{J}^{\mu\nu}_{(c)} + {}^2\hat{S}_\lambda{}^{\mu\nu}\xi^\lambda &(3..63)\\
&= \frac{1}{\kappa}\hat{l}^{\rho[\mu}\overline{D}_\rho\xi^{\nu]} + \hat{\mathcal{P}}^{\mu\nu}{}_\rho\xi^\rho &(3..64)\\
&= \frac{1}{\kappa}\left(\hat{l}^{\rho[\mu}\overline{D}_\rho\xi^{\nu]} + \xi^{[\mu}\overline{D}_\sigma\hat{l}^{\nu]\sigma} - \bar{D}^{[\mu}\hat{l}^{\nu]}_\sigma\xi^\sigma\right). &(3..65)
\end{aligned}$$

For $\overline{g}_{\mu\nu} \to \eta_{\mu\nu}$ and $\xi^\rho \to \lambda^\rho = \delta^\rho_{(\alpha)}$ the superpotential $\hat{J}^{\mu\nu}_{(s)}$ transforms into the the Papapetrou superpotential (3..15) (see the form (3..64)). The last term on the left hand side of

Eq. (3..62) is connected with the expression (3..52) as

$$\begin{aligned}2\kappa\hat{Z}^{\mu}_{(s)} &\equiv 2\kappa\left({}^{2}\hat{S}_{\tau}^{\;\mu\nu}\overline{g}^{\tau\lambda}+\frac{\partial\hat{\mathcal{L}}_G}{\partial\left(\bar{D}_\mu g_{\rho\sigma}\right)}\,g_{\rho\sigma}|^{\nu}_{\tau}\overline{g}^{\lambda\tau}\right)\zeta_{\nu\lambda}+2\kappa\hat{Z}^{\mu}_{(c)}\\ &= 2\left(\zeta^{\rho\sigma}\overline{D}_\rho\hat{l}^{\mu}_{\sigma}-\hat{l}^{\rho\sigma}\overline{D}_\rho\zeta^{\mu}_{\sigma}\right)-\left(\zeta_{\rho\sigma}\overline{D}^{\mu}\hat{l}^{\rho\sigma}-\hat{l}^{\rho\sigma}\overline{D}^{\mu}\zeta_{\rho\sigma}\right)\\ &+ \left(\hat{l}^{\mu\nu}\overline{D}_\nu\zeta^{\rho}_{\rho}-\zeta^{\rho}_{\rho}\overline{D}_\nu\hat{l}^{\mu\nu}\right). \end{aligned} \qquad (3..66)$$

The way of constructing (3..56) - (3..62) is natural, but a cumbersome one. Formally, one concludes that the final identity (3..62) is a result of a difference between identities (3..43) and (3..57). Thus, the Nœther procedure could be applied directly to the difference $\hat{\mathcal{L}}_G-\hat{\mathcal{L}}_{G2}=\hat{\mathcal{L}}_{G1}$. Indeed, recalculating the identity

$$\pounds_\xi\hat{\mathcal{L}}_{G1}+\partial_\mu(\xi^\mu\hat{\mathcal{L}}_{G1})\equiv 0 \qquad (3..67)$$

one obtains directly the identity (3..62). The result is not degenerated because $\hat{\mathcal{L}}_{G1}$ contains derivatives of the background metric up to the second order.

After substituting the field equations (2..24) into the identity (3..62) one obtains the conservation law in the form:

$$\hat{J}^{\mu}_{(s)}\equiv\hat{\Theta}^{\mu}_{(s)\nu}\xi^\nu+\hat{Z}^{\mu}_{(s)}=\partial_\nu\hat{J}^{\mu\nu}_{(s)} \qquad (3..68)$$

where the conserved current $\hat{J}^{\mu}_{(s)}$ is expressed through the superpotential $\hat{J}^{\mu\nu}_{(s)}$. The generalized total energy-momentum tensor density is

$$\hat{\Theta}^{\mu}_{(s)\nu}=\hat{t}^{(tot)\mu}_{\nu}+\frac{1}{\kappa}\left(\hat{l}^{\mu\lambda}\overline{R}_{\lambda\nu}-\hat{\Phi}^{L\mu}_{\nu}\right). \qquad (3..69)$$

The role of interaction with the background is played by the term $(\kappa)^{-1}(\hat{l}^{\mu\lambda}\overline{R}_{\lambda\nu}-\hat{\Phi}^{L\mu}_{\nu})$. Next, if one uses the form of the field equations (2..32), then (3..69) looks as

$$\hat{\Theta}^{\mu}_{(s)\nu}=\hat{t}^{(eff)\mu}_{\nu}+\frac{1}{\kappa}\hat{l}^{\mu\lambda}\overline{R}_{\lambda\nu}, \qquad (3..70)$$

and one needs to consider $\hat{t}^{(eff)\mu}_{\nu}$ as the energy-momentum of perturbations and $(\kappa)^{-1}\hat{l}^{\mu\lambda}\overline{R}_{\lambda\nu}$ as the interacting term. It is interesting to note that the form of the energy-momentum (3..70) is more close to the form of the KBL energy-momentum (3..49). The other point is that the general term $\sim\hat{l}^{\mu\lambda}\overline{R}_{\lambda\nu}$ destructs the symmetry of $\hat{\Theta}^{\mu\nu}_{(s)}$.

Let us present also gauge invariance properties of the presented current. Substituting (2..49) and (2..50) into (3..68) and using the identity (3..62) one obtains for $\xi^\alpha=\lambda^\alpha$:

$$\begin{aligned}&\hat{J}'^{\mu}_{(s)}(\lambda)=\\ &\hat{J}^{\mu}_{(s)}(\lambda)+\frac{1}{\kappa}\partial_\nu\left[(\hat{l}'^{\rho[\mu}-l^{\rho[\mu})\overline{D}_\rho\lambda^{\nu]}+\lambda^{[\mu}\overline{D}_\rho(\hat{l}'^{\nu]\rho}-\hat{l}^{\nu]\rho})-\overline{D}^{[\mu}(\hat{l}'^{\nu]\rho}-\hat{l}^{\nu]\rho})\lambda_\rho\right]\\ &-\frac{1}{\kappa}\lambda^\nu\overline{g}^{\mu\lambda}\frac{\partial\hat{\overline{g}}^{\rho\sigma}}{\partial\overline{g}^{\lambda\nu}}\sum_{k=1}^{\infty}\frac{1}{k!}\pounds_\xi{}^{k}\left[\frac{\partial\overline{g}^{\delta\pi}}{\partial\hat{\overline{g}}^{\rho\sigma}}\left(\hat{G}^{L}_{\delta\pi}+\hat{\Phi}^{L}_{\delta\pi}-\kappa\hat{t}^{(tot)}_{\delta\pi}\right)-2\kappa\frac{\delta\overline{\hat{\mathcal{L}}^{E}}}{\delta\hat{\overline{g}}^{\rho\sigma}}\right],\end{aligned} \qquad (3..71)$$

thus $\hat{J}^{\mu}_{(s)}$ is gauge invariant up to a divergence on the dynamic and background equations.

Conserved quantities in Eq. (3..68), like the KBL quantities, satisfy all the requirements of subsection 3.2.. Besides, the field-theoretical derivation has some advantages. First, the quantities in Eq. (3..68) do not depend on divergences in the dynamical Lagrangian. Second, the current in Eq. (3..68) is defined by the unique quantity, the generalized energy-momentum $\Theta^{\mu}_{(s)\nu}$, without a spin term. Third, the conservation law (3..68) explicitly describes perturbations.

However, the currents and superpotentials in the framework of the field approach bring in themselves the Boulware-Deser ambiguity defined for the total energy-momentum tensor in subsection 2.8.. Let us outline this problem in detail. Basing on the Lagrangian (2..78) one can write out the identity:

$$\frac{1}{\kappa}\hat{G}^{L(a)\mu}_{\nu}\xi^{\nu} + \frac{1}{\kappa}\hat{l}^{\mu\lambda}_{(a)}\overline{R}_{\lambda\nu}\xi^{\nu} + \hat{Z}^{\mu}_{(sa)} \equiv \partial_{\nu}\hat{J}^{\mu\nu}_{(sa)}\,. \tag{3..72}$$

Substituting the field equations (2..80) one obtains

$$\hat{J}^{\mu}_{(sa)} = \left[\hat{t}^{(tot\ a)\mu}_{\nu} + \kappa^{-1}\left(\hat{l}^{\mu\lambda}_{(a)}\overline{R}_{\lambda\nu} - \hat{\Phi}^{L\mu}_{(a)\nu}\right)\right]\xi^{\nu} + \hat{Z}^{\mu}_{(sa)} = \partial_{\nu}\hat{J}^{\mu\nu}_{(sa)} \tag{3..73}$$

where $\hat{Z}^{\mu}_{(sa)}$ is defined in (3..66) with exchanging $\hat{l}^{\mu\nu}$ by $\hat{l}^{\mu\nu}_{(a)}$ (see the definition (2..81)). The generalized superpotential is

$$\hat{J}^{\mu\nu}_{(sa)} = \frac{1}{\kappa}\left(\hat{l}^{\rho[\mu}_{(a)}\overline{D}_{\rho}\xi^{\nu]} + \xi^{[\mu}\overline{D}_{\sigma}\hat{l}^{\nu]\sigma}_{(a)} - \bar{D}^{[\mu}\hat{l}^{\nu]}_{(a)\sigma}\xi^{\sigma}\right). \tag{3..74}$$

Thus, taking into account the difference in perturbations (2..82) one obtains an analog of the Boulware-Deser ambiguity in the definition of the superpotentials:

$$\Delta\hat{J}^{\mu\nu}_{(sa)12} = \frac{1}{\kappa}\left(\hat{\beta}^{\rho[\mu}_{(a)12}\overline{D}_{\rho}\xi^{\nu]} + \xi^{[\mu}\overline{D}_{\sigma}\hat{\beta}^{\nu]\sigma}_{(a)12} - \bar{D}^{[\mu}\hat{\beta}^{\nu]\sigma}_{(a)12}\xi_{\sigma}\right). \tag{3..75}$$

In the next subsection we present arguments to resolve this ambiguity.

3.5. The Belinfante Procedure on Curved Backgrounds

Returning to the problems of the canonical approach (see the end of subsection 3.3.) we remark that the classical Belinfante method resolves analogous problems in a canonical field theory in Minkowski space. Indeed, the Belinfante corrected current (3..29), unlike the current (3..25), is presented without a spin term. At the same time the Belinfante symmetrized energy-momentum is equal to the symmetrical (metric) energy momentum (3..30), and thus it does not depend on divergences in Lagrangian. Of course, a perturbed system on a curved background in GR is a more complicated case. However, the KBL model [176] looks an appropriate one for an application of the Belinfante procedure. Indeed, for a Killing vector the current $\hat{J}^{\mu}_{(c)}$ in (3..48) has the form (3..25) with the correspondent energy-momentum $\hat{\Theta}^{\mu}_{(c)\nu}$ and spin term $\hat{\sigma}^{\mu\rho\sigma}$. Thus, it is anticipated that the Belinfante procedure could 1) resolve the problems of the KBL approach, 2) connect it with the field-theoretical approach and, consequently, 3) resolve problems of the last also. Therefore, in this subsection, we generalize the Belinfante method for constructing conservation laws in the perturbed GR on arbitrary curved backgrounds and following the papers [115, 116] apply it to the KBL model.

Thus, for the spin tensor density (3..51) with the rules (3..27) we construct the quantity:

$$\hat{S}^{\mu\nu\rho} = -\hat{S}^{\nu\mu\rho} = \hat{\sigma}^{\rho[\mu\nu]} + \hat{\sigma}^{\mu[\rho\nu]} - \hat{\sigma}^{\nu[\rho\mu]} \tag{3..76}$$

and present the KBL conservation law (3..53) in the equivalent form:

$$\hat{J}^{\mu}_{(c)} + \partial_\nu \left(\hat{S}^{\mu\nu\rho}\xi_\rho\right) = \partial_\nu \left(\hat{J}^{\mu\nu}_{(c)} + \hat{S}^{\mu\nu\rho}\xi_\rho\right). \tag{3..77}$$

Defining the left hand side as a symmetrized current $\hat{J}^{\mu}_{(B)}$ and the right hand side as a divergence of a superpotential $\hat{J}^{\mu\nu}_{(B)}$, we rewrite (3..77) in the form:

$$\hat{J}^{\mu}_{(B)} = \hat{\Theta}^{\mu}_{(B)\nu}\xi^\nu + \hat{Z}^{\mu}_{(B)} = \partial_\nu \hat{J}^{\mu\nu}_{(B)}. \tag{3..78}$$

Here, the spin term is absent; if ξ^μ is a Killing vector of the background Z-term disappears; the symmetrized energy-momentum tensor density has the form:

$$\hat{\Theta}^{\mu}_{(B)\nu} = \hat{\Theta}^{\mu}_{(c)\nu} + \overline{D}_\rho \hat{S}^{\mu\rho}{}_{\nu}\,. \tag{3..79}$$

Now, consider properties of Belinfante symmetrized quantities. The symmetrized energy-momentum tensor density (3..79) has the structure

$$\hat{\Theta}^{\mu\nu}_{(B)} = \hat{t}^{\mu\nu}_{B} + (\hat{T}^{(\mu}_{\rho}\overline{g}^{\nu)\rho} - \hat{\overline{T}}^{\mu\nu}) + \frac{1}{2\kappa}\hat{l}^{\rho\sigma}\overline{R}_{\rho\sigma}\overline{g}^{\mu\nu} + \frac{1}{\kappa}\hat{l}^{\lambda[\mu}\overline{R}^{\nu]}_{\lambda}. \tag{3..80}$$

Here, the first term is the symmetrical energy-momentum tensor density for the free gravitational field:

$$\begin{aligned}
\kappa\hat{t}^{\mu\nu}_{B} &= \tfrac{1}{2}\left(\hat{l}^{\mu\nu}\overline{g}^{\rho\sigma} - \overline{g}^{\mu\nu}\hat{l}^{\rho\sigma}\right)\overline{D}_\sigma\Delta^{\lambda}_{\rho\lambda}\\
&+ \left(\hat{l}^{\rho\sigma}\overline{g}^{\lambda(\mu} - \overline{g}^{\rho\sigma}\hat{l}^{\lambda(\mu}\right)\overline{D}_\sigma\Delta^{\nu)}_{\lambda\rho}\\
&+ \overline{g}^{\rho\sigma}\left(\tfrac{1}{2}\hat{g}^{\mu\nu}\Delta^{\lambda}_{\rho\lambda}\Delta^{\eta}_{\sigma\eta} + \hat{g}^{\lambda\eta}\Delta^{(\mu}_{\lambda\rho}\Delta^{\nu)}_{\eta\sigma}\right)\\
&+ \overline{g}^{\rho\sigma}\left(\Delta^{\lambda}_{\sigma\eta}\Delta^{(\mu}_{\lambda\rho}\hat{g}^{\nu)\eta} - 2\Delta^{\lambda}_{\sigma\lambda}\Delta^{(\mu}_{\eta\rho}\hat{g}^{\nu)\eta}\right)\\
&+ \tfrac{1}{2}\hat{g}^{\lambda\eta}\overline{g}^{\mu\nu}\Delta^{\sigma}_{\rho\lambda}\Delta^{\rho}_{\sigma\eta}\\
&+ \hat{g}^{\lambda\eta}\left(\Delta^{\sigma}_{\rho\sigma}\Delta^{(\mu}_{\lambda\eta} - \Delta^{\sigma}_{\lambda\eta}\Delta^{(\mu}_{\rho\sigma} - \Delta^{\sigma}_{\lambda\rho}\Delta^{(\mu}_{\eta\sigma}\right)\overline{g}^{\nu)\rho}.
\end{aligned} \tag{3..81}$$

The second term in (3..80) is the perturbation of the matter energy-momentum, the third and fourth terms describe the interaction with the background geometry.

Because the fourth term in (3..80) is antisymmetric, the energy-momentum is symmetric: $\hat{\Theta}^{\mu\nu}_{(B)} = \hat{\Theta}^{\nu\mu}_{(B)}$ if and only if $\overline{R}_{\mu\nu} = \overline{\Lambda}\overline{g}_{\mu\nu}$, i.e., for the cases when backgrounds are Einstein spaces in Petrov's terminology [143]. Thus, the Belinfante symmetrization is restricted in the sense of constructing really *symmetric* quantities. Similary, the differential conservation law $\overline{D}_\nu\hat{\Theta}^{\mu\nu}_{(B)} = 0$ has also a place if and only if backgrounds are Einstein spaces. However, even on arbitrary curved backgrounds for Killing vectors λ^μ the current is defined, like in (3..29): $\hat{J}^{\mu}_{(B)} = \hat{\Theta}^{\mu}_{(B)\nu}\lambda^\nu$, and is conserved: $\partial_\mu\left(\hat{\Theta}^{\mu}_{(B)\nu}\lambda^\nu\right) = 0$. Thus both "defects" compensate one another. This could be useful, e.g., for constructing angular momenta of relativistic astrophysical objects on a FRW background, which is not an Einstein space, but which has rotating Killing vectors.

Unlike the KBL quantities, second derivatives of $g_{\mu\nu}$ appear in the energy-momentum tensor density (3..81). This needs some comments also. The canonical energy-momentum $\hat{t}^\mu_\nu$, see (3..50), is quadratic in first order derivatives, and this is the normal behaviour of a conserved quantity relating to initial conditions. Consider the local quantities $\hat{J}^\mu_{(B)}$ in (3..77) and (3..78). Suppose that initial conditions are defined on a hypersurface at a given time coordinate $x^0 = \mathrm{const}$. Then it is important to examine initial conditions for zero's component $\hat{J}^0_{(B)} = \hat{J}^0_{(c)} + \partial_k(\hat{S}^{0k\sigma}\xi_\sigma)$ only, see Eq. (3..33). Such a form is because $\hat{S}^{\mu\nu\sigma}$ is anti-symmetric in the first two indices, see (3..76). Thus since $\hat{J}^0_{(c)}$ and $\hat{S}^{0k\sigma}\xi_\sigma$ contain only first order *time* derivatives, $\hat{J}^0_{(B)}$ itself contains only first order *time* derivatives of the metric and therefore has the normal behaviour with respect to initial conditions. Thus, this requirement may be unnecessarily restrictive.

Now let us discuss the properties of the quantities in (3..78). Comparing the expressions in Eqs. (3..60) and (3..76) and definitions in Eqs. (3..63) and (3..77) we find out that the superpotentials are identical. The same conclusion is related to Z-terms, thus

$$\hat{J}^{\mu\nu}_{(B)} = \hat{J}^{\mu\nu}_{(s)}, \tag{3..82}$$

$$\hat{Z}^{\mu}_{(B)} = \hat{Z}^{\mu}_{(s)}. \tag{3..83}$$

Next, taking into account these relations and comparing (3..68) with (3..78) we conclude that $\hat{J}^\mu_{(B)} = \hat{J}^\mu_{(s)}$ and, consequently it has to be

$$\hat{\Theta}^\mu_{(B)\nu} = \hat{\Theta}^\mu_{(s)\nu}. \tag{3..84}$$

Although the structure of these generalized energy-momenta are different, direct calculations with the use of the field equations also assert the claim (3..84). Thus, on arbitrary curved background one cannot use separately the energy-momentum of the free gravitational field neither $\hat{t}^B_{\mu\nu}$, nor $\hat{t}^g_{\mu\nu}$. One needs to use the *total* energy-momentum tensor density, which is the same, as is seen from in (3..84).

From all of these one concludes

- The construction of the conservation laws for perturbations on arbitrary curved backgrounds in GR with the use of the generalized Belinfante symmetrization transforms the conservation laws of the canonical system into the ones of the field-theoretical approach. Thus, the Belinfante method is a "bridge" between the canonical Nœther and the field-theoretical approaches.

In spite of that Eq. (3..84) is an analog of the relation (3..30) in a simple field theory (3..18), this conclusion is not so evident. Indeed, we cannot apply the Belinfante method *only* in the framework of the KBL bimetric system (3..40) because we cannot define a symmetrical energy-momentum analogously to (3..19). Really, the variation of the Lagrangian (3..40) with respect to the background metric gives only the background Einstein equations, whereas the variation of the Lagrangian (3..41) gives only the non-dynamical background

Einstein tensor. Only a non-trivial connection with the field approach turns out fruitful leading to (3..82) - (3..84).

Relations (3..82) - (3..84) solve the aforementioned problems of the canonical construction. Indeed the quantities $\hat{\Theta}^{\mu}_{(B)\nu}$, $\hat{Z}^{\mu}_{(B)}$ and $\hat{J}^{\mu\nu}_{(B)}$ are, in fact, the quantities in the field-theoretical frame and therefore they do not depend on divergences in the Lagrangian. A direct mathematical substantiation is given in subsection 4.1.. Thus we have arrived a one of desirable results.

Now, return to the ambiguity of the field approach in currents and superpotentials noted at the end of subsection 3.4.. Remark that the KBL approach and the Belinfante symmetrization do not depend on a choice of the variables, like $g^{\mu\nu}$, $g_{\mu\nu}$, $\sqrt{-g}g^{\mu\nu}$, ..., and *uniquely* lead to the conservation law (3..78). On the other hand, it coincides *uniquely* with conservation law (3..68) defined by the second decomposition in (2..73). Thus it is a theoretical argument in favor of a choice $\hat{l}^{\mu\nu}_{(a)} = \hat{l}^{\mu\nu}$ in (2..81). To support this theoretical conclusion below we present a useful test. Recall that differences between superpotentials in the family (3..84) appear in the second order. Just the second order is crucial, e.g., in calculations of energy and its flux [185] at null infinity. As it was checked in [116], only the superpotential (3..65) from the family (3..75) gives the standard result [185], whereas all the others, e.g., the superpotential corresponding to the first decomposition in (2..73) and defined in [68], do not.

3.6. Applications

The above presented formalism, its technique and expressions (in particular, the energy-momentum tensor $\hat{t}^{(tot)}_{\mu\nu}$ and its properties, including gauge invariance ones) were used in various applications in GR. Shortly we note that $\hat{t}^{(tot)}_{\mu\nu}$ was used in development of quantum mechanics with non-classical gravitational self-interaction [108, 109], in the framework of which inflation scenario was analyzed [110, 111]. Some other applications are reviewed below in more detail.

3.6.1. The Weakest Falloff at Spatial Infinity

The proof of the positive-energy theorem in [70] - [73] has renewed a great interest to asymptotically flat spacetimes in GR. In particular, these investigations showed that the standard asymptotic conditions at spatial infinity ($1/r$ falloff in metric) can be significantly relaxed (see, e.g., [186] - [189] and references there in). These models are very appropriate for applications in the framework of the field approach. Indeed, asymptotically a background is just chosen naturally, it is a Minkowski space at spatial infinity. Gravitational field at spatial infinity is weak metrical perturbations with respect to this background. We have done such applications: both in Lagrangian [112] and in Hamiltonian [113] derivation we examine the weakest asymptotic behaviour at spatial infinity. We consider so-called *real* isolated systems, for which all the physical fields are effectively concentrated in a confined space at finite time intervals, and a falloff of which is invariant under asymptotic Poincaré transformations.

As an initial falloff we use the standard one: $1/r$. Then choosing a background spacetime as the Minkowski space we evaluate a field configuration in the framework of the field approach. For simplicity it is assumed that a manifold, which globally supports the physical metric, supports also the auxiliary flat metric. To search for the weakest falloff global conserved quantities of the isolated system (integrals of motion) are examined. We construct them in the form of the surface integrals (3..39) with the use of the superpotential defined in (3..65) and the ten Killing vectors of the Minkowski space.

To find the weakest falloff conditions one uses the gauge invariance properties of the field formulation of GR in section 2.6.. Taking into account the general formula (3..71) for the current (3..68) we conclude that the integrals of motion are gauge invariant up to surface term on the solutions of the field equations. Then we search for the weakest falloff for the gauge potentials ξ^α and their derivatives which ensure vanishing the non-invariant terms. After that we find out the weakest falloff for the gravitational potentials:

$$l^{\mu\nu} = O^+(r^{-\varepsilon}) + O^-(r^{-\delta})\,; \qquad \varepsilon + \delta > 2\,,\ 1 \geq \varepsilon > \frac{1}{2}\,,\ \delta > 1 \tag{3..85}$$

where $(+)$ and $(-)$ mean even and odd functions with respect to changing sign of the 3-vector $\vec{n} = x^k/r$. To obtain the simple conclusion that the falloff can be weakened up to $r^{-1/2}$ one needs to examine energy and momentum only (see, e.g., [187, 188]). To obtain the more detailed falloff (3..85) one has to study all the ten integrals of motion. In this relation we remark also the works [186, 189]. In [186] the algebra of asymptotic Poincaré generators is studied, and in [189] the finiteness of the ADM definition of angular momentum is examined. The result (3..85) based on the gauge invariance requirement coincides with the results in [186] corrected not significantly, and with the results in [189], if the last are adopted by some natural requirements (see [112, 113]). Thus the three different approaches give the same weakest falloff.

Historically, among the ten integrals of motion, not enough attention was payed to the Lorentzian momentum. In this relation the paper by Regge and Teitelboim [190] who included the falloff of odd and even parity, and the paper by Beig and ÓMurchadha [191] who improved the results of the work [190] were the most important earlier works. Currently this gap is closing [117], [192] - [194]. In [117] we demonstrate that the quasi-local Brown-York [79] center of mass integral asymptotically agrees with the one given in [191] and with the one expressed by the curvature integrals, e.g., in [195]. In [192] the falloff of the gravitational potentials is analyzed in $n+1$ dimensions with special attention to the Lorenzian momentum and matter variables. In [117] and [192] (for $n = 3$) the falloff exploited by Beig and ÓMurchadha [191] is used. It is r^{-1} for the even parity first term and $r^{-\delta}$, $\delta > 1$, for the next of arbitrary parity term. Comparing with the spectrum of the conditions (3..85) one finds that this condition corresponds to $\varepsilon = 1$ and $\delta > 1$ in the full spectrum (3..85). As an example, the other end of the spectrum is $\varepsilon > 1/2$ and $\delta \geq 3/2$. In [193] and [194], in the framework of the covariant Hamiltonian approach [94] the *quasi-local* center of mass integral in the teleparallel gravity and GR is considered. This approach gives *quasi-local* 4-momentum, angular momentum and center of mass integral in the framework of an unique relativistic invariant (Poincaré invariant) description.

3.6.2. Closed Worlds as Gravitational Fields

In the paper [101], we show how a closed-world geometry can be viewed as a gravitational field configuration, superposed on a topologically trivial geometrically flat spacetime. As seems, it is not very natural. However, we do this without contradictions. The space components of this configuration corresponds to a so-called stereographic projection of 3-sphere S^3 onto a flat 3-space E^3, when the "south" pole corresponds to the origin of the coordinate system, whereas the "north" pole is identified with all the points at infinity of E^3. Thus, to make the full identification the "north" pole is "knocked out" of the sphere S^3 to simplify its topology.

The background Minkowski space has an auxiliary character, as it has to be. Formally, the field configuration is specified in an infinite volume. However, with the use of the physically reasonable measurements examining light signals in gravitational fields one finds the standard volume for S^3 (see [1]). By considering the metric relations established through the physical measurements an observer will infer that homogeneity, an equivalence of observers at all points of space, has a place.

A picture when topological and geometrical properties are replaced by the properties of an effective field propagated in a trivial spacetime could be interesting and useful. Thus, Rubakov and Shaposhnikov [40] have shown that not too energetic scalar particles could become effectively trapped in a potential well even in a topologically trivial universe, although nontrivial classical solutions would have to be present to play a role of an external field.

The field configuration presenting the closed world is static and has an asymptotic behaviour $1/r^2$. This ensures zero energy, momentum and angular momentum of the system, which then could be treated as a microuniverse. Indeed, such characteristics are true for a Minkowski vacuum where classical fields and particles are absent, and thus an idea of quantum birth of the universe could be supported.

3.6.3. A Point Particle in GR

The Schwarzschild solution is the one of the most popular models in GR. Frequently, in spite of a complicated geometry, it is treated as a point mass solution in GR [1]; currently the interest to this viewpoint is renewed [196] - [199]. However, if one considers GR in the usual geometrical description, then this interpretation meets conceptual difficulties (see, for example, [200] and also [114]).

Except a pure theoretical interest the point particle description could be interesting and useful for experimental gravity problems. Gravitational wave detectors such as LIGO and VIRGO will definitely discover gravitational waves from coalescing binary systems comprising of compact relativistic objects. Therefore it is necessary to derive equations of motion of such components, e.g., two black holes. As a rule, at an *initial* step the black holes are modeled by point-like particles presented by Dirac's δ-function. Then consequent post-Newtonian approximations are used (see, for example, [201, 202] and references therein).

The aforementioned difficulties [200], at least, do not appear in Newtonian gravity. To describe a point particle one has to assume that the mass distribution has the form $\rho(\mathbf{r}) = m\delta(\mathbf{r})$ where δ-function satisfies the ordinary Poisson equation, which in spherical coordinates is

$$\nabla^2 \left(\frac{1}{r}\right) \equiv \left(\frac{d^2}{dr^2} + \frac{2}{r}\frac{d}{dr}\right)\frac{1}{r} = -4\pi\delta(\mathbf{r})\,. \tag{3..86}$$

Then, the Newtonian potential $\phi = m/r$ satisfies the Newtonian equation in the whole space including the point $r = 0$. Besides, the whole mass of the system is defined by the *unique* expression $\int_\Sigma dx^3\rho(\mathbf{r})$ where both the usual regular distribution and the point particle density can be integrated providing the standard mass m.

In [114]) we show that, analogously to the Newtonian prescription, the point mass in GR can be described in a non-contradictory manner in the framework of the field formulation. The Schwarzschild solution was presented as a gravitational field configuration in a background Minkowski space described in the standard spherical Schwarzschild (*static*) coordinates. The concept of Minkowski space was extended from spatial infinity (frame of reference of a distant observer) up to the horizon $r = r_g$, and even under the horizon including the worldline $r = 0$ of the true singularity. The configuration satisfies the Einstein equations at all the points of the Minkowski space, including $r = 0$, if the operations with the generalized functions [203] are valid. The energy-momentum tensor was constructed. Its components are presented by expressions proportional to $\delta(\mathbf{r})$ and by free gravitational field outside $r = 0$. The picture is clearly interpreted as a point particle distribution in GR. Indeed, the configuration is essentially presented by δ-function, one can use the volume integration over the *whole* Minkowski space and obtain the total energy mc^2 in the natural way. However at $r = r_g$ the energy density and other characteristics have discontinuities. A"visible" boundary between the regions outside and inside the horizon exists and does not allow to consider an evolution of events continuously.

However, it is not a real singularity. In the field formulation this situation is interpreted as a "bad" fixing of gauge freedom, which can be improved. That is the break at $r = r_g$ can be countered with the use of an appropriate choice of a flat background, which is determined by related coordinates for the Schwarzschild solution. At least, the use of the coordinates without singularities at the horizon could resolve the problem locally at neighborhood of $r = r_g$. Next, it would be natural to describe the true singularity by the world line $r = 0$ of the chosen polar coordinates. Besides, it is desirable to have appropriate coordinates, in which the Schwarzschild solution conserves the form of an asymptotically flat spacetime.

Recently Pitts and Schieve [130] defined and studied properties of a so-called "η-causality". Its fulfilment means that the physical light cone is inside the flat light cone at all the points of the Minkowski space. It is necessary to avoid interpretation difficulties under the field-theoretical presentation of GR. By this requirement all the causally connected events in the physical spacetime are described by the right causal structure of the Minkowski space. A related position of the light cones is not gauge invariant. We consider this requirement *only* to construct a more convenient in applications and interpretation field configuration for the Schwarzschild solution. The requirement of the η-causality can be strengthened by the requirement of a "stable η-causality" [130]. The last means that the physical light cone has to be *strictly* inside the flat light cone, and this is important when quantization problems are under consideration. Indeed, in the case of tangency a field is on the verge of η-causality violation. Returning to the presentation in the Schwarzschild coordinates in [114] we note that it does not satisfy the η-causality requirement.

In the paper [120], taking into account the above requirements we have found a desirable description. A more appropriate gauge fixing corresponds to the *stationary* (not static) coordinates presented in [105, 106] and recently improved in [130]. We consider also the contracting Eddingtom-Finkenstein coordinates in stationary form [3]. These two coordinate systems belong to a parameterized family where all of systems satisfies all the above

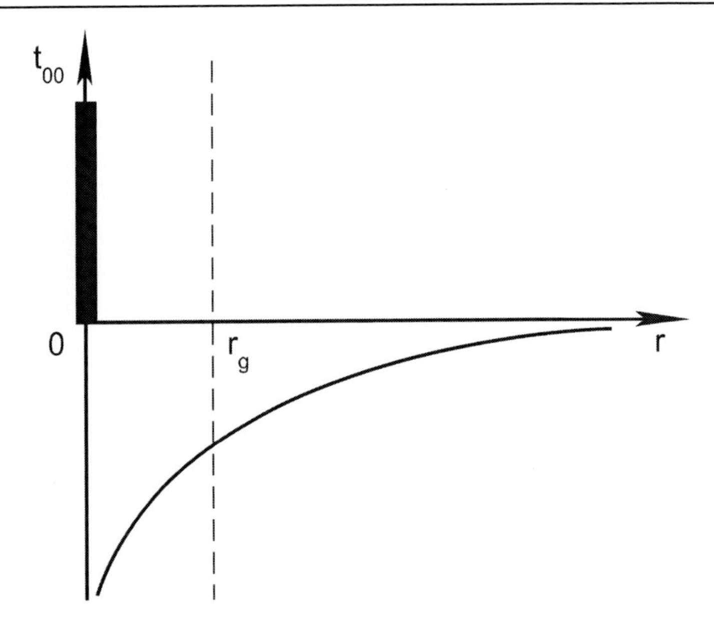

Figure 4. The energy distribution for the field configuration associated with the Schwarzschild solution. The case: $\alpha \in (0,\ 2]$.

requirements. The transformation

$$ct' = ct + r_g \ln \left| \left(\frac{r}{r_g} - 1 \right) \left(\frac{r_g}{r} \right)^{\alpha} \right| \tag{3..87}$$

gives just the parameterized by $\alpha \in [0, 2]$ family of metrics. The cases $\alpha = 1$ and $\alpha = 0$ correspond to the first and to the second examples above. The *stable* η-causality *is not* satisfied with $\alpha = 0$ at $0 \le r \le \infty$. Thus, all the configurations $\alpha \in (0, 2]$ are appropriate for the study both classical and quantum problems, whereas the case $\alpha = 0$ could not be useful for the study quantized fields. Properties of field configurations corresponding to $\alpha \in (0, 2]$ qualitatively are the same as for $\alpha = 1$. In the terms of the field approach [100], all the field configurations for $\alpha \in [0, 2]$ are connected by gauge transformations and are physically equivalent.

Both the gravitational potentials and the components of the total energy-momentum tensor for the field configurations with $\alpha \in [0, 2]$ have no a break at $r = r_g$. The energy distribution is described by the 00-component of the energy-momentum tensor, and for $\alpha \in (0, 2]$ qualitatively is presented on the figure 4. Then the total energy of the system can be calculated both by the volume integration and by the surface integration over the 2-sphere with $r \to \infty$ giving mc^2. For the case $\alpha = 0$ the components of the total energy-momentum tensor for the field configuration have the simplest form:

$$\begin{aligned} t_{00}^{tot} &= mc^2\delta(\mathbf{r})\,, \\ t_{11}^{tot} &= -mc^2\delta(\mathbf{r})\,, \\ t_{AB}^{tot} &= -\tfrac{1}{2}\overline{g}_{AB}\, mc^2\delta(\mathbf{r})\,. \end{aligned} \tag{3..88}$$

This energy-momentum is concentrated *only* at $r = 0$, see the energy distribution on the

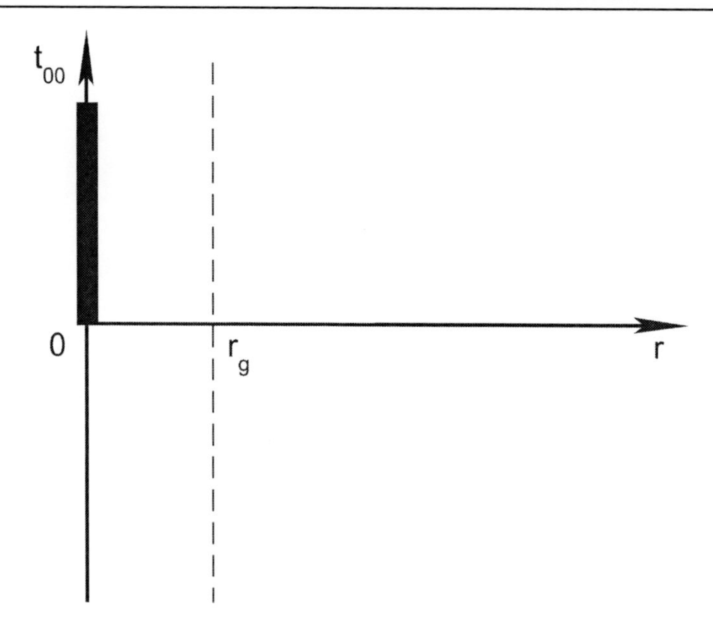

Figure 5. The energy distribution for the field configuration associated with the Schwarzschild solution. The case: $\alpha = 0$.

figure 5. The other component t^{tot}_{11} and the angular ones t^{tot}_{AB} formally could be interpreted as related to the "inner" properties of the point. Indeed, they are proportional only to $\delta(\mathbf{r})$ and, thus, describe the point "inner radial" and "inner tangent" pressure. At last, again the total energy for (3..88) is mc^2.

The field configurations for $\alpha \in [0, 2]$ satisfy the Einstein equations (2..66) at all the points of the Minkowski space including $r = 0$. Then, keeping in mind the presentations on the figures 4 and 5 one can conclude that an appropriate description of a point particle in GR is approached. It is directly and simply continues the Newtonian derivation and, at the same time, does not contradict GR corresponding its principles.

3.6.4. Integral Constraints in FRW Models

In this subsection we consider the linear perturbations on FRW backgrounds from the point view of the conservation laws (3..78), or the same (3..68), for the detail calculations see [116]. Write the background metric $d\overline{s}^2$ in dimensionless coordinates $x^0 = \eta, x^k$ as follows [153]:

$$d\overline{s}^2 = \overline{g}_{\mu\nu}dx^\mu dx^\nu = a^2(\eta)(d\eta^2 - f_{kl}dx^k dx^l) = a^2(\eta)e_{\mu\nu}dx^\mu dx^\nu, \qquad (3..89)$$

$a(\eta)$ is the scale factor, $f_{kl} = \delta_{kl} + k\delta_{km}\delta_{ln}x^m x^n(1 - kr^2)^{-1}$, $f = det\, f_{kl} = (1 - kr^2)^{-1}$, $k = 0$ or ± 1, and $r^2 = \delta_{kl}x^k x^l$. We define the perturbations by $g_{\mu\nu} = a^2(e_{\mu\nu} + \tilde{h}_{\mu\nu})$ as often used in cosmology (see, e.g., [126, 204]), which are connected with the the usual definition (2..13) in linear approximation:

$$\hat{l}^{\mu\nu} = a^2\sqrt{f}(-e^{\mu\rho}e^{\nu\sigma} + \tfrac{1}{2}e^{\mu\nu}e^{\rho\sigma})\tilde{h}_{\rho\sigma}\,. \qquad (3..90)$$

Because the FRW spacetimes are conformal to Minkowski space the conformal Killing vectors of FRW spacetimes are thus those of Minkowski space written in η, x^k coordinates (see [205] - [207]). There are 15 conformal Killing vectors, ξ_A, $A = 1, 2, \ldots, 15$, some of which are pure Killing ones.

These vectors we use as displacement vectors in the expression (3..78). Next, we make linear combinations of ξ_A's with functions of η, say $c^A(\eta)\xi_A$. In general the $c^A\xi_A$'s are not conformal Killing vectors but some linear combinations turn out to be rather simple with clear physical interpretations in appropriate gauge conditions. Integrating the conservation law (3..78) we construct the integral relations connecting perturbations inside of a volume with perturbations at its surface. Thus we consider a sphere $r = \mathrm{const}$ at constant time η and integrate zero components of (3..78) $\hat{J}^0_{(B)} = \partial_l \hat{J}^{0l}_{(B)}$ over such a ball. After cumbersome calculations for perturbations (3..90) one has 15 volume integrals connected with surface integrals.

Besides the matter perturbations δT^0_μ the volume integrands contain two field quantities $\mathcal{Q} = \delta(-D_\mu n^\mu)$ — the perturbation of the external curvature scalar of $\eta = \mathrm{const}$, and $\tilde{h}^m_m$ — the space trace of the metric petrurbations. There are four linear combinations which do not involve $\tilde{h}^m_m$; they are associated with some linear combinations of the conformal Killing vectors, which are just Traschen's [172] "integral constraint vectors".

Of course, we must still impose four gauge conditions to fix the mapping of the perturbed spacetime onto the background. One gauge condition that simplifies almost all volume integrands is the "uniform Hubble expansion" gauge $\mathcal{Q} = 0$ discussed, e.g., by Bardeen [10]. Then, 14 of the 15 volume integrands reduce to linear combinations of δT^0_μ *only*, and thus the 14 relations present "integral constraints", among which the 4 are new. The integrals represent momenta of the matter energy-momentum tensor of order 0, 1 and 2 in powers of x^a when $k = 0$ and with similar interpretations when $k = \pm 1$. There remains one integral that contains both δT^0_μ and $\tilde{h}^m_m$: it is related to so-called conformal time translations for $k = \pm 1$ or to so-called time accelerations for $k = 0$.

Another gauge condition often used (see, e.g., [204]) is $\nabla_l \underset{T}{\tilde{h}^l_k} = 0$ in which $\underset{T}{\tilde{h}^l_k}$ is the traceless part of $\tilde{h}^l_k$; these gauges remove the longitudinal modes of the gravitational waves. Combining $\nabla_l \underset{T}{\tilde{h}^l_k} = 0$ with $\mathcal{Q} = 0$ one finds four relations that are independent of the gravitational radiation.

4. D-Dimensional Metric Theories of Gravity

As was discussed in Introduction, multidimensional gravitational models become more and more popular, their solutions (for example, generalized p-branes and brane-world black hole solutions [208] - [211]) induce an arising interest. Other generalizations, say, scalar-tensor theories of gravity [212], are developed intensively also. In this section, keeping in mind such generalizations we construct a perturbed scheme and conservation laws for perturbations in a generic metric theory. The presentation is based on the works [103, 104, 119, 121]. Besides, here the approach is presented in the united scheme, and more details are given.

4.1. The Main Identities

In this subsection, we present necessary identities and stress their important properties. Let the system of fields, set of tensor densities Q^A, be described by the Lagrangian

$$\hat{\mathcal{L}} = \hat{\mathcal{L}}(Q^A; Q^A{}_{,\alpha}; Q^A{}_{,\alpha\beta}) \tag{4..1}$$

including derivatives up to the second order. We assume that it is an *arbitrary metric* theory of gravity in D dimensions with the generalized gravitational variables (g^a, Ψ^B) and the matter sources Φ^C. Thus $Q^A = \{(g^a, \Psi^B), \Phi^C\}$; the variables Ψ^B are included keeping in mind, say, scalar-tensor or vector-tensor theories; the metric variables g^a are thought as defined in (2..74), only in D dimensions. Then we include $\overline{g}_{\mu\nu}$ as usual. The ordinary derivatives ∂_α are rewritten over the covariant $\overline{D}_\alpha$ ones by changing $Q_{B,\tau} \equiv \overline{D}_\tau Q_B - \overline{\Gamma}^\sigma_{\tau\rho}\, Q_B|^\rho_\sigma$. Then the Lagrangian (4..1) takes an explicitly covariant form:

$$\hat{\mathcal{L}} = \hat{\mathcal{L}}_c = \hat{\mathcal{L}}_c(Q_B; \overline{D}_\alpha Q_B; \overline{D}_\beta \overline{D}_\alpha Q_B)\,. \tag{4..2}$$

After that indexes are shifted by $\overline{g}_{\mu\nu}$ and $\overline{g}^{\mu\nu}$.

Now we use the standard technique (see, for example, [140]). For the Lagrangian (4..2), as a scalar density, we write out again the identity: $\pounds_\xi \hat{\mathcal{L}}_c + (\xi^\alpha \hat{\mathcal{L}}_c)_{,\alpha} \equiv 0$, which can be rewritten in the form:

$$-\left[\frac{\delta \hat{\mathcal{L}}_c}{\delta Q_B}\overline{D}_\alpha Q_B + \overline{D}_\beta\left(\frac{\delta \hat{\mathcal{L}}_c}{\delta Q_B}\, Q_B|^\beta_\alpha\right)\right]\xi^\alpha + \overline{D}_\alpha\left[\hat{u}^\alpha_\sigma \xi^\sigma + \hat{m}^{\alpha\tau}_\sigma \overline{D}_\tau \xi^\sigma + \hat{n}^{\alpha\tau\beta}_\sigma \overline{D}_\beta \overline{D}_\tau \xi^\sigma\right] \equiv 0\,. \tag{4..3}$$

In (4..3), the coefficients are defined by the Lagrangian in unique way:

$$\begin{aligned}\hat{u}^\alpha_\sigma \;\equiv\;& \hat{\mathcal{L}}_c \delta^\alpha_\sigma + \frac{\delta \hat{\mathcal{L}}_c}{\delta Q_B}\, Q_B|^\alpha_\sigma - \left[\frac{\partial \hat{\mathcal{L}}_c}{\partial(\overline{D}_\alpha Q_B)} - \overline{D}_\beta\left(\frac{\partial \hat{\mathcal{L}}_c}{\partial(\overline{D}_\beta \overline{D}_\alpha Q_B)}\right)\right]\overline{D}_\sigma Q_B\\ -\;& \frac{\partial \hat{\mathcal{L}}_c}{\partial(\overline{D}_\beta \overline{D}_\alpha Q_B)}\overline{D}_\sigma \overline{D}_\beta Q_B - \hat{n}^{\alpha\tau\beta}_\lambda \overline{R}^\lambda{}_{\tau\beta\sigma}\,,\end{aligned} \tag{4..4}$$

$$\begin{aligned}\hat{m}^{\alpha\tau}_\sigma \;\equiv\;& \left[\frac{\partial \hat{\mathcal{L}}_c}{\partial(\overline{D}_\alpha Q_B)} - \overline{D}_\beta\left(\frac{\partial \hat{\mathcal{L}}_c}{\partial(\overline{D}_\beta \overline{D}_\alpha Q_B)}\right)\right] Q_B|^\tau_\sigma\\ -\;& \frac{\partial \hat{\mathcal{L}}_c}{\partial(\overline{D}_\tau \overline{D}_\alpha Q_B)}\overline{D}_\sigma Q_B + \frac{\partial \hat{\mathcal{L}}_c}{\partial(\overline{D}_\beta \overline{D}_\alpha Q_B)}\overline{D}_\beta(Q_B|^\tau_\sigma)\,,\end{aligned} \tag{4..5}$$

$$\hat{n}^{\alpha\tau\beta}_\sigma \equiv \tfrac{1}{2}\left[\frac{\partial \hat{\mathcal{L}}_c}{\partial(\overline{D}_\beta \overline{D}_\alpha Q_B)}\, Q_B|^\tau_\sigma + \frac{\partial \hat{\mathcal{L}}_c}{\partial(\overline{D}_\tau \overline{D}_\alpha Q_B)}\, Q_B|^\beta_\sigma\right]. \tag{4..6}$$

The coefficient at ξ^σ in the first term in (4..3) is identically equal to zero (generalized Bianchi identity). Thus the identity (4..3) has the form of the differential conservation law:

$$\overline{D}_\alpha \hat{\imath}^\alpha \equiv \partial_\alpha \hat{\imath}^\alpha \equiv 0 \tag{4..7}$$

with a generalized current

$$\hat{\imath}^\alpha \equiv -\left[\hat{u}^\alpha_\sigma \xi^\sigma + \hat{m}^{\alpha\tau}_\sigma \overline{D}_\tau \xi^\sigma + \hat{n}^{\alpha\tau\beta}_\sigma \overline{D}_\beta \overline{D}_\tau \xi^\sigma\right]. \tag{4..8}$$

We also use the helpful form:

$$\hat{\imath}^\alpha \equiv - \left[(\hat{u}^\alpha_\sigma + \hat{n}^{\alpha\beta\gamma}_\lambda \overline{R}^\lambda{}_{\beta\gamma\sigma})\xi^\sigma + \hat{m}^{\alpha\beta}_\sigma \bar{g}^{\sigma\rho} \partial_{[\beta}\xi_{\rho]} + \hat{z}^\alpha \right] \tag{4..9}$$

with z-term defined as

$$\hat{z}^\alpha(\xi) \equiv \hat{m}^{\alpha\beta}_\sigma \zeta^\sigma_\beta + \hat{n}^{\alpha\beta\gamma}_\sigma \bar{g}^{\sigma\rho} \left(2\overline{D}_\gamma \zeta_{\beta\rho} - \overline{D}_\rho \zeta_{\beta\gamma} \right) . \tag{4..10}$$

Again, if $\xi^\alpha = \lambda^\alpha$, then $\hat{z}^\alpha = 0$ and the current (4..9) is determined by the energy-momentum $(u + n\overline{R})$-term and the spin m-term. Opening the identity (4..7) and, since ξ^σ, $\partial_\alpha \xi^\sigma$, $\partial_{\beta\alpha}\xi^\sigma$ and $\partial_{\gamma\beta\alpha}\xi^\sigma$ are arbitrary at every world point, equating independently to zero the coefficients at ξ^σ, $\overline{D}_\alpha \xi^\sigma$, $\overline{D}_{(\beta\alpha)}\xi^\sigma$ and $\overline{D}_{(\gamma\beta\alpha)}\xi^\sigma$ we get a "cascade" (in terminology by Julia and Silva [183]) of identities:

$$\begin{aligned} &\overline{D}_\alpha \hat{u}^\alpha_\sigma + \tfrac{1}{2}\hat{m}^{\alpha\rho}_\lambda \overline{R}_\sigma{}^\lambda{}_{\rho\alpha} + \tfrac{1}{3}\hat{n}^{\alpha\rho\gamma}_\lambda \overline{D}_\gamma \overline{R}_\sigma{}^\lambda{}_{\rho\alpha} \equiv 0, \\ &\hat{u}^\alpha_\sigma + \overline{D}_\lambda \hat{m}^{\lambda\alpha}_\sigma + \hat{n}^{\tau\alpha\rho}_\lambda \overline{R}_\sigma{}^\lambda{}_{\rho\tau} + \tfrac{2}{3}\hat{n}^{\lambda\tau\rho}_\sigma \overline{R}^\alpha{}_{\tau\rho\lambda} \equiv 0, \\ &\hat{m}^{(\alpha\beta)}_\sigma + \overline{D}_\lambda \hat{n}^{\lambda(\alpha\beta)}_\sigma \equiv 0, \\ &\hat{n}^{(\alpha\beta\gamma)}_\sigma \equiv 0. \end{aligned} \tag{4..11}$$

The above constructions are the generalization to arbitrary curved backgrounds of the expressions given by Mitzkevich [140].

Since Eq. (4..7) is identically satisfied, the current (4..9) must be a divergence of a superpotential $\hat{\imath}^{\alpha\beta}$, for which $\partial_{\beta\alpha}\hat{\imath}^{\alpha\beta} \equiv 0$, that is

$$\hat{\imath}^\alpha \equiv \overline{D}_\beta \hat{\imath}^{\alpha\beta} \equiv \partial_\beta \hat{\imath}^{\alpha\beta}. \tag{4..12}$$

Indeed, substituting $\hat{u}^\alpha_\sigma$ and $\hat{m}^{\alpha\beta}_\sigma$ from Eqs. (4..11) *directly* into the current and using algebraic properties of $n^{\alpha\beta\gamma}_\sigma$ and $\overline{R}^\alpha{}_{\beta\rho\sigma}$, and the third identity in Eqs. (4..11) we reconstruct (4..9) into the form (4..12) where the superpotential is

$$\hat{\imath}^{\alpha\beta} \equiv \left(\tfrac{2}{3}\overline{D}_\lambda \hat{n}^{[\alpha\beta]\lambda}_\sigma - \hat{m}^{[\alpha\beta]}_\sigma \right) \xi^\sigma - \tfrac{4}{3}\hat{n}^{[\alpha\beta]\lambda}_\sigma \overline{D}_\lambda \xi^\sigma . \tag{4..13}$$

It is explicitly antisymmetric in α and β. Of course, Eq. (4..12) has also a sense of the differential conservation law Eq. (4..7).

Let us find contributions into currents and superpotentials from a divergence in the Lagrangian $\delta_d \hat{\mathcal{L}}_c = div = \hat{d}^\nu{}_{,\nu}$. For the scalar density $\hat{d}^\nu{}_{,\nu}$ one has the identity $(\pounds_\xi \hat{d}^\alpha + \xi^\alpha \hat{d}^\nu{}_{,\nu})_{,\alpha} \equiv 0$, which gives contributions $\delta_d \hat{\imath}^\alpha$ into the current (4..8) and $\delta_d \hat{\imath}^{\alpha\beta}$ into the superpotential (4..13). The additional quantities are

$$\delta_d \hat{u}^\alpha_\sigma = 2\overline{D}_\beta(\delta^{[\alpha}_\sigma \hat{d}^{\beta]}) , \tag{4..14}$$

$$\delta_d \hat{m}^{\alpha\beta}_\sigma = 2\delta^{[\alpha}_\sigma \hat{d}^{\beta]} , \tag{4..15}$$

$$\delta_d \hat{n}^{\alpha\beta\gamma}_\sigma = 0 , \tag{4..16}$$

$$\delta_d \hat{\imath}^{\alpha\beta} = -2\xi^{[\alpha}\hat{d}^{\beta]} . \tag{4..17}$$

Then, for $\hat{\mathcal{L}}_c \to \hat{\mathcal{L}}_c + \delta_d \hat{\mathcal{L}}_c$ Eq. (4..12) changes as

$$\hat{\imath}^\alpha + \delta_d \hat{\imath}^\alpha \equiv \overline{D}_\beta \left(\hat{\imath}^{\alpha\beta} + \delta_d \hat{\imath}^{\alpha\beta} \right) , \tag{4..18}$$

where the changes do not depend on a structure of $\hat{d}^\nu$.

Next, using the general Belinfante rule (3..27) we define a tensor density

$$\hat{s}^{\alpha\beta\sigma} \equiv -\hat{s}^{\beta\alpha\sigma} \equiv -\hat{m}_\lambda^{\sigma[\alpha}\bar{g}^{\beta]\lambda} - \hat{m}_\lambda^{\alpha[\sigma}\bar{g}^{\beta]\lambda} + \hat{m}_\lambda^{\beta[\sigma}\bar{g}^{\alpha]\lambda}, \qquad (4..19)$$

add $\overline{D}_\beta(\hat{s}^{\alpha\beta\sigma}\xi_\sigma)$ to both sides of (4..12), and obtain a new identity:

$$\hat{\imath}_B^\alpha \equiv \overline{D}_\beta \hat{\imath}_B^{\alpha\beta} \equiv \partial_\beta \hat{\imath}_B^{\alpha\beta}. \qquad (4..20)$$

This modification cancels the spin term from the current (4..9):

$$\hat{\imath}_B^\alpha \equiv \left(-\hat{u}_\sigma^\alpha - \hat{n}_\lambda^{\alpha\beta\gamma}\overline{R}^\lambda{}_{\beta\gamma\sigma} + \overline{D}_\beta \hat{s}^{\alpha\beta}{}_\sigma\right)\xi^\sigma + \hat{z}_B^\alpha(\xi) \equiv \hat{u}_{B\sigma}^\alpha \xi^\sigma + \hat{z}_B^\alpha(\xi), \qquad (4..21)$$

and a new z-term disappears also on Killing vectors of the background:

$$\hat{z}_B^\alpha(\xi) = \left(\bar{g}^{\lambda\tau}\hat{m}_\lambda^{\beta\alpha} + \hat{m}_\lambda^{\alpha\tau}\bar{g}^{\beta\lambda} - \hat{m}_\lambda^{\tau\beta}\bar{g}^{\alpha\lambda}\right)\zeta_{\tau\beta} + \hat{n}_\lambda^{\alpha\tau\beta}\left(2\overline{D}_{(\beta}\zeta_{\tau)}^\lambda - \overline{D}_\sigma\zeta_{\beta\tau}\bar{g}^{\lambda\sigma}\right). \qquad (4..22)$$

Thus, the current $\hat{\imath}_B^\alpha$ is defined, in fact, by the modified energy-momentum tensor density $\hat{u}_{B\sigma}^\alpha$. Because the new superpotential depends on the n-coefficients only:

$$\hat{\imath}_B^{\alpha\beta} \equiv 2\left(\tfrac{1}{3}\overline{D}_\rho\hat{n}_\sigma^{[\alpha\beta]\rho} + \overline{D}_\tau\hat{n}_\lambda^{\tau\rho[\alpha}\bar{g}^{\beta]\lambda}\bar{g}_{\rho\sigma}\right)\xi^\sigma - \tfrac{4}{3}\hat{n}_\sigma^{[\alpha\beta]\lambda}\overline{D}_\lambda\xi^\sigma, \qquad (4..23)$$

then due to the definition (4..6) it vanishes for Lagrangians with only the first order derivatives. On the other hand, the superpotential (4..23) is well adapted to theories with second derivatives in Lagrangians, like GR or the Einstein-Gauss-Bonnet gravity. It is also important to note that for the superpotentials (4..13) and (4..23) their forms do not depend *explicitly* on a dimension D; the same property is related to the Deser and Tekin superpotentials [75].

It is important to note that the Belinfante procedure applied to (4..18) cancels the quantities $\delta_d\hat{\imath}^\alpha$ and $\delta_d\hat{\imath}^{\alpha\beta}$ and gives again (4..20). Let us show this. The quantity (4..19) constructed for $\delta_d\hat{m}_\sigma^{\alpha\beta}$ in (4..15) gives

$$\delta_d\hat{s}^{\alpha\beta\sigma}\xi_\sigma = 2\xi^{[\alpha}\hat{d}^{\beta]}. \qquad (4..24)$$

Then, adding $\overline{D}_\beta(\hat{s}^{\alpha\beta\sigma}\xi_\sigma + \delta_d\hat{s}^{\alpha\beta\sigma}\xi_\sigma)$ to Eq. (4..18) one cancels completely the spin term and suppresses $\delta_d\hat{u}_\sigma^\alpha$ and $\delta_d\hat{\imath}^{\alpha\beta}$. Indeed, combining (4..14) and (4..17) with (4..24) one has $\delta_d\hat{u}_\sigma^\alpha - \overline{D}_\beta(\delta_d\hat{s}^{\alpha\beta\rho}\bar{g}_{\rho\sigma}) = 0$ and $\delta_d\hat{\imath}^{\alpha\beta} + \delta_d\hat{s}^{\alpha\beta\sigma}\xi_\sigma = 0$ for an arbitrary $\hat{d}^\nu$. This just supports the claim that $\hat{J}^\mu_{(B)}$, $\hat{\Theta}^\mu{}_{(B)\nu}$ and $\hat{J}^{\mu\nu}_{(B)}$ in (3..78) are independent on divergences.

4.2. The Field-Theoretical Formulation for Perturbations

In this subsection, following the method of section 2. we develop a perturbed derivation. The field equations corresponding to the system (4..1) are derived as usual:

$$\delta\hat{\mathcal{L}}/\delta Q^A = 0. \qquad (4..25)$$

Let us decompose Q^A onto the background part $\overline{Q}^A$ and the dynamic part q^A:

$$Q^A = \overline{Q}^A + q^A. \qquad (4..26)$$

The background fields satisfy the background equations

$$\delta\overline{\hat{\mathcal{L}}}/\delta\overline{Q}^A = 0 \tag{4..27}$$

where $\overline{\hat{\mathcal{L}}} = \hat{\mathcal{L}}(\overline{Q})$. Taking into account the form of the Lagrangian (2..19) (or (2..78)) we will describe the perturbed system by the Lagrangian

$$\hat{\mathcal{L}}^{dyn}(\overline{Q}, q) = \hat{\mathcal{L}}(\overline{Q} + q) - q^A \frac{\delta\overline{\hat{\mathcal{L}}}}{\delta\overline{Q}^A} - \overline{\hat{\mathcal{L}}} + div \tag{4..28}$$

instead of the original Lagrangian $\hat{\mathcal{L}}(Q) = \hat{\mathcal{L}}(\overline{Q}+q)$. As before, the background equations should not be taken into account before variation of $\hat{\mathcal{L}}^{dyn}(\overline{Q}, q)$ with respect to $\overline{Q}^A$. Using the evident property $\delta\hat{\mathcal{L}}(\overline{Q} + q)/\delta\overline{Q}^A = \delta\hat{\mathcal{L}}(\overline{Q} + q)/\delta q^A$, the equations of motion related to the Lagrangian (4..28) are presented as

$$\frac{\delta\hat{\mathcal{L}}^{dyn}}{\delta q^A} = \frac{\delta}{\delta\overline{Q}^A}\left[\hat{\mathcal{L}}(\overline{Q} + q) - \overline{\hat{\mathcal{L}}}\right] = 0\,. \tag{4..29}$$

It is clear that they are equivalent to the equations (4..25) if the background equations (4..27) hold.

Defining the "background current"

$$t^q_A \equiv \frac{\delta\hat{\mathcal{L}}^{dyn}}{\delta\overline{Q}^A} = \frac{\delta\hat{\mathcal{L}}^{dyn}}{\delta q^A} - \frac{\delta}{\delta\overline{Q}^A} q^B \frac{\delta\overline{\hat{\mathcal{L}}}}{\delta\overline{Q}^B} \tag{4..30}$$

and combining this expression with (4..29) one obtains another form for the equation (4..29):

$$G^{Lq}_A + \Phi^{Lq}_A \equiv -\frac{\delta}{\delta\overline{Q}^A} q^B \frac{\delta\overline{\hat{\mathcal{L}}}}{\delta\overline{Q}^B} = t^q_A\,. \tag{4..31}$$

The left hand side in (4..31) is a linear perturbation of the expression $\delta\hat{\mathcal{L}}/\delta Q^A$ in (4..25), G^L_A is a pure gravitational part. Really gravitational variables g^a and Ψ^B have a different nature. Conservation laws are connected with symmetries of a geometry related directly to g^a, not with Ψ^B. Thus, we will consider Ψ^B as included in Φ^B, setting thus in calculations $\Psi^B = 0$ and $Q^A = \{g^a,\ \Phi^B\}$. Later we will consider the lagrangian

$$\hat{\mathcal{L}} = \hat{\mathcal{L}}_{Dg}(g^a) + \hat{\mathcal{L}}_{Dm}(g^a, \Phi^B) \tag{4..32}$$

with the pure metric gravitational part $\hat{\mathcal{L}}_{Dg}$. Then the equations (4..30) are separated into gravitational and matter parts as follows

$$G^{Lq}_a + \Phi^{Lq}_a \equiv -\frac{\delta}{\delta\overline{g}^a} q^B \frac{\delta\overline{\hat{\mathcal{L}}}}{\delta\overline{Q}^B} = t^q_a\,, \tag{4..33}$$

$$\Phi^{Lq}_C \equiv -\frac{\delta}{\delta\overline{\Phi}^C} q^B \frac{\delta\overline{\hat{\mathcal{L}}}_{Dm}}{\delta\overline{Q}^B} = t^q_C\,. \tag{4..34}$$

As is seen, the form of the perturbed equations (4..31), (4..33) and (4..34) is a quite universal form with the "background current" as a source on the right hand side. Contracting (4..33) with $2\partial\overline{g}^{a}/\overline{g}^{\mu\nu}$ one obtains

$$\hat{G}^{Lq}_{\mu\nu} + \hat{\Phi}^{Lq}_{\mu\nu} = \hat{t}^{q}_{\mu\nu} \tag{4..35}$$

where the linear operator is defined as

$$\hat{G}^{Lq}_{\mu\nu} + \hat{\Phi}^{Lq}_{\mu\nu} \equiv -2\frac{\delta}{\delta\overline{g}^{\mu\nu}}\left(\hat{l}^{\alpha\beta}_{(a)}\frac{\delta\overline{\hat{\mathcal{L}}}_{Dg}}{\delta\overline{g}^{\alpha\beta}} + q^{B}\frac{\delta\overline{\hat{\mathcal{L}}}_{Dm}}{\delta\overline{Q}^{B}}\right) \tag{4..36}$$

with independent gravitational variables $\hat{l}^{\alpha\beta}_{(a)}$ defined as in (2..81). The right hand side in (4..35) is defined by variation of (4..28)

$$\hat{t}^{q}_{\mu\nu} \equiv 2\frac{\delta\hat{\mathcal{L}}^{dyn}(\overline{Q}, q)}{\delta\overline{g}^{\mu\nu}} \tag{4..37}$$

and is the generalized symmetric energy-momentum.

The equations (4..35) generalize the equations of GR (2..24), they generalize also the Deser-Tekin equations [75, 213, 214] constructed for the quadratic theories by direct calculations. Below we will show that for the vacuum backgrounds (when $\hat{\Phi}^{Lq}_{\mu\nu} = 0$) $\overline{D}^{\mu}\hat{G}^{Lq}_{\mu\nu} \equiv 0$, and thus the energy momentum tensor $\hat{t}^{q}_{\mu\nu}$ is differentially conserved, like in the Petrov spaces (2..67) in GR. The presented here model has the same properties for expansions as was described in subsection 2.5. and the same gauge properties as was described in detail in subsection 2.6..

4.3. Currents and Superpotentials in the Field Formulation

In the recent series of the current works [75, 76, 213, 214] Deser with coauthors develop a construction of conserved charges for perturbations about vacua in metric quadratic (in curvature) gravity theories in D dimensions. They apply the Abbott and Deser procedure [68] and develop it. The aim of the present subsection is to suggest an approach, which generalizes the Deser and Tekin constructions. We construct conserved currents and superpotentials corresponding to the equations (4..35) derived for an arbitrary metrical gravitational theory, and not only on a vacuum background. Below taking into account the definition of the gravitational part of linear operator (4..36) we demonstrate that conserved quantities of the system and their properties can be obtained and described analyzing *only* the scalar density $\hat{\mathcal{L}}_1 \equiv \hat{l}^{\alpha\beta}_{(a)}(\delta\overline{\hat{\mathcal{L}}}_{Dg}/\delta\overline{g}^{\alpha\beta})$, which is the gravitational part of second term in the Lagrangian (4..28). As an important case we consider explicitly only such theories where $\hat{\mathcal{L}}_1$ has derivatives not higher than of second order, like the Einstein-Gauss-Bonnet gravity. In principle, our results can be repeated when $\hat{\mathcal{L}}_1$ has derivatives of higher orders, like in [75].

Keeping in mind that $\hat{\mathcal{L}}_1$ is the scalar density we again follow the standard technique and use the results of the subsection 4.1., which are universal. One transforms the identity $\pounds_\xi\hat{\mathcal{L}}_1 + \partial_\alpha(\xi^\alpha\hat{\mathcal{L}}_1) \equiv 0$ into the identity:

$$\overline{D}_\mu\hat{\imath}^{\mu}_1 \equiv \partial_\mu\hat{\imath}^{\mu}_1 \equiv 0 \tag{4..38}$$

where

$$\hat{\imath}_1^\mu \equiv -\left(\hat{\mathcal{L}}_1\xi^\mu + \xi^\nu\, \hat{l}^{\rho\sigma}_{(a)}\Big|^\mu_\nu \frac{\delta\overline{\hat{\mathcal{L}}}_{Dg}}{\delta\overline{\hat{g}}^{\rho\sigma}} + 2\xi^\sigma \frac{\delta\hat{\mathcal{L}}_1}{\delta\overline{g}^{\rho\sigma}}\overline{g}^{\rho\mu}\right) + \hat{\mathcal{Z}}^\mu_{(s)}\,, \tag{4..39}$$

$$\hat{\mathcal{Z}}^\mu_{(s)} \equiv 2\frac{\partial\hat{\mathcal{L}}_1}{\partial\overline{g}_{\rho\sigma,\mu\nu}}\overline{D}_\nu\zeta_{\rho\sigma} - 2\zeta_{\rho\sigma}\overline{D}_\nu\frac{\partial\hat{\mathcal{L}}_1}{\partial\overline{g}_{\rho\sigma,\mu\nu}}\,. \tag{4..40}$$

It is just the current (4..8) for the Lagrangian $\hat{\mathcal{L}}_1$.

In the case of a vacuum background one has $\delta\overline{\hat{\mathcal{L}}}_{Dg}/\delta\overline{\hat{g}}^{\rho\sigma} = 0$ and $\hat{\mathcal{L}}_1 = 0$. Then assuming arbitrary Killing vectors $\xi^\alpha = \lambda^\alpha$ one transforms the identity (4..38) into

$$\overline{D}_\mu\left(\frac{\delta\hat{\mathcal{L}}_1}{\delta\overline{g}_{\mu\nu}}\right) \equiv \overline{D}_\mu\left(\frac{\delta}{\delta\overline{g}_{\mu\nu}}\hat{l}^{\alpha\beta}_{(a)}\frac{\delta\overline{\hat{\mathcal{L}}}_{Dg}}{\delta\overline{\hat{g}}^{\alpha\beta}}\right) \equiv 0\,. \tag{4..41}$$

Recalling that for the vacuum case $\hat{\Phi}^{Lq}_{\mu\nu} = 0$ and taking into account the identity (4..41) with the definition (4..36) in the equations (4..35) one gets $\overline{D}_\mu\hat{t}^{\mu\nu}_q = 0$. This generalizes the results in [213, 75] for the quadratic theories. Recall also that in the Lagrangian (2..69) Λ-term was interpreted as "degenerated" matter, then with this assumption the identity (2..72) has been approached. However, the Lagrangian (2..69) fully is, of course, a kind of metric Lagrangians, therefore the identity (2..72) also is interpreted in the terms of the general formula (4..41).

Again, because (4..38) is the identity the current $\hat{\imath}_1^\mu$ has to be presented through a divergence of a superpotential. To construct the last we use the results of the subsection 4.1. adopted for the Lagrangian $\hat{\mathcal{L}}_1$. We set $\overline{g}_{\mu\nu} \to g_{\mu\nu}$ and construct the coefficients (4..5) - (4..6) with $\hat{\mathcal{L}}_1 = \hat{\mathcal{L}}_1(Q;\, \partial_\mu Q;\, \partial_{\mu\nu}Q) \equiv \hat{\mathcal{L}}_1^c(Q;\, \overline{D}_\mu Q;\, \overline{D}_{\mu\nu}Q)$, where $Q^A = \{\hat{l}^{\mu\nu}_a,\, g_{\mu\nu}\}$. Then we go back, $g_{\mu\nu} \to \overline{g}_{\mu\nu}$, and obtain simple expressions

$$\hat{m}_{1\sigma}{}^{\mu\nu} = 2\overline{D}_\lambda\left(\frac{\partial\hat{\mathcal{L}}_1}{\partial\overline{g}_{\rho\nu,\mu\lambda}}\right)\overline{g}_{\rho\sigma}\,, \qquad \hat{n}^{\lambda\mu\nu}_{1\sigma} = -2\frac{\partial\hat{\mathcal{L}}_1}{\partial\overline{g}_{\rho(\mu,\nu)\lambda}}\overline{g}_{\rho\sigma}\,. \tag{4..42}$$

Then, substituting these into the expression (4..13) we can define the superpotential $\hat{\mathcal{I}}^{\mu\nu}_{(s)}$, which is evidently linear in $\hat{l}^{\mu\nu}_{(a)}$. In the case of the Einstein gravity it is the superpotential (3..74). Thus, we have the identity

$$\hat{\imath}_1^\mu \equiv \partial_\nu\hat{\mathcal{I}}^{\mu\nu}_{(s)}\,, \tag{4..43}$$

which generalizes the identity (3..72) and the superpotential has the form:

$$\hat{\mathcal{I}}^{\alpha\beta}_{(s)} \equiv \left(\tfrac{2}{3}\overline{D}_\lambda\hat{n}^{[\alpha\beta]\lambda}_{1\sigma} - \hat{m}^{[\alpha\beta]}_{1\sigma}\right)\xi^\sigma - \tfrac{4}{3}\hat{n}^{[\alpha\beta]\lambda}_{1\sigma}\overline{D}_\lambda\xi^\sigma. \tag{4..44}$$

Using in (4..39) the field equations in the form (4..35) and the definition (4..36) we write out the physically real current: $\hat{\imath}_1^\mu \to \hat{\mathcal{I}}^\mu_{(s)}$:

$$\hat{\mathcal{I}}^\mu_{(s)} \equiv \hat{\mathcal{T}}^\mu_{(s)\nu}\xi^\nu + \hat{\mathcal{Z}}^\mu_{(s)} \tag{4..45}$$

where the generalized energy-momentum is

$$\hat{\mathcal{I}}^{\mu}_{(s)\nu} \equiv \left(\hat{t}^{\mu}_{q\nu} - \hat{\Phi}^{\mu}_{Lq\nu}\right) - \left(\delta^{\mu}_{\nu}\hat{l}^{\rho\sigma}_{(a)}\frac{\delta\overline{\hat{\mathcal{L}}_{Dg}}}{\delta\overline{g}^{\rho\sigma}} + \hat{l}^{\rho\sigma}_{(a)}\Big|^{\mu}_{\nu}\frac{\delta\overline{\hat{\mathcal{L}}_{Dg}}}{\delta\overline{g}^{\rho\sigma}}\right). \tag{4..46}$$

Thus finally the identity (4..43) transforms to the conservation law:

$$\hat{\mathcal{I}}^{\mu}_{(s)} = \partial_{\nu}\hat{\mathcal{I}}^{\mu\nu}_{(s)}. \tag{4..47}$$

The expressions (4..45) and (4..46) are based on the symmetrical energy-momentum (4..37) and generalize the correspondent expressions in GR: (3..68) and (3..69), therefore we choose the subscript "(s)". Thus, this subsection, generalizing the results of the subsection 3.4. develop the idea to use the identity (3..67) in GR. The expression (4..47), on the one hand, generalizes the conservation law (3..73) in GR, on the other hand, it generalizes the Deser-Tekin expressions [75, 213, 214]. In the case of a vacuum background and using the Killing vectors $\xi^{\alpha} = \lambda^{\alpha}$ one has for the current

$$\hat{\mathcal{I}}^{\mu}_{(s)} = \hat{t}^{\mu}_{q\nu}\lambda^{\nu}, \tag{4..48}$$

that also is in correspondence with the general definition (3..23) in a field theory.

4.4. Canonical Nœther and Belinfante Symmetrized Currents and Superpotentials

The expressions presented in subsection 4.1. are maximally adopted to construct both Noether canonical conserved quantities in the framework of the bimetric formulation and Belinfante corrected quantities. To construct such quantities one has to consider a pure metric part $\hat{\mathcal{L}}_{Dg}$ of the Lagrangian (4..32). At the beginning we construct the Nœther canonical quantities. We construct the corresponding to $\hat{\mathcal{L}}_{Dg}$ coefficients (4..4) - (4..6), $\hat{u}^{\alpha}_{g\sigma}$, $\hat{m}^{\alpha\beta}_{g\sigma}$, $\hat{n}^{\alpha\beta\gamma}_{g\sigma}$, with the use of which we present the related identity (4..12):

$$\hat{\imath}^{\alpha}_{g} \equiv \partial_{\beta}\hat{\imath}^{\alpha\beta}_{g}. \tag{4..49}$$

Next, following to the KBL ideology (3..41) we construct a metric Lagrangian for the perturbed system:

$$\hat{\mathcal{L}}_{DG} = \hat{\mathcal{L}}_{Dg} - \overline{\hat{\mathcal{L}}}_{Dg} + \partial_{\alpha}\hat{d}^{\alpha}. \tag{4..50}$$

Then, we apply the barred procedure to the identity (4..49) and take into account the divergence keeping in mind (4..14) - (4..18). As a result we obtain the identity corresponding to (4..50):

$$\hat{\imath}^{\alpha}_{g} - \overline{\hat{\imath}^{\alpha}_{g}} + \delta_{d}\hat{\imath}^{\alpha}_{g} \equiv \overline{D}_{\beta}\left[\hat{\imath}^{\alpha\beta}_{g}(\xi) - \overline{\hat{\imath}^{\alpha\beta}_{g}}(\xi) + \delta_{d}\hat{\imath}^{\alpha\beta}_{g}(\xi)\right]. \tag{4..51}$$

After substituting the dynamical equations (4..25) and the background equations (4..27) in the form

$$\frac{\delta\hat{\mathcal{L}}_{Dg}}{\delta g^{a}} = -\frac{\delta\hat{\mathcal{L}}_{Dm}}{\delta g^{a}}, \tag{4..52}$$

$$\frac{\delta\overline{\hat{\mathcal{L}}}_{Dg}}{\delta\overline{g}^{a}} = -\frac{\delta\overline{\hat{\mathcal{L}}}_{Dm}}{\delta\overline{g}^{a}} \tag{4..53}$$

into $\hat{u}^\alpha_{g\sigma}$ and $\overline{\hat{u}}^\alpha_{g\sigma}$ one obtains $\hat{U}^\alpha_{g\sigma}$ and $\overline{\hat{U}}^\alpha_{g\sigma}$, respectively. Then the identity (4..51) transforms into a real conservation law:

$$\hat{\mathcal{I}}^\alpha_{(c)}(\xi) = \partial_\beta \hat{\mathcal{I}}^{\alpha\beta}_{(c)}(\xi) \,. \tag{4..54}$$

The left hand side (current), in correspondence with (4..9), is

$$\hat{\mathcal{I}}^\alpha_{(c)}(\xi) \equiv \hat{\mathcal{T}}^\alpha_{\sigma(c)}\xi^\sigma + \hat{\mathcal{S}}^{\alpha\beta\rho}_{(c)}\partial_{[\beta}\xi_{\rho]} + \hat{\mathcal{Z}}^\alpha_{(c)}(\xi) \tag{4..55}$$

where the generalized canonical energy-momentum, spin and Z-term are

$$\hat{\mathcal{T}}^\alpha_{\sigma(c)} \equiv -\left[\left(\delta\hat{U}^\alpha_{g\sigma} + 2\overline{D}_\beta(\delta^{[\alpha}_\sigma \hat{d}^{\beta]})\right) + \delta\hat{n}^{\alpha\beta\gamma}_{g\lambda}\overline{R}^\lambda{}_{\beta\gamma\sigma}\right] \tag{4..56}$$

$$\hat{\mathcal{S}}^{\alpha\rho\beta}_{(c)} \equiv \delta\hat{m}^{\alpha\beta}_{g\sigma} + 2\delta^{[\alpha}_\sigma \hat{d}^{\beta]}\overline{g}^{\sigma\rho} \tag{4..57}$$

$$\hat{\mathcal{Z}}^\alpha_{(c)}(\xi) \equiv -\left(\delta\hat{z}^\alpha_g + 2\delta^{[\alpha}_\sigma \hat{d}^{\beta]}\zeta^\sigma_\beta\right) . \tag{4..58}$$

A more detailed expression for the superpotential is

$$\begin{aligned} \hat{\mathcal{I}}^{\alpha\beta}_{(c)}(\xi) &\equiv \hat{\imath}^{\alpha\beta}_g(\xi) - \overline{\hat{\imath}^{\alpha\beta}_g}(\xi) + \delta_d \hat{\imath}^{\alpha\beta}_g(\xi) \\ &= \left(\tfrac{2}{3}\overline{D}_\lambda \delta\hat{n}^{[\alpha\beta]\lambda}_\sigma - \delta\hat{m}^{[\alpha\beta]}_\sigma\right)\xi^\sigma - \tfrac{4}{3}\delta\hat{n}^{[\alpha\beta]\lambda}_{g\sigma}\overline{D}_\lambda\xi^\sigma - 2\xi^{[\alpha}\hat{d}^{\beta]} . \end{aligned} \tag{4..59}$$

Here the perturbed expressions are used:

$$\delta\hat{U}^\alpha_{g\sigma} = \hat{U}^\alpha_{g\sigma} - \overline{\hat{U}}^\alpha_{g\sigma} , \tag{4..60}$$

$$\delta\hat{m}^{\alpha\beta}_{g\sigma} = \hat{m}^{\alpha\beta}_{g\sigma} - \overline{\hat{m}}^{\alpha\beta}_{g\sigma} , \tag{4..61}$$

$$\delta\hat{n}^{\lambda\alpha\beta}_{g\sigma} = \hat{n}^{\lambda\alpha\beta}_{g\sigma} - \overline{\hat{n}}^{\lambda\alpha\beta}_{g\sigma} . \tag{4..62}$$

The perturbation $\delta\hat{z}^\alpha_g$ is defined by the the definition (4..10) and by the perturbations (4..61) and (4..62). Of course, if a displacement vector is a Killing vector in the background spacetime then $\hat{\mathcal{Z}}^\alpha_{(c)}$ disappears. In the case of GR the conservation law (4..54) goes to the KBL conservation law (3..55), if $\hat{d}^\alpha$ is defined as $\hat{k}^\alpha$ in (2..20).

The presented here procedure gives well defined currents and superpotentials in the following sense. Without changing the identity (4..7) one can add to the current an arbitrary quantity $\Delta\hat{\imath}^\alpha(\xi)$ satisfying $[\Delta\hat{\imath}^\alpha(\xi)]_{,\alpha} \equiv 0$. Analogously, without changing $\hat{\imath}^\alpha$ in (4..12) the superpotential can be added by $\Delta\hat{\imath}^{\alpha\beta}(\xi)$ with the property $[\Delta\hat{\imath}^{\alpha\beta}(\xi)]_{,\beta} \equiv 0$. However, the "broken" current and superpotential can be "restored" by the same way because the quantities $\Delta\hat{\imath}^\alpha(\xi)$ and $\Delta\hat{\imath}^{\alpha\beta}(\xi)$ are not connected *at all* with the procedure applied to the given Lagrangian in a non-explicit form. Whereas, in the sense of the procedure, the current and the superpotential in Eq. (4..12) are given by the coefficients (4..4) - (4..6) *uniquely* defined by the Lagrangian. This claim develops also the criteria by Szabados [168] who suggested to consider a connection of pseudotensors with Lagrangians "as a selection rule to choose from the mathematically possible pseudotensors". Thus one can assert that the current and superpotential in (4..54) are defined by the unique way in the above sense. Of course, the same claim is related to the KBL quantities in (3..55). They *uniquely* are defined by the Lagrangian (3..41) in the sense of the Nœther procedure.

To construct the Belinfante corrected conserved quantities for the perturbed system (4..50) we again turn to subsection 4.1.. For the coefficients $\hat{u}^{\alpha}_{g\sigma}$, $\hat{m}^{\alpha\beta}_{g\sigma}$ and $\hat{n}^{\alpha\beta\gamma}_{g\sigma}$ we construct the identity (4..20) and subtract the same barred identity

$$\hat{\imath}^{\alpha}_{gB} - \overline{\hat{\imath}^{\alpha}_{gB}} \equiv \overline{D}_{\beta}\left[\hat{\imath}^{\alpha\beta}_{gB}(\xi) - \overline{\hat{\imath}^{\alpha\beta}_{gB}}(\xi)\right]. \tag{4..63}$$

After substitution the equations (4..52) and (4..53) into the left hand side of (4..63) one gets the conservation law:

$$\hat{\mathcal{I}}^{\alpha}_{(B)}(\xi) = \partial_{\beta}\hat{\mathcal{I}}^{\alpha\beta}_{(B)}(\xi). \tag{4..64}$$

The current is

$$\hat{\mathcal{I}}^{\alpha}_{(B)} \equiv \hat{\mathcal{T}}^{\alpha}_{\sigma(B)}\xi^{\sigma} + \hat{\mathcal{Z}}^{\alpha}_{(B)}(\xi). \tag{4..65}$$

where the Belinfante corrected energy-momentum and Z-term are

$$\hat{\mathcal{T}}^{\alpha}_{\sigma(B)} \equiv -\delta\hat{U}^{\alpha}_{g\sigma} - \delta\hat{n}^{\alpha\beta\gamma}_{g\lambda}\overline{R}^{\lambda}{}_{\beta\gamma\sigma} + \overline{D}_{\beta}\delta\hat{s}^{\alpha\beta}_{g}{}_{\sigma}, \tag{4..66}$$

$$\hat{\mathcal{Z}}^{\alpha}_{(B)}(\xi) \equiv \delta\hat{z}^{\alpha}_{gB}(\xi). \tag{4..67}$$

The perturbations $\delta\hat{z}^{\alpha}_{gB}$ and $\delta\hat{s}^{\alpha\beta}_{g}{}_{\sigma}$ in (4..65) are defined by the definitions (4..22) and (4..19) and by the perturbations (4..61) and (4..62). A detailed expression for the superpotential is

$$\begin{aligned}\hat{\mathcal{I}}^{\alpha\beta}_{(B)}(\xi) &\equiv \hat{\imath}^{\alpha\beta}_{gB}(\xi) - \overline{\hat{\imath}^{\alpha\beta}_{gB}}(\xi)\\ &= 2\left(\tfrac{1}{3}\overline{D}_{\rho}\delta\hat{n}^{[\alpha\beta]\rho}_{g\sigma} + \overline{D}_{\tau}\delta\hat{n}^{\tau\rho[\alpha}_{g\lambda}\bar{g}^{\beta]\lambda}\bar{g}_{\rho\sigma}\right)\xi^{\sigma} - \tfrac{4}{3}\delta\hat{n}^{[\alpha\beta]\lambda}_{g\sigma}\overline{D}_{\lambda}\xi^{\sigma}.\end{aligned} \tag{4..68}$$

In the case of GR the conservation law (4..64) goes to the Belinfante corrected conservation law (3..78). We conclude also that in the sense of the united Nœther-Belinfante procedure the current and the superpotential in (4..64) are defined in unique way by the Lagrangian (4..50).

Now, rewrite the conservation laws (4..47), (4..54) and (4..64) in the united form:

$$\hat{\mathcal{I}}^{\alpha}_{D}(\xi) = \partial_{\beta}\hat{\mathcal{I}}^{\alpha\beta}_{D}(\xi). \tag{4..69}$$

This allows us to construct the conserved charges in generalized form in D-dimensions:

$$\mathcal{P}(\xi) = \int_{\Sigma} d^{D-1}x\,\hat{\mathcal{I}}^{0}_{D}(\xi) = \oint_{\partial\Sigma} dS_i\,\hat{\mathcal{I}}^{0i}_{D}(\xi) \tag{4..70}$$

where Σ is a spatial $(D-1)$ hypersurface $x^0 = \text{const}$ and $\partial\Sigma$ is its $(D-2)$ dimensional boundary.

4.5. The Einstein-Gauss-Bonnet Gravity. The Mass of the Schwarzschild-anti-de Sitter Black Hole

To illustrate the above theoretical results we apply them to the Einstein-Gauss-Bonnet (EGB) gravity. The action of the Einstein D-dimensional theory with a bare cosmological term Λ_0 corrected by the Gauss-Bonnet term (see, for example, [75]) is

$$\begin{aligned}S &= \int d^D x\left\{\hat{\mathcal{L}}_{EGB} + \mathcal{L}_{Dm}\right\}\\ &= \int d^D x\left\{-\frac{\sqrt{-g}}{2}\left[\kappa^{-1}(R - 2\Lambda_0) + \gamma\left(R^2_{\mu\nu\rho\sigma} - 4R^2_{\mu\nu} + R^2\right)\right] + \mathcal{L}_{Dm}\right\}\end{aligned} \tag{4..71}$$

where $\kappa = 2\Omega_{D-2}G_D > 0$ and $\gamma > 0$; G_D is the D-dimension Newton's constant. The equations of motion that follow from (4..71) are

$$\hat{\mathcal{E}}^{\mu\nu} = \hat{T}^{\mu\nu}\,. \tag{4..72}$$

The metric part is defined as

$$\begin{aligned}\hat{\mathcal{E}}^{\mu\nu} \equiv \frac{\delta}{\delta g_{\mu\nu}}\hat{\mathcal{L}}_{EGB} &\equiv \frac{\sqrt{-g}}{2}\left\{\frac{1}{\kappa}\left(R^{\mu\nu} - \tfrac{1}{2}g^{\mu\nu}R + g^{\mu\nu}\Lambda_0\right)\right.\\ &+ 2\gamma\left[RR^{\mu\nu} - 2R^{\mu}{}_{\sigma}{}^{\nu}{}_{\rho}R^{\sigma\rho} + R^{\mu}{}_{\sigma\rho\tau}R^{\sigma\nu\rho\tau} - 2R^{\mu}{}_{\sigma}R^{\sigma\nu}\right.\\ &- \left.\left.\tfrac{1}{4}g^{\mu\nu}\left(R^2_{\tau\lambda\rho\sigma} - 4R^2_{\rho\sigma} + R^2\right)\right]\right\}, \end{aligned} \tag{4..73}$$

In the vacuum case, $T^{\mu\nu} = 0$, and the equations are

$$\hat{\mathcal{E}}^{\mu\nu} = 0\,. \tag{4..74}$$

In the present subsection, as a background we consider the AdS solution, which is a solution to the equations (4..74) and is described by the metric:

$$d\overline{s}^2 = -(1+\overline{f})dt^2 + (1+\overline{f})^{-1}dr^2 + r^2\sum_{a,b}^{D-2} q_{ab}dx^a dx^b\,. \tag{4..75}$$

The last term describes $(D-2)$-dimensional sphere of the radius r, and q_{ab} depends on coordinates on the sphere only; for the other components $\overline{g}_{00} = -(1+\overline{f})$ and $\overline{g}_{11} = (1+\overline{f})^{-1}$ where

$$\overline{f}(r) = -r^2\frac{2\Lambda_{eff}}{(D-1)(D-2)}\,. \tag{4..76}$$

The background Christoffel symbols corresponding (4..75) are

$$\begin{aligned}\overline{\Gamma}^1_{00} &= \tfrac{1}{2}(1+\overline{f})\,\overline{f}', \qquad \overline{\Gamma}^0_{10} = \frac{\overline{f}'}{2(1+\overline{f})}, \qquad \overline{\Gamma}^1_{11} = -\frac{\overline{f}'}{2(1+\overline{f})},\\ \overline{\Gamma}^a_{1b} &= \frac{1}{r}\,\delta^a_b, \qquad \overline{\Gamma}^1_{ab} = -r\,(1+\overline{f})q_{ab}\,. \end{aligned} \tag{4..77}$$

The effective cosmological constant (see [75]):

$$\Lambda_{eff} = \frac{\Lambda_{EGB}}{2}\left(1 \pm \sqrt{1 - \frac{4\Lambda_0}{\Lambda_{EGB}}}\right) \tag{4..78}$$

is the solution of the equation $\Lambda^2_{eff} - \Lambda_{eff}\Lambda_{EGB} + \Lambda_{EGB}\Lambda_0 = 0$, where

$$\Lambda_{EGB} = -\frac{(D-2)(D-1)}{2\kappa\gamma(D-4)(D-3)} \tag{4..79}$$

is defined *only* by the Gauss-Bonnet term. Thus Λ_{eff} is negative, and the background Riemannian, Ricci tensors and curvature scalar are

$$\overline{R}_{\mu\alpha\nu\beta} = 2\Lambda_{eff}\frac{(\overline{g}_{\mu\nu}\overline{g}_{\alpha\beta} - \overline{g}_{\mu\beta}\overline{g}_{\nu\alpha})}{(D-2)(D-1)}, \qquad \overline{R}_{\mu\nu} = 2\Lambda_{eff}\frac{\overline{g}_{\mu\nu}}{D-2}, \qquad \overline{R} = 2\Lambda_{eff}\frac{D}{D-2}\,. \tag{4..80}$$

Then equations (4..74) are transformed into

$$\overline{R}_{\mu\nu} - \tfrac{1}{2}\overline{g}_{\mu\nu}\overline{R} + \Lambda_{eff}\overline{g}_{\mu\nu} = 0\,. \tag{4..81}$$

The Schwarzschild-AdS (S-AdS) solution [215] can be also considered as a solution to (4..74):

$$ds^2 = -(1+f)dt^2 + (1+f)^{-1}dr^2 + r^2\sum_{a,b}^{D-2} q_{ab}dx^a dx^b \tag{4..82}$$

where

$$f(r) = \frac{r^2}{2\kappa\gamma(D-3)(D-4)}\left\{1 \pm \sqrt{1 - \frac{4\Lambda_0}{\Lambda_{EGB}} + 4\kappa\gamma(D-3)(D-4)\frac{r_0^{D-3}}{r^{D-1}}}\right\}. \tag{4..83}$$

For the metrics (4..75) and (4..82) the relation $-g = -\overline{g} = r^{D-2}\det q_{ab}$ has a place and is important for future calculations. The Riemannian, Ricci tensors and curvature scalar corresponding to (4..82) are

$$\begin{aligned}
R_{0101} &= \tfrac{1}{2}f'', \qquad R_{0a0b} = \tfrac{1}{2}r(1+f)f'q_{ab}\,, \qquad R_{1a1b} = -\frac{rf'}{2(1+f)}q_{ab}\,,\\
R_{abcd} &= -r^2 f(q_{ac}q_{bd} - q_{ad}q_{bc})\,; \qquad (4..84)\\
R_{00} &= \frac{1+f}{2}\left(f'' + f'\frac{D-2}{r}\right), \qquad R_{11} = -\frac{1}{2(1+f)}\left(f'' + f'\frac{D-2}{r}\right),\\
R_{ab} &= -\left[f(D-3) + rf'\right]q_{ab}\,; \qquad (4..85)\\
R &= -\left(f'' + f'\frac{D-2}{r}\right) - \frac{D-2}{r^2}\left[f(D-3) + rf'\right]. \qquad (4..86)
\end{aligned}$$

Of course, the barred solution (4..82) goes to (4..75), and the barred expressions (4..84) - (4..86) go to (4..80). It is evidently, in the case of the solutions (4..75) and (4..82) perturbations can be described only by $\Delta f = f - \overline{f}$. In linear approximation it is

$$\Delta f = \pm\left(\sqrt{1 - \frac{4\Lambda_0}{\Lambda_{EGB}}}\right)^{-1}\left(\frac{r_0}{r}\right)^{D-3}. \tag{4..87}$$

4.5.1. The Field-Theoretical Prescription

Here, we turn to the results of subsection 4.3.. To concretize and to have a possibility to compare with [75], we define the gravitational perturbations from the set h^a, like in (2..77), as $h_{\alpha\beta} = g_{\alpha\beta} - \overline{g}_{\alpha\beta}$. Thus $\hat{\mathcal{L}}_1 = \hat{\mathcal{L}}_1^{EGB} = h_{\alpha\beta}(\delta\overline{\hat{\mathcal{L}}}_{EGB}/\delta\overline{g}_{\alpha\beta}) = h_{\alpha\beta}\overline{\hat{\mathcal{E}}}^{\alpha\beta}$, where $\overline{\hat{\mathcal{E}}}^{\alpha\beta}$ is the barred expression (4..73). Then for the AdS background (4..75) the equations (4..72) can be rewritten in the form of the equations (4..35):

$$G^{\mu\nu}_{Lq} \equiv \left[1_{(E)} - \left(1 \pm \sqrt{1 - \frac{4\Lambda_0}{\Lambda_{EGB}}}\right)_{(GB)}\right] G^{\mu\nu}_L \equiv \mp\sqrt{1 - \frac{4\Lambda_0}{\Lambda_{EGB}}}\; G^{\mu\nu}_L = \kappa t^{\mu\nu}_q \tag{4..88}$$

where subscripts $_{(E)}$ and $_{(GB)}$ are related to the Einstein and the Gauss-Bonnet part in (4..71), which are with the coefficients "κ" and "γ", respectively. The left hand side in (4..88) is calculated using (4..36) and

$$\begin{aligned} 2G_L^{\mu\nu} \equiv & - \overline{D}_\sigma \overline{D}^\sigma h^{\mu\nu} - \overline{D}^\mu \overline{D}^\nu h^\sigma_\sigma + \overline{D}_\sigma \overline{D}^\nu h^{\sigma\mu} + \overline{D}_\sigma \overline{D}^\mu h^{\sigma\nu} - \frac{4\Lambda_{eff}}{D-2} h^{\mu\nu} \\ & - \overline{g}^{\mu\nu} \left(-\overline{D}_\sigma \overline{D}^\sigma h^\rho_\rho + \overline{D}_\sigma \overline{D}_\rho h^{\sigma\rho} - \frac{2\Lambda_{eff}}{D-2} h^\sigma_\sigma \right). \end{aligned} \quad (4..89)$$

The right hand side of (4..88) generalizes the energy-momentum in (2..71). It could be presented as

$$t^q_{\mu\nu} \equiv -\frac{2}{\sqrt{-\overline{g}}} \left[\hat{\mathcal{E}}_{\mu\nu}(\overline{g}_{\alpha\beta} + h_{\alpha\beta}) - \hat{T}_{\mu\nu}(\overline{g}_{\alpha\beta} + h_{\alpha\beta}) \right] + \kappa^{-1} G^{Lq}_{\mu\nu} \quad (4..90)$$

where indexes for $\hat{\mathcal{E}}_{\mu\nu}$ and $\hat{T}_{\mu\nu}$ were lowered by $g_{\mu\nu}$, and they are thought as depending on the sum $\overline{g}_{\alpha\beta} + h_{\alpha\beta}$.

Constructing the charges for the EGB system one has to use the generalized expression (4..70). The current could be used. However, looking at (4..48), (4..73) and (4..90) one can see that it is very complicated. Evidently that the superpotential expression (4..44) is significantly simpler. We set $\hat{\mathcal{L}}_1 = \hat{\mathcal{L}}_1^{EGB}$ in (4..42) and obtain

$$\begin{aligned} \hat{m}_{1\sigma}{}^{\mu\nu} &= \mp \frac{\sqrt{-\overline{g}}}{\kappa} \sqrt{1 - \frac{4\Lambda_0}{\Lambda_{EGB}}} \overline{g}_{\sigma\rho} \overline{D}_\lambda H^{\mu(\nu\rho)\lambda}, \\ \hat{n}^{\rho\mu\nu}_{1\sigma} &= \mp \frac{\sqrt{-\overline{g}}}{\kappa} \sqrt{1 - \frac{4\Lambda_0}{\Lambda_{EGB}}} \overline{g}_{\sigma(\lambda} \delta^{(\mu}_{\pi)} H^{\nu)\pi\rho\lambda}; \\ H^{\mu\nu\rho\lambda} &\equiv h^{\mu\rho} \overline{g}^{\nu\lambda} + h^{\lambda\nu} \overline{g}^{\rho\mu} - h^{\mu\lambda} \overline{g}^{\rho\nu} - h^{\rho\nu} \overline{g}^{\mu\lambda} + h^\sigma_\sigma \left(\overline{g}^{\mu\lambda} \overline{g}^{\rho\nu} - \overline{g}^{\mu\rho} \overline{g}^{\nu\lambda} \right). \end{aligned} \quad (4..91)$$

The substitution of the quantities (4..91) into the expression (4..44) gives the superpotential related to the general EGB case:

$$\begin{aligned} \hat{\mathcal{I}}^{\mu\rho}_{(s)} \equiv & \pm \frac{\sqrt{-\overline{g}}}{\kappa} \sqrt{1 - \frac{4\Lambda_0}{\Lambda_{EGB}}} \times \\ & \times \left(\xi^{[\mu} \overline{D}_\nu h^{\rho]\nu} - \xi_\nu \overline{D}^{[\mu} h^{\rho]\nu} - \xi^{[\mu} \overline{D}^{\rho]} h - h^{\nu[\mu} \overline{D}^{\rho]} \xi_\nu - \tfrac{1}{2} h \overline{D}^{[\mu} \xi^{\rho]} \right). \end{aligned} \quad (4..92)$$

It is expressed through the Abbott-Deser superpotential in the Einstein theory [68, 75], $\hat{I}^{\mu\rho}_{AD}$, as $\hat{\mathcal{I}}^{\mu\rho}_{(s)} = \mp\sqrt{1 - 4\Lambda_0/\Lambda_{EGB}}\, \hat{I}^{\mu\rho}_{AD}$. Recently, developing the results [75] Deser with co-authors [76] have reached the same expression. Paddila [216] reffering to [75] also analyzed it. Keeping in mind that different definitions for the metric perturbations h^a can be used we find that in fact the superpotential (4..92) belongs to the set

$$\hat{\mathcal{I}}^{\mu\rho}_{(s)} \equiv \frac{1}{\kappa} \left[1_{(E)} - \left(1 \pm \sqrt{1 - \frac{4\Lambda_0}{\Lambda_{EGB}}} \right)_{(GB)} \right] \left(\hat{l}^{\sigma[\mu}_{(a)} \overline{D}_\sigma \xi^{\rho]} + \xi^{[\mu} \overline{D}_\sigma \hat{l}^{\rho]\sigma}_{(a)} - \bar{D}^{[\mu} \hat{l}^{\rho]}_{(a)\sigma} \xi^\sigma \right) \quad (4..93)$$

where the Einstein part defined in D dimensions formally coincides with (3..74) in GR.

For the S-AdS solution (4..82) considered as perturbations with respect to the AdS spacetime (4..75) one has in linear approximation:

$$h_{00} = h^{11} = -\Delta f \approx \mp \left(\sqrt{1 - \frac{4\Lambda_0}{\Lambda_{EGB}}} \right)^{-1} \left(\frac{r_0}{r} \right)^{D-3} . \tag{4..94}$$

To calculate the total conserved energy of the S-AdS solution we use the formula (4..70) where we substitute (01)-component of the superpotential (4..92). The last is calculated with (4..94) and the time-like Killing vector $\lambda^\mu = (-1, \mathbf{0})$; covariant derivatives are defined by (4..77). Finally (4..70) gives:

$$E = \frac{(D-2) r_0^{D-3}}{4G_D} \tag{4..95}$$

that is the result of [75], and is the standard accepted result obtained with using the various approaches (see [216] - [225] and references therein).

4.5.2. The Canonical Nœther Mass

The results of this subsubsection, in fact, repeat the results by Deruelle, Katz and Ogushi [182]. However, there is a difference between the approach in [182] and our approach in [119]. In [182] conserved quantities are constructed in the way, when the charges for the physics system and for the background system are constructed separately. Only after that they are compared and differences are interpreted as charges for perturbations. On the other hand, in [119] the conserved charges from the starting are constructed in the perturbed form. This approach is in the spirit of the present paper and is presented explicitly here in significantly more detail than in [119].

To construct global conserved quantities in the above prescription we again turn to the generalized integral (4..70), only this time with the superpotential (4..59). Then it is necessary to calculate the perturbations (4..61) and (4..62). For this one has to use the general formulae (4..5) and (4..6) with the metric Lagrangian of the EGB gravity in (4..71). Also the expressions in D-dimensions:

$$\Delta^\alpha_{\mu\nu} = \Gamma^\alpha_{\mu\nu} - \overline{\Gamma}^\alpha_{\mu\nu} = \tfrac{1}{2} g^{\alpha\rho} \left(\overline{D}_\mu g_{\rho\nu} + \overline{D}_\nu g_{\rho\mu} - \overline{D}_\rho g_{\mu\nu} \right) , \tag{4..96}$$

$$\overline{D}_\rho g_{\mu\nu} = g_{\tau\mu} \Delta^\tau_{\rho\nu} + g_{\tau\nu} \Delta^\tau_{\rho\mu} , \tag{4..97}$$

$$R^\lambda{}_{\tau\rho\sigma} = \overline{D}_\rho \Delta^\lambda_{\tau\sigma} - \overline{D}_\sigma \Delta^\lambda_{\tau\rho} + \Delta^\lambda_{\rho\eta} \Delta^\eta_{\tau\sigma} - \Delta^\lambda_{\eta\sigma} \Delta^\eta_{\tau\rho} + \overline{R}^\lambda{}_{\tau\rho\sigma} \tag{4..98}$$

are useful. For calculation of the superpotential (4..59) we need in antisymmetric part of (4..61) only. Thus after prolonged calculations we have

$$\begin{aligned}
& \hat{m}^{[\alpha\beta]}_{g\sigma} = {}_{(E)}\hat{m}^{[\alpha\beta]}_\sigma + {}_{(GB)}\hat{m}^{[\alpha\beta]}_\sigma \\
= \; & \frac{\sqrt{-g}}{2\kappa} \left\{ 2\Delta^{[\alpha}_{\sigma\rho} g^{\beta]\rho} + \delta^{[\alpha}_\sigma \Delta^{\beta]}_{\rho\tau} g^{\rho\tau} \right\} \\
+ \; & \gamma\sqrt{-g} \left\{ 2R_\sigma{}^{\rho\tau[\alpha} \Delta^{\beta]}_{\rho\tau} - 6R_\tau{}^{[\alpha\beta]\rho} \Delta^\tau_{\sigma\rho} + 6R^{\rho[\alpha\beta]}{}_\sigma \Delta^\tau_{\tau\rho} - 4R^{\rho[\alpha} \Delta^{\beta]}_{\sigma\rho} + 2R^\rho_\sigma \Delta^{[\alpha}_{\rho\tau} g^{\beta]\tau} \right. \\
+ \; & 4R^{\rho\tau} \delta^{[\alpha}_\sigma \Delta^{\beta]}_{\rho\tau} - 6R^{\rho[\alpha} g^{\beta]\tau} \Delta^\pi_{\rho\tau} g_{\sigma\pi} - 12R^{[\alpha}_\rho g^{\beta]\tau} \Delta^\rho_{\sigma\tau} - 8R^{[\alpha}_\sigma g^{\beta]\rho} \Delta^\tau_{\tau\rho} + 2R\Delta^{[\alpha}_{\sigma\rho} g^{\beta]\rho} \\
+ \; & \left. R\Delta^{[\alpha}_{\rho\tau} \delta^{\beta]}_\sigma g^{\rho\tau} + 3R\Delta^\tau_{\tau\rho} \delta^{[\alpha}_\sigma g^{\beta]\rho} - 2g_{\sigma\rho} g^{\tau[\alpha} \overline{D}_\tau R^{\beta]\rho} - \delta^{[\alpha}_\sigma g^{\beta]\rho} \overline{D}_\rho R \right\} .
\end{aligned} \tag{4..99}$$

The expression for (4..6) is more simple:

$$
\begin{aligned}
\hat{n}^{\lambda\alpha\beta}_{g\sigma} &= {}_{(E)}\hat{n}^{\lambda\alpha\beta}_{\sigma} + {}_{(GB)}\hat{n}^{\lambda\alpha\beta}_{\sigma} = \\
&= \frac{\sqrt{-g}}{2\kappa}\left\{g^{\alpha\beta}\delta^{\lambda}_{\sigma} - g^{\lambda(\alpha}\delta^{\beta)}_{\sigma}\right\} \\
&+ \gamma\sqrt{-g}\left\{-2R_{\sigma}{}^{(\alpha\beta)\lambda} - 4R^{\lambda}_{\sigma}g^{\alpha\beta} + 4R^{(\alpha}_{\sigma}g^{\beta)\lambda} + R\left(g^{\alpha\beta}\delta^{\lambda}_{\sigma} - g^{\lambda(\alpha}\delta^{\beta)}_{\sigma}\right)\right\} \quad (4..100)
\end{aligned}
$$

For calculation of the superpotential (4..59) we need in the next part of (4..100) [1]:

$$
\begin{aligned}
\hat{n}^{[\alpha\beta]\lambda}_{g\sigma} &= {}_{(E)}\hat{n}^{[\alpha\beta]\lambda}_{\sigma} + {}_{(GB)}\hat{n}^{[\alpha\beta]\lambda}_{\sigma} \\
&= \frac{3\sqrt{-g}}{4\kappa}\delta^{[\alpha}_{\sigma}g^{\beta]\lambda} + \frac{3\gamma\sqrt{-g}}{2}\left\{R_{\sigma}{}^{\lambda\alpha\beta} + 4g^{\lambda[\alpha}R^{\beta]}_{\sigma} + \delta^{[\alpha}_{\sigma}g^{\beta]\lambda}R\right\}. \quad (4..101)
\end{aligned}
$$

Perturbations of (4..99) and (4..101) are obtained following the recommendation given in (4..61) and (4..62):

$$
\delta\,\hat{m}^{[\alpha\beta]}_{g\sigma} = \delta[{}_{(E)}\hat{m}^{[\alpha\beta]}_{\sigma}] + \delta[{}_{(GB)}\hat{m}^{[\alpha\beta]}_{\sigma}]\,, \quad (4..102)
$$

$$
\delta\,\hat{n}^{[\alpha\beta]\lambda}_{g\sigma} = \delta[{}_{(E)}\hat{n}^{[\alpha\beta]\lambda}_{\sigma}] + \delta[{}_{(GB)}\hat{n}^{[\alpha\beta]\lambda}_{\sigma}]\,. \quad (4..103)
$$

We do not present explicit expressions for these perturbations because they are evident due to (4..99) and (4..101). We note that only the two last terms in (4..99) contribute into $\overline{m}^{[\alpha\beta]}_{g\sigma}$ because $\overline{\hat{\Delta}\gamma}_{\alpha\beta} \equiv 0$.

Recall that in the canonical prescription one has to define a divergence in the Lagrangian (4..50). We exactly follow the recommendation in [182]. In our notations it is the divergence of $\hat{d}^{\lambda} = \hat{n}^{\lambda\alpha\beta}_{g\sigma}\Delta^{\sigma}_{\alpha\beta}$. In the GR case this leads to the choice of KBL, that is to $\hat{k}^{\alpha}$ in (3..40) that leads to variation of the action under Dirichlet boundary conditions. A more detail discussion on a choice of a divirgence one can found in [182]. Thus, we define

$$
\begin{aligned}
\hat{d}^{\lambda} &= \left({}_{(E)}\hat{n}^{\lambda\alpha\beta}_{\sigma} + {}_{(GB)}\hat{n}^{\lambda\alpha\beta}_{\sigma}\right)\Delta^{\sigma}_{\alpha\beta} = \frac{\sqrt{-g}}{2\kappa}\left(\Delta^{\lambda}_{\alpha\beta}g^{\alpha\beta} - \Delta^{\alpha}_{\alpha\beta}g^{\lambda\beta}\right) \\
&+ \gamma\sqrt{-g}\left(-2R_{\sigma}{}^{\alpha\beta\lambda} + 4R^{\alpha}_{\sigma}g^{\beta\lambda} - 4R^{\lambda}_{\sigma}g^{\alpha\beta} + \delta^{\lambda}_{\sigma}g^{\alpha\beta} - \delta^{\alpha}_{\sigma}g^{\beta\lambda}\right)\Delta^{\sigma}_{\alpha\beta}\,. \quad (4..104)
\end{aligned}
$$

Substitution of (4..102), (4..103) and (4..104) into (4..59) gives the canonical Nœther superpotential $\hat{\mathcal{I}}^{\alpha\beta}_{(c)} = \hat{\mathcal{I}}^{\alpha\beta}_{(c)E} + \hat{\mathcal{I}}^{\alpha\beta}_{(c)GB}$ for perturbations in EGB gravity. Its Einstein part formally coincides with the KBL superpotential (3..54):

$$
\hat{\mathcal{I}}^{\alpha\beta}_{(c)E} = \frac{1}{\kappa}(\hat{g}^{\rho[\alpha}\overline{D}_{\rho}\xi^{\beta]} + \hat{g}^{\rho[\alpha}\Delta^{\beta]}_{\rho\sigma}\xi^{\sigma} - \overline{D^{[\alpha}\hat{\xi}^{\beta]}}) - 2\xi^{[\alpha}\hat{d}^{\beta]}_{E}\,. \quad (4..105)
$$

only one has to keep in mind that it is presented in D-dimensions [182].

To calculate the mass of the S-AdS black hole we use the formulae (4..75) - (4..87) for background and perturbed systems. We use the component $\hat{\mathcal{I}}^{01}_{(c)}$ in the general formula (4..70) under the requirement $r \to \infty$. For this it is enough to calculate $\hat{\mathcal{I}}^{01}_{(c)}$ in linear approximation with respect to the perturbation Δf in (4..87). Next, for calculating the mass

[1] In [119] an analogous to (4..100) expression has a missprint.

we again need in the Killing vector $\lambda^\alpha = \{-1, \mathbf{0}\}$. In linear approximation the symbols (4..96) are

$$\Delta^1_{00} = -\frac{1}{2r}\Delta f \overline{f}(D-5)\,, \;\; \Delta^0_{10} = -\frac{1}{2r}\frac{\Delta f}{\overline{f}}(D-1)\,,$$
$$\Delta^1_{11} = \frac{1}{2r}\frac{\Delta f}{\overline{f}}(D-1)\,, \;\; \Delta^1_{ab} = -r\,\Delta f q_{ab}\,. \tag{4..106}$$

In calculations the simple relations $(\Delta f)' = -(D-3)\Delta f/r$, $(\overline{f})' = 2\overline{f}/r$ are used. Then in linear approximation the Einstein part (4..105) gives

$$\hat{\mathcal{I}}^{01}_{(c)E} = -\frac{\sqrt{-\overline{g}}}{2\kappa r}\Delta f(D-2)\,. \tag{4..107}$$

The contribution from the Gauss-Bonnet part of (4..59) we write out in part. Thus in linear approximation

$$-\delta[_{(GB)}\hat{m}^{[01]}_\sigma]\lambda^\sigma = \frac{\gamma\sqrt{-\overline{g}}}{2r^3}\Delta f \overline{f}(D-2)(D-3)\,[3+2(D-5)]\,. \tag{4..108}$$

The contribution from (4..103) is

$$\tfrac{2}{3}\lambda^\sigma \overline{D}_\lambda \delta[_{(GB)}\hat{n}^{[01]\lambda}_\sigma] - \tfrac{4}{3}\delta[_{(GB)}\hat{n}^{[01]\lambda}_\sigma]\overline{D}_\lambda\lambda^\sigma = -\frac{3\gamma\sqrt{-\overline{g}}}{2r^3}\Delta f \overline{f}(D-2)(D-3)\,. \tag{4..109}$$

At last, the contribution from the Gauss-Bonnet part of (4..104) is

$$-2\lambda^{[0}\hat{d}^{1]}_{GB} = \frac{\gamma\sqrt{-\overline{g}}}{r^3}\Delta f \overline{f}(D-2)(D-3)\,. \tag{4..110}$$

Summing (4..108) - (4..110) we obtain the Gauss-Bonnet part of the superpotential (4..59) in linear approximation

$$\hat{\mathcal{I}}^{01}_{(c)GB} = \frac{\sqrt{-\overline{g}}}{2\kappa r}\Delta f(D-2)\left[1 \pm \sqrt{1-\frac{4\Lambda_0}{\Lambda_{EGB}}}\right]\,. \tag{4..111}$$

The definitions (4..76) - (4..79) were used. Summing (4..107) and (4..111) we obtain the full 01-component of the EGB canonical superpotential necessary for calculating the mass of the S-AdS black hole:

$$\hat{\mathcal{I}}^{01}_{(c)} = \frac{\sqrt{-\overline{g}}}{2\kappa r}\left(\frac{r_0}{r}\right)^{D-3}(D-2). \tag{4..112}$$

The definition (4..87) was used. Substitution of (4..112) into (4..70) gives again the standard result (4..95).

4.5.3. The Belinfante Corrected Mass

Comparing the Belinfante corrected superpotential (4..68) and the canonical one (4..59) we find that for constructing the first we need (together with (4..101)) in the expression

$\hat{n}_{g\lambda}^{\tau\rho[\alpha}\overline{g}^{\beta]\lambda}\overline{g}_{\rho\sigma}$ instead of (4..99) and (4..104). Thus in EGB gravity

$$\begin{aligned}
&\hat{n}_{g\lambda}^{\tau\rho[\alpha}\overline{g}^{\beta]\lambda}\overline{g}_{\rho\sigma} = \\
&\frac{\sqrt{-g}}{4\kappa}\left(2g^{\rho[\alpha}\overline{g}^{\beta]\tau}\overline{g}_{\rho\sigma} - g^{\tau[\alpha}\delta^{\beta]}_{\sigma}\right) + \frac{\gamma\sqrt{-g}}{2}\left\{2R_\lambda{}^{\rho\tau[\alpha} - 2R^{\rho\tau}{}_\lambda{}^{[\alpha} - 8R^\tau_\lambda g^{\rho[\alpha}\right.\\
+ \quad & \left. 4R^\rho_\lambda g^{\tau[\alpha} + 4g^{\rho\tau}R^{[\alpha}_\lambda + R\left(2\delta^\tau_\lambda g^{\rho[\alpha} - \delta^\rho_\lambda g^{\tau[\alpha}\right)\right\}\overline{g}^{\beta]\lambda}\overline{g}_{\rho\sigma}\,.
\end{aligned} \tag{4..113}$$

As is seen, it significantly simplify the expression (4..68) with respect to (4..59). Perturbation of (4..113) is again obtained following the recommendation given in (4..62):

$$\delta[\hat{n}_{g\lambda}^{\tau\rho[\alpha}\overline{g}^{\beta]\lambda}\overline{g}_{\rho\sigma}] = \delta[_{(E)}\hat{n}_{\lambda}^{\tau\rho[\alpha}\overline{g}^{\beta]\lambda}\overline{g}_{\rho\sigma}] + \delta[_{(GB)}\hat{n}_{\lambda}^{\tau\rho[\alpha}\overline{g}^{\beta]\lambda}\overline{g}_{\rho\sigma}]\,. \tag{4..114}$$

Combining (4..103) and (4..114) in the definition (4..68) we obtain the Belinfante corrected superpotential $\hat{\mathcal{I}}^{\alpha\beta}_{(B)} = \hat{\mathcal{I}}^{\alpha\beta}_{(B)E} + \hat{\mathcal{I}}^{\alpha\beta}_{(B)GB}$ in EBG gravity. Its Einstein part

$$\hat{\mathcal{I}}^{\alpha\beta}_{(B)E} = \frac{1}{\kappa}\left(\hat{l}^{\rho[\alpha}\overline{D}_\rho\xi^{\beta]} + \xi^{[\alpha}\overline{D}_\rho\hat{l}^{\beta]\rho} - \bar{D}^{[\alpha}\hat{l}^{\beta]}_{\rho}\xi^\rho\right) \tag{4..115}$$

formally coincides with the superpotential (3..65), only one has to keep in mind that it is presented in D-dimensions.

To calculate the mass of the S-AdS black hole we use again the formulae (4..75) - (4..87) for background and perturbed systems. We use the component $\hat{\mathcal{I}}^{01}_{(B)}$ in the general formula (4..70) under the requirement $r \to \infty$ and for $\lambda^\alpha = \{-1, \mathbf{0}\}$. We calculate a linear approximation of $\hat{\mathcal{I}}^{01}_{(B)}$ in Δf. The (01)-component of the Einstein part (4..115) gives the same result as in (4..107). The contribution from the Gauss-Bonnet part of (4..68) we again write out in part. Thus in linear approximation

$$2\lambda^\sigma\overline{D}_\tau\delta[_{(GB)}\hat{n}_{\lambda}^{\tau\rho[0}\overline{g}^{1]\lambda}\overline{g}_{\rho\sigma}] = \frac{\gamma\sqrt{-\overline{g}}}{2r^3}\Delta f\overline{f}(D-2)(D-3)\left[3 + 2(D-4)\right]\,. \tag{4..116}$$

Summing (4..109) and (4..116) we obtain the Gauss-Bonnet part of the (01)-component of the superpotential (4..68) in linear approximation that is exactly (4..111). Finally, thus, adding the Einstein part we again approach the standard result (4..95).

4.6. Discussion

The results of subsections 4.2. and 4.3., first, generalize the corresponding results in GR presented in sections 2. and 3., second, they generalize the Deser and Tekin approach [75] in quadratic theories that is particulary presented in the framework of EGB gravity in subsection 4.5.1.. Thus really we answer the criticism in [216] related to the Deser and Tekin approach. Indeed, in subsections 4.2. and 4.3. we have extended the approach in [75] to perfectly arbitrary backgrounds, which can be without symmetries at all, and arbitrary displacement vectors (no necessity in Killing vectors) can be used.

The results of subsection 4.4. generalize the corresponding results in GR presented in section 3.. Also, our canonical Nœther expressions generalize the Deruelle, Katz and Ogushi results [182] in EGB gravity, which in fact are presented in subsection 4.5.2., but only in a preferred here perturbed form. The Belinfante corrected expressions both in a general form in subsection 4.4. and in application to EGB gravity in subsection 4.5.3. are quite new.

A choice for more suitable and successful expressions for conserved quantities is not so simple problem even in the framework of GR. Therefore, only we discuss general proposals for the generalized superpotentials (4..44), (4..59) and (4..68) from a general viewpoint. However, a relation to GR, and to EGB gravity (subsection 4.5.), of course, could be useful.

Let us turn to the superpotentials of the family (4..44) with the substitution of (4..42). They differ from each other starting from the second order for different $\hat{l}^{\mu\nu}_{(a)}$, like the superpotentials (3..74) in GR. The choice $h^a \to \hat{l}^{\mu\nu} = \hat{g}^{\mu\nu} - \overline{\hat{g}}^{\mu\nu}$ in GR was based on the two points (see the end of subsection 3.5.). First, only for this choice the superpotential from (3..74) gives the standard energy-momentum and its flux at null infinity for the Bondi-Sachs solution. Second, only the superpotential related to $\hat{l}^{\mu\nu}$ from the family (3..74) coincides with the Belinfante correction of the KBL superpotential. We do not know a test model for a choice of perturbations h^a in the family (4..44) in EBG gravity. For example, an arbitrary choice h^a for the solution (4..82) leads to the same result (4..95) if we use a correspondent superpotential from the family (4..93) instead of (4..92). Indeed, for this calculations only the linear order is crucial, but it is the same for all of them. Considering the Belinfante corrected expression (4..68) and the family (4..44) we see that, in general, they cannot be identified. Indeed, the superpotentials (4..44) are linear in $\hat{l}^{\mu\nu}_{(a)}$ for an arbitrary metric theory, whereas (4..68) does not. However many modified D-dimensional gravities, as a rule, are the Einstein theories with corrections (quadratic and others, like string and M-theory corrections). Among such theories the Lanczos-Lovelock theory (in particulary, EGB gravity) plays an important role. Then, comparing the Einstein parts *only* (in the EGB gravity they are (4..115) and the one in (4..93)) we again can prefer $\hat{l}^{\mu\nu}$ from the set $\hat{l}^{\mu\nu}_{(a)}$. Of course, it is not an absolute choice, but could be a recommendation.

Comparing (4..44), (4..59) and (4..68) we remark that the first is in the framework of the field-theoretical approach, where perturbations are examined explicitly. At the same time, the expressions (4..59) and (4..68) are presented rather as bimetric ones. Moreover, the expression (4..44), being a linear one, is significantly simpler. Next, the superpotential (4..59), as a canonical quantity, essentially depends on a choice of a divergence in the Lagrangian. But for an arbitrary metric theory, as we know, there is no a crucial principle for the definition of such a divergence; there are only reasonable recommendations (see [182] and subsection 4.5.). On the other hand, expressions (4..44) and (4..68) do not depend on divergences at all, that could be an advantage. Comparing (4..59) and (4..68) we can also note that the Belinfante corrected expressions are simpler significantly. This is demonstrated clearly by the corresponding expression in EGB gravity and by the calculation of mass of the S-AdS black hole in subsection 4.5.3..

The applications in subsection 4.5. related to calculation of mass of the S-AdS black hole in EGB gravity are, in fact, the test of the presented here approach. Recalling the papers [216] - [225] where the S-AdS mass was calculated we note that all of them are in agreement giving the acceptable result, although they use different approaches. In the most of the papers the definitions are based on the first law of the black hole thermodynamics and the connection with the surface terms of Hamiltonian dynamics. The surface terms are defined following the recommendation by Regge and Teitelboim [190]. On the other hand, considering the AdS background as an arena for perturbations one can connect conserved charges with AdS symmetries expressed by Killing vectors and define them in the quite classical form. Such approach was used in [182] and in just the present paper. In [224] the charges are associated to the diffeomorphism symmetries of Lovelock gravities in any odd dimensions. In [225] asymptotic symmetries of AdS spacetime were used with employing the conformal completion technique.

Sometimes it is important to interpret some special situations with the use of developing approaches. In the works [76, 216, 221, 226, 227] such a case, when

$$\Lambda_{EGB} = 4\Lambda_0 , \qquad (4..117)$$

is remarked. Indeed, this case looks as degenerated one. The reason is that the condition (4..117) leads to a zero coefficient at the linear approximation of the EGB gravity equations around the AdS background. Applying the generalized KBL approach in EGB gravity [182] Deruelle and Morisava [226, 227] have found that mass and angular momentum expressions for the Kerr-AdS solution in EGB gravity have the same coefficient and have to be defined only as zero. In the framework of the field-theoretical formalism we approach the analogous conclusion. Turn to the equations (4..88). Under the condition (4..117) their linear left hand side disappears: the Einstein part is canceled by the Gauss-Bonnet part. Consequently the total energy-momentum for perturbations at the right hand side disappears also. This means that the matter energy-momentum is compensated by the energy-momentum of the metric perturbations. In the vacuum case the energy-momentum of the metric perturbations is equal to zero. Thus the conserved integrals have to be equal to zero, like the charges in [226, 227]. This situation on the AdS background is similar to the definition of zero energy of the S-dS black hole in a so-called Nariari space in [222]. The last is interpreted as a real ground state whose energy is lower than energy of the pure dS space. To escape the difficulties with the situation (4..117) Paddila [216] using his approach suggests to use a modified background without all the symmetries, unlike AdS one. All three the approaches presented in the present paper have this possibility also.

In future we plan to check the formulae of the subsection 4.5. applying them to the Kerr-AdS solution in EGB gravity keeping in mind, firstly, like in [226, 227], that asymptotically in linear approximation it has to go to the solution in the Einstein D-dimensional theory [228, 229, 230], secondly, checking the just now appeared exact solutions [231, 232]. Moreover, the expressions related both to the field-theoretical and to the canonical Nœther prescriptions already were used and discussed for examination of higher D Kerr-AdS spacetimes [76, 226, 227, 230] including EGB gravity. Planning to check the formulae in EGB gravity of subsections 4.5.1. and 4.5.2. for the Kerr-AdS solution we anticipate that the results will be identical with the ones in [76, 226, 227, 230]. Applications of the Belinfante corrected expressions of subsection 4.5.3. for calculating mass and angular momentum for the Kerr-AdS black hole in EGB gravity have to be testable.

Gravitational theories in D dimensions are very intensively developed. New solutions with interesting properties appear following one by one (see, for example, [233, 234]). To understand and describe their properties, which could be quite unusual, many of new solutions need in calculating conserved charges. Therefore all of known and new possibilities to define and calculate conserved quantities, if they are non-contradictive and satisfy all the acceptable tests, could be interesting for future studies.

Acknowledgments

Author is very grateful to Chiang-Mei Chen, Nathalie Deruelle, Stanley Deser, Jacek Jezierski, Sergei Kopeikin, James Nester, László Szabados, Brian Pitts, Bayram Tekin and Shwetketu Vibhadra for their useful conversations, comments, remarks related to the problems of this review and explanations of their works. Author expresses a special gratitude to Leonid Grishchuk and Alla Popova, the earlier collaboration with whom became a basis of the review; and to coauthors Joseph Katz, Stephen Lau and Deepak Baskaran for very fruitful collaboration in last years. Author also is grateful to Deepak Baskaran for a big help in checking English, and to Maria Grigorian and Michael Prokhorov for help in preparing figures.

References

[1] Landau L D and Lifshitz E M 1975 *The Classical Theory of Fields*, (Oxford, Pergamon)

[2] Fock V A 1964 *The Theory of Space, Time and Gravitation* (Oxford, Pergamon)

[3] Misner Ch W, Thorne K S and Wheller J A 1973 *Gravitation* (San Francisco, Freeman)

[4] DeWitt B S 1965 *Dynamical Theory of Groups and Fields* (New York, Gordon and Breach)

[5] Lifshitz E M 1946 On a gravitational stability of an expanding world *J. of Phys., Moscow* 10 116 [*Zh. Eksp. Teor. Fiz.* 16 587 (1946)]

[6] Mukhanov V F, Feldman H A and Brandenberger R H 1992 Theory of cosmological perturbations *Phys. Rep.* 215 203

[7] Lukash V N 2006 On the relation between tensor and scalar perturbation modes in Friedmann cosmology *Phys. Usp.* 49 103 [2006 *Usp. Fiz. Nauk* 176 113] *(Preprint astro-ph/0610312)*

[8] Grishchuk L P 1974 An amplification of gravitational waves in isotropic world *Zh. Eksp. Teor. Fiz.* 67 825

[9] Lukash V N 1980 The birth of sound waves in an isotropic universe *Zh. Eksp. Teor. Fiz.* 79 1601

[10] Bardeen J M 1980 Gauge invariant cosmological perturbations *Phys. Rev.* D22 1882

[11] Khalatnikov I M, Kamenshchik A Yu and Starobinsky A A 2002 Comment about quasi-isotropic solution of Einstein equations near cosmological singularity *Class. Quantum Grav.* 19, 3845 *(Preprint gr-qc/0204045)*

[12] Khalatnikov I M, Kamenshchik A Yu, Martellini M and Starobinsky A A 2003 Quasi-isotropic solution of the Einstein equations near a cosmological singularity for a two-fluid cosmological model *JCAP*, 0303 001 *(Preprint gr-qc/0301119)*

[13] Chernin A D 2002 Physical vacuum and cosmic coincidence problem *New Astron.*, 7, 113 *(Preprint astro-ph/0107071)*

[14] Bartolo N, Komatsu E, Matarrese S and Riotto A 2004 Non-Gaussianity from inflation: Theory and observations *Phys. Rep.*, 402 103 *(Preprint astro-ph/0406398)*

[15] Grishchuk L P 2001 Relic gravitational waves and their detection *Lect. Notes Phys.* 562, 167 *(Preprint gr-qc/0002035)*

[16] Grishchuk L P 2005 Relic gravitational waves and cosmology *Phys.Usp.* 48, 1235 [2005 *Usp. Fiz. Nauk* 175 1289] *(Preprint gr-qc/0504018)*

[17] Dolgov A D, Sazhin M V and Zel'divich Ya B 1990 *Basics of Modern Cosmology* (Editions Frontiéres, France).

[18] Kiefer C, Polarski D and Starobinsky A A 2000 Entropy of gravitons produced in the early Universe *Phys. Rev. D* 62 043518 *(Preprint gr-qc/9910065)*

[19] Starobinsky A A, Tsujikawa S and Yokoyama J 2001 Cosmological perturbations from multi-field inflation in generalized Einstein theories *Nucl. Phys. B* 610 383 *(Preprint astro-ph/0107555)*

[20] Kiefer C, Polarski D and Starobinsky A A 2006 Pointer states for primordial fluctuations in inflationary cosmology *(Preprint astro-ph/0610700)*

[21] Lukash V N 1996 The very early Universe, in Australian: *Cosmology: The physics of the universe* eds. Robson B A et al. (World Scientific) 213 [in Barsilian: *Cosmology and gravitation, II.* ed. Novello M (Editions Frontieres) 288] *(Preprint astro-ph/9910009)*

[22] Linde A D 2005 Inflation and String Cosmology *J. Phys. Conf. Ser.* 24 151 *(Preprint hep-th/0503195)*

[23] Linde A D 2005 Inflation and String Cosmology *Contemp. Concepts Phys.* 5 1 *(Preprint hep-th/0503203)*

[24] Chernin A D 2001 Cosmic vacuum *Uspekhi Fiz. Nauk* 171 1153

[25] Chernin A D, Santiago D I and Silbergleit A S 2002 Interplay between gravity and quintessence: A set of new GR solutions *Phys. Lett. A* 294 79 *(Preprint astro-ph/0106144)*

[26] Chandrasekhar S 1983 *The Mathematical Theory of Black Holes* (Oxfrod Univ. Press, New York)

[27] Kokkotas K D and Schmidt B G 1999 Quasi-Normal Modes of Stars and Black Holes *Living Rev. Relativity* 2 2 *(Preprint gr-qc/9909058)*

[28] Sarbach O and Tiglio M 2001 Gauge invariant perturbations of Schwarzschild black holes in horizon-penetrating coordinates *Phys. Rev. D* 64 084016 *(Preprint gr-qc/0104061)*

[29] Glampedakis K and Andersson N 2001 Scattering of scalar waves by rotating black holes *Class. Quant. Grav.* 18 1939 *(Preprint gr-qc/0102100)*

[30] Karkowski J, Malec E and Świerczyński Z 2002 Backscattering of electromagnetic and gravitational waves off Schwarzschild geometry *Class. Quantum Grav.* 19 953 (2002) *(Preprint gr-qc/0105042)*

[31] Karlovini M 2002 Axial perturbations of general spherically symmetric spacetimes *Class. Quantum Grav.* 19 2125 *(Preprint gr-qc/0111066)*

[32] Karkowski J, Roszkowski K, Świerczyński Z and Malec E 2003 Waves in Schwarzschild spacetimes: How strong can imprints of the spacetime curvature be *Phys. Rev. D* 67 064024 *(Preprint gr-qc/0210041)*

[33] Frolov V P and Lee H K 2005 Observable form of pulses emitted from relativistic collapsing objects *Phys. Rev. D* 71 044002 *(Preprint gr-qc/0412124)*

[34] Grishchuk L P, Lipunov V M, Postnov K A, Prokhorov M E and Sathyaprakash B S 2001 Gravitational wave astronomy: In anticipation of first sources to be detected *Phys. Usp.* 44, 1 [2001 *Usp. Fiz. Nauk* 171, 3] *(Preprint astro-ph/0008481)*

[35] Alekseev G A 1987 The exact solutions in general relativity *Transactions of the Steklov institute* 176, 211, in Russian

[36] Alekseev G A 2005 Monodromy-data parameterization of spaces of local solutions of integrable reductions of Einstein's field equations *Theor. Math. Phys.* 143 720 [2005 *Teor. Mat. Fiz.* 143 278] *(Preprint gr-qc/0503043)*

[37] Alekseev G A and Griffiths J B 2000 Infinite hierarchies of exact solutions of the Einstein and Einstein-Maxwell equations for interacting waves and inhomogeneous cosmologies *Phys. Rev. Lett.* 84 5247 *(Preprint gr-qc/0004034)*

[38] Alekseev G A and Griffiths J B 2001 Solving the characteristic initial value problem for colliding plane gravitational and electromagnetic waves *Phys. Rev. Lett.* 87, 221101 *(Preprint gr-qc/0105029)*

[39] Alekseev G A and Griffiths J B 2004 Collision of plane gravitational and electromagnetic waves in a Minkowski background: solution of the characteristic initial value problem *Class. Quantum Grav.* 21 5623 *(Preprint gr-qc/0410047)*

[40] Rubakov V A and Shaposhnikov M E 1983 Do we live inside a domain wall? *Phys. Lett. B* 125 136

[41] Akama K 2000 An early proposal of "Brane World" *(Preprint hep-ph/0001113)*

[42] Randall L and Sundrum R 1999 A large mass hierarchy from a small extra dimension *Phys. Rev. Lett.* 83 3370 *(Preprint hep-ph/9905221)*

[43] Randall L and Sundrum R 1999 An alternative to compactification *Phys. Rev. Lett.* 83 4690 *(Preprint hep-th/9906064)*

[44] Rubakov V A 2001 Large and infinite extra dimensions *Phys. Usp.* 44 871 [2001 *Usp. Fiz. Nauk* 171 913] *(Preprint hep-ph/0104152)*

[45] Kodama H, Ishibashi A and Seto O 2000 Brane world cosmology — gauge-invariant formalism for perturbation *Phys. Rev. D* 62 064022 *(Preprint hep-th/0004160)*

[46] Van de Bruck C, Dorca M, Brandenberger R H and Lukas A 2000 Cosmological perturbations in brane-world theories: Formalism *Phys. Rev. D* 62 123515 *(Preprint hep-th/0005032)*

[47] Deruelle N, Dolezel T and Katz J 2001 Perturbations of brane worlds *Phys. Rev. D* 63 083513 *(Preprint hep-th/0010215)*

[48] Gorbunov D S, Rubakov V A and Sibiryakov S M 2001 Gravity waves from inflating brane or mirrors moving in AdS_5 *JHEP* 0110 015 *(Preprint hep-th/0108017)*

[49] Nojiri S, Odintsov S D and Ogushi S 2002 Friedmann-Robertson-Walker brane cosmological equations from the five-dimensional bulk (A)dS black hole *Int. J. Mod. Phys. A* 17 4809 (*Preprint* hep-th/0205187)

[50] Binetruy P, Bucher M and Carvalho C 2004 Models for the brane-bulk interaction: Toward understanding braneworld cosmological perturbation *Phys. Rev. D* 70 043509 *(Preprint hep-th/0403154)*

[51] Libanov M V and Rubakov V A 2005 Lorentz-violating brane worlds and cosmological perturbations *Phys. Rev. D* 72 123503 *(Preprint hep-th/0509148)*

[52] Einstein A 1915 On general theory of relativity *Sitzungsber. preuss. Akad. Wiss.* 44 778

[53] Einstein A 1915 On general theory of relativity (Comment) *Sitzungsber. preuss. Akad. Wiss.* 46 799

[54] Einstein A 1915 Equations of gravitational field *Sitzungsber. preuss. Akad. Wiss.* 48 844

[55] Einstein A 1916 The Hamiltonian principle and general theory of relativity *Sitzungsber. preuss. Akad. Wiss.* 2 1111

[56] Einstein A 1916 Aproximate integration of equations of gravitational field *Sitzungsber. preuss. Akad. Wiss.* 1 688

[57] Einstein A 1918 On gravitational waves *Sitzungsber. preuss. Acad. Wiss.* 1 154

[58] Einstein A 1918 An energy conservation law in general theory of relativity *Sitzungsber. preuss. Akad. Wiss.* 1 448

[59] Einstein A 1918 A remark to the work by E Schrödinger "Components of energy of gravitational field" *Phys. Z.* 19 115

[60] Szabados L B 2004 Quasi-local energy-momentum and angular momentum in GR: A review article *Living Rev. Relativity* 7 4 [Online article: cited on 7 June 2004, http://www.livingreviews.org/lrr-2004-4]

[61] York J W 1980 Energy and momentum of the gravitational field *Essays in General Relativity* ed. Tipler F J (Academic, New York, 1980) 39

[62] Ashtekar A 1980 Asymptotic structure of the gravitational field at spatial infinity *General Relativity and Gravitation* ed. Held A (Plenum, New York) 2 37

[63] Ashtekar A, Bicak J and Schmidt B G 1997 Asymptotic structure of symmetry reduced general relativity *Phys. Rev. D* 55 669 *(Preprint gr-qc/9608042)*

[64] Chruściel P T, Jezierski J and MacCallum M A H 1998 Uniqueness of the Trautman-Bondi mass *Phys. Rev. D* 58 084001 *(Preprint gr-qc/9803010)*

[65] Dain S and Friedrich H 2001 Asymptotically flat initial data with prescribed regularity at infinity *Commun. Math. Phys.* 222 569 *(Preprint gr-qc/0102047)*

[66] Beig R and Schmidt B G 2000 Time-independent gravitational fields *Lect. Notes Phys.* 540 325 *(Preprint gr-qc/0005047)*

[67] Chruściel P T, Jezierski J and Kijowski J 2001 Hamiltonian field theory in the radiating regime *Lect. Notes Phys.* 570 (Springer, Berlin, Heidelberg, New York)

[68] Abbott L F and Deser S 1982 Stability of gravity with a cosmological constant *Nucl. Phys. B* 195 76

[69] Ashtekar A and Das S 2000 Asymptotically anti-de Sitter space-times: Conserved quantities *Class. Quantum Grav.* 17 L17 *(Preprint hep-th/9911230)*

[70] Schoen R and Yau S-T 1979 On the proof of the positive mass conjecture in general relativity *Commun. Math. Phys.* 65 45

[71] Schoen R and Yau S-T 1981 Proof of the positive mass theorem. II *Commun. Math. Phys.* 79 231

[72] Witten E 1981 A new proof of the positive energy theorem *Commun. Math. Phys.* 80 381

[73] Nester J M 1981 A new gravitational energy expression with a simple positive proof *Phys. Lett. A* 83 241

[74] Faddeev L D 1982 The problem of energy in the Einstein gravitational theory *Uspekhi Fiz. Nauk* 136 433

[75] Deser S and Tekin B 2003 Energy in generic higher curvature gravity theories *Phys. Rev. D* 67 084009 *(Preprint hep-th/0212292)*

[76] Deser S, Kanik I and Tekin B 2005 Conserved charges in higher D Kerr-AdS spacetimes *Class. Quantum Grav.* 22 3383 *(Preprint gr-qc/0506057)*

[77] Kofinas G and Olea R 2006 Vacuum energy in Einstein-Gauss-Bonnet AdS gravity *Phys. Rev. D* 74 084035 *(Preprint hep-th/0606253)*

[78] Miskovic O and Olea R 2006 On boundary conditions in three-dimensional AdS gravity *Phys. Lett. B* 640 101 *(Preprint hep-th/0603092)*

[79] Brown J D and York J W 1993 Quasilocal energy and conserved charges derived from the gravitational action *Phys. Rev. D* 47 1407 *(Preprint gr-qc/9209012)*

[80] Lau S R 1993 Canonical variables and quasilocal energy in general relativity *Class. Quantum Grav.* 10 2379 *(Preprint gr-qc/9307026)*

[81] Lau S R 1995 Spinors and the reference point of quasilocal energy *Class. Quantum Grav.* 12 1063 *(Preprint gr-qc/9409022)*

[82] Lau S R 1996 On the canonical reduction of spherically symmetric gravity *Class. Quantum Grav.* 13 1509 *(Preprint gr-qc/9508028)*

[83] Brown J D, Lau S R and York J W 1997 Energy of isolated systems at retarded times as the null limit of quasilocal energy *Phys. Rev. D* 55 1977 *(Preprint gr-qc/9609057)*

[84] Brown J D, Lau S R and York J W 1999 Canonical quasilocal energy and small spheres *Phys. Rev. D* 59 064028 *(Preprint gr-qc/9810003)*

[85] Lau S R 1999 Lightcone reference for total gravitational energy *Phys. Rev. D* 60 104034 *(Preprint gr-qc/9903038)*

[86] Brown J D, Lau S R and York J W 2002 Action and energy of the gravitational field *Ann. Phys.* 297 175 *(Preprint gr-qc/0010024)*

[87] Arnowitt R, Deser S and Misner C W 1962 The dynamics of general relativity *Gravitation: an Introduction to Current Research* ed. Witten L (Wiley, New York) 227

[88] Kijowski J and Tulczyiew W M 1979 A symplectic framework for field theories *Lect. Notes Phys.* 107 (Springer, Berlin)

[89] Jezierski J and Kijowski J 1990 The localization of energy in gauge field theories and in linear gravitation *Gen. Relat. Grav.* 22 1283

[90] Kijowski J 1997 A simple derivation of canonical structure in quasilocal Hamiltonian in general relativity *Gen. Relat. Grav.* 29 307

[91] Jezierski J 1999 Energy and angular momentum of the weak gravitational waves on the Schwarzschild background — quasi-local gauge-invariant formulation *Gen. Relat. Grav.* 31 1855 *(Preprint gr-qc/9801068)*

[92] Nester J M 1991 Covariant Hamiltonian for gravity theories *Mod. Phys. Lett. A* 6 2655

[93] Chen C-M, Nester J M and Tung R-S 1995 Quasilocal energy-momentum for geometric gravity theories *Phys. Lett. A* 203, 5 *(Preprint gr-qc/9411048)*

[94] Chen C-M and Nester J M 1999 Quasilocal quabtities fo GR and other gravity theories *Class. Quantum Grav.* 16 1279 *(Preprint gr-qc/9809020)*

[95] Chang C-C, Nester J M and Chen C-M 1999 Pseudotensors and quasilocal energy-momentum *Phys. Rev. Lett.* 83 1897 *(Preprint gr-qc/9809040)*

[96] Chen C-M and Nester J M 2000 A simplectic Hamiltonian derivation of quasilocal energy-momentum for GR *Grav. Cosmol.* 6 257 *(Preprint gr-qc/0001088)*

[97] Nester J M 2004 General pseudotensors amd quasilocal quantities *Class. Quantum Grav.* 21 S261

[98] Chen C-M, Nester J M and Tung R-S 2005 The Hamiltonian boundary term and quasilocal energy flux *Phys. Rev. D* 72 104020 *(Preprint gr-qc/0508026)*

[99] Deser S 1970 Self-interaction and gauge invariance *Gen. Relat. Grav.* 1 9 *(Preprint gr-qc/0411023)*

[100] Grishchuk L P, Petrov A N and Popova A D 1984 Exact theory of the (Einstein) gravitational field in an arbitrary background space-time *Commun. Math. Phys.* 94 379

[101] Grishchuk L P and Petrov A N 1986 Closed worlds as gravitational fields *Sov. Astron. Lett.* 12 179 [1986 *Pis'ma Astron. Zh.* 12, 429]

[102] Grishchuk L P and Petrov A N 1987 The Hamiltonian description of the gravitational field and gauge symmertries *Sov. Phys.: JETP* 65 5 [1986 *Zh. Eksp. Teor. Fiz.* 92, 9]

[103] Petrov A N and Popova AD 1987 On exact dynamic theories acting on a given background *Vestnik Mosk. Univ. Fiz. Astron.* 28 No 6 13

[104] Popova A D and Petrov A N 1988 The dynamic theories on a fixed background in gravitation *Int. J. Mod. Phys. A* 3 2651

[105] Petrov A N 1990 New harmonic coordinates for the Schwarzschild geometry *Vestnik Mosk. Univ. Fiz. Astron.* 31 No 5 88

[106] Petrov A N 1992 New harmonic coordinates for the Schwarzschild geometry and the field approach *Astronom. Astrophys. Trans.* 1 195

[107] Petrov A N 1993 General relativity from 'localization' of Killing vector fields *Class. Quantum Grav.* 10 2663

[108] Popova A D and Petrov A N 1993 Nonlinear quantum mechanics with nonclassical gravitational self-interaction. II. Nonstationary situation *Intern. J. Mod. Phys. A* 8 2683

[109] Popova A D and Petrov A N 1993 Nonlinear quantum mechanics with nonclassical gravitational self-interaction. III. Related topics *Intern. J. Mod. Phys. A* 8 2709

[110] Petrov A N and Popova A D 1994 Associated length and inflation in quantum mechanics with gravitational self-interaction *Intern. J. Mod. Phys. D* 3 461

[111] Petrov A N and Popova A D 1994 The associated length and inflation in quantum mechanics with gravitational coupling *Gen. Relat. Grav.* 26 1153

[112] Petrov A N 1995 Asymptotically flat spacetimes at spatial infinity: The field approach and the Lagrangian description *Int. J. Mod. Phys. D* 4 451

[113] Petrov A N 1997 Asymptotically flat spacetimes at spatial infinity: II. Gauge invariance of the integrals of motion in the field approach *Int. J. Mod. Phys. D* 6 239

[114] Petrov A N and Narlikar J V 1996 The energy distribution for a spherically symmetric isolated system in general relativity *Found. Phys.* 26 1201; Erratum 1998 *Found. Phys.* 28 1023

[115] Petrov A N and Katz J 1999 Conservation laws for large perturbations on curved backgrounds *Fundamental Interactions: From Symmetries to Black Holes* eds. Frere J M et al. (Universite de Bruxelles, Belgium) 147 *(Preprint gr-qc/9905088)*

[116] Petrov A N and Katz J 2002 Conserved currents, superpotentials and cosmological perturbations *Proc. R. Soc. A, London* 458 319 *(Preprint gr-qc/9911025)*

[117] Baskaran D, Lau S R and Petrov A N 2003 Center of mass integral in canonical general relativity *Ann. Phys.* 307 90 *(Preprint gr-qc/0301069)*

[118] Petrov A N 2004 Perturbations in the Einstein theory of gravity: Conserved currents *Mosc. Univ. Phys. Bull.* 59 No 1 24 [2004 *Vestnik Mosk. Univ. Fiz. Astron.* No 1 18] *(Preprint gr-qc/0402090)*

[119] Petrov A N 2004 Conserved currents in D-dimensional gravity and brane cosmology *Mosc. Univ. Phys. Bull.* 59 No 2 11 [2004 *Vestnik Mosk. Univ. Fiz. Astron.* No 2 10] *(Preprint gr-qc/0401085)*

[120] Petrov A N 2005 The Schwarzschild black hole as a point particle *Found. Phys. Lett.* 18 477 *(Preprint gr-qc/0503082)*

[121] Petrov A N 2005 A note on the Deser-Tekin charges *Class. Quantum Grav.* 22 L83 *(Preprint gr-qc/0504058)*

[122] Babak S V and Grichshuk L P 2000 The energy-momentum tensor for the gravitational field *Phys. Rev. D* 61 24038 *(Preprint gr-qc/9907027)*

[123] Babak S V and Grichshuk L P 2003 Finite-range gravity and its role in gravitational waves, black holes and cosmology *Int. J. Mod. Phys. D* 12 1905 *(Preprint gr-qc/0209006)*

[124] Petrov A N 2004 The field theoretical formulation of general relativity and gravity with non-zero masses of gravitons *Searches for a mechanism of gravity* eds. Ivanov M A and Savrov L A (Nizhny Novgorod, Nickolaev Publisher) 230 [a compressed

version: 2004 *Proceedings to PIRT-IX, London* ed. Duffy M 2 433] *(Preprint gr-qc/0505058)*

[125] Pinto-Neto N and Silva R R 2000 Generalized field theoretical approach to general relativity and conserved quantities in anti-de Sitter spacetimes *Phys. Rev. D* 61 104002 *(Preprint gr-qc/0005101)*

[126] Kopeikin S, Ramirez J, Mashhoon B and Sazhin M 2001 Cosmological perturbations: A new gauge-invariant approach *Phys. Lett. A* 292 173 *(Preprint gr-qc/0106064)*

[127] Ramirez J and Kopeikin S 2002 A decoupled system of hyperbolic equations for linearized cosmological perturbations *Phys. Lett. B* 532, 1 *(Preprint gr-qc/0110071)*

[128] Pitts J B and Schieve W C 2001 Slightly Bimetric Gravitation *Gen. Relat. Grav.* 33 1319 *(Preprint gr-qc/0101058)*

[129] Pitts J B and Schieve W C 2001 Null cones in Lorentz-covariant general relativity *(Preprint gr-qc/0111004)*

[130] Pitts J B and Schieve W C 2004 Null cones and Einstein's equations in Minkowski spacetime *Found. Phys.* 34 211 *(Preprint gr-qc/0406102)*

[131] Pitts J B and Schieve W C 2003 Nonsingularity of flat Robertson-Walker models in the special relativistic approach to Einstein's equations *Found. Phys.* 33 1315 *(Preprint gr-qc/0406103)*

[132] Zeldovich Ya B and Grishchuk L P 1986 Gravity, general relativity and altrernative theories *Sov. Phys. Usp.* 29 780 [1986 *Usp. Fiz. Nauk* 149 695]

[133] Grishchuk L P 1990 The general theory of relativity: Familiar and unfamiliar *Sov. Phys. Usp.* 33 669 [1990 *Usp. Fiz. Nauk* 160, 147]

[134] Grishchuck L P 1992 Gravity-wave astronomy: Some mathematical aspects *Carrent Topics in Astrofundamental Physics* eds. Sanches N et al. (World Scientific) 435

[135] Barnebey T A 1974 Gravitational waves: the nonlinearized theory *Phys. Rev D* 10 1741

[136] Rosen N 1940 General relativity and flat space *Phys. Rev.* 57 147

[137] Katz J 1985 A note on Komar's anomalous factor *Class. Quantum Grav.* 2 423

[138] Eisenkhart L P 1933 *Continuous Groups of Transformations* (Princeton Univ. Press, Princeton)

[139] Schouten J A 1951 *Tensor Analysis for Physicists* (Claredon Press, Oxford)

[140] Mitzkevich N V 1969 *Physical Fields in General Theory of Relativity* (Nauka, Moscow), in Russian.

[141] Bruni M, Matarrese S, Mollerqash S and Sonego S 1997 Perturbations of spacetime: Gauge transformations and gauge invariance at second order and beyond *Class. Quantum Grav.* 14 2585 *(Preprint gr-qc/9609040)*

[142] Abramo L R W, Branderberg R H and Mukhanov V F 1997 Energy-momentum tensor for cosmological perturbations *Phys. Rev. D* 56 3248 *(Preprint gr-qc/9704037)*

[143] Petrov A Z 1969 *Einstein Spaces* (Pergamon Press, London)

[144] Deser S 1987 Gravity from self-interaction in a curved background *Class. Quantum Grav.* 4 99

[145] Boulware D C and Deser S 1975 Classical general relativity derived from quantum gravity *Ann. Phys.* 89 193

[146] Trautman A 1962 Conservation laws in general relativity *Gravitation: an Introduction to Current Research*, ed. Witten L (Wiley, New York)

[147] Tolman R C 1934 *Relativity, Thermodynamics and Cosmology* (Oxford, Oxford Univ. Press)

[148] Von Freud Ph 1939 Über die ausdrücke der gesamtenergie und des gesamtimpulses eins materiellen systems in der allgemeinen relativitätstheorie *Ann. of Math.* 40 417

[149] Bergmann P G 1949 Non-linear field theories *Phys. Rev.* 75 680

[150] Goldberg J N 1958 Conservation laws in general relativity *Phys. Rev.* 111 315

[151] Møller C 1958 On the localization of the energy of a physical system in the general theory of relativity *Ann. Phys.* 4 347

[152] Papapetrou A 1948 Einstein's theory of gravitation and flat space *Proc. R. Irish Ac.* 52 11

[153] Weinberg S 1972 *Gravitation and Cosmology* (Wiley, New York)

[154] Komar A 1959 Covariant conservation laws in general relativity *Phys. Rev.* 113 934

[155] Garecki J 2001 Remarks on the Bergmann-Thomsom expression on angular momentum in general relativity *Grav. Cosmol.* 7 131 *(Preprint gr-qc/0102091)*

[156] Rosen N and Virbhadra K S 1993 Energy and momentum of cylindrical gravitational waves *Gen. Relat. Grav.* 25 429

[157] Virbhadra K S 1995 Energy and momentum of cylindrical gravitational waves. II *Pramana J. Physics* 45 215 *(Preprint gr-qc/9509034)*

[158] Favata M 2001 Energy localization invariance of tidal work in general relativity *Phys. Rev. D* 63 064013 *(Preprint gr-qc/0008061)*

[159] Aguirregabiria J M, Chammorro A and Virbhadra K S 1996 Energy and angular momentum of charged rotating black holes *Gen. Relat. Grav.* 28 1393 *(Preprint gr-qc/9501002)*

[160] Virbhadra K S 1999 Naked singularities and Seifert's conjecture *Phys. Rev. D* 60 104041 *(Preprint gr-qc/9809077)*

[161] Radinschi I 2000 Energy of a conformal scalar dyon black hole *Mod. Phys. Lett. A* 15 2171 *(Preprint gr-qc/0010094)*

[162] Radinschi I 2000 The energy distribution of the Bianchi type I universe *Acta. Phys. Slov.* 50 609 *(Preprint gr-qc/0008034)*

[163] Radinschi I 2001 Energy distribution of a charged regular black hole *Mod. Phys. Lett. A* 16, 673 *(Preprint gr-qc/0011066)*

[164] Radinschi I 2005 On the Moller energy-momentum complex of the Melvin magnetic universe *Fizika B* 14 311 *(Preprint gr-qc/0202075)*

[165] Xulu S S 2000 Moller energy for the Kerr-Newman metric *Mod. Phys. Lett. A* 15 1511 *(Preprint gr-qc/0010062)*

[166] Xulu S S 2003 Moller energy of the nonstatic spherically symmetric metrics *Astrophys. Space Sci.* 283 23 *(Preprint gr-qc/0010068)*

[167] Belinfante F J 1939 On the spin angular momentum and mesons *Physica* 6 887

[168] Szabados L B 1991 Canonical pseudotensor Sparling's form and Noether currents *Preprint:* KFKI-1991-29/B.

[169] Berezin V T 1992 Phenomenological foundations for a theory of gravity *Teor. Mat. Fiz.* 93 154

[170] Szabados L B 1992 On canonical pseudotensor Sparling's form and Noether currents *Class. Quantum Grav.* 9 2521

[171] Borokhov V 2002 Belinfante tensors induced by matter-gravity couplings *Phys. Rev. D* 65 125022 *(Preprint hep-th/0201043)*

[172] Traschen J 1985 Constraints on stress-energy perturbations in general relativity *Phys. Rev. D* 31 283

[173] Traschen J and Eardley D M 1986 Large-scale anisotropy of the cosmic background radiation in Friedmann universes *Phys. Rev. D* 34 1665

[174] Veeraraghavan S and Stebbin A 1990 Causal compensated perturbations in cosmology *Ap. J.* 365 37

[175] Uzan J P, Deruelle N and Turok N 1998 Conservation laws and cosmological perturbations in curved universes *Phys. Rev. D* 57 7192 *(Preprint gr-qc/9805020)*

[176] Katz J, Bičák J and Lynden-Bell D 1997 Relativistic conservation laws and integral constraints for large cosmological perturbations *Phys. Rev. D* 55 5957 *(Preprint gr-qc/0504041)*

[177] Chruściel P T 1985 On the relation between the Einstein and the Komar expressions for the energy of the gravitational field *Ann. Inst. Henri Poincaré* 42 267

[178] Katz J, Lynden-Bell D and Israel W 1988 Quasilocal energy in static gravitational fields *Class. Quantum Grav.* 5 971

[179] Katz J and Ori A 1990 Localization of field energy *Class. Quantum Grav.* 7 787

[180] Katz J and Lerer D 1997 On global conservation laws at null infinity *Class. Quantum Grav.* 14 2297 *(Preprint gr-qc/9612025)*

[181] Deruelle N, Katz J and Uzan J P 1997 Integral constraints on cosmological perturbations and their energy *Class. Quantum Grav.* 14 421 *(Preprint gr-qc/9608046)*

[182] Deruelle N, Katz J and Ogushi S 2004 Conserved charges in Einstein Gauss-Bonnet theory *Class. Quantum Grav.* 21 1971 *(Preprint gr-qc/0310098)*

[183] Julia B and Silva S 1998 Currents and superpotentials in classical invariant theories. I. Local results with applications to perfect fluids and general relativity *Class. Quantum Grav.* 15 2173 *(Preprint gr-qc/9804029)*

[184] Silva S 1999 On superpotentials and charge algebras of gauge theories *Nucl. Phys. B* 558, 391 *(Preprint hep-th/9809109)*

[185] Bondi H, Metzner A W K and Van der Berg M J C 1962 Gravitational waves in general relativity. VII. Waves from axi-symmetrical isolated systems *Proc. R. Soc. A London* 269 21

[186] Soloviev V O 1985 The generator algebra of asymptotic Poincré group in general relativity *Teor. Mat. Fiz.* 65 400

[187] Bartnic R 1986 The mass of an asymptotically flat manifold *Commun. Pure Appl. Math.* 39 661

[188] ÓMurchadha N 1986 Total energy-momentum in general realtivity *J. Math. Phys.* 27 2111

[189] Chruściel P T 1987 On angular momentum at spatial infinity *Class. Quantum Grav.* 4 L205

[190] Regge T and Teitelboim T 1974 Role of surface integrals in the Hamiltonian formulation *Ann. Phys.* 88 286

[191] Beig R and ÓMurchadha N 1987 The Poincaré group as the symmetry group of canonical general relativity *Ann. Phys.* 174 463

[192] Szabados L B 2003 On the roots of the Poincaré structure of asymptotically flat spacetimes *Class. Quantum Grav.* 20 2627

[193] Nester J M, Ho F-H and Chen C-M 2003 Quasilocal center-of-mass for teleparallel gravity *In the proceedings of the 10th Marcel Grossman meeting* (Rio de Janeiro, 2003) *(Preprint gr-qc/0403101)*

[194] Nester J M, Meng F-F and Chen C-M 2004 Quasilocal center-of-mass *J. Korean Phys. Soc.* 45 S22 *(Preprint gr-qc/0403103)*

[195] Ashtekar A and Hansen R O 1978 A unified treatment of null and spatial infinity in general relativity. I. Universal structure of asymptotic symmetries and conserved quantities at spatial infinity *J. Math. Phys.* 19 1542

[196] Joshi P S, Vaz C and Witten L 2004 A time-like naked singularity *Phys. Rev. D* 70 084038 *(Preprint gr-qc/0410041)*

[197] Fiziev P The gravitational field of massive non-charged point source in general relativity *(Preprint gr-qc/0412131)*

[198] Goswami R and Joshi P S A resolution of spacetime singularity and black hole paradoxes through avoidance of trapped surface formation in Einstein gravity *(Preprint gr-qc/0504019)*

[199] Golubev M B and Kelner S R 2005 Point charge self-energy in the general relativity *Int. J. Mod. Phys. A* 20 2288 *(Preprint gr-qc/0504097)*

[200] Narlikar J V 1985 Some conceptual problems in general relativity and cosmology *In A Random Walk in Relativity and Cosmology* eds N Dadhich et al. (Viley Eastern Limited, New Delhi, 1985) 171

[201] Damour T Jaranowski P and Schäfer G 2001 Dimensional regularization of the gravitational interaction of point masses *Phys. Lett. B* 513 147 *(Preprint gr-qc/0105038)*

[202] Schaefer G 2003 Binary black holes and gravitational wave production: Post-newtonian analytic treatment *Current Trends in Relativistic Astrophysics*, eds: Fernendez-Jambrina L and Gonzolez-Romero L M (Springer's *Lecture Notes in Physics*, Vol. 617 2003) 195

[203] Gelfand I M and Shilov G E 1964 *Generalized functions. Vol. 1. Properties and Operations* (Academic Press, New York)

[204] Bertschinger E 1996 Cosmological dynamics *In Cosmology and Large Scale Structures* eds Schaefer R, Silk J, Spiro M and zin-Justin J (North Holland, Amsterdam) 273

[205] Fulton Y, Rohrlich F and and Witten L 1976 Conformal invariance in physics *Rev. Mod. Phys.* 34 442

[206] Penrose R and Rindler W 1988 *Spinors and Space-Time: Spinor and Twistor Methods in Space-Time Geometry* (Cambridge Univ. Press, Cambridge)

[207] Keane A J and Barrett R K 2000 *Class. Quantum Grav.* 17 201 *(Preprint gr-qc/9907002)*

[208] Bronnikov K A, Grebeniuk M A, Ivashchuk V D and Melnikov V N 1997 Integrable multidimensional cosmology for intersecting p-branes *Grav. Cosmol.* 3 105 *(Preprint gr-qc/9709006)*

[209] Grebeniuk M A, Ivashchuk V D and Melnikov V N 1998 Multidimensional cosmology for intersecting p-branes with static internal spaces *Grav. Cosmol.* 4 145 *(Preprint gr-qc/9804042)*

[210] Ivashchuk V D and Melnikov V N 2000 Billiard representation for multidimensional cosmology with intersecting p-branes near the singularity *J. Math. Phys.* 41 6341 *(Preprint hep-th/9904077)*

[211] Bronnikov K A, Dehnen H and Melnikov V N 2003 On a general class of brane-world black holes *Phys. Rev. D*, 68 024025 *(Preprint gr-qc/0304068)*

[212] Kopeikin S and Vlasov I 2004 Parametrized post-Newtonian theory of reference frames, multipolar expansions and equations of motion in the N-body problem *Phys. Rep.* 400 209 *(Preprint gr-qc/0403068)*

[213] Deser S and Tekin B 2002 Gravitational energy in quadratic curvature gravities *Phys. Rev. Lett.* 89 101101 *(Preprint hep-th/0205318)*

[214] Deser S and Tekin B 2003 Energy in topologically massive gravity *Class. Quantum Grav.* 20 L259 *(Preprint gr-qc/0307073)*

[215] Boulware D C and Deser S 1985 String-generated gravity models *Phys. Rev. Lett.* 55 2656

[216] Paddila A 2003 Surface terms and Gauss-Bonnet Hamiltonian *Class. Quantum Grav.* 20 3129 (*Preprint* gr-qc/0303082)

[217] Jacobson T and Myers R C 1993 Entropy of Lovelock black holes *Phys. Rev. Lett.* 70 3684 *(Preprint hep-th/9305016)*

[218] Banados M, Teitelboim C and Zanelli J 1993 Black hole entropy and the dimensional continuation of the Gauss-Bonnet theorem *(Preprint gr-qc/9309026)*

[219] Louko J, Simon J Z and Winter-Hilt S N 1996 Hamiltonian thermodynamics of a Lovelock black hole *(Preprint gr-qc/9610071)*

[220] Nojiri S, Odintsov S D and Ogushi S 2002 Cosmological and black hole brane-world Universes in higher derivative gravity *Phys. Rev. D* 65 023521 *(Preprint hep-th/0108172)*

[221] Cai R-G 2002 Gauss-Bonnet black holes in AdS spaces *Phys. Rev. D* 65 084014 *(Preprint hep-th/01092133)*

[222] Cvetic M, Nojiri S and Odintzov S D 2002 Black hole thermodynamics and negative entropy in de Sitter and anti-de Sitter Einstein-Gauss-Bonnet gravity *Nucl. Phys. B* 628 295 *(Preprint hep-th/0112045)*

[223] Cho Y M and Neupane I P 2002 Anti-de Sitter black holes, thermal phase trnsition and holography in higher curvature gravity *Phys. Rev. D* 66 024044 *(Preprint hep-th/0202140)*

[224] Allemandi G, Francaviglia M and Raiteri M Charges and energy in Chern-Simons theories and Lovelock gravity *(Preprint gr-qc/0308019)*

[225] Okuyama N and Koga J-I 2005 Asymptotically anti-de Sitter spacetimes and conserved quantities in higher curvature gravitational theories *Phys. Rev. D* 71 084009 *(Preprint hep-th/0501044)*

[226] Deruelle N and Morisawa 2005 Mass and angular momenta of Kerr anti-de Sitter spacetimes in Einstein-Gauss-Bonnet theory *Class. Quantum Grav.* 22 933 (*Preprint* gr-qc/0411135)

[227] Deruelle N 2005 Mass and angular momentum of a Kerr-anti-de Sitter spacetimes (*Preprint* gr-qc/0502072)

[228] Gibbons G W, Lu H, Page D N and Pope C N 2005 The General Kerr-de Sitter Metrics in All Dimensions *J. Geom. Phys.* 53 49 *(Preprint hep-th/0404008)*

[229] Gibbons G W, Perry M J and Pope C N 2005 The first law of thermodynamics for Kerr-anti-de Sitter black holes *Class. Quantum Grav.* 22 1503 *(Preprint hep-th/0408217)*

[230] Deruelle N and Katz J 2005 On the mass of a Kerr-anti-de Sitter spacetime in *D* dimensions *Class. Quantum Grav.* 22 421 (*Preprint* gr-qc/0411035)

[231] Alexeyev S, Popov N, Startseva M, Barrau and Grain J 2007 Kerr-Gauss-Bonnet black holes: Exact analitical solution (*Preprint* arXiv:0712.3546[gr-qc])

[232] Brihaye Y and Radu E 2008 Five-dimensional rotating black holes in Einstein-Gauss-Bonnet theory (*Preprint* arXiv:0801.1021[hep-th])

[233] Aliev A N 2006 A slowly rotating charged black hole in five dimensions *Mod. Phys. Lett. A* 21 751 (*Preprint* gr-qc/0505003)

[234] Aliev A N 2007 Electromagnetic Properties of Kerr-Anti-de Sitter Black Holes (*Preprint* hep-th/0702129)

In: Classical and Quantum Gravity Research
Editors: M.N. Christiansen et al, pp. 161-168
ISBN 978-1-60456-366-5

Chapter 3

THE EUCLIDEAN PATH INTEGRAL IN QUANTUM GRAVITY

Arundhati Dasgupta
Department of Mathematics and Statistics,
University of New Brunswick.
Fredericton E3B 5A3

Abstract

We discuss the progress in the path-integral approach to quantum gravity, particularly in the context of the divergence of the Euclidean classical action which is unbounded from below. We show that the effective action in the path-integral can be made positive definite by isolating the trace part from the measure.

1. Introduction

One of the ways of 'quantising' any classical theory is the 'path-integral' approach, originally introduced by Feynman [1] for quantum mechanics of a particle. The integral is over all possible paths a particle can take, given the boundary points. The paths in the integral can be absolutely different from the classical path predicted by the Euler Lagrange's equation. Each path in the integral is weighted with the exponential of the 'action' for the path times the square root of -1. The path-integral $G(x_0 t_0, x_1 t_1) = \int \mathcal{D}x \exp(\frac{\iota}{h} S)$ (S is the action, h the Planck's constant which we set to 1) is a function of the boundary points, and is interpreted as the quantum amplitude of a particle to propagate from (x_0, t_0) to (x_1, t_1) and is known as the propagator. Surprisingly in the case the initial point is collapsed, the propagator reduces to the Schrodinger wave-function, whose modulus square is the probability of a particle to be exist at (x_1, t_1). This rather interesting 'quantisation' is naturally extended to 'quantum field theories', where one sums over all possible field configurations of the system given the boundary data, with each field configuration weighted with the exponential of square root -1 times the action. Thus the 'path integral' quantisation of gravity is simply given as

$$G(\Sigma_0 h_0, \Sigma_1 h_1) = \int \mathcal{D}g_{\mu\nu} \exp(\frac{\iota}{\hbar} S) \tag{1}$$

where $g_{\mu\nu}$ represents the metric or the 'field' for gravity given the boundary metrics $h_{0,1}$ at $\Sigma_{0,1}$, and $S = \frac{-1}{16\pi G}\int d^4x\sqrt{g}R$ is the Einstein's action for each metric configuration which exists given the boundary data. As expected, the path-integral suffers from two very obvious problems
(i)The measure and the weights are complex and the integral does not converge
(ii)The measure in the path integral is infinite, and one has to use a suitable discretisation to regularise the infinity and extract a finite answer.
both of these problems are suitably solved for a particle by
(i)Wick rotation from Lorentzian time to Euclidean time makes the weight and the measure real. The effect is evident in $\iota S \rightarrow -S_E$, where the added advantage which ensures convergence for most actions is that the Euclidean action S_E is positive definite.
(ii)The 'paths' can be broken into discrete time steps with each 'sub-path' being piecewise analytic.

However both of these problems appear difficult to implement for gravity as (i) Time is not unique for gravity, and hence 'Wick rotating' which time is a question which has no easy answer, and moreover Wick rotating time in a particular coordinate frame leads to complex metrics. Further, [2] ab-initio a sum over Euclidean metrics, weighted with the Euclidean action, the usual advantage of 'positive definite' Euclidean action for gravity does not exist, the action for gravity can be arbitrarily negative for certain field configurations or *gravitational Euclidean action is unbounded from below.*
(ii)The regularisation of the 'geometries' can be done using lattices which are piecewise flat, with the curvature concentrated at the vertex. But these discrete sums have proved difficult to completely evaluate as they yield infinities, like the continuum path-integral. However, recent results are emerging from dynamical triangulations in certain approximations, which are very promising [3].

In this article I shall talk about problem (i) and whether there are any resolutions of this. In the next section I shall talk about 'Wick rotation' for gravity, in the third section I shall introduce the Euclidean action, and discuss how the 'effective' action for quantum gravity might eventually prove to be positive definite, and finally I shall conclude.

2. Wick Rotation in Quantum Gravity

The space-time we live in has Lorentzian signature, and the metric of space-time is indeed Lorentzian. So a Lorentzian path-integral is the fundamental 'quantum propagator' for gravity. The Euclidean sum is just a tool used for computational purposes (similar to contour integrals where the complex plane techniques are used to extract answers for real integrals). In usual field theory, Wick rotation is a analytic continuation of time to the imaginary axis $t \rightarrow it$, however, in gravity, which is invariant under general coordinate transformation, this does not make sense. As simply a coordinate transformation $t \rightarrow e^t$ will end up taking the new Euclidean time to be e^{it}, which is complex. Thus a coordinate invariant 'Wick rotation' hasn't been defined. Infact, this is a second order problem, as identifying a globally defined time for gravity, or a observable which can be measured using clocks is still an open problem. But lets say, just at the basic level, if we take a metric which has time dependence, lets say the de Sitter metric (cosmological constant Λ).

$$ds^2 = -dt^2 + e^{\sqrt{\Lambda/3}\,t}\left\{dr^2 + r^2(d\theta^2 + \sin^2\theta d\phi^2)\right\} \tag{2}$$

time is identified here, and if one does the usual analytic continuation $t \to it$, the resultant complex metric is not a part of the 'real positive definite metrics' we would like to sum over in a 'Euclidean' path-integral. One could ask why use 'Wick rotation' from a Lorentzian path-integral to a Euclidean path-integral. One could just sum over Euclidean metrics, and analytically continue back to get a 'Lorentzian propagator'. But herein is the interesting observation from two dimensional gravity: A ab-initio Euclidean path-integral doesn't have any causality restrictions, and the sum results in 'branching polymer like behaviour' which has no 'Lorentzian analog'. Thus a path-integral for gravity should be defined in Lorentzian space-time, and one should find a way to identify a corresponding 'Euclidean path-integral' corresponding to the Lorentzian one. This was done using a 'non-perturbative' Wick rotation defined in [5, 6], where a particular gauge was chosen in which time is distinguished and causality is manifest. In this gauge, the metric is written in a particular coordinates τ, x_i, where τ is the time coordinate and x_i are the spatial coordinates

$$ds^2 = -d\tau^2 + g_{ij}dx^i dx^j \tag{3}$$

This proper-time gauge ofcourse inherits the problems of lack of 'global time' in quantum gravity. This is due to the fact that cusps or caustics develop very fast in this gauge, the proper-time lines converge and the determinant of the three metric is singular. However, one could divide the entire history of the metric into a set of time intervals, for each of which a proper-time can be identified. As we know, the path-integral is a product of these discrete intervals, as obtained initially by Feynman. Now, in this, once the path-integral is defined, it is very easy to identify the corresponding 'Euclidean' set of metrics, and that is by simply introducing a discrete parameter $\epsilon = \pm 1$ such that

$$ds^2 = \epsilon d\tau^2 + g_{ij}dx^i dx^j \tag{4}$$

and perform the path-integral in the $\epsilon = 1$ sector, and then analytically continue back in the final answer. Infact this is what was attempted in [6], where indeed the path-integral gave a answer (similar to that obtained in [3] and [7]) which was different from a Euclidean path-integral obtained in [8]. A proper-time quantisation of 2-dimensional gravity rules out topology change unlike [8] but gives a propagator as a function of the total proper-time elapsed between the boundary data, and a Lorentzian propagator is obtained by analytically continuing back in that proper-time. Thus one could take this 'Wick rotation' in proper-time gauge and apply it to a four dimensional path-integral. However, as we found in [5] a four dimesnional computation meets with the first hurdle, that of the unboundedness of the Euclidean action. In the next section I shall discuss this and plausible resolutions of the same.

3. The Unbounded Euclidean Action

The Euclidean gravitational action or the action for positive definite metric comprises of

$$S = -\frac{1}{16\pi G}\int \sqrt{\det g}\; R\, d^4x \tag{5}$$

In [2] Hawking showed that if there is a particular decomposition of the metric of the form;

$$g_{\mu\nu} = e^{2\phi}\bar{g}_{\mu\nu} \tag{6}$$

where $e^{2\phi}$ is a conformal factor associated with the metric, and $\bar{g}_{\mu\nu}$ has a constant Ricci curvature, then the gravity action reduces to

$$S = -\frac{1}{16\pi G}\int d^4xe^{2\phi}\sqrt{\bar{g}}\left[\bar{R} + 6(\nabla\phi)^2\right] \tag{7}$$

The kinetic term of the conformal mode is positive definite, and hence the Euclidean action can assume as negative values as possible. This pathology can be assumed to be a signature of redundant degrees of freedom, existing in the theory, and indeed the diffeomorphisms leave the action invariant, and they have to be factored out. This procedure, known as gauge fixing, leads to an effective action, re-written only in terms of the physical degrees of freedom. For perturbative gravity, it has been shown that [9], the action is indeed positive definite. For non-perturbative gravity, this gauge fixing is non-trivial, and it is difficult to find the 'physical degrees of freedom'. However I use the reasoning that if the trace part of the diffeomorphism is isolated and its contribution to the action investigated, then that is enough to argue for the positive definiteness of the effective action. The details of the gauge fixing are used in the traceless part, and hence this derivation of the positive definite action is independent of the details of gauge fixing. The gauge transformation of the tangent space vectors in the space of metrics is written as:

$$h_{\mu\nu} = h^{\perp}_{\mu\nu} + (L\xi)_{\mu\nu} + \left(2\phi + \frac{1}{2}\nabla\xi\right)g_{\mu\nu} \tag{8}$$

Here $h^{\perp}_{\mu\nu}$ is the traceless part of the gauge invariant metric, and the

$$(L\xi)_{\mu\nu} = \nabla_\mu\xi_\nu + \nabla_\nu\xi_\mu - \frac{1}{2}\nabla\xi g_{\mu\nu} \tag{9}$$

is an operator which maps vectors to traceless tensors.

The trace part of the diffeomorphisms, contributes to the scale factor or the conformal trajectories in the metric space. The gauge fixing procedure is very similar to other field theories, and leads to a Fadeev-Popov determinant remaining in the measure. To resolve these problems, we suggest a redefinition of the generators of diffeomorphisms, which isolates the trace part of the diffeomorphism, just as the conformal mode is isolated in the metric. The traceless part of the metric remains where one can implement the gauge fixing.

The trace part can be isolated in the diffeomorphisms by writing the generator of the diffeomorphisms as comprised of two parts one arising from the gradient of a scalar σ, and the other from a divergenceless vector. This will imply that the trace part of the diffeomorphisms will be generated by the scalar σ. Thus

$$\xi \equiv \hat{\xi} + \nabla\sigma \Rightarrow \nabla.\xi = \nabla^\mu\nabla_\mu\sigma \tag{10}$$

Note that at this stage the σ directions are not orthogonal to the tracefree directions, as the conformal killing vectors are, as clearly $L\xi$ has a dependence in the σ directions.

The new 'generators' of diffeomorphisms, are a vector $\hat{\xi}_\mu$ and a scalar σ. The vector which generates the diffemorphism is

$$\xi_\mu = \hat{\xi}_\mu + \nabla_\mu \sigma \tag{11}$$

$$\nabla^\mu \xi_\mu = \nabla^\mu \nabla_\mu \sigma \tag{12}$$

The interesting aspect of this decomposition is that one can write $\hat{\xi}_\mu$ in terms of the divergence of a two tensor $B_{\mu\nu}$. Thus

$$\hat{\xi}_\mu = \epsilon_{\mu\nu\rho\lambda} \nabla^\nu B^{\rho\lambda} \tag{13}$$

Clearly,

$$\begin{aligned} \nabla^\mu \xi_\mu &= \epsilon_{\mu\nu\rho\lambda} \nabla^\mu \nabla^\nu B^{\rho\lambda} && (14) \\ &= \frac{1}{2}\epsilon_{\mu\nu\rho\lambda}[\nabla^\mu, \nabla^\nu] B^{\rho\lambda} && (15) \\ &= \frac{1}{2}\epsilon_{\mu\nu\rho\lambda}(R^{\mu\nu\rho}_{\tau} B^{\tau\lambda} + R^{\mu\nu\lambda}_{\tau} B^{\rho\tau}) && (16) \\ &= 0 \quad \text{(by Bianchi identity)} && (17) \end{aligned}$$

The diffeomorphism transformations written purely in terms of the antisymmetric two tensor and the scalar field can be easily shown to form a group structure.

Thus to decide the uniqueness of the new decomposition, one finds:

$$\nabla^\mu \xi_\mu = \nabla^2 \sigma \tag{18}$$

The above equation determines σ upto a scalar (whose divergence is zero), in terms of the trace of the diffeomorphism. The very interesting aspect of this breakdown is the fact that 'orthogonal' orbits of the σ field are not the 'conformal killing' orbits, but solutions to the equation

$$\left(\nabla_\mu \nabla_\nu - \frac{1}{2} g_{\mu\nu} \nabla^2 \right) \sigma = 0 \tag{19}$$

This does not reduce to

$$L\xi = 0 \tag{20}$$

which is a much restricted equation for ξ, and for this particular way of redefining the diffeomorphisms, the $\hat{\xi}_\mu$ directions remain. If however, the $\hat{\xi}_\mu$ is vanishing, then Equation (19) will reduce to the conformal killing vectors, as the diffeomorphisms will be only along the scalar directions. Similarly if $\hat{\xi}_\mu$ is zero, but Equation (19), does not hold, then it is not a pure scalar transformation, and σ can be absolutely arbitrary for this. We proceed to find the effective action as obtained from this particular method.

The effective action includes the classical action and the terms from the Faddeev-Popov determinant. The interesting aspect of this new calculation is the separation of the trace part of the diffeomorphisms and evaluation of the Faddeev-Popov determinant for this separately. The path-integral is defined to be

$$Z = \int \mathcal{D}\phi \, \mathcal{D}\bar{g}_{\mu\nu} \exp\left(\frac{1}{16\pi G} \int d^4x e^{2\phi} \sqrt{\bar{g}}[\bar{R} + 6(\nabla\phi)^2] \right) \tag{21}$$

(we subsequently set $16\pi G$=1) Further the gauge transformations are implemented, the Faddeev-Popov determinant appears in the measure. To identify the correct measure one writes in the cotangent space of the De-Witt super space [10] or Equation (8), a coordinate transformation to the traceless gauge fixed part (denoted by a $\perp$) and a trace part and the pure diffeomorphsims generated by $\xi_\mu = \hat{\xi}_\mu + \nabla_\mu \sigma$. The Jacobian of the coordinate transformations is the Faddeev-Popov determinant and written as (detM) subsequently.

$$Z = \int \mathcal{D}\phi\, \mathcal{D}g^{\perp}_{\mu\nu}\, \mathcal{D}\hat{\xi}_\mu\, \mathcal{D}\sigma \;\mathrm{detM}\; \exp\left(\int d^4x\sqrt{\bar{g}}\, e^{2\phi}[\bar{R} + 6(\nabla\phi)^2] \right) \tag{22}$$

This way of gauge fixing is completely non-perturbative, and is not specific to a gauge fixing. The det M is shown to have a scalar determinant times a vector and a tensor determinant.

$$\mathrm{detM} = \mathrm{det}_S[8(1+2C)(-2(\nabla)^4 + 4\nabla_\mu\nabla^2\nabla^\mu + 4\nabla_\mu\nabla_\rho\nabla^\mu\nabla^\rho)]^{1/2}\mathrm{det_V}\tilde{\mathrm{V}}\mathrm{det_T}\mathrm{T} \tag{23}$$

The operators V_μ and $T_{\mu\nu}$ which are vector and tensor operators. The gauge volume or the integrations over the $\hat{\xi}_\mu$ and the σ can be taken out of the path-integral, and the Faddeev-Popov determinant contributes to the effective action. For the purpose of the calculation of this paper, we concentrate on the scalar determinant and find the contribution to the effective action, in the perturbative case.

Note that since this isolation of the trace part is independent of the gauge fixing condition, our results will be true in any gauge.

3.1. The Resolution of the Conformal Mode Problem: Fluctuations about a Classical Solution

We begin by taking the perturbative case and give the resolution of the conformal mode problem in some known cases. In the perturbative situation ϕ is taken to be very small, thus $e^{2\phi} = 1 + 2\phi$ and hence the action gives

$$\begin{aligned} S &= -\int d^4x\sqrt{\bar{g}}(1 + 2\phi + 4\phi^2)[\bar{R} + 6(\nabla\phi)^2] & (24)\\ &= -\int d^4x\sqrt{\bar{g}}[6(\nabla\phi)^2 + 2\phi\bar{R} + 4\phi^2\bar{R} + \bar{R}] & (25)\\ &= -\int d^4x\sqrt{\bar{g}}\left\{2\phi'[-3\nabla^2 + 2\bar{R}]\phi' + \bar{R} - \bar{R}(\Delta^{-1})\bar{R}\right\} & (26) \end{aligned}$$

Where the square in ϕ' is completed and one obtains a non-local term in that process.

The first we consider is perturbing about Minkowski space-time. Clearly, $R_{\mu\nu\lambda\rho} = 0$ in the first approximation, and one obtaines the scalar determinant (23) as

$$\mathrm{det_S}[48(1 + 2\mathrm{C})\nabla^4]^{1/2} \tag{27}$$

Under suitable boundary conditions (e.g. for the eigenfunction of the Laplacian, one fixes the boundary condition for ψ, ψ' and then one takes ψ'', ψ''' consistently with those), the determinant splits into

$$\mathrm{det_S}[-48(1 + 2\mathrm{C})\nabla^2]\mathrm{det_S}[-\nabla^2] \tag{28}$$

Thus the partition function acquires the following form

$$\int \mathcal{D}\phi \mathrm{dets}[-48(1+2\mathrm{C})\nabla^2]^{1/2}\mathrm{dets}[-\nabla^2]^{1/2}\exp(-\int \mathrm{d}^4\mathrm{x}\phi'[-6\nabla^2]\phi') \tag{29}$$

Clearly the ϕ integral can be written formally as a determinant, which is divergent.

$$\mathrm{dets}[-48(1+2\mathrm{C})\nabla^2]^{1/2}\mathrm{dets}[-\nabla^2]^{1/2}\frac{1}{\mathrm{dets}[6\nabla^2]^{1/2}} \tag{30}$$

The C is a one-parameter ambiguity which exists in the definition of the measure. For Einstein gravity $C = -2$ and thus the factor $1 + 2C$ is a negative factor, whcih makes the determinant divergent. Thus the divergent determinants from the Faddeev-Popov and the ϕ integral can be cancelled leaving a finite determinant (with the correct sign) and this can be obtained by a suitable regularisation method. Thus at the perturbative level, the conformal mode is a unphysical mode consistent with the fact that it is a longitudinal mode of the graviton.

Next the de-sitter and anti de Sitter backgrounds which are constant curvature metrics and hence have $R_{\mu\nu} = \Lambda g_{\mu\nu}$ are considered (Λ is a cosmological constant), and the conformal mode is treated as a fluctuation over that.

The scalar determinant (23) factorises rather neatly as fourth order determinant is

$$6\nabla^4 + 4[\nabla_\mu, \nabla^2]\nabla^\mu + 4\nabla_\mu[\nabla_\rho, \nabla^\mu]\nabla^\rho = 6\nabla^4 + 8\Lambda\nabla^2 \tag{31}$$

Thus the determinant can be factorised for the above operator using $R = 4\Lambda$, as

$$\mathrm{dets}(-\nabla^2)(8(1+2\mathrm{C})(-3\nabla^2 - \mathrm{R}) = \mathrm{dets}(-\nabla^2)\mathrm{dets}(8(1+2\mathrm{C})(-3\nabla^2 - \mathrm{R})). \tag{32}$$

The second factor of this is clearly what one obtaines by integrating the conformal mode including the cosmological factor.

So, for most Ricci flat and constant curvature metrics, we seem to have correctly identified the resolution.

For generic conformal perturbations about a arbitrary solution of Einstein's equation, we use the Heat Kernel method, to estimate the effective action [12].

4. Conclusion

In this paper, we discuss the Euclidean path-integral for gravity, and show that the effective action in the perturbative case does not have the conformal mode divergence associated with it. The measure in the path integral constrains the classical action. What is new in this calculation is that this is obtained independent of gauge fixing conditions which are imposed in the traceless part of the metric. The isolation of the trace part of the diffeomorphism simply takes care of the divergence due to the conformal mode. The non-perturbative aspect of this calculation will be obtained in a publication to appear [12].

Acknowledgment

I would like to thank A. Ghosh and P. Majumdar for suggesting the symmetric two tensor in the decomposition of the diffeomorphism.

References

[1] R. P. Feynman, *Rev. Mod. Phys.* 20 367 (1948).

[2] G. W. Gibbons, S. W. Hawking and M. J. Perry, *Nucl. Phys. B* 138 141-150 (1978).

[3] J. Ambjorn and R. Loll, *Nucl. Phys. B* 536 407 (1998).

[4] J. Ambjorn, J. Jurkiewicz, R. Loll, *Phys. Rev. D* 72 064014 (2005).

[5] A. Dasgupta and R. Loll, *Nucl. Phys. B* 606 357 (2001).

[6] A. Dasgupta, *JHEP* 0207, 062 (2002).

[7] R. Nakayama, *Phys. Lett. B* 325 347-353 (1994).

[8] A. M. Polyakov, *Phys. Lett. B* 103 207 (1981).

[9] K. Schleich, *Phys. Rev. D* 36 2342 (1987).

[10] Z. Bern, S. K. Blau, E. Mottola, *Phys. Rev. D* 43 1212 (1991).

[11] E. Mottola, *J. Math. Phys.* 36 2470-2511 (1995).

[12] A. Dasgupta, *The gravitational path-integral and the trace of the diffeomorphisms* arXiv:0801.4770 [gr-qc].

In: Classical and Quantum Gravity Research
Editors: M.N. Christiansen et al, pp. 169-243
ISBN 978-1-60456-366-5

Chapter 4

OPEN QUANTUM RELATIVITY

***Giuseppe Basini*[1] *and Salvatore Capozziello*[2*]**
[1] Laboratori Nazionali di Frascati, INFN,
Via E. Fermi C.P. 13, I-0044 Frascati, Italy,
[2] Dipartimento di Scienze Fisiche,
Universitá di Napoli "Federico II" and INFN Sez. di Napoli
Complesso Universitario di Monte S. Angelo,
Via Cinthia, I-80126 Napoli Italy.

Abstract

Open Quantum Relativity is a theory based on two fundamental features: the assumption of a General Conservation Principle which states that the conservation laws can never be violated and the achievement that both General Relativity and Quantum Mechanics can be described under the standard of a covariant symplectic formalism. These facts lead to some important consequences. First of all the existence of a dynamical unification scheme of fundamental interactions achieved by assuming a $5D$ space which allows that the conservation laws are always and absolutely valid as a natural necessity. Then what we usually describe as violations of conservation laws can be described by a process of topology change, embedding and dimensional reduction, which gives rise to an induced-gravity-matter theory in the $4D$ space-time by which the usual masses, spins and charges of particles, naturally spring out. As results the theory leads to a dynamical explanation of several problems of modern physics (e.g. entanglement of quantum states, quantum teleportation, gamma ray bursts origin, black hole singularities, cosmic primary antimatter absence). Moreover the theory provides a self-consistent picture of the observed accelerated cosmological behavior with the correct reproduction of experimental cosmological parameters and new predictions as gravitationally induced neutrino oscillations and further scalar modes in gravitational waves. A fundamental role in this approach is the link between the geodesic structure and the field equations of the theory before and after the dimensional reduction process. The emergence of an extra force term in the reduction process and the possibility to recover the masses of particles, allow to reinterpret the Equivalence Principle as a dynamical consequence which naturally "selects" geodesics from metric structure and vice-versa the metric structure from the geodesics. It is worth noting that, in the Einstein General

[*]E-mail address: capozziello@na.infn.it

Relativity, geodesic structure is "imposed" by choosing a Levi-Civita connection and this fact can be criticized considering a more general completely "affine" approach like in the Palatini formalism (in agreement with our covariant symplectic approach). As we will show, the dimensional reduction process gives rise to the generation of the masses of particles which emerge both from the field equations and the embedded geodesics. Due to this result, the coincidence of chronological and geodesic structures is derived from the embedding and a new dynamical formulation of the Equivalence Principle is the direct consequence of dimensional reduction. The dynamically derived structure becomes more general since two time arrows and closed time-like paths naturally emerge, so opening the doors to even more fundamental consequences, first of all a reinterpretation of the standard notion of causality which can be, in this way, always recovered, even in the case in which it is questioned (like in entanglement phenomena and quantum teleportation), since it is generalized to a "forward" and a "backward" causation.

1. Introduction

Modern physics presents, in several aspects, a contradictory situation, since the great achievements of XX century, aimed to realize a unified picture of physical world, are still far from this goal and gave rise to many new paradoxes. The successes of experimental microscopic physics, astrophysics and cosmology led very far from standard human environment (*e.g.* primordial particles, quasar, last scattering surface), have brought surprising results very difficult to be settled in standard schemes and, together with the persistent failure in describing General Relativity and Quantum Mechanics under the same standard theory, are at the basis of the lack of this unified vision.

In order to achieve this goal, a covariant symplectic structure can be introduced [1, 2] where the covariant formalism of General Relativity and the symplectic features of Quantum Mechanics are unified by taking into account affine connections instead of space-time metric as fundamental fields. This in the light of a General Conservation Principle, which, differently from the current standard hypotheses, find a natural place in a covariant symplectic formalism. As we will see, such an approach can be applied to frame some shortcomings of modern physics in a comprehensive picture.

In fact, in our epoch of very rapid progress, and due to this rapidity, several new problems were substantially removed and confined outside the mainstream of physics, but they are no longer forgettable if we want to carry on in this progress. Let us remember some of them: (i) the contradiction pointed out by the Einstein-Podolsky-Rosen (EPR) paradox [3], *i.e.* the possibility, under particular conditions, to perturb a physical object without interacting with it in any known way; (ii) the existence of objects, like the black holes, which, despite several serious attempts (Refs. [4, 5]), seem to violate the energy conservation [6]; (iii) the consideration of quarks as elementary constituents even though it is generally hypothesized that probably they can never get their individuality as single particles [7]; (iv) the up to now observed absence of primary antimatter in our universe, despite the standard symmetric creation of matter and antimatter pairs [8, 9, 10, 11]; (v) the Big Bang theory, in both standard and inflationary cosmology, which is not yet satisfactory solved in the initial singularity [12]; (vi) the experimental results of quantum teleportation [13, 14] which suggest a $\Delta t = 0$ in transferring information, in that way questioning the Relativity and the Causality; (vii) the lack of a consistent unitary description of all fundamental interactions [7]; $(viii)$ and last but not least, the fundamental question of the absence of a general approach [15, 16] connecting Quantum Mechanics and Relativity (the core of this paper). Now, from our point of view, several contradictions of today physics arise from a principle, implicitly or explicitly contained in every scientific formulation: The assumption that the geometry is always given in the evolution of physical systems, assumption which determines that any possible topology change is considered as a "singularity". In other words, the topology of manifolds where systems evolve, is given from the beginning and never considered as a dynamical structure. In our opinion, this implicit principle is hiding a possible dynamical explanation and is leading to singularities, symmetry breakings and violations of conservation laws, and is responsible for the most significant contradictions of today physics, since it seems that symmetry breakings and violations are a sort of *ad hoc* hypotheses invoked as soon as a new phenomenon cannot be included in standard schemes. As we will show, a straightforward way to take into account the above issues is deeply related to conservation laws, so then we propose to investigate what happens if conservation laws are *always valid and symmetries are always maintained*. In other words, we propose a new approach in which several contradictions of modern physics can be framed and solved since conservation laws are always conserved. This is the starting point of the *Open Quantum Relativity*, the subject of this paper.

Directly due to the fact that conservation laws can never be violated, the symmetry of the theory leads to the general consequence that backward and forward time evolution (and causation) are both allowed. Then, as we shall see, the necessity of a generalization to five physical dimensions leads to the derivation of particle features as mass, spin, and charge as result of an embedding process. In this paper, we start with a discussion on the EPR paradox (the initial point of the considerations involving Quantum Mechanics and Relativity); we continue with the consideration of the absolute validity of conservation laws and the existence, as a consequence, of two time arrows which make possible the entanglement of physical systems; then, in order to extend the time definition, we generalize the approach to five dimensions and we show how this leads to the generation of the features of particles. The theoretical basis of such an approach is the covariant symplectic structure by which all the standard tensor, vector and scalar fields can be framed into a unified picture.

This review is organized as follows. In Sec.II, we introduce a discussion on Quantum Mechanics and General Relativity, describing the role of conservation laws, the necessity of a $5D$-space, the emergence of two time arrows, together with masses, spins and charges of particles as consequence of a dimensional reduction process. Sec.III is devoted to the discussion of the role of time in modern physics, considering the topology changes and the concept of causality which can be, in general, extended due to *forward* and *backward* causation. In Sec.IV, we present the methodological approach of Open Quantum Relativity: the covariant symplectic structure by which gravity and the other interactions can be dealt under the same standard considering conserved Hamiltonian invariant constructed by tensor, vector and spinor fields. Sec.V is devoted to the applications of Open Quantum Relativity to quantum and classical cosmology. In Sec. VI, we discuss further observational and experimental evidences in the framework of Open Quantum Relativity like gamma ray bursts, galactic structures, gravitationally induced neutrino oscillations, stochastic background of gravitational waves. Sec.VII is devoted to discussion and conclusions.

2. Conservation laws in Quantum Mechanics and General Relativity

2.1. Time Arrows vs. EPR Paradox

The foundations of Quantum Mechanics, especially in relations to Relativity, are always largely discussed, since the intrinsic characteristic of Quantum Mechanics is the existence of systems that have not definite values of measurable quantities, unless one does not measure them. The most significant point is that this is due to fact that the state of a system is a superposition of different states, and the only possibility left to the observation consists in interacting in an irreversible way with the system, by the measure process. This is also described as the collapse of the wave function by the measure, and then the evolution of the system, after the measure, has the collapsed state as initial wave function. This description, mainly due to Bohr, is known as Copenhagen Interpretation (CI), and following it, the irreversibility of the measure process, and consequentially the irreversibility of the time direction, plays a key role. Moreover the superposition of quantum states, that in CI is characteristic of the microscopic description of Nature, may have macroscopic effects and this is at the base of the EPR paradox. Let us take into account (considering the Bohm's [17] very clear example) a spin zero particle, or in general a bound state, which decays into two particles each of spin 1/2. As far as the spin of the particles in a definite direction is concerned, the state of the system is described by a state vector of the following form :

$$|0\rangle = |\uparrow\rangle|\downarrow\rangle - |\downarrow\rangle|\uparrow\rangle\,, \tag{1}$$

where we must stress that every single particle is not a state by itself and the evolution of the state vector, with respect to the spatial distribution, of the wave function is not yet specified. We know that at the time $t = t_0 = 0$, in which the decay occurs, the spatial support of the variable $\vec{x}_1 \equiv \{x_{1,1}, x_{2,1}, x_{3,1}\}$, (position of the particle 1) which appears in the wave function $\Psi(\vec{x}_1, \vec{x}_2, \sigma_1, \sigma_2, t_0)$ (where σ_1 and σ_2 are the spins of the particles) coincides with the spatial support of the variable $\vec{x}_2$ (position of the particle 2). Now we let the system evolve with an Hamiltonian $\mathcal{H}$ such that, after a time $\triangle t = t_m - t_0$ (where $t_m$is the time of the measurement), the wave function of the system

$$\Psi(\vec{x}_1, \vec{x}_2, \sigma_1, \sigma_2, t_m) = e^{\frac{i}{\hbar}\mathcal{H}\triangle t}\Psi(\vec{x}_1, \vec{x}_2, \sigma_1, \sigma_2, t_0)\,, \tag{2}$$

is such that the spatial supports in x_1 and x_2 are separated, *i.e.* the particles coming from the decay get their individuality. Note that, in the CI, the evolution of the system after the measure is given by (if, for instance, the measure of σ_1 gives the value + 1/2)

$$\Psi(\vec{x}_1, \vec{x}_2, 1/2, -1/2, t) = e^{\frac{i}{\hbar}\mathcal{H}(t-t_m)}\Psi(\vec{x}_1, \vec{x}_2, 1/2, -1/2, t_m) \qquad t \geq t_m\,. \tag{3}$$

Note that we do not perform measurement at time t such that $t_0 < t < t_m$ and therefore we do not know anything about the state of the system in that range of time. Now, at time t_m, we measure the spin σ_1 of the particle 1. The spin state of the system is still determined by Eq.(1), but now the measure operation causes the collapse of the wave function in one of the two states $|\uparrow\rangle|\downarrow\rangle$ or $|\downarrow\rangle|\uparrow\rangle$. This implies that, once the spin of particle 1 is measured, also the spin of particle 2 is instantaneously acquired, although the particles are far apart. Therefore we have the following paradox: we have two particles in absence of direct interaction, but the state of the system is such that, if we fix the spin state of one particle, also the spin state of the other particle is instantaneously fixed. The nonlocal behavior (connected with the Bell's inequalities) of this kind of systems has been tested by Aspect et al. [18] and the experimental results show that the non-locality is an actual feature of Nature. The experiments were performed with a slightly different device, *i.e.* the products of the decay were photons, but, even if this is a particular case, the formal description of the experiment remains the one described above. Let us summarize the results of this experiment (and of the others made by several groups of physicists) using Sakurai's words (in a book revised by J. Bell himself [16]): "All the experiments made, have conclusively shown that Bell's inequality (which comes from the "locality prescription" of Einstein) is violated and violated in a way which is compatible, within the errors, with Quantum Mechanic's prediction". In order to explain this paradox, so crucial since at the intersection between Quantum Mechanics and Relativity, we develop a new approach, starting from the general remark that the Noether Theorem [19] states that for every conservation law of Nature a symmetry must exist. From this statement, it comes out directly (as we will see later) the consideration of a backward time evolution of wave function, since dynamics, if derived from a variational principle, is always symmetric under time reversal transformations. This general remark indicates that conservation laws intrinsically contain forward and backward causation, even if against common sense and local realism, which instead assume just one time arrow. Below we show, by very general arguments, that Bianchi's identities, which are geometric identities directly connected to conservation laws, contain symmetric dynamics. From such a dynamics, it is therefore possible to describe backward and forward evolution of the wave function, basing our considerations on the fact that quantum matter can be described by a scalar field ϕ [20, 21]. Let us start from a phenomenological definition of entanglement as the phenomenon which takes place when two or more physical objects, despite being spatially disconnected, are subjected to an inter-relation for which the effect of a perturbation on one of them induces a perturbation on the other one, without any direct interaction on each other. This definition will be generalized below, starting from this new approach [22]). Let us now recall that, in a previous paper [23], it has been proposed a solution of the EPR paradox which does not change formally the scheme of Bohr interpretation. This means that all the features of CI are preserved, but the time flow of the wave function can be different from the time flow of the observer. As we outline, this possibility, although not intuitive and not referable to our macroscopic perception of reality, is compatible with Quantum Mechanics, since our intrinsic ignorance of the state of the system in the time interval $t_0 < t < t_m$.

Dynamics of a scalar field ϕ, describing quantum matter on a (curved) space-time, is given by the stress-energy tensor

$$T_{\mu\nu} = \nabla_\mu\phi\nabla_\nu\phi - \frac{1}{2}g_{\mu\nu}\nabla_\mu\phi\nabla^\mu\phi + g_{\mu\nu}V(\phi)\,, \tag{4}$$

which is a completely symmetric object, being

$$T_{[\mu,\nu]} = T_{\mu\nu} - T_{\nu\mu} = 0\,. \tag{5}$$

$V(\phi)$ is a self-interacting potential specified below; ∇_μ is the covariant derivative on a given spacetime manifold (as standard in General Relativity, we assume a Riemannian manifold with a metric tensor $g_{\mu\nu}$ defined on it); the indexes are $\mu, \nu = 0, 1, 2, 3$ and the signature of the metric is $(+ - - -)$. Since ϕ is a scalar field, the covariant derivative coincides with the partial derivative so that

$$\nabla_\mu \phi = \partial_\mu \phi = \phi_\mu \,. \tag{6}$$

Such a tensor has to satisfy the conservation laws

$$\nabla_\mu T^\mu_\nu = 0 \,, \tag{7}$$

which are nothing else but the contracted Bianchi identities for $T_{\mu\nu}$. Sending to [23] for the full derivation, it is straightforward to show that

$$\nabla_\mu T^\mu_\nu = \phi_\nu \Box \phi + V_\nu = \phi_\nu \left[\Box \phi + \frac{dV}{d\phi} \right] \,, \tag{8}$$

where $\Box$ is the d'Alembert operator and $\nabla_\mu \phi^\mu = \Box \phi$.

The result is that the validity of the contracted Bianchi identities, implies the Klein-Gordon equation which gives the dynamics of ϕ, that is

$$\nabla_\mu T^\mu_\nu = 0 \qquad \Longleftrightarrow \qquad \Box \phi + \frac{dV}{d\phi} = 0 \,. \tag{9}$$

Obviously we are assuming $\phi_\nu \neq 0$.

It is interesting to note the full symmetry of the result, *i.e.* the Klein-Gordon operator is symmetric.

Specifying the problem to the case of a self-interacting massive particle, we can write

$$V(\phi) = \frac{1}{2} m^2 \phi^2 \,, \qquad \text{and then} \qquad \frac{dV}{d\phi} = m^2 \phi \,, \tag{10}$$

so that we obtain the Klein-Gordon equation

$$\left(\Box + m^2 \right) \phi = 0 \,. \tag{11}$$

Being ϕ the scalar field, we want to stress that it can be interpreted as the product of two conjugate complex numbers

$$\phi = \psi^* \psi \,. \tag{12}$$

Now, let us take into account, as an example, the case of a Minkowski space-time, and develop the Klein-Gordon operator (suppressing the indexes)

$$\Box + m^2 = \partial_\alpha \partial^\alpha + m^2 = (\partial - i\,m)(\partial + i\,m) \,. \tag{13}$$

For consistency, the Klein-Gordon equation gives

$$\left(\Box + m^2 \right) \phi = \left(\partial_\alpha \partial^\alpha + m^2 \right) (\psi^* \psi) = (\partial - i\,m)(\partial + i\,m)(\psi^* \psi) = 0 \,, \tag{14}$$

and Eq.(14) can be split, for massive particles, in the cases

$$(\partial - i\,m)\,\Psi = 0 \qquad\qquad (\partial + i\,m)\,\Psi^* = 0 \tag{15}$$

$$(\partial - i\,m)\,\Psi^* = 0 \qquad\qquad (\partial + i\,m)\,\Psi = 0\,. \tag{16}$$

Instead, for massless particles, the four conditions reduce to two

$$\partial\psi = \partial\psi^* = 0\,. \tag{17}$$

Taking into account spinors, we can write down

$$(i\gamma^\mu\partial_\mu - m)\,\psi = 0\,, \tag{18}$$

and analogous equations for the other cases, where γ^μ are the standard Dirac matrices [19]

$$\gamma^0 = \begin{pmatrix} I & 0 \\ 0 & -I \end{pmatrix}, \qquad \gamma^i = \begin{pmatrix} 0 & \sigma^i \\ -\sigma^i & 0 \end{pmatrix}, \tag{19}$$

with I the (2×2) identity matrix and σ^i the three Pauli matrices [20].

We can see that Eqs.(15) can be considered as a "forward" and "backward" propagators and viceversa for the other two.

In terms of 4−momenta, we have

$$(\partial - i\,m) \quad\longrightarrow\quad (k - i\,m)\,, \tag{20}$$

$$(\partial + i\,m) \quad\longrightarrow\quad (k + i\,m)\,, \tag{21}$$

so the general solutions have the forms

$$\psi(x) = e^{-ikx}u(k)\,, \qquad \psi^*(x) = e^{ikx}u^*(k) \tag{22}$$

which can be interpreted, respectively, as *progressive* and *regressive* solutions. It comes out that a function (a superposition) of the form

$$\varphi = \alpha_1\psi + \alpha_2\psi^*\,, \tag{23}$$

where $\alpha_{1,2}$ are arbitrary constants, is a general solution of the dynamics and the states ψ and ψ^* can be considered, in our scheme, as *entangled* since they can influence each other also when they are disconnected. In other words, the absolute validity of conservation laws gives rise to a symmetric dynamics (backward and forward evolution of the system) and the entanglement of states is naturally determined. We want to stress that we found the four conditions (15)–(16) which satisfy Eq.(14), and this fact implies that backward and forward evolutions exist both for the ψ-field and the conjugate ψ^*-field. In some sense, it seems that all the folds of light–cone, in Minkowski spacetime, have the same dignity, but we have been confined to investigate (at least macroscopically) just the fold toward the future (the arrow of time which we can usually perceive). Finally, by writing the wave function in polar representation

$$\psi = e^{-i\theta}\rho \tag{24}$$

it is immediate to see that backward or forward evolution depends on the sign of the phase θ, and then a pure geometric representation is allowed for time evolution. We conclude with some remarks about the pictures presented. In order to explain the fact that a measurement made on a particle seems able to affect the state of another particle, entangled but spatially disconnected, it is important to notice that the statement can be re-expressed in terms of an effect of a measurement of a particle able to affect the system in the past: *i.e.* the relation of interference among variables known as Bell inequalities can be obtained if the particles decay in a state that depend on what will be measured [23].

The above considerations are completely general and can be developed in the framework of curved space-times as we will discuss below in the framework of the covariant symplectic structure (see Sec.IV).

2.2. The General Conservation Principle. From the $5D$-Space Extension to the $4D$ Reduction

Let us continue the description of Open Quantum Relativity showing how and why it is able to take in account and solve several contradictions in the relations between Quantum Mechanics and Relativity [25, 26, 27]. It is relevant now the introduction of a *General Conservation Principle*: "The conservation laws have an absolute meaning and maintain always their validity. This is the reason why, when it would be otherwise impossible to preserve this validity, they determine the entanglement phenomenon, which allows in any case their recovery, thanks to topology changes with related time arrow inversions. So, the conservation laws are invariant while the topologies are not" [22, 27]. This principle, thanks to the fact that mathematical formalism does not prevent backward and forward evolution in time, $i)$ allows an explanation in terms of non-local physical behavior for entangled systems, $ii)$ agrees with the experimentally observed violation of Bell inequalities and consequently of Einstein locality principle [15], $iii)$ and open a discussion on the first principles leading to such a symmetric time behavior.

For the "traditional" causality principle [15] and Einstein-Podolsky-Rosen point of view [3], such a behavior is a paradox. Nevertheless, using backward and forward evolution, entangled systems are naturally explained [23]. In other words, if we maintain conservation laws absolutely valid, we need two time arrows, and backward evolution turns out to be a feature of Nature, which we cannot ordinarily feel, but which emerges as soon as a conservation law has no other way to maintain its validity. As a consequence of this view, strictly related to the concept of entanglement, there are topology changes taking place in order to preserve a conservation law, like matter-energy in black hole dynamics, or quantum numbers in EPR effects [23]. In the first case, as we will see in the following, the entangled system is constituted by a black hole which dynamically evolves in a white hole through a topology change (the basic essence of a worm hole). The general feature of such a result is that the conservation laws are invariant, while topologies are not invariant and so they can be considered dynamical quantities. The existence of backward and forward evolution leads however to the necessity of a fifth dimension. Let us sketch a very simple picture, imagining a hypothetical one-dimensional universe. In such a universe, a point placed to the left of another one could never invert its position exchanging the left with the right, while if we pass to a 2D–universe, it becomes obviously possible to exchange the relative position of the two. This is conceptually true also passing from four to five dimensions, allowing the evolution in both directions of the time axis [25, 26, 27]. But this is not, in our conception, just a technicality, because the $5D$ universe has a real physical meaning, as we will show below. In our approach, as the fourth dimension is related to time in Relativity, the fifth dimension can be related to the mass of particles, as we will see [27]. In other words, we can deal with a generalized $5D$ mass-chronotope, every point of which is labelled by space, time and mass. It is relevant to note the fact that we do not perceive the fourth time-like dimension in the same way of space-like dimensions and the situation is analogue for this fifth mass-like dimension. The set of equations (15) and (16) are fundamental in our approach, since they give rise to forward and backward dynamics. From a relativistic point of view, in the cases (15) and (16) we are inside one of the folds of the light-cone, in the case (17) we are, instead, on the null surface. A role in distinguishing dynamics is played by the mass of particles which was introduced by the self-interacting potential (10). In the following, we find, instead, a general approach through which the masses of particles are generated by a geometrical procedure. The distinction of backward and forward dynamics depends on the dimension of the space-time we are dealing with. In a $5D$-manifold, all particles are moving without preference of backward and forward evolution, while it is a reduction procedure from $5D$-space to $4D$-spacetime which gives rise to two dynamics (*i.e.* two arrows of time), and conservation laws are preserved in any case. In the light of these considerations, the concepts of entanglement and topology change are features of the theory which emerge in order to preserve conservation laws [22].

In previous papers, we discussed the approach for different cases. In [23], we showed that EPR paradox can be interpreted under the standard of conservation laws since the paradox is solved by an entangled superposition of forward and backward solutions. Then, in [43], we have taken into account an astrophysical system like a black hole. Assuming as natural law, that the collapsing matter-energy is totally conserved, such a black hole evolves, trough two topology changes (at the two junctions of the worm hole) in a white hole. The black hole and the white hole (in fact a white fountain) are so two entangled objects which can live in two different causally disconnected regions of spacetime and, again, entanglement and topology changes are dynamically generated by the same request of absolute validity of conservation laws. All these indications suggest that the conservation laws exhibit the same general validity, so they seem all related by a unique general and fundamental conservation principle. Finally, in this approach, the possibility to evolve backward to the past (*i.e.* in what we conceive as a past) is a general feature of Nature, but in our "forward" fold of light-cone, we cannot sensorially perceive it. The only way to experience the backward evolution is under the extreme conditions in which a conservation law, being otherwise violated, determines a spacetime breaking via a topology change connecting entangled systems and this phenomenon is allowed by the presence of a fifth dimension of real physical meaning. As we will see, it seems that several shortcomings of current physics can be solved assuming this general conservation principle always valid. We want to outline that this is a general result, because the connection among conservation laws, symmetries and first integrals of motion have a deep physical meaning.

In view of what we discussed above, a $5D$-theory can be the *minimal* dimensional unification scheme which, by a reduction process to $4D$, can give rise to the physics which we experimentally know (see also [28]).

In [30], beside a dynamical unification scheme of all the interactions, we develop an approach by which, starting from a pure theory of gravitation formulated in $5D$, we are capable of obtaining $4D$-physics where the ordinary features of particles, as mass, spin and charge, are the product of an embedding process as well as the same structure of spacetime, which results richer than the standard $4D$-one, since backward and closed timelike solutions naturally emerges. The major feature of the approach is that conservation laws are always and absolutely valid (*i.e.* $5D$-Bianchi identities always hold, see also [27]) and what we consider as different symmetries and interactions can be all reconducted to a $5D$-supergroup $\mathcal{G}_5$, whose a possible $4D$-decomposition, including the Standard Model, is [30]

$$\mathcal{G}_5 \supset IO(3,1) \otimes SU(3) \otimes SU(2) \otimes U(1)\,. \tag{25}$$

Due to this fact, we are going to deal with the degrees of freedom of the spacetime and of the particles under the same standard and when $\mathcal{G}_5 = SL(5)$, this is the minimal group which is capable of including all the standard fundamental interactions and the 10 generators of inhomogeneous Lorentz group [30].

2.3. From $5D$- to $4D$- Field Equations: Conservation Laws, Extra Force and Geodesics

Let us now outline the main features of this $5D$-theory discussing, in particular, the $4D$-reduction procedure which induces a scalar-tensor theory of gravity where conservation laws (*i.e.* Bianchi identities) play a fundamental role into dynamics. After we develop the related quantum cosmology showing that, using these results, it is always possible to select classical trajectories, *i.e.* observable universes.

The $5D$-manifold which we are taking into account is a Riemannian space provided with a $5D$-metric of the form

$$dS^2 = g_{AB}dx^A dx^B\,, \tag{26}$$

where the Latin indexes are $A, B = 0, 1, 2, 3, 4$. We do not specify the signature yet, since it can be dynamically fixed by the reduction procedure in $4D$. Canonically, it is a pseudo-Lorentzian metric.

Let us define the curvature invariants, the field equations and the conservation laws in the $5D$-space. In general, we ask for a space which is a smooth manifold, singularity free because defined in such a way that every conservation law on it has to be always and absolutely valid [30]. The $5D$-Riemann tensor is

$$R^D_{ABC} = \partial_B \Gamma^D_{AC} - \partial_C \Gamma^D_{AB} + \Gamma^D_{EB}\Gamma^E_{AC} - \Gamma^D_{EC}\Gamma^E_{AB}\,. \tag{27}$$

The number of independent components of such a tensor, after the full derivation and thanks to the Petrov classification [29], is $\frac{1}{12}N^2(N^2-1) = 50$ for $N = 5$. The Ricci tensor and scalar are derived from the contractions

$$R_{AB} = R^C_{ACB}\,, \qquad {}^{(5)}R = R^A_A\,. \tag{28}$$

The field equations can be derived from the $5D$ Hilbert–Einstein action

$${}^{(5)}\mathcal{A} = -\frac{1}{16\pi\,{}^{(5)}G}\int d^5x\sqrt{-g^{(5)}}\left[{}^{(5)}R\right]\,, \tag{29}$$

where ${}^{(5)}G$ is the $5D$-gravitational coupling and $g^{(5)}$ is the determinant of the $5D$-metric. The variational principle

$$\delta \int d^5x \sqrt{-g^{(5)}} \left[{}^{(5)}R \right] = 0 \,, \tag{30}$$

gives the $5D$-field equations which are

$$G_{AB} = R_{AB} - \frac{1}{2} g_{AB} \, {}^{(5)}R = 0 \,, \tag{31}$$

so that at least the Ricci-flat space is always a solution. Let us define now the $5D$-stress-energy tensor:

$$T_{AB} = \nabla_A \Phi \nabla_B \Phi - \frac{1}{2} g_{AB} \nabla_C \Phi \nabla^C \Phi \,, \tag{32}$$

where, because of the General Conservation Principle and so the presence of symmetries, only the kinetic terms are considered. As standard, such a tensor can be derived from a variational principle

$$T^{AB} = \frac{2}{\sqrt{-g^{(5)}}} \frac{\delta \left(\sqrt{-g^{(5)}} \mathcal{L}_\Phi \right)}{\delta g_{AB}} \,, \tag{33}$$

where $\mathcal{L}_\Phi$ is a Lagrangian density connected with the scalar field Φ. Because of the definition of $5D$ space itself [27], it is important to stress now that no self-interaction potential $V(\Phi)$ has been taken into account so that T_{AB} is a completely symmetric object and Φ is, by definition, a cyclic variable. This fact guarantees that Noether theorem always holds for T_{AB} and a conservation law intrinsically exists. With these considerations in mind, the field equations can now assume the form

$$R_{AB} = \chi \left(T_{AB} - \frac{1}{2} g_{AB} T \right) \,, \tag{34}$$

where T is the trace of T_{AB} and $\chi = 8\pi \, {}^{(5)}G$, being $\hbar = c = 1$. The form (34) of field equations is useful in order to put in evidence the role of the scalar field Φ, if we are not simply assuming Ricci-flat $5D$-spaces. As we said, T_{AB} is a symmetric tensor for which the relation

$$T_{[A,B]} = T_{AB} - T_{BA} = 0 \,, \tag{35}$$

holds. Due to the choice of the metric and to the symmetric nature of the stress-energy tensor T_{AB} and the Einstein field equations G_{AB}, the contracted Bianchi identities

$$\nabla_A T^A_B = 0 \,, \qquad \nabla_A G^A_B = 0 \,, \tag{36}$$

must always hold. Developing the stress-energy tensor, we have

$$\begin{aligned} \nabla_A T^A_B &= \nabla_A \left(\partial_B \Phi \partial^A \Phi - \frac{1}{2} \delta^A_B \partial_C \Phi \partial^C \Phi \right) \\ &= \left(\nabla_A \Phi_B \right) \Phi^A + \Phi_B \left(\nabla_A \Phi^A \right) - \frac{1}{2} \left(\nabla_B \Phi_C \right) \Phi^C - \frac{1}{2} \Phi_C \left(\nabla_B \Phi^C \right) \\ &= \left(\nabla_A \Phi_B \right) \Phi^A + \Phi_B \left(\nabla_A \Phi^A \right) - \Phi_C \left(\nabla_B \Phi^C \right) . \end{aligned} \tag{37}$$

Since our $5D$-space is a Riemannian manifold, it is

$$\nabla_A \Phi_B = \nabla_B \Phi_A \tag{38}$$

and then

$$\Phi^A \left(\nabla_A \Phi_B\right) - \Phi_C \left(\nabla_B \Phi^C\right) = \Phi^A \left(\nabla_B \Phi_A\right) - \Phi_C \left(\nabla_B \Phi^C\right) = 0 \,. \tag{39}$$

In this case, partial and covariant derivatives coincide for the scalar field Φ. Finally

$$\nabla_A T^A_B = \Phi_B \, {}^{(5)}\Box \Phi \,, \tag{40}$$

where ${}^{(5)}\Box$ is the $5D$ d'Alembert operator defined as $\nabla_A \Phi^A \equiv g^{AB} \Phi_{,A;B} \equiv {}^{(5)}\Box \Phi$. The general result is that the conservation of the stress-energy tensor T_{AB} (*i.e.* the contracted Bianchi identities) implies the Klein-Gordon equation which assigns the dynamics of Φ, that is

$$\nabla_A T^A_B = 0 \qquad \Longleftrightarrow \qquad {}^{(5)}\Box \Phi = 0 \,, \tag{41}$$

assuming $\Phi_B \neq 0$ since we are dealing with a real physical field. Let us note again the absence of self-interaction (*i.e.* potential) terms. The relations (41), being field equations, are giving a physical meaning to the fifth dimension. Splitting the $5D$-problem in a $(4+1)$-problem, from Eq.(41), it is possible to generate the mass of particles in $4D$ [26, 27, 30], as we shall see below.

Let us now derive and discuss the geodesic equation in this $5D$-manifold. The action for geodesics is

$$\mathcal{A} = \int dS \left(g_{AB} \frac{dx^A}{dS} \frac{dx^B}{dS} \right)^{1/2} \,, \tag{42}$$

the Euler-Lagrange equations give the geodesic equation

$$\frac{d^2 x^A}{dS^2} + \Gamma^A_{BC} \frac{dx^B}{dS} \frac{dx^C}{dS} = 0 \,, \tag{43}$$

where Γ^A_{BC} are the $5D$-Christoffel symbols. Eq.(43) can be split in the $(4+1)$ form

$$2 g_{\alpha\mu} \left(\frac{dx^\alpha}{ds} \right) \left(\frac{d^2 x^\mu}{ds^2} + \Gamma^\mu_{\beta\gamma} \frac{dx^\beta}{ds} \frac{dx^\gamma}{ds} \right) + \frac{\partial g_{\alpha\beta}}{dx^4} \frac{dx^4}{ds} \frac{dx^\alpha}{ds} \frac{dx^\beta}{ds} = 0 \,, \tag{44}$$

where the Greek indexes are $\mu, \nu = 0, 1, 2, 3$ and $ds^2 = g_{\alpha\beta} dx^\alpha dx^\beta$. Clearly, in the $4D$ reduction (*i.e.* in the usual spacetime) we ordinarily experience only the standard geodesics of General Relativity, *i.e.* the $4D$-component of Eq.(44)

$$\frac{d^2 x^\mu}{ds^2} + \Gamma^\mu_{\beta\gamma} \frac{dx^\beta}{ds} \frac{dx^\gamma}{ds} = 0 \,, \tag{45}$$

so that, under these conditions, the last part of the representation given by Eq.(44) has to vanish in $4D$. In other words, for standard laws of physics, the metric $g_{\alpha\beta}$ does not depend on x^4 in the embedded $4D$-manifold. On the other hand, the last component of Eq.(44) can be read as an extra "force" which gives the motion of a $4D$-frame with respect to the fifth coordinate x^4. This fact shows us that the fifth dimension has a *real physical meaning* and any embedding procedure scaling up in $5D$-manifold (or reducing to $4D$-spacetime) has a dynamical description [31]; the quantity

$$\mathcal{F} = \frac{\partial g_{\alpha\beta}}{dx^4}\frac{dx^4}{ds}\frac{dx^\alpha}{ds}\frac{dx^\beta}{ds}, \tag{46}$$

has to be related to the mass of moving particles being the extra-force related to the embedding procedure. The emergence of this term in Eq.(44), leaving the $5D$-geodesic equation verified, means that the Equivalence Principle in $4D$ can be seen as a dynamical result [28]. In other words, the quantity $\mathcal{F}$, which distinguishes the masses of particles, gives a dynamical base of the Equivalence Principle. Finally *all* particles are represented as massless in $5D$ while, for the physical meaning of the fifth coordinate, they acquire mass in $4D$ thanks to Eq.(46).

Another important consideration has to be done on the line element dS^2. Let us take into account a $5D$-null path. It is given by

$$dS^2 = g_{AB}dx^A dx^B = 0\,. \tag{47}$$

A splitting of Eq.(47), made considering the $4D$-part of the metric and the homogeneous $5D$-component, is

$$dS^2 = ds^2 + g_{44}(dx^4)^2 = 0\,, \tag{48}$$

so that a null path in $5D$ can result a pure time-like or a space-like path in $4D$ depending on the sign of g_{44}. Let us consider now the $5D$-generalized velocity $u^A = dx^A/dS$. It can be split as a velocity in the ordinary 3D-space v, a velocity along the ordinary time axis w and a velocity along the fifth dimension z. In general, for null paths, we can have $v^2 = w^2+z^2$ and this should lead to super-luminal speed, explicitly overcoming the Lorentz transformations. The problem is solved if we consider the $5D$-motion as *a-luminal*, because all particles and fields have the same speed (being massless) and the distinction among (an eventual) super-luminal motion, luminal motion and sub-luminal (or canonically causal motion for massive particles) emerges only *after* the dynamical reduction from $5D$-space to $4D$-spacetime. In this way, the fifth dimension is the entity which, by assigning the masses, is able to generate the different dynamics which we conceive in $4D$. Consequently, it is the process of mass generation which sets the particles in the $4D$-light-cone. Below, we will discuss this mechanism.

2.4. Scalar-Tensor Gravity as the Result of $4D$-Reduction. Two Time Arrows and Closed Time-Like Paths

The above results can be reduced to a $4D$-dynamics taking into account the Campbell theorem [32]. This theorem states that it is always possible to consider a $4D$-Riemannian manifold, defined by the line element $ds^2 = g_{\alpha\beta}dx^\alpha dx^\beta$, in a $5D$ one with $dS^2 = g_{AB}dx^A dx^B$. We have $g_{AB} = g_{AB}(x^\alpha, x^4)$ with x^4 the yet unspecified extra coordinate. The metric g_{AB} is covariant under the group of $5D$ coordinate transformations $x^A \to \overline{x}^A(x^B)$, *but not* under the (restricted) group of $4D$ transformations $x^\alpha \to \overline{x}^\alpha(x^\beta)$. This relevant fact has, as a consequence, that the choice of $5D$ coordinates results as the *gauge* necessary to specify the $4D$ physics also in non-standard aspects (*e.g.* entanglement [22]). Vice-versa, in specifying the $4D$ physics, the bijective embedding process in $5D$ gives physical meaning to the fifth coordinate x^4 [30]. Furthermore the signature of the fifth coordinate results from the dynamics generated by the physical quantities which we observe in $4D$ (mass, spin, charge). Such a process gives rise to fundamental features (*e.g.* two time arrows, closed time-like paths) capable, in principle, of solving several paradoxes of modern physics (*e.g.* entanglement of quantum systems [22], or black holes [25]).

Let us start our considerations by replacing the variational principle (30) with

$$\delta\int d^{(5)}x\sqrt{-g^{(5)}}\left[{}^{(5)}R+\lambda(g_{44}-\epsilon\Phi^2)\right]=0\,, \tag{49}$$

where λ is a Lagrange multiplier, Φ a generic scalar field and $\epsilon=\pm1$. This approach is completely general and used in theoretical physics when we want to put in evidence some specific feature. In this case, we need it in order to derive the physical gauge for the $5D$ metric. Starting from Eq.(49), we can write down the metric as

$$dS^2=g_{AB}dx^Adx^B=g_{\alpha\beta}dx^\alpha dx^\beta+g_{44}(dx^4)^2=g_{\alpha\beta}dx^\alpha dx^\beta+\epsilon\Phi^2(dx^4)^2 \tag{50}$$

from which we obtain directly particle-like solutions ($\epsilon=-1$) or wave-like solutions ($\epsilon=+1$) in the $4D$-reduction procedure. The standard signature of $4D$-component of the metric is $(+\;-\;-\,-)$ and $\alpha,\beta=0,1,2,3$. Furthermore, the $5D$-metric can be written as the matrix

$$g_{AB}=\begin{pmatrix} g_{\alpha\beta} & 0\\ 0 & \epsilon\Phi^2\end{pmatrix}, \tag{51}$$

and the $5D$-curvature Ricci tensor, one time fully developed, is expressed as

$${}^{(5)}R_{\alpha\beta}=R_{\alpha\beta}-\frac{\Phi_{,\alpha;\beta}}{\Phi}+\frac{\epsilon}{2\Phi^2}\left(\frac{\Phi_{,4}g_{\alpha\beta,4}}{\Phi}-g_{\alpha\beta,44}+g^{\lambda\mu}g_{\alpha\lambda,4}g_{\beta\mu,4}-\frac{g^{\mu\nu}g_{\mu\nu,4}g_{\alpha\beta,4}}{2}\right), \tag{52}$$

where $R_{\alpha\beta}$ is the $4D$-Ricci tensor. The expressions for ${}^{(5)}R_{44}$ and ${}^{(5)}R_{4\alpha}$ can be analogously derived [28]. After the projection from $5D$ to $4D$, $g_{\alpha\beta}$, derived from g_{AB}, no longer explicitly depends on x^4, so, from Eq.(52), a very useful expression for the Ricci scalar can be derived:

$${}^{(5)}R=R-\frac{1}{\Phi}\Box\Phi\,, \tag{53}$$

where the dependence on ϵ is explicitly disappeared and $\Box$ is the $4D$-d'Alembert operator which gives $\Box\Phi\equiv g^{\mu\nu}\Phi_{,\mu;\nu}$. The action in Eq.(49) can be recast in a $4D$-reduced Brans-Dicke-like action of the form

$$\mathcal{A}=-\frac{1}{16\pi G_N}\int d^4x\sqrt{-g}\left[\Phi R+\mathcal{L}_\Phi\right], \tag{54}$$

where the Newton constant is given by

$$G_N=\frac{{}^{(5)}G}{2\pi l} \tag{55}$$

where it is important to notice that l is a characteristic length in $5D$ which can be the Planck length as we shall discuss below. Defining a generic function of a $4D$-scalar field ϕ as

$$-\frac{\Phi}{16\pi G_N} = F(\phi) \tag{56}$$

we get, in $4D$, the most general action in which gravity is nonminimally coupled to a scalar field:

$$\mathcal{A} = \int_{\mathcal{M}} d^4x\sqrt{-g}\left[F(\phi)R + \frac{1}{2}g^{\mu\nu}\phi_{;\mu}\phi_{;\nu} - V(\phi) + \mathcal{L}_m\right] + \int_{\partial\mathcal{M}} d^3x\sqrt{-h}K\,, \tag{57}$$

(see Basini, Capozziello [30]) where the form and the role of $V(\phi)$ are still general and $\mathcal{L}_m$ represents the standard fluid matter content of the theory generated by the splitting of the $5D$-Klein-Gordon equation (41). The state equation of fluid matter is $p = \gamma\rho$ and $0 \leq \gamma \leq 1$ where p and ρ are, respectively, the ordinary pressure and density. The second integral is the boundary term where $K \equiv h^{ij}K_{ij}$ is the trace of the extrinsic curvature tensor K_{ij} of the hypersurface $\partial\mathcal{M}$ which is embedded in the $4D$-manifold $\mathcal{M}$; h is the metric determinant of the 3D-manifold.

The field equations can be derived by varying with respect to the $4D$-metric $g_{\mu\nu}$

$$G_{\mu\nu} = \tilde{T}_{\mu\nu}\,, \tag{58}$$

where

$$G_{\mu\nu} = R_{\mu\nu} - \frac{1}{2}g_{\mu\nu}R \tag{59}$$

is the Einstein tensor, while

$$\tilde{T}_{\mu\nu} = \frac{1}{F(\phi)}\left\{-\frac{1}{2}\phi_{;\mu}\phi_{;\nu} + \frac{1}{4}g_{\mu\nu}\phi_{;\alpha}\phi^{;\alpha} - \frac{1}{2}g_{\mu\nu}V(\phi) - g_{\mu\nu}\Box F(\phi) + F(\phi)_{;\mu\nu} + T^{m}_{\mu\nu}\right\}, \tag{60}$$

is the *effective stress–energy tensor* containing the nonminimal coupling contributions, the kinetic terms, the potential of the scalar field ϕ, and the ordinary matter stress-energy tensor. In the case in which $F(\phi)$ is a constant F_0 (in our units, $F_0 = -1/(16\pi G_N)$), we get the usual stress–energy tensor of a scalar field, minimally coupled to gravity, that is

$$T^{\phi}_{\mu\nu} = \phi_{;\mu}\phi_{;\nu} - \frac{1}{2}g_{\mu\nu}\phi_{;\alpha}\phi^{;\alpha} + g_{\mu\nu}V(\phi)\,. \tag{61}$$

By varying with respect to ϕ, we get the $4D$-Klein–Gordon equation

$$\Box\phi - RF'(\phi) + V'(\phi) = 0\,, \tag{62}$$

where $F'(\phi) = dF(\phi)/d\phi$ and $V'(\phi) = dV(\phi)/d\phi$. At this point it is important noting that Eq.(62) is nothing else but the contracted Bianchi identity so the above results hold demonstrating the consistency of the scheme.

This feature shows that the effective stress–energy tensor at right hand side of (58) is a zero–divergence tensor and this fact is fully compatible with Einstein theory of gravity also if we started from a $5D$-space. Specifically, the reduction procedure, which we have used, preserves the features of standard General Relativity. Now we have the task of the physical identification of the fifth dimension. To this goal, let us recast the generalized Klein-Gordon equation (62) as

$$\left(\Box + m_{eff}^2\right)\phi = 0\,, \tag{63}$$

where

$$m_{eff}^2 = \left[V'(\phi) - RF'(\phi)\right]\phi^{-1} \tag{64}$$

is the effective mass, *i.e.* a function of ϕ, where self-gravity contributions $RF'(\phi)$ and scalar field self-interactions $V'(\phi)$ are taken into account. In any quantum field theory formulated on curved space-times, these contributions, at one-loop level, have the "same weight" [21]. We will show that a natural way to generate the masses of particles can be achieved starting from a $5D$ picture and the concept of *mass* can be recovered as a geometric derivation.

Finally, as we have seen, the reduction mechanism can select also $\epsilon = 1$ in the metric (50). In this case, the $5D$-Klein Gordon equation (41), and the $5D$-field equations (31) have wave-like solutions of the form

$$dS^2 = dt^2 - f(t,x_1)(dx^1)^2 - f(t,x_2)(dx^2)^2 - f(t,x_3)(dx^3)^2 + (dx^4)^2\,, \tag{65}$$

where

$$f(t,x_j) = \exp i(\omega t + k_j x^j)\,, \qquad j = 1,2,3\,. \tag{66}$$

It is very interesting to note that intrinsically such a solution has two times and, as a direct consequence, due to the structure of the functions $f(t,x_j)$, closed time-like paths are allowed. The existence of closed time-like paths means that Anti-De Sitter [33] and Gödel [34] solutions are naturally allowed possibilities in the dynamics.

2.5. Mass, Spin and Charge of Particles in $4D$

Let us consider a flat $5D$-space. The $5D$ d'Alembert operator can be split, following the metric definition (50) for particle-like solutions, as:

$$^{(5)}\Box = \Box - \partial_4{}^2\,, \tag{67}$$

so selecting the value $\epsilon = -1$ in the metric. Introducing the scalar field Φ, we have

$$^{(5)}\Box\Phi = \left[\Box - \partial_4{}^2\right]\Phi = 0\,, \tag{68}$$

and then

$$\Box\Phi = \partial_4{}^2\Phi\,. \tag{69}$$

The problem is solvable by separation of variables and then we split the scalar field Φ into two functions

$$\Phi = \phi(t,\vec{x})\chi(x_4)\,, \tag{70}$$

where the field ϕ depends on the ordinary space-time coordinates, while χ is a function of the fifth coordinate x_4. Inserting (70) into Eq.(69), we get

$$\frac{\Box\phi}{\phi} = \frac{1}{\chi}\left[\frac{d^2\chi}{dx_4^2}\right] = -k_n^2 \tag{71}$$

where k_n must be a constant for consistency. From Eq.(71), we obtain the two equations of motion

$$\left(\Box + k_n^2\right)\phi = 0\,, \tag{72}$$

and

$$\frac{d^2\chi}{dx_4^2} + k_n^2\chi = 0\,. \tag{73}$$

Eq.(73) describes a harmonic oscillator whose general solution is

$$\chi(x_4) = c_1 e^{-ik_n x_4} + c_2 e^{ik_n x_4}\,. \tag{74}$$

The constant k_n has the physical dimension of the inverse of a length and, assigning the boundary conditions for consistency with (73), we can derive the eigenvalue relation

$$k_n = \frac{2\pi}{l}n\,, \tag{75}$$

where n is an integer and l a length which we have previously defined in Eq.(55). As a result, in standard units, we can recover the physical lengths through the Compton lengths

$$\lambda_n = \frac{\hbar}{2\pi m_n c} = \frac{1}{k_n} \tag{76}$$

which always assign the mass of a particle. It has to be emphasized that, the eigenvalues of Eq.(73) are the masses of particles which are generated by the process of reduction [Eqs.(68),(69)] from $5D$ to $4D$. The solution (74) is the superposition of two mass eigenstates. The $4D$-evolution is given by Eq.(63) [or equivalently by (72)]. An important point to be noticed is that, due to the fact that we have a harmonic oscillator [Eq.(73)], we have that the solutions of this one give us the associated Compton lengths from which the effective physical masses are derived. More in detail, different values of n fix the families of particles, while, for any given value n, different values of parameters $c_{1,2}$ distinguish the different particles within a family.

Furthermore, the effective mass can be *geometrically* derived as

$$m_{eff} \equiv \int |\Phi| dx^4 = \int |\Phi(dx^4/ds)| ds \tag{77}$$

where ds is the affine parameter used in the above derivation of geodesic equation. Eq.(64) is an effective mass definition, based on gravitational and scalar field self-interactions. The above reduction procedure from $5D$ to $4D$ tells us that, due to the coincidence of the descriptions (74) and (77), this scheme is an effective mechanism for mass generation. Substituting $m_{eff}^2 = m^2$, the splitting of Klein-Gordon equation developed in Sec.IIA (from Eq.(8 to Eq.(23)) gives rise to forward and backward evolution of (entangled) particles, *i.e.* it yields two time arrows together with the other features of particles. In fact, let us rewrite Eq.(22), which are also solutions of Eq.(72)

$$\psi(x) = e^{-ikx}u(k)\,, \qquad \psi^*(x) = e^{ikx}u^*(k) \tag{78}$$

which can be interpreted, respectively, as *progressive* and *regressive* solutions. From the above calculations, it is straightforward that they contains all the information on the spin and the charge of the particle due to the fact that they derive *directly* from conjugate Dirac equations. A function (a superposition) of the form

$$\varphi(x) = \alpha_1 \psi(x) + \alpha_2 \psi^*(x)\,, \tag{79}$$

where $\alpha_{1,2}$ are constants, is a general solution of the $4D$-dynamics and the states ψ and ψ^* can be interpreted as *entangled* since they can influence each other also when they are disconnected (see for the demonstration [23]). In other words, the absolute validity of conservation laws gives rise to a symmetric dynamics (backward and forward evolution of the system) and the entanglement of states is determined without arbitrary violations [22, 27], argument completely general and developed in the framework of any curved space-time.

At this point, remembering the condition (70), we can *recombine* all the solutions which we have obtained through Eqs.(74) and (78). The field Φ can be expanded as a Fourier series so that

$$\Phi(x^\alpha, x^4) = \sum_{n=-\infty}^{+\infty} \left[\psi_n(x^\alpha) e^{-ik_n x_4} + \psi_n^*(x^\alpha) e^{ik_n x_4}\right]\,, \tag{80}$$

where it is important to notice that ψ and ψ^* are the $4D$-solutions, while $e^{\pm ik_n x_4}$ are the embedded components coming from the reduction procedure. In general, every particle mass can be selected by solutions of type (74) while other particle features, as charge and spin, are selected by the other $4D$ solutions (78). It is worth noticing that the number $k_n x_4$, *i.e.* the ratio between the two lengths x_4/λ_n, fixes the interaction scale. Geometrically, such a scale is related to the curvature radius of the embedded $4D$-space where particles can be identified. In this sense, our approach is an *induced-matter* theory, where the extra dimension cannot be simply classified as "compactified" since it yields all the $4D$ dynamics giving origin to the masses. Moreover, Eq.(80) is not a simple "tower of mass states" but a spectrum capable of explaining the hierarchy problem. Finally also gravitational interaction scale can be discussed in this framework considering as fundamental the scale in Eq.(55), so then the Planck length

$$\lambda_P = l = \left(\frac{\hbar G_N}{c^3}\right)^{1/2}\,, \tag{81}$$

instead of the above Compton length. It fixes the vacuum state of the system since the masses of all particles can be considered negligible if compared with the Planck scales.

These results deserve a further discussion due to the fact that the mass of a given particle is fixed by the embedding process while the further splitting of $4D$-Klein Gordon equation in two (forward and backward) Dirac equations gives rise to the emergence of other features of particle as the spin and the charge which are *automatically* conjugated and conserved by the occurrence of two time arrows. In terms of a group description, the splitting process defines the $SU(2)$ (spin) and $U(1)$ (charge) components of the above considered supergroup $\mathcal{G}_5$.

3. Time in Modern Physics

3.1. Causality, Entanglement and Topology Changes

Einstein, Podolsky and Rosen [3] stated that Quantum Mechanics, contradicting the locality principle, brings to results which violates the causality, so the EPR paradox, apparently showing the incompatibility between Causality and Quantum Mechanics, leads to a deep crisis of the concept of Causality as a Principle of general validity. This fundamental question has become even more urgent, because, after the Bell's Inequalities and the related experiments [18], Quantum Mechanics is confirmed as a description fundamentally correct, so leaving the problem open and unresolved, despite many attempts and exotic hypotheses like the "hidden variables" , by definition impossible to be revealed and so then epistemologically unsatisfactory, since not solving the problem, but only bringing it farer. In the new approach, the redefinition of the entanglement concept, based on the impossibility of violation of conservation laws and on the generalization to a five dimensional space, leads instead to the recovery of the classical and fundamental idea of Causality. The overall validity of the conservation laws induces, in the cases where they would be otherwise violated, topology changes which make possible "tunnels" (like wormholes in relativistic astrophysics) connecting separated spacetime regions. This phenomenon turns out to be describable as *a-luminal* and able to open the door, without overcoming the light speed limit, to conceivable time-tunnels. In fact, Quantum Mechanics states that an interaction, even if acting only on a single part of a quantum system, determines a dependent evolution of the correlated quantities of the other part, also when these parts are placed in regions "causally disconnected" of spacetime, (*i.e.* when they are so far that no direct interaction between the part can occur, in the light speed limit). From the beginning, Schrödinger tried to overcome the problem by the "qualitative concept " of entanglement, described as a sort of deep connection, not yet dynamically specified, able to link two (apparently) causally disconnected, but quantum related, objects. Starting from the qualitative concept by Schrödinger, later accepted and elaborated by many authors [35], in a previous paper [22], we recovered the causality giving, in the light of the new approach started from the EPR effect, this new entanglement definition : "Two states, spatially separated and causally disconnected in the four dimensional space that we ordinarily perceive, are entangled if an interaction with one of them can influence the other one, without in any way directly interact with it, because a four dimensional entanglement means that it exists a causality nexus in a larger five dimensional physics space ". The reasoning which leads to this definition seems a real necessity, since the entanglement, unavoidable concept to explain why Quantum Mechanics works, would be otherwise simply impossible, without violating the causality principle and also the pure logic, which states that: "It is impossible an interaction with an object without in some way interacting with it ". If we do not hypothesize another physical dimension (the fifth one), in which the two states are causally connected, so then restoring the causality principle, what we call entanglement would remain a "necessary but inexplicable" phenomenon. In our theory, the conservation laws are the first principle which determines all the following evolution, since the fact that they can never be violated leads to a mechanism to avoid such a violation, also in the cases in which, for the standard interpretations, the violation should occur. This mechanism is the topology change. It is the topology change which provides the dynamics allowing the very particular interaction between two otherwise causally disconnected states, which we call entanglement, it is the topology change which makes possible the "hole" connecting the top and the bottom of the Fig.1, or, as an example, the formation of a wormhole connecting a black hole with its entangled white hole [43]. Entanglement, in this picture, maintains the meaning of the underlying mechanism allowing interactions otherwise impossible (which initially Schrödinger gave to it), but with a endowed dynamics providing an explanation of the phenomenon which emerges in $4D$-spacetime thanks to the existence of a larger dimensional space.

All these considerations and results always hold for Quantum Mechanics. In general, conservation laws are defined in Quantum Mechanics through the Heisenberg relations

$$[\Sigma_j, \mathcal{H}] = 0\,. \tag{82}$$

They means that j momenta of the n-dimensional phase space are conserved. Clearly, a conservation law is not admitted if

$$i\frac{\partial \Sigma_j}{\partial t} = [\Sigma_j, \mathcal{H}] \neq 0,\,, \tag{83}$$

where we are assuming Planck units with $\hbar = c = k_B = 1$. By the Dirac canonical quantization procedure, we have by definition

$$\pi_j \longrightarrow (\hat{\pi}_j = -i\partial_j)\,, \qquad \mathcal{H} \longrightarrow \hat{\mathcal{H}}(q^j, -i\partial_{q^j})\,, \tag{84}$$

where π_j are conjugate momenta. If $|\Psi\rangle$ is a *state* of the system (*i.e.* the wave function of a particle), dynamics is given by the Schrödinger eigenvalue equation

$$\hat{\mathcal{H}}|\Psi\rangle = E|\Psi\rangle\,, \tag{85}$$

where the whole wave-function is given by $|\varphi(t,x)\rangle = e^{iEt/\hbar}|\Psi\rangle$. If conserved quantities exist, we get classically

$$\pi_1 \equiv \frac{\partial \mathcal{L}}{\partial \dot{Q}^1} = i_{X_1}\theta_{\mathcal{L}} = \Sigma_1\,, \quad \pi_2 \equiv \frac{\partial \mathcal{L}}{\partial \dot{Q}^2} = i_{X_2}\theta_{\mathcal{L}} = \Sigma_2\,, \quad \ldots \quad \pi_j \equiv \ldots\,, \tag{86}$$

where $\mathcal{L}$ is a canonical Lagrangian, i_{X_j} is a contraction and $\theta_{\mathcal{L}}$ is the Cartan one-form, where the index j depends on the number of symmetry vectors.

After Dirac quantization, we get

$$-i\partial_1|\Psi\rangle = \Sigma_1|\Psi\rangle\,, \quad -i\partial_2|\Psi\rangle = \Sigma_2|\Psi\rangle\,, \quad \ldots \quad -i\partial_j|\Psi\rangle = \Sigma_j|\Psi\rangle \tag{87}$$

which are nothing else but translations along the Q^j axis singled out by the corresponding conserved quantity.

Eqs. (87) can be immediately integrated and, being Σ_j real constants, we obtain oscillatory behaviors for $|\Psi\rangle$ in the directions of such quantities, *i.e.*

$$|\Psi\rangle = \sum_{j=1}^{m} e^{i\Sigma_j Q^j}|\chi(Q^l)\rangle\,, \quad m < l \leq n\,, \tag{88}$$

where m is the number of symmetries, l are the directions where symmetries do not exist and n is the number of dimensions of configuration space. Vice-versa, dynamics given by (85) can be reduced if, and only if, it is possible to define constant conjugate momenta as in (86), *i.e.* the oscillatory behaviors of a subset of solutions $|\Psi\rangle$ exist always as a consequence of the fact that conserved quantities are present in the dynamics. The m symmetries give first integrals of motion and then the possibility to select exact solutions for particles. Moreover, if $m = n$, the problem is completely solvable and a symmetry exists for every variable of configuration space.

From a physical point of view, we can say that the existence of conservation laws, determines also the structure of the configuration space (a vector space in the case of Hilbert) where the physical system is set. Vice-versa, the degree of solvability and separability of a system is deeply related to the existence of conservation laws and, moreover, all physical quantities are conserved in a completely integrable system. In other words in Quantum Mechanics, we must stress that Eq.(82) assumes the meaning of Bianchi identities. It is straightforward to show that entangled states like

$$|\Psi\rangle = \frac{1}{\sqrt{2}}\{|\chi\rangle_1|\eta\rangle_2 + |\chi\rangle_2|\eta\rangle_1\} \tag{89}$$

rise up when the superposition (88) is not possible, i.e. when the Heisenberg relations (82) do not hold. Such a mechanism of entanglement is necessary in a $4D$-spacetime, but not in a $5D$-one, where the Hamiltonian is always conserved. Finally, we must stress that the entanglement of quantum states (i.e. the necessity to overcame the impossibility to write down the wave function of a system as a superposition) comes out in order to preserve in any case the conservation laws.

In the light of the new approach, another very important question, coming always from the EPR effect finds a possible solution. The question of the hypothesized contemporaneousness of the effects of an interaction on one part of a system, with the induced effect on another part (placed far and non directly interacting, but entangled). This question has become even more important after recent results claiming for instantaneous quantum teleportation [13, 14]. This issue is a problem also in Relativity, since immediately out of the limits of uncertainty principle, it suggests the possibility to travel faster than light speed. This would be a major problem, since in clear contradiction with the basis of Relativity itself, if we continue to treat it in terms of super-luminality, while it is no longer a puzzle if treated in terms of a-luminality.

Looking at Fig.1, the question finds a possible solution.

Let us take into account a bounded surface: it is evident that it would take time to go from a point at the upper side to the corresponding image point at the lower side of such a surface, but this time is reduced to zero, if a mechanism exists to make an hole and get directly the bottom from the top. It is straightforward to see that, once defined a standard Lorentz transformation for time and space intervals $\triangle t'$ and $\triangle t$:

$$\triangle t' = \triangle t(1 - v^2/c^2)^{1/2}, \qquad \triangle x' = \triangle x(1 - v^2/c^2)^{1/2}, \tag{90}$$

the reduction to zero of the space interval $\triangle x'$ implies, in the second of (90), that $v = c$, so then the time interval is $\triangle t' = 0$. The situation is that depicted in Fig.1: the travel is from A to B (or viceversa), the mechanism is the topology change and the deep

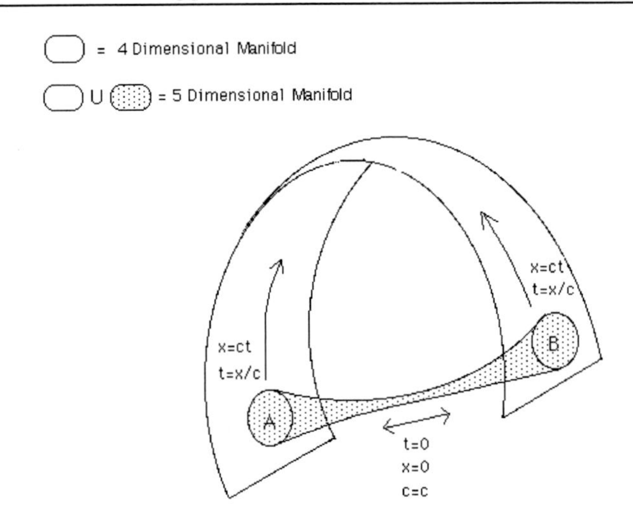

Figure 1. A graphic representation of an a-luminal travel involving a topology change.

reason of this is the physical necessity to save the conservation laws, without overcoming the light speed. This type of travel is a-luminal. In our theory, the conservation laws are the first principle which determines all the following evolution, since the fact that they can never be violated leads, under the particular condition in which for the standard physics the violation should be a necessity, to a mechanism to avoid such a violation. This mechanism is the topology change. It is the topology change, as we will see in the next subsection, which provides the dynamics allowing the very particular interaction between two otherwise causally disconnected states, that we call entanglement, it is the topology change which makes possible the "hole" connecting the top and the bottom of the above example, or, more specifically, the formation of a wormhole connecting a black hole with its entangled withe hole, via a Gödel type change of topology [30, 34]. The fundamental concept which has to be stressed here, is the "*a-luminality*", since it saves both conservation laws and causality principle, and it seems moreover the only mechanism capable to be compatible with the necessity to extend, but not contradict, the Relativity.

3.2. The Time in Relativitic Theories of Gravitation

Even if it can be considered a final synthesis of classical physics, General Relativity is, above all, the cornerstone of a new way to interpret Nature and it opened the door to a different way to see and define also old concepts, especially the time. In fact, if the possibility of "relative" time travels in the relative future is possible already in the General Relativity framework, it is in the post-relativistic theories that the time travel hypothesis takes a more general meaning, mainly if such theories are taking into account also effects coming from Quantum Mechanics. If we fix now our attention on the possibility (induced by conservation laws) of topology changes, we can get, as an example, a Gödel-type condition of closed time-like curve (CTC) geometry, recovering the hypothesis to travel even back in time. In fact a closed time line, which Gödel first presented as purely formal possibility in an unconventional solution of Einstein equations (that we can sketch as a cylindrical coordinates choice where the longitudinal-one is spatial and the circular is temporal, instead of the usual contrary) makes it possible re-track back in time the path done, so then "to pass again" in the past. Now, what for Gödel was a mathematical picture, here becomes a physical hypothesis, because based on a dynamics (the forward and backward evolution related to the splitting of Klein-Gordon equation) on a "necessity" (the preservation of conservation laws and the causality principle), on a conceivable path (the induced topology change defined in an appropriate mathematical domain) and finally on a theory able to propose an explanation of several physics problem (e.g. EPR paradox, entangled teleportation, black hole existence, gamma rays bursts and a good fit with the observed cosmological parameters). After Gödel, many authors showed that, far to be an exception, the possible existence of natural CTCs in the Universe should be an infinity, unless the existence of some rules (like, for instance, the Hawking Chronology Protection) based on semi-classical considerations, which are no longer holding in the framework of quantum gravity, since unappropriate combination of quantum matter fields and classical spacetime devices" [25, 36]. Another important point is the possibility to transfer information, instantaneous and impossible to be detected outside the entangled emitter-receiver system. The long series of Bob and Alice (receiver and emitter) papers is the most famous example of it [13] and a special relevance deserves the very important claim by Bouwmeester et al. [14], where they presented experimental evidences of an effect of polarization, given on only one of a couple of entangled photons, transferred on the other one without any direct standard interaction between them. Following Bennett et al. suggestion [13]: "It is possible to transfer the quantum state of a particle into another particle, provided one does not get any information about the state in the course of the transformation". It is also possible to realize an "entanglement swapping", where photon 1 is no longer available in the original state, but it exists a photon 3 which is now in that state and this as a result of teleportation.

It remains the fact that it is necessary to send to Bob, in standard way, the information of the state of the entanglement (one possibility among four) between photon 1 and photon 2, and this fact opens a big discussion on the real meaning of teleportation. Anyway, besides the possible and very important applications in computing sciences, these experimental results deserve some important comments : i) it seems possible to determine the entanglement also between particles (and very probably also between more complicated objects) without the same origin, so opening the doors to a conceivable general technique; ii) it seems possible to send information, via entanglement, without any destroying influence of the environment and as far as one wants; iii) it seems possible to send information instantaneously, even if a conventional message is necessary to inform and check. The last two points seem to indicate a $\triangle t = 0$ in the transfer operation, the only one way to send information instantaneously (and as far as we want) and without any environmental influence. This really seems in contradiction with Special Relativity, even if some authors try to avoid such a contradiction saying that, due to the necessity to send also a conventional message, this would restrict again the phenomenon in the limits of Relativity, because this could not be used to send signals faster than light. Let us now enter into some general properties of spacetimes with CTCs (see [36, 37, 38, 39]). Solutions of Einstein's equations which allow CTCs have been known for a long time. The earliest example of such a spacetime is a solution obtained by Van Stockum in 1937 [40], which describes an infinitely long cylinder of rigidly and rapidly rotating dust. Another very important example and perhaps the most famous, is Gödel solution [34], discussed above, representing a stationary homogeneous universe with non zero cosmological constant, filled with rotating dust. CTCs are also present in the interior of Kerr black hole in the vicinity of its ring singularity, and other examples of spacetimes with CTCs[1] were discussed by de Felice [42]. In the general case, a spacetime can be divided into chronal regions, without CTCs, and achronal regions which contain CTCs. The boundaries between the chronal and achronal regions are formed by chronology horizons, precisely chronal regions end and achronal regions begin at a future chronology horizon, while achronal regions end and chronal regions begin at past chronology horizon. Thus, achronal regions are intersections of the regions bounded by both of these horizons. A future chronology horizon is a special type of Cauchy horizon, and it is subject to all the properties of such horizons. In particular, it is generated by null geodesics that have no past endpoints, but can leave the horizon when followed into the future. If the generators, monitored into the past, enter one or more compact regions of spacetime and never thereafter leave them, the future chronology horizon is said to be compactly generated. In a wormhole model with CTCs, the future chronology horizon is compactly generated. The inner horizon of a Kerr-Newman solution is an example of a Cauchy horizon that is not compactly generated. A compactly generated chronology horizon cannot form in a spacetime developed from a spacelike non-compact surface, without boundary, if the null energy condition holds [39]. The past-directed generators of the compactly generated future chronology horizon have no past end points. They will enter and remain in a compact region C. Hawking, in 1992 [39], showed that there exists a non empty set E of generators, each of them remains in the compact set C in the future direction, as well as in the past direction.

The sets E generically contain at least one closed null geodesic. More precisely, Hawking showed that: 1) if E contains such a closed null geodesic, small variations of the metric preserve this property. 2) If E does not contain a closed null geodesic, then in geometries obtained by small variation of the metric such curves do exist. In the framework of General Relativity, in order to create a time machine (which, in this context, means in general a region with CTCs) by using a wormhole, one needs to assume that there exists the possibility, in principle, to make them long living and traversable, which, in standard view, needs the violation of the averaged null energy condition.

3.3. A New Concept of Time in Open Quantum Relativity

Let us now deal with the new approach, which starts from the main stream of standard General Relativity and Quantum Mechanics, but providing another point of view in the light of a new General Conservation Principle [27]. The first characteristic of this new scheme is that it is not violating Relativity principles but it is extending their validity to include

[1] Frolov and Novikov [41] demonstrated that the interaction of a wormhole with classical matter generates a non-potential component of the gravitational field. That is why a locally static wormhole is generically unstable with respect to the processes which could transform it into a "time machine". Relative motion of the wormhole's mouths also generates a time gap for clock synchronization as discussed in [10]. One can interpret the above results in the following way. There exist inner relations between the topological and causal properties of a spacetime. The existence of CTCs is a general property of multiply connected locally static spacetimes [41].

backward time solutions as a "necessity" to preserve always the conservation laws and the causality principle. In this sense, we speak in terms of an Extended (or Open) Relativity [25]. The Many Worlds interpretation of Quantum Mechanics [46] is another necessary ingredient of this new approach, since it comes out as a final point of a process which starts from the very basic principles. Namely: the solution of the EPR paradox as a conflict between Relativity and Quantum Mechanics [23]; the generalization of entanglement concept as a gate, through topology changes, recovering backward causation [22]; and finally a dynamical scheme for the unification of the different interactions [30]. We should stress immediately that the whole question of time, and of time machine hypothesis, is completely changed by this new picture, because "moving in time" is no longer an extreme possibility, only due to the relative motion of two frameworks on the same time arrow, but a more general possibility related also to the existence of backward directed time arrow [25]. In the "conventional" time machine, "backward in time" means only a relative or circular "past", while, in the new scheme, there is also another general possibility and the real question is how to get "physically" a backward time arrow, that we cannot ordinarily perceive, but which has to exist (see [22, 23, 25]). In order to substantiate the previous statement, let us recall the equations (22) describing quantum matter, that is $\psi(x) = e^{-ikx}u(k)$, and $\psi^*(x) = e^{ikx}u^*(k)$, which, as we have seen, can be interpreted respectively, as progressive an regressive solutions in four combination. It comes out that a function (a superposition) of the form $\varphi(x) = \alpha_1\psi(x) + \alpha_2\psi^*(x)$, is a general solution of the dynamics and the states ψ and ψ^* can be interpreted as *entangled* since they can influence each other also when they are, in standard picture, considered disconnected. The absolute validity of conservation laws yields a symmetric dynamics (backward and forward evolution of the system) and the entanglement of states is naturally determined without any arbitrary violation. This starting point, based on first principles, is evidently of very general meaning, because it is suggesting that the existence of two time arrows is a necessary feature of Nature, independently of our capability of perceiving them.

The necessity to take into account two arrows of time is implying the consideration of a fifth dimension to open both directions in time and, in the new picture the fifth dimension is not a pure mathematical tool, but a feature of real physical meaning, because, from the embedding of this fifth dimension, it has been shown that is possible to get mass, charge and spin of all particles [30]. The resulting scheme seems properly to work in explaining many paradoxes and shortcomings of modern physics (for instance, it seems capable, beside the EPR paradox, of explaining highly questioning problems, like the existence of black holes and gamma ray bursts [43], rotation curves of galaxies [44]) and also in recovering basic principle, like conservation and causality, with the minimal necessary number of parameters, in comparison to all existing theories (like Strings, or Supergravity). Particularly interesting, in the context which we are discussing here, is the fact that entangled gravitational systems constituted by black holes, wormholes and white holes naturally emerge, through topology changes, starting from the request that the mass-energy of collapsing systems is conserved in the framework of the General Conservation Principle. The main point of this result is that such systems can be stable so that (due to conservations which avoid spontaneous symmetry breakings) time travels and time machine become (at least on a theoretical ground) a real possibility [43], and moreover, starting from $5D$ dynamics, CTCs are ordinary solutions of field equations. Due to this point, entangled gravitational systems (in $4D$) are a straightforward necessity of the theory [22, 43]. These results deserve a further discussion due to the fact that the mass of a given particle is fixed by the embedding process, while the splitting of $4D$-Klein Gordon equation in two Dirac equations (forward and backward) gives rise to the emergence of other features of particle as the spin and the charge which are automatically conjugated and conserved by the occurrence of two time arrows. In terms of a group description, the splitting process defines the $SU(2)$ (spin) and $U(1)$ (charge) components of the supergroup $\mathcal{G}_5$ whose features was discussed in [30]. Finally, as we have seen, from the reduction mechanism we get solutions of the form (65), (66), *i.e.* $dS^2 = dt^2 - f(t, x_1)(dx^1)^2 - f(t, x_2)(dx^2)^2 - f(t, x_3)(dx^3)^2 + (dx^4)^2$, where $f(t, x_j) = \exp i(\omega t + k_j x^j)$, with $j = 1, 2, 3$. Intrinsically such a solution has two times and, due to the structure of the functions $f(t, x_j)$ closed time-like paths are naturally allowed possibilities starting from $5D$–dynamics. These results, in the context of time machine, have a deep meaning, since they are not "anomalies" in the framework of the standard theory (General Relativity) but are "ordinary" outputs in the framework of this "Open Relativity". While the concept of time travels in General Relativity is related only to the relativistic effects among frameworks in relative motion on the same time arrow, here, in the new approach, the relative motion can be taken into account on two time arrows, this means that a closed time-like path does not need to a topology change imposed "by hand", because the projected motion along a circumference is the combination of two linear motions in opposite sense, so then the topology change is the natural result of an underlying feature of Nature [30, 25]. This feature comes out every time a topology change is necessary, and this occurs if a conservation law would be otherwise violated. It is evident, even if starting from Quantum Mechanics considerations, that there is the full recovering of Special and General Relativity principles (because we do not need to travel faster than light) but the mechanism of relative motion between two frameworks, which is generating the hypothesis of time travels, is clearly not the same if applied on one or two time arrows. In this sense, we speak in terms of Open Quantum Relativity [25]. The new scheme, even not entering in any suggestion for the technical devices of a hypothetical time machine, provides nevertheless first principles on which such a machine should be based, because it is fixing the limits in which the phenomenon would be eventually possible. In fact, only in a situation of impossibility to avoid a violation of a conservation law, it happens that the Nature reacts changing the topology, so then we should first of all find such a situation in the universe or be able to recreate it. In the first case (up today), the "laboratory" for such an experiment should be a black hole, in the second case, an "entanglement machine". This would be a kind of machine not possible to design yet, but which we cannot in principle exclude. In fact, there are works of authors [45] which demonstrate that an entanglement of two macroscopic systems does not need the entanglement of every component of them in correspondence one-to-one (which will be probably impossible to obtain) but only a correspondence at least under the limit of the uncertainty principle (certainly not easy, but perhaps not impossible). In this framework, the Many Worlds Theory should be taken not as a possibility, but as a necessity, in order to avoid new and deeper paradoxes induced by backward time travels, like the famous hypothesis of a time traveller killing his grandmother when she was young (before his own birth) so destroying himself and making impossible to kill grandmother. Only the Many Worlds Theory, in its full version [46] of an infinity of universes (each one representing a virtual possibility of evolution which can become a reality), can settle this puzzle, saving time travel generalized hypothesis, causality and pure logic together. In this picture, every universe can be reduced to a "local fact". It is very difficult, today, to estimate all the consequences of the above considerations, mainly concerning a "time machine", nevertheless one should try to derive some conclusions to open an useful debate. Time travel hypothesis in which two time arrows (and CTCs) are general features of Nature, represents a break point with the more traditional view of time machine but able, in principle, to reconcile General Relativity and Causality in a framework in which Relativity and Quantum Mechanics are deeply connected. Due to the fact that there are experimental evidences, not only for the well established standard effects (e.g. the lengthening of the life of accelerated elementary particles) but also for new ones (like the instantaneous teleportation via entanglement effects), it seems impossible to simply close the doors to any time travelling hypothesis and so the necessity for a new theoretical framework to include this possibility seems unavoidable. We are in the middle of a ford, we should accomplish this crossing and find a new synthesis of the scientific discoveries of last century, which are tesserae of a mosaic still incomplete. In any case, we do not believe that any semi-classical approximation will be enough to settle this modern puzzle. Like many other people, we see the necessity of a Quantum Relativity as an issue which cannot be postponed any longer as solution of modern paradoxes (from EPR to twins) because they are coming out directly from first principles of physics.

4. The Covariant Symplectic Structure of Open Quantum Relativity

The above discussion gives evidences that it has to be a deep link between General Relativity and Quantum Mechanics, moreover, such a link seems necessary to overcome the shortcomings and the paradoxes of modern physics. To seek for this connection, a fundamental point is that an intrinsic symplectic structure is present in General Relativity and this can be recovered from conservation laws in Hamiltonian formulation. Before entering into the formalism of this derivation, we have to recall what it means a symplectic structure and how it can be recovered in General Relativity. Furthermore, we have to enter into some details of canonical quantization procedure.

A self-consistent quantum field theory of space-time has not been achieved, up to date, using standard quantization approaches. In fact, the request of general coordinate invariance of General Relativity gives rise to several shortcomings in quantum field theory. For a generic physical field, one has to assign the field amplitudes and their first time derivatives, in order to determine the time evolution of such a field considered as a dynamical entity and in General Relativity, these quantities are not suitable for dynamical determination, since the metric $g_{\alpha\beta}$ can evolve at any time by a general coordinate transformation. Physical observables do not change and the consequence of such an operation is a pure relabelling which leaves the theory invariant. This means that it is necessary a separation of metric degrees of freedom into a part related to the true dynamical information and a part related to the coordinate system. General Relativity, from this point of view, has common features with classical Electromagnetism: the coordinate invariance plays a role analogous to the electromagnetic gauge invariance and in both cases gives rise to redundant variables to insure the transformation properties. Shortcomings arise when one is going to distinguish dynamical from gauge variables. This operational approach works in Electromagnetism while it fails in General Relativity, where dynamics is highly non-linear. Independent dynamical modes of gravitational field are recovered when theory is formulated in a canonical form and the canonical formalism is preliminary in quantization program, since it gives rise directly to Poisson brackets. To realize such a program, one needs first order field equations and a $(3+1)$-form of dynamics. In General Relativity, the ADM formalism [80] has led to the definition of gravitational Hamiltonian and time as a conjugate couple of variables. In any case, the main feature of General Relativity, the covariance, is impaired and the full quantization of gravity has not been achieved up to now. The main problems are related to the lack of Hilbert space and of a quantum measure for the metric.

As we will show below, [1, 2], a prominent role, in the identification of a covariant symplectic structure, can be played by bilinear Hamiltonians which have to be conserved. In fact, taking into account generic Hamiltonian invariants, constructed by covariant vectors, bivectors or tensors, it is possible to show that a symplectic structure can be achieved in any case. By specifying the nature of such vector fields (or, in general, tensor invariants), it gives rise to intrinsically symplectic structure which is always related to Hamilton-like equations (and a Hamilton-Jacobi-like approach is always found). This works for curvature invariants, Maxwell and Dirac fields. In any case, the two main basic assumptions are that conservation laws always have to be identified and moreover that the structure is completely affine and not metric. Such a program of "generalized" quantization works for every generic fields, once that bilinear Hamiltonian invariants are correctly identified starting from suitable vectors, tensors or spinors. From a theoretical point of view, such a covariant symplectic structure constitutes the core of Open Quantum Relativity.

4.1. The Symplectic Structure and the Canonical Formulation of Mechanics

As we said, the identification of a syplectic structure has a prominent role in the formulation of a physical theory at fundamental level. In this section, we outline the main features of a generic symplectic structure and relate it to the Hamiltonian mechanics, with the goal to achieve a covariant formulation.

In general, a symplectic structure [1] is achieved if the couple

$$\{\mathbf{E_{2n}}, \mathbf{w}\}, \tag{91}$$

is defined, where $\mathbf{E_{2n}}$ is a vector space and the tensor $\mathbf{w}$, on $\mathbf{E_{2n}}$, associates scalar functions to pairs of vectors, that is

$$[\mathbf{x}, \mathbf{y}] = \mathbf{w}(\mathbf{x}, \mathbf{y}), \tag{92}$$

which is the *antiscalar* product. Such an operation satisfies the following properties

$$[\mathbf{x}, \mathbf{y}] = -[\mathbf{y}, \mathbf{x}] \qquad \forall \mathbf{x}, \mathbf{y} \in \mathbf{E_{2n}} \tag{93}$$

$$[\mathbf{x}, \mathbf{y}+\mathbf{z}] = [\mathbf{x}, \mathbf{y}] + [\mathbf{x}\mathbf{z}] \quad \forall \mathbf{x}, \mathbf{y}, \mathbf{z} \in \mathbf{E_{2n}}, \tag{94}$$

$$\mathrm{a}[\mathbf{x}, \mathbf{y}] = [\mathrm{a}\mathbf{x}, \mathbf{y}] \qquad \forall \mathrm{a} \in R, \quad \mathbf{x}, \mathbf{y} \in \mathbf{E_{2n}} \tag{95}$$

$$[\mathbf{x}, \mathbf{y}] = 0 \qquad \forall \mathbf{y} \in \mathbf{E_{2n}} \Rightarrow \mathbf{x} = 0 \tag{96}$$

$$[\mathbf{x}, [\mathbf{y}, \mathbf{z}]] + [\mathbf{y}, [\mathbf{z}, \mathbf{x}]] + [\mathbf{z}, [\mathbf{x}, \mathbf{y}]] = 0 \tag{97}$$

The last one is the Jacobi cyclic identity.

If $\{\mathbf{e}_i\}$ is a vector basis in $\mathbf{E_{2n}}$, the antiscalar product is completely singled out by the matrix elements

$$w_{ij} = [\mathbf{e}_i, \mathbf{e}_j], \tag{98}$$

where $\mathbf{w}$ is an antisymmetric matrix with determinant different from zero. Every antiscalar product between two vectors can be expressed as

$$[\mathbf{x}, \mathbf{y}] = w_{ij} x^i y^j, \tag{99}$$

where x^i and y^j are the vector components in the given basis.

The form of the matrix $\mathbf{w}$ and the relation (99) become considerably simpler if a canonical basis is taken into account for $\mathbf{w}$. Since $\mathbf{w}$ is an antisymmetric non-degenerate tensor, it is always possible to represent it through the matrix

$$J = \begin{pmatrix} 0 & I \\ -I & 0 \end{pmatrix}, \tag{100}$$

where I is a $(n \times n)$ unit matrix. Every basis where $\mathbf{w}$ can be represented through the form (100) is a *symplectic basis*. In other words, the symplectic bases are the canonical bases for any antisymmetric non-degenerate tensor $\mathbf{w}$ and can be characterized by the following conditions:

$$[\mathbf{e}_i, \mathbf{e}_j] = 0, \quad [\mathbf{e}_{n+i}, \mathbf{e}_{n+j}] = 0, \quad [\mathbf{e}_i, \mathbf{e}_{n+j}] = \delta_{ij}, \tag{101}$$

which have to be verified for every pair of values i and j ranging from 1 to n.

Finally, the expression of the antiscalar product between two vectors, in a symplectic basis, is

$$[\mathbf{x}, \mathbf{y}] = \sum_{i=1}^{n} \left(x^{n+i} y^i - x^i y^{n+i}\right), \tag{102}$$

and a symplectic transformation in $\mathbf{E_{2n}}$ leaves invariant the antiscalar product

$$\mathbf{S}[\mathbf{x}, \mathbf{y}] = [\mathbf{S}(\mathbf{x}), \mathbf{S}(\mathbf{y})] = [x, y]. \tag{103}$$

It is easy to see that standard Quantum Mechanics satisfies such properties and so it is endowed with a symplectic structure.

On the other hand a standard canonical description can be sketched as follows. For example, the relativistic Lagrangian of a charged particle interacting with a vector field $A(q; s)$ is

$$\mathcal{L}(q, u; s) = \frac{mu^2}{2} - eu \cdot A(q; s), \tag{104}$$

where the scalar product is defined as

$$z \cdot w = z_\mu w^\mu = \eta_{\mu\nu} z^\mu w^\nu, \tag{105}$$

and the signature of the Minkowski spacetime is the usual one with

$$z_\mu = \eta_{\mu\nu} z^\nu, \qquad \widehat{\eta} = \mathrm{diag}(1, -1, -1, -1). \tag{106}$$

Furthermore, the contravariant vector u^μ with components $u = (u^0, u^1, u^2, u^3)$ is the four-velocity

$$u^\mu = \frac{\mathrm{d}q^\mu}{\mathrm{d}s}. \tag{107}$$

The canonical conjugate momentum π^μ is defined as

$$\pi^\mu = \eta^{\mu\nu}\frac{\partial \mathcal{L}}{\partial u^\nu} = mu^\mu - eA^\mu, \tag{108}$$

so that the relativistic Hamiltonian can be written in the form

$$\mathcal{H}(q, \pi; s) = \pi \cdot u - \mathcal{L}(q, u; s). \tag{109}$$

Suppose now that we wish to use any other coordinate system x^α as Cartesian, curvilinear, accelerated or rotating one. Then the coordinates q^μ are functions of the x^α, which can be written explicitly as

$$q^\mu = q^\mu(x^\alpha). \tag{110}$$

The four-vector of particle velocity u^μ is transformed according to the expression

$$u^\mu = \frac{\partial q^\mu}{\partial x^\alpha}\frac{\mathrm{d}x^\alpha}{\mathrm{d}s} = \frac{\partial q^\mu}{\partial x^\alpha}v^\alpha, \tag{111}$$

where

$$v^\mu = \frac{\mathrm{d}x^\mu}{\mathrm{d}s}, \tag{112}$$

is the transformed four-velocity expressed in terms of the new coordinates. The vector field A^μ is also transformed as a vector

$$\mathcal{A}^\mu = \frac{\partial x^\mu}{\partial q^\alpha}A^\alpha. \tag{113}$$

In the new coordinate system x^α the Lagrangian (104) becomes

$$\mathcal{L}(x, v; s) = g_{\mu\nu}\left[\frac{m}{2}v^\mu v^\nu - ev^\mu \mathcal{A}^\nu(x; s)\right], \tag{114}$$

where

$$g_{\alpha\beta} = \eta_{\mu\nu}\frac{\partial q^\mu}{\partial x^\alpha}\frac{\partial q^\nu}{\partial x^\beta}. \tag{115}$$

The Lagrange equations can be written in the usual form

$$\frac{\mathrm{d}}{\mathrm{d}s}\left(\frac{\partial \mathcal{L}}{\partial v^\lambda}\right) - \frac{\partial \mathcal{L}}{\partial x^\lambda} = 0. \tag{116}$$

In the case of a free particle (no interaction with an external vector field), we have

$$\frac{\mathrm{d}}{\mathrm{d}s}(g_{\lambda\mu}v^\mu) - \frac{1}{2}\frac{\partial g_{\mu\nu}}{\partial x^\lambda}v^\mu v^\nu = 0. \tag{117}$$

Specifying the covariant velocity v_λ as

$$v_\lambda = g_{\lambda\mu}v^\mu, \tag{118}$$

and using the well-known identity for connections $\Gamma^{\alpha}_{\mu\nu}$

$$\frac{\partial g_{\mu\nu}}{\partial x^{\lambda}} = \Gamma^{\alpha}_{\lambda\mu} g_{\alpha\nu} + \Gamma^{\alpha}_{\lambda\nu} g_{\alpha\mu}, \tag{119}$$

we obtain

$$\frac{\mathrm{D}v_{\lambda}}{\mathrm{D}s} = \frac{\mathrm{d}v_{\lambda}}{\mathrm{d}s} - \Gamma^{\mu}_{\lambda\nu} v^{\nu} v_{\mu} = 0. \tag{120}$$

Here $\mathrm{D}v_{\lambda}/\mathrm{D}s$ denotes the covariant derivative of the covariant velocity v_{λ} along the curve $x^{\nu}(s)$. Using Eqs. (118) and (119) and the fact that the affine connection $\Gamma^{\lambda}_{\mu\nu}$ is symmetric in the indices μ and ν, we obtain the equation of motion for the contravariant vector v^{λ}

$$\frac{\mathrm{D}v^{\lambda}}{\mathrm{D}s} = \frac{\mathrm{d}v^{\lambda}}{\mathrm{d}s} + \Gamma^{\lambda}_{\mu\nu} v^{\mu} v^{\nu} = 0. \tag{121}$$

Before taking into account the Hamiltonian description, let us note that the generalized momentum p_{μ} is defined as

$$p_{\mu} = \frac{\partial \mathcal{L}}{\partial v^{\mu}} = m g_{\mu\nu} v^{\nu}, \tag{122}$$

while, from Lagrange equations of motion, we obtain

$$\frac{\mathrm{d}p_{\mu}}{\mathrm{d}s} = \frac{\partial \mathcal{L}}{\partial x^{\mu}}. \tag{123}$$

The transformation from $(x^{\mu}, v^{\mu}; s)$ to $(x^{\mu}, p_{\mu}; s)$ can be accomplished by means of a Legendre transformation, and instead of the Lagrangian (114), we consider the Hamilton function

$$\mathcal{H}(x, p; s) = p_{\mu} v^{\mu} - \mathcal{L}(x, v; s). \tag{124}$$

The differential of the Hamiltonian in terms of x, p and s is given by

$$\mathrm{d}\mathcal{H} = \frac{\partial \mathcal{H}}{\partial x^{\mu}} \mathrm{d}x^{\mu} + \frac{\partial \mathcal{H}}{\partial p_{\mu}} \mathrm{d}p_{\mu} + \frac{\partial \mathcal{H}}{\partial s} \mathrm{d}s. \tag{125}$$

On the other hand, from Eq.(124), we have

$$\mathrm{d}\mathcal{H} = v^{\mu} \mathrm{d}p_{\mu} + p_{\mu} \mathrm{d}v^{\mu} - \frac{\partial \mathcal{L}}{\partial v^{\mu}} \mathrm{d}v^{\mu} - \frac{\partial \mathcal{L}}{\partial x^{\mu}} \mathrm{d}x^{\mu} - \frac{\partial \mathcal{L}}{\partial s} \mathrm{d}s. \tag{126}$$

Taking into account Eq.(122), the second and the third term on the right-hand-side of Eq.(126) cancel out. Eq.(123) can be further used to cast Eq.(126) into the form

$$\mathrm{d}\mathcal{H} = v^{\mu} \mathrm{d}p_{\mu} - \frac{\mathrm{d}p_{\mu}}{\mathrm{d}s} \mathrm{d}x^{\mu} - \frac{\partial \mathcal{L}}{\partial s} \mathrm{d}s, \tag{127}$$

Comparison between Eqs.(125) and (127) yields the Hamilton equations of motion

$$\frac{\mathrm{d}x^{\mu}}{\mathrm{d}s} = \frac{\partial \mathcal{H}}{\partial p_{\mu}}, \qquad \frac{\mathrm{d}p_{\mu}}{\mathrm{d}s} = -\frac{\partial \mathcal{H}}{\partial x^{\mu}}, \tag{128}$$

where the Hamiltonian is given by

$$\mathcal{H}(x, p; s) = \frac{g^{\mu\nu}}{2m} p_{\mu} p_{\nu} + \frac{e}{m} p_{\mu} \mathcal{A}^{\mu}. \tag{129}$$

In the case of a free particle, the Hamilton equations can be written explicitly as

$$\frac{\mathrm{d}x^{\mu}}{\mathrm{d}s} = \frac{g^{\mu\nu}}{m}p_{\nu}, \qquad \frac{\mathrm{d}p_{\lambda}}{\mathrm{d}s} = -\frac{1}{2m}\frac{\partial g^{\mu\nu}}{\partial x^{\lambda}}p_{\mu}p_{\nu}. \tag{130}$$

To obtain the equations of motion we need the expression

$$\frac{\partial g^{\mu\nu}}{\partial x^{\lambda}} = -\Gamma^{\mu}_{\lambda\alpha}g^{\alpha\nu} - \Gamma^{\nu}_{\lambda\alpha}g^{\alpha\mu}, \tag{131}$$

which can be derived from the obvious identity

$$\frac{\partial}{\partial x^{\lambda}}(g^{\mu\alpha}g_{\alpha\nu}) = 0, \tag{132}$$

and Eq.(119). From the second of Eqs. (130), we obtain

$$\frac{\mathrm{D}p_{\lambda}}{\mathrm{D}s} = \frac{\mathrm{d}p_{\lambda}}{\mathrm{d}s} - \Gamma^{\mu}_{\lambda\nu}v^{\nu}p_{\mu} = 0, \tag{133}$$

similar to equation (120). Differentiating the first of the Hamilton equations (130) with respect to s and taking into account equations (131) and (133), it is worthwhile stressing that we again arrive to the equation for the geodesics (121).

Let us now show that on a generic curved manifolds the Poisson brackets are conserved so that such a symplectic structure is, in any case, related to a general conservation principle [81]. To achieve this result, we need the following identities

$$g^{\mu\nu} = g^{\nu\mu} = \eta^{\alpha\beta}\frac{\partial x^{\mu}}{\partial q^{\alpha}}\frac{\partial x^{\nu}}{\partial q^{\beta}}, \tag{134}$$

$$\frac{\partial^2 x^{\lambda}}{\partial q^{\alpha}\partial q^{\beta}} = -\Gamma^{\lambda}_{\mu\nu}\frac{\partial x^{\mu}}{\partial q^{\alpha}}\frac{\partial x^{\nu}}{\partial q^{\beta}}, \tag{135}$$

To prove (135), we differentiate the obvious identity

$$\frac{\partial x^{\lambda}}{\partial q^{\rho}}\frac{\partial q^{\rho}}{\partial x^{\nu}} = \delta^{\lambda}_{\nu}. \tag{136}$$

As a result, we find

$$\Gamma^{\lambda}_{\mu\nu} = \frac{\partial x^{\lambda}}{\partial q^{\rho}}\frac{\partial^2 q^{\rho}}{\partial x^{\mu}\partial x^{\nu}} = -\frac{\partial q^{\rho}}{\partial x^{\nu}}\frac{\partial q^{\sigma}}{\partial x^{\mu}}\frac{\partial^2 x^{\lambda}}{\partial q^{\rho}\partial q^{\sigma}}. \tag{137}$$

The next step is to calculate the fundamental Poisson brackets in terms of the variables (x^{μ}, p_{ν}), initially defined using the canonical variables (q^{μ}, π_{ν}) according to the relation

$$[U, V] = \frac{\partial U}{\partial q^{\mu}}\frac{\partial V}{\partial \pi_{\mu}} - \frac{\partial V}{\partial q^{\mu}}\frac{\partial U}{\partial \pi_{\mu}}, \tag{138}$$

where $U(q^{\mu}, \pi_{\nu})$ and $V(q^{\mu}, \pi_{\nu})$ are arbitrary functions. Making use of Eqs.(108) and (111), we know that the variables

$$q^{\mu} \quad \Leftrightarrow \quad \pi_{\mu} = mu_{\mu} = m\eta_{\mu\nu}u^{\nu} = m\eta_{\mu\nu}\frac{\partial q^{\nu}}{\partial x^{\alpha}}v^{\alpha}, \tag{139}$$

give rise to a canonical conjugate pair. Using Eq.(122), we would like to check whether the variables

$$x^{\mu} \quad \Leftrightarrow \quad p_{\mu} = m g_{\mu\nu} v^{\nu} = g_{\mu\nu} \eta^{\alpha\lambda} \pi_{\lambda} \frac{\partial x^{\nu}}{\partial q^{\alpha}}, \tag{140}$$

form a canonical conjugate pair. We have

$$\begin{aligned}
[U, V] = & \left[\frac{\partial U}{\partial x^{\alpha}} \frac{\partial x^{\alpha}}{\partial q^{\mu}} + \frac{\partial U}{\partial p_{\sigma}} \eta^{\beta\lambda} \pi_{\lambda} \frac{\partial}{\partial q^{\mu}} \left(g_{\sigma\nu} \frac{\partial x^{\nu}}{\partial q^{\beta}} \right) \right] \\
& \times \frac{\partial V}{\partial p_{\alpha}} g_{\alpha\chi} \eta^{\rho\mu} \frac{\partial x^{\chi}}{\partial q^{\rho}} \\
& - \left[\frac{\partial V}{\partial x^{\alpha}} \frac{\partial x^{\alpha}}{\partial q^{\mu}} + \frac{\partial V}{\partial p_{\sigma}} \eta^{\beta\lambda} \pi_{\lambda} \frac{\partial}{\partial q^{\mu}} \left(g_{\sigma\nu} \frac{\partial x^{\nu}}{\partial q^{\beta}} \right) \right] \\
& \times \frac{\partial U}{\partial p_{\alpha}} g_{\alpha\chi} \eta^{\rho\mu} \frac{\partial x^{\chi}}{\partial q^{\rho}}.
\end{aligned} \tag{141}$$

The first and the third term on the right-hand-side of Eq.(141) can be similarly manipulated as follows

$$\begin{aligned}
\mathbf{I} - \text{st term} &= \frac{\partial U}{\partial x^{\alpha}} \frac{\partial V}{\partial p_{\beta}} g_{\beta\chi} \eta^{\rho\mu} \frac{\partial x^{\chi}}{\partial q^{\rho}} \frac{\partial x^{\alpha}}{\partial q^{\mu}} \\
&= g_{\beta\chi} g^{\chi\alpha} \frac{\partial U}{\partial x^{\alpha}} \frac{\partial V}{\partial p_{\beta}} = \frac{\partial U}{\partial x^{\alpha}} \frac{\partial V}{\partial p_{\alpha}},
\end{aligned} \tag{142}$$

$$\mathbf{III} - \text{rd term} = -\frac{\partial V}{\partial x^{\alpha}} \frac{\partial U}{\partial p_{\alpha}}. \tag{143}$$

Next, we manipulate the second term on the right-hand-side of Eq.(141). We obtain

$$\begin{aligned}
\mathbf{II} - \text{nd term} &= \frac{\partial U}{\partial p_{\sigma}} \frac{\partial V}{\partial p_{\alpha}} g_{\alpha\chi} \eta^{\rho\mu} \frac{\partial x^{\chi}}{\partial q^{\rho}} \eta^{\beta\lambda} \pi_{\lambda} \\
& \times \left[g_{\sigma\nu} \frac{\partial^2 x^{\nu}}{\partial q^{\mu} \partial q^{\beta}} + \frac{\partial x^{\nu}}{\partial q^{\beta}} \frac{\partial g_{\sigma\nu}}{\partial x^{\gamma}} \frac{\partial x^{\gamma}}{\partial q^{\mu}} \right] \\
&= \frac{\partial U}{\partial p_{\sigma}} \frac{\partial V}{\partial p_{\alpha}} g_{\alpha\chi} \eta^{\rho\mu} \frac{\partial x^{\chi}}{\partial q^{\rho}} \eta^{\beta\lambda} \pi_{\lambda} \\
\times & \left[-g_{\sigma\nu} \Gamma^{\nu}_{\gamma\delta} \frac{\partial x^{\gamma}}{\partial q^{\mu}} \frac{\partial x^{\delta}}{\partial q^{\beta}} + \frac{\partial x^{\delta}}{\partial q^{\beta}} \frac{\partial x^{\gamma}}{\partial q^{\mu}} \left(\Gamma^{\nu}_{\gamma\sigma} g_{\nu\delta} + \Gamma^{\nu}_{\gamma\delta} g_{\nu\sigma} \right) \right] \\
&= \frac{\partial U}{\partial p_{\sigma}} \frac{\partial V}{\partial p_{\alpha}} g_{\alpha\chi} g^{\chi\gamma} \eta^{\beta\lambda} \pi_{\lambda} \frac{\partial x^{\delta}}{\partial q^{\beta}} g_{\nu\delta} \Gamma^{\nu}_{\gamma\sigma} \\
&= \frac{\partial U}{\partial p_{\sigma}} \frac{\partial V}{\partial p_{\beta}} g_{\mu\nu} \eta^{\alpha\lambda} \pi_{\lambda} \frac{\partial x^{\nu}}{\partial q^{\alpha}} \Gamma^{\mu}_{\beta\sigma} \\
&= \Gamma^{\lambda}_{\mu\nu} p_{\lambda} \frac{\partial U}{\partial p_{\nu}} \frac{\partial V}{\partial p_{\mu}}.
\end{aligned} \tag{144}$$

The fourth term is similar to the second one but with U and V interchanged

$$\mathbf{IV} - \text{th term} = -\Gamma^{\lambda}_{\mu\nu} p_{\lambda} \frac{\partial U}{\partial p_{\mu}} \frac{\partial V}{\partial p_{\nu}}. \tag{145}$$

In the absence of torsion, the affine connection $\Gamma^{\lambda}_{\mu\nu}$ is symmetric with respect to the lower indices, so that the second and the fourth term on the right-hand-side of Eq.(141) cancel each other. Therefore,

$$[U, V] = \frac{\partial U}{\partial x^{\mu}}\frac{\partial V}{\partial p_{\mu}} - \frac{\partial V}{\partial x^{\mu}}\frac{\partial U}{\partial p_{\mu}}, \tag{146}$$

which means, at the end, that the fundamental Poisson brackets are always conserved. On the other hand, this implies that the variables $\{x^{\mu}, p_{\nu}\}$ are a canonical conjugate pair.

As a final remark, we have to say that a generic metric $g_{\alpha\beta}$ and a connection $\Gamma^{\alpha}_{\mu\nu}$ are related to the fact that we are passing from a Minkowski-flat spacetime (local inertial reference frame) to an accelerated reference frame (curved spacetime). In what follows, we want to show that a generic bilinear Hamiltonian invariant, which is conformally conserved, is always based on a canonical symplectic structure. The specific theory is assigned by the vector (or tensor) fields which define the Hamiltonian invariant.

4.2. The Symplectic Structure Compatible with the General Covariance

The above considerations can be linked together leading to a more general scheme where a covariant symplectic structure is achieved. Summarizing, the main points which we need are: $i)$ an even-dimensional vector space $\mathbf{E_{2n}}$ equipped with an antiscalar product satisfying the algebra (93)-(97); $ii)$ generic vector fields defined on such a space which have to satisfy the Poisson brackets; $iii)$ first-order equations of motion which can be read as Hamilton-like equations; $iv)$ and the feature that general covariance has to be preserved.

Such a program can be pursued by taking into account covariant and contravariant vector fields. In fact, it is always possible to construct the Hamiltonian invariant

$$\mathcal{H} = V^{\alpha}V_{\alpha}, \tag{147}$$

which satisfies the relation

$$\delta\mathcal{H} = \delta(V^{\alpha}V_{\alpha}) = 0, \tag{148}$$

being δ a spurious variation due to the transport. It is worth stressing that the vectors V^{α} and V_{α} are not specified and the following considerations are completely general. Eq.(147) is an already parameterized invariant which can be seen as the density of a parameterized action principle [82, 83] where the time coordinate is not distinguished *a priori* from the other coordinates.

The intrinsic variation of V^{α} is

$$DV^{\alpha} = dV^{\alpha} - \delta V^{\alpha} = \partial_{\beta}V^{\alpha}dx^{\beta} - \delta V^{\alpha}, \tag{149}$$

where d is the total variation and δ the spurious variation. In General Relativity, if the spurious variation of a given quantity is equal to zero, means that the quantity is conserved. From the definition of covariant derivative, we have

$$DV^{\alpha} = \partial_{\beta}V^{\alpha}dx^{\beta} + \Gamma^{\alpha}_{\sigma\beta}V^{\sigma}dx^{\beta}, \tag{150}$$

and

$$\nabla_{\beta}V^{\alpha} = \partial_{\beta}V^{\alpha} + \Gamma^{\alpha}_{\sigma\beta}V^{\sigma}, \tag{151}$$

and then

$$\delta V^\alpha = -\Gamma^\alpha_{\sigma\beta} V^\sigma dx^\beta. \tag{152}$$

Analogously,

$$DV_\alpha = dV_\alpha - \delta V_\alpha = \partial_\beta V_\alpha dx^\beta - \delta V_\alpha, \tag{153}$$

and then

$$DV_\alpha = \partial_\beta V_\alpha dx^\beta - \Gamma^\sigma_{\alpha\beta} V_\sigma dx^\beta, \tag{154}$$

and

$$\nabla_\beta V_\alpha = \partial_\beta V_\alpha - \Gamma^\sigma_{\alpha\beta} V_\sigma. \tag{155}$$

The spurious variation is now

$$\delta V_\alpha = \Gamma^\sigma_{\alpha\beta} V_\sigma dx^\beta. \tag{156}$$

Developing the variation (148), we have

$$\delta\mathcal{H} = V_\alpha \delta V^\alpha + V^\alpha \delta V_\alpha, \tag{157}$$

and

$$\frac{\delta\mathcal{H}}{dx^\beta} = V_\alpha \frac{\delta V^\alpha}{dx^\beta} + V^\alpha \frac{\delta V_\alpha}{dx^\beta}, \tag{158}$$

which becomes

$$\frac{\delta\mathcal{H}}{dx^\beta} = \frac{\delta V^\alpha}{dx^\beta}\frac{\partial\mathcal{H}}{\partial V^\alpha} + \frac{\delta V_\alpha}{dx^\beta}\frac{\partial\mathcal{H}}{\partial V_\alpha}, \tag{159}$$

being

$$\frac{\partial\mathcal{H}}{\partial V^\alpha} = V_\alpha, \qquad \frac{\partial\mathcal{H}}{\partial V_\alpha} = V^\alpha. \tag{160}$$

From Eqs.(152) and (156), it is

$$\frac{\delta V^\alpha}{dx^\beta} = -\Gamma^\alpha_{\sigma\beta} V^\sigma = -\Gamma^\alpha_{\sigma\beta}\left(\frac{\partial\mathcal{H}}{\partial V_\sigma}\right), \tag{161}$$

$$\frac{\delta V_\alpha}{dx^\beta} = \Gamma^\sigma_{\alpha\beta} V_\sigma = \Gamma^\sigma_{\alpha\beta}\left(\frac{\partial\mathcal{H}}{\partial V^\sigma}\right), \tag{162}$$

and substituting into Eq.(159), we have

$$\frac{\delta\mathcal{H}}{dx^\beta} = -\Gamma^\alpha_{\sigma\beta}\left(\frac{\partial\mathcal{H}}{\partial V_\sigma}\right)\left(\frac{\partial\mathcal{H}}{\partial V^\alpha}\right) + \Gamma^\sigma_{\alpha\beta}\left(\frac{\partial\mathcal{H}}{\partial V_\alpha}\right)\left(\frac{\partial\mathcal{H}}{\partial V^\sigma}\right), \tag{163}$$

and then, since α and σ are mute indexes, the expression

$$\frac{\delta\mathcal{H}}{dx^\beta} = (\Gamma^\alpha_{\sigma\beta} - \Gamma^\alpha_{\sigma\beta})\left(\frac{\partial\mathcal{H}}{\partial V_\sigma}\right)\left(\frac{\partial\mathcal{H}}{\partial V^\alpha}\right) \equiv 0, \tag{164}$$

is identically equal to zero. In other words, $\mathcal{H}$ is absolutely conserved, and the analogy with a canonical Hamiltonian structure is straightforward. In fact, if, as above,

$$\mathcal{H} = \mathcal{H}(p, q) \tag{165}$$

is a classical generic Hamiltonian function, expressed in the canonical phase-space variables $\{p, q\}$, the total variation, in a vector space $\mathbf{E_{2n}}$ whose dimensions are generically given by p_i and q_j with $i, j = 1, ..., n$, is

$$d\mathcal{H} = \frac{\partial \mathcal{H}}{\partial q} dq + \frac{\partial \mathcal{H}}{\partial p} dp, \tag{166}$$

and

$$\begin{aligned} \frac{d\mathcal{H}}{dt} &= \frac{\partial \mathcal{H}}{\partial q}\dot{q} + \frac{\partial \mathcal{H}}{\partial p}\dot{p} \\ &= \frac{\partial \mathcal{H}}{\partial q}\frac{\partial \mathcal{H}}{\partial p} - \frac{\partial \mathcal{H}}{\partial p}\frac{\partial \mathcal{H}}{\partial q} \equiv 0, \end{aligned} \tag{167}$$

thanks to the Hamilton canonical equations

$$\dot{q} = \frac{\partial \mathcal{H}}{\partial p}, \qquad \dot{p} = -\frac{\partial \mathcal{H}}{\partial q}. \tag{168}$$

Such a canonical approach holds also in our covariant case if we operate the substitutions

$$V^{\alpha} \longleftrightarrow p \qquad V_{\alpha} \longleftrightarrow q \tag{169}$$

and the canonical equations are

$$\frac{\delta V^{\alpha}}{dx^{\beta}} = -\Gamma^{\alpha}_{\sigma\beta}\left(\frac{\partial \mathcal{H}}{\partial V_{\sigma}}\right) \quad \longleftrightarrow \quad \frac{dp}{dt} = -\frac{\partial \mathcal{H}}{\partial q}, \tag{170}$$

$$\frac{\delta V_{\alpha}}{dx^{\beta}} = \Gamma^{\sigma}_{\alpha\beta}\left(\frac{\partial \mathcal{H}}{\partial V^{\sigma}}\right) \quad \longleftrightarrow \quad \frac{dq}{dt} = \frac{\partial \mathcal{H}}{\partial p}. \tag{171}$$

In other words, starting from the (Hamiltonian) invariant (147), we have recovered a covariant canonical symplectic structure. The variation (157) may be seen as the generating function $\mathcal{G}$ of canonical transformations where the generators of q, p and t changes are dealt under the same standard. The covariant and contravariant vector fields can be also of different nature so that the above fundamental Hamiltonian invariant can be generalized as

$$\mathcal{H} = W^{\alpha} V_{\alpha}, \tag{172}$$

or, considering scalar smooth and regular functions, as

$$\mathcal{H} = f(W^{\alpha} V_{\alpha}), \tag{173}$$

or, in general

$$\mathcal{H} = f\Big(W^{\alpha} V_{\alpha}, B^{\alpha\beta} C_{\alpha\beta}, B^{\alpha\beta} V_{\alpha} V'_{\beta}, \ldots\Big), \tag{174}$$

where the invariant can be constructed by covariant vectors, bivectors and tensors. Clearly, as above, the identifications

$$W^{\alpha} \longleftrightarrow p \qquad V_{\alpha} \longleftrightarrow q \tag{175}$$

hold and the canonical equations are

$$\frac{\delta W^\alpha}{dx^\beta} = -\Gamma^\alpha_{\sigma\beta}\left(\frac{\partial \mathcal{H}}{\partial V_\sigma}\right) \qquad \frac{\delta V_\alpha}{dx^\beta} = \Gamma^\sigma_{\alpha\beta}\left(\frac{\partial \mathcal{H}}{\partial W^\sigma}\right). \tag{176}$$

Finally, conservation laws are given by

$$\frac{\delta \mathcal{H}}{dx^\beta} = (\Gamma^\alpha_{\sigma\beta} - \Gamma^\alpha_{\sigma\beta})\left(\frac{\partial \mathcal{H}}{\partial V_\sigma}\right)\left(\frac{\partial \mathcal{H}}{\partial W^\alpha}\right) \equiv 0. \tag{177}$$

In our picture, this means that the canonical symplectic structure is assigned in the way in which covariant and contravariant vector fields are related. If the Hamiltonian invariant is constructed by bivectors and tensors, Eqs. (176) and (177) have to be generalized but the structure is the same. It is worth noticing that, in our derivation, we never used the metric field but only connections considering a completely affine Palatini approach [47].

These considerations can be made independent of the reference frame, if we define a suitable system of unitary vectors by which we can pass from holonomic to anholonomic description and viceversa. We can define the reference frame on the event manifold $\mathcal{M}$ as vector fields $e_{(k)}$ in event space and dual forms $e^{(k)}$ such that vector fields $e_{(k)}$ define an orthogonal frame at each point and the relations

$$e^{(k)}\left(e_{(l)}\right) = \delta^{(k)}_{(l)}, \tag{178}$$

hold. If these vectors are unitary, in a Riemann 4-spacetime, they are the standard *vierbeins* [24].

If we do not limit this definition of reference frame by orthogonality, we can introduce a *coordinate reference frame* $(\partial_\alpha, ds^\alpha)$ based on vector fields tangent to line $x^\alpha = const.$ Both reference frames are linked by the relations

$$e_{(k)} = e^\alpha_{(k)}\partial_\alpha; \qquad e^{(k)} = e^{(k)}_\alpha dx^\alpha. \tag{179}$$

Greek indices indicate holonomic coordinates while Latin indices between brackets, the anholonomic coordinates (*vierbein* indices in 4-spacetimes). We can prove the existence of a reference frame using the orthogonalization procedure at every point of spacetime. From the same procedure, we get that coordinates of frame smoothly depend on the point. The statement about the existence of a global reference frame follows from this. A smooth field on time-like vectors of each frame defines congruence of lines that are tangent to this field. We say that each line is a world line of an observer or a *local reference frame*. Therefore a reference frame is a set of local reference frames. The *Lorentz transformation* can be defined as a transformation of a reference frame

$$x'^\alpha = f\left(x^0, x^1, x^2, x^3, \dots, x^n\right), \tag{180}$$

$$e'^\alpha_{(k)} = A^\alpha_\beta B^{(l)}_{(k)} e^\beta_{(l)}, \tag{181}$$

where

$$A^\alpha_\beta = \frac{\partial x'^\alpha}{\partial x'^\beta}, \qquad \delta_{(i)(l)} B^{(i)}_{(j)} B^{(l)}_{(k)} = \delta_{(j)(k)}. \tag{182}$$

We call the transformation A^{α}_{β} the holonomic part and transformation $B^{(l)}_{(k)}$ the anholonomic part.

A vector field V has two types of coordinates: *holonomic coordinates* V^{α} relative to a coordinate reference frame and *anholonomic coordinates* $V^{(k)}$ relative to a reference frame. For these two kinds of coordinates, the relation

$$V^{(k)} = e^{(k)}_{\alpha} V^{\alpha}, \tag{183}$$

holds. We can study parallel transport of vector fields using any form of coordinates. Because Eqs. (180) and (181) are linear transformations, we expect that parallel transport in anholonomic coordinates has the same form as in holonomic coordinates. Hence we write

$$DV^{\alpha} = dV^{\alpha} + \Gamma^{\alpha}_{\beta\gamma} V^{\beta} dx^{\gamma}, \tag{184}$$

$$DV^{(k)} = dV^{(k)} + \Gamma^{(k)}_{(l)(p)} V^{(l)} dx^{(p)}. \tag{185}$$

Because DV^{α} is also a tensor, we get

$$\Gamma^{(k)}_{(l)(p)} = e^{\alpha}_{(l)} e^{\beta}_{(p)} e^{(k)}_{\gamma} \Gamma^{\gamma}_{\alpha\beta} + e^{\alpha}_{(l)} e^{\beta}_{(p)} \frac{\partial e^{(k)}_{\alpha}}{\partial x^{\beta}}. \tag{186}$$

Eq.(186) shows the similarity between holonomic and anholonomic coordinates. Let us introduce the symbol $\partial_{(k)}$ for the derivative along the vector field $e_{(k)}$

$$\partial_{(k)} = e^{\alpha}_{(k)} \partial_{\alpha}. \tag{187}$$

Then Eq.(186) takes the form

$$\Gamma^{(k)}_{(l)(p)} = e^{\alpha}_{(l)} e^{\beta}_{(p)} e^{(k)}_{\gamma} \Gamma^{\gamma}_{\alpha\beta} + e^{\alpha}_{(l)} \partial_{(p)} e^{(k)}_{\alpha}. \tag{188}$$

Therefore, when we move from holonomic coordinates to anholonomic ones, we must underline that the connection also transforms in the the same way to when we move from one coordinate system to another. This leads us to the model of anholonomic coordinates. The vector field $e_{(k)}$ generates lines defined by the differential equations

$$e^{\alpha}_{(l)} \frac{\partial \tau}{\partial x^{\alpha}} = \delta^{(k)}_{(l)}, \tag{189}$$

or the symbolic system

$$\frac{\partial \tau}{\partial x^{(l)}} = \delta^{(k)}_{(l)}. \tag{190}$$

Keeping in mind the symbolic system (190), we denote also the functional τ as $x^{(k)}$ and call it the anholonomic coordinate. Then we can find derivatives and get

$$\frac{\partial x^{(k)}}{\partial x^{\alpha}} = \delta^{(k)}_{\alpha}. \tag{191}$$

The necessary and sufficient conditions to complete the integrability of system (191) are

$$\omega^{(i)}_{(k)(l)} = e^{\alpha}_{(k)} e^{\beta}_{(l)} \left(\frac{\partial e^{(i)}_{\alpha}}{\partial x^{\beta}} - \frac{\partial e^{(i)}_{\beta}}{\partial x^{\alpha}} \right) = 0, \tag{192}$$

where we introduced the anholonomic object $\omega^{(i)}_{(k)(l)}$. Therefore each reference frame has the vector fields

$$\partial_{(k)} = \frac{\partial}{\partial x^{(k)}} = e^{\alpha}_{(k)}\partial_{\alpha}, \tag{193}$$

which have the commutators

$$\begin{aligned} \left[\partial_{(i)}, \partial_{(j)}\right] &= \left(e^{\alpha}_{(i)}\partial_{\alpha}e^{\beta}_{(j)} - e^{\alpha}_{(j)}\partial_{\alpha}e^{\beta}_{(i)}\right)e^{(m)}_{\beta}\partial_{(m)} \\ &= e^{\alpha}_{(i)}e^{\beta}_{(j)}\left(-\partial_{\alpha}e^{(m)}_{\beta} + \partial_{\beta}e^{(m)}_{\beta}\right)\partial_{(m)} = \omega^{(m)}_{(i)(j)}\partial_{(m)}. \end{aligned} \tag{194}$$

For the same reason, we introduce the forms

$$dx^{(k)} = e^{(k)} = e^{(k)}_{\beta}dx^{\beta}, \tag{195}$$

and a differential of this form is

$$\begin{aligned} d^2x^{(k)} = d\left(e^{(k)}_{\alpha}dx^{\alpha}\right) &= \left(\partial_{\beta}e^{(k)}_{\alpha} - \partial_{\alpha}e^{(k)}_{\beta}\right)dx^{\alpha} \wedge dx^{\beta} \\ &= -\omega^{(m)}_{(k)(l)}dx^{(k)} \wedge dx^{(l)}. \end{aligned} \tag{196}$$

Therefore when $\omega^{(i)}_{(k)(l)} \neq 0$, the differential $dx^{(k)}$ is not an exact differential and the system (191), in general, cannot be integrated. However, the problem can be settled with meaningful objects which model the solution. We can study how the functions $x^{(i)}$ changes along different lines, and due to the fact that the functions $x^{(i)}$ are natural parameters along a flow line of vector fields $e_{(i)}$, they are defined along any line.

Finally, all the above results can be immediately achieved in holonomic and anholonomic formalism considering the equation

$$\mathcal{H} = W^{\alpha}V_{\alpha} = W^{(k)}V_{(k)}, \tag{197}$$

and the analogous ones. This means, as a conclusion, that the results are independent of the reference frame and the symplectic covariant structure always holds.

4.3. The Hydrodynamic Picture Compatible with the Covariant Symplectic Structure

In order to further check the validity of the above approach, we can prove that it is always consistent with the hydrodinamic picture of mechanics (see also [84] for details on hydrodynamic covariant formalism). After, we will take into account remarkable applications of such an approach.

Let us define a phase space density $f(x, p; s)$ which evolves according to the Liouville equation

$$\frac{\partial f}{\partial s} + \frac{1}{m}\frac{\partial}{\partial x^{\mu}}(g^{\mu\nu}p_{\nu}f) - \frac{1}{2m}\frac{\partial}{\partial p_{\lambda}}\left(\frac{\partial g^{\mu\nu}}{\partial x^{\lambda}}p_{\mu}p_{\nu}f\right) = 0. \tag{198}$$

Next we define the density $\varrho(x; s)$, the covariant current velocity $v_{\mu}(x; s)$ and the covariant stress tensor $\mathcal{P}_{\mu\nu}(x; s)$ according to the relations

$$\varrho(x; s) = mn\int \mathrm{d}^4 p f(x, p; s), \tag{199}$$

$$\varrho(x;s)v_\mu(x;s) = n\int \mathrm{d}^4p p_\mu f(x,p;s), \tag{200}$$

$$\mathcal{P}_{\mu\nu}(x;s) = \frac{n}{m}\int \mathrm{d}^4p p_\mu p_\nu f(x,p;s). \tag{201}$$

It can be verified, by direct substitution, that a solution of the Liouville Eq.(198) of the form

$$f(x,p;s) = \frac{1}{mn}\varrho(x;s)\delta^4[p_\mu - mv_\mu(x;s)], \tag{202}$$

leads to the equation of continuity

$$\frac{\partial\varrho}{\partial s} + \frac{\partial}{\partial x^\mu}(g^{\mu\nu}v_\nu\varrho) = 0, \tag{203}$$

and to the equation for the balance of momentum

$$\frac{\partial}{\partial s}(\varrho v_\mu) + \frac{\partial}{\partial x^\lambda}\left(g^{\lambda\alpha}\mathcal{P}_{\alpha\mu}\right) + \frac{1}{2}\frac{\partial g^{\alpha\beta}}{\partial x^\mu}\mathcal{P}_{\alpha\beta} = 0. \tag{204}$$

Taking into account the fact that for the particular solution (202), the stress tensor, as defined by Eq.(201), is given by the expression

$$\mathcal{P}_{\mu\nu}(x;s) = \varrho v_\mu v_\nu, \tag{205}$$

we obtain the final form of the hydrodynamic equations

$$\frac{\partial\varrho}{\partial s} + \frac{\partial}{\partial x^\mu}(\varrho v^\mu) = 0, \tag{206}$$

$$\frac{\partial v_\mu}{\partial s} + v^\lambda\left(\frac{\partial v_\mu}{\partial x^\lambda} - \Gamma^\nu_{\mu\lambda}v_\nu\right) = \frac{\partial v_\mu}{\partial s} + v^\lambda\nabla_\lambda v_\mu = 0. \tag{207}$$

It is straightforward to see that, through the substitution $v_\mu \rightarrow V_\mu$, Eq.(162) is immediately recovered along a geodesic, that is our covariant symplectic structure is consistent with a hydrodynamic picture. It is worth noting that if $\dfrac{\partial v_\mu}{\partial s} \neq 0$ in Eq.(207), the motion is not geodesic. The meaning of this term different from zero is that an extra force is acting on the system.

4.4. Some Relevant Applications: Gravitational, Electromagnetic and Dirac Fields

Many applications of the previous results can be achieved specifying the nature of vector (or tensor) fields which define the conserved Hamiltonian invariant $\mathcal{H}$. Considerations in General Relativity and Electromagnetism are particularly interesting at this point. Let us take into account the Riemann tensor $R^\rho_{\sigma\mu\nu}$. It comes out when a given vector V^ρ is transported along a closed path on a generic curved manifold. It is

$$[\nabla_\mu, \nabla_\nu]V^\rho = R^\rho_{\sigma\mu\nu}V^\sigma, \tag{208}$$

where ∇_μ is the covariant derivative. We are assuming a Riemannian $\mathbf{V}_n$ manifold as standard in General Relativity. If connection is not symmetric, one has to take into account an additive torsion field coming out from the parallel transport.

Clearly, the Riemann tensor results from the commutation of covariant derivatives and it can be expressed as the sum of two commutators

$$R^\rho_{\sigma\mu\nu} = \partial_{[\mu}, \Gamma^\rho_{\nu]\sigma} + \Gamma^\rho_{\lambda[\mu}, \Gamma^\lambda_{\nu]\sigma}. \tag{209}$$

Furthermore, (anti) commutation relations and cyclic identities (in particular Bianchi's identities) hold for the Riemann tensor [24].

All these straightforward considerations suggest immediately the presence of a symplectic structure whose elements are covariant and contravariant vector fields, V^α and V_α, satisfying the properties (93)-(97). In this case, the dimensions of vector space $\mathbf{E}_{2n}$ are assigned by V^α and V_α. It is important to notice that such properties imply the connections (Christoffel symbols) and not the metric tensor.

The invariant (147) is a generic conserved quantity specified by the choice of V^α and V_α. If

$$V^\alpha = \frac{dx^\alpha}{ds}, \tag{210}$$

is a 4-velocity, with $\alpha = 0, 1, 2, 3$, from Eq.(170), we obtain the equation of geodesics of General Relativity,

$$\frac{d^2x^\alpha}{ds^2} + \Gamma^\alpha_{\mu\nu}\frac{dx^\mu}{ds}\frac{dx^\nu}{ds} = 0. \tag{211}$$

On the other hand, being

$$\delta V^\alpha = R^\alpha_{\beta\mu\nu} V^\beta dx_1^\mu dx_2^\nu, \tag{212}$$

the result of the transport along a closed path, it is easy to recover the geodesic deviation considering the geodesic (211) and the infinitesimal variation ξ^α with respect to it, i.e.

$$\frac{d^2(x^\alpha+\xi^\alpha)}{ds^2} + \Gamma^\alpha_{\mu\nu}(x+\xi)\frac{d(x^\mu+\xi^\mu)}{ds}\frac{d(x^\nu+\xi^\nu)}{ds} = 0, \tag{213}$$

which gives, through Eq.(209),

$$\frac{d^2\xi^\alpha}{ds^2} = R^\alpha_{\mu\lambda\nu}\frac{dx^\mu}{ds}\frac{dx^\nu}{ds}\xi^\lambda. \tag{214}$$

The symplectic structure is due to the fact that the Riemann tensor is derived from covariant derivatives either as

$$[\nabla_\mu, \nabla_\nu]V^\rho = R^\rho_{\sigma\mu\nu}V^\sigma, \tag{215}$$

or

$$[\nabla_\mu, \nabla_\nu]V_\rho = R^\sigma_{\mu\nu\rho}V_\sigma. \tag{216}$$

In other words, fundamental equations of General Relativity are recovered from our covariant symplectic formalism.

Analogously, another choice allows to recover the standard Electromagnetism. If $V^\alpha = A^\alpha$, with A^α the vector potential, the Hamiltonian invariant is

$$\mathcal{H} = A^\alpha A_\alpha. \tag{217}$$

Following the above procedure, we obtain, from the covariant Hamilton equations, the electromagnetic tensor field

$$F_{\alpha\beta}=\nabla_{\alpha}A_{\beta}-\nabla_{\beta}A_{\alpha}=\nabla_{[\alpha}A_{\beta]}, \tag{218}$$

and the Maxwell equations (in a generic empty curved spacetime)

$$\nabla^{\alpha}F_{\alpha\beta}=0, \qquad \nabla_{[\alpha}F_{\lambda\beta]}=0. \tag{219}$$

The standard Lorentz gauge is

$$\nabla^{\alpha}A_{\alpha}=0, \tag{220}$$

and electromagnetic wave equation is recovered.

Another Hamiltonian invariant can be constructed by taking into account the "*mass-shell condition*" of quantum field theory [20], i.e.

$$\mathcal{H}=p^{\alpha}p_{\alpha}=m^{2}. \tag{221}$$

where, obviously, $p^{\alpha}p_{\alpha}=E^2-\vec{p}^2$, E the particle energy and $\vec{p}$ the momentum. By standard canonical quantization procedure we can recast such quantities as operators

$$E\ \rightarrow\ -i\partial_t\,, \qquad \vec{p}\ \rightarrow\ -i\vec{\nabla}\,, \tag{222}$$

obtaining the above standard Klein-Gordon equation for a scalar field which we rewrite for completeness

$$(\Box+m^2)\phi=0\,. \tag{223}$$

Being ϕ a scalar field, as above, it can be interpreted as the product of two conjugate complex fields [30, 85] $\phi=\psi^{\dagger}\psi$ and the Klein-Gordon operator can be split, as discussed above

$$\left(\Box+m^2\right)\phi=\left(\partial_{\alpha}\partial^{\alpha}+m^2\right)\left(\psi^{\dagger}\psi\right)=\left(\partial_{\alpha}-i\,m\right)\left(\partial^{\alpha}+i\,m\right)\left(\psi^{\dagger}\psi\right)=0\,. \tag{224}$$

This means that the heuristic considerations of Sec.II can be framed into the self-consistent scheme of the covariant symplectic structure. Then, taking into account spinors, we can immediately write

$$\left(i\gamma^{\alpha}\partial_{\alpha}-m\right)\left(i\gamma_{\alpha}\partial^{\alpha}+m\right)\left(\psi^{\dagger}\psi\right)=0\,. \tag{225}$$

Taking into account the general considerations developed in Sec.II, this is a natural symplectic basis. The discussion can be further generalized introducing the matrix

$$\sigma_{\mu\nu}=\frac{i}{2}[\gamma_{\mu},\gamma_{\nu}]\,, \tag{226}$$

by which it is straightforward to demonstrate that the quantity $\bar{\psi}\sigma_{\mu\nu}\psi$ transforms as an antisymmetric second rank tensor under the Lorentz group, where $\bar{\psi}$ is a generic spinor defined starting from the Hermitian conjugate of ψ, i.e. $\bar{\psi}\equiv\psi^{\dagger}\gamma^{0}$. In order to find all the Lorentz tensors that can be represented as bilinear spinors (and then capable of being represented by our symplectic structure), let us introduce the matrix:

$$\gamma_5 = \gamma^5 = i\gamma^0\gamma^1\gamma^2\gamma^3 = -\frac{i}{4!}\epsilon_{\mu\nu\sigma\rho}\gamma^\mu\gamma^\nu\gamma^\sigma\gamma^\rho\,, \tag{227}$$

where $\epsilon^{\mu\nu\sigma\rho} = -\epsilon_{\mu\nu\sigma\rho}$ and $\epsilon^{0123} = +1$. Because γ^5 transforms like $\epsilon^{\mu\nu\sigma\rho}$, it is a pseudoscalar, that is it changes sign under a parity transformation. Thus, $\bar{\psi}\gamma_5\psi$ is a pseudoscalar. By analogous considerations, the complete set of bilinear spinors can be achieved:

$$\text{Scalar:} \quad \bar{\psi}\psi \tag{228}$$

$$\text{Vector:} \quad \bar{\psi}\gamma^\mu\psi \tag{229}$$

$$\text{Tensor:} \quad \bar{\psi}\sigma^{\mu\nu}\psi \tag{230}$$

$$\text{Pseudovector:} \quad \bar{\psi}\gamma_5\gamma^\mu\psi \tag{231}$$

$$\text{Pseudoscalar:} \quad \bar{\psi}\gamma_5\psi\,. \tag{232}$$

All these bilinear forms undergoes symplectic transformations as discussed above. In particular, the Dirac Lagrangian

$$\mathcal{L} = \bar{\psi}\left(i\gamma^\mu\partial_\mu - m\right)\psi \tag{233}$$

is invariant under the Lorentz group and variations with respect to both ψ and $\bar{\psi}$ give rise to the two versions of the Dirac equation[2]

$$\left(i\gamma^\mu\partial_\mu - m\right)\psi = 0 \tag{234}$$

and

$$\bar{\psi}\left(i\gamma_\mu\partial^\mu + m\right) = 0\,. \tag{235}$$

The above arguments are completely general and can be developed in the framework of any curved space-time by defining suitable bilinear Hamiltonian invariants. Due to the curved space-time, partial derivatives must be substituted with covariant derivatives obtaining, at the end, the Dirac equation in curved space-times which is

$$\left[i\gamma^\mu(x)\partial_\mu - i\gamma^\mu(x)\Gamma_\mu(x)\right]\psi = m\psi\,, \tag{236}$$

and the analogous one for the Hermitian conjugate quantities $\bar{\psi}$ and γ_μ. Here $\gamma^\mu(x)$ are curvature dependent Dirac matrices and $\Gamma_\mu(x)$ are the spin connections. The relation among them is given by the commutators:

$$\Gamma_\mu = -\frac{1}{8}\left[\gamma^\nu(x), \gamma_\nu(x)_{;\mu}\right]\,, \tag{237}$$

where semicolon inside the square brackets indicates the covariant derivative. The relations between curvature-dependent Dirac matrices $\gamma_\mu(x)$ and curvature-independent Dirac matrices $\gamma_{(a)}$, using the anticommutation relations, are

$$\gamma_\mu\gamma_\nu + \gamma_\nu\gamma_\mu = 2g_{\mu\nu} \tag{238}$$

[2] In order to write the Dirac equation correctly, we have to consider how the Hermitian conjugation works on γ^μ matrices and spinors. As we said $\bar{\psi} \equiv \psi^\dagger\gamma^0$, and $(\gamma^0)^\dagger = \gamma^0$, $(\gamma^i)^\dagger = -\gamma^i$, where γ^i is anti-Hermitian and γ^0 is Hermitian. Another way to express this property is $\gamma^{\mu\dagger} = \gamma_\mu$.

and

$$\gamma_{(a)}\gamma_{(b)}+\gamma_{(b)}\gamma_{(a)}=2\eta_{(a)(b)} \tag{239}$$

defined through the above *vierbein* fields $e_{\mu}^{(a)}$ by the equations

$$\gamma_{\mu}(x)=e_{\mu}^{(a)}\gamma_{(a)}\,, \tag{240}$$

so that the relations between the holonomic and anholonomic coordinates are fully recovered. In conclusion, a general covariant symplectic structure works also for spinor fields.

4.5. Remarks on the Covariant Symplectic Structure

We have shown that a covariant, symplectic structure can be found for every bilinear Hamiltonian invariant, which can be constructed by covariant vector, tensor and spinor fields. The basic request for such invariants is that they have to be conserved, so the approach can be always related to a General Conservation Principle [81]. In particular, the Poisson brackets are always covariantly conserved for generic vector fields and the approach is fully compatible with the hydrodynamic picture of mechanics. Besides, the approach can be formulated in a holonomic and anholonomic representations, once vector, tensor and spinor fields are represented both in *vierbein* and coordinate-frames. This feature is essential to be sure that general covariance and symplectic structure are always conserved. Another very important point is the role of affine connections which are considered independently from metric. Due to this feature, the approach is genuinely Palatini-like since no relation (i.e. Levi-Civita connection) is requested between $g_{\mu\nu}$ and $\Gamma^{\lambda}_{\mu\nu}$. Furthermore, covariant and contravariant vector, tensor and spinor fields can be read as the configurations q^i and the momenta p_i of standard Hamiltonian dynamics in such a way that specific Hamilton field equations are recovered from the covariant derivatives of such covariant and contravariant vector fields. Specifying the nature of vector fields, we select the particular theory. We have applied the method to gravity, electromagnetism and Dirac fields. For example, the curvature invariants of General Relativity can be recovered from Hamiltonian invariants opportunely defined, so that we must stress that this theory is essentially *covariant* and *symplectic*. If the vector fields are the 4-velocity in a generic curved space-time, we obtain the equation of geodesic motion and the geodesic deviation, which can be read as covariant Hamilton equations. If we consider the vector potential of Electromagnetism, Maxwell equations and Lorentz gauge are recovered and also in this case, these two sets of equations (Maxwell ones and Lorentz gauge) can be read as two sets of conjugate Hamilton equations. Finally, we discussed the mass-shell conditions of relativistic particle mechanics, which can be intrinsically considered as a bilinear Hamilton invariant. We showed that it gives rise to Klein-Gordon equation and Dirac equation when scalar or spinor fields are considered. The procedure can be generalized to any space-time, once the Dirac matrices or generic Dirac invariant quantities are defined for holonomic and anholonomic coordinate reference frames. This is the core of Open Quantum Relativity to which can be reduced all the previous considerations.

5. Application of Open Quantum Relativity to Cosmology

From Open Quantum Relativity, we can construct self-consistent cosmological models. In general, conservation laws assume a prominent role in cosmology thanks to the fact that they can give rise to singularity free, exactly integrable, cosmological models. In this section, we discuss this issue first of all in connection to quantum cosmology. After, starting from exact cosmological solutions, we show that observed cosmological parameters can be recovered in the framework of Open Quantum Relativity.

5.1. Conservation Laws in Quantum Cosmology

In the last thirty years, strong efforts have been directed to the quantization of gravity and then to achieve a comprehensive unification theory [48, 49]. The goal is to obtain a coherent scheme in which the gravity is treated under the same standard of the other interaction of Nature. In order to test these theories, we have to work at Planck's scale, so the cosmology is the most reasonable area for the application of the observable predictions for every theory of quantum gravity. Moreover, it is possible to define a *quantum cosmology*, that is the quantization of dynamical systems which are "*universes*". In this context, it is supposed that the Universe as a whole (the ensemble of all the possible universes as in the Many Worlds picture) has a quantum mechanical nature and that a *classical universe* is only a limit concept valid in particular regions of a manifold (*superspace*) composed of all the possible space-like 3-geometries and local configurations of the matter fields. The task of quantum cosmology is then to relate all the observable quantities of a classical universe to the assigned boundary conditions for a wave function in the superspace. This wave function should be connected with the probability of obtaining standard universes (even if, in the common approach, it is not a proper probability amplitude since a Hilbert space does not exist in the canonical formulation of quantum gravity) [50].

Quantum cosmology can solve, in principle, the problem of the initial conditions of the classical cosmology: *i.e.* it has to explain the very peculiar aspect of the universe which we see today, simply specifying the physical meaning of the boundary conditions for the wave function which describes it. Then the main question of quantum cosmology is to search for boundary conditions in agreement with the astronomical observations and these conditions have to be compatible with the wave function of the universe $|\Psi\rangle$. The dynamical behavior of $|\Psi\rangle$ in the superspace is described by the Wheeler-DeWitt (WDW) equation [48] that is a second order functional differential equation hard to handle because it has infinite degrees of freedom. Usually attention has been concentrated on finite dimensional models in which the metrics and the matter fields are restricted to particular forms (*minisuperspace models*), like homogeneous and isotropic models. With these choices, the WDW equation becomes a second order partial differential equation which can be, in principle, exactly integrated.

An interpretative scheme for the solutions of the WDW equation is the *Hartle criterion* [50]. Hartle proposed to look for peaks of the wave function of the universe: if it is strongly peaked, we have correlations among the geometrical and matter degrees of freedom; if it is not peaked, correlations are lost. In the first case, the emergence of classical relativistic trajectories (*i.e.* universes) is expected. Consequently, the wave function is peaked on a subset of the general solution. In this sense, the boundary conditions on the wave function imply initial conditions for the classical solutions. It is possible to show that the Hartle criterion is always connected to the presence of a conservation law and then to the emergence of classical trajectories [51] which are *classical relativistic universes* where cosmological observations are possible[3].

Let us now take into account a minisuperspace where the wave function of the universe can be consistently defined. The Hamiltonian constraint (corresponding to the $(0,0)$-Einstein energy equation) gives the WDW equation, so that if $|\Psi>$ is a *state* of the system (*i.e.* the wave function of the universe), dynamics is given by

$$\mathcal{H}|\Psi>=0\,. \tag{241}$$

If conservation laws exist, the reduction procedure, outlined in Sec.III, can be applied and then we get

$$\pi_1 \equiv \frac{\partial\mathcal{L}}{\partial\dot{Q}^1} = i_{X_1}\theta_{\mathcal{L}} = \Sigma_1\,, \qquad \pi_2 \equiv \frac{\partial\mathcal{L}}{\partial\dot{Q}^2} = i_{X_2}\theta_{\mathcal{L}} = \Sigma_2\,, \qquad \ldots \tag{242}$$

depending on the number of symmetries. After quantization, we get

$$-i\partial_1|\Psi>=\Sigma_1|\Psi>\,, \qquad -i\partial_2|\Psi>=\Sigma_2|\Psi>\,, \qquad \ldots \tag{243}$$

which are nothing else but generalized translations along the Q^j axis singled out by the corresponding symmetry (*i.e.* a conserved physical quantity). Eqs. (243) can be immediately integrated and, being Σ_j real constants, we obtain oscillatory behaviors for $|\Psi>$ in the directions of symmetries, *i.e.* the wave function of the universe is given by the superposition (88), that is

$$|\Psi>=\sum_{j=1}^{m} e^{i\Sigma_j Q^j}|\chi(Q^l)>\,, \quad m<l\leq n\,, \tag{244}$$

where m is the number of symmetries, l are the directions where symmetries do not exist, n is the total dimension of minisuperspace.

Viceversa, dynamics given by (241) can be reduced by (243) if, and only if, it is possible to define constant conjugate momenta as in (242), that is an oscillatory behavior of a subset of solutions $|\Psi>$ exists only if conservation laws are valid for dynamics.

The m symmetries give first integrals of motion and then the possibility to select classical relativistic trajectories. In conclusion, we can set out the following statement

In minisuperspace quantum cosmology, the existence of conservation laws yields a reduction procedure of dynamics which allows to find oscillatory behaviors for the general solution of WDW equation. Viceversa, if a subset of solutions of WDW equation has an oscillatory behavior, conserved momenta have to exist and conservation laws are present. If a conservation law exists for every configuration variable, the dynamical system is completely integrable and the general solution of WDW equation is a superposition of oscillatory behaviors. In other words, conservation laws allow and select observable universes.

This fact, in the framework of the Hartle interpretative criterion of the wave function of the universe, gives conserved momenta and trajectories which can be interpreted as classical

[3]An operative definition of "classical relativistic universe" could be a universe where cosmological parameters as the Hubble one H_0, the deceleration parameter q_0, the density parameters Ω_M, Ω_Λ, Ω_k and the age t_0 can, in principle, be evaluated [52]. It is not possible to have "quantum universes" with an ill posed concept of metric in which the "cosmological principle" becomes a nonsense.

relativistic cosmological solutions. The major result of the approach is the fact that the quantum cosmology corresponding to a theory where conservation laws are absolutely valid *must* generate classical universe. From this point of view, what was a criterion to select classical universes (the Hartle one) becomes a necessity.

5.2. Exact Cosmological Solutions and Observational Constraints

A cosmological model directly related to the previous results can be derived by the action (57) through the transformations:

$$\phi = \exp[-\varphi]\,, \quad F(\phi) = \frac{1}{8}\exp[-2\varphi]\,, \quad V(\phi) = U(\varphi)\exp[-2\varphi]\,, \tag{245}$$

Let us take into consideration a $4D$–FRW metric. In this case, we derive the pointlike Lagrangian

$$\mathcal{L} = \frac{1}{8}a^3 e^{-2\varphi}\left[6\left(\frac{\dot{a}}{a}\right)^2 - 12\dot{\varphi}\left(\frac{\dot{a}}{a}\right) - 6\frac{k}{a^2} + 4\dot{\varphi}^2 - 8U(\varphi)\right] + a^3\tilde{\mathcal{L}}_m\,. \tag{246}$$

The scale–factor duality symmetry arises if the transformation of the scale factor of a homogeneous and isotropic space-time metric, $a(t) \to a^{-1}(-t)$, leaves the model invariant, following the form of the potential U. Provided the transformations

$$\psi = \varphi - \frac{3}{2}\ln a\,, \quad Z = \ln a\,, \tag{247}$$

the Lagrangian (246) becomes:

$$\mathcal{L} = \frac{1}{2}e^{-2\psi}\left[4\dot{\Psi}^2 - 3\dot{Z}^2 - 6ke^{-2Z} - 8W(\psi, Z)\right] + De^{-3\gamma Z}\,, \tag{248}$$

where the potential $U(\varphi) \to W(\psi, Z)$. We note that in such new variables, the duality invariance is establishing an equivalence with the parity invariance since Z and $-Z$ are solutions depending on time and have to be considered entangled. The emergence of this feature is related to the presence of nonminimal coupling; it allows that cosmological solutions can be extended for $t \to -\infty$ without singularities.

The cosmological equations (Friedmann, energy condition and Klein-Gordon equations), considering also the continuity and the state equations, are

$$4\dot{\psi}^2 - 3\dot{Z}^2 + 6ke^{-2Z} + 8W - 2De^{-3\gamma Z}e^{2\psi} = 0\,, \tag{249}$$

$$3\ddot{Z} - 6\dot{\psi}\dot{Z} + 6ke^{-2Z} - 4\frac{\partial W}{\partial Z} + 3(1-\gamma)De^{-3\gamma Z}e^{2\psi} = 0\,, \tag{250}$$

$$\ddot{\psi} - \dot{\psi}^2 - \frac{3}{4}\dot{Z}^2 - \frac{3k)}{2}e^{-2Z} - 2W + \frac{\partial W}{\partial \psi} = 0\,, \tag{251}$$

$$\dot{\rho} + 3H(p+\rho) = 0\,, \tag{252}$$

$$p = \gamma\rho\,, \tag{253}$$

Equations (249)-(251) are a parametric system depending on the form of the potential W, on the density constant of fluid matter D, on the thermodynamic state of fluid γ and on the spatial curvature k.

Dynamics due to the potential W is related to standard fluid matter, being:

$$W(Z,\psi) = \frac{D}{4}e^{-3\gamma Z}e^{2\psi}\,, \qquad W(Z,\psi) = \Lambda\,, \tag{254}$$

where Λ =const. This scenario leads to a class of exact cosmological solutions [54] which can be selected thanks to the presence of conservation laws [52, 54]. Examples are

$$a(t) = a_0 \exp\left\{\mp\frac{1}{\sqrt{6}}\arctan\left[\frac{1-2e^{4\lambda\tau}}{2e^{2\lambda\tau}\sqrt{1-e^{4\lambda\tau}}}\right]\right\}, \tag{255}$$

$$\varphi(t) = \frac{1}{4}\ln\left[\frac{2\lambda^2 e^{4\lambda\tau}}{(1-e^{4\lambda\tau})}\right] \mp \frac{1}{\sqrt{6}}\arctan\left[\frac{1-2e^{4\lambda\tau}}{2e^{2\lambda\tau}\sqrt{1-e^{4\lambda\tau}}}\right] + \varphi_0, \tag{256}$$

where $\tau = \pm t$, $\lambda^2 = \Lambda/2$ and duality is evident. These solutions agree with data coming from recent observational surveys (see *e.g.* [55]) since they are consistent with all the cosmological parameters in a universe accelerated by Ω_Λ as indicated by the Cosmic Microwave Background Radiation (CMBR) observations, recently published by the collaborations BOOMERanG, MAXIMA [56] and WMAP [57] which give $\Omega_M \simeq 0.3$, $\Omega_\Lambda \simeq 0.7$, and $\Omega_k \simeq 0.0$ for the density parameters of matter, dark energy and spatial curvature respectively. A quantitative evaluation of these features, can be achieved using the SNe Ia surveys [55]. Let us apply the method to the above solutions. It is well known that the use of astrophysical standard candles provides a fundamental mean of measuring the cosmological parameters. Type Ia supernovae (SNe Ia) are the best candidates for this aim since their luminosity can be accurately calibrated and they can be detected at enough high red-shift. This fact allows to discriminate among cosmological models. To this aim, one can fit a given model to the observed magnitude - redshift relation, conveniently expressed as:

$$\mu(z) = 5\log\frac{c}{H_0}d_L(z) + 25 \tag{257}$$

being μ the distance modulus and $d_L(z)$ the dimensionless luminosity distance. The distance in the model we are considering is completely equivalent to the one in a spatially flat universe with a non-zero cosmological constant. Thus $d_L(z)$ is simply given as:

$$d_L(z) = (1+z)\int_0^z dz'[\Omega_M(1+z')^3 + \Omega_\varphi]^{-1/2}\,. \tag{258}$$

where $\Omega_\varphi = 1 - \Omega_M$ plays the same role as the usual Ω_Λ. Our model can be fully characterized by two parameters : the today Hubble constant H_0 and the matter density Ω_M. We find their best fit values minimizing the χ^2 defined as :

$$\chi^2(H_0,\Omega_M) = \sum_i \frac{[\mu_i^{theor}(z_i|H_0,\Omega_M) - \mu_i^{obs}]^2}{\sigma^2_{\mu_0,i} + \sigma^2_{mz,i}} \tag{259}$$

where the sum is over the data points [58]. The results of the fit are presented in Fig.2 where we show the 1,2 and 3 σ confidence regions in the (Ω_M, H_0) plane. The best fit values (with

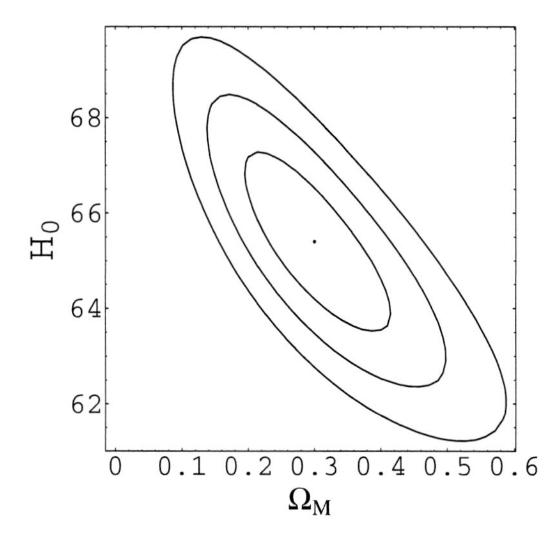

Figure 2. 1, 2 and 3σ confidence regions in the (Ω_M, H_0) plane. The central dot represents the best fit : $\Omega_M = 0.3, H_0 = 65.4\,\mathrm{km\,s^{-1}\,Mpc^{-1}}$.

1σ error) turn out to be :

$$\Omega_M = 0.30 \pm 0.08 \ , \ H_0 = 65.4 \pm 1.2\, km\ s^{-1}\ Mpc^{-1}\ .$$

which allow to conclude that a Ω_ϕ, deduced from the above singularity free solutions, can explain SNe Ia observations very well. Besides, we can discuss how H_0 and Ω_ϕ can be alternatively constrained also by the angular diameter distance D_A as measured using the Sunyaev-Zeldovich effect (SZE) and the thermal bremsstrahlung (X-ray brightness data) for galaxy clusters (see [59] and references therein). The results are shown in the Fig.3, where we see the contours corresponding to the 68.5% and 98% confidence levels: The best fit values (at 1σ) turn out to be :

$$\Omega_M = 0.30 \pm 0.3 \ , \ H_0 = 67 \pm 6\, km\ s^{-1}\ Mpc^{-1}\ .$$

in good agreement with the above fit derived from SNeIa data. Finally, the age of the universe can be directly obtained, by definition, if one knows the value of the Hubble parameter, that is $t_0 = \alpha H_0^{-1}$ where α is a constant depending on the model. For a matter-dominated, flat model $(k = 0)$, it is $\alpha = 2/3$. We evaluate the age taking into account the above solutions in the 3σ-range of variability of the Hubble parameter deduced by the Supernovae fit and SZE method. We get age estimates included between $13\,Gyr$ and $15Gyr$. A further check for the allowed values of t_0 is to verify if the considered range provides also accelerated expansion rates. We get a negative deceleration parameter, converging to $q_0 \simeq -1/2$, for the above singularity free solutions evolving toward the future, which is perfectly in agreement with observations. A further test of the model can be performed by the age estimate obtained by the WMAP campaign [57]. Using the above approach, with the only difference to restrict the age estimates between $13.5Gyr$ and $13.9Gyr$, the WMAP results fall perfectly in the ranges theoretically calculated, in the limit of errors. As

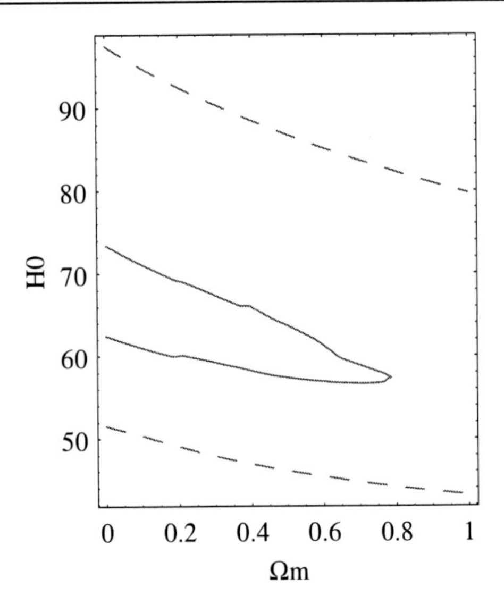

Figure 3. 1, 68.5% and 98% confidence regions in the (Ω_M, H_0) plane. By this further method, the best fit values are : $\Omega_M = 0.3$, $H_0 = 67$ km s^{-1} Mpc^{-1}.

a last remark, it is worth stressing the fact that the results of recent observational campaigns can be matched with our singularity free solutions, and the set of cosmological parameters $\{\Omega_M, \Omega_\Lambda (= \Omega_\phi), H_0, q_0, t_0\}$ can be entirely reproduced in ranges of physical interest. Another important feature is that the solutions match with different sets of data (*e.g.* SNeIa and SZE taken at different redshift regimes and epochs) so that the model is consistent from early to late eras of the universe [52, 54].

6. Further Observational and Experimental Evidences

Beside cosmology, the above effective theory derived from Open Quantum Relativity could have several observational and experimental evidences which we are going to outline below. The final picture, as we will see, results extremely coherent since coming from the same dynamical approach.

6.1. Gamma-ray Bursts as a Signature for Black-White Hole Entangled Systems

Gamma Ray Bursts (GRBs) are among the most interesting and debated issues of modern astrophysics. It has to be noted, however, that the large amount of observational data available in literature, while providing a quite acceptable scenario in which GRBs take place, it leaves nevertheless unsolved the problem of the first origin, and of the nature of the engine which produces the GRBs. First of all, we recall that there is a general consensus on the fact that GRBs are isotropically distributed on the sky and are located at cosmological distances. The overall observed fluences range from 10^{-7} erg/cm^2 to 10^{-4} erg/cm^2. At a redshift z in the range $2 \div 3$, this implies a corresponding isotropic emission range 10^{51} to

10^{54} erg. The emission is non-thermal and with a peculiar slope; the number of photons per unit of energy ϵ follows a power law $N(\epsilon) \propto \epsilon^{-\alpha}$ where $\alpha \sim 1$ for low energy photons and $\sim 1 \div 2$ for energies greater than $0.1 \div 1.0$ MeV up to several GeV. Light curves are extremely irregular and present short term fluctuations (up to a few milliseconds) which, together with the usual causality arguments, lead to the conclusion that the emitting region must be incredibly small (*i.e.* $c\Delta t < 100$ Km in diameter). The duration of the bursts ranges from less than 0.01 sec to more than 100 sec.

A further characteristic of GRBs is polarization. It was soon proposed that the observed polarization could arise from observing a collimated fireball with a slightly off axis line of sight [66]. As a consequence, it was also realized that the degree of polarization could be connected with the light curve expected from the jets of fireballs. In other words, the ultra-relativistic motion of matter ejected is the source of polarization.

Considering specifically the most accredited model, the *Fireball Model* [65], the intense burst of energy originates the formation of a e^+, e^-, γ fireball relativistically propagating into the surrounding medium and giving rise to external and/or internal shock waves [67].

However, a dichotomy exists between the observed phenomenology and the possible energy sources. This fact leads to the shortcoming of a rather large number of different mechanisms invoked as progenitors. Black holes of few solar masses, considered, in conventional way, as the primary sources of GRBs, seem, up to now, the most probable candidates as GRB progenitors among the hypotheses presented in literature. Such black holes are assumed either surrounded by a short living torus of debris (a remnant of the progenitor star) or not. Two large reservoirs of energy are available in black hole systems: i) the binding energy of the orbiting debris which, for a maximally rotating black hole, can convert into energy up to 42% of the rest mass of the disk, and ii) the spin energy of the black hole itself which can release up to 29% of the rest mass of the black hole itself. Mechanisms capable to extract this energy pose, however, major problems which are not yet solved. Other exotic processes can be invoked as colliding neutron stars or vacuum phase transitions but, at the end, no one of them can be considered fully satisfactory [65]. We propose an alternative approach [43] to the GRBs puzzle considering them, essentially, as the signature of a process which, in order to preserve the mass-energy of a dynamically evolving black hole gives rise to a white hole as the ending point of mass-energy flux. The peak of GRBs flux can be connected to the birth of such a structure, the afterglows are connected to the late and the less energetic story of the system. Our starting point is the key hypothesis that a black hole does not give rise to a ill-defined *singularity*, where the laws of physics do not work, but instead that its mass–energy is conserved in any case and, dynamically, through a worm hole, results in a white hole. Every object falling into black hole (acting as a sink) is so converted in energy emerging from a white fountain into another region of space-time. Black hole and white hole result naturally "entangled" since every perturbation of one of the two systems "necessarily" perturbs the other one without any standard direct interaction between them. From the GRB-point of view, a black hole, worm hole, white hole system (from now on BWWH) result a "natural engine". In fact the "birth cries" of white holes are nothing else but the typical energy release of extremely evolved stars which, exceeding the Chandrasekhar limit, give rise to black holes. The flux peak of GRB is connected to this birth act, while the following luminosity could be connected with the following dynamics around both the edges of black hole and white hole. The mechanism is completely compatible with the *Fireball model* [65] which asserts that GRBs are produced when kinetic energy (or Poynting flux) of a relativistic flow is dissipated by shocks. These shocks accelerate electrons and generate strong magnetic fields. The relativistic electrons emit the observed γ-rays via synchrotron or self-Compton effects (in any case with a non-thermal spectrum). The Fireball model is, in some sense independent of the nature of the generating engine (and this is its limit): the only very stringent request is that the matter, in form of beams or jets, is emitted from the central source ultra-relativistically.

Conversely, in the framework of our model, the problem of the origin seems solved because not only the energetics, but also the beaming, the polarization and the cosmological distribution of GRBs can be explained [43]. In fact, the energy flux is naturally beamed by the throat of the worm hole, whose width and length give rise also to the polarization of the burst. The cosmological distribution is a consequence of the topology change in BWWH system (occurring at the two junctions of the worm hole), due to the fact that this process occurs only in a final stage of star life. As we shall see, the path along the worm hole throat (*i.e.* the topology change) can result in a connection between far regions of the universe. The black holes (due to the late evolution of their progenitor stars) form relatively near to us ($z \simeq 0$), while the white holes emerge far from us at high red shifts. More precisely, the worm hole throat is defined as

$$r = \frac{2G_N M}{c^2} \tag{260}$$

which is nothing else but the Schwarzschild radius. We can define the *proper length*, which goes to zero as soon as the radial coordinate reaches its minimum. We obtain [43]

$$l_{WH} = \pm\left[\sqrt{r\left(r - \frac{2G_N M}{c^2}\right)} + \frac{2G_N M}{c^2}\ln\left(\sqrt{\frac{rc^2}{2G_N M}} + \sqrt{\frac{rc^2}{2G_N M} - 1}\right)\right]. \tag{261}$$

The positive component of formula (261) is the proper distance from the black hole while the negative component is the proper distance from the white hole. Such distances directly depend on the mass of the collapsing object. The situation is the following: for $l > 0$, we have the proper distance from the black hole; $l = 0$ implies $r = 2G_N M/c^2$, *i.e.* the worm hole throat; $l < 0$ gives the proper distance from the white hole. The proper "length" of the whole system is $2|l|$. This is a mechanism which generally rules the topology in the worm hole transition. The conservation of mass–energy, once the system is stabilized against perturbations [43], is obtained by considering that the mass–energy falling into the black hole, outcomes from the white hole after passing through the worm hole. Let us now show how GRBs can be considered as the signature of BWWH systems acting as source-engines. The aspects of GRBs phenomenology which can be naturally explained by our model are the energetics, the beaming, the polarization and the fact that GRBs are observed only at high red shifts. Let us take into account the first feature: the energetics. Black holes are created naturally in the aging of massive stars. Some simple dimensional considerations are due now. A solar mass is $M_\odot \simeq 1.9 \times 10^{33}$g $\simeq 1.1 \times 10^{57}$GeV and 1 GeV$\simeq 1.6 \times 10^{-3}$erg so that the mass-energy converted in radiant energy could be of the order of 10^{54}erg, which characterizes GRBs. It is interesting to note that this is the "natural" energy emission corresponding to the birth act of white holes, where no signal or object can enter but everything has to come out from the event anti-horizon. In other words, the *simple* request of conservation of mass-energy of a black hole, through a white hole, justifies the huge amount of GRBs emission. Besides, if BWWH system is stable and does not evaporate [43], a part of mass-energy falling into the black hole emerges at the white hole edge as γ-photons of wave length $\simeq 1.33 \times 10^{-13}$ cm, *i.e.* a γ ray of energy ~ 1 GeV.

Beaming and polarization are the consequence of the BWWH system dynamics and it is compatible with the Fireball Model [65]. In fact, beaming is considered due to the ultra-relativistic motion of source matter, so then the collapse of a massive star gives rise to beamed jets of energy and cannot be isotropic, due to the macroscopic matter flows with Lorentz factors $\Gamma \geq 100$ [65, 67]. Polarization is another natural consequence and it is connected with the achromatic break in the light curve due to the jets produced by the relativistic source [68]. In the case of a BWWH system, we have a *longitudinal* polarization and a *circular* polarization due, respectively, to the length and the width of the worm hole throat which generates an extremely anisotropic relativistic beam. Considering the mass-energy involved ($\sim M_{\odot}$, 10^{54}erg) and the relativistic motions, BWWH model fits with the observations [62, 63].

The last aspect of GRBs phenomenology which could be explained by Open Quantum Relativity with the BWWH model is the observed high red shift. Our working hypothesis is that black holes are the result of late evolution of stars so that they form relatively nearby to us. Furthermore, no signal can come out from such "sinks" so that we can "observe" them only by secondary effects as accretion disks, matter infalls toward the event horizon and so on. Topology changes through a tunnelling allow to span large geodesic intervals and, at the end of these, white hole "fountains" could form. We perceive space-time distances in the FRW metric measuring the red shift z of a given object, while we cannot perceive a topology change by standard measurement techniques (see Fig.1). Let l_{WH} (the worm hole length derived above) be the distance due to the topology change, which in our model depends on the mass M of the progenitor black hole, and l_{FRW} be the distance computed without taking into account the topology change, *i.e.* the length of the geodesic arc between the black hole and the point where the white hole forms. A proper distance in a FRW metric is

$$l_{FRW} = \int_0^r \frac{adr}{\sqrt{1-kr^2}} = a(t)f(r)\,. \tag{262}$$

In agreement with recent observations, we shall assume $k = 0$ and therefore $l_{FRW}(t) = a(t)r$.

The proper distance $l_{FRW}(t_0)$ at the epoch t_0 (*i.e.* where the black hole is) is related to the proper distance at time t (*i.e.* where the white hole is) by the relation:

$$l_{FRW}(t_0) = a_0 f(r) = \frac{a_0}{a(t)} l_{FRW}(t) = a_0^2 \frac{r}{a(t)}\,, \tag{263}$$

which can be read as a proper time, so that the comparison can be performed at two different red shifts (*i.e.* epochs). Being

$$\frac{a_0}{a(t)} = 1 + z(t)\,, \tag{264}$$

the proper distance at t becomes

$$l_{FRW} = \frac{c}{H_0}\left[z + \frac{1}{2}(1-q_0)z^2 + \cdots\right]. \tag{265}$$

In the case of the BWWH system, we have that the proper length l_{WH} is given by (261), from which the distance between the event horizons of the black hole and of the white hole is $\tilde{l}_{WH} = 2|l_{WH}|$. By comparing the FRW and the wormhole proper length in this new scheme, substituting $r = cz/(H_0 a_0) = \alpha_0 z$, we get the relation

$$\alpha_0 z = 2\left[\sqrt{z(z-\tilde{M})} + \tilde{M}\ln\left(\sqrt{\frac{z}{\tilde{M}}} + \sqrt{\frac{z}{\tilde{M}} - 1}\right)\right]. \tag{266}$$

Considering the first order approximation and solving for the mass, we get a relation between M and z, that is

$$M = \frac{c^2}{2G_N}\left[\frac{(\alpha_0^2 z^2 - a_0\alpha_0 z - 2\alpha_0 z) \pm \sqrt{(\alpha_0^2 z^2 - a_0\alpha_0 z - 2\alpha_0 z) - 4\alpha_0^3 z}}{2\alpha_0 z}\right], \tag{267}$$

where it is easy to obtain the order of magnitude of the solar mass in the range $0.5 < z < 5$, in agreement with observations. From this point of view, the cosmological origin of GRBs can be suggested and the topology changes (*i.e.* the worm holes) acquire a fundamental role in the game. It is important to stress the fact that the distribution of BWWH systems has to follow the star formation and then the black hole production: in this sense the isotropic distribution in the sky of GRBs is recovered.

6.2. The Post-Newtonian Limit and the Galactic Dynamics

The weak-field limit of the theory (57) could give rise other interesting experimental tests. As we will see, the corrections to the Newtonian potential are directly related to astrophysical sizes which, in this case, could be explained without invoking the standard huge amount of dark matter (see also [44]).

In order to work out the post-Newtonian limit of the theory, we need an operative choice of the coupling and the potential in the action (57), as

$$F(\phi) = \xi\phi^m, \qquad V(\phi) = \lambda\phi^n, \tag{268}$$

where ξ is a coupling constant, λ gives the self–interaction potential strength, m and n are arbitrary, up to now, parameters. At first order, the metric tensor and the scalar field can be written as

$$g_{\mu\nu} = \eta_{\mu\nu} + h_{\mu\nu}, \qquad \phi = \varphi_0 + \psi, \tag{269}$$

where $\eta_{\mu\nu}$ is the Minkoskwi metric while $h_{\mu\nu}$ and ψ are small corrections; φ_0 is a constant of order unit. For $\varphi_0 = 1$ and $\psi = 0$ Einstein General Relativity is recovered.

Inserting these positions into the field equations [44], we get, respectively, the time-time component, the spatial components and the scalar field perturbation

$$\begin{aligned} h_{00} \simeq\ & \left[(4\pi\tilde{G})\frac{\varphi_0^m}{\xi}\right]\frac{M}{r} - \left[\frac{4\pi\lambda\varphi_0^{m+n}}{\xi}\right]r^2 \\ & - \left[(4\pi\tilde{G})\frac{m^2\varphi_0^{2m-2}M}{1-3\,\xi m^2\varphi_0^{m-2}}\right]\frac{e^{-c_1 r}}{r} - \left[\frac{4\pi m\varphi_0^{2m}}{n-1}\right]\cosh(c_1 r), \end{aligned}$$

$$
\begin{aligned}
h_{il} &\simeq \delta_{il}\left\{\left[(4\pi\tilde{G})\frac{\varphi_0^m}{\xi}\right]\frac{M}{r}+\left[\frac{4\pi\lambda\varphi_0^{m+n}}{\xi}\right]r^2\right.\\
&+\left.\left[(4\pi\tilde{G})\frac{m^2\varphi_0^{2m-2}M}{1-3\,\xi m^2\varphi_0^{m-2}}\right]\frac{e^{-c_1 r}}{r}\right\}-\delta_{il}\left[\frac{4\pi\varphi_0^{2m}m}{n-1}\right]\cosh(c_1 r)\,,\\
\psi &\simeq \left[(4\pi\tilde{G})\frac{mM}{1-3\,\xi m^2\varphi_0}\right]\frac{e^{-c_1 r}}{r}-\left[\frac{4\pi\varphi_0}{n-1}\right]\cosh(c_1 r)\,,
\end{aligned}
\tag{270}
$$

where the parameter c_1 is given by the combination of the physical parameters of the system

$$
c_1^2=\frac{\lambda(n-2m)(n-1)\varphi_0^{m-2}}{1-3\,\xi m^2\varphi_0^{m-2}}\,. \tag{271}
$$

and the Newton constant is defined as

$$
G_N=-\frac{\varphi_0^m}{2\,\xi}\left(\frac{1-4\,\xi m^2\varphi_0^{m-2}}{1-3\,\xi m^2\varphi_0^{m-2}}\right)\tilde{G}\,. \tag{272}
$$

The solutions (270) can be recast in the form

$$
h_{00}\simeq-\frac{2G_NM}{r}(1-e^{-c_1 r})+c_2r^2+c_3\cosh(c_1 r)\,, \tag{273}
$$

$$
h_{ii}\simeq-\frac{2G_NM}{r}(1+e^{-c_1 r})-c_2r^2-c_3\cosh(c_1 r)\,, \tag{274}
$$

$$
\psi\simeq\frac{2G_NM}{r}e^{-c_1 r}+c_3\cosh(c_1 r)\,, \tag{275}
$$

where $c_1,\ c_2,\ c_3$ are combinations of ξ, m, n, λ and φ_0. The solution (273), defining $h_{00}=2U$, can be read as a Newtonian potential with exponential and quadratic corrections *i.e.*

$$
U(r)\simeq-\frac{G_NM}{r}(1-e^{-c_1 r})+\frac{c_2}{2}r^2+\frac{c_3}{2}\cosh(c_1 r)\,. \tag{276}
$$

In general, it can be shown [69, 71, 73] that most of the extended theories of gravity has a weak field limit of similar form, *i.e.*

$$
U(r)=-\frac{G_NM}{r}\left[1+\sum_{k=1}^{n}\alpha_k e^{-r/r_k}\right]\,, \tag{277}
$$

where G_N is the value of the gravitational constant considered at infinity, r_k is the interaction length of the k-th component of non-Newtonian corrections. The amplitude α_k of each component is normalized to the standard Newtonian term; the sign of α_k tells us if the corrections are attractive or repulsive (see [74] for details). Clearly the parameters α_k and r_k are functions of ξ, m, n, λ. As an example, let us take into account only the first term of the series in (277) which is usually considered the leading term. We have

$$
U(r)=-\frac{G_NM}{r}\left[1+\alpha_1 e^{-r/r_1}\right]\,. \tag{278}
$$

The effect of non-Newtonian term can be parameterized by $(\alpha_1,\ r_1)$. For large distances, at which $r\gg r_1$, the exponential term vanishes and the gravitational coupling is G_N. If

$r \ll r_1$, the exponential becomes 1 and, by differentiating Eq.(278) and comparing with the gravitational force measured in laboratory, we get

$$G_{lab} = G_N \left[1 + \alpha_1 \left(1 + \frac{r}{r_1} \right) e^{-r/r_1} \right] \simeq G_N(1 + \alpha_1) , \tag{279}$$

where $G_{lab} = 6.67 \times 10^{-8}$ $g^{-1}cm^3s^{-2}$ is the usual Newton constant measured by Cavendish-like experiments. Of course, G_N and G_{lab} coincide in the standard gravity. It is worthwhile noting that, asymptotically, the inverse square law holds but the measured coupling constant differs by a factor $(1 + \alpha_1)$. In general, any correction introduces a characteristic length that acts at a certain scale for the self-gravitating systems. The range of r_k of the kth-component of non-Newtonian force can be identified with the mass m_k of a pseudo-particle whose Compton's length is

$$r_k = \frac{\hbar}{m_k c} . \tag{280}$$

The interpretation of this fact is that, in the weak energy limit, fundamental relativistic theories which attempt to unify gravity with the other forces introduce, in addition to the massless graviton, particles *with mass* which also carry the gravitational force [30, 75]. These masses introduce length scales which are

$$r_k = 2 \times 10^{-5} \left(\frac{1\,\text{eV}}{m_k} \right) \text{cm} . \tag{281}$$

There have been several attempts to constrain r_k and α_k (and then m_k) by experiments on scales in the range $1\,\text{cm} < r < 1000\,\text{km}$, using very different techniques [76, 77, 78]. The expected masses for particles which should carry the additional gravitational force are in the range $10^{-13}\text{eV} < m_k < 10^{-5}\,\text{eV}$. The general outcome of these experiments, even retaining only the term $k = 1$, is that a "geophysical window" between the laboratory and the astronomical scales has to be taken into account. In fact, the range

$$|\alpha_1| \sim 10^{-2} , \qquad r_1 \sim 10^2 \div 10^3 \text{ m} , \tag{282}$$

is not excluded at all in this window. The astrophysical phenomena corresponding of these non-Newtonian corrections seemed ruled out till some years ago, due to the fact that experimental tests of General Relativity predict "exactly" the Newtonian potential in the weak energy limit, "inside" the Solar System. Recently, indications of an anomalous long–range acceleration, revealed from the data analysis of Pioneer 10/11, Galileo, and Ulysses spacecrafts (which are now almost outside the Solar System) makes these Yukawa–like corrections come again into the game [72]. Furthermore, Sanders [70] reproduced phenomenologically the flat rotation curves of spiral galaxies by using

$$\alpha_1 = -0.92 , \qquad r_1 \sim 40 \text{ kpc} . \tag{283}$$

His main hypothesis is that the additional gravitational interaction is carried by an ultra-soft boson whose range of mass is $m_1 \sim 10^{-27} \div 10^{-28}$eV. The action of this boson becomes significant at galactic scales without the request of enormous amounts of dark matter to stabilize the systems. Eckhardt [79] uses a combination of two exponential terms and gives a detailed explanation of the kinematics of galaxies and galaxy clusters, again without dark matter models. It is worthwhile noting that both the spacecrafts measurements and galactic rotation curves observational indications come from "outside" the usual Solar System boundaries used up to now to test General Relativity. However, the above authors do not start from any fundamental theory in order to explain the outcome of Yukawa corrections. In their contexts, these terms are phenomenological while, in the framework of Open Quantum Relativity, we have dynamically derived them from first principles.

6.3. Gravitationally Induced Neutrino Oscillations from Open Quantum Relativity

Neutrinos act at all length scales, ranging from nuclei [86], to molecular structures [87], up to galaxies [29, 88] and whole universe [89]. In particular, current hypotheses as *dark matter* and *dark energy* are related to the issue that neutrinos have masses and that such mass eigenstates mix and/or superimpose. In order to observe such a mixing, it is important to select constraints applicable to observables sensitive to the absolute neutrino masses, as the effective neutrino mass in Tritium-beta decay, the sum of neutrino masses in cosmology and even the effective Majorana neutrino mass in neutrinoless double-beta decay (for a recent review, see [90]). A key role in this context is played by the neutrino oscillations which allow the transition among the three weak flavor eigenstates e, μ, and τ. It is well known that such a problem is still open and the research of new effects, in which the oscillations could manifest, is one of the main task of the today physics. For this reason, the quantum mechanical phase of neutrinos propagating in gravitational field has been discussed by several authors, also in view of the astrophysical consequences. More controversial is the debate concerning the red-shift of flavor oscillation clocks, in the framework of the weak gravitational field of a star [91]. It has also been suggested that the gravitational oscillation phase might have a significant effect in supernova explosions due to the extremely large fluxes of neutrinos produced with different energies and flavors. This result has been confirmed in [92], and it has been also derived under the assumption that the radial momentum of neutrinos is constant along the trajectory of the neutrino itself [93]. Besides, neutrino oscillations, in particular the gravitational part of the oscillation phase, could straightforwardly come into the debate, which is recently risen, to select what is the *correct* theory of gravity, due to the well known experimental and theoretical shortcomings of General Relativity [74]. Scalar-Tensor Gravity theories provide the most natural generalizations of General Relativity. They can be thought as minimal extensions of Einstein theory in which Mach's principle and Dirac's large number hypothesis are properly accommodated by means of a nonminimal coupling between the geometry and a generic scalar field. The scalar field rules dynamics together with geometry and, furthermore, induces a variation of the gravitational *coupling* in time and space. The gravitational constant G_N is recovered in the limit $\phi \rightarrow constant$. Some recent experiments [95] seem to confirm a variation of the Newton coupling on astrophysical and cosmological scales. The consequences of Scalar-Tensor Gravity have been analyzed for the light deflection, the relativistic perihelion rotation of Mercury, and the time delay experiment, resulting in reasonable agreement with all available observations [74]. On the other hand, bounds on the anisotropy of the microwave background radiation give the upper limit on the variable gravitational coupling

[96]. Scalar-Tensor Gravity can arise from phenomenological considerations or can be the effective counterpart of some fundamental theory. In this last case, the scalar field achieves the role of a prominent dynamical ingredient related to gravity which assumes the meaning of an *induced* interaction as it is the case of Open Quantum Relativity. In fact, the effective action (57) is the result of a well-defined physical process [30] where the scalar field has the role to lead the reduction dynamics. In other words, it is not only a phenomenological ingredient. Now, we want to show how the quantum mechanical phase of neutrino oscillations can be affected by such a nonminimal coupling. In this framework, gravitationally induced neutrino oscillations are the result of a mechanism which can be led back to a fundamental theory. On the other hand, these shifts in phase, if observed, could be a further test for gravitational interaction capable of selecting the correct theory of gravity eventually related to Open Quantum Relativity [30].

Specifically, let us recast the effective action (57) as a Brans-Dicke theory in order to put in evidence the role of nonminimal coupling in the gravitationally induced neutrino oscillation phase. We have

$$\varphi = F(\phi)\,, \quad \omega(\phi) = -\frac{F(\phi)}{2F'(\phi)^2}\,, \quad V(\phi) = 0\,. \tag{284}$$

where we are using physical units. A key role is played by the parameter ω. Considering nonminimal couplings physically motivated as $F(\phi) = \xi\phi^2$, ω is a constant which can be determined by observations. In particular, its value can be constrained by classical tests of General Relativity. In fact, the Einstein theory is recovered if $\varphi \rightarrow$const for $t \rightarrow \infty$ (that is, in our units, $F(\phi) \rightarrow -1/2$). This implies $F'(\phi) \rightarrow 0$ and then $\omega \rightarrow \infty$. Furthermore, the Solar System tests require $(d\omega/d\varphi)\omega^{-3} \rightarrow 0$, which is $(d\omega/d\varphi)\omega^{-3} = 4F'(\phi)^4/F(\phi)^3 \rightarrow 0$. This condition is satisfied when $F(\phi)$ approaches to a constant without asymptotic variations in the first derivative. We are assuming the simplest case where $V(\phi)$ is zero in order to parameterize the results just in terms of the coupling (in particular ω) but the following discussion and results can be enlarged also to cases in which self-interactions are present. The above $4D$-field equations, assuming such $\{\varphi, \omega\}$ - representation, are

$$R_{\mu\nu} - \frac{1}{2}g_{\mu\nu}R = \frac{8\pi}{c^4\varphi}T_{\mu\nu} + \frac{\omega}{\varphi^2}\left(\varphi_{,\mu}\varphi_{,\nu} - \frac{1}{2}g_{\mu\nu}\varphi_{,\alpha}\varphi^{,\alpha}\right) + \frac{1}{\varphi}\left(\varphi_{,\mu;\nu} - g_{\mu\nu}\Box\varphi\right) \tag{285}$$

for the geometric part, and

$$\frac{2\omega}{\varphi}\Box\varphi - \frac{\omega}{\varphi^2}\varphi_{,\mu}\varphi^{,\mu} + R = 0 \tag{286}$$

for the scalar field (Klein-Gordon equation). $T_{\mu\nu}$, as above, is the energy–momentum tensor of standard matter. In order to see how the gravitational field (and then the nonminimal coupling) affects the neutrino oscillation phase, let us assume a massive body M (e.g. a star) as source; then the line element describing a static and isotropic geometry related to such a source is

$$ds^2 = e^{2\alpha}dt^2 - e^{2\beta}[dr^2 + r^2(d\theta^2 + \sin^2\theta d\xi^2)]\,, \tag{287}$$

where the functions α and β depend on the radial coordinate r. The general solution is

$$e^{2\alpha} = e^{2\alpha_0}\left[\frac{1-B/r}{1+B/r}\right]^{2/\lambda}, \tag{288}$$

$$e^{2\beta} = e^{2\beta_0}\left(1+\frac{B}{r}\right)^4\left[\frac{1-B/r}{1+B/r}\right]^{2(\lambda-C-1)/\lambda}, \tag{289}$$

$$\varphi = \varphi_0\left[\frac{1-B/r}{1+B/r}\right]^{-C/\lambda}, \tag{290}$$

where the constants, appropriately chosen, are given by

$$\lambda=\sqrt{\frac{2\omega+3}{2(\omega+2)}}, \quad C=-\frac{1}{2+\omega}, \quad \alpha_0=\beta_0=0\,, \quad B=\frac{M}{2c^2\varphi_0}\sqrt{\frac{2\omega+4}{2\omega+3}}. \tag{291}$$

Splitting the metric tensor as $g_{\mu\nu}\simeq\eta_{\mu\nu}+h_{\mu\nu}$, the above solution gives

$$g_{00} \simeq 1-\frac{2M\phi_0^{-1}}{c^2r}\frac{4+2\omega}{3+2\omega}, \tag{292}$$

$$g_{ii} \sim 1+\frac{2M\phi_0^{-1}}{c^2r}\frac{2+2\omega}{3+2\omega}, \quad i=1,2,3, \tag{293}$$

$$g_{0i} = 0\,, \quad g_{ij}=0\,, \quad i\neq j\,, \tag{294}$$

$$\varphi = \varphi_0+\frac{2M}{c^2r}\frac{1}{3+2\omega}, \tag{295}$$

where an effective gravitational coupling

$$G_{eff}=\frac{4+2\omega}{\varphi_0(3+2\omega)}, \tag{296}$$

varying with ω has to be taken into account in the Newtonian potential. Clearly, for $\omega\to\infty$, $G_{eff}=\varphi_0^{-1}=G_N$. Our task is now to investigate the consequences on the oscillation phase for neutrinos propagating in such a geometry.

Before considering the consequences of nonminimal coupling, let us discuss, briefly, how the gravitational field contributes to neutrino oscillations. If R_A is a physical region where neutrinos are created, a neutrino energy eigenstate E_ν can be denoted by $\mid\nu_l, R_A\rangle$ (where $l=e,\mu,\tau$ represents the weak flavor eigenstates). The three neutrino mass eigenstates can be represented by $\mid\nu_i\rangle$ with $i=1,2,3$ corresponding to the masses m_1, m_2, m_3. The mixing between mass and flavor eigenstates is achieved by the transformation

$$\mid\nu_{l'}, R_A\rangle=\sum_{i=1,2,3}U_{li}\mid\nu_i\rangle\,, \tag{297}$$

where

$$U(\theta,\beta,\psi)=\begin{pmatrix} c_\theta c_\beta & s_\theta c_\beta & s_\beta \\ -c_\theta s_\beta s_\psi - s_\theta c_\psi & c_\theta c_\psi - s_\theta s_\beta s_\psi & c_\beta s_\psi \\ -c_\theta s_\beta c_\psi + s_\theta s_\psi & -s_\theta s_\beta c_\psi - c_\theta s_\psi & c_\beta c_\psi \end{pmatrix}, \tag{298}$$

is a 3×3 unitary matrix parametrized by the three mixing angles $\eta = \theta, \beta, \psi$ with $c_\eta = \cos(\eta)$ and $s_\eta = \sin(\eta)$. At time $t = t_B > t_A$, the weak flavor eigenstates can be detected in a region R_B and, in general, the evolution is given by

$$\mid \nu_l, R_B\rangle = \exp\left(-\frac{i}{\hbar}\int_{t_A}^{t_B} \mathcal{H}dt + \frac{i}{\hbar}\int_{r_A}^{r_B} \vec{P}\cdot d\vec{x}\right) \mid \nu_l, R_A\rangle\,, \tag{299}$$

where $\mathcal{H}$ is the Hamiltonian operator associated to the system (the time translation operator) and $\vec{P}$ the momentum operator (the spatial translation operator). The change in phase in Eq.(299), i.e. the argument of the exponential function, can be recast in the form

$$\Phi = \frac{1}{\hbar}\int_{r_A}^{r_B}\left[E\frac{dt}{dr} - p_r\right] dr\,, \tag{300}$$

which will be useful below. Let us consider now a covariant formulation as

$$\Phi = \frac{1}{\hbar}\int_A^B m ds = \frac{1}{\hbar}\int_A^B p_\mu dx^\mu\,, \tag{301}$$

where $p_\mu = m g_{\mu\nu}(dx^\nu/ds)$ is the 4-momentum of the particle. The effect of the gravitational field is given by $g_{\mu\nu}$ and, in general, the neutrino oscillation probability from a state $\mid \nu_l, R_A\rangle$ to another state $\mid \nu_{l'}, R_B\rangle$ is given by

$$\begin{aligned}\mathcal{P}\left[\mid \nu_l, R_A\rangle \rightarrow \mid \nu_{l'}, R_B\rangle\right] &= \delta_{ll'} - 4U_{l'1}U_{l1}U_{l'2}U_{l2}\sin^2[\Phi_0^{21} + \Phi_G^{21}] \\ &\quad -4U_{l'1}U_{l1}U_{l'3}U_{l3}\sin^2[\Phi_0^{31} + \Phi_G^{31}] \\ &\quad -4U_{l'2}U_{l2}U_{l'3}U_{l3}\sin^2[\Phi_0^{32} + \Phi_G^{32}]\end{aligned} \tag{302}$$

where Φ_0^{ij} are the usual kinematic phases while Φ_G^{ij} are the gravitational contributions. It is easy to show that, in a flat space-time, the Φ_G^{ij} contributions are zero. In fact, a particle passing nearby a point mass feels a Schwarzschild geometry so that the trajectories is given by

$$dx \simeq \left[1 - \frac{2G_N M}{c^2 r}\right] cdt\,. \tag{303}$$

If the effects of gravitational field are vanishing, Eq.(303) reduces to $dx \simeq cdt$. Considering two generic neutrino mass eigenstates in a Schwarzschild geometry, the total gravitational phase shift is

$$\Phi_G = \frac{G_N \Delta m^2 M}{4\hbar E}\log\frac{r_B}{r_A}\,, \tag{304}$$

as shown in [91], where Δm^2 is the mass squared difference, $\Delta m^2 = \mid m_2^2 - m_1^2 \mid$, E the neutrino energy, r_A and r_B the points where neutrinos are created and detected, respectively. Nevertheless, assuming that the neutrino energy is constant along the trajectory, it has been shown that the term (304) is cancelled out [94] so several authors are wondering if gravitational corrections to neutrino oscillation phase are true and detectable effects. Also the mere extension of the result in (304) to the standard Brans-Dicke theory does not give appreciable corrections to the quantum dynamical phase. In fact the corrective factor is of the form $(3 + 2\omega)(4 + 2\omega)^{-1}$ and it is of the order ~ 1 in the interesting limits $\omega \geq 500$ and $\omega \leq 30$ which we will discuss below.

If we assume the so called *covariant form* of the quantum phase, it is possible to solve this controversy and show why the gravitational contribution could be not observable without considering the effects of nonminimal coupling on the gravitationally induced neutrino oscillation phase. As we will see, the result could allow to discriminate among relativistic theories of gravity.

Let us consider now the representation (300) of quantum mechanical phase and the solution (288)–(290) of the scalar–tensor field equations (285)–(286). Inserting the momentum of the particle, coming from the shell–condition, one gets

$$p_r = e^{\beta-\alpha}\sqrt{E^2 - m^2 e^{2\alpha}} \tag{305}$$

into Eq. (300), and using the fact that $dt/dr = e^{\beta-\alpha}$, one gets the phase Φ given by

$$\Phi = \frac{1}{\hbar E}\int_{r_A}^{r_B}\left(1+\frac{B}{r}\right)^2\left(\frac{1-B/r}{1+B/r}\right)^{(\lambda-C/2)/\lambda}\left[E-\sqrt{E^2-m^2\left(\frac{1-B/r}{1+B/r}\right)^{2/\lambda}}\right]dr. \tag{306}$$

By using Eqs. (292)–(294), one can separate out the scalar-tensor gravitational contribution to the neutrino oscillation phase, so that Eq. (306) can be cast in the form

$$\Phi = \Phi_0 + \Phi_\omega\,, \tag{307}$$

where (restoring the constants c and $\hbar$)

$$\Phi_0 = \frac{\Delta m^2 c^3}{2E\hbar}(r_B - r_A)\,, \tag{308}$$

which represents the standard phase of neutrino oscillations, and

$$\Phi_\omega = \frac{\Delta m^2 c\, G_N M}{2\hbar E\;\; 2+\omega}\log\frac{r_B}{r_A}\,, \tag{309}$$

which clearly reproduces Eq.(304) for $\omega \to 0$ and disappears in the GR limit $\omega \to \infty$. In principle, this result could solve the above mentioned controversy [94].

In deriving Eqs.(308) and (309), we have considered ultra–relativistic neutrinos, $E >> mc^2$, where E is interpreted as the energy at the infinite. The integration has been performed along the trajectory where E is constant.

It is convenient to rewrite the phases (308) and (309) in the following way

$$\Phi_0 \approx 2.5\cdot 10^3 \frac{\Delta m^2}{\text{eV}^2/c^4}\frac{\text{MeV}}{E}\frac{r_B - r_A}{\text{Km}}\,, \tag{310}$$

and

$$\Phi_\omega \approx 3.5\cdot 10^3\,\frac{1}{2+\omega}\frac{\Delta m^2}{\text{eV}^2/c^4}\frac{\text{MeV}}{E}\frac{M}{M_\odot}\log\frac{r_B}{r_A}\,, \tag{311}$$

where $M_\odot$ is the solar mass. Estimations of the difference phases (310) and (311) are carried out for solar, atmospheric and astrophysical neutrinos. To this end, we will introduce the ratio q defined as

$$q = \frac{\Phi_\omega}{\Phi_0} \approx 1.5\,\frac{1}{2+\omega}\frac{M}{M_\odot}\frac{\log(r_B/r_A)}{(r_B - r_A)/\text{Km}}\,. \tag{312}$$

q does not depend on the squared–mass difference Δm^2 and on the neutrino energy E. For solar neutrinos, we use the following values: $M \sim M_\odot$, $r_A \sim r_{Earth} \sim 6.3 \cdot 10^3$Km, and $r_B \sim r_A + D$, where $D \sim 1.5 \cdot 10^8$Km is the Sun–Earth distance. Eq. (312) gives the result

$$q \sim 10^{-8} \frac{1}{2+\omega}, \tag{313}$$

which is an irrelevant correction to the difference phase (310). Analogous conclusion holds for atmospheric neutrinos.

Concerning the astrophysical neutrinos, the effect could be more relevant and could be measured by terrestrial experiments. In fact, setting $r_B = \alpha r_A$, $1 < \alpha \leq \infty$ and using the typical values of neutron stars, $M \sim 1.4 M_\odot$ and radius $r_A \sim 10$ Km, we get

$$q \sim \frac{0.2}{2+\omega} \frac{\log \alpha}{\alpha - 1}. \tag{314}$$

This analysis has been done for radially propagating neutrinos. In the case of motion transverse to the radial propagation and near to the detection point r_A we have

$$\Phi_\omega^\perp = \frac{\Delta m^2 c\, G_N M}{2\hbar E} \frac{r_B - r_A}{2+\omega} \frac{}{r_A} \approx 3.5 \cdot 10^3 \frac{1}{2+\omega} \frac{\Delta m^2}{\text{eV}^2} \frac{\text{MeV}}{E} \frac{M}{M_\odot} \frac{r_B - r_A}{r_A}. \tag{315}$$

Then, the ratio between the difference phases (315) and (310) is

$$q^\perp = \frac{\Phi_\omega^\perp}{\Phi_0} \approx 1.5 \frac{1}{2+\omega} \frac{M}{M_\odot} \frac{\text{Km}}{r_A}. \tag{316}$$

For the numerical constants corresponding to Sun and Earth, we have

$$q_{Sun}^\perp \sim \frac{1.5 \cdot 10^{-5}}{2+\omega}, \qquad q_{Earth}^\perp \sim \frac{5 \cdot 10^{-10}}{2+\omega}. \tag{317}$$

Using the above values for a neutron star, Eq.(316) gives the result

$$q^\perp \sim \frac{0.2}{2+\omega}. \tag{318}$$

Experimental data from the Solar System imply that the parameter ω can assume the value $\omega \geq 500$. For the lower limit, one gets from Eqs. (314) and (318),

$$q \sim 4 \cdot 10^{-4} \frac{\log \alpha}{\alpha - 1}, \qquad q^\perp \sim 4 \cdot 10^{-4}, \tag{319}$$

giving a correction of the 0.01 percent. For the sake of completeness, we have to consider also values $\omega \leq 30$, coming from the anisotropy of microwave background radiation [96]. In this case, we get corrections of few percents, as one can derive from Eqs.(314) and (318). Obviously, the limit $\omega \leq 30$ does not confirm the limit obtained from the Solar System measurements and it has to be discussed. The shortcoming, as reported in [96], comes out when one studies the prescriptions to obtain a successful extended inflationary stage. In fact, in order to remove the defects of the old inflationary model, one has to introduce an effective gravitational coupling, function of a scalar field φ, which allows a successful inflationary phase transition. The model is consistent with cosmological bounds from CMBR, if the Brans-Dicke parameter $\omega \leq 30$, limit which is clearly in contradiction to the Solar System measurements (see also [97]). The shortcoming is solved by introducing a non-trivial self-interaction potential $V(\varphi)$ (so then extending the original Brans-Dicke model where $V(\varphi) = 0$ and ω =const) which allows an evolution of the scalar field capable of reconciling the two limits at different scales [96]. So then, in this sense, we are dealing with a general scalar-tensor theory as in the action (57) where Brans-Dicke theory is only a particular case. Going back to our aims, using the Eqs.(314) and (318), we can get significant corrections to neutrino oscillation phase in both limits and this fact could be another experimental evidence in support of Open Quantum Relativity.

6.4. The Stochastic Background of Gravitational Waves and Open Quantum Relativity

Another interesting "signature" for Open Quantum Relativity could come from the gravitational waves (GW) detection. In fact, the design and the construction of a number of sensitive detectors for GWs is underway today. There are some laser interferometers like the VIRGO detector, built in Cascina, near Pisa by a joint Italian-French collaboration, the GEO 600 detector built in Hannover, Germany, by a joint Anglo-German collaboration, the two LIGO detectors built in the United States (one in Hanford, Washington, and the other in Livingston, Louisiana) by a joint Caltech-MIT collaboration, and the TAMA 300 detector, in Tokyo, Japan. Many bar detectors are currently in operation too, and several interferometers and bars are in a phase of planning and proposal stages (for the current status of gravitational waves experiments see [100, 101]). The results of these detectors will have a fundamental impact on astrophysics and gravitation physics. There will be a huge amount of experimental data to be analyzed, and theorists would face new aspects of physics from such a data stream. Furthermore, gravitational wave detectors will be of fundamental importance to probe the General Relativity or every new extended theory of gravitation [102, 103, 104]. A possible target of these experiments is the so called stochastic background of gravitational waves [105, 106, 107, 108]. The production of the primordial part of this stochastic background (relic GWs) is well known in literature [105, 108], where, using the so called adiabatically-amplified zero-point fluctuations process, several authors have shown in two different ways how the inflationary scenario for the early universe can, in principle, provide the signature for the spectrum of relic GWs.

Here we want to show that the effective scalar-tensor theory coming from Open Quantum Relativity can admit, in principle, the existence of GW scalar modes. Such a result can be confronted with the WMAP data [109, 110] and then it is possible to achieve a possible upper limit for the scalar part of relic GWs.

As above, we can recast the theory in a Brans-Dicke-like form and then taking into account the weak field limit in vacuum ($T^{(m)}_{\mu\nu} = 0$) up to first-order in perturbations. This means

$$g_{\mu\nu} = \eta_{\mu\nu} + h_{\mu\nu}\,, \qquad \varphi = \varphi_0 + \delta\varphi \tag{320}$$

and

$$V \simeq \frac{1}{2}\alpha\delta\varphi^2 \Rightarrow V' \simeq \alpha\delta\varphi \tag{321}$$

for the self-interacting, scalar-field potential. These assumptions allow to derive the "linearized" curvature invariants $\widetilde{R}_{\mu\nu\rho\sigma}$, $\widetilde{R}_{\mu\nu}$ and $\widetilde{R}$ which correspond to $R_{\mu\nu\rho\sigma}$, $R_{\mu\nu}$ and R, and then the linearized field equations [80]

$$\begin{array}{c} \widetilde{R}_{\mu\nu} - \frac{\widetilde{R}}{2}\eta_{\mu\nu} = -\partial_\mu\partial_\nu\Phi + \eta_{\mu\nu}\Box\Phi \\ \Box\Phi = m^2\Phi, \end{array} \tag{322}$$

where

$$\Phi \equiv -\frac{\delta\varphi}{\varphi_0}, \qquad m^2 \equiv \frac{\alpha\varphi_0}{2\omega + 3}. \tag{323}$$

The case $\omega = const$ and $W = 0$ can be analyzed in considering the so-called "canonical" linearization [80]. In particular, the transverse-traceless (TT) gauge (see [80]) can be generalized to scalar-tensor gravity obtaining the total perturbation due to a GW incoming in the $z+$ direction in this gauge as

$$h_{\mu\nu}(t - z) = A^+(t - z)e^{(+)}_{\mu\nu} + A^\times(t - z)e^{(\times)}_{\mu\nu} + \Phi(t - z)e^{(s)}_{\mu\nu}. \tag{324}$$

The term $A^+(t - z)e^{(+)}_{\mu\nu} + A^\times(t - z)e^{(\times)}_{\mu\nu}$ describes the two standard (i.e. tensorial) polarizations of a gravitational wave arising from General Relativity in the TT gauge [80], while the term $\Phi(t-z)e^{(s)}_{\mu\nu}$ is the extension of the TT gauge to the scalar case. This means that, in a Scalar-Tensor Gravity theory, the scalar field generates a third component for the tensor polarization of GWs. This is because three different degrees of freedom are present, while only two are present in standard General Relativity. In this sense, such the detection of such a new component could be a signature for the Open Quantum Relativity.

Let us now take into account the primordial physical process (where this effect is specially enhanced and so detectable with appropriate detector configurations) which gave rise to a characteristic spectrum Ω_{sgw} for the early stochastic background of relic scalar GWs. The production physical process has been analyzed, for example, in [105, 108, 113] but only for the first two tensorial components of eq. (324) due to standard General Relativity. Actually the process can be improved considering also such a third scalar-tensor component.

Before starting with the analysis, it has to be emphasized that, considering a stochastic background of scalar GWs, it can be described in terms of the scalar field Φ and characterized by a dimensionless spectrum (see the analogous definition for tensorial waves in [105, 106, 107, 108])

$$\Omega_{sgw}(f) = \frac{1}{\rho_c}\frac{d\rho_{sgw}}{d\ln f}, \tag{325}$$

where

$$\rho_c \equiv \frac{3H_0^2}{8\pi G} \tag{326}$$

is the (actual) critical energy density of the Universe, H_0 the today observed Hubble expansion rate, and $d\rho_{sgw}$ is the energy density of the scalar part of the gravitational radiation contained in the frequency range f to $f + df$. We are considering now standard units.

The existence of a relic stochastic background of scalar GWs is a consequence of very general assumptions. Essentially, in the framework of Open Quantum Relativity, it derives from basic principles of Quantum Field Theory and General Relativity. The strong variations of gravitational field in the early Universe amplifies the zero-point quantum fluctuations and produces relic GWs. It is well known that the detection of relic GWs is a good way to learn about the evolution of the very early universe [105, 106, 107, 108]. It is very important to stress the unavoidable and fundamental character of such a mechanism. It directly derives from the inflationary scenario [114, 115], which well fit the WMAP data in particular good agreement with almost exponential inflation and spectral index ≈ 1, [109, 110].

A remarkable fact about the inflationary scenario is that it contains a natural mechanism which gives rise to perturbations for any field. It is important for our aims that such a mechanism provides also a distinctive spectrum for relic scalar GWs. These perturbations in inflationary cosmology arise from the most basic quantum mechanical effect: the uncertainty principle. In this way, the spectrum of relic GWs that we could detect today is nothing else but the adiabatically-amplified zero-point fluctuations [105, 108]. The calculation for a simple inflationary model can be performed for the scalar field component of eq. (324). Let us assume that the early universe is described by an inflationary de Sitter phase emerging in a radiation dominated phase [105, 106, 108]. The conformal metric element is

$$ds^2 = a^2(\eta)[d\eta^2 - d\overrightarrow{x}^2 - h_{\mu\nu}(\eta, \overrightarrow{x})dx^\mu dx^\nu], \tag{327}$$

where, for a purely scalar GW the metric perturbation (324) reduces to

$$h_{\mu\nu} = \Phi e^{(s)}_{\mu\nu}, \tag{328}$$

Following [105, 108], if we express the scale factor in terms of comoving time $cdt = a(t)d\eta$, we have

$$a(t) \propto \exp(H_{ds}t)\,, \qquad a(t) \propto \sqrt{t} \tag{329}$$

for the de Sitter and radiation phases respectively. In order to solve the horizon and flatness problems, the condition $\frac{a(\eta_0)}{a(\eta_1)} > 10^{27}$ has to be satisfied, where η_1 is the inflation-radiation transition conformal time and η_0 is the value of conformal time today. The relic scalar-tensor GWs are the weak perturbations $h_{\mu\nu}(\eta, \overrightarrow{x})$ of the metric (328) which can be written in the form

$$h_{\mu\nu} = e^{(s)}_{\mu\nu}(\hat{k})X(\eta)\exp(\overrightarrow{k}\cdot\overrightarrow{x}), \tag{330}$$

in terms of the conformal time η where $\overrightarrow{k}$ is a constant wavevector. From Eq.(330), the scalar component is

$$\Phi(\eta, \overrightarrow{k}, \overrightarrow{x}) = X(\eta)\exp(\overrightarrow{k}\cdot\overrightarrow{x}). \tag{331}$$

Assuming $Y(\eta) = a(\eta)X(\eta)$, from the Klein-Gordon equation in the FRW metric, one gets

$$Y'' + (|\overrightarrow{k}|^2 - \frac{a''}{a})Y = 0 \tag{332}$$

where the prime $'$ denotes derivative with respect to the conformal time. The solutions of Eq. (332) can be expressed in terms of Hankel functions in both the inflationary and radiation dominated eras, that is:
For $\eta < \eta_1$

$$X(\eta) = \frac{a(\eta_1)}{a(\eta)}[1 + H_{ds}\omega^{-1}] \exp -ik(\eta - \eta_1), \tag{333}$$

for $\eta > \eta_1$

$$X(\eta) = \frac{a(\eta_1)}{a(\eta)}[\alpha \exp -ik(\eta - \eta_1) + \beta \exp ik(\eta - \eta_1), \tag{334}$$

where $\omega = ck/a$ is the angular frequency of the wave (which is function of the time being $k = |\vec{k}|$ constant), α and β are time-independent constants which we can obtain demanding that both X and $dX/d\eta$ are continuous at the boundary $\eta = \eta_1$ between the inflationary and the radiation dominated eras. By this constraint, we obtain

$$\alpha = 1 + i\frac{\sqrt{H_{ds}H_0}}{\omega} - \frac{H_{ds}H_0}{2\omega^2}, \qquad \beta = \frac{H_{ds}H_0}{2\omega^2} \tag{335}$$

In Eqs. (335), $\omega = ck/a(\eta_0)$ is the angular frequency as observed today, $H_0 = c/\eta_0$ is the Hubble expansion rate as observed today. Such calculations are referred in literature as the Bogoliubov coefficient methods [105, 108].

In an inflationary scenario, every classical or macroscopic perturbation is damped out by the inflation, i.e. the allowed level of fluctuations is that required by the uncertainty principle. The solution (333) corresponds to a de Sitter vacuum state. If the period of inflation is long enough, the today observable properties of the Universe should be indistinguishable from the properties of a Universe started in the de Sitter vacuum state. In the radiation dominated phase, the eigenmodes which describe particles are the coefficients of α, and these which describe antiparticles are the coefficients of β [116]. Thus, the number of particles created at angular frequency ω in the radiation dominated phase is given by

$$N_\omega = |\beta_\omega|^2 = \left(\frac{H_{ds}H_0}{2\omega^2}\right)^2. \tag{336}$$

Now it is possible to write an expression for the energy density of the stochastic scalar relic gravitons background in the frequency interval $(\omega, \omega + d\omega)$ as

$$d\rho_{sgw} = 2\hbar\omega\left(\frac{\omega^2 d\omega}{2\pi^2 c^3}\right)N_\omega = \frac{\hbar H_{ds}^2 H_0^2}{4\pi^2 c^3}\frac{d\omega}{\omega} = \frac{\hbar H_{ds}^2 H_0^2}{4\pi^2 c^3}\frac{df}{f}, \tag{337}$$

where f, as above, is the frequency in standard comoving time. Eq. (337) can be rewritten in terms of the today and de Sitter value of energy density being

$$H_0 = \frac{8\pi G\rho_c}{3c^2}, \qquad H_{ds} = \frac{8\pi G\rho_{ds}}{3c^2}. \tag{338}$$

Introducing the Planck density $\rho_{Planck} = \dfrac{c^7}{\hbar G^2}$ the spectrum is given by

$$\Omega_{sgw}(f) = \frac{1}{\rho_c}\frac{d\rho_{sgw}}{d\ln f} = \frac{f}{\rho_c}\frac{d\rho_{sgw}}{df} = \frac{16}{9}\frac{\rho_{ds}}{\rho_{Planck}}. \tag{339}$$

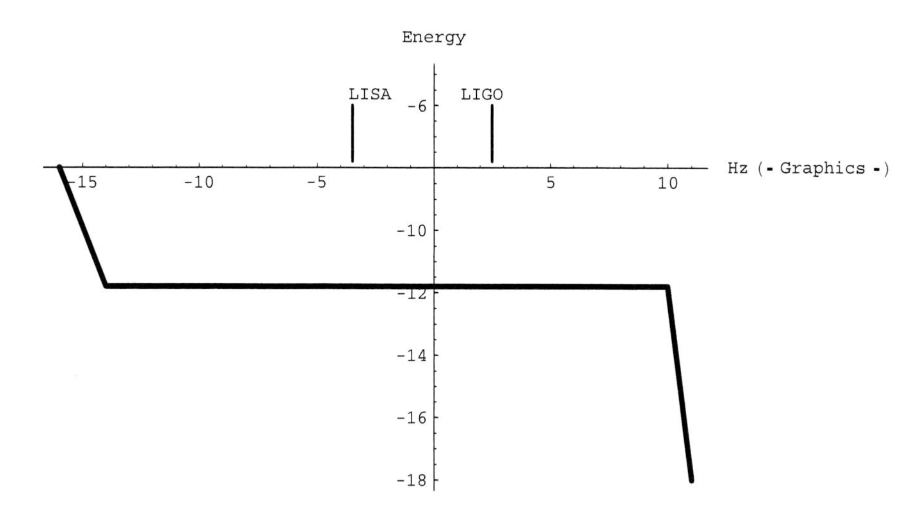

Figure 4. The spectrum of relic scalar GWs in inflationary models is flat over a wide range of frequencies. The horizontal axis is $\log_{10}$ of frequency, in Hz. The vertical axis is $\log_{10}\Omega_{gsw}$. The inflationary spectrum rises quickly at low frequencies (wave which re-entered in the Hubble sphere after the Universe became matter dominated) and falls off above the (appropriately redshifted) frequency scale f_{max} associated with the fastest characteristic time of the phase transition at the end of inflation. The amplitude of the flat region depends only on the energy density during the inflationary stage; we have chosen the largest amplitude consistent with the WMAP constrains on scalar perturbations. This means that, at LIGO and LISA frequencies, one has $\Omega_{sgw} < 2.3 * 10^{-12}$.

At this point, some comments are in order. First of all, such a calculation works for a simplified model that does not include the matter dominated era. If such an era is also included, the redshift at equivalence epoch has to be considered. Taking into account also results in [113], we get

$$\Omega_{sgw}(f) = \frac{16}{9}\frac{\rho_{ds}}{\rho_{Planck}}(1+z_{eq})^{-1}, \tag{340}$$

for the waves which, at the epoch in which the Universe becomes matter dominated, have a frequency higher than H_{eq}, the Hubble parameter at the equivalence era. This situation corresponds to frequencies $f > (1+z_{eq})^{1/2}H_0$. The redshift correction in Eq.(340) is needed since the today observed Hubble parameter H_0 would result different without a matter dominated contribution. At lower frequencies, the spectrum is given by [105, 108]

$$\Omega_{sgw}(f) \propto f^{-2}. \tag{341}$$

As a further consideration, let us note that the results (339) and (340), which are not frequency dependent, does not work correctly in all the range of physical frequencies. For waves with frequencies less than today observed H_0, the notion of energy density has no sense, since the wavelength becomes longer than the Hubble scale of the Universe. In analogous way, at high frequencies, there is a maximal frequency above which the spectrum rapidly drops to zero. In the above calculation, the simple assumption that the phase transition from the inflationary to the radiation dominated epoch is instantaneous has been made. In the physical Universe, this process occurs over some time scale $\Delta\tau$, being

$$f_{max} = \frac{a(t_1)}{a(t_0)}\frac{1}{\Delta\tau}, \tag{342}$$

which is the redshifted rate of the transition. In any case, Ω_{sgw} drops rapidly. The two cutoffs at low and high frequencies for the spectrum guarantee that the total energy density of the relic scalar gravitons is finite. For GUT energy-scale inflation it is of the order [105]

$$\frac{\rho_{ds}}{\rho_{Planck}} \approx 10^{-12}. \tag{343}$$

These results can be quantitatively constrained considering the recent WMAP release. In fact, it is well known that WMAP observations put strongly severe restrictions on the spectrum. In Fig. 4 the spectrum Ω_{sgw} is mapped : considering the ratio ρ_{ds}/ρ_{Planck}, the relic scalar GW spectrum seems consistent with the WMAP constraints on scalar perturbations. Nevertheless, since the spectrum falls off $\propto f^{-2}$ at low frequencies, this means that today, at LIGO-VIRGO and LISA frequencies (indicated in fig. 4), one gets

$$\Omega_{sgw}(f)h_{100}^2 < 2.3 \times 10^{-12}. \tag{344}$$

It is interesting to calculate the corresponding strain at $\approx 100Hz$, where interferometers like VIRGO and LIGO reach a maximum in sensitivity. The well known equation for the characteristic amplitude [105, 108] adapted to the scalar component of GWs can be used:

$$\Phi_c(f) \simeq 1.26 \times 10^{-18}(\frac{1Hz}{f})\sqrt{h_{100}^2\Omega_{sgw}(f)}, \tag{345}$$

and then we obtain

$$\Phi_c(100Hz) < 2 \times 10^{-26}. \tag{346}$$

Then, since we expect a sensitivity of the order of 10^{-22} for the above interferometers at $\approx 100Hz$, we need to gain four order of magnitude. Let us analyze the situation also at smaller frequencies. The sensitivity of the VIRGO interferometer is of the order of 10^{-21} at $\approx 10Hz$ and in that case it is

$$\Phi_c(100Hz) < 2 \times 10^{-25}. \tag{347}$$

The sensitivity of the LISA interferometer will be of the order of 10^{-22} at $\approx 10^{-3}Hz$ and in that case it is

$$\Phi_c(100Hz) < 2 \times 10^{-21}. \tag{348}$$

This means that a stochastic background of relic scalar GWs could be, even if difficult in principle, detected by the forthcoming LISA interferometer.

In summary, the above results point out that a further scalar component of GWs, coming from an the effective scalar-tensor theory derived from Open Quantum Relativity, could be seriously considered in the signal detection of interferometers at least in the future. In general, this fact could constitute an independent test for alternative theories of gravity or a further probe for General Relativity.

7. Discussion and Conclusions

Modern physics presents several shortcomings and paradoxes and an apparent dichotomy between Relativity and Quantum Mechanics. Moreover, many of the attempts made to settle this situation have led to an ever increasing number of dimensions and free parameters but leaving the problems still unsolved. Such issues, instead, could be framed, and solved, considering the lowest dimensional extension and assuming a *General Conservation Principle* by which the conservation laws can never be violated. This principle leads, in fact, to some basic features: $i)$ a unifying dynamical scheme which can be formulated in a $5D$ space; $ii)$ a covariant symplectic structure for tensor, vector and spinor fields by which it is possible to construct suitable Hamiltonian scalar invariant. These two ingredients constitute the basic structure of Open Quantum Relativity. The reduction process of dynamics from $5D$ to $4D$ gives rise to several new features and some of them have observable effects.

In $5D$, it is possible to construct a description based on a stress-energy tensor where Bianchi identities must hold in any case, since there are only kinetic terms without self-interacting potentials [30]. In this scheme, dynamics is singularity free and the reduction procedure to a $4D$-spacetime gives rise to a scalar-tensor theory of gravity where gravitational coupling is a function of a scalar field. A very significant consequence of this general approach is that, in $4D$, two time arrows and closed time-curves are exact solutions of dynamics, allowing a new interpretation of several phenomena, among them the entanglement.

In fact, entangled systems are not "hidden features" of Quantum Mechanics but superpositions of progressive and regressive solutions in a dynamical framework were backward and forward causation is treated under the same standard without any violation of conservation laws [22].

In this sense, topology change is the mechanism capable of guaranteeing the conservation not only in EPR-type phenomena but also, for instance, in collapsing astrophysical systems (*e.g.* mass-energy for black holes) [43]. This feature is fully compatible with Relativity since it can be dynamically realized in the $5D$–$4D$ reduction process considering a-luminal geodesics, and, moreover, it allows the full recovery of the essential concept of causality.

Furthermore, this scheme leads to an *induced-gravity-matter* theory where standard features of particles as mass, spin and charge (and also their hierarchy) come out as a direct consequence of the reduction procedure.

Finally, we have to stress that a covariant symplectic structure can be found for every Hamiltonian invariant. In fact, following the same scheme of Quantum Mechanics, any theory of physics has to be endowed with a symplectic structure in order to be formulated at a fundamental level.

We pointed out that curvature invariants of General Relativity can show such a feature and, furthermore, they can be recovered from Hamiltonian invariants opportunely defined. Another interesting remark deserves the fact that, starting from such invariants, covariant and contravariant vector fields can be read as the configurations q^i and the momenta p_i of classical Hamiltonian dynamics so then the Hamilton-like equations of motion are recovered from the application of covariant derivative to both these vector fields. Specifying the nature of vector fields, we select the particular theory. For example, if the vector field is the

4-velocity, we obtain geodesic motion and geodesic deviation. If the vector is the vector potential of electromagnetism, Maxwell equations and Lorentz gauge are recovered. The scheme is independent of the nature of vector field and it lead to a unifying view of basic interactions, gravity included.

Our approach is supported by matching with several observational and experimental evidences. In particular, starting from it, we can construct self-consistent cosmological models. This fact allows to reproduce the observed cosmological parameters as measured by the most recent observational campaigns as SNeIa project [55], WMAP [57], and SZE observations [59]. It is worth stressing that such a cosmology [54] is not phenomenological but it dynamically derives from first principles.

Another significant result is related to GRBs, considering an entangled gravitational system constituted by a black hole which, through a worm hole, emerges in a white hole as GRB source [43]. In this way, it is possible to explain the main observed features of these bursts [43]; specifically: 1) the order of magnitude of released energy, 2) the red-shift range in which such phenomena are observed, 3) the polarization and beaming.

Furthermore, since the $4D$-dynamics results in an effective scalar-tensor theory of gravity, the weak field limit gives rise to Yukawa-like corrections in the Newtonian potential. This feature could account for sizes of astrophysical systems, in particular the flat rotation curves of galaxies, without the necessity of huge amounts of dark matter, whose definitive experimental evidences are still lacking [44].

Finally the effective scalar tensor theory emerging from Open Quantum Relativity, could lead to new effects, like a gravitationally induced neutrino oscillation phase and a scalar component in the gravitational wave spectrum, which could be detected in the stochastic background by the forthcoming space experiments.

In conclusion, asking for a *General Conservation Principle*, and a covariant symplectic structure, we have achieved a self-consistent theory with the minimal request of free parameters and dimensions: the Open Quantum Relativity. As a consequence of the theory, the role of time has to be revised toward a full recover of causality: this means that the absolute validity of conservation laws needs backward and forward causation and then two time arrows. This is the core of our scheme which results in a unified view of modern physics (no longer sharply separated in "quantum" and "classical") and allows to frame in a comprehensive picture several shortcomings and paradoxes, maybe opening new perspectives.

References

[1] G. Basini, S. Capozziello, *Mod. Phys. Lett.* A 20, 251 (2005).

[2] G. Basini, S. Capozziello, *Int. Journ. Mod. Phys.* D 15, 583 (2006).

[3] A. Einstein, B. Podolsky, N. Rosen, *Phys. Rev.* **47** (1935) 777.

[4] S.O. Alexeyev, M.V. Sazhin, *Gen. Relativ. Gravit.* **30** (1998),1187.

[5] S.Alexeyev,A.Barrau,G.Boudoul,M.Sazhin,*Cl.Q.Gravity* **19** (2002)4431.

[6] J.D. Bekenstein, Black Hole Thermodynamics, *Physics Today*,1980.

[7] C. Quigg, *Gauge Theories of Strong, Weak, and Electromagnetic Interactions*, Addison-Wesley, Reading, MA, 1983; P.D.B. Collins, A.D. Martin, E.I. Squires, Particle Physics and Cosmology, Wiley, New York, 1991.

[8] G. Basini, A. Morselli, M. Ricci, *La Riv. del N. Cimento* 12 (1989) 4.

[9] A. Zichichi, *La Riv. del N. Cimento* **24** (2001) 12.

[10] A.D. Dolgov, M.V. Sazhin, Ya.B. Zeldovich, *Basic of Modern Cosmology*, Editions Frontieres, Gif-sur-Yvette, 1990.

[11] V.A. Rubakov, M.E. Shaposhnikov, *Usp. Fiz. Nauk.* 166 (1996) 493.

[12] E.W.Kolb, M.S. Turner, *The Early Universe*, Addison-Wesley, Redwood, 1990.

[13] C.H. Bennett et al. *Phys. Rev. Lett.* 70, 1985 (1993); M. Zukowski et al., *Phys. Rev. Lett.* 71, 4287 (1993); H. Weinfurter, *Europhys. Lett.* 25, 559 (1994).

[14] D. Bouwmeester et al. *Nature* 390, 575 (1997).

[15] F. Selleri (Ed.), *Conceptual Foundations of Quantum Mechanics*, Benjamin, Menlo Park, CA, 1988; F. Selleri (Ed.), *Quantum Mechanics versus Local Realism*,Plenum, London, 1988; J.Bell, *Rev. Mod. Phys.* 38 (1966) 447; J.Bell, *Physics* 1 (1965) 195.

[16] J.J.Sakurai, *Modern Quantum Mechanics*, Revised Edition, Addison-Wesley Publ. Co., New York, 1994.

[17] D. Bohm, *Quantum Mechanics*, Prentice-Hall, Englewood, N.J. (1951).

[18] A. Aspect, P. Grangier, G. Roger, *Phys. Rev. Lett.* 47, 460 (1981); A. Aspect, P. Grangier, G. Roger, *Phys. Rev. Lett.* 49, 91 (1982); A. Aspect, J. Dalibard, G. Roger, *Phys. Rev. Lett.* 49, 1804 (1982).

[19] C. Itzykson, J.B. Zuber, *Quantum Field Theory*, McGraw-Hill, Singapore, 1980.

[20] M. Kaku, *Quantum Field Theory*, Oxford Univ. Press, Oxford, 1993.

[21] N.Birrell, P.C.Davies, *Quantum Fields in Curved Space*, Cambridge Univ.1984.

[22] G. Basini S. Capozziello, *Europhys. Lett.* 63, 166 (2003).

[23] G. Basini, S. Capozziello, G.Longo, *Gen. Relativ. Grav.* 35 (2003) 189.

[24] L.D. Landau, E.M. Lifshitz, *Theorie du Champs*, Mir, Moscow, 1960.

[25] G. Basini, S. Capozziello, *La Riv. del N. Cimento* 27 N. 11 (2004).

[26] G. Basini S. Capozziello, *Europhys. Lett.* 63, 635 (2003).

[27] G. Basini, S. Capozziello, G.Longo *Phys. Lett.* 311 A, 465 (2003).

[28] G. Basini and S. Capozziello, *Prog. in Phys.* 3, 36 (2007).

[29] S. Weinberg, *Gravitation and Cosmology*, Wiley, New York (1972).

[30] G. Basini, S. Capozziello, *Gen. Relativ. Grav.* 35, 2217 (2003).

[31] D. Youm, *Phys. Rev.* D 62, 084002 (2000).

[32] J.E. Campbell *A Course of Differential Geometry*, Clarendon, Oxford (1926).

[33] D.H. Satinger, O.L.Weaver, *Lie Groups and Algebras with Applications to Physics, Geometry and Mechanics*, Springer-Verlag, Berlin (1986).

[34] K. Gödel, *Rev. Mod. Phys.*, 21, 447 (1949).

[35] C. Brukner,M.Zukowski,A.Zeilinger, quantum-ph/0106119 (2001).

[36] M. Visser, *Lorentzian Wormholes: From Einstein to Hawking*, American Institute of Physics Press (1995).

[37] K. S. Thorne, in GRG13: *General Relativity and Gravitation 1992* - Proceedings of the 13th International Conference on General Relativity and Gravitation, Cordoba, Argentina, 1992, Bristol Institute of Physics, p.295.

[38] J.L. Friedman et al. *Phys. Rev.* D 42, 1915 (1990).

[39] S.W. Hawking *Phys. Rev.* D 46 2 (1992).

[40] W.J. van Stockum, *Proc. Roy. Soc. Edin.* 57, 135 (1937).

[41] V.P. Frolov and I.D. Novikov, *Phys. Rev.* D 42, 1057 (1990).

[42] F. de Felice, *Il N. Cimento* 65, 224 (1981).

[43] G. Basini, S. Capozziello, G. Longo *Astrop. Phys.* 20, 457 (2004).

[44] G. Basini, S. Capozziello, M.Ricci, F.Bongiorno *Int. Journ. Mod.Phys.* D 13, 359 (2004).

[45] B. Julsgaard, A. Kozhekin, and E.S. Polzik, *Nature* 413, 400 (2001).

[46] J.J. Halliwell, *Nucl. Phys.* B 266, 228 (1986); J.J. Halliwell, *Phys. Rev.* D 36, 3626 (1987); J.J. Halliwell, in *Quantum Cosmology and Baby Universes*, Eds. S. Coleman S., Hartle J.B., Piran T., Weinberg S., World Scientioc, Singapore, (1991).

[47] A. Palatini, *Rend. Circ. Mat. Palermo* 43(1919), 203.

[48] B.S. DeWitt *Phys. Rev.* 160 (1967) 1113; C.W. Misner in *Relativity* eds. Carmeli, Fickler and Witten (Plenum Pub. Co., San Francisco, 1970); C.W. Misner in *Magic Without Magic* ed. J. Klauder, W.H. Freeman, San Francisco, (1972).

[49] A. Ashtekar *Phys. Rev. Lett.* 57 (1986) 2244; A. Ashtekar *Phys. Rev.* D 36 (1987) 1587; A. Ashtekar in *Proceedings of Banff Workshop on Grav. Phys.* (1990) and references therein.

[50] J.B. Hartle in *Gravitation in Astrophysics* Cargese, 1986, eds. B. Carter and J.B. Hartle, Plenum, New York, (1986); S.W. Hawking and D.N. Page *Nucl. Phys.* B 264 (1986) 185.

[51] S. Capozziello and G. Lambiase *Gen. Relativ. Grav.* 32, 673, 2000.

[52] G. Basini, S. Capozziello, F. Bongiorno, *Int. Jou. Mod. Phys.* D 13, 717 (2004).

[53] S.M. Caroll, W.H. Press, E.L. Turner, *Ann. Rev. Astr. Ap.* 30, 499 (1992).

[54] G. Basini, S. Capozziello, *Astrop. Phys.*, 21, 543 (2004).

[55] B.P. Schmidt et al. *Ap. J.* 507, 46 (1998); A.G. Riess et al. *Ap. J.* 116, 1009 (1998); S. Perlmutter et al. *Ap. J.* 483, 565 (1997); S. Perlmutter et al. *Nature* 391, 51 (1998); S. Perlmutter et al., *Ap. J.* 517, 565 (1999).

[56] P. de Bernardis et al. Nature 404, 955 (2000); R. Stompor et al. *Ap. J.* 561, L7 (2001); D.N. Spergel et al. astro-ph/0302209 (2003).

[57] M. Tegmark et al. astro-ph/0310723 (2003).

[58] Y. Wang, *Ap. J.* 536,531 (2000).

[59] M. Birkinshaw *Phys. Rep.* 310(1999) 97.

[60] M. Schmidt, *Ap. J.* 535, 117 (1999).

[61] J.I. Katz and L.M. Canel, *Ap. J.* 471, 915 (1996).

[62] S. Covino *et al*, *Astron. Astrophys.* 348, L1 (1999).

[63] R.A.M.J. Wijers *et al*, *Ap. J.* 523, 33L (1999).

[64] J. van Paradijs *et al*, *Nature*, 386, 686 (1997).

[65] T. Piran *Phys. Rep.*, 314, 575 (1999); T. Piran *Phys. Rep.*, 333, 529 (2000).

[66] G. Ghisellini, D. Lazzati, *Mon. Not. R. Ast. Soc.* 309, L7 (1999).

[67] M.J. Rees, P. Meszaros, *Ap. J.* 430, L93 (1994).

[68] R. Sari, *Ap. J.* 524, L43 (1999).

[69] K. Stelle, *Gen. Rel. Grav.* 9 (1978) 353.

[70] R.H. Sanders, *Ann. Rev. Astr. Ap.* 2 (1990) 1.

[71] M. Kenmoku, Y. Okamoto, and K. Shigemoto, *Phys. Rev.* 48 D (1993) 578.

[72] J.D. Anderson et al. *Phys. Rev. Lett.* 81 (1998) 2858.

[73] I. Quant and H.-J. Schmidt, *Astron. Nachr.* 312 (1991) 97.

[74] C.M. Will, *Theory and Experiments in Gravitational Physics* (1993) Cambridge Univ. Press, Cambridge.

[75] G.W. Gibbons and B.F. Whiting, *Nature* 291 (1981) 636.

[76] E. Fischbach, D. Sudarsky, A. Szafer, C. Talmadge, and S.H. Aroson, *Phys. Rev. Lett.* 56, 3 (1986) .

[77] C.C. Speake and T.J. Quinn, *Phys. Rev. Lett.* 61, 1340 (1988).

[78] D.H. Eckhardt, C. Jekeli, A.R. Lazarewicz, A.J. Romaides, *Phys. Rev. Lett.* 60, 2567 (1988).

[79] D.H. Eckhardt, *Phys. Rev.* 48 D (1993) 3762.

[80] C.W. Misner, K.S. Thorne, and J.W. Wheeler, *Gravitation*, W.H. Freeman and Company, New York (1970).

[81] G. Basini, S. Capozziello, G. Longo *Phys. Lett.* A 311, 465 (2003).

[82] C. Lanczos, *The Variational Principles of Mechanics*, Toronto Univ. Press (1949).

[83] E. Schwinger, *Phys. Rev.* 82, 914 (1951), *Phys. Rev.* 91, 713 (1953).

[84] E.V. Ferapontov, M.V. Pavlov, arXiv:nlin.SI/0212026 (2002).

[85] G. Basini and S. Capozziello, *Gen. Rel. Grav* 37, 115 (2005).

[86] F. Reines and C.L. Cowan Jr., *Phys. Rev.* 92, 830 (1953).

[87] J.I. Collar, *Phys. Rev. Lett.* 76, 999 (1996).

[88] S. Capozziello, G. Lambiase, G. Iovane *Mod. Phys. Lett.* A 18, 905 (2003).

[89] M. Blasone, A. Capolupo, S. Capozziello, S. Carloni, G. Vitiello, *Phys. Lett.* A 323, 182 (2004).

[90] G.L. Fogli et al. hep-ph/0408045 (2004).

[91] D.V. Ahluwalia and C. Burgard, *Gen. Rel. Grav.* 28, 1161 (1996); *Phys. Rev.* D57, 4724 (1998).

[92] Y. Grossman and H.J. Lipkin, *Phys. Rev.* D55, 2760 (1997).

[93] K. Konno and M. Kasai, *Progr. Theor. Phys.* 100, 1145 (1998).

[94] T. Bhattacharya, S. Habib and E. Mottola *Phys. Rev.* D59, 067301 (1999).

[95] J.D. Anderson, et al., *Phys. Rev.* D 65 (2002) 082004.

[96] D. La, P.J. Steinhard, and E.W. Bertschinger, *Phys. Lett.* B220, 375 (1989).

[97] D. La and P.J. Steinhard *Phys. Rev. Lett.* 62, 376 (1989).

[98] R.V. Pound and G.A. Rebka, *Phys. Rev. Lett.* 4, 337 (1960).

[99] R. Colella, A.W. Overhauser, S.A. Werner *Phys. Rev. Lett.* 34, 1472 (1975).

[100] http://www.ligo.org/pdf_public/camp.pdf.

[101] http://www.ligo.org/pdf_public/hough02.pdf.

[102] S. Capozziello and A. Troisi, *Phys. Rev. D* **72**, 044022 (2005).

[103] S. Capozziello - *Newtonian Limit of Extended Theories of Gravity* in *Quantum Gravity Research Trends* Ed. A. Reimer, pp. 227-276 Nova Science Publishers Inc., NY (2005).

[104] M. E. Tobar , T. Suzuki and K. Kuroda *Phys. Rev. D* **59**, 102002 (1999).

[105] B. Allen - *Proceedings of the Les Houches School on Astrophysical Sources of Gravitational Waves*, eds. Jean-Alain Marck and Jean-Pierre Lasota (Cambridge University Press, Cambridge, England 1998).

[106] B. Allen and A.C. Ottewill - *Phys. Rev. D* **56**, 545-563 (1997).

[107] M. Maggiore - *Phys. Rep.* **331**, 283-367 (2000).

[108] L. Grishchuk et al. - *Phys. Usp.* **44** 1-51 (2001); *Usp. Fiz. Nauk* 171 3-59 (2001).

[109] C.L. Bennet et al. - *ApJS* **148**, 1 (2003).

[110] D.N. Spergel et al. - *ApJS* **148**, 195 (2003).

[111] S. Nojiri and S.D. Odintsov *Int. J. Geom. Meth. Mod. Phys.* 4, 115 (2007).

[112] C. Brans and R. H. Dicke - *Phys. Rev.* **124**, 925 (1961).

[113] B. Allen - *Phys. Rev. D* **3-7**,2078 (1988).

[114] G.S. Watson - *"An exposition on inflationary cosmology"* - North Carolina University Press (2000).

[115] A. Guth - *Phys. Rev.* **23**, 347 (1981).

[116] S. Capozziello, Ch. Corda, M. De Laurentis, *Mod. Phys. Lett.* A 22, 1097 (2007).

In: Classical and Quantum Gravity Research
Editors: M.N. Christiansen et al, pp. 245-269
ISBN: 978-1-60456-366-5

Chapter 5

CONDITIONS FOR STIMULATED EMISSION IN ANOMALOUS GRAVITY-SUPERCONDUCTORS INTERACTIONS

***Giovanni Modanese*[1], *Timo Junker*[2] *and Göde Wissenschaftsstiftung*[2]**
Am Heerbach 5, 63857 Waldaschaff, Germany
[1] University of Bolzano – Logistics and Production Engineering; Via Sernesi 1, 39100 Bolzano, Italy

Abstract

Several authors have studied the generation of gravitational fields by condensed-matter systems in non-extreme density conditions (i.e., conditions not like those of collapsed stars, but such to be possibly obtained in a laboratory). General Relativity and lowest-order perturbative Quantum Gravity predict in this case an extremely small emission rate, so these phenomena can become relevant only if some strong quantum effect occurs. Quantum aspects of gravity are still poorly understood. It is believed that they could play a role in systems which exhibit macroscopic quantum coherence, like superconductors and superfluids, leading to an "anomalous" coupling between matter and field. We mention here recent work in this field by Woods, Chiao, Becker, Agop et al., Ummarino, Kiefer and Weber. Many of these theoretical works were stimulated by the experimental claims of Podkletnov. His results have not yet been confirmed, but the published replication attempts have admittedly been incomplete. Recently, Tajmar claimed to have detected a gravitomagnetic field generated by a spinning superconductor. Chiao also made some attempts at the construction of a gravity/e.m. transducer based on quantum effects. In our previous theoretical work, we sought an interpretation of the anomalous emission reported by Podkletnov as a consequence of the local modification of the vacuum energy density in the superconductor. We hypothesed that the vacuum energy density term could interfere with a set of strong gravitational fluctuations called "dipolar fluctuations". In this chapter we improve our earlier model and also present new results concerning anomalous stimulated gravitational emission in a layered superconductor like YBCO. We model the superconductor as an array of intrinsic Josephson junctions. The superconducting parameters are defined by our preliminary measurements with melt-textured samples. Coherent e.m. emission by synchronized Josephson junctions arrays was first reported by Barbara et al. in 1999. We write explicitly and solve numerically the

Josephson equations which give the normal and super components of the total current in the superconductor, and derive from this the total available power $P=IV$. Then, assuming that the coefficients A and B for spontaneous and stimulated gravitational emission are known, we apply to this case the Frantz-Nodvik equation for a laser amplifier. The equation is suitably modified in order to allow for a "continuous pumping" given by an oscillating transport current. The conclusions are relevant for the evaluation of gravitational emission from superconductors. We find that even if the A and B coefficients are anomalously large (possibly because of the Quantum Gravity effects mentioned above), the conditions for stimulated emission are quite strict and the emission rate strongly limited by the IV value, for reasons intrinsic to the nature of the superconductor.

1. Introduction

Over the last decades several authors, mainly from the General Relativity community, were intrigued by the idea that the interaction between gravity and superconductors might be somehow peculiar. The simplest proposals were about using superconductors as sensitive field detectors. Some also speculated, however, that superconductors could act as effective emitting antennas of gravitational waves. This is clearly outside the orthodoxy of General Relativity, which "weighs" any gravitational source only with regard to its energy-momentum, independently from its microscopic structure or composition.

But if gravitation has to be eventually reconciled with quantum mechanics, the macroscopic quantum character of superconductors might actually matter. In a recent authoritative review on the "Interaction of gravity with mesoscopic systems" [1], Kiefer and Weber recall that the interaction of gravitational fields with quantum fluids has been extensively studied. They mention work published by De Witt in 1966, Papini in 1967, Anandan and Chiao in 1981-84, Peng and Torr in 1990-91. Then they focus on the ideas which describe generation and detection of gravitational waves via the use of quantum fluids. They investigate the arguments suggesting that quantum fluids should be better interaction partners of gravitational waves than classical materials, and discuss proposed coupling schemes, including those by De Matos and Tajmar [2] and by Chiao et al. [3].

The "HFGW conferences" (High Frequency Gravitational Waves) held in 2003 and 2007 collected numerous conservative and speculative works. Conservative works typically involve General Relativity estimates of (very low) gravitational waves power emitted by laboratory devices with high-frequency vibrations. Speculative works (for instance, [4]) hypotesized that the graviton emission amplitude is somehow amplified by quantum properties of matter. Woods [5] discusses impedance mismatch at superconductor-air interfaces in the propagation of HFGWs. In the case of type-II superconductors with variable internal magnetization, he shows that this amounts to a sizeable interaction between the gravitational wave and the magnetic field ("enhanced Gertsenshtein effect"). He argues that this may be exploited for the design of a novel type of lens for HFGWs, using a magnetic field to adjust the focal length.

Agop et al. [6] write equations for a generalized Meissner effect, which take into account the gravito-magnetic and gravito-electric fields in the Maxwell-Einstein approximation. They find a very large "gravitational screening length", in accordance with previous authors, and yet their screening equation also involves the short length $\lambda_e \approx 10^{-8}$ m. Ummarino [7] calculates the possible alteration of the gravitational field in a superconductor using the time-dependent Ginzburg-Landau equations and compares the behaviour of a high-T_c superconductor like YBCO with a classical low-T_c superconductor like Pb.

Many of these theoretical works were stimulated by the experimental claims of Podkletnov [8,9]. His results have not yet been confirmed, but the published replication attempts have admittedly been incomplete [10].

One of the problems of current models of gravity-superconductors interaction is the over-simplified representation of the superconductors. Type-II superconductors and anisotropic ceramic superconductors have a complex microscopic structure, far from the ideal fluid model suitable for type-I superconductors with long-range coherent wave functions. In this work we shall model a melt-textured YBCO emitter according to the intrinsic Josephson junctions picture which has been firmly established starting from the '90.

In general terms, we believe that gravitational emission from superconductors is limited by three main factors.

1. The fundamental coupling. A quantum mechanism is needed, which escapes the severe limitations of the standard General Relativity coupling. We have proposed earlier such a dynamical mechanism, based on the vacuum fluctuations of Quantum Gravity. Here we recall it briefly and add some new remarks. This mechanism is only able to generate a strong virtual radiation, ie off-shell gravitons with $\lambda f \ll c$ and finite propagation range (but also with spin 0 and 1 components).
2. Energetic efficiency. This is a key limitation even in electromagnetic Josephson emission from superconductors. The maximum available power $P=IV$ is usually small, due to very small voltage drops in good superconductors. In addition, any current injected from the outside causes considerable dissipation at the superconducting-normal contacts.
3. Stimulated emission. It has been found experimentally [26] that electromagnetic emission from Josephson junctions arrays can be amplified by stimulated emission. In the absence of any resonant cavity suitable for gravitational radiation, this can only occur in a single-pass mode, like in an optical or maser amplifier governed by the Frantz-Nodvik rate equation. The non-standard dispersion relation $\lambda f \ll c$ implies that real stimulated electromagnetic emission does not compete with virtual graviton emission.

We shall address the Points (1), (2) and (3) above in Sec. 2, 3, 4, respectively. The techniques employed vary. Sec. 2 mainly involves Quantum Gravity considerations and a model of gravitational vacuum fluctuations. We recall previous work and prove, or a least justify, a chain of equivalences: (i) the presence of a superconductor amounts to a local variation in the vacuum energy density Λ; (ii) a time-variable Λ is equivalent to an oscillating virtual mass $M_{\Lambda,\mathrm{eff}}$; (iii) this means in turn that in a Josephson junction under high-frequency current, sizeable coefficients of spontaneous and stimulated graviton emission can be defined. The weakest link in this chain is the proof of the "amplification" $M_{\Lambda,\mathrm{eff}} \gg M_\Lambda=\Lambda V/G$, where V is the spatial volume of the region where Λ is present, and the vacuum energy density Λ/G associated with a superconductor is typically 10^6-10^8 J/m^3.

Sec. 3 employs notions and techniques from the theory and experimental practice of ceramic superconductors and Josephson junctions. On the base of data from our preliminary measurements, an YBCO emitter is analysed as a series of intrinsic Josephson junctions, whose behaviour is numerically simulated within the RSJ model. The main outcome is an estimate for the maximum available emission power (see also the Conclusion Section for a summary). Several details related to the superconducting properties of the emitter are discussed: emitter inductance, plasma frequency, dampening parameter, normal resistance of the intrinsic junctions and resistive shunts, synchronization of the junctions, effect of an external magnetic field, contact resistance and heating.

In Sec. 4 we compute the probability of stimulated emission through a specific rate equation, derived from the Frantz-Nodvik equation, but with three important modifications: (a) A spontaneous emission term, absent in the original equation, is introduced; this is responsible for the start of the emission. (b) Correspondingly, the initial conditions for the solution of the differential equation are different: there is no incoming beam, since the beam is generated inside the active material by spontaneous emission. (c) The population inversion and pumping occur via an external oscillating current, which generates a voltage on the intrinsic junctions, as computed in Sec. 3.

2. Anomalous Emission as a Consequence of the Local Modification of Vacuum Energy Density in Superconductors

Throughout this Section we use units in which $\hbar=c=1$.

2.1. Previous Work in Perturbation Theory

In General Relativity and related models (including modifications of Einstein's theory and quantization attempts) the coupling between matter and the gravitational field is described by the tensor $GT_{\mu\nu}$ in the field equations, or equivalently by an interaction term $GT_{\mu\nu}g^{\mu\nu}$ in the Lagrangian. In general, the field equations can also contain a term $\Lambda g_{\mu\nu}$, corresponding to $\sqrt{g}\Lambda$ in the Lagrangian. This term, traditionally named "cosmological term", describes the coupling of the field with the so-called vacuum energy density. By definition, the vacuum energy density is Lorentz invariant, ie it looks the same for any observer in relative uniform motion. It follows that the energy-momentum tensor of vacuum energy density must have the form $const\cdot g_{\mu\nu}$. In the vacuum energy density are usually included the zero-point energies of the quantized fields (including the gravitational field itself). Further contributions to Λ originate from the non-vanishing vacuum expectation values of quantum fields in the presence of spontaneous symmetry breaking.

For instance, for a scalar field with vacuum expectation value φ_0 and Lagrangian $L(\varphi)$, the cosmological term is $-8\pi GL(\varphi_0)$, because the energy-momentum tensor has the form $T_{\mu\nu}=\partial_\mu\varphi\partial_\nu\varphi-g_{\mu\nu}L$. For the electromagnetic field, the part of $T_{\mu\nu}$ proportional to $g_{\mu\nu}$ is (B^2-E^2). Possible contributions of this kind to the global vacuum energy density are supposed to define a uniform background present in the whole universe. Unless there is an exact cancellation of the various contributions, the curvature of the universe should be very large; but this is not observed, and that is the origin of the well-known "cosmological constant problem".

A *local* contribution to the vacuum energy density can arise when the state of a localized physical system is described by a classical field comparable with the vacuum expectation value of a quantum field. We are interested into cases of this kind occurring in condensed matter physics. In this context, the physical systems properly described by *continuous* classical-like fields (also at microscopic level, not just in a macroscopic-average sense as for fluids) are basically: (1) the electromagnetic field in the low-frequency limit, in states where the photons number uncertainty is much larger than the phase uncertainty; (2) systems with macroscopic quantum coherence, described by "order parameters", like superfluids, superconductors and spin systems.

Suppose that, in one of these systems, the field has a constant value φ_0 in a bounded region and is zero outside. (Consider for instance a container with superfluid helium of constant density.) We can speak of a contribution of the field to the cosmological constant in this region, equal to $-8\pi GL(\varphi_0)$, if it is a scalar field, or the analogous quantity for other fields.

From the classical point of view, one can correctly object that the description of this situation in terms of a local cosmological constant is purely formal, because the gravitational field present is just that due to the superfluid regarded as an energy-momentum source. Moreover, there is no distinction, still at the classical level, between a truly continuous source, like the superfluid wave function, and an incoherent fluid.

The perspective changes if one takes into account short-scale gravitational quantum fluctuations. Suppose to describe gravity with the covariant perturbation theory in the weak-field approximation. The action contains some parameters, and one of these is the effective Λ in the considered region. The Λ term in superconductors turns out to be much larger than the cosmological background: one typically has $\Lambda/G=10^6$-10^8 J/m^3 in superconductors, depending on the type, while the currently accepted value for the cosmological background is of the order of $\Lambda/G=10^{-9}$ J/m^3 [11]. This is interesting in principle, because it implies in any case a peculiar dynamical condition. At the classical level, however, such a small mass-energy density is irrelevant. As we said, it should be treated quantum-mechanically, being microscopically uniform throughout the superconductor.

In perturbation theory a negative Λ (in our conventions) gives gravitons a small real mass, while a positive Λ gives them an imaginary mass, ie creates an instability. This seems to suggest a non-trivial role of the Λ term in quantum gravity, especially in situations with positive local Λ, like the unusual case of local density maxima in superconductors [12]. Any supposed instability, however, takes us outside the validity range of perturbation theory. Furthermore, there is no evidence that anomalous effects only occur in situations with positive local Λ. More often, there appears to be a local Λ oscillating in time (e.g., in layered superconductors with high-frequency supercurrents).

The idea, then, is to look for a more fundamental mechanism. We hypothesized earlier [13] that the Λ term can be particularly relevant for field configurations with zero scalar curvature. We found a large novel class of off-shell weak field configurations (gravitational vacuum fluctuations) having this property, and studied their modification under the effect of a Λ term. Although the fluctuations themselves are very strong and bear a large virtual mass, we found that the effect of the Λ term upon them is small, corresponding only to the mass-energy equivalent of Λ itself.

In this approximation there is no amplification, ie the Λ term does not cause any appreciable variations of the virtual mass density. In general in quantum mechanics an oscillating charge [mass] source emits photons [gravitons], the emission probability being proportional to the square of the source. Then for gravitons the probability is proportional to

Λ^2 and very small, because it also contains the small coupling G [14]. Note that we are talking here of source in a virtual sense, as intermediate state in a quantum process. Such sources would generate the virtual gravitons we called for as possible explanation of the anomalous effects [9].

2.2. Vacuum Fluctuations with Large Virtual Mass, in Strong-Field Regime

In subsequent work [15] we thus extended the concept of "dipolar" virtual mass fluctuations to the strong-field case. The $\sqrt{g}$ volume factor in the gravitational action is relevant in this case, and the fluctuations are not exactly dipolar any more. We obtained a wider set of vacuum field configurations, with large virtual masses, present at any length scale. We also showed that this set has finite functional measure, so that the fluctuation effects indeed enter physical averages.

And yet an "amplification" effect of the Λ term is still missing. If we insert a static Λ into the zero-mode equation

$$\frac{rg_{rr}'}{g_{rr}^2} - \frac{1}{g_{rr}} + 1 = 0, \tag{2.1}$$

the solution will be of the same form as for zero-curvature, but with a slightly different virtual mass. The mass variations will be again of the order of the space integral of the mass density equivalent of Λ/G. Let us show this in detail. The Λ term appears in the zero-mode equation as a constant source term. We already considered in [15] source terms with spatial oscillations, leading to "excited" zero-modes, for instance with the sources sin(ns). In that case, the explicit solution of the zero-mode equation was only approximate, although we knew that the exact solution satisfies the zero-mode condition exactly. Here, since the Λ term in the Lagrangian includes a factor $\sqrt{g}$, this factor is eliminated from the equation and we obtain an explicit exact solution. The Λ term is first supposed constant in time.

Apart from factors 4π, writing A=g_{rr}, B=g_{00}, the zero-mode equation becomes

$$\sqrt{|AB|}\left(\frac{rA'}{A^2} + 1 - \frac{1}{A}\right) = \sqrt{g}\frac{\Lambda}{G} \tag{2.2}$$

but $\sqrt{g} = \sqrt{|AB|}r^2$, therefore the equation is simplified and the solution is

$$\alpha(s) = \frac{1}{s}\left[\int\left(1 + t^2\widetilde{\Lambda}\right)dt + k\right] \tag{2.3}$$

where $\alpha=1/A$, k is an integration constant and $\widetilde{\Lambda}$ is the adimensional value of Λ after re-scaling to the size $r_{\rm ext}$ of the fluctuation: $\widetilde{\Lambda}=r_{ext}^{\ 2}\Lambda$. We obtain, denoting by r_0 the size of the region where Λ is not zero and by $s_0=r_0/r_{\rm ext}$ the corresponding adimensional quantity:

$$\alpha(s)=1+\frac{k}{s}+\frac{1}{s}\frac{\widetilde{\Lambda}s_0^{\ 3}}{3} \tag{2.4}$$

By correspondence, k must be the virtual adimensional mass $\widetilde{M}$ in the absence of Λ. The factor $\frac{\widetilde{\Lambda}s_0^{\ 3}}{3}$ gives the additional mass due to the cosmological term. Remembering that adimensional masses are scaled according to the rule $M=\frac{\widetilde{M}r_{ext}}{2G}$, we find that the additional mass is of the order of $M_\Lambda=\frac{\Lambda r_0^{\ 3}}{G}$. This is the mass resulting from the energy-density Λ/G integrated over the volume $r_0^{\ 3}$ where Λ is not zero.

From this point on, we offer a new conjecture which has a definite intuitive justification but still lacks a proof. We try to show that an oscillating Λ term generates synchronous oscillations in the virtual mass density.

2.3. A Possible Connection between an Oscillating Local a Term and Virtual Mass Fluctuations

We have seen that strong vacuum fluctuations with large virtual mass exist in quantum gravity. We would like now to display a connection between an oscillating Λ and virtual mass fluctuations. The main idea is, that a local time-variable Λ term causes changes in the fluctuations spectrum. These changes have observable physical effects. Like in the Casimir effect, there exists a definite interface between the real and the virtual world. For the Casimir effect, the interface are the metal plates which cut the virtual electromagnetic modes. Here Λ is not a static cut-off, but oscillates in the superconductor with a certain frequency. Λ appears in the "state" equation of the vacuum fluctuations as an external source; when it is oscillating, it could be regarded as a forcing term. We know that Λ is very small and we have seen in the previous section that its static effects are minimal, namely tiny shifts in the virtual masses. If we model vacuum fluctuations by a collection of harmonic oscillators, then a static Λ term would shift slightly their equilibrium positions. Even as a forcing term, Λ would be inefficient, causing small oscillation amplitudes proportional only to Λ itself.

In gravitation the vacuum fluctuations are much stronger than in QED, and virtual Casimir-like forces can be much stronger, too. We think that the Λ term could "attract" to its own frequency some of the natural virtual amplitude present at nearby frequencies. This change of spectrum would be observable, and its amplitude proportional both to Λ and to the pre-existing natural virtual amplitude.

This frequency change of an oscillator in response to an external, possibly weak "pilot" signal is a phenomenon also known as "entrainment" of the oscillator [16]. When many weakly interacting oscillators with different proper frequencies are involved, one often speaks of "synchronization" of the oscillators. It is a phenomenon widespread in nature and well described by general mathematical models like for instance the Kuramoto model [31]. A large ensemble of coupled oscillators can be in synchronized or non-synchronized phases, depending on the values of the coupling parameter, on the natural frequency spread and on the amplitude of the pilot signal. Our dipolar vacuum fluctuations are not exactly an ensemble of oscillators, but a definite analogy can be drawn.

In the functional integral all these fluctuations have the same probability, because their action is zero. (A slightly different weight for the configurations could originate from the functional measure, which is however unknown.) We model the evolution of the fluctuations as a random process, like in a Montecarlo simulation: the system makes frequent attempts at transitions to different states, and the transition probability is given by $\exp(-\beta\delta m)$, where β is the analogue of an inverse temperature $\beta=1/kT$ and $\delta m=|m-m'|$ is the mass difference between the initial and final state. Transitions are more likely when the mass difference is small, because in that case the functional change in the configuration is also small.

Transitions between fluctuations of different mass effectively appear as mass oscillations. Let us consider a set of fluctuations with mass distributed within a certain interval σ_m about a reference mass M which can be very large, up to $M \approx 10^{57}$ cm^{-1} $\approx 10^{22}$ g at the "condensed matter" scale $r_0=10^{-9}$ m, ie the scale of the local Λ [15].

Let the mass take discrete values m_i. With a large attempt frequency f_0, the system makes attempts at transitions, with probability $\exp(-\beta\delta m)$. The result is a temporal sequence of mass values $m_i(t_j)$. By Fourier-analyzing this sequence, we should obtain a spectrum with a broad maximum around some multiple n of the attempt period $T_0=1/f_0$. (Of course, the parameters f_0, β and δm all have to be normalized at the same time.) This means that the average transition time is $\approx nT_0$. The average is made with transitions with different δm and different probabilities; the dominant transitions are those with smaller δm.

We know that a static Λ term changes the masses m_i by an amount $M_\Lambda=\Lambda V/G$, because it enters the "state equation" of the fluctuations as a source term (eq. 2.2). Call f_Λ the oscillation frequency of a varying $\Lambda(t)$ (with $f_\Lambda<f_0$). Can we expect that the transition probability is affected, and thus the spectrum of $m_i(t_j)$?

The mass transitions are not strictly periodic, while the external signal $\Lambda(t)$ is, so we expect that in some case the transition will be easier, when $\Lambda(t)$ decreases δm, or viceversa. Since the amplitudes of the mass oscillations are much larger than the amplitude of M_Λ, we expect that the variations in the spectral amplitudes of the fluctuations are proportional to M_Λ but also to the initial amplitude of the fluctuations. This is the analogue of the frequency entrainment of an oscillator by an external pilot signal, except that the fluctuations are not exactly like an oscillator; they are random transitions without a sharp proper frequency, and so we speak of a change in their spectrum, instead of entrainment of their frequency.

A typical feature of phenomena of frequency entrainment is, that the amplitude of the entrained signal is inversely proportional to the difference between the pilot signal and the proper frequency of the oscillation [17]. We can check that this is also the case in our simple model of transitions. If we consider a transition with probability close to 1, which means that the inverse temperature parameter β is small (ie, the temperature large, and transitions quite probable), we can write the probability as $P \approx 1-\beta\delta m+(\beta\delta m)^2/2$. If δm changes by an amount M_Λ, the transition can be favoured or hindered and the average probability change ΔP is of order $(\beta\delta m)^2$. The change in the average transition frequency is $\Delta f_{av} \approx \Delta P/T_0$ and so finally $\Delta f_{av}/f_0 \approx (\beta M_\Lambda)^2$.

Since in the current approximation the transition probability is of order 1, then βM_Λ is much smaller than 1, and of the order of $M_\Lambda/\delta m$. Therefore $\Delta f_{av}/f_0 \approx (M_\Lambda/\delta m)^2$. This gives a relation between the frequency shift, or entrainment of the fluctuations, and the fluctuation amplitude which is shifted in frequency. We are interested into frequency shifts of fluctuations with an amplitude δm which is much larger than M_Λ. The squared inverse proportionality between the frequency shift and the amplitude sets a limit on the deformation of the fluctuations spectrum.

In order to assess the final effect of the Λ term, and judge which deformations of the spectrum are observable, we still need to know how the "natural" spectrum is. Presently we do not have any information about the shape of this spectrum, but just about the single modes. A further missing piece of information is the number of fluctuations or mass oscillations present in a given region of space. We can not know this until we know the upper cut-off on their mass or some other UV cut-off. The total change of amplitude in mass fluctuations depends on the number of field oscillators; the amplitudes displaced from their original frequency add coherently (because they all follow the oscillations of the pilot Λ) and therefore this sum process between modes is crucial.

Also in systems of weakly interacting classical oscillators (and our fluctuations are certainly weakly interacting) the emergence of phenomena of synchronization or collective entrainment by an external pilot signal depends on the presence of a large number of oscillators. We shall see an example of this behaviour in the Josephson junctions arrays: their synchronization is observed only above a certain number of junctions.

2.4. How to Define the *A* and *B* Coefficients for Virtual Graviton Emission?

The last logical step required is: how to relate the occurrence of virtual mass oscillations at the same frequency of $\Lambda(t)$ (but with larger amplitude) to the emission of virtual gravitons with Einstein coefficients A and B of spontaneous and stimulated emission? Such a relation has been previously demonstrated by Rogovin and Scully [27] for the electromagnetic emission in a Josephson junction under finite voltage. In that case, the macroscopic classical picture of the oscillating dipole is well complemented by the quantum-mechanical picture of electrons in a collective wave function, undergoing quantum tunnelling between two states. In the case of the Josephson junction, the oscillating charge/current is that of the Cooper pairs. In the gravitational case, the oscillating mass is not directly the mass of the Cooper pairs, it is $M_{\Lambda(t)}=\Lambda(t)V/G$, amplified by "entrained" vacuum fluctuations. $\Lambda(t)$ has a definite expression in terms of the pairs density ρ [12]:

$$\Lambda = -\frac{1}{2m}\left[\hbar^2(\nabla\rho)^2 + \hbar^2\rho\nabla^2\rho - m\beta\rho^4\right], \tag{2.5}$$

where β is the second Ginzburg-Landau coefficient and m is the Cooper pair mass.

3. Ceramic Superconducting "Emitters" Modelled as Series of Intrinsic Josephson Junctions

We are currently performing precise measurements of the behaviour of melt-textured ceramic superconductors under the conditions of Podkletnov's latest experiment [9]. The superconductors are subjected to powerful high-frequency current or current pulses. Since the superconductors are expected to emit Josephson radiation under these conditions, and possibly (according to [9]) also anomalous gravitational-like radiation, we shall call them "emitters".

Our emitters are made of melt-textured YBCO and are reduced in size (diameter 5 cm). They can be modelled as a stack of intrinsic Josephson junctions, in agreement with all modern studies of the conduction of cuprates in the *c* direction [18]. Alternatively, one can describe the material as a homogeneous superconductor with complex conductivity $\sigma=\sigma_1+i\sigma_2$; the conclusions are compatible with the Josephson junctions model, but the information available in the literature about σ_1 and σ_2 in the cuprates in dependence on T and other parameters is scarce (one should extrapolate from Bardeen theory [19]). The intrinsic Josephson junctions model is furthermore better suited to describe the electromagnetic emission of the material, which is important because partly related to the anomalous emission.

The appropriate parameters of the intrinsic Josephson junctions will be discussed in detail below. For now we observe that, as confirmed by numerical simulations, being the emitter inductance and capacitance L_E and C_E much smaller than those of the external circuit, they do not substantially affect the oscillation frequency. The simulations also show, as can be expected since the material "must" anyway conduct with excellent values of σ at MHz frequency, that the normal current in the junctions adjusts itself to a value I_n ($<<I_s$; I_s supercurrent) such that the voltage-per-plane corresponds to an AC Josephson frequency equal to the external frequency. This AC Josephson frequency is in turn the same of the Cooper pairs interplane tunnelling, and so the same of the anomalous emission.

3.1. Emitter Inductance, Capacitance, Plasma Frequency, Dampening Parameter

Let us find the inductance of the emitter as a series of Josephson junctions. There are $\approx 10^7$ junctions, since the emitter thickness is about 1 cm and the inter-plane spacing about 1 nm. The inductance of a single junction is $L \approx \phi_0/I_J$, with $\phi_0=h/2e\approx 2\cdot 10^{-15}$ Wb and I_J critical current (at least 10^4 A in our case). We find $L \approx 10^{-19}$ H. With 10^7 crystal planes, the total inductance is: $L_E = 10^{-12}$ H, to be compared with $L_L\approx 10^{-6}$ H of the external circuit. Each layer is seen simply as a very wide junction in this model; in fact there will be distinct coherence regions in the layer, each in parallel with the others, but the final result is the same.

Next we find the capacitance of the emitter as a series of junctions. For a couple of crystal planes (cross-section $S\approx 20$ cm^2, distance $d\approx 1$ nm, relative dielectric constant ε of the order of 10), we have $C = \varepsilon_0\varepsilon S/d \approx 10^{-4}$ F. Dividing by 10^7, the total capacitance is found to be $C_E\approx 10^{-11}$ F, to be compared with $C_L\approx 10^{-8}$ F of the external circuit.

Therefore the proper frequency of the Josephson junctions, also called plasma frequency f_P, is $f_P = 1/2\pi\sqrt{(L_L C_L)} \approx 10^{11}$ Hz. It is natural to expect that this proper oscillation does not influence the behaviour of the system under the effect of an external forcing frequency typically smaller, of the order of 0.1-10 MHz. Note that f_P is the same for the emitter and for any single junction, because the capacitances and inductances in series scale in the opposite way. The formula for f_P can be easily re-written as follows, with reference to a single junction [18]:

$$f_P = \left(\frac{I_J}{2\pi\phi_0 C}\right)^{1/2} = \left(\frac{j_J d}{2\pi\phi_0 \varepsilon\varepsilon_0}\right)^{1/2} \tag{3.1}$$

The McCumber parameter of a junction β_c is defined by $\sqrt{\beta_c} = 2\pi f_P RC$. This is connected to the hysteresis of the I-V curve of the junction, because $\sqrt{\beta_c} = (4/\pi) I_J / I_r$, where I_r is the so-called return current. For $\beta_c < 1$ we have over-dampened, non-hysteretic junctions; for $\beta_c > 1$ we have under-damped, hysteretic junctions. With the data above, one finds for the single junction $\sqrt{\beta_c} \approx 10^8 R$. Therefore our junctions are strongly over-dampened, because R is less than 10^{-10} Ω for the single junction.

3.2. Normal Resistance of the Intrinsic Josephson Junctions in YBCO and "Resistive Shunts"

The intrinsic Josephson effect has been observed in YBCO as clearly as in BSCCO. Kawae et al. [20] give evidence of I-V curves in YBCO with multiple branches and hysteresis, very similar to those reported by Kleiner and Muller for BSCCO [18]. This can only be seen, however, in very small samples, with area about 0.25 μm^2. In larger samples, grain borders or other defects act as low-resistance shunts. The total resistance is essentially determined by these shunts, and so depends on the micro-structure and not just on the material. The junctions become non-hysteretic, because the McCumber parameter β_c is proportional to R, and a small β_c means no hysteresis. For this reason, the presence of the single junctions can not be seen in the I-V curves of "large" samples. From the practical point of view, all this does not disturb much, except that it is impossible to know in advance the resistance of our material, it depends on the micro-structure. The CRC data [25] is only an indication: $\rho = 5 \cdot 10^{-5}$ Ωm at room temperature, implying $R_E = 10^{-4}$ Ω for our emitter. According to CRC, there is only a small variation in the normal resistance between room temperature and 100 K.

Ref.s [20] and [21] also allow to estimate the normal resistance of the employed samples. For a stack of 80 junctions, ref. [20] gives at 4.2 K a critical current of 0.1 mA (40 kA/cm^2); the return voltage is 0.2 V and the slope of the I-V characteristic in the single normal branch is 800-1000 Ω. The $I_c R$ product (I_c critical current of YBCO material) is therefore 2-3 mV per junction; the material has T_c=43 K, so the BCS prediction is about 10 mV. The resistivity computed from the data above is $3 \cdot 10^{-3}$ Ωm. In ref. [21] the junctions have size 0.65×0.85 mm^2 and R=2 μΩ per crystal plane at the peak resistance value (84K; T_c=93 K). This gives $\rho = 1.1 \cdot 10^{-3}$ Ωm.

The measured resistance of our emitter, including contacts, is $3{\cdot}10^{-4}$ Ω at room temperature and $5{\cdot}10^{-6}$ - $12{\cdot}10^{-6}$ Ω at 77 K. For noble metals the resistivity varies by a factor 5-10 between 300 K and 80 K; for iron more than 10 times. Alloys with noble metals show smaller variations, typically a factor 2-3. At 77 K the DC resistance of YBCO alone vanishes, therefore the residual resistance is that of the metal layer and of the contact. On the contrary, the $3{\cdot}10^{-4}$ Ω at room temperature are essentially due to the YBCO, in agreement with the CRC data above.

The aim of works like [20] is to see the features of the microscopic junctions (their resistance, capacitance, impedance, McCumber parameter), in view of possible applications for fast electronics, microwave collective emission or detection etcetera. In our case we will be satisfied to know that the intrinsic Josephson junctions are active and syncronized. Seeing their individual signature is not essential, actually impossible in a bulk sample. We believe they have to be syncronized, because otherwise they could not sustain the oscillating current; they would instead pass into a purely resistive state and the oscillating current would be normal current. The voltage on the emitter would then be of the order of 10^4 A $\cdot 10^{-4}$ $\Omega \approx 1$ V.

3.3. First Estimate of the Total Anomalous Radiation Energy U_{max} and of the Dissipation in the Emitter

We have found a simple way to make an order-of-magnitude estimate of U_{max}, which agrees with the results of the detailed simulations (see below). Assume that the voltage-per-plane in the emitter is given by the Josephson relation $V=hf/2e=\phi_0 f$. Here f is the frequency of the *external* circuit. This is necessary in order that the external current flows in the emitter as supercurrent (except for the small normal component I_n which gives the finite voltage). The numerical simulations confirm this coincidence of external frequency and Josephson frequency.

Take, for instance, an external circuit with frequency f=10 MHz, current 1 kA and dampening time $\tau=10^{-4}$ s. The voltage over a single junction is $V = 2{\cdot}10^{-15} f = 2{\cdot}10^{-8}$ V. The total number of junctions in 1 cm thickness is $\approx 10^7$ (each is 1.17 nm). Thus the total voltage on the emitter is 0.2 V. The IV product in the emitter, also called DC-power P_{DC} is $P_{DC}=IV$=200 W. In the time τ this makes available in the emitter an energy of the order of 20 mJ.

The electromagnetic emission generated in the AC Josephson effect has an energetic efficiency which is typically of the order of 10% [22]. We suppose that the anomalous emission is associated with the electromagnetic emission, ie a graviton is emitted together with the photon at each Cooper pair tunnelling, at the same frequency and with an emission probability of the same magnitude order. So the total energy U_{max} of the anomalous radiation is ≈ 2 mJ.

Approximately the 80% of the IV product is wasted for the emission and dissipated as heat. This does not cause any serious temperature increase. A reasonable estimate for the thermal capacity of our emitter is 200 mJ/K; the temperature increase of the bulk is therefore negligible. This thermal capacity can be obtained as follows, using, for instance, data from Tilley [23] for the specific heat of YBCO as a function of temperature. The density of YBCO depends on the cell parameters, which are variable. Taking for instance a=0.38 nm, b=0.39,

c=1.17 nm (Waldram [24], $YBCO_{7-x}$, x=0.4), one finds a unit cell volume of $1.7\cdot10^{-28}$ m^3. The unit cell has a total mass of 560 a.m.u. This gives a density of 6300 kg/m^3 (measured value: about 6000 kg/m^3). Therefore the thermal capacity of the emitter, supposed it has volume 20 cm^3, is 200 mJ/K. 10 mJ/Kcm^3 corresponds to 1.5 mJ/Kg.

Note that the IV product on the emitter does not depend on the external load resistance R_L. If we can reduce R_L, we will increase τ and so proportionally increase the target energy. There is a practical problem with a small R_L, however: increased dissipation at the S/N contacts and possible damage to the external capacitors (see below).

The normal current I_n in the emitter can be computed through the relation $V=I_nR_E$. The resistance R_E can roughly be guessed from the data in the literature. For instance, if $R_E\approx10^{-4}$ Ω (CRC) one finds $I_n\approx10^2$ A; if $R_E\approx10^{-3}$ Ω [21], $I_n\approx10$ A. The pure ohmic heating is therefore irrelevant. The figure for I_n gives the *total* normal current. The density of normal current varies locally, as more current flows in the shunts with smaller resistance.

3.4. Synchronization of the Emission. Magnetic Field. Literature Review

Emission of laser-like, coherent radiation from intrinsic Josephson junctions has been observed by Barbara et al. [26] when the junctions are enclosed in a microwave cavity, which serves to impose a definite common oscillation frequency to the junctions. In our case the common frequency is set by the external circuit. The superconductor just follows the external oscillation. The general response of a superconductor to an AC voltage in the KHz-MHz range is to exhibit a small impedance, with small resistive and inductive components (related to the σ_1 and σ_2 mentioned above). For cuprates this is still true, independently from their intrinsic-Josephson structure. We are assuming that in large samples with resistive shunts the intrinsic Josephson structure, un-observable in the I-V curves, is nevertheless active for coherent electromagnetic and anomalous emission.

In the cited works, the array size is comparable or larger than the free-space radiation wavelength λ=2 mm. The emission frequency corresponds to a high-Q resonance in the structure formed by the array and the resonator ground plane. The power coupled to the detector is actually transmitted through a non-linear transmission line, so λ is not exactly that of free space. The detector is itself made of junctions, and is very close.

At MHz frequency, λ is clearly much larger than the system's size. In our case, λ is comparable to the system size, but the anomalous radiation is supposed to be only virtual (compare also Sect. 4).

The transition of the junctions array to a coherent state was predicted by Bonifacio et al. [29] on the basis of the formal analogy between Josephson junctions arrays and free electron lasers. Earlier, such a quantum coupling mechanism was predicted by Tilley [30]. Jain [22], on the contrary, describes classical synchronization of over-dampened Josephson junctions with resistive shunts and low efficiency, about 1%. Note that our junctions are over-dampened because $\omega_cRC<<1$, due to resistive shunts and low resistance. Our junctions are driven from an external AC current, however, so they keep oscillating in spite of being over-dampened.

An extensive literature search and study about the onset of synchronization in arrays of artificial and intrinsic Josephson junctions confirmed that an external load causes synchronization. In many experiments and simulations, the external load is just an RLC circuit as in our case. Our simulations with few junctions clearly exhibit synchronization. Note that in our case, not only has the external circuit a definite proper frequency, but the initial conditions are such that the circuit oscillates from the beginning, while in several other experiments and simulations the junctions are DC biased and coupled to a resonant circuit which is initially passive.

In general, a magnetic field in the *ab* direction should increase the inter-plane coupling (Kleiner et al., [18]). It does so, however, at the price of decreasing the critical current I_J. Other coupling mechanisms should be more effective in our case, in particular the external driving frequency and the normal current ("quasi-particles" current). The numerical simulations predict that, paradoxically, a lower I_J (provided still larger than the external current I_0) increases the emitter voltage and thus the DC Josephson power. Therefore it can be helpful to apply an uniform magnetic field to the emitter. The dependence of I_J on the field is not strong and V_E is inversely proportional to I_J, so the power gain is not expected to be dramatic. Also, it is difficult to define the degree of uniformity needed.

In the computations by Rogovin and Scully [27] the magnetic field appears explicitly. Sometimes it couples e.m. normal modes with different polarizations (see also the work by Almaas and Stroud below). In principle, certain modes of the Josephson junctions would be decoupled from the radiation in the absence of a static magnetic field.

Acebron et al. [31] and K. Wiesenfeld et al. [32] consider several synchronization problems, among them the one we are concerned with. They find that Josephson-junction arrays connected in series through a load exhibit "all-to-all" (that is, global) coupling. A schematic circuit is given, with ideal junctions in series coupled through a resistance-inductance-capacitance load; in parallel to both is a bias current generator.

A model for a large number of Josephson junctions coupled to a cavity and an attempt at an explanation of the experiment by Barbara et al. for 2D arrays was given in [33]. The synchronization behavior was reproduced. Junctions are under-dampened, with non-zero C. A bias current is taken such that each junction is in the hysteretic regime. Depending on the intial conditions, the junctions may work in each of two possible states, with zero or non-zero voltage. In the latter case, the phases vary with time and the junctions are called "active".

Filatrella and Pedersen [34] find that the transition from a state where the junctions are essentially oscillating at the unperturbed frequencies to one where they oscillate at the same frequency occurs above a threshold number of active junctions, in agreement with the experimental results by Barbara et al. A subsequent work [35] studied conditions when there is no threshold.

In general, the model employed in the papers above includes a global, "classical" coupling (external oscillating circuit), while Barbara et al. in their PRL article stressed the fact that the coupling is local and typically quantum-mechanical, with stimulated emission. For this reason, probably, is the threshold behaviour not properly reproduced.

In our simulations we feed an oscillating current into the junctions, instead of a DC bias.The authors of [35] suppose that all junctions have the same R (for us, not necessarily, because variations are compensated by variations in the normal current I_n), and let instead the critical current I_J vary (our simulation also supports this; the voltage on the junctions depends on I_c; in any case we have $I<I_J$).

Almaas and Stroud [37] give a theory of 2D Josephson arrays in a resonant cavity. They consider the dynamics of a 2D array of under-dampened junctions placed in a single-mode resonant cavity, in the limit of many photons. The numerical results show many features similar to the experiment by Barbara et al., namely: (1) self-induced resonant steps; (2) a threshold number of active rows; (3) a time-averaged cavity energy which is quadratic in the number of the active junctions. They predict a strong polarization effect: if the cavity mode is polarised perpendicular to the direction of current injection in a square array, then it does not couple to the array and no power is radiated into the cavity. In the presence of an applied magnetic field, however, a mode with this polarisation would couple to an applied current.

3.5. Contact Resistance and Heating

In experiments on intrinsic Josephson junctions there is usually a transport current along the c axis, fed in from a generator through special contacts on the top and bottom of the samples. At the contacts, most of the external current is converted into super-current. The same should happen in our case, but our current is large and there is the problem of contacts over-heating. (The current is well below I_c, but this is so because melt-textured materials have especially large I_c.) If the material is driven normal near the contacts, all the mechanism of superconducting conduction and Josephson tunnelling is lost. The material can be driven normal also because contact is not uniform and local current density exceeds J_c.

We have seen that dissipation and heating in the bulk of the emitter can be disregarded. We must then check heating at the contacts. In work by Takeya et al. [18] the heat diffusion in BSCCO is taken into account. A heat diffusion length l can be defined, both in the ab and c directions. Heat delivered at one point spreads over a volume of approximate size $l^2_{ab} \cdot l_c$. Knowing the specific heat of the material, one can compute the temperature increase of that volume. We are interested only in l_c, since heat is generated at planar contacts.

Next we need a guess for the surface resistance of the contacts R_c. Take for instance $R_c = 10^{-5}$ Ω (IoP Handbook [19], Sect. B.5), ie $\rho_c = 2 \cdot 10^{-4}$ Ωcm^2; then for larger or smaller values all scales in proportion. Heat generated at the contacts is of the order of the total energy (10^2 J) multiplied by the ratio R_c/R_{load}. For instance, with $R_{load} = 0.1$ Ω we find a dissipation of 10^{-2} J.

The heat diffusion length is given by the formula $l = 2\sqrt{(Kt/c)}$, where K is the heat conductivity, c the specific heat, t the duration of the pulse. For BSCCO, Takeya et al. give $K = 0.25$ W/mK, $c = 2$ kJ/Km^3. For YBCO, $K = 15$ W/mK along ab ([24], p. 254), $c = 10$ kJ/Km^3 (see above). For instance, with a pulse duration t=0.5 μs, we find $l = 1.7$ mm and the interested volume is $3 \cdot 10^{-6}$ m^3; its thermal capacity is 30 mJ/K. The thermal capacity of a copper feeding electrode, however, is bigger, so it takes much of the heat. (One should also take into account a possible indium layer between YBCO and copper.) The temperature increase would then be negligible.

So, supposed a surface resistivity of the order of 10^{-4} Ωcm^2 can be obtained, there would be room for a reduction of the external resistance R_{load}, admitted this is possible in practice. If R_{load} is smaller, then the dampening time τ is larger and the total energy U_{max} of the anomalous radiation increases in proportion. Dissipation in the bulk of the emitter and in the contacts also increases in proportion to τ.

3.6. Simulation of a Josephson Junction Inserted in a RLC Circuit, in the RSJ Model

According to the RSJ model (resistively-shunted junction), a Josephson junction can be represented as a non-linear circuit element obeying the Josephson effect equations below, plus an ohmic resistance R in parallel. In the purely Josephson element flows only supercurrent while in the resistance flows normal current. We have seen that when the Josephson junction is placed in an external oscillating circuit with large C and L, it should not influence the external current. This is confirmed by the simulation below and is true also for many junctions in series. Therefore we first simulate one single junction and then we shall consider the synchronization of several junctions.

The two fundamental equations of the Josephson effect are

$$I_s = I_J \sin\phi , \tag{3.2}$$

where I_s is the supercurrent in the junction, I_J is the critical current and ϕ the phase difference over the link, and

$$\phi' = \frac{2e}{\hbar} V , \tag{3.3}$$

where the prime denotes time derivative and V is the voltage applied to the junction. According to the RSJ model, $V=RI_n$, where I_n is the normal current flowing in the normal resistance R of the junction, parallel to I_s.

Only I_n generates a voltage in the emitter, but both I_n and I_s flow in the external resistance and inductance (I_s after conversion to normal) and discharge the capacitor.

Denote $a = \frac{2e}{\hbar} R$ and rewrite (3.3) and the second derivative of (3.2) as follows

$$\begin{cases} \phi' = aI_n \\ I_s'' = aI_J (I_n' \cos\phi - aI_n^2 \sin\phi) \end{cases} . \tag{3.4}$$

These are the first two equations of a system, whose unknowns are the functions of time $\phi(t)$, $I_s(t)$, $I_n(t)$.

Write the derivative of the Kirchoff equation over the loop including the external load (L_L, C_L, R_L) and the junction

$$\frac{1}{C_L}(I_s + I_n) + L_L (I_s'' + I_n'') + R_L (I_s' + I_n') + RI_n' = 0 . \tag{3.5}$$

This is going to be the third equation of the system. Divide by L_L and note that the proper frequency of the external circuit is $\omega = 1/\sqrt{L_L C_L}$. Disregard the last term because R is about 10^{10} times smaller than R_L. Replace I_s" with the second equation in (3.4), where I_J is denoted g. Finally define $b=R_L/L_L$. We find

$$\omega^2(I_s + I_n) + ag(I_n'\cos\phi - aI_n^2 \sin\phi) + I_n'' + b(I_s' + I_n') = 0. \tag{3.6}$$

Isolating I_n", we obtain the final complete non-linear system, where the currents are denoted simply by s and n:

$$\begin{cases} \phi' = an \\ s'' = ag(n'\cos\phi - an^2 \sin\phi) \\ n'' = -\omega^2(s+n) - ag(n'\cos\phi - an^2 \sin\phi) - b(s'+n') \end{cases} \tag{3.7}$$

Summarizing, the symbols and typical magnitude orders of the parameters are, in SI units,

$$\begin{aligned} n &= I_n(t) \\ s &= I_s(t) \\ a &= \frac{2e}{\hbar} R = 3 \cdot 10^4 \\ b &= \frac{R_L}{L_L} = 1.5 \cdot 10^4 \end{aligned} \tag{3.8a}$$

$$\begin{aligned} R &= R_E / N = 10^{-11} \\ \omega &= 3.3 \cdot 10^6 \\ g &= I_J = 5 \cdot 10^4 \end{aligned} \tag{3.8b}$$

The initial conditions at time t=0 (when the external circuit is closed) are the following:

$$\begin{aligned} \phi(0) &= 0 \\ I_n(0) &= 0 \\ I_s(0) &= 0 \\ I_s'(0) &= 0 \\ I_n'(0) &= I_0\omega \end{aligned} \tag{3.9}$$

At the time t=0 the external circuit begins to oscillate, starting from a state in which the capacitor is fully loaded. The initial value for $I'_n(0)$ is standard for an RLC circuit. I_0 is the maximum external current, which depends on V, C_L and L_L as

$$I_0 \approx V\sqrt{\frac{C_L}{L_L}} \tag{3.10}$$

Note that I_s' is initially zero due to eq. (3.2) and (3.3), since V is initially zero. It is interesting to note that in spite of this, I_s rapidly grows and becomes almost equal to I_0 in the emitter, where I_n stays small (see below).

With these initial conditions the equation system (3.7) can be solved numerically through the Runge-Kutta method. The result is clear: for $I_0<I_J$ (which is usually the case) all functions oscillate with the external frequency. With the parameters above, ϕ oscillates between 0.1 and -0.1, while $I_n \approx 10$ A. For $I_0=I_J$, the phase makes a complete oscillation in the period of the circuit oscillation.

A refined version of the RSJ model includes a junction capacitance C_J in parallel to the resistance [28]. For high frequency, the capacitive channel can become important. We have seen that $C_J \approx 10^{-4}$ F, so the impedance of the C-channel at $\omega \approx 1$ MHz is of the order of 10^{-2} Ω, much larger than $R_E \approx 10^{-11}$ Ω. It should therefore be legitimate to disregard C_J. For a check, we included a capacitive channel in the numerical simulation. The results are at first sight puzzling, because in this case the capacitance of the Josephson junction affects the circuit behaviour much more than the (smaller) external capacitance C_L; but this is an artefact, because eventually we want to simulate a large number of junctions in series, and in that case their total capacitance will be small, so C_L will actually dominate and the C-channels of the junction carry very little current.

Let us write the equation for 2 junctions in series:

$$\begin{aligned}
&s_1 = g\sin\phi_1 \\
&s_2 = g\sin\phi_2 \\
&\phi_1' = an_1 \\
&\phi_2' = an_2 \\
&s_1'' = ac(n_1'\cos\phi_1 - an_1^2\sin\phi_1) \\
&s_2'' = ac(n_2'\cos\phi_2 - an_2^2\sin\phi_2) \\
&\frac{1}{C_L}(n_1+s_1) + L_L(n_1''+s_1'') + R_L(n_1'+s_1') = 0
\end{aligned} \tag{3.11}$$

In the last equation one isolates n_1'' and replaces s_1''. In order to find n_2", we note that $n_1+s_1=n_2+s_2 \rightarrow n_1''+s_1''=n_2''+s_2'' \rightarrow n_2''=\ldots$ We so have 6 equations with unknown ϕ_1, s_1, n_1, f_2, s_2, n_2. The initial condition is the same, as is easily obtained differentiating the equation $n_1+s_1=n_2+s_2$.

For 3 junctions: the equations for ϕ_3' and s_3" are simple. Then current conservation gives $n_3''=n_1''+s_1''-s_3''$. This also holds for the first derivatives, and for the initial condition, which is just the same. And so on.

In this way we check directly the synchronization, at least for few junctions, and regarding a whole crystal layer (with surface of the order of square centimetres!) as a single junction. Phase, voltage and normal current are synchronized. The synchronization also occurs for higher frequency (larger than 10 MHz).

The simulations allow to compute the emitter voltage V_E by multiplying I_n and R_E. This voltage turns out to be smaller (typically 10 times smaller, with the parameters above) than the simple estimate based on the relation $V=(h/2e)f$. This relation holds rigorously for a constant voltage, while in our case we have $V=RI_n$, and I_n oscillates.

In addition, the emitter voltage depends on the critical Josephson current I_J, and is larger when I_J is smaller (inversely proportional, see below). This could not have been predicted without the simulations, but the qualitative reason is clear. The supercurrent is fixed (approximately equal to the total current) and $I_s=I_J\sin\phi$. If I_J is larger, then the oscillations of ϕ are smaller, and so ϕ' is smaller too; and V is proportional to ϕ'. A possible way to depress I_J is to apply a magnetic field. So the magnetic field may be not needed for synchronization, but improves the IV power and U_{max}.

It is not easy to understand intuitively how an oscillating I_s is obtained when the voltage itself oscillates. Mathematically, the point is that ϕ does not evolve linearly in time, but oscillates in turn, therefore I_s is not perfectly harmonic while I_0 is harmonic, and the difference $I_n=I_0-I_s$ oscillates.

On a short time scale, the simulations show that after $t=0$ the normal current, starting from zero, rises quickly and then begins to oscillate from its maximum. The Josephson junctions are very quick (10^{-11}-10^{-12} s) to adapt to the least energy configuration, in which most external current is converted into super-current.

Some simulations were run to look for the dependence of the normal current upon I_J. It turns out that there is an inverse proportionality. For instance, with $a_1=a_2=3\cdot10^4$ one finds the following values:

Critical current (kA)	**Normal current (A)**
20	25
40	12.5
80	6.1
160	3.1

4. Stimulated Emission

We have analysed the behaviour of a superconducting emitter, modelled as a series of Josephson junctions, when it is inserted into an oscillating circuit with proper frequency much smaller than the Josephson plasma frequency. We concluded that the junctions are synchronized with the external circuit and we evaluated the normal- and super-current components I_n and I_s. It is interesting to compare the situation with the experiments of Ref.s [26,34]. In that case, the junctions are individually biased with a DC, and are synchronized by a passive external cavity [26] or by a passive external circuit [34,35]. In our case the external current, in the MHz frequency range, serves at the same time as bias and coupling device.

The main question now is: does stimulated emission occur, like in [26]? In the absence of a resonant cavity, this can only occur in a single-pass mode, like in optical or maser amplifiers. Each junction is "pumped" and first emits spontaneously. The emitted photons/gravitons (the model applies to both) propagate and stimulate further emission. A representation of a Josephson current as an ensemble of Cooper pairs tunnellings or as an oscillating macroscopic quantum dipole as been given earlier, as mentioned, by Rogovin and Scully. A rate equation appropriate for a single-pass linear photon amplifier has been given by Frantz and Nodvik [36], and subsequently applied to several cases. It has the form

$$\frac{dn}{dt}+c\frac{dn}{dx}=\sigma cn(N_2-N_1) \tag{4.1}$$

The Frantz-Nodvik equation has the typical structure of a conservation equation describing the longitudinal propagation of a particles beam with light velocity. The equation takes into account the possibility that particles are absorbed or generated at any point. The l.h.s. represents the net variation of the particles density $n(x,t)$ along the beam.

The equation was originally written for photons, σ being the "resonant photon absorption cross section". In other words, the product σc gives the transition probability per unit time. $N_1(x,t)$ and $N_2(x,t)$ give the density of atoms in the levels 1 and 2. The incoming beam is supposed to be monochromatic.

We would like to consider an incident beam with a frequency spread and to make a connection with the basic equation describing spontaneous and stimulated emission, namely

$$\frac{dN_2}{dt}=-B\rho N_2-AN_2\ , \tag{4.2}$$

where ρ is the energy density per volume and frequency:

$$\rho=\frac{dE}{Vdf}=\frac{nhf}{df}. \tag{4.3}$$

In our case the bandwith ratio f/df is fixed by the merit factor Q of the external oscillating circuit. Therefore

$$\rho=hQn=const\cdot n. \tag{4.4}$$

For our purposes, the FN equation needs to be modified and adapted as follows. (a) A spontaneous emission term is added. This is important at early times, because in our case there is no in-going beam in the initial conditions, but the initial photon density $n(x,0)$ is zero everythere. (b) In the FN equation, the population difference (N_2-N_1) is a function of x and t, albeit one which is eliminated in the final solution. We replace that difference with a constant N giving the number of tunnelling processes ("transitions") of Cooper pairs in the intrinsic Josephson junctions per unit time and volume.

The constant N depends on the super-current, which together with the emitter voltage V_E defines the maximum pumping power $P\approx V_E I_S$. At each transition, there is a certain probability of spontaneous or stimulated emission of a graviton by each Cooper pair. Competing electromagnetic emission can also occur, but being on-shell it does not have the right wavelength for amplification (see below; in [26] the size of the junctions is much larger, and the emission frequency too, f=150 GHz).

In the end, we shall mainly be interested into the saturation condition, when stimulated emission dominates and n grows rapidly to the maximum value $n \approx N$ allowed by the pumping. In these conditions, the dominant emission is necessarily longitudinal, because only the longitudinal mode is amplified, while spontaneous emission might be preferentially transverse (like for electromagnetic Josephson emission), also possibly depending on the applied magnetic field. "Longitudinal" means here along the c crystal axis, orthogonal to the junctions and parallel to the super-current.

N is the number of Josephson transitions per unit time and volume, ie the number of layers per volume in the emitter multiplied by $I/2e$. Each layer is seen simply as a "giant" junction in the model, though more realistically there will be in the layer distinct coherence regions, each in parallel with the others. If S is the superconductor cross section, δ its thickness and d the thickness of a single layer, we have

$$N = \frac{j}{2ed} \tag{4.5}$$

Eq. (4.1) becomes

$$\frac{dn}{dt} + c\frac{dn}{dx} = \gamma nN + AN, \tag{4.6}$$

where γ is a pure number, which we can express in terms of the other parameters by comparison with (4.2), (4.3) and (4.5). We find the relation

$$\gamma nN = \frac{Bh}{2ed} nQj. \tag{4.7}$$

The factor $(Bh/2ed)$ contains only fundamental constants or fixed experimental parameters; the factor Qj can be tuned in a certain range. Defining the constants

$$\begin{aligned} \alpha &= \frac{Aj}{2ed} \\ \beta &= \frac{Bh}{2ed} Qj \end{aligned}, \tag{4.8}$$

with ratio $\alpha/\beta = A/(BhQ)$, we obtain the rate equation in final form:

$$\frac{dn}{dt} + c\frac{dn}{dx} = \beta n + \alpha. \tag{4.9}$$

In order to solve it, we define as usual the auxiliary variables

$$\xi = \frac{x}{c} \qquad \rho = t - \frac{x}{c} . \tag{4.10}$$

The equation then becomes

$$\frac{dn}{d\xi} = \beta n + \alpha , \tag{4.11}$$

with solution

$$n = k(\rho)e^{\beta\xi} - \frac{\alpha}{\beta} . \tag{4.12}$$

Here $k(\rho)$ is an arbitrary function which we determine returning to the original variables and imposing the initial condition $n(x,0)=0$ for any x. The final solution is remarkably independent from x:

$$n(x,t) = \frac{\alpha}{\beta}\left(e^{\beta t} - 1\right). \tag{4.13}$$

Note that the emission is supposed to be in the positive x direction (this is implicit in the definition of the variable ρ).

We can see here a necessary condition for saturation. When the current oscillates, saturation can only occur if the characteristic time $t_c=1/\beta$ of the exponential growth is smaller than the oscillation time. For a magnitude order estimate, consider for instance Q=10-100, $j=5\cdot 10^6$ A/m^2, d=1.17 nm; we find that $t_c<10^{-7}$ if $B>10^4$ m^3/Js^2. This is much smaller than the B-coefficient of atomic optical transitions. (In principle, we could estimate B from Podkletnov's data, but the uncertainties on his parameters j and Q are too large.)

The overall amplitude α/β depends on the known ratio between the Einstein coefficients $A/B=8\pi f^3 h/c^3$. We find $\alpha/\beta=(\omega/c)^3/Q$. It is known that ω enters the A/B ratio because of phase-space considerations based on the formula $p=h/\lambda$ for the photon momentum. Therefore re-inserting λ we have $\alpha/\beta=1/(\lambda^3 Q)$. It follows that the ratio α/β, and thus $n(x,t)$, is large when λ is small, as happens for the virtual anomalous radiation: we know from the experiment [9] that $\lambda f\approx 1$ m/s; then for instance with $f=10^7$ Hz we have $\lambda\approx 10^{-7}$ m. At the same time, this shows that the corresponding real electromagnetic radiation with $f=10^7$ Hz and λ=30 m is strongly suppressed.

Now taking Q=10 we find $\alpha/\beta=10^{20}$. The Cooper pairs density in YBCO is at least 10^{26} m^{-3}, therefore for saturation the exponential factor must be 10^6, or the exponent $t/t_c\approx 10$.

5. Conclusion

The first necessary condition for a sizeable gravitational emission from superconductors subjected to high-frequency currents is that the standard matter/field coupling must be amplified by some microscopic quantum mechanism. We have pointed out the possible existence and nature of such a dynamical mechanism, but a rigorous proof is not yet available. We call "anomalous" any kind of gravitational emission due to this fundamental anomalous coupling, which can only occur when matter is in a coherent state. According to our model, the coupling is mediated by the vacuum energy density Λ, and amplified by gravitational vacuum fluctuations. The emitted gravitons can only be off-shell, with $\lambda f \ll c$ and have finite propagation range (but can have spin 0 and 1 components). They are virtual particles, ie they exist only as intermediate states of quantum processes.

The second condition is, that the overall energetic balance must be respected. The emitting transitions are Josephson tunnellings of Cooper pairs in the intrinsic junctions of oriented (melt-textured) anisotropic ceramic superconductors. The pumping occurs via a high-frequency current in the *c* crystal direction. The current has super- and normal-components, I_n and I_s. I_n is much smaller than I_s, typically 10^3 times smaller, and is necessary in order to establish the voltage $V_E=R_nI_n$ which supports the oscillations of I_s. The maximum available power is $P=IV_E\approx I_sV_E=I_sI_nR_n$. The effective normal resistance R_n of the material depends on its micro-structure (resistive micro-shunts), but numerical simulations show that I_n adapts to R_n, keeping V_E constant; therefore the shunts micro-structure is not critical. The emitter voltage depends on the external frequency and on the critical Josephson current I_J.

In our laboratory trials, the estimated maximum pumping power is of the order of 10^2 W, but the emitter voltage V_E could not yet be measured, due to powerful disturbances generated by the external circuit. An increase in the maximum available power is technically difficult, and in any case there are stringent theoretical limits on the maximum voltage present in the superconducting emitter. Further technical problems which emerged from our preliminary trials and planning concern thermal dissipation in the bulk and at normal-superconducting contacts, and the application of a proper magnetic field.

The third necessary condition is, that the pumping power must be exploited as much as possible, and for this it is crucial that a cascade process of stimulated emission is activated. The cascade can occur in a single passage, because the emitting layers are very numerous ($\approx 10^7$), and is governed by the rate equation (4.9) which takes into account the Einstein coefficients A and B, the current density j in the emitter and the merit factor Q of the external circuit. A compromise between these parameters should be found, such that the stimulated emission cascade can fully develop in a high-frequency cycle or in short pulses.

Acknowledgment

This work was supported by the Göde Wissenschaftsstiftung (Goede Science Foundation).

References

[1] C. Kiefer, C. Weber, *Annalen Phys.* 14 (2005) 253-278.

[2] C. De Matos and M. Tajmar, *Physica C* 385, 551 (2003). M. Tajmar, F. Plesescu, B. Seifert, R. Schnitzer, I. Vasiljevich, *Search for Frame-Dragging in the Vicinity of Spinning Superconductors,* gr-qc/0707.3806.

[3] R.Y. Chiao, Conceptual tensions between quantum mechanics and general relativity: Are there experimental consequences? In: *Science and ultimate reality: Quantum theory, cosmology and complexity*, ed. J.D. Barrow et al., Cambridge Univ. Press, Cambridge, 2003. R.Y. Chiao, W.J. Fitelson and A.D. Speliotopoulos, Search for quantum transducers between electromagnetic and gravitational radiation: A measurement of an upper limit on the transducer conversion efficiency of yttrium barium copper oxide, gr-qc/0304026.

[4] Robert M. L. Baker, Jr., Precursor Proof-of-Concept Experiments for Various Categories of High-Frequency Gravitational Wave (HFGW) Generators, STAIF 2004, AIP Conference Proceedings 699, 1093-1097. G. Fontana, *Design of a Quantum Source of High-Frequency Gravitational Waves (HFGW) and Test Methodology,* physics/0410022.

[5] R.C. Woods, *Physica C* 433 (2005) 101-107; *Physica C* 442 (2006) 85-90; *Physica C* 453 (2007) 31-36.

[6] M. Agop et al., *Physica C* 339 (2000) 120-128.

[7] G.A. Ummarino, *Possible alterations of the gravitational field in a superconductor*, cond-mat/0010399.

[8] Podkletnov E. and Nieminen R., *Physica C* 203 (1992) 441. E. Podkletnov, Weak gravitation shielding properties of composite bulk $YBa_2Cu_3O_{\{7-x\}}$ superconductor below 70 K under e.m. field, cond-mat/9701074.

[9] E. Podkletnov and G. Modanese, *J. Low Temp. Phys.* 132 (2003) 239-259.

[10] Ning Li, David Noever, Tony Robertson, Ron Koczor and Whitt Brantley, Physica C 281 (1997) 260-267. G. Hathaway, B. Cleveland and Y. Bao, *Physica C.* 385 (2003) 488-500.

[11] D.N. Spergel et al., Astrophys. *J. Supp. Series* 148 (2003) 148. C. Beck and M.C. Mackey, *Physica A* 379, 101 (2007).

[12] G. Modanese, *Mod. Phys. Lett.* A18 (2003) 683-690.

[13] G. Modanese, *Europhys. Lett.* 35 (1996) 413-418; *Nucl. Phys. B* 588 (2000) 419-435. *Phys. Rev.* D62 (2000) 087502.

[14] L. Halpern and B. Laurent, *Nuovo Cim.* 33, 728-751 (1964). E.G. Harris, On the possibility of gravitational wave amplification by stimulated emission of radiation, unpublished, 2000.

[15] G. Modanese, *Class. Quantum Grav.* 24 (2007) 1899-1909.

[16] M. Zalalutdinov et al., *Appl. Phys. Lett.* 83 (2003) 3281-3283 and refs. P.S. Landa, *Nonlinear oscillations and waves in dynamical systems*, Kluwer, Dordrecht, 1996.

[17] P.M. Varangis et al., *Phys. Rev. Lett.* 78 (1997) 2353-2356.

[18] R. Kleiner and P. *Muller, Phys. Rev.* B 49 (1994) 1327; *Physica C* 293 (1997) 156. R. Kleiner et al., *Phys. Rev. B* 50 (1994) 3942. J. Takeya et al., *Physica C* 293 (1997) 220.

Y. Mizugaki et al., *J. Appl. Phys.* 94 (2003) 2534. M. Sakai et al., *Physica C* 299 (1998) 31.
[19] Porch, High frequency electromagnetic properties, in *Handbook of SC materials*, ed.s D.A. Cardwell, D.S. Ginley, *IoP Bristol.* 2003, Sect. A2.5, p. 99.
[20] T. Kawae et al., *Physica C* 426-431 (2005) 1479.
[21] D.C. Ling et al., *Phys. Rev. Lett.* 75 (1995) 2011.
[22] A.K. Jain et al., *Phys. Rep.* 109 (1984) 309.
[23] D. Tilley and J. Tilley, *Superfluidity and Superconductivity*, 3rd edition, IOP, London, 1990.
[24] J. Waldram, *Superconductivity of metals and cuprates*, IoP, London, 1996.
[25] *Handbook of Chemistry and Physics*, 85th edition, D.R. Lide Editor-in-Chief, CRC Press, Boca Raton, 2004.
[26] P. Barbara, A.B. Cawthorne, S.V. Shitov and C.J. Lobb, *Phys. Rev. Lett.* 82 (1999) 1963. Vasilic et al., *Appl. Phys. Lett.* 78 (2001) 1137. Vasilic, B., Barbara, P., Shitov, S.V., Lobb, C.J., *IEEE Trans. on Appl. Superconductivity*, 11 (2001) 1188-1190. B. Vasilic, P. Barbara, S.V. Shitov, and C.J. Lobb, *Phys. Rev. B* 65, 180503 (2002).
[27] D. Rogovin and M. Scully, *Phys. Rep.* 25 C (1976) 175.
[28] M. Tinkham, *Introduction to superconductivity*, McGraw-Hill, New York, 1996.
[29] Bonifacio et al., *Lett. Nuovo Cim.* 34 (1982) 520.
[30] D.R. Tilley, *Phys. Lett.* 33A (1970) 205-206.
[31] Acebron et al., *Rev. Mod. Phys.* 77 (2005).
[32] K. Wiesenfeld et al., *Phys. Rev. Lett.* 76 (1996) 404.
[33] Daniels et al., *Phys. Rev*. E 67 (2003) 026216.
[34] G. Filatrella et al., *Phys. Rev.* E 61 (2000) 2513.
[35] Filatrella, G., Pedersen, N.F., *Physica C, Vol.* 372-376, Suppl. Part 1 (2002) 11-13.
[36] L.M. Frantz, J.S. Nodvik, J. *Appl. Phys.* 34 (1963) 2346-2349.
[37] E. Almaas and D. Stroud, *Phys. Rev. B* 67 (2003) 064511.

In: Classical and Quantum Gravity Research
Editors: M.N. Christiansen et al, pp. 271-307
ISBN 978-1-60456-366-5

Chapter 6

TOPOLOGICAL ORIGIN OF THE COUPLING CONSTANTS HIERARCHY

***Vladimir N. Efremov*[1]*, *Nikolai V. Mitskievich*[2],**
***Alfonso M. Hernandez Magdaleno*[1] and *Claudia Moreno Gonzalez*[1]**
[1]Mathematics Department, University of Guadalajara, México
[2]Physics Department, University of Guadalajara, México

Abstract

We construct a class of cosmological models with the adequately defined topological field configurations (Blau-Thompson-Horowitz-Baez BF-systems) on a specific collection of plumbed V-cobordisms (pV-cobordisms) and corresponding plumbed V-manifolds (pV-manifolds), modeling a four-dimensional space-time in the Euclidean regime. The explicit expressions for transition amplitudes (partition functions) are written in these BF-models and it is shown that the basic topological invariants of the pV-cobordisms and pV-manifolds (intersection matrices) play the role of coupling constants between the formal analogues of electric and magnetic fluxes quantized à la Dirac, using Poicaré–Lefschetz duality. The Hartle-Hawking quantum amplitudes (in semi-classical approximation) for the BF-systems on pV-manifolds are calculated, they turn out to be the space-time topological invariants and correspond to the top-down approach to cosmology. The diagonal elements and eigenvalues of the intersection matrix for a definite pV-cobordism and corresponding pV-manifold reproduce the hierarchy of dimensionless low-energy coupling constants of the fundamental interactions acting in the real universe at the present time.

PACS 0420G, 0240, 0460.

1. Introduction

The main hypothesis we advance in this paper is that the spacetime topology determines (*via* an Abelian topological BF-type field theory) the number and hierarchy of coupling constants of the fundamental (pre-)interactions which are adequate to the topological structure of the real universe. Thus we continue to develop the principal idea of [1] that the

*E-mail address: efremov@cencar.udg.mx

primary values of coupling constants (which correspond to a vacuum without any local excitations) are topological invariants of four-dimensional spacetime manifold. The topological field theories [2, 3, 4, 5] are in fact exercises in calculation of topological invariants (the "theory of nothing") [6] accompanied by certain physical interpretations. In particular, in our $U(1)$ gauge BF-model (in the vein of works on global aspects of electric-magnetic duality [7, 8, 9, 10, 11, 12, 13]) a determination of the complete system of the gauge classes of BF system solutions (phase space) is equivalent to specification of topology of spacetime M because of the existence of isomorphisms of type

$$\text{Princ}\,(M) \overset{c}{\cong} H^2(M, \mathbb{Z})$$

between the group of principal $U(1)$-bundles over M and the group of cohomology classes of 2-cycles of M (see the formulae (2.10) – (2.13)) and since the groups of cohomological classes of cocycles with dimensions 1 and 3 are trivial for the set of four-dimensional manifolds under consideration, namely pV-cobordisms and pV-manifolds, which are intensively studied by mathematicians in recent years [14, 15, 16, 17].

The problem is set as in self-consistent cosmological model building: to construct a spacetime manifold which admits fields configurations with non-zero fluxes through the collection of homologically non-trivial 2-cycles [7], *i.e.* closed surfaces that do not correspond to the boundary of a three-dimensional submanifold in M. A generalization of Dirac's quantization conditions [7, 8, 10, 11, 13] (formulated on the basis of the Poincaré–Lefschetz duality) implies that these fluxes must be quantized. We show that the coupling constant matrix describing interactions between the quantized fluxes, coincides up to a scale factor (which also is found to be a topological invariant) with the basic topological invariant of spacetime M, *i.e.* with its intersection matrix Q. For the specially constructed pV-cobordism $M_D^+(0)$ and pV-manifold $M_V^+(0)$ (see section 4), diagonal elements of the intersection matrix $Q^+(0)$ reproduce rather exactly the hierarchy of dimensionless low-energy coupling (DLEC) constants of the universal physical interactions acting in our universe. The other pV-cobordisms $M_D^+(t)$ and pV-manifolds $M_V^+(t)$, $t = -1, -2, -3, -4$, constructed in the same section, probably describe earlier stages of cosmological evolution, which are distinguished from each other rather by the spacetime topology than by gauge groups of physical fields (the gauge group we use is $U(1)$). Thus our model demonstrates that topological invariants may codify information about the physical interactions, which can be naturally introduced on a non-trivial topological space forming the spacetime background of the self-consistent cosmological model which involves qualitatively different evolutionary phases connected one to another by topology changes.

The relations between the topology fluctuations and the problem of fixing fundamental coupling constants are discussed widely, see for example [18, 19]. In particular, the solution of the cosmological-constant problem in terms of spontaneous topology changes is treated in [20, 21, 22, 23]. We propose a fairly different approach to these old problems by means of using cobordisms of sufficiently specific type to model the spacetime manifolds.

We remind that if the boundary of a compact four-dimensional manifold M, ∂M, is a disjoint sum of two three-dimensional manifolds Σ_{in} and Σ_{out}, the 3-tuple $(M, \Sigma_{\text{in}}, \Sigma_{\text{out}})$ is called a cobordism [24]. Both Σ_{in} and Σ_{out} are often merely implied, thus the very manifold M is called a cobordism. A quantum amplitude for topology change between the boundary components Σ_{in} and Σ_{out} is constructed by integrating and/or summing over all physically distinct histories that satisfy the appropriate boundary conditions weighted by the Euclidean action. In quantum gravity such history consists of a four-dimensional manifold (or its generalization [25]), Euclidean metric g and a matter field configuration ϕ on the manifold. In terms of such histories, an amplitude (partition sum) is heuristically

$$Z(h,\varphi,\partial M)=\sum_{\{M\}}w(M)\int[dg][d\phi]\exp[-S_E(g,M)], \tag{1.1}$$

where h and φ are the metric and the matter field configuration induced on the boundary ∂M by g and ϕ respectively. The sum is over all the 4-manifolds such that $\partial M=(-\Sigma_{\text{in}})\sqcup\Sigma_{\text{out}}$ with a measure (probability) $w(M)$; the integral is the quantum gravitational path integral (see detailed discussion, for example, in [26]). In particular, the Hartle-Hawking "no boundary" (one-boundary) proposal [27] is that there should be no initial component of the boundary (*i.e* $\Sigma_{\text{in}}=\varnothing$) and the path integral (which defines the Hartle-Hawking wave function) should be evaluated for a compact manifold M with only a single (final) boundary component Σ_{out}. The quantum state of the universe is determined by a certain specification of the spacetime class $\{M\}$ which is involved in the summing.

In our purely topological model, a quantum amplitude is constructed by summing over all physically distinct histories that satisfy a certain boundary condition (formulated on the level second cohomology groups) weighted by the BF-theory Euclidean action. Such a history consists of both a pV-cobordism (or pV-manifold) and a principal $U(1)$-bundle P over it, with the specified (induced) principal $U(1)$-bundle P_∂ over the boundary ∂M. This corresponds to generalizations of path integral which include conifolds [28, 25] and orbifolds [29, 30] for definition of path histories.

The paper is organized as follows. The section 2. contains the basic mathematical concepts which are used to construct a topological field theory on pV-cobordisms and pV-manifolds. In the subsection 2.1. we reproduce the definitions of spicing and plumbing operations which are necessary to glue pV-cobordisms and pV-manifolds according to algorithms codified by means of splice diagrams. These concepts are unusual for physicists, but are well developed by topologists [31, 32, 33]. The subsection 2.2. contains a review of the basic notions connected with the principal $U(1)$-bundles on four manifolds with nonempty boundaries in the style of [13]. Here we also discuss the non-trivial (but rather simple) (co-)homological properties of pV-cobordisms. The following subsection 2.3. is dedicated to a brief survey of Poincaré–Lefschetz duality [34, 33] which we use instead of the common Hodge duality to define the formal analogues of electric and magnetic fluxes in our version of BF-theory. Moreover we remind the definition of the main topological invariant of four-dimensional manifolds, namely intersection form [31, 32] which serves for the determination of coupling constant matrices characterizing interactions of these "electric" and "magnetic" fluxes.

In the section 3. we construct the simple version of the Abelian BF-theory on pV-cobordisms, also known as graph cobordisms [15]. In analogy with the electrodynamics including theta term [7, 8, 13], the "electric" and "magnetic" fluxes are defined as linear combinations of the first Chern classes of pV-cobordism M and its boundary ∂M. Then the transition amplitudes (partition functions) are expressed as functionals of these fluxes, intersection matrices and the BF scale factor λ. These transition amplitudes are topological invariants and represent something resembling to the theta function. They also support a certain form of strong-weak coupling duality.

In the section 4. we present numerical calculations of intersection matrices for a specific sequence of pV-cobordisms and pV-manifolds which can be interpreted as a series of the topology changes leading to a certain state of the universe. This state can be identified with the contemporary one by means of the elementary interactions between "electric" fluxes, since the coupling constants hierarchy of these fluxes reproduces the hierarchy of the fundamental interactions in the real universe. At the end of this section we give interpretation of the obtained results on the base of the Hartle-Hawking "no-boundary" proposal modified for pV-manifolds. The section 5. contains the main conclusions.

The standard notations $\mathbb{Z}$, $\mathbb{R}$ and $\mathbb{C}$ are used for the sets of integer, real and complex numbers, respectively. The symbol $\mathbb{N}$ denotes the set of positive integers and $\mathbb{Z}_p = \mathbb{Z}/p\mathbb{Z}$ is a cyclic group of the order $p \in \mathbb{N}, p > 1$.

2. Mathematical Concepts

As this was said in the Introduction, we build here a rather simple topological gauge (BF-) model on sufficiently complicated topological spaces belonging to the class of pV-cobordisms imitating tunnelling topological changes. These four-dimensional smooth manifolds with the Euclidean signature possess non-trivial (co-)homological characteristics. This leads to a specific generalization of the Dirac quantization conditions [7, 8, 10] and enables us to explicitly express transition amplitudes (partition functions) in terms of the topological invariants (intersection forms) of the pV-cobordisms. The boundary components of pV-cobordisms are disjoint sums of lens spaces and $\mathbb{Z}$-homology spheres. First we give some necessary definitions following the works of Saveliev [33, 15], see also [32].

2.1. $\mathbb{Z}$-Homology Spheres, Splicing, Plumbing, pV-cobordisms and pV-manifolds

Let $a_1,\ a_2,\ a_3$ be pairwise relatively prime positive numbers. The Brieskorn homology sphere (Bh-sphere) $\Sigma(\underline{a}) := \Sigma(a_1, a_2, a_3)$ is defined as the link of (Brieskorn) singularity

$$\Sigma(\underline{a}) := \Sigma(a_1, a_2, a_3) := \{{z_1}^{a_1} + {z_2}^{a_2} + {z_3}^{a_3} = 0\} \cap S^5 \tag{2.1}$$

where $z_i \in \mathbb{C}_i$, and S^5 is the unit five-dimensional sphere $|z_1|^2 + |z_2|^2 + |z_3|^2 = 1$. The singular complex algebraic surface ${z_1}^{a_1} + {z_2}^{a_2} + {z_3}^{a_3} = 0$ has the canonical orientation which induces the canonical orientation of the link $\Sigma(\underline{a})$. If any of a_i is equal to 1, the manifold $\Sigma(\underline{a})$ is homeomorphic to the ordinary S^3. Bh-spheres belong to the class of Seifert fibered homology (Sfh-) spheres [35]. On this manifold, there exists a unique Seifert fibration which has unnormalized Seifert invariants [36] (a_i, b_i) subject to $e(\Sigma(\underline{a})) = \sum_{i=1}^{3} b_i/a_i = 1/a$, where $a = a_1a_2a_3$ and $e(\Sigma(\underline{a}))$ is its Euler number (the well known topological invariant of a Bh-sphere). This Seifert fibration is defined by the S^1-action which reads $t(z_1, z_2, z_3) = (t^{\sigma_1} z_1, t^{\sigma_2} z_2, t^{\sigma_3} z_3)$, where $t \in S^1$, and $\sigma_i = a/a_i$. This action is fixed-point-free. The only points of $\Sigma(a_1, a_2, a_3)$ which have non-trivial isotropy group $\mathbb{Z}_{a_i}$ are those with one coordinate z_i equal to 0 ($i = 1, 2, 3$). The fiber through such a point is called an exceptional (singular) fiber of degree a_i. All other fibers are called regular (non-singular). In general, Sfh-spheres $\Sigma(a_1, ..., a_n)$ have n different exceptional fibers and represent special cases of $\mathbb{Z}$-homology spheres [32].

By a $\mathbb{Z}$-homology sphere we mean a closed three-manifold Σ such that all homology groups of Σ with integer coefficients are isomorphic to homology groups of the ordinary three-sphere S^3 over $\mathbb{Z}$. All $\mathbb{Z}$-homology spheres used in this paper can be obtained from Bh-spheres by the splicing operation. This operation is defined for any Sfh-sphere as follows: First we define [32] a Seifert link as a pair $(\Sigma, S) = (\Sigma, S_1 \bigcup \cdots \bigcup S_m)$ consisting of oriented $\mathbb{Z}$-homology sphere Σ and a collection S of Seifert fibers (exceptional or regular) $S_1, \ldots, S_m$ in Σ. Note that the links (S^3, S) where S^3 is an ordinary three-sphere, are also allowed. Let (Σ, S) and (Σ', S') be links and choose components $S_i \in S$ and $S'_j \in S'$. Let also $N(S_i)$ and $N(S'_j)$ be their tubular neighbourhoods, while $m, l \subset \partial N(S_i)$ and $m', l' \subset \partial N(S'_j)$ be standard meridians and longitudes. The manifold $\Sigma'' = (\Sigma \setminus \text{int}N(S_i)) \bigcup (\Sigma' \setminus \text{int}N(S'_j))$ obtained by pasting along the torus boundaries by matching m to l' and m' to l, is a $\mathbb{Z}$-homology sphere. The link $\left(\Sigma'', (S \setminus S_i) \bigcup (S' \setminus S'_j)\right)$ is called the splice (splicing) of (Σ, S) and (Σ', S') along S_i and S'_j. We shall use the standard notation $\Sigma'' = \Sigma \frac{\quad}{S_i \; S'_j} \Sigma'$ or simply $\Sigma'' = \Sigma - \Sigma'$. Any link which can be obtained from a finite number of Seifert links by splicing is called a graph link. Empty graph links are precisely the (graph) $\mathbb{Z}$-homology spheres. All graph links are classified as in [32] by their splice diagrams.

A splice diagram Δ is a finite tree graph with vertices of three types. Vertices with at least three adjacent edges are called *nodes*. Each node with n adjacent edges corresponds to a Sfh-sphere $\Sigma(a_1, ..., a_n)$. The edges adjacent to a node correspond to exceptional fibers and are weighted by $a_1, ..., a_n$, respectively. The very node carries a sign plus if $\Sigma(a_1, ..., a_n)$ is oriented as a link of singularity of the type (2.1) and minus otherwise. Two other types of vertices have only one adjacent edge and are called either *leaves* or *arrowheads*. The former ones represent singular fibers in a Sfh-sphere, while the latter ones, components of a graph link. We shall use splice diagrams Δ_r with nodes of valence $n = 3$ only, see figure 1 . This type of splice diagrams corresponds to splicing of r Bh-spheres. We just consider the splice diagrams with pairwise coprime positive weights around each node. In this case the concepts of integer ($\mathbb{Z}$-) homology spheres and graph homology spheres are equivalent.

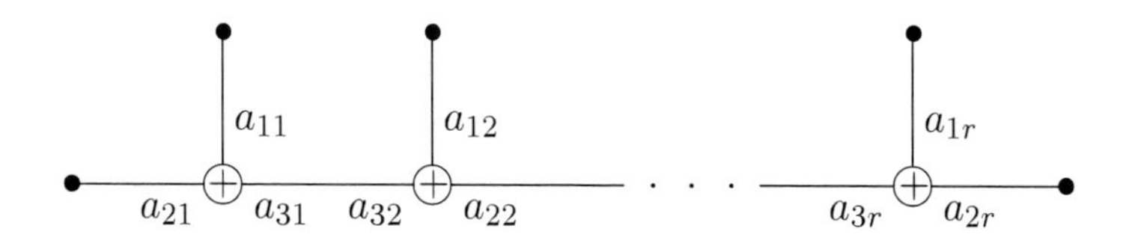

Figure 1. A splice diagram Δ_r.

In general, graph homology spheres can be conveniently described by plumbing. Plumbing graphs are required for introduction of four-dimensional manifolds with graph homology spheres as boundaries. The plumbing representation make it also possible to define intersection forms of these four-manifolds and to pass to the definition of pV-cobordisms as well as pV-manifolds.

A *plumbing graph* Γ is a graph with no cycles (a finite tree) each of whose vertices v_i carries an integer weight e_i, $i = 1, \ldots, r$. To each vertex v_i a D^2-bundle $Y(e_i)$ over S^2 is associated, whose Euler class (self-intersection number of zero-section) is e_i. If the vertex v_i has d_i edges connected to it on the graph Γ, choose d_i disjoint discs in the base S^2 of $Y(e_i)$ and call the disc bundle over the jth disc $B_{ij} = \left(D^2_j \times D^2\right)_i$. When two vertices v_i and v_k are connected by an edge, the disc bundles B_{ij} and B_{kl} should be identified by exchanging the base and fiber coordinates [37]. This pasting operation is called *plumbing*, and the resulting smooth four-manifold $P(\Gamma)$ is known as *graph manifold* (*plumbed four-manifold*). Its boundary $\Sigma(\Gamma) = \partial P(\Gamma)$ is referred to as a *plumbed three-manifold*.

Since the homology group $H_1(P(\Gamma), \mathbb{Z}) = 0$, the unique non-trivial homology characteristic is $H_2(P(\Gamma), \mathbb{Z})$ which has a natural basis (set of generators) represented by the zero-sections of the plumbed bundles. All these sections are embedded 2-spheres z_i where $i = 1, \ldots, r =$ rank $H_2(P(\Gamma), \mathbb{Z})$, and they can be oriented in such a way that the intersection (bilinear) form [33]

$$Q: \; H_2(P(\Gamma), \mathbb{Z}) \otimes H_2(P(\Gamma), \mathbb{Z}) \to \mathbb{Z} \tag{2.2}$$

will be represented by the $r \times r$-matrix $Q(\Gamma) = (q_{ij})$ with the entries: $q_{ij} = e_i$ if $i = j$; $q_{ij} = 1$ if the vertex v_i is connected to v_j by an edge; and $q_{ij} = 0$ otherwise. The three-manifold $\Sigma(\Gamma)$ is $\mathbb{Z}$-homology sphere iff the matrix $Q(\Gamma)$ is unimodular, that is det $Q(\Gamma) = 1$.

In order to construct a plumbing representation for a $\mathbb{Z}$-homology sphere given by a splice diagram Δ, we need two things:

1. Plumbing graphs for the basic building blocks, *i.e.* Sfh-spheres (in our case, Bh-spheres).

2. A procedure to splice together plumbing graphs.

First, let Σ be a Sfh-sphere with unnormalized Seifert invariants (a_i, b_i), $i = 1, ..., n$, and the splice diagram of figure 2. It can be obtained as a boundary of the graph manifold $P(\Gamma)$ where Γ is a star graph shown in figure 3 [37]. The integer weights t_{ij} in this graph are found from continued fractions $a_i/b_i = [t_{i1}, \ldots, t_{im_i}]$; here

$$[t_1, \ldots, t_k] = t_1 - \cfrac{1}{t_2 - \cfrac{1}{\cdots - \cfrac{1}{t_k}}}.$$

Lens spaces represent a special case of Seifert fibered manifolds. Expanding $-p/q = [t_1, \ldots, t_n]$ into a continued fraction, we encounter $L(p, q)$ as a boundary of the 4-manifold obtained by plumbing on the chain Γ^{ch}_i shown in figure 4.

Notice that this plumbing graph simultaneously represents the lens space $L(p, q^*)$ with $-p/q^* = [t_n, \ldots, t_1]$ where $qq^* = 1 \bmod p$. This reflects the fact that $L(p, q)$ and $L(p, q^*)$

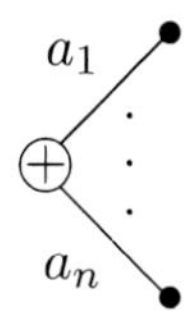

Figure 2. The splice diagram of an Sfh-sphere Σ.

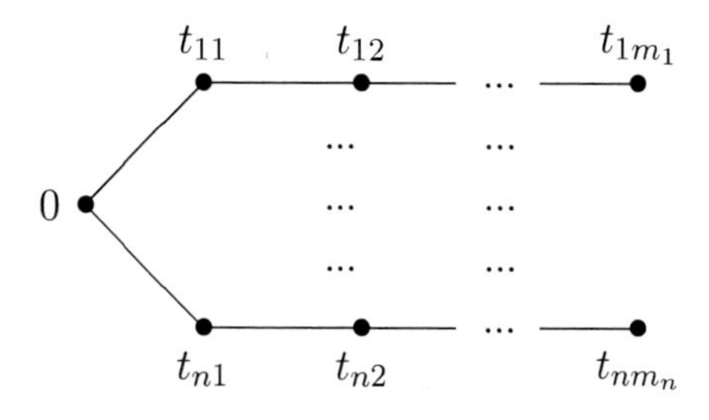

Figure 3. The star graph Γ.

are homeomorphic. Moreover, to the continuous fraction $a_i/b_i = [t_{i1}, \ldots, t_{im_i}]$ there correspond both the subgraph Γ_i^{ch} of Γ shown in figure 4, and the lens space $L(-a_i, b_i)$ (also called *leaf lens space*).

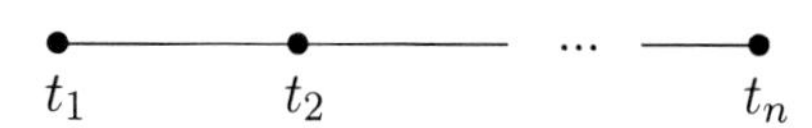

Figure 4. One-dimensional chain Γ_i^{ch}.

Next, we address the problem of splicing together plumbing graphs. It can be described as follows. Suppose that two graph links are represented by their plumbing diagrams $\overline{\Gamma}$ and $\overline{\Gamma'}$ (see figure 5) with arrows attached to vertices e_n and e'_m, respectively. The corresponding plumbing diagram for a spliced link is shown in figure 6 where $a = \det Q(\Gamma_0)/\det Q(\Gamma)$, while Γ is the plumbing graph $\overline{\Gamma}$ with the arrow deleted, and Γ_0 is a portion of Γ obtained by removing the nth vertex weighted by e_n as well as all its adjacent edges. Another integer a' is similarly obtained from the graph Γ' (examples see in [32, 15]). The above description of splicing in terms of plumbing graphs makes it possible to treat splicing as an operation on the corresponding plumbed 4-manifold; moreover, $\Sigma(\Gamma) = \partial P(\Gamma)$.

The important problem in calculation with plumbed graphs is to identify the *extra lens space* arising between nodes in the course of splicing. See figure 7 where the lens space in question is decorated by oval. The resulting lens space $L(p, q)$ $(-p/q = [t_1, \ldots, t_k])$ is

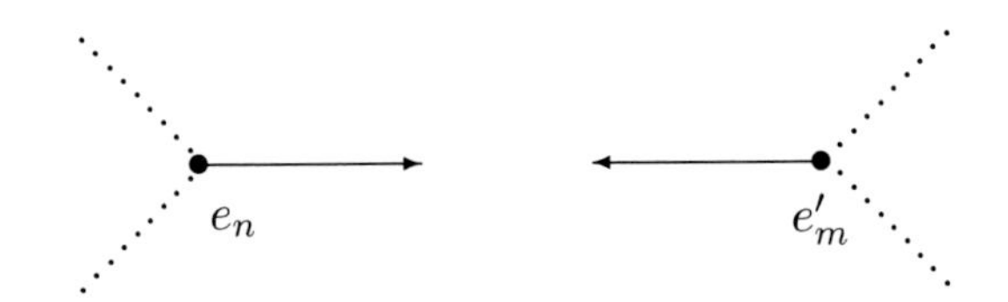

Figure 5. Plumbing diagrams $\overline{\Gamma}$ and $\overline{\Gamma'}$ prepared for splicing.

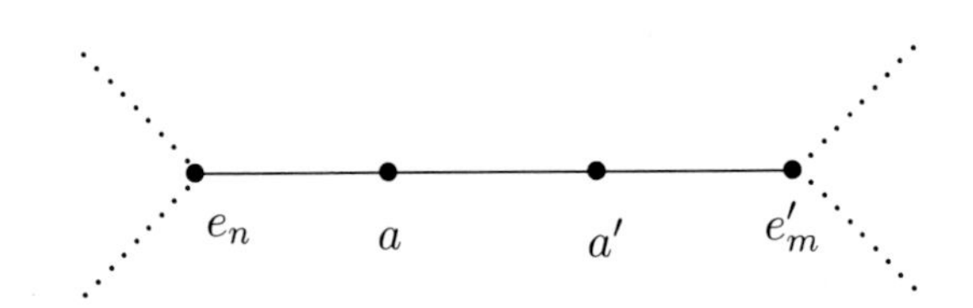

Figure 6. Plumbing diagram obtained by splicing $\overline{\Gamma}$ and $\overline{\Gamma'}$.

characterized by the following parameters,

$$p = a_1 \cdots a_{n-1} \alpha_1 \cdots \alpha_{m-1} - a_n \alpha_m, \tag{2.3}$$

$$q = -a_1 \cdots a_{n-1} \alpha_1 \cdots \alpha_{m-1} \sum_{i=1}^{n-1} \frac{b_i}{a_i} - b_n \alpha_m. \tag{2.4}$$

This lens space depends only on the spliced Seifert links $(\Sigma(a_1, ..., a_n), S_n)$ and $(\Sigma(\alpha_1, ..., \alpha_m), S_m)$, but does not depend on the rest of the splice diagram [15, 32].

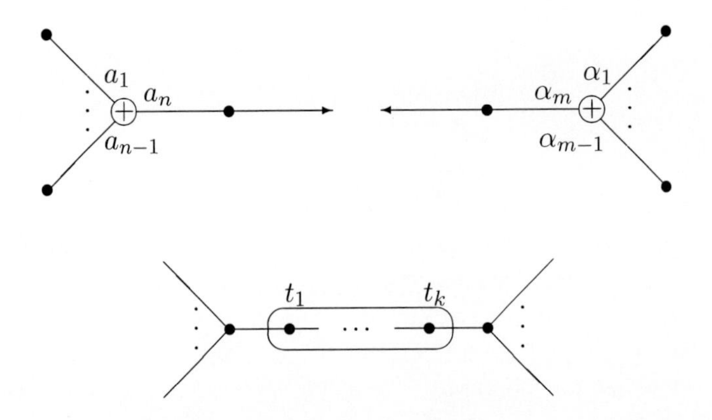

Figure 7. The extra lens space arising as a result of splicing.

If all extra lens spaces (which we also call "lens spaces between nodes") $L(p'_k, q'_k)$, $k = 1, ..., N_{\text{extra}}$, are subject to the condition $p'_k < 0$, then $\partial P(\Gamma) = \Sigma_{\text{alg}}$ is an *algebraic link* [32] which consequently bounds the graph manifold $P(\Gamma)$ having a negative defined intersection form $Q(\Gamma)$.

Now we are ready to define topological spaces of most importance in this paper, the pV-cobordisms also known as graph cobordisms [15] related to the decorated plumbed graphs. Let $P(\Gamma)$ be a plumbed four-manifold corresponding to graph Γ, and Γ^{ch} be a chain in Γ of the form shown in Figure 4. Plumbing on Γ^{ch} yields a submanifold $P(\Gamma^{\text{ch}})$ of $P(\Gamma)$ whose boundary is a lens space $L(p', q')$. The closure of $P(\Gamma) \setminus P(\Gamma^{\text{ch}})$ is a smooth compact 4-manifold with oriented boundary $-L(p', q') \bigsqcup \partial P(\Gamma)$ where $\bigsqcup$ denotes the disjoint sum operation. Starting with several chains Γ^{ch}_s ($s = \overline{1, N}$) in Γ (where $\overline{1, N}$ is the integer numbers interval from 1 to N), one can introduce a cobordism $P(\Gamma_D)$ between $\Sigma(\Gamma) := \partial P(\Gamma)$ and the disjoint union $L = \bigsqcup_{s=1}^{N} L(p'_s, q'_s)$, *i.e.*

$$\partial P(\Gamma_D) = \left(- \bigsqcup_{s=1}^{N} L(p'_s, q'_s) \right) \bigsqcup \Sigma(\Gamma).$$

Such a cobordism will be called *pV-cobordism*. Here naturally appears the concept of a *decorated graph* Γ_D shown as an ordinary graph Γ but with ovals or circles, each enclosing exactly one chain Γ^{ch}_s (see, *e.g.* figure 8).

Observation 2.1. The chains $\{\Gamma^{\text{ch}}_s\}$ must be disjoint in the following sense: No two chains should have a common vertex, and no edges of Γ should have one endpoint on one chain and another, on any other chain [15].

Observation 2.2. Consider the algebraic case when all extra lens spaces $L(p'_k, q'_k)$ are subject to the condition $p'_k < 0$, and there are decorated (by ovals) N_{extra} extra lens spaces and N_{leaf} leaf lens spaces $L(-a_i, b_i) \equiv L(p'_i, q'_i)$. In this case the cobordism $P(\Gamma_D)$ has the boundary

$$\partial P(\Gamma_D) = \left(- \bigsqcup_{k=1}^{N_{\text{extra}}} L(p'_k, q'_k) \right) \left(- \bigsqcup_{i=1}^{N_{\text{leaf}}} L(p'_i, q'_i) \right) \bigsqcup \Sigma_{\text{alg}}, \tag{2.5}$$

and its intersection matrix $Q(\Gamma_D)$ is negative defined. The series defining transition amplitudes (partition functions) in section 3. converge if the intersection matrix of the four-dimensional cobordism is positive defined. Hence the orientation of the spacetime cobordism describing a topological tunnelling should be inverse to that of $P(\Gamma_D)$. We define the pV-cobordism as $M_D = -P(\Gamma_D) = P(-\Gamma_D)$ where $-\Gamma_D$ is the decorated graph Γ_D with all weights being sign-inverse. Thus

$$\partial M_D = \left(- \bigsqcup_{k=1}^{N_{\text{extra}}} L(|p'_k|, q'_k) \right) \left(- \bigsqcup_{i=1}^{N_{\text{leaf}}} L(a_i, b_i) \right) \bigsqcup \Sigma \tag{2.6}$$

where $\Sigma = \Sigma(-\Gamma) = -\Sigma_{\text{alg}}$, and it was taken into account that $p'_k < 0$ and $a_i = -p'_i > 0$. Since $N = N_{\text{extra}} + N_{\text{leaf}}$, (2.6) now reads

$$\partial M_D = \left(- \bigsqcup_{s=1}^{N} L(|p'_s|, q'_s) \right) \bigsqcup \Sigma. \tag{2.7}$$

The resulting orientation of the four-manifold M_D coincides with that introduced by Hirzebruch in [31] for plumbed manifolds. The pV-cobordisms M_D have positive defined intersection forms. In section 4. we build examples of such cobordisms as a spacetime basis for cosmological models.

Observation 2.3. A pV-cobordism is always a smooth manifold. The notation 'V-' refers to the fact that each lens space $L(|p'_s|, q'_s)$ on the boundary of M_D may be eliminated by pasting a cone $c_s L(|p'_s|, q'_s)$ over the lens space (c_s is the vertex of this cone). This yields the well known V-manifold [38]

$$M_V = M_D \bigsqcup_{s=1}^{N} c_s L(|p'_s|, q'_s) \qquad (2.8)$$

with isolated singular points. V-manifolds of the type M_V are called *plumbed V-manifolds* (pV-manifolds). These topological spaces are also known as *pseudofree orbifolds* [29].

From the viewpoint of physics, the pV-cobordism M_D with boundary $\partial M_D = -L \bigsqcup \Sigma$ can be treated as a topology change from the disjoint union L of lens spaces to the $\mathbb{Z}$-homology sphere Σ. Then the pV-manifold M_V may be interpreted as a topology change from the finite set $\{c_s | s \in \overline{1, N}\}$ of singular points (say, big bangs [39]) to the $\mathbb{Z}$-homology sphere Σ. It means that to the end of creating of a universe with $\mathbb{Z}$-homology sphere as its spatial section, one needs N (at least three) 'big bangs' described by cones of the type $c_s L(|p'_s|, q'_s)$.

2.2. Principal $U(1)$-Bundles, Connections and Cohomologies

Let M be a four-manifold with boundary ∂M. We denote by $i_\partial : \partial M \to M$ the natural inclusion map and by d, the de Rham differential on M. By $H^p(M, \mathbb{Z})$ we further denote the absolute pth cohomology group with integer coefficients, and by $H^p(M, \partial M, \mathbb{Z})$, the relative pth cohomology group modulo ∂M. We also denote by $\Omega^p(M) = C^p(M, \mathbb{R})$ the space of p-forms on M, and by $\Omega^p(M, \partial M) = C^p(M, \partial M, \mathbb{R})$ the space of relative p-forms on M. The subscript $\mathbb{Z}$ is attached for corresponding subsets of p-forms with integer (relative) periods:

$$\int_\Sigma f \in \mathbb{Z} \qquad (2.9)$$

where $f \in \Omega^p_{\mathbb{Z}}(M)$ (or $\Omega^p_{\mathbb{Z}}(M, \partial M)$, while Σ is (relative) closed p-dimensional surface (cycle).

The quantization of BF-theory and evaluation of the transition amplitude corresponding to the cobordism M involve summation over the topological classes of gauge fields. Mathematically, these classes can be identified with the isomorphism classes of principal $U(1)$-bundles.

Let Princ (M) be the group of principal $U(1)$-bundles over M. It is well known that there exists the isomorphism

$$\text{Princ}\,(M) \overset{c}{\cong} H^2(M, \mathbb{Z}) \qquad (2.10)$$

which assigns to a bundle P its first Chern class $c(P)$, see, *e.g.*, [12, 13]. Since $H^2(M, \mathbb{Z})$ is an Abelian group, the subset of elements of finite order is a subgroup, Tor$H^2(M, \mathbb{Z})$, called the torsion subgroup of $H^2(M,\mathbb{Z})$. The preimage by the map c of the torsion subgroup is the subgroup $\mathrm{Princ}_0(M)$ of $\mathrm{Princ}\,(M)$, the elements of which are called flat principal $U(1)$-bundles, *i.e.*

$$\mathrm{Princ}_0(M) \overset{c}{\cong} \mathrm{Tor}H^2(M,\mathbb{Z}). \tag{2.11}$$

A relative principal $U(1)$-bundle (P,t) on M consists both of principal $U(1)$-bundle P on M such that its restriction $i_\partial^* P$ on ∂M is trivial, and of trivialization $t : i_\partial^* P \to \partial M \times U(1)$. The relative principal $U(1)$-bundles form a group $\mathrm{Princ}(M,\partial M)$ which is isomorphic to the 2nd relative cohomology group,

$$\mathrm{Princ}(M,\partial M) \overset{c_{\mathrm{rel}}}{\cong} H^2(M,\partial M,\mathbb{Z}), \tag{2.12}$$

where the map c_{rel} assigns to a relative bundle (P,t) its relative first Chern class $c(P,t) = c_{\mathrm{rel}}(P)$.

Of course, we can describe the group $\mathrm{Princ}(\partial M)$ of principal $U(1)$-bundles on ∂M in the same way as we did for the group $\mathrm{Princ}\,(M)$. Thus the isomorphism (2.10) holds in the form

$$\mathrm{Princ}(\partial M) \overset{c_\partial}{\cong} H^2(\partial M,\mathbb{Z}). \tag{2.13}$$

The preimage of $\mathrm{Tor}H^2(\partial M,\mathbb{Z})$ relative to the map c_∂ is the subgroup $\mathrm{Princ}_0(\partial M)$ of flat principal $U(1)$-bundles on ∂M.

Consider now the problem of extendability of principal bundles on ∂M to M following Zucchini [13]. This is important since the gauge theory transition amplitude (partition function) involves a sum over the set of the bundles $P \in \mathrm{Princ}(M)$ such that their restrictions $i_\partial^* P$ on the boundary ∂M coincide with a fixed bundle $P_\partial \in \mathrm{Princ}(\partial M)$. Every bundle $P \in \mathrm{Princ}(M)$ yields by pull-back $i_\partial^* : P \to P_\partial$ (induced by the natural inclusion $i_\partial : \partial M \to M$) a bundle $P_\partial \in \mathrm{Princ}(\partial M)$. But the converse is in general false: not every bundle P_∂ is a pull-back of some bundle P. When this does indeed happen, one says that P_∂ is extendable to M. We shall show that in the case of four-dimensional pV-cobordisms (considered in this paper as a model of spacetime) any $P_\partial \in \mathrm{Princ}(\partial M)$ is extendable. To this end, consider the absolute/relative cohomology exact sequence:

$$\ldots \to H^p(\partial M,\mathbb{Z}) \to H^{p+1}(M,\partial M,\mathbb{Z}) \to H^{p+1}(M,\mathbb{Z}) \to H^{p+1}(\partial M,\mathbb{Z}) \to \ldots. \tag{2.14}$$

We can now use the isomorphisms (2.10), (2.12) and (2.13) to draw the commutative diagram

$$\begin{array}{ccccccccc} H^1(\partial M,\mathbb{Z}) \to & H^2(M,\partial M,\mathbb{Z}) & \overset{j^*}{\to} & H^2(M,\mathbb{Z}) & \overset{i_\partial^*}{\to} & H^2(\partial M,\mathbb{Z}) & \overset{\delta^*}{\to} & H^3(M,\partial M,\mathbb{Z}) \\ & c_{\mathrm{rel}} \uparrow & & c \uparrow & & c_\partial \uparrow & & \\ & \mathrm{Princ}(M,\partial M) & \overset{j^*}{\to} & \mathrm{Princ}(M) & \overset{i_\partial^*}{\to} & \mathrm{Princ}(\partial M) & & \end{array} \tag{2.15}$$

in which the lines are exact and vertical maps are isomorphisms. In the case of pV-cobordisms under consideration it is true that [14, 15]

$$H^3(M,\partial M,\mathbb{Z}) = 0, \tag{2.16}$$

$$H^1(\partial M,\mathbb{Z}) = 0. \tag{2.17}$$

Interpretation of the second line in (2.15) is quite simple: the mapping j^* associates with every relative bundle $(P, t) \in \mathrm{Princ}(M, \partial M)$ the underlying bundle $P \in \mathrm{Princ}(M)$, and the mapping i_∂^* associates with every bundle P its pull-back bundle $P_\partial = i_\partial^* P \in \mathrm{Princ}(\partial M)$. This interpretation then applies to the first line too.

By the exactness of lines in (2.15) the bundle $P_\partial \in \mathrm{Princ}(\partial M)$ is the pull-back of the bundle $P \in \mathrm{Princ}(M)$ iff

$$\delta^*(c_\partial(P_\partial)) = 0. \tag{2.18}$$

Consequently, the obstruction to the extendability of P_∂ is a class of $H^3(M, \partial M, \mathbb{Z})$. Since (2.16) is true for all our pV-cobordisms, every principal $U(1)$-bundles on ∂M are extendable to M. Each P_∂ has several extensions to M. Again, by the exactness of (2.15), its extensions are parametrized by the group of relative bundles $\mathrm{Princ}(M, \partial M)$. This parametrization is one-to-one due to (2.17) (the mapping j^* is injective).

Connections

Let $P \in \mathrm{Princ}(M)$ be a principal $U(1)$-bundle. We denote by $\mathrm{Conn}(P)$ the affine space of connections of P. Let $A \in \mathrm{Conn}(P)$ be a connection on P. We can fix a trivializing cover $\{U_\alpha\}$ of M ($M = \cup_\alpha U_\alpha$) and assign to each open set U_α a vector potential A^α. The connection $\{A^\alpha\}$ is a Čech 0-cochain with values in 1-forms $A^\alpha = A_i^\alpha dx^i$ [40]. The curvature F_A of a connection A is defined by

$$F_A = dA. \tag{2.19}$$

This is a brief expression of the local relations $F_A^\alpha = F_A|_{U_\alpha} = dA^\alpha$. The gauge transformation properties of A [9]

$$A^\alpha - A^\beta = d\chi^{\alpha\beta} \tag{2.20}$$

ensure that F_A does not depend on the chosen local trivialization of P, *i.e.* F is gauge invariant, $F^\alpha = F^\beta$ in $U_\alpha \cap U_\beta$, thus $F_A \in \Omega^2(M)$ is a 2-form, F_A obviously being closed:

$$dF_A = 0. \tag{2.21}$$

2.3. Intersection Forms and the Poincaré–Lefschetz Duality

Now let M be a pV-cobordism corresponding to a decorated graph Γ_D (see subsection 2.1. for M_D; we suppress the subindex $_D$ below in this section). Then for any elements $f, f' \in H^2(M, \mathbb{Z})$ the rational intersection number $\langle f, f' \rangle_\mathbb{Q}$ is defined as follows [14]: We start with the part

$$0 \to H^2(M, \partial M, \mathbb{Z}) \xrightarrow{j^*} H^2(M, \mathbb{Z}) \xrightarrow{i_\partial^*} H^2(\partial M, \mathbb{Z}) \to 0 \tag{2.22}$$

of the exact sequence (2.14). Since $H^2(\partial M, \mathbb{Z})$ is a pure torsion (finite Abelian group), we see that for any $f \in H^2(M, \mathbb{Z})$ there exists $p \in \mathbb{Z}$ with $i_\partial^*(pf) = 0$, hence $pf = j^*(b)$ for unique $b \in H^2(M, \partial M, \mathbb{Z})$. Then we put

$$\langle f, f' \rangle_\mathbb{Q} := \frac{1}{p} \langle b, f' \rangle_\mathbb{Z} \in \mathbb{Q}, \tag{2.23}$$

where $\langle\ ,\ \rangle_{\mathbb{Z}}$ on the right-hand side is the usual integer intersection number well defined due to the Poincaré–Lefschetz duality [33]

$$H^2(M,\mathbb{Z}) \cong H_2(M,\partial M,\mathbb{Z}), \tag{2.24}$$

$$H^2(M,\partial M,\mathbb{Z}) \cong H_2(M,\mathbb{Z}). \tag{2.25}$$

The Poincaré–Lefschetz duality pairing (PL-pairing)

$$\langle\ ,\ \rangle_{\mathbb{Z}} : H^2(M,\partial M,\mathbb{Z}) \times H^2(M,\mathbb{Z}) \to \mathbb{Z} \tag{2.26}$$

can be written in the de Rham representation as

$$\langle b, f' \rangle = \int_M b \wedge f' \in \mathbb{Z}. \tag{2.27}$$

This is true since $\Lambda = H^2(M,\partial M,\mathbb{Z})$ and $\Lambda^{\#} = H^2(M,\mathbb{Z})$ are the integer cohomology lattices in the de Rham cohomology space $H^2(M,\mathbb{R})$. Note that if $H^2(M,\partial M,\mathbb{Z})$ and $H^2(M,\mathbb{Z})$ had torsion, this inclusion would be impossible, but in the case of pV-cobordisms these groups are finitely generated free Abelian ones [14, 16]. Moreover,

$$\text{rank}\, H^2(M,\partial M,\mathbb{Z}) = \text{rank}\, H^2(M,\mathbb{Z}) = \text{rank}\, H^2(M,\mathbb{R})$$

since $H^2(\partial M,\mathbb{Z})$ is pure torsion [41].

From exactness of the sequence (2.22) it follows that $H^2(M,\partial M,\mathbb{Z})$ is the subgroup of $H^2(M,\mathbb{Z})$ (the mapping j^* is monomorphism). Since both groups are torsion-free, the group $H^2(M,\mathbb{Z})$ can be represented as a homomorphism group

$$H^2(M,\mathbb{Z}) = \text{Hom}(H_2(M,\mathbb{Z}),\mathbb{Z}) \cong \text{Hom}(H^2(M,\partial M,\mathbb{Z}),\mathbb{Z}), \tag{2.28}$$

i.e. as the lattice $\Lambda^{\#}$ dual to $\Lambda = H^2(M,\partial M,\mathbb{Z})$ with respect to the scalar product (PL-paiting)

$$\Lambda^{\#} = H^2(M,\mathbb{Z}) := \{f \in H^2(M,\mathbb{R}) | \langle b, f \rangle \in \mathbb{Z}, \forall b \in \Lambda\} \tag{2.29}$$

where $\langle b, f \rangle$ is defined in de Rhamian representation by (2.27). Note that $\Lambda \subset \Lambda^{\#}$, thus Λ is not unimodular, and it is possible to introduce the nontrivial discriminant group

$$T(\Lambda) := \Lambda^{\#}/\Lambda = H^2(M,\mathbb{Z})/H^2(M,\partial M,\mathbb{Z}) = H^2(\partial M,\mathbb{Z}) \tag{2.30}$$

which is the finite Abelian group. (The last equality follows from exactness of the sequence (2.22)).

For our purposes there will be useful the following proposition, which is the cohomological version of the Proposition 4 from [14]. Let M be a pV-cobordism. If we choose a certain basis b_I of $\Lambda = H^2(M,\partial M,\mathbb{Z})$ and the dual basis f^I of $\Lambda^{\#} = H^2(M,\mathbb{Z})$ ($I = 1,...,r = \text{rank}\, H^2(M,\mathbb{Z})$), dual in the sense that $\langle b_I, f^J \rangle_{\mathbb{Z}} = \delta_I^J$, then the integral intersection matrix

$$Q_{IJ} = \langle b_I, b_J \rangle_{\mathbb{Z}} \tag{2.31}$$

for $H^2(M,\partial M,\mathbb{Z})$, is inverse of the rational intersection matrix

$$Q^{IJ} = \langle f^I, f^J \rangle_{\mathbb{Q}}. \tag{2.32}$$

Note that order of the discriminant group (2.30) is [10]

$$|T(\Lambda)| = \det \langle b_I, b_J \rangle_{\mathbb{Z}} = \det Q_{IJ}. \tag{2.33}$$

Let us now calculate the discriminant group $T(\Lambda)$ defined in (2.30). The exactness of the cohomology sequence (2.22) results in existence of such a basis $\{f^I \,|I \in \overline{1,r}\}$ in the group $H^2(M,\mathbb{Z})$ that $i^*_\partial(f^I) = t^I$ are generators of $H^2(\partial M,\mathbb{Z})$ [29]. Due to finiteness of the group $T(\Lambda) = H^2(\partial M,\mathbb{Z})$ there exist minimal integers $p(I) > 1$ such that $p(I)t^I = 0$ (without summation in I). Since the mapping i^*_∂ is linear, $i^*_\partial(p(I)f^I) = p(I)t^I = 0$. Moreover, the monomorphism property of j^* in (2.22) yields existence of the unique element $\tilde{f}_I \in H^2(M,\partial M,\mathbb{Z})$ such that $j^*(\tilde{f}_I) = p(I)f^I$. The class $j^*(\tilde{f}_I)$ can be considered as an element $\tilde{f}^I$ in the subgroup $H^2(M,\partial M,\mathbb{Z})$ of $H^2(M,\mathbb{Z})$. Due to rank $H^2(M,\partial M,\mathbb{Z}) =$ rank $H^2(M,\mathbb{Z}) = r$ and to the minimality of the integers $p(I)$, the set of classes $\left\{\tilde{f}^I\right\} = \{p(I)f^I\}$ forms the basis in $H^2(M,\partial M,\mathbb{Z})$ [14]. The theorems **5.1.1** and **5.1.3** in [41] then lead to the conclusion that the discriminant group $T(\Lambda)$ reads as

$$T(\Lambda) = \bigoplus_{I=1}^{r} \mathbb{Z}_{p(I)}. \tag{2.34}$$

Orders $p(I)$ of cyclic groups $\mathbb{Z}_{p(I)}$ can be calculated from the characteristics of the decorated graph Γ_D which determines topology of the cobordism M, *i.e.* from the topological invariants of this pV-cobordism. To this end note that the number of elements in the discriminant group $T(\Lambda)$ is

$$|T(\Lambda)| = p(1)\cdots p(r) = \det Q_{IJ} = |p'_1 \cdots p'_{2r+1}|. \tag{2.35}$$

The last equality follows from juxtaposition of the generators $\{f^I \,|I \in \overline{1,r}\}$ of $H^2(M,\mathbb{Z})$ to the vertices $\{v_I \,|I \in \overline{1,r}\}$ of the graph Γ_D outside of all the decorated ovals [15, 17]. In this case

$$\det Q_{IJ} = |p'_1 \cdots p'_{2r+1}| \tag{2.36}$$

where p'_s are characteristics of decorated ovals (decorated linear chains) of the graph Γ_D, see (2.7) with $N = 2r+1$ and the first decorated graph Γ^r_D in figure 8.

Now consider sublattices Λ_{r-1} and $\Lambda^{\#}_{r-1}$ (of the lattices Λ_r and $\Lambda^{\#}_r$) generated by the subbases $\left\{\tilde{f}^I \,|I \in \overline{1,r-1}\right\}$ and $\{f^I \,|I \in \overline{1,r-1}\}$, respectively. This singles out from the graph $\Gamma_D \equiv \Gamma^r_D$, the subgraph Γ^{r-1}_D which consists of the vertices $\{v_I \,|I \in \overline{1,r-1}\}$ and decorated ovals adjusted to these vertices, *i.e.* decorated chains with characteristics $p'_1, ..., p'_{2(r-1)+1} = p'_{2r-1}$, see figure 8. The discriminant group corresponding to this subgraph is

$$T(\Lambda_{r-1}) = \Lambda^{\#}_{r-1}/\Lambda_{r-1} = \bigoplus_{I=1}^{r-1} \mathbb{Z}_{p(I)} \tag{2.37}$$

and possesses the order

$$|T(\Lambda_{r-1})| = p(1)\cdots p(r-1) = |p'_1 \cdots p'_{2r-1}|. \tag{2.38}$$

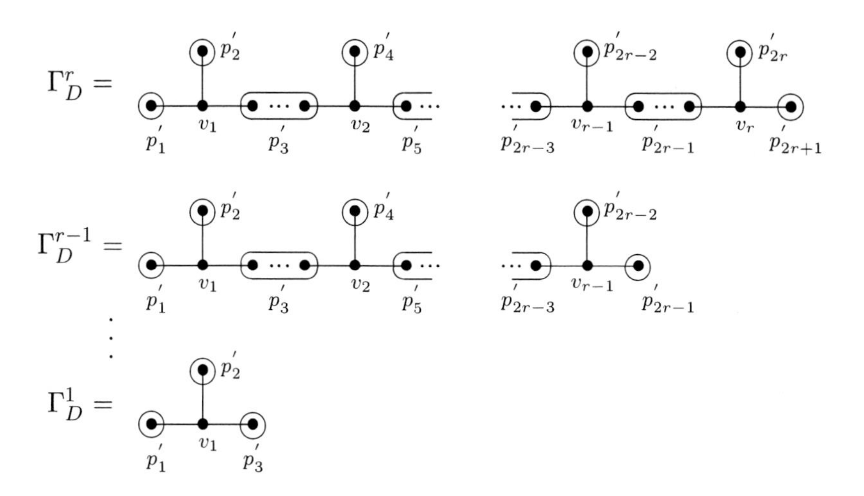

Figure 8. The decorated graphs Γ_D^I.

A comparison of (2.38) and (2.35) shows that

$$p(r) = |p'_{2r}p'_{2r+1}|. \tag{2.39}$$

After a finite number of steps, we encounter the following sequence of relations:

$$p(I) = |p'_{2I}p'_{2I+1}| \text{ for } 2 \leq I \leq r \text{ and finally } p(1) = |p'_1 p'_2 p'_3| \tag{2.40}$$

(the last expression is related to the final graph Γ_D^1).

3. Topological Gauge Theory

3.1. The Classical Abelian BF-Model

Let $P \in \text{Princ}\,(M)$ be a principal $U(1)$-bundle on M and $P_\partial := i_\partial^* P \in \text{Princ}\,(\partial M)$ be the induced principal $U(1)$-bundle on ∂M. The Abelian BF gauge theory action is a functional of the connection $A \in \text{Conn}\,(P)$ and of the auxiliary 2-cochain $B \in C^2(M, \mathbb{R})$ (2-form in the de Rham representation) [4]

$$S = \frac{1}{2\pi}\int_M \left(B \wedge F - \frac{\lambda}{2} B \wedge B\right) \tag{3.1}$$

where $F = dA$ is the curvature of the connection A (see subsection 2.2.), λ being a scale factor analogous to the cosmological constant [5].

The dynamical equations following from (3.1) are quite simple,

$$F = \lambda B, \tag{3.2}$$

$$dB = 0, \tag{3.3}$$

if the normal boundary condition [13] on the variation of the connection δA is accepted,

$$i_{\partial}^{*}(\delta A) = 0. \tag{3.4}$$

Moreover the equation (3.3) is a consequence of (3.2) and the Bianchi identity $dF = 0$. The action (3.1) is actually invariant under very large gauge transformations

$$\delta A = w, \tag{3.5}$$

$$\delta B = \frac{1}{\lambda} dw \tag{3.6}$$

where w is an arbitrary 1-form. The gauge-inequivalent classical solutions of the BF-system (3.1) are thus characterized by a 2-cocycle $F_{\rm cl}$ modulo coboundary dw (*i.e.* $F_{\rm cl} \in H^2(M,\mathbb{R})$) and by a 2-cocycle $B_{\rm cl}$ modulo coboundary $\frac{1}{\lambda}dw$ (*i.e.* $B_{\rm cl} \in H^2(M,\mathbb{R})$).

Quantization of the BF-theory involves a summation over topological classes of gauge fields. Formally, these classes may be identified with the isomorphism classes of principal $U(1)$-bundles Princ (M) and of relative principal $U(1)$-bundles Princ $(M, \partial M)$ which were described in subsection 2.2. by means of isomorphisms (2.10) and (2.12). Thus it is natural to introduce the BF analogue of the generalized Dirac quantization conditions (*cf.* [7]) for a pV-cobordism M as follows:

$$c_{\rm rel}(P) = \frac{B_{\rm cl}}{2\pi} \in \Lambda \cong H^2(M,\partial M,\mathbb{Z}) \subset H^2(M,\mathbb{R}), \tag{3.7}$$

$$c(P) = \frac{F_{\rm cl}}{2\pi} \in \Lambda^{\#} \cong H^2(M,\mathbb{Z}) \subset H^2(M,\mathbb{R}). \tag{3.8}$$

Namely these quantization conditions lead to an automatic quantization of the scale factor λ (see subsection 3.3).

3.2. Transition Amplitudes

In an analogy with the partition function of Abelian gauge theory [7, 8] we construct the transition amplitude for purely topological gauge (BF-) theory over a pV-cobordism M. As usual [13], this transition function (in semi-classical approximation) involves both summation over the set of principal $U(1)$ bundles $P \in \mathrm{Princ}(M) \cong H^2(M,\mathbb{Z})$ such that the principal bundle $P_{\partial} = i_{\partial}^{*}P$ is fixed, and functional integration over bulk quantum fluctuations v of the connection on P which satisfies the ordinary normal boundary condition [42]

$$i_{\partial}^{*}v = 0. \tag{3.9}$$

We apply the customary classical-background-quantum-splitting method [43] which can be realized as follows. We write a general representative of cohomology class in $H^2(M,\mathbb{Z})$ as

$$\frac{F}{2\pi} = n_I f^I + dv = m^I b_I + l_I f^I + dv \tag{3.10}$$

(a cohomology class and its representative we usually denote by the same symbol). This expression needs a more detailed comments: v is a proper 1-form describing quantum fluctuations. The first two terms of the last right-hand-side corresponding to non-trivial cohomology classes, describe classical (background) solution of the BF-theory,

$$\frac{F_{\rm cl}}{2\pi} = n_I f^I = m^I b_I + l_I f^I. \tag{3.11}$$

In this formula $\{b_I\}$ is basis of lattice (group) $\Lambda = H^2(M, \partial M, \mathbb{Z})$ while $\{f^I\}$, basis of the dual lattice (group) $\Lambda^{\#} = H^2(M, \mathbb{Z})$, see subsection 2.3.. Since Λ is a sublattice of $\Lambda^{\#}$ and since $i_\partial^*(f^I) = t^I$ are generators of the subgroup $\mathbb{Z}_{p(I)}$ of the discriminant group $T(\Lambda) = \Lambda^{\#}/\Lambda = H^2(M, \mathbb{Z})/H^2(M, \partial M, \mathbb{Z})$, an arbitrary element of $\Lambda^{\#}$ can be represented as a sum of some elements $m^I b_I$ of the lattice Λ ($m^I \in \mathbb{Z}$) and of such a linear combination $l_I f^I$, so that $l_I \in \overline{0, p(I) - 1}$. The restriction on the values of the coefficients l_I is due to $p(I) f^I \in \Lambda$ (there is no summation in I).

Now note that $i_\partial^*(l_I f^I) = l_I t^I \in H^2(\partial M, \mathbb{Z})$, then due to the isomorphism (2.13) the set of numbers $l_I \in \overline{0, p(I) - 1}$ determines the bundle P_∂ on the boundary ∂M which should be fixed in the calculation of a transition amplitude. Thus in this calculation the summation has to be done over the sets $m^I \in \mathbb{Z}$, *i.e.* the summation over $P \in \text{Princ}(M) \cong H^2(M, \mathbb{Z})$ reduces to that over $P \in \text{Princ}(M, \partial M) \cong H^2(M, \partial M, \mathbb{Z})$ which coincides with the result in [13].

Passing to the procedure of calculation of the transition amplitude we see that the equation of motion for B (3.2) is in fact an algebraic constraint which we can substitute back to (3.1) in order to obtain a more usual form of the BF-action [6]

$$S = \frac{1}{4\pi\lambda} \int_M F \wedge F. \tag{3.12}$$

Inserting the expression (3.10) into (3.12), we find

$$S(\lambda, \bar{m}, \underline{l}) = \frac{\pi}{\lambda} \left(m^I + Q^{IJ} l_J\right) Q_{IK} \left(m^K + Q^{KL} l_L\right) \tag{3.13}$$

where $\bar{m} = \{m^I | I \in \overline{1, r}\}$, $\underline{l} = \{l_I | I \in \overline{1, r}\}$, and $Q_{IJ} = \langle b_I, b_J \rangle_{\mathbb{Z}} = \int_M b_I \wedge b_J$, $Q^{IJ} = \langle f^I, f^J \rangle_{\mathbb{Q}} = \int_M f^I \wedge f^J$ are integer and rational intersection matrices defined in subsection 2.3.; remember that $Q_{IJ} Q^{JK} = \delta_I^K$. Note that the quantum fluctuations v give zero contributions in (3.13) due to the Stokes theorem

$$\int_M dv \wedge dv = \int_{\partial M} v \wedge dv = 0 \tag{3.14}$$

and to the normal boundary conditions (3.9). Thus we see that the transition amplitude (partition function corresponding to the pV-cobordism M in Euclidean regime) reads

$$Z(\lambda, \underline{l}) = \frac{1}{c} \sum_{m^I \in \mathbb{Z}} \exp[-S(\lambda, \bar{m}, \underline{l})]$$

where the constant $c = 1$ due to (3.14), so that

$$Z(\lambda, \underline{l}) = \sum_{m^I \in \mathbb{Z}} \exp\left[-\frac{\pi}{\lambda} \left(m^I + Q^{IJ} l_J\right) Q_{IK} \left(m^K + Q^{KL} l_L\right)\right]. \tag{3.15}$$

This is recognizable as a sort of theta function associated with the flux lattice $\Lambda + Q\underline{l}$ since $m^I b_I \in \Lambda$. To clarify this question, we define the analogues of electric and magnetic fluxes related to the field strength $\frac{F_{\text{cl}}}{2\pi} \in H^2(M, \mathbb{Z})$ through a set of homologically non-trivial 2-cycles. First note that in the case of a pV-cobordism M both integer intersection form (2.26) and rational intersection form (2.23) are rigorously determined as

$$\langle\ ,\ \rangle_{\mathbb{Z}} : H^2(M, \mathbb{Z}) \times H^2(M, \partial M, \mathbb{Z}) \to \mathbb{Z} \tag{3.16}$$

and

$$\langle\ ,\ \rangle_{\mathbb{Q}} : H^2(M,\mathbb{Z}) \times H^2(M,\mathbb{Z}) \to \mathbb{Q}, \tag{3.17}$$

respectively. Due to the Poincaré–Lefschetz duality in the form (2.24) and (2.25), there are two induced pairings

$$\langle\ ,\ \rangle_{\mathbb{Z}} : H^2(M,\mathbb{Z}) \times H_2(M,\mathbb{Z}) \to \mathbb{Z} \tag{3.18}$$

and

$$\langle\ ,\ \rangle_{\mathbb{Q}} : H^2(M,\mathbb{Z}) \times H_2(M,\partial M,\mathbb{Z}) \to \mathbb{Q} \tag{3.19}$$

which enable one to determine the fluxes of the field strength $\frac{F_{\mathrm{cl}}}{2\pi}$ through non-trivial absolute and relative 2-cycles, respectively.

Let us introduce a basis $\{\phi_I | I \in \overline{1,r}\}$ of homologically non-trivial 2-cycles in $\Lambda \cong H_2(M,\mathbb{Z})$ dual to the basis $\{f^I\}$ in $\Lambda^{\#} \cong H^2(M,\mathbb{Z})$ with respect to the pairing (3.18) in the sense that

$$\langle f^I, \phi_J \rangle_{\mathbb{Z}} = \delta^I_J. \tag{3.20}$$

Moreover, using once again the Poincaré–Lefschetz duality (2.24), we come from (3.18) to another induced pairing

$$\langle\ ,\ \rangle^{\mathrm{hom}}_{\mathbb{Z}} : H_2(M,\partial M,\mathbb{Z}) \times H_2(M,\mathbb{Z}) \to \mathbb{Z} \tag{3.21}$$

which gives the usual intersection numbers between the 2-cycles of $H_2(M,\partial M,\mathbb{Z})$ and $H_2(M,\mathbb{Z})$ [14]. Now, we can define a basis $\{\beta^I | I \in \overline{1,r}\}$ of homologically non-trivial relative 2-cycles in $\Lambda^{\#} \cong H_2(M,\partial M,\mathbb{Z})$ dual to the basis $\{\phi_I\}$ with respect to the pairing (3.21) in the sense that

$$\langle \beta^I, \phi_J \rangle^{\mathrm{hom}}_{\mathbb{Z}} = \delta^I_J, \tag{3.22}$$

i.e. dual in the sense of Poincaré–Lefschetz. In the definition of analogues of the electric and magnetic fluxes we apply the Poincaré–Lefschetz duality instead of the Hodge duality used in the ordinary electrodynamics with a theta term [7, 8, 10, 11].

We define the "electric" fluxes of field strength as fluxes through the homologically non-trivial 2-cycles $\phi_I \in \Lambda$ using the scalar product (3.18) as

$$\Phi^{(\mathrm{el})}_I(\bar{m},\underline{l}) := \left\langle \frac{F_{\mathrm{cl}}}{2\pi}, \phi_I \right\rangle_{\mathbb{Z}} = Q_{IJ} m^J + l_I \in \mathbb{Z}. \tag{3.23}$$

We analogously define "magnetic" fluxes of field strength as fluxes through the homologically non-trivial relative 2-cycles $\beta^I \in \Lambda^{\#}$ using the scalar product (3.19) as

$$\Phi^I_{(\mathrm{mag})}(\bar{m},\underline{l}) := \left\langle \frac{F_{\mathrm{cl}}}{2\pi}, \beta^I \right\rangle_{\mathbb{Q}} = m^I + Q^{IJ} l_J \in \mathbb{Q}. \tag{3.24}$$

The expressions of the fluxes in terms of "quantum numbers" m^I and l_I in formulae (3.23) and (3.24) follow from the relation (3.11) and from the duality of corresponding bases. Thus the transition amplitude (3.15) can be rewritten as

$$Z(\lambda,\underline{l}) = \sum_{\bar{m}\in\Lambda} \exp\left[-\frac{\pi}{\lambda}\Phi^I_{(\mathrm{mag})}(\bar{m},\underline{l}) Q_{IK} \Phi^K_{(\mathrm{mag})}(\bar{m},\underline{l})\right]. \tag{3.25}$$

Consequently, this transition amplitude describes the strong coupling of "magnetic" fluxes (3.24) since the product $\frac{1}{\lambda}Q_{IJ}$ plays the rôle of coupling constants' matrix. Here Q_{IJ} is the integer intersection matrix of the cobordism M (with all non-zero elements being > 1), and the scale factor $\frac{1}{\lambda} \geq 1$ (to be shown in subsection 3.3.).

Observation 3.1. Note that if the lattice Λ is self-dual ($\Lambda \cong \Lambda^{\#}$), the "electric" and "magnetic" fluxes (in the sense of our definition) mutually coincide, thus one has to introduce the Hodge operator, which presumes existence of metric (and we are trying to avoid this), to distinguish between these two types of fluxes. Thus in the consideration of a closed 4-manifold ($\partial M = 0$) our model becomes trivial, and the non-triviality of our approach is due to the substantiveness of the exact cohomological sequence (2.22) which degenerates into isomorphism,

$$0 \to H^2(M, \partial M, \mathbb{Z}) \overset{j^*}{\cong} H^2(M, \mathbb{Z}) \to 0,$$

if $\partial M = 0$ or even $H^2(\partial M, \mathbb{Z}) = 0$.

To determine the behaviour of $Z(\lambda, \underline{l})$ under the transformation $\lambda \to 1/\lambda$ we use the same trick as Olive and Alvarez [10]. Using the Poisson summation formula in matrix form [44]

$$\sum_{\bar{m} \in \Lambda} \exp\left[-\pi(\bar{m} + \bar{x}) \cdot A \cdot (\bar{m} + \bar{x})\right] =$$

$$(\det A)^{-1/2} \sum_{\underline{n} \in \Lambda^{\#}} \exp\left[-\pi \underline{n} \cdot A^{-1} \cdot \underline{n} + 2\pi i \underline{n} \cdot \bar{x}\right],$$

we can rewrite the transition amplitude (3.15) in terms of the weak coupling between "electric" fluxes (3.23) as

$$\left.\begin{array}{l} Z(\lambda, \underline{l}) = \lambda^{r/2} (\det Q_{IJ})^{-1/2} \times \\ \sum\limits_{n_I \in \mathbb{Z}} \exp\left[\pi\lambda\left(2 i n_I Q^{IJ} l_J - n_I Q^{IJ} n_J\right)\right] = \\ \\ \lambda^{r/2} (\det Q_{IJ})^{-1/2} \sum\limits_{\underline{l}' \in T(\Lambda)} \exp\left(2\pi i \lambda l'_I Q^{IJ} l_J\right) \times \\ \sum\limits_{\bar{m} \in \Lambda} \exp\left[-\pi\lambda\left(Q_{IJ} m^J + l'_I\right) Q^{IK} \left(Q_{KL} m^L + l'_K\right)\right] = \\ \\ \lambda^{r/2} (\det Q_{IJ})^{-1/2} \sum\limits_{\underline{l}' \in T(\Lambda)} \exp\left(2\pi i \lambda l'_I Q^{IJ} l_J\right) \times \\ \sum\limits_{\bar{m} \in \Lambda} \exp\left[-\pi\lambda \Phi_I^{(\mathrm{el})}(\bar{m}, \underline{l}') Q^{IK} \Phi_K^{(\mathrm{el})}(\bar{m}, \underline{l}')\right]. \end{array}\right\} \quad (3.26)$$

In the second and third parts of this formula we used the expressions (3.11) in the form

$$\frac{F_{\mathrm{cl}}}{2\pi} = n_I f^I = m^I b_I + l'_I f^I = \left(Q_{IJ} m^J + l'_I\right) f^I = \Phi_I^{(\mathrm{el})}(\bar{m}, \underline{l}') f^I \quad (3.27)$$

and $b_I = Q_{IJ} f^J$ [14]. The prime in l'_I (with respect to which the summation is performed) distinguishes it from the fixed l_I, while the symbolic notation $\underline{l}' \in T(\Lambda)$ means that the summation runs over all collections $\{l'_I\}$ which determine the elements $l'_I t^I$ of the discriminant group $T(\Lambda)$, *i.e.* $l'_I \in \overline{1, p(I) - 1}$.

The transition amplitudes $Z(\lambda, \underline{l})$ can be considered as $|T(\Lambda)|$ modifications of generalized theta functions. Comparing the expressions (3.25) and (3.26), one finds the relation

$$Z(\lambda, \underline{l}) = \frac{\lambda^{r/2}}{\sqrt{|T(\Lambda)|}} \sum_{\underline{l}' \in T(\Lambda)} \exp\left(2\pi i \lambda l'_I Q^{IJ} l_J\right) Z\left(\frac{1}{\lambda}, \underline{l}'\right) \tag{3.28}$$

which can be regarded as action of a variant of the Montonen–Olive duality transformation [45], that is, a $(\lambda \to \frac{1}{\lambda})$-analogue of the S-transformation $\tau \to -\frac{1}{\tau}$ of electric-magnetic duality [7, 8] applied to the transition amplitude in the BF-model (*cf.* section 6 in [10]). Thus the transition amplitudes written as (3.25) and (3.26), are expressed by means of $|T(\Lambda)|$ "theta functions" depending on the principal $U(1)$-bundles P_∂ (on the boundary ∂M) classified by Chern classes $c_\partial = l_I t^I \in T(\Lambda)$ where $\{t^I\}$ are generators of $H^2(\partial M, \mathbb{Z})$. Since for our pV-cobordisms $T(\Lambda) = H^2(\partial M, \mathbb{Z}) = \oplus_{I=1}^r \mathbb{Z}_{p(I)}$ (pure torsion), all principal $U(1)$-bundles on ∂M are flat [12]:

$$\text{Princ}_0(\partial M) \overset{c_\partial}{\cong} H^2(\partial M, \mathbb{Z}) = \bigoplus_{I=1}^{r} \mathbb{Z}_{p(I)}. \tag{3.29}$$

Due to (2.16) for our case, all flat principal $U(1)$-bundles on ∂M (fixed by the sets $\left\{l_I \in \overline{0, p(I) - 1} | I \in \overline{1, r}\right\}$ also known as *rotation numbers* [29, 33]) are extendable to M. These extensions are parametrized by the group of relative bundles $\text{Princ}(M, \partial M) \overset{c_{\text{rel}}}{\cong} H^2(M, \partial M, \mathbb{Z})$ (or equivalently by the set of integer parts $\left\{m^I \in \mathbb{Z} | I \in \overline{1, r}\right\}$ of rational "magnetic" fluxes $\Phi^I_{(\text{mag})}$) in the one-to-one manner due to (2.17).

Observation 3.2. The partition sums (3.25) and (3.26) converge if the intersection matrices Q_{IJ} and Q^{IJ} are positive definite. In section 4. examples of pV-cobordisms and pV-manifolds satisfying this condition will be given.

Observation 3.3. The passage from (3.25) to (3.26) for the transition amplitudes corresponds to an interchange of strong and weak couplings: (3.25) involves the coupling constants matrix $\frac{1}{\lambda} Q_{IJ}$ whose non-zero elements are > 1 (strong coupling); this formula is related to interaction of "magnetic" fluxes passing through homologically non-trivial closed 2-surfaces β^I, while the expression (3.26) contains the coupling constants matrix λQ^{IJ} whose non-zero elements are < 1 (weak coupling), and it is related to interaction of "electric" fluxes $\Phi^{(\text{el})}_I(\bar{m}, \underline{l})$ (through homologically non-trivial closed 2-cycles ϕ_I) which mimic presence of quantized "electric charges" always being integer according to (3.23). In the same manner, the "magnetic" fluxes captured by homologically non-trivial 2-cycles β^I can be interpreted as effective quantized "magnetic charges" possessing rational values since $\Phi^I_{(\text{mag})}(\bar{m}, \underline{l}) \in \mathbb{Q}$. Note that these effective "magnetic charges" pertain to a specific fixed subset of rational numbers with a finite collection of different denominators composed only by products of the topological invariants $p'_1, \ldots, p'_{2r+1}$ of the pV-cobordism M (see subsection 2.3.). Thus from the BF-analogue of the generalized Dirac quantization conditions (3.7) and (3.8) it follows that the "electric charges" come to be integer, while the "magnetic charges" are found to be rational.

Observation 3.4. It is interesting to note that the same formulae (3.23) and (3.24) can be obtained from fluxes of basic 2-cocycles $b_I \in H^2(M, \partial M, \mathbb{Z}) = \Lambda$ and $f^I \in H^2(M, \mathbb{Z}) = \Lambda^\#$ through a general 2-cycle Z_2 which is possible to represent in the form

$$Z_2 = m^J \phi_J + l_J \beta^J, \tag{3.30}$$

using the same ideas as for writing the expression (3.11). Then the flux of the basic field b_I through Z_2 reads

$$< b_I, m^J \phi_J + l_J \beta^J >= m^J Q_{IJ} + l_J \delta_I^J = \Phi_I^{(\text{el})}(\bar{m}, \underline{l}) \in \mathbb{Z} \tag{3.31}$$

which coincides with the "electric" flux (3.23). Thus the matrix element $Q_{IJ} =< b_I, \phi_J >\in \mathbb{Z}$ may be interpreted as an elementary "electric charge" imitated by the flux of basic "electric field" b_I through the basic 2-cycle $\phi_J \in \Lambda$. Analogously, $\delta_I^J =< b_I, \beta^J >$ can be understood as an elementary "electric charge" simulated by the flux of basic 2-cocycle b_I through the basic 2-cycle $\beta^J \in \Lambda^\#$. In the same way the flux of the dual basic field $f^I \in \Lambda^\#$ through Z_2 is

$$< f^I, m^J \phi_J + l_J \beta^J >= m^J \delta_J^I + l_J Q^{IJ} = \Phi_{(\text{mag})}^I(\bar{m}, \underline{l}) \in \mathbb{Q}, \tag{3.32}$$

being the same as the "magnetic" flux (3.24). This leads to the interpretation of matrix element $Q^{IJ} =< f^I, \beta^J >\in \mathbb{Q}$ as an elementary "magnetic charge" imitated by the flux of basic "magnetic field" f^I captured by the basic 2-cycle $\beta^J \in \Lambda^\#$, and of matrix element $\delta_J^I =< f^I, \phi_J >$ as an elementary "magnetic charge" imitated by the flux of basic 2-cocycle f^I through the basic 2-cycle $\phi_J \in \Lambda$.

This picture resembles well aged ideas of Wheeler and Misner [46, 47] about "charges without charges" when the field strength lines are captured by topological handles (wormholes = topological non-trivialities of the spacetime manifold). It occurs that the spacetime topology has to be unexpectedly complex when one is trying to reproduce certain characteristic features of the real universe. In section 4. we propose a concrete model exemplifying the possibility of dealing with the problems of the number of fundamental interactions in the universe as well as the hierarchy of their coupling constants on the purely topological level, but using rather complicated four-manifolds (pV-cobordisms).

Observation 3.5. There are no obstacles for extension of the transition amplitudes (3.25) and (3.26) from a pV-cobordism $M = M_D$ to a pV-manifold M_V defined in (2.8). This follows from the existence of the homeomorphisms [33, 29, 14]

$$H^2(M_D, \mathbb{Z}) \cong H_2(M_D, \partial M_D, \mathbb{Z}) \cong H_2(M_V, \mathbb{Z}),$$

$$H^2(M_D, \partial M_D, \mathbb{Z}) \cong H_2(M_D, \mathbb{Z}) \cong H^2(M_V, \mathbb{Z}).$$

Due to these homeomorphisms the exact sequence (2.22) with $M = M_D$ it is possible to represent in the form

$$0 \to H^2(M_V, \mathbb{Z}) \xrightarrow{j^*} H_2(M_V, \mathbb{Z}) \xrightarrow{i_\partial^*} \bigoplus_{I=1}^{r} \mathbb{Z}_{p(I)} \to 0.$$

Therefore the fluxes $\Phi_I^{(\text{el})}(\bar{m}, \underline{l})$ and $\Phi_{(\text{mag})}^I(\bar{m}, \underline{l})$ as well as the intersection matrix Q_{IJ} can be determined on a pV-manifold M_V and they coincide with the corresponding topological invariants defined for the pV-cobordism M_D connecting with M_V by (2.8). Since the pV-manifold possesses only the outgoing boundary $\Sigma_{\text{out}} = \partial M_V$, it is natural to interpret the partition functions (3.25) and (3.26) as Hartle-Hawking-type wave functions [27], *i.e* in terms of the "no-boundary" proposal, generalized to the conifold case [25, 28] (see also the Introduction).

3.3. Upper Bounds of the Scale Factor

In our BF-model, the generalized Dirac quantization conditions (3.7) and (3.8), together with the exactness of cohomological sequence (2.22), give upper bounds of the scale factor λ introduced in the action (3.1). These bounds are determined in terms of the topological invariants of the pV-cobordism M, but they also depend on the Chern classes $c_\partial = l_I t^I$ of the principal $U(1)$-bundles P_∂ which are fixed on the boundary ∂M. Note that the parameter λ does determine the scale factor of the coupling constants matrix as λQ^{IJ} for weak coupling in (3.26) and $\frac{1}{\lambda}Q_{IJ}$ for strong coupling in (3.25) where the rational and integer intersection matrices Q^{IJ} and Q_{IJ} give the hierarchy of the corresponding coupling constants. Due to exactness of the cohomological sequence (2.22) and since $H^2(\partial M, \mathbb{Z})$ is a pure torsion for any class $\frac{F_{\text{cl}}}{2\pi} \in H^2(M, \mathbb{Z})$ (3.11), there exists such a minimal positive integer q_0 that $q_0 \frac{F_{\text{cl}}}{2\pi}$ is a certain element $\frac{B_{\text{cl}}}{2\pi}$ of the group $H^2(M, \partial M, \mathbb{Z})$ [29, 14], *i.e*

$$q_0 F_{\text{cl}}/2\pi = B_{\text{cl}}. \tag{3.33}$$

(It is obvious that for any positive integer k the class $kq_0 F_{\text{cl}}/2\pi$ will certainly belong to the group $H^2(M, \partial M, \mathbb{Z})$.) Comparing the relation (3.33) and the classical constraint equation (3.2), $F = \lambda B$, we find the upper bound of λ, namely $\lambda_0 = 1/q_0$. Moreover, the scale factor λ is quantized in the sense that $\lambda_{k-1} = 1/(kq_0)$, $k \in \mathbb{N}$. The value of q_0 can be found from the expansion of solutions of the equation (3.33) with respect to the bases $\{b_I\}$ and $\{f^I\}$ of the groups $H^2(M, \partial M, \mathbb{Z})$ and $H^2(M, \mathbb{Z})$, respectively,

$$\frac{B_{\text{cl}}}{2\pi} = k^I b_I, \;\; k^I \in \mathbb{Z}, \tag{3.34}$$

$$\frac{F_{\text{cl}}}{2\pi} = m^I b_I + l_I f^I, \;\; m^I \in \mathbb{Z}, l_I \in \overline{0, p(I) - 1} \tag{3.35}$$

(generalized Dirac quantization conditions). A substitution of these solutions into (3.33) yields

$$k^I b_I = q_0 \left(m^I b_I + l_I f^I \right). \tag{3.36}$$

This condition would be satisfied if the term $q_0 l_I f^I$ pertained to the group $H^2(M, \partial M, \mathbb{Z})$, *i.e.* if such a collection $\{s^I | I \in \overline{1, r}\}$ of integers could be found that

$$q_0 l_I f^I = s^I b_I. \tag{3.37}$$

A pairing of the last equation with the basis elements f^J of the group $H^2(M, \mathbb{Z})$ yields

$$s^I = q_0 Q^{IJ} l_J. \tag{3.38}$$

Here the problem consists of rationality of Q^{IJ} while $\{s^I\}$ is a collection of integers. Note that the cobordisms under consideration correspond to graphs shown in figure 8 thus having the only non-zero elements Q^{II} and $Q^{I\,I\pm1}$. If for some value of J the rotation number $l_J = 0$, the matrix elements Q^{JJ} and $Q^{J\,J\pm1}$ do not enter (3.38). If $l_J \neq 0$, such elements give non-zero contribution. These elements have the common denominator $\tilde{P}_J = \mathrm{LCM}\,(p'_{2J-1}, p'_{2J}, p'_{2J+1})$ where LCM means Least Common Multiple, while p'_s are the positive integers characterizing the decorated graph Γ_D in figure 8 (see subsections 2.1., 2.3.). Thus all terms in the right-hand side of (3.38) will be integers, if

$$q_0 \equiv q_0(\underline{l}) = \mathrm{LCM}\left(\frac{\tilde{P}_J}{\mathrm{GCD}\,(\tilde{P}_J, l_J)}, \text{ over all } J \text{ such that } l_J \neq 0\right) \tag{3.39}$$

where GCD means Greatest Common Divisor, and the notation $q_0(\underline{l})$ takes into account dependence on the rotation numbers l_J. It is worth being emphasized that the upper bound $\lambda_0(\underline{l}) = 1/q_0(\underline{l})$ of the scale factor in our BF-model depends not only on topological invariants p'_s of the pV-cobordism M, but also on the Chern class $c_\partial = l_I t^I$ which fixes the principal bundle P_∂ on the boundary ∂M. Note that $\lambda_0(\underline{l}) = 1/q_0(\underline{l})$ itself is the upper bound of the sequence of admissible scale factors $\lambda_{k-1}(\underline{l}) = 1/(kq_0(\underline{l}))$, $k \in \mathbb{N}$.
Observation 3.5 From (3.39) one sees that the quantity $q_0(\underline{l})$ takes its maximum value when all $l_J \neq 0$ and $\mathrm{GCD}\,(\tilde{P}_J, l_J) = 1$. Then

$$\bar{q}_0 := \max_{\underline{l}} q_0(\underline{l}) = \mathrm{LCM}\,(\tilde{P}_J, J = \overline{1, r}) = \mathrm{LCM}\,(p'_s, s = \overline{1, 2r+1}). \tag{3.40}$$

The quantity $q_0(\underline{l})$ takes its minimum value when all $l_J = 0$. It is obvious that in this case $\underline{q}_0 := \min_{\underline{l}} q_0(\underline{l}) = 1$. Thus the upper bounds $\lambda_0(\underline{l})$ of the scale factor in our BF-model take discrete values in the interval

$$\frac{1}{\bar{q}_0} \leq \lambda_0(\underline{l}) \leq 1. \tag{3.41}$$

In section 4. we shall build a cosmological model in which the present-stage universe is characterized by an integer $\bar{q}_0$ having the order of magnitude $3.28 \cdot 10^{177}$.

4. The Family of pV-cobordisms and pV-manifolds as a Sequence of Cosmological Models

In this section we construct a collection of pV-cobordisms and pV-manifolds interpretable as sequences of topological changes finally resulting in the state of universe which we identify as its contemporary stage by the number of fundamental interactions and the hierarchy of their coupling constants. We proposed a similar type of model in recent papers [1, 48]. The construction we realize now differs by an additional condition on the four-dimensional topological space playing the rôle of the spacetime manifold: its intersection matrix is demanded to be positive defined (see observation 2.2). This guarantees convergence of partition sums (3.25) and (3.26) in the topological gauge theories built on the pV-cobordisms and pV-manifolds (see section 3.).

4.1. The Basic Family of Seifert Fibred Homology Spheres

The basic structure elements of pV-cobordisms and pV-manifolds used in this paper are simple graph four-manifolds with Seifert fibred Brieskorn homology (Bh-) spheres

$\Sigma(a_1, a_2, a_3)$ as boundaries (see subsection 2.1.). (Compact locally homogeneous universes with spatial sections homeomorphic to Seifert fibrations were considered at length in [49, 50, 51].) We use only a specific bi-parametric family of Bh-spheres which is defined as follows: First, we introduce the *primary sequence* of Bh-spheres (see [1] for more details). Let p_i be the ith prime number in the set of positive integers $\mathbb{N}$, *e.g.* $p_1 = 2,\ p_2 = 3, \ldots, p_9 = 23, \ldots$. Then the primary sequence is defined as

$$\{\Sigma(p_{2n}, p_{2n+1}, q_{2n-1}) | n \in \mathbb{Z}^+\} \tag{4.1}$$

where $q_i := p_1 \cdots p_i$, $\mathbb{Z}^+$ is the set of non-negative integers [1]. The first terms in this sequence with $n > 0$ (which we really use) are $\Sigma(2, 3, 5)$ (the Poincaré homology sphere), $\Sigma(7, 11, 30)$, $\Sigma(13, 17, 2310)$, and $\Sigma(19, 23, 510510)$. We also include in this sequence as its first term ($n = 0$) the usual three-dimensional sphere S^3 (Sf-sphere) with Seifert fibration determined by the mapping $h_{pq} : S^3 \to S^2$, in its turn defined as $h_{pq}(z_1, z_2) = z_1^p / z_2^q$ [52]. Recall that $S^3 = \{(z_1, z_2) | |z_1|^2 + |z_2|^2 = 1\}$ and $z_1^p / z_2^q \in \mathbb{C} \cup \{\infty\} \cong S^2$. In this paper we consider the case $p = 1$, $q = 2$ and denote this Sf-sphere as $\Sigma(1, 2, 1)$, *i.e.* $p_0 = q_{-1} = 1$ in (4.1). In this notation we use two additional units which correspond to two arbitrary regular fibers. This will enable us to operate with $\Sigma(1, 2, 1)$ in the same manner as with other members of the sequence (4.1).

Second, we define $k^{\pm}$-operations for each of Bh-spheres $\Sigma(a_1, a_2, a_3)$. To start with, we renumber Seifert's invariants so that $a_1 < a_2 < a_3$; this is always possible since a_1, a_2 and a_3 are pairwise coprime (in the case of the Sf-sphere, we take the order $\Sigma(1, 1, 2)$). The result of $k^{\pm}$-operation acting on $\Sigma(a_1, a_2, a_3)$ is another Bh-sphere

$$\Sigma^{\pm}_{k_1}(a_1^{(1)}, a_2^{(1)}, a_3^{(1)}) = \Sigma(a_1, a_2 a_3, k_1 a \pm 1), \tag{4.2}$$

i.e. it is the Bh-sphere with Seifert invariants

$$a_1^{(1)} = a_1, a_2^{(1)} = a_2 a_3, a_3^{(1)} = k_1 a \pm 1 \tag{4.3}$$

where $a = a_1 a_2 a_3$, $k_1 \in \mathbb{N}$. The upper index in the parentheses means a single application of the $k^{\pm}$-operation. A repeated application of this operation yields still another Bh-sphere

$$\Sigma^{\pm}_{k_1 k_2}(a_1^{(2)}, a_2^{(2)}, a_3^{(2)}) = \Sigma(a_1, a_2 a_3(k_1 a \pm 1), k_2 a(k_1 a \pm 1) \pm 1) \tag{4.4}$$

where $k_2 \in \mathbb{N}$; in general, $k_2 \neq k_1$. The l-fold application of the $k^{\pm}$-operation again gives an Bh-sphere, $\Sigma^{\pm}_{k_1 \ldots k_l}(a_1^{(l)}, a_2^{(l)}, a_3^{(l)})$ whose invariants are found by induction from the invariants $a_1^{(l-1)}, a_2^{(l-1)}, a_3^{(l-1)}$, with arbitrary $k_l \in \mathbb{N}$. Note that the least Seifert invariant does not change under $k^{\pm}$-operations ($a_1^{(l)} = a_1$ for any $l = 1, 2, \ldots$) while the two other Seifert invariants depend both on the order (multiplicity) of the $k^{\pm}$-operation fulfilment and on which (k^+ or k^-)-operation is applied. A hint of such an operation can be found in Saveliev's paper [53].

In [54, 1] we defined only the k^+-operation in the special case $k_1 = k_2 = \cdots = k_l = 1$ and named it (not quite aptly) "derivative of Bh-sphere". In our new terminology this is the

[1] If we change the definition of the primary sequence of Bh-spheres then the hierarchy of coupling constants that is predicted by our model will be other than the experimental one; see the table 2 for numerical justifications and the subsection 4.4 for the correspondence with a top down approach to cosmology.

Table 1. Euler number of (n, t)-family of Sf- and Bh-spheres.

$n \backslash t$	-4	-3	-2	-1	0	1	2	3	4
0					$\mathbf{5.0\times 10^{-1}}$	1.7×10^{-1}	2.3×10^{-2}	5.5×10^{-4}	3.1×10^{-7}
1				3.3×10^{-2}	$\mathbf{1.1\times10^{-3}}$	1.2×10^{-6}	1.3×10^{-12}	1.8×10^{-24}	
2			4.3×10^{-4}	1.9×10^{-7}	$\mathbf{3.5\times10^{-14}}$	1.2×10^{-27}	1.5×10^{-54}		
3		2.0×10^{-6}	3.8×10^{-12}	1.5×10^{-23}	$\mathbf{2.2\times10^{-46}}$	4.7×10^{-92}			
4	4.5×10^{-9}	2.0×10^{-17}	4.0×10^{-34}	1.6×10^{-67}	$\mathbf{2.7\times10^{-134}}$				

1^+-operation; its l-fold application gives the Bh-sphere denoted in [1] as $\Sigma(a_1^{(l)}, a_2^{(l)}, a_3^{(l)})$. In the same paper we showed that the application of this operation to the primary sequence (4.1) yields a bi-parametric family of Bh-spheres whose Euler numbers reproduce fairly well the experimental hierarchy of *dimensionless low-energy coupling* (DLEC) constants of the fundamental interactions in the real universe. For the reader's convenience we concisely reiterate here some results obtained in [54, 1].

This bi-parametric family of Bh-spheres is

$$\left\{\Sigma(a_{1n}^{(l)}, a_{2n}^{(l)}, a_{3n}^{(l)}) = \Sigma(p_{2n}^{(l)}, p_{2n+1}^{(l)}, q_{2n-1}^{(l)})|n, l \in \mathbb{Z}^+\right\}. \tag{4.5}$$

(Note that the $k^{\pm}$-operation involves a renumbering of Seifert's invariants such that the inequalities $a_{1n}^{(l)} < a_{2n}^{(l)} < a_{3n}^{(l)}$ become valid. Thus the collections of Seifert's invariants $\left\{a_{1n}^{(l)}, a_{2n}^{(l)}, a_{3n}^{(l)}\right\}$ and $\left\{p_{2n}^{(l)}, p_{2n+1}^{(l)}, q_{2n-1}^{(l)}\right\}$ are equivalent up to ordering.) In [54, 1] it was shown that to reproduce the hierarchy of the DLEC constants of the known five fundamental interactions (including the cosmological one) it is sufficient to restrict values of parameters as $n, l \in \overline{0, 4}$. With this restriction, the Euler numbers of the Bh-spheres family are given in table 1 (the revised tables 1 and 3 of [54]).

To make the comparison with the experimental hierarchy of DLEC constants (see table 2) easier, we introduced instead of l a new parameter $t := l - n$ which plays the rôle of "discrete cosmological time".

Just at $t = 0$ $(l = n)$ the experimental hierarchy of DLEC constants is reproduced properly. This enables us to consider the ensemble of Bh-spheres (4.5) at $l = n$

$$E_0 = \left\{\Sigma(a_{1n}^{(n)}, a_{2n}^{(n)}, a_{3n}^{(n)}) = \Sigma(p_{2n}^{(n)}, p_{2n+1}^{(n)}, q_{2n-1}^{(n)})|n \in \overline{0, 4}\right\} \tag{4.6}$$

as the basis elements used in constructing the spatial section Σ_0 of the contemporary universe by means of the splicing operation. The key factor in this (at first glance, exotic) hypothesis is the fact that the diagonal elements (and eigenvalues) of the rational intersection matrix for the corresponding pV-cobordism M_0 (that is, its $\partial M_0 = \Sigma_0 \bigsqcup_{s=1}^{N} (-L(|p_s'|, q_s')))$ show the same hierarchy as the Euler numbers, thus reproducing the DLEC constants' hierarchy (see two last columns in table 2). Then it is natural to suppose that at $t \in \overline{-4, -1}$ the ensembles

$$E_t = \left\{\Sigma(a_{1n}^{(n+t)}, a_{2n}^{(n+t)}, a_{3n}^{(n+t)}) = \Sigma(p_{2n}^{(n+t)}, p_{2n+1}^{(n+t)}, q_{2n-1}^{(n+t)})|n \in \overline{-t, 4}\right\} \tag{4.7}$$

of Bh-spheres forming basic elements for gluing (by splicing) spatial sections of the universe on earlier stages characterized, in particular, by a diminishing of the number of fundamental interactions from five at $t = 0$ to one at $t = -4$.

Table 2. Euler numbers *vs.* experimental DLEC constants *vs.* diagonal elements of intersection matrix $Q^{+}(0)$ (see subsection 4.3.).

n	$e\left(\Sigma_n^{(n)}\right)$	Interaction	α_{exper}	$Q^{+II}(0)$	I
0	0.5	strong	1	9.69×10^{-1}	1
1	1.07×10^{-3}	electromagnetic	7.30×10^{-3}	7.21×10^{-3}	2
2	3.51×10^{-14}	weak	3.04×10^{-12}	1.76×10^{-12}	3
3	2.17×10^{-46}	gravitational	2.73×10^{-46}	3.68×10^{-44}	4
4	2.70×10^{-134}	cosmological	$< 10^{-120}$	2.66×10^{-134}	5

Notes: 1. The dimensionless strong interaction constant is $\alpha_{\text{st}} = G/\hbar c$, G characterizes the strength of the coupling of the meson field to the nucleon. **2.** The fine structure (electromagnetic) constant is $\alpha_{\text{em}} = e^2/\hbar c$. **3.** The dimensionless weak interaction constant is $\alpha_{\text{weak}} = (G_{\text{F}}/\hbar c)(m_{\text{e}}c/\hbar)^2$, G_{F} being the Fermi constant (m_{e} is mass of electron). **4.** The dimensionless gravitational coupling constant is $\alpha_{\text{gr}} = G_N m_{\text{e}}^2/\hbar c$, G_{N} being the Newtonian gravitational constant. **5.** The cosmological constant Λ multiplied by the squared Planckian length is $\alpha_{\text{cosm}} = \Lambda G_{\text{N}}\hbar/c^3$. We select five coupling constants (and their combinations) of 31 dimensionless physical constants required by particle physics and cosmology that were indicated in [55], which characterize the low energy approximation to the Standard Model.

Observation 4.1. In this paper we shall not consider splice diagrams occurring for $t > 0$. Just note that in accordance with table 1 the number of interactions should diminish from five to one with the increase of the parameter t from 0 to 4, but the intersection matrices (related to the coupling constants ones) are different in ascending and descending stages.
Observation 4.2. It is worth being observed that in our scheme the five (low energy) interactions are related to the first nine prime numbers as (1,2), (3,5), (7,11), (13,17), (19,23). To obtain any new interaction, one has to attach a new pair of prime numbers to the preceding set. For example, taking the next pair (29,31), we come with the same algorithm to a new coupling constant of the order of magnitude $\alpha_6 \approx 10^{-361}$. Thus our model answers the question we did not even put: How many fundamental interactions may really exist in the universe? Our model predicts an infinite number of interactions due to the infinite succession of prime numbers. We simply cannot detect too weak interactions beginning with α_6 since all subsequent are even weaker: $\alpha_7 \approx 10^{-916}$, *etc.* [56].

4.2. The Construction of pV-cobordisms

The splice diagrams corresponding to states of the universe at the cosmological time $t \in \overline{-4, 0}$ are shown in figure 9 where we consider them as subdiagrams being parts of a disjoint total splice diagram. To any of these subdiagrams one associates (in accordance with the well-known algorithm, [32, 15, 17]) the pV-cobordism M_D (constructed in Observation 2.2, subsection 2.1.) whose intersection form is positive definite iff for each edge joining two nodes the edge determinant is positive. The *edge determinant* $\det(e_{mn})$ of an edge joining two nodes is the product of the two weights on the edge minus the product of the weights adjacent to the edge [17]. In our case this criterion means that in any portion of

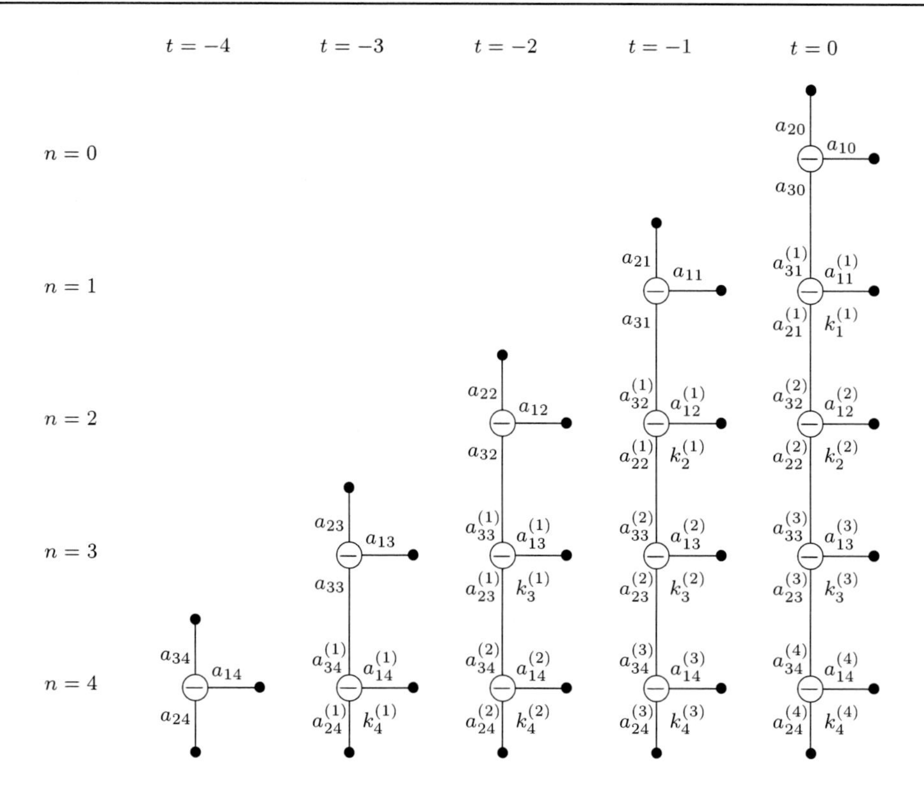

Figure 9. The splice diagram of different states of the universe.

splice diagram shown in figure 10, *i.e.* for any edge e_{mn} one has

$$\det(e_{mn}) = a_{2n}a_{3m} - a_{1n}a_{3n}a_{1m}a_{2m} > 0, \tag{4.8}$$

cf. subsection 2.1. where a more general form of the edge determinant is given, $\det(e_{mn}) = -p$, p being defined in (2.3). Taking certain collections of integers $k_1, \ldots, k_4$ (which participate in $k^{\pm}$-operations), one gets positive definite intersection forms for all splice diagrams shown in figure 9. Note that when $k_1 = k_2 = k_3 = k_4 = 1$, all intersection matrices have indefinite signature, thus convergence of transition amplitudes (3.25) and (3.26) for corresponding graph cobordisms is not ensured. Below we confine ourselves to the k^+-operation (for the k^--operation we shall only give the final result in the Appendix).

The total disconnected diagram in figure 9 consists of five connected splice subdiagrams. Naturally, there exists an ambiguity in the splice operation (related to this type of diagrams). The diagram in figure 9 contains fifteen nodes, each of them having three adjacent edges. Thus one can glue 3^{15} different graph cobordisms. Moreover, there is an infinite set of integers k_i which guarantee positive definiteness of the intersection matrices of respective cobordisms. It is however possible to fix a unique gluing procedure imposing a minimality condition on the coefficients k_i at each level of realization of the k^+-operation. In particular, this condition immediately yields a conclusion that the vertices (leaves) corresponding to minimal Seifert invariants ($a^{(l)}_1$) remain free (not subjected

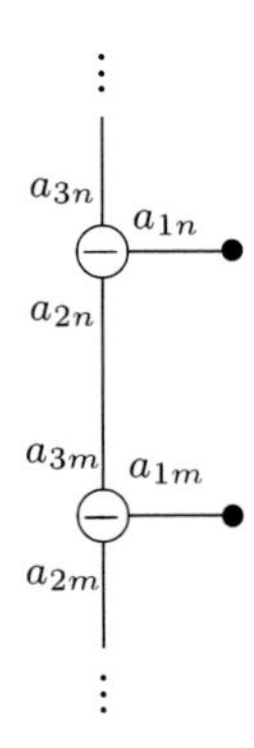

Figure 10. A portion of a splice diagram.

to slicing). Applying the k^+-operation to all Bh-spheres of the primary sequence

$$\left\{\Sigma(a_{1n}, a_{2n}, a_{3n}) | n \in \overline{1,4}\right\}, \tag{4.9}$$

we find the minimal $k_n^{(1)}$ for which the conditions $a_{3n}a_{3,n+1}^{(1)} - a_{1n}a_{2n}a_{1,n+1}^{(1)}a_{2,n+1}^{(1)} > 0,\ n \in \overline{0,3}$ are satisfied. Thus we unambiguously fixed the collection of the first-level Bh-spheres, *i.e.* those with the parameter $l = 1$:

$$\left\{\Sigma_{k_n^{(1)}}(a_{1n}^{(1)}, a_{2n}^{(1)}, a_{3n}^{(1)}) | n \in \overline{1,4}\right\}. \tag{4.10}$$

Now we execute the first splicing procedure (along the upper vertical edges between nodes in figure 9):

$$\Sigma(a_{1n}, a_{2n}, a_{3n}) \overline{\underset{S\ S'}{\quad}} \Sigma_{k_{n+1}^{(1)}}(a_{1,n+1}^{(1)}, a_{2,n+1}^{(1)}, a_{3,n+1}^{(1)}) \tag{4.11}$$

where $n \in \overline{0,3}$, $S = S_{a_{3n}}$, $S' = S_{a_{3,n+1}}$. The same algorithm is applied to determine the collection of the second-level Bh-spheres, and further by induction the lth-level Bh-spheres:

$$\left\{\Sigma_{k_n^{(1)}\ldots k_n^{(l)}}(a_{1n}^{(l)}, a_{2n}^{(l)}, a_{3n}^{(l)}) | n \in \overline{l,4}\right\}. \tag{4.12}$$

In each step, there is executed the splice operation according to the diagram in figure 9.

Consequently, we obtain the five connected subdiagrams $\Delta^+(t)$ where superscript $^+$ corresponds to the use of k^+-operation with minimization of parameters $k_n^{(l)}$ at each step. According to the procedure described in the subsection 2.1., the corresponding decorated plumbed graphs $\Gamma_D^+(t)$ are constructed. These decorated graphs codify the definite pV-cobordisms $M_D^+(t)$, $t \in \overline{-4,0}$, which are interpreted as the spacetime manifolds corresponding to different values of cosmological time parameter t. The boundaries of these cobordosms are represented as follows:

$$\partial M_D^+(t) = \left(-\bigsqcup_{s=1}^{N(t)} L(|p_s'(t)|, q_s'(t))\right) \bigsqcup \Sigma^+(t). \tag{4.13}$$

(see the expression (2.7)). It is worth being underlined that both $\mathbb{Z}$-homology sphere $\Sigma^{+}(t)$ and the collection of lens spaces $L(|p'_s(t)|, q'_s(t))$ depend on the cosmological time t. Note that among the lens spaces forming the boundaries of different cobordisms $M_D^{+}(t)$ there exist mutually homeomorphic, namely $L(a_1^{(l)}, b_1^{(l)})$, since $a_1^{(l)} = a_1$ for $\forall l \in \mathbb{N}$. By means of successive pairwise gluing together these lens spaces, it is possible to form a cobordism M_{total} which connects the initial state of universe, with a spatial section $\Sigma^{+}(-4)$, to the final one with a spatial section $\Sigma^{+}(0)$. The cobordism M_{total} will include all intermediate stages with the following sequence of spatial sections

$$\Sigma^{+}(-4) \to \Sigma^{+}(-3) \to \Sigma^{+}(-2) \to \Sigma^{+}(-1) \to \Sigma^{+}(0). \tag{4.14}$$

This is accompanied by creation and annihilation of a certain set of disjoint lens spaces of the type $L(|p'_s(t)|, q'_s(t))$ which have no homeomorphic counterparts. This procedure is outlined in [1].

4.3. Coupling Constants of Fundamental Interactions as Cosmological Circumstances

In order to deepen our physical discussion, we give below the calculations results for rational intersection matrices $Q^{+IJ}(t)$ $(I, J = \overline{1, 5+t})$, their eigenvalues, and determinants, corresponding to pV-cobordisms $M_D^{+}(t)$:

$$Q^{+IJ}(0) = \begin{pmatrix} \mathbf{9.7 \times 10^{-1}} & 3.1 \times 10^{-2} & 0 & 0 & 0 \\ 3.1 \times 10^{-2} & \mathbf{7.2 \times 10^{-3}} & 1.4 \times 10^{-8} & 0 & 0 \\ 0 & 1.4 \times 10^{-8} & \mathbf{1.8 \times 10^{-12}} & 1.9 \times 10^{-29} & 0 \\ 0 & 0 & 1.9 \times 10^{-29} & \mathbf{3.7 \times 10^{-44}} & 3.1 \times 10^{-89} \\ 0 & 0 & 0 & 3.1 \times 10^{-89} & \mathbf{2.7 \times 10^{-134}} \end{pmatrix},$$

$$\lambda_I^{+}(0) = \left\{9.7 \times 10^{-1}, 6.2 \times 10^{-3}, 1.7 \times 10^{-12}, 3.7 \times 10^{-44}, 6.4 \times 10^{-139}\right\},$$

$$\det Q^{+IJ}(0) = 2.4 \times 10^{-196};$$

$$Q^{+IJ}(-1) = \begin{pmatrix} 8.3 \times 10^{-2} & 1.1 \times 10^{-4} & 0 & 0 \\ 1.1 \times 10^{-4} & 9.6 \times 10^{-6} & 1.2 \times 10^{-14} & 0 \\ 0 & 1.2 \times 10^{-14} & 2.5 \times 10^{-21} & 2.0 \times 10^{-44} \\ 0 & 0 & 2.0 \times 10^{-44} & 1.6 \times 10^{-67} \end{pmatrix},$$

$$\lambda_I^{+}(-1) = \left\{8.3 \times 10^{-2}, 9.5 \times 10^{-6}, 2.5 \times 10^{-21}, 9.7 \times 10^{-72}\right\},$$

$$\det Q^{+IJ}(-1) = 1.9 \times 10^{-98};$$

$$Q^{+IJ}(-2) = \begin{pmatrix} 3.0 \times 10^{-3} & 1.5 \times 10^{-7} & 0 \\ 1.5 \times 10^{-7} & 6.6 \times 10^{-10} & 5.1 \times 10^{-22} \\ 0 & 5.1 \times 10^{-22} & 4.0 \times 10^{-34} \end{pmatrix},$$

$$\lambda_I^{+}(-2) = \left\{3.0 \times 10^{-3}, 6.5 \times 10^{-10}, 7.9 \times 10^{-37}\right\},$$

$$\det Q^{+IJ}(-2) = 1.5 \times 10^{-48};$$

$$Q^{+IJ}(-3) = \begin{pmatrix} 2.2 \times 10^{-6} & 2.1 \times 10^{-12} \\ 2.1 \times 10^{-12} & 2.2 \times 10^{-17} \end{pmatrix},$$

$$\lambda_I^+(-3) = \{2.2 \times 10^{-6}, 2.0 \times 10^{-17}\},$$

$$\det Q^{+IJ}(-3) = 4.4 \times 10^{-23};$$

$$Q^{+IJ}(-4) = (4.5 \times 10^{-9}).$$

(In Appendix we shall give the rational intersection matrices $Q^{-IJ}(t)$ corresponding to the cobordisms $M_D^-(t)$ obtained from the splice diagrams in figure 9 by an application of the k^--operation while minimizing the parameter k^- at each step.)

Note that all elements $Q^{+IJ}(t)$ in these matrices are rational; they are given here up to two significant digits. The inverse matrices $Q_{IJ}^+(t)$ are integer. The inversion of the rational intersection matrices with the help of the MAPLE program is an excellent test of the correctness of their calculation according to algorithms described in [15, 32], since any error leads to non-integer elements in resulting matrices $Q_{IJ}^+(t)$. Recalling the interpretation of rational intersection matrices $\lambda Q^{+IJ}(t)$ as the coupling constants of "electric" fluxes proposed in subsection 3.2. (see *Observation 3.3*) we observe that the diagonal elements of 5×5 matrix $Q^{+IJ}(0)$ (see boldface numbers) reproduce rather exactly the hierarchy of DLEC constants for the well known five fundamental interactions (see the fifth column in the table 2). The eigenvalues of this matrix reveal the same hierarchy. This enables us to consider the interactions between "electric" fluxes $\Phi_I^{(\mathrm{el})}(\bar{m}, \underline{l})$ defined in (3.23), see also their interpretation after (3.31), as "pre-images" of the real fundamental interactions (or elementary pre-interactions [1]). Then in accordance with table 2 we shall relate the matrix elements $Q^{+II}(0)$ to strong (for $I = 1$), electromagnetic ($I = 2$), weak ($I = 3$), gravitational ($I = 4$), and cosmological ($I = 5$) pre-interactions. In this sense the "electric" fluxes in the BF-model acquire the status of quantized pre-fields bearing these names, *e.g.*, $\Phi^{(\mathrm{strong})}(\bar{m}, \underline{l}) := \Phi_1^{(\mathrm{el})}(\bar{m}, \underline{l})$, $\Phi^{(\mathrm{electromagnetic})}(\bar{m}, \underline{l}) := \Phi_2^{(\mathrm{el})}(\bar{m}, \underline{l})$, and so on.

It is natural to suppose that diagonal elements of the other rational intersection matrices $Q^{+IJ}(t)$, $t \in \overline{-4, -1}$ have hierarchy of the vacuum-level coupling constants of the fundamental interactions (pre-interactions) acting at earlier phases of cosmological evolution (which correspond to the spacetime manifolds modeled by cobordisms $M_D^+(t)$). Thus our model includes a certain unification scheme of pre-interactions. So the intersection matrix $Q^{+IJ}(-1)$ has the rank 4 and hence it describes the stage of universe with four fundamental pre-interactions. This stage can be associated with higher density of vacuum energy under which the topological structure of the universe is reconstructed. But it would be too speculative to directly connect this "unification" with the electroweak unification theory, since in our model five pre-interactions (between"electric" fluxes) are replaced by rather different (at least in the sense of hierarchy) four pre-interactions.

With the same reservations one can relate the 3×3 matrix $Q^{+IJ}(-2)$ to grand unified theories (GUT) in ordinary gauge terms. The next 2×2 matrix $Q^{+IJ}(-3)$ may be associated with a supersymmetric unification including the gravitation, since out of five low-energy (for $t = 0$) pre-interactions there survive only two of them which correspond to gravitational and cosmological pre-interactions. In this case the cobordism $M_D^+(-3)$ should pertain to the Planck scales. Then the 1×1 matrix (one rational number) $Q^{+IJ}(-3)$ might belong to the sub-Planckian level where only one pre-interaction (pre-image of the cosmological one) remains. It is obvious that in order these interrelations might have some sense, we should first introduce metric structures on the pV-cobordisms $M_D^+(t)$, then constructing over them field theories with local degrees of freedoms. But we do not pose such a vast problem in this paper.

4.4. Modified Hartle-Hawking Proposal: "No-boundary Conditions for pV-manifold

Now we shall give a brief description of another approach which allows us to an interpret the topological partition functions (3.25) and (3.26) in terms of quantum wave functions of the universe in a semi-classical approximation. This approach realizes a generalization [25] of the Hartle-Hawking [27] "no-boundary" proposal in quantum cosmology. We start with the so called top down approach to cosmology [58]. In the top down approach one computes amplitudes for alternative histories of universe with final boundary conditions only. These boundary conditions act as late time constraints and provide information that is supplementary to the dynamical laws, which selects a *subclass* of histories and enables one to identify alternatives that (within this subclass) have probabilities (measures $w(M)$ in (1.1)) near one. As a final (contemporary) boundary conditions we suggest the set of coupling constants corresponding to low energy approximations to the Standard Model for elementary particle physics and the upper bound of the cosmological constant (see, *e.g.* tables 1 and 2 in [55], and the experimental column of the table 2 in this paper). Assuming the hypothesis that these fundamental coupling constants should have the topological origin, we select as a subclass of histories the set of pV-manifolds (2.8). Within this subclass we consider as the most probable the pV-manifolds [39]

$$M_V^+(t) = M_D^+(t) \bigsqcup_{s=1}^{N(t)} c_s L(|p'_s(t)|, q'_s(t)), \tag{4.15}$$

that are built on the base of the primary sequence (4.1) (see the footnote near this formula) and correspond to pV-cobordisms $M_D^+(t)$ constructed in subsection 4.2. This probability conjecture finds its justification in the fact that the basic topological invariants of these pV-manifolds, namely intersection matrices $Q^{+IJ}(t)$, contain a specific information about the hierarchy of coupling constants (see subsection 4.3).

Using the expression (3.25), an analogue of semi-classical approximation to the Hartle-Hawking wave functions for the universe (corresponding to the pV-manifold $M_V^+(t)$) can be written as

$$\Psi(\lambda(t), \underline{l}(t)) = \sum_{\bar{m}\in\Lambda^{r(t)}} \exp\left[-\frac{\pi}{\lambda(t)} \Phi^I_{(\mathrm{mag})}(\bar{m}, \underline{l}(t)) Q^+_{IK}(t) \Phi^K_{(\mathrm{mag})}(\bar{m}, \underline{l}(t))\right], \tag{4.16}$$

where $t \in \overline{-4, 0}$; $r(t) = \operatorname{rank} H_2(M_V^+(t), \mathbb{Z})$; $\overline{m}(t) = \{m^I \in \mathbb{Z} | I \in \overline{1, r(t)}\}$; $\underline{l}(t) = \{l_I \in \mathbb{Z}_{p(I,t)} | I \in \overline{1, r(t)}\}$ (integer numbers $p(I)$ which were calculated in subsection 2.3, see (2.40) , now depend on t). The expression (4.16) gives us a semi-classical approximation since the summing is over all classes of BF-theory classical solutions expressed in terms of magnetic fluxes $\Phi^I_{(\mathrm{mag})}(\bar{m}, \underline{l}(t)) = m^I + Q^{+IJ}(t) l_J(t)$, with Dirac quantization conditions of type (3.7) and (3.8). The collection of integer numbers $\underline{l}(t)$ describes the final boundary conditions on $\mathbb{Z}$-homology sphere $\partial M^+_V(t)$ and have to be fixed.

If a quantum state of the universe is mixed one so the Hartle-Hawking wave function reads as a direct sum

$$\Psi = \bigoplus_{t=-4}^{0} w(t)\Psi(\lambda(t), \underline{l}(t)).$$

This wave function should describe a "multiverse" in the sense of [59], *i.e.* a disjoint union $\bigsqcup_{t=-4}^{0} M^+_V(t)$ of "universes". The corresponding quantum cosmology in the many-world interpretation will be discussed elsewhere.

4.5. Linear Scales of $\mathbb{Z}$-homology Spheres

Now, let us see why manifestations of the presence of the exceptional orbits in $\mathbb{Z}$-homology spheres (which are spatial sections of our cosmological model), could be unobservable by astronomical means. The idea is essentially the same as in the inflation theory: to show that the linear scales of the present-epoch $\mathbb{Z}$-homology sphere $\Sigma^+(0)$, are by many orders of magnitude larger than the characteristic size of the observable part of the universe ($L_0 \sim 10^{28}$cm). To evaluate the universe scales we take the following presumption. Let the four-dimensional volume of universe $M^+(t)$ be proportional to $\det Q^+_{IJ}(t)$ while the "minimal volume" in this universe be proportional to $\det Q^{+IJ}(t)$. Then the universe volume $V^+(t)$ expressed in terms of the "minimal volume" should be $\det Q^+_{IJ}(t)/\det Q^{+IJ}(t) = \left(\det Q^{+IJ}(t)\right)^{-2}$ which yields the expression for the linear size of the universe as

$$L^+(t) \simeq \sqrt[4]{V^+(t)} = 1/\sqrt{\det Q^{+IJ}(t)}.$$

Numerical estimates give the following results:

$$\left.\begin{array}{r} L^+(-4) \sim 1.5 \times 10^4 \\ L^+(-3) \sim 1.5 \times 10^{11} \\ L^+(-2) \sim 8.0 \times 10^{23} \\ L^+(-1) \sim 7.2 \times 10^{48} \\ L^+(0) \sim 6.4 \times 10^{97} \end{array}\right\}. \tag{4.17}$$

As it was mentioned in subsection 4.3, the state of universe corresponding to $t = -3$ may be associated with a supersymmetric unification which includes the gravitational pre-interaction. If linear scales of the universe in this state might be considered as Planckian ones ($L_{Pl} \simeq 1.6 \times 10^{-33}$cm), then the hierarchy (4.17) would be expressed in centimeters:

$$\left.\begin{array}{r} L^+(-4) \sim 1.6 \times 10^{-40} \text{ cm} \\ L^+(-3) \sim 1.6 \times 10^{-33} \text{ cm (normalization)} \\ L^+(-2) \sim 8.6 \times 10^{-21} \text{ cm} \\ L^+(-1) \sim 7.7 \times 10^{4} \text{ cm} \\ L^+(0) \sim 6.8 \times 10^{53} \text{ cm} \end{array}\right\}. \tag{4.18}$$

These estimates give a plausible picture of expansion of the universe in the course of cosmological evolution. Four periods of moderate inflation take place,

$$1.6 \times 10^{-40} \to 1.6 \times 10^{-33} \to 8.6 \times 10^{-21} \to 7.7 \times 10^4 \to 6.8 \times 10^{53},$$

which correspond to the sequence of topology changes (4.14). The size of the universe after the last "inflation" ($\sim 6.8 \times 10^{53}$ cm) occurs to be 25 orders of magnitude greater then the size of its part which is observed now by means of the most sophisticated astronomical devices. These evaluations coincide with those obtained in T_0-discrete cosmological model [57] except for the last one.

Consequently, all that we astronomically observe is a three-dimensional almost flat disk about 10^{28} cm in diameter cut out of the $\mathbb{Z}$-homology sphere whose characteristic size amounts 6.8×10^{53} cm. But while astronomical observations then have nothing to do with spacetime topology, the local experiments providing information about the hierarchy of fundamental interactions (in contrast to ordinary inflation models) tell in our model sufficiently much about non-trivial topological structure of the spacetime.

5. Conclusion

Let us now summarize the basic features of the model proposed in this paper.

- Our model contains a unification scheme of fundamental (pre-)interactions that is based on the primary sequence of Seifert fibered homology spheres (4.1) and on the corresponding pV-manifolds $M_V^+(t)$ and pV-cobordisms $M_D^+(t)$, constructed according to the splice diagrams shown in figure 9. Instead of a symmetry breaking that is utilized in usual gauge unification theories our approach connects the unification principle with the simplification of the spacetime topological structures. Thus, from this point of view the main information is contained no in the group-theoretic structure of field configurations but in the topological complexity of pV-manifolds and pV-cobordisms which model the spacetime relations in a top-down approach to cosmology (à la Hartle-Hawking-Hertog) in terms of BF-model that we have proposed in this article.

- BF systems on pV-cobordisms and pV-manifolds hint that the hierarchy of physical interactions originates at the global level (the utmost topological generalization of the Mach principle), so that the background vacuum (excitations-free) coupling constants naturally occur to coincide with basic topological invariants (intersection matrices) of the spacetime manifold.

- It is clear that the pre-interactions between "electric" fluxes in the framework of Abelian BF-model considered in this paper, cannot comprehensively express specific characteristics of the real fundamental interactions. However our model (in spite of exotic structure of the spacetime manifold or even due to these exotica) heuristically circumscribes certain properties of Nature.

Acknowledgments

We are grateful to Nikolai Saveliev for fruitful discussions in our daily meetings during his visit to the University of Guadalajara organized in the framework of our Proyecto de Posgrado en Ciencias en Física. We thank Gustavo López Velázquez for his interest and stimulating questions.

Appendix

In this appendix we present the rational intersection matrices $Q^{-}(t)$ corresponding to the cobordisms $M_{\bar{D}}^{-}(t)$ obtained from the splice diagrams in figure 9 applying the k^{-}-operation:

$$Q^{-}(0) = \begin{pmatrix} \mathbf{1} & 3.6\times10^{-2} & 0 & 0 & 0 \\ 3.6\times10^{-2} & \mathbf{3.6\times10^{-2}} & 9.4\times10^{-8} & 0 & 0 \\ 0 & 9.4\times10^{-8} & \mathbf{1.9\times10^{-7}} & 2.1\times10^{-24} & 0 \\ 0 & 0 & 2.1\times10^{-24} & \mathbf{1.5\times10^{-23}} & 1.3\times10^{-68} \\ 0 & 0 & 0 & 1.3\times10^{-68} & \mathbf{1.1\times10^{-113}} \end{pmatrix},$$

$$\lambda_{\bar{I}}^{-}(0) = \{1., 3.4\times10^{-2}, 1.9\times10^{-7}, 1.5\times10^{-23}, 8.\times10^{-139}\};$$

$$Q^{-}(-1) = \begin{pmatrix} 8.3\times10^{-2} & 1.1\times10^{-4} & 0 & 0 \\ 1.1\times10^{-4} & 4.3\times10^{-4} & 5.5\times10^{-13} & 0 \\ 0 & 5.5\times10^{-13} & 3.8\times10^{-12} & 3.1\times10^{-35} \\ 0 & 0 & 3.1\times10^{-35} & 2.5\times10^{-58} \end{pmatrix},$$

$$\lambda_{\bar{I}}^{-}(-1) = \{8.3\times10^{-2}, 4.3\times10^{-4}, 3.8\times10^{-12}, 9.8\times10^{-72}\};$$

$$Q^{-}(-2) = \begin{pmatrix} 3.0\times10^{-3} & 1.5\times10^{-7} & 0 \\ 1.5\times10^{-7} & 2.0\times10^{-6} & 1.5\times10^{-18} \\ 0 & 1.5\times10^{-18} & 1.2\times10^{-30} \end{pmatrix},$$

$$\lambda_{\bar{I}}^{-}(-2) = \{3.0\times10^{-3}, 2.0\times10^{-6}, 8.0\times10^{-37}\};$$

$$Q^{-}(-3) = \begin{pmatrix} 2.2\times10^{-6} & 2.1\times10^{-12} \\ 2.1\times10^{-12} & 2.2\times10^{-17} \end{pmatrix},$$

$$\lambda_{\bar{I}}^{-}(-3) = \{2.2\times10^{-6}, 2.0\times10^{-17}\};$$

$$Q^{-}(-4) = (4.5\times10^{-9}).$$

It is easy to note that both diagonal elements of the matrix $Q^{-}(0)$ and its eigenvalues show another hierarchy then that of the DLEC constants of the fundamental interactions (compare the boldface numbers in this matrix with the "experimental" column of the table 2 and with the corresponding characteristics of the matrix $Q^{+}(0)$). So the matrices $Q^{-}(t)$ may be related to some other universe, not with our one.

References

[1] Efremov, V. N.; Mitskievich, N. V.; Hernández Magdaleno, A. M. and Serrano Bautista, R. *Class. Quantum Grav.* 2005, 22, 3725.

[2] Witten, E. *Commun. Math. Phys.* 1988, 117, 325.

[3] Blau, M. and Thompson, G. *Ann. Phys., NY*, 1991, 205, 130.

[4] Horowitz, G. T. *Commun. Math. Phys.* 1989, 125, 417.

[5] Baez, J. in *Proc. 7th Marcel Grossman Meeting on General Relativity*, 1996, (World Scientific; Singapore pp 779-798

[6] Thompson, G.(1995) New Results in Topological Field Theory and Abelian Gauge Theory. arXiv:hep-th/9511038.

[7] Verlinde, E. *Nucl. Phys.*, 1995, B455, 211.

[8] Witten, E. *Selecta Math. (NS)*, 1995, 1, 383.

[9] Alvarez, M. and Olive, D. I. *Commun. Math. Phys.*, 2000, 210, 13.

[10] Alvarez, M. and Olive, D. I. *Commun. Math. Phys.*, 2001, 217, 331.

[11] Alvarez, M. and Olive, D. I. *Commun. Math. Phys.*, 2006, 267, 279.

[12] Zucchini, R. *Commun. Math. Phys.*, 2003, 242, 473.

[13] Zucchini, R. *Adv. Theor. Math. Phys.*, 2005, 8, 895.

[14] Fukumota, Y.; Furuta, M. and Ue, M. *Topology and its Applications*, 2001, 116, 333.

[15] Saveliev, N. *Pacific J. Math.*, 2002, 205, 465.

[16] Nemethi, A. and Nicolaescu, L. I. *Geometry and Topology*, 2002, 6, 269.

[17] Neumann, W. D. and Wahl, J. *Geometry and Topology*, 2005, 9, 757.

[18] Preskill, J. *Nucl. Phys.*, 1989, B323, 141.

[19] Weinberg S 1989 *Rev. Modern Phys.* 61 1

[20] Hawking, S. *Phys. Rev.*, 1988, D37, 904.

[21] Coleman, S. *Nucl. Phys.*, 1988, B307, 864.

[22] Giddings, S. and Strominger, A. *Nucl. Phys.*, 1988, B307, 854.

[23] Klebanov, I.; Susskind, L. and Banks, T. *Nucl. Phys.*, 1989, B317, 665.

[24] Rourke, S. P. and Sanderson, B. J. *Introduction to Piecewise-lineal Topology*; Springer-Verlag: Berlin, New York, 1972.

[25] Schleich K. and Witt D.M. *Nucl. Phys.* 1993, B402, 411.

[26] Kiefer C. *Quantum Gravity;* Oxford Univ. Press: New York, 2004.

[27] Hartle J.B. and Hawking S.W. *Phys. Rev.* 1983, D28, 2960.

[28] Ghoshal D. and Vafa C. *Nucl. Phys.* 1995, B453, 121.

[29] Fintushel, R. and Stern, R, *Ann. Math.*, 1985, 122, 335.

[30] Kachru S. and Silverstein E. *Phys. Rev. Lett.* 1998, 80, 4855.

[31] Hirzebruh, F. *Differentiable Manifolds and Quadratic Forms* ; Marcel Dekker: New York, 1971.

[32] Eisenbud, D. and Neumann, W. *Three-dimensional Link Theory and Invariants of Plane Curve Singularities*; Princeton Univ. Press: Princeton, 1985.

[33] Saveliev, N. *Invariants for Homology 3-Spheres* Springer: Berlin, 2002.

[34] Saveliev, N. *Lectures on the Topology of 3-Manifolds. An Introduction to the Casson Invariant*; Walter de Gruyter: Berlin, 1999.

[35] Neumann, W. *Inventions Math.*, 1977, 42, 285.

[36] Neumann, W. and Raymond, F. *Lect. Notes Math.*, 1978, 664, 163.

[37] Orlik, P. *Lect. Notes Math.*, 1972, 291, 464.

[38] Stake, I, *J. Math. Soc. Japan*, 1957, 9, 464.

[39] Efremov, V.N. *Intern. J. Theor. Phys.*, 1996, 35, 63.

[40] Alvarez, O. *Commun. Math. Phys.*, 1985, 100, 279.

[41] Hilton, P. J. and Wylie, S. *Homology Theory: An Introduction to Algebraic Topology* ; Cambridge Univ. Press: Cambridge, 1960.

[42] Witten, E.(2003) $SL(2,\mathbb{Z})$ action on three-dimensional conformal field theories with Abelian Symmetry. arXiv:hep-th/0307041.

[43] Dijkgraaf, R. (1997) Les Houches Lestures on Fields, Strings and Duality. arXiv:hep-th/9703136

[44] Green, M. B.; Schwarz, J. H. Witten, E. *Superstring Theory*; Cambridge Univ. Press: Cambridge, 1988 Vol.2.

[45] Montonen, C. and Olive, D. *Phys. Lett.*, 1977, 72B, 117.

[46] Wheeler, J. A. *Geometrodynamics*; Academic Press: New York, 1962.

[47] Misner, C. W. and Wheeler, J. A. *Ann. Phys., USA*, 1957, 2, 525.

[48] Efremov, V. N.; Mitskievich, N. V. and Hernández Magdaleno, A. M. *Gravitation and Cosmology*, 2006, 12, 199.

[49] Koike, T.; Tanimoto, M. and Hosoya, A. *J. Math. Phys.*, 1994, 35, 4855.

[50] Koike, T.; Tanimoto, M. and Hosoya, A. *J. Math. Phys.*, 1997, 38, 350.

[51] Fagundes, H. V. *Gen. Rel. Grav.*, 1992, 24, 199.

[52] Scott, P. *Bull. London Math. Soc.*, 1983, 15, 401.

[53] Saveliev, N. *Matemat. Sbornik*, 1992, 183, 125 (in Russian).

[54] Efremov, V. N. and Mitskievich, N.V. In *Quantum Cosmology Research Trends*; Reimer, A.; Ed,; Horizons in World Physics; Nova Science Publishers, Inc.: New York, 2005; Vol. 246, pp 1-48.

[55] Tegmark M.; Aguirre A.; Rees J. and Wilczek F. *Phys. Rev.* 2006, D73, 023505.

[56] Mitskievich, N. V.; Efremov, V. N. and Hernández Magdaleno A. M. (2007) Topological gravitation on graph manifolds *Preprint* arXiv:0706.2736v1 [gr-qc]

[57] Efremov, V. N.; Mitskievich, N. V. and Hernández Magdaleno, A. M. *Gravitation and Cosmology*, 2004, 10, 201.

[58] Hawking S.W. and Hertog T. *Phys. Rev.* 2006, D73, 123527.

[59] Gibbons G.W.; Turok N. (2007). The Measure Problem in Cosmology. arXiv:hep-th/0609095v2.

In: Classical and Quantum Gravity Research
Editors: M.N. Christiansen et al, pp. 309-317
ISBN 978-1-60456-366-5

Chapter 7

SEMICLASSICAL DYNAMICS OF BLACK HOLES

Arundhati Dasgupta
Department of Mathematics and Statistics,
University of New Brunswick

Abstract

We derive semiclassical physics of black holes using coherent states in loop quantum gravity (LQG). We find an explanation for the origin of horizon entropy in this framework by tracing over the wavefunction within the horizon. The coherent state within the horizon is shown to be correlated with the coherent state outside the horizon, and the physics for the outside observer is described by a reduced density matrix which yields the entropy. We then examine the next order quantum fluctuations as measured in the coherent state, and show that information emerges from behind the horizon. This appears to be the origin of Hawking radiation.

1. Introduction

Black hole thermodynamics has been the subject of research for decades and the resolution appears in sight with various approaches to quantum gravity maturing to candidate theories. One such promising approach has been loop quantum gravity (LQG), where the search has been for a ab-initio background independent non-perturbative formulation of gravity. The LQG approach is particularly suited to answer questions of black hole thermodynamics, as black holes are strongly gravitating objects and cannot be obtained in perturbative gravity, or weak gravitational perturbations of flat space. Further, the semi-classical sector of this theory has been obtained, and black hole thermodynamics is essentially semiclassical in origin. The laws of black hole mechanics [1] and the subsequent interpretation of that by Bekenstein, as a set of thermodynamical laws with horizon area as entropy is semiclassical in origin. Hawking confirmed the thermal behaviour of black holes by quantising scalar fields near a black hole horizon, at energy regimes far below the Planck scale, where gravity is classical, but all other fields are quantum. Thus, starting from a quantum theory of gravity, which describes physics at Planck scales, one should be able to take the limit to semi-classical physics, and investigate the origin of entropy and Hawking radiation. The semi-classical sector of a quantum theory is recovered in certain approximations, but a very interesting construct, is the coherent state wavefunction, a 'wavepacket' which assigns the classical configuration maximum probability. The states are peaked both in the momentum and position representations and are minimum uncertainty states. Using this, one can access a classical solution, and obtain quantum fluctuations around the system. I shall describe a LQG coherent state wavefunction for a black hole, and show how the thermodynamics of black holes can be recovered from first principles, and the singularity at the center of the black hole appears to be smoothened. I shall also show how Hawking radiation can emerge from this formulation.

In the next section I describe the coherent states defined in the LQG phase space variables, and describe how a black hole space-time is described by such a system. The next section describes a derivation of the area operator spectrum, and a singularity resolution. The third section describes black hole entropy, and the fourth concludes with a discussion about how the next order corrections from the coherent states, seem to indicate the origin of Hawking radiation.

2. Coherent States

A physical system is described using the position and momentum variables, (x,p), which comprise the phase space-variables of the system. For gravity, the field variables in a canonical slicing of space-time, are identified with the intrinsic metric of the spatial slice q_{ab} and the corresponding momentum, π_{ab}, is a function of the extrinsic curvature with which the slice is embedded in the given space-time. A redefinition of the variables in terms of tangent space densitised triads E_a^I and a corresponding gauge connection A_a^I where I represents the tangent space SO(3) index simplifies the quantisation considerably. The quantisation of the Poisson Algebra of these variables is done by smearing the connection along one dimensional edges e to get holonomies $h_e(A)$ and the triads in a set of 2-surface decomposition of the three dimensional spatial slice to get the corresponding momentum P_e^I. The algebra is then represented in a kinematic 'Hilbert space', in which the physical constraints have been 'formally' realised [2], but are yet to be concretised for a solvable system. Once the phase space variables have been identified, one can write a coherent state for these [3] i.e. minimum uncertainty states peaked at classical values of h_e, P_e^I. Whether these are physical coherent states, or have appropriate behaviour under the action of the constraints has to be examined carefully [4, 5]. The coherent state for one edge is defined to be

$$\psi^t(g_e h_e^{-1}) = \sum_j e^{-tj(j+1)/2} \chi_j(g_e h_e^{-1}) \tag{1}$$

where $\chi_j(g_e h_e^{-1})$ corresponds to the character of the SU(2) element $g_e h_e^{-1}$ in the jth representation. The element g_e corresponds to a complexified classical phase space element $e^{iT^I P_e^{I\text{cl}}/2} h_e^{\text{cl}}$. The coherent state is precisely peaked with maximum probability at the h_e^{cl} for the variable h_e. The fluctuations about the classical value are controlled by the parameter t (the semiclassicality parameter). The coherent state for a entire slice can be obtained by taking the tensor product of the coherent state for each edge.

$$\Psi = \prod_e \psi^t \tag{2}$$

What remains to be understood is what the classical variables g_e mean for a physical space-time. Thus to implement this, one regularises even the classical space time in terms of holonomy and momentum defined for graphs and 2-surface decomposition of the three slices. In [6], this was done for the Schwarzschild black hole, for a particular graph which had the edges along the coordinate lines of a sphere. This simplistic graph, was very useful in obtaining the description of the space-time in terms of discretised holonomy and momenta. A particular interesting consequence of this was that the phase space variables were defined even including the otherwise singular black hole center. However, the graph in actuality was good for description of the space-time near the center, but away at the asymptotics, the density of vertices decreased to almost zero. We however can cure this problem by taking a different graph, which samples vertices uniformly distributed as a function of distance from the center. This is obtained by taking the following example, let us take the planar example, we take the circles at different radii, each a integer distance in terms of a fundamental length δ. Thus we can label the circles at different radii by just integers 1,2,3. We then proceed thus, we put n vertices at circle 1, connecting it to the center. We connect these to n vertices at circle 2. But at cricle 2, we bisect each circular edge with one more vertex, and thus have 2n vertices at circle 2 which will end at circle 3. At circle 3, we further introduce n vertices by bisecting each alternate edge. Thus at circle 3, there are 3n vertices. In this way, the number of vertices will grow with the radius, and the sampled points will be equally dense throughout the space-time. The price to pay for this is of course that the edge lengths are also functions of the radius. Sticking to the original graph taken in [6], this is also the set of basic blocs used for describing classical Schwarzschild black holes in Regge calculus. A classical holonomy along the radial edge of a Schwarzschild black hole with flat slicing is

$$h_{e_r} = \cos\left(\tau'\left\{\frac{1}{r_2^{1/2}} - \frac{1}{r_1^{1/2}}\right\}\right) - \imath\gamma^1 \sin\left(\tau'\left\{\frac{1}{r_2^{1/2}} - \frac{1}{r_1^{1/2}}\right\}\right) \tag{3}$$

where r_1, r_2 are the begining and end points of the edge, and γ^1 is a Pauli Matrix, $\tau' = r_g/2$, where r_g is the Schwarzschild radius for the black hole. Clearly as is evident taking $r_2 \to 0$ coincides with the black hole center, but the holonomy is confined to $-1..1$, which is finite. The momenta $P^I_{e_r}$ will have three independent components, and the 2-surface is centered at r, θ_0 and has extent $2\theta'$ in the radial direction (the ϕ direction has been suppressed) and are listed as:

$$P^1_r = \frac{X_1(r)}{r_g^2} \qquad P^2_r = \frac{X_3}{r_g^2\sqrt{\alpha^2+1}}\left[\alpha\sin\left(\gamma'\alpha^3\right) + \cos\left(\gamma'\alpha^3\right)\right] \tag{4}$$

$$P^3_r = \frac{X_3}{r_g^2\sqrt{\alpha^2+1}}\left[-\alpha\cos\left(\gamma'\alpha^3\right) + \sin\left(\gamma'\alpha^3\right)\right]$$

($\gamma' = \delta/2r_g$ is radial edge length divided by twice the Schwarzschild radius). where

$$X_1(r) = \frac{r_g^2}{\alpha^4}\sin\theta_0\left[\frac{\sin(1-\alpha')\theta'}{1-\alpha'} + \frac{\sin(1+\alpha')\theta'}{(1+\alpha')}\right] \tag{5}$$

$$X_3(r) = \frac{r_g^2}{\alpha^4}\cos\theta_0\left[\frac{\sin(1-\alpha')\theta'}{(1-\alpha')} - \frac{\sin(1+\alpha')\theta'}{(1+\alpha')}\right] \tag{6}$$

$$X(r) = X_1(r)\gamma^1 + X_3(r)\frac{1}{\sqrt{\alpha^2+1}}\gamma^2 + X_3\frac{\alpha}{\sqrt{\alpha^2+1}}\gamma^3 \quad (7)$$

(γ^I are Pauli matrices $\alpha = \sqrt{\frac{r_g}{r}}$ $\alpha' = \sqrt{\frac{\alpha^2+1}{2}}$) The detailed derivation of these can be found in [6]. The complexified phase space-variable is then simply $g_{e_r} = \exp(i\gamma^I P^I/2)h_{e_r}$

Armed with this, we try to investigate the area operator and the curvature operator as a function of these variables, and find their expectation values in the coherent states.

3. Area and the Curvature Operator

Given that the area of a surface in gravity is measured as the integral of the square root of the metric over the surface, the area operator can be written simply as $\hat{A} = \sqrt{P_e^I P_e^I}$. The expectation value of the area operator in the coherent state emerges as

$$< \psi|\hat{A}|\psi >= (j + \frac{1}{2})t \quad (8)$$

Where note, j is a particular eigenvalue of the Casimir. This is really the expectation value of the area operator in a coherent state, and is not a eigenvalue equation. This operator is of course corrected away from the classical limit $t \to 0$, and the coherent state is not a eigenstate of the area operator, What is however, clear is that there exists a minimum of energy, which is infinitesimal in the $t \to 0$, but nevertheless the area of a 'point' would still have a finite extent as measured. This obviously brings us to the question of whether the 'point' singularity at the center of the black hole is smoothened, and this results in a upper bound to the curvature operator. To verify this we measure the curvature operator as regularised in terms of the loop quantum gravity phase space variables. For this we take the Kretschmann scalar $R^2 = R_{\mu\nu\tau\sigma}R^{\mu\nu\tau\sigma}$, and since we work in the particular slicing of the black hole where the intrinsic curvature of the space-time is flat, this operator has contribution from the extrinsic curvature of the slice.

$$\sqrt{g}R^2 = 2\sqrt{q}\left[K^4 - K_{ab}K_{cd}K^{ac}K^{bd}\right] \quad (9)$$

In [7], I chose a regularisation of the extrinsic curvature in terms of the holonomy and the volume operator as given below

$$K_{ab} = \frac{C_{e_b}}{2(e_a(0) - e_a(1))}\mathrm{Tr}\left[h_{e_b}^{-1}\left[h_{e_b}, V\right]h_{e_a}\right] \quad (10)$$

As pointed out in [7], the operator ordering ambiguities would be order (t^2) corrections to the expression for the expectation value of the numbers. In equation (10), one takes $e_a(0,1)$ to be the begining and end point of the edges, and C_{e_b} is due to the volume operator, and fixed by edge lengths etc. The expectation value

$$< \psi|\sqrt{g}R^2|\psi >\propto \frac{V}{e^3 r_g}\frac{1}{t^{10}} \quad (11)$$

Where V is the volume of the region e is the typical edge length, and r_g is the Schwarzschild radius. Clearly for large black holes, the upper bound for the curvature is obtained, and for zero t the classical singularity re-appears as it should.

4. Entropy

The coherent state is defined for a graph which when embedded in the classical space-time comprises of edges along the coordinate lines of spherically symmetric axis (r, θ, ϕ). The apparent horizon equation $\nabla_a S^a - K_{ab} S^a S^b - K = 0$ (S^a is the normal to the apparent horizon,K_{ab} the extrinsic curvature of the spatial surface and K, the trace of the extrinsic curvature) is re-written in the regularised 'variables', and encodes correlation across the horizon [7]. Written in terms of the holonomies at a vertex v_O outside the horizon, and v_I inside the horizon, one obtaines

$$4P_{e_\theta}^2 \left[\mathrm{Tr}\left(T^J h_{e_\theta}^{-1} V^{1/2} h_{e_\theta}\right)_{v_O} - \mathrm{Tr}\left(T^J h_{e_\theta}^{-1} V^{1/2} h_{e_\theta}\right)_{v_I} \right] \mathrm{Tr}\left(T^J h_{e_\theta}^{-1} V^{1/2} h_{e_\theta}\right)_{v_O}$$
$$- \frac{1}{\sqrt{\beta}} \frac{\partial}{\partial \beta} \mathrm{Tr}\left(T^I \,{}^{\beta} h_{e_\theta}\right) P_{e_\theta\ v_O}^I = 0 \tag{12}$$

e_θ denotes a edge along the coordinate lines of θ, β denotes the Immirzi parameter, h_{e_θ} and $P_{e_\theta}^I$ denote the holonomy and momenta along the edge e_θ, one set for a edge beginning/ending at vetex v_O another set for a edge beginning/ending at vertex v_I. Symbolically it has the following form

$$\hat{A}[\hat{B}^I(v_O) - \hat{B}^I(v_I)]\hat{C}^I(v_O) - \hat{D} = 0 \tag{13}$$

where $\hat{A}, \hat{B}, \hat{C}, \hat{D}$ are operators given as functions of h_e, P_e^I of the angular edges which begin/end at a vertex v_O outside the horizon and those angular edges which begin/end at a vertex inside the horizon. In the coherent states this can be written as

$$\mathrm{Limit}_{t \to 0} < \Psi | \hat{H} | \Psi >= 0, \tag{14}$$

The expectation value of the horizon operator is thus vanishing only in the semi-classical limit, realised as $t \to 0$. where $\hat{H} = \hat{A}\hat{B}^I(v_O)\hat{C}^I(v_I) - \hat{A}\hat{B}^I(v_O)\hat{C}^I(v_I) - \hat{D}$.

Note that this set of equations show that the coherent state wavefunction which is a function of the classical holonomy and the momentum within the horizon must be correlated with the coherent state outside the horizon. The correlation was encoded in a conditional probability function to derive what the zeroeth order entropy should be, but the actual computation of the correlations and the density function should be obtained using physical coherent states, a first step towards which has been taken in a recent paper by Thiemann et al [5].

In the above, the graph at the horizon is taken to simply comprise of radial edges linking vertices inside the horizon to those outside the horizon. Thus the coherent state at the horizon is written thus

$$|\Psi >= \prod f(v_O, v_I) |\psi_{v_O} > |\psi_H > |\psi_{v_I} > \tag{15}$$

where $|\psi_{v_{(0,I)}} >$ represent coherent states for edges inside and outside the horizon, and $|\psi_H >$ is the coherent state for the horizon radial edge linking the inside vertices with the outside ones. A tracing over of the edges inside the horizon (or the set of coherent states in the above tensor product state labelled by v_I), gives a mixed density matrix ρ, which in the

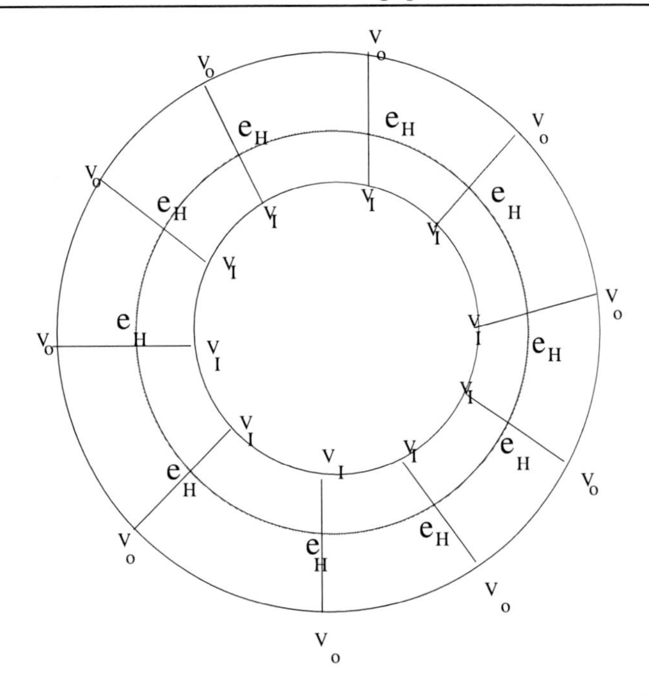

classical limit is diagonal [7]. The entropy obtained as $-\mathrm{Tr}(\rho\ln\rho)$ of this density matrix, is the number of ways to induce the horizon area as the degeneracy due to the horizon state $|\psi_H>$ contributes to the trace (The degeneracy of a horizon state which induces a area $(j_e+1/2)t$ is $2j_e+1$. Thus given a set of radial edges inducing the horizon with area, one simply obtaines the degeneracy associated with the coherent state for those edges, given the total area of the horizon, and a log of that gives the entropy.
Having only one graph, with the edges each carrying spin j_e, with the area equally distributed on the horizon sphere one can count the entropy as simply by summing the degeneracy $(2j_e+1)$ associated with each edge. The constraint is that if N is the number of edges, $(j_e+1/2)N=\frac{A_H}{l_p^2}$. The log of the degeneracy is thus

$$S_{BH}=\frac{A_H}{l_p^2}\ln(2j_e+1) \tag{16}$$

The entropy calculation is exact, and there are no corrections to the Bekenstein-Hawking term.

In the situation one fixes the number of edges a priori, to be a number N and these edges are allowed to be distributed asymmetrically, i.e. the spins j_e of the edges need not be equal, the constraint is $\sum_{j_e}(j_e+1/2)=\frac{A_H}{l_p^2}$ the entropy is

$$S_{BH}=\frac{A_H}{l_p^2}\left(\frac{3}{2}\ln 3-\ln 2\right)-\frac{1}{2}\ln(\frac{A_H}{l_p^2})+.. \tag{17}$$

As seen above, the log area correction appears here with the coefficient $1/2$, and this type of correction has been obtained in other derivations of entropy [10]. However as we show below, this is not unique, and entropy corrections will differ if the number of edges is allowed to vary.

We then generalised the case of one graph coherent state to the sum over graphs [9] 'generalised coherent state'. Keeping the spherical symmetry in place, the graph at the horizon can vary as per (i) the number of edges crossing the horizon (ii) the distribution of the edges across the horizon. These different graphs are labeled as 'minimal graphs', as one graph *cannot* be obtained by subdividing the edges of the other graph. The generalised LQG Hilbert space can be written as a direct sum of Hilbert spaces corresponding to each minimal graph.

$$H = \oplus_{\Gamma} H_{\Gamma} \tag{18}$$

In a sum over graphs situation, where the number of edges is not fixed a priori, and the (j_e) are arbitrary, subject only to the constraint that $\sum_{je}(j_e + 1/2) = A_H/l_p^2$, the entropy is Bekenstein-Hawking with corrections. Two different answers are obtained, as per the two restrictions in the tracing procedure to obtain the generalised density matrix. (i)The generalised density matrix is a tensor sum over the density matrices for each Hilbert space, the entropy is given by

$$S_{BH} = (2\frac{A_H}{l_p^2} - 1)\ln 2 + \exp(-(2\frac{A_H}{l_p^2} - 1)\ln 2)\ln(\frac{A_H}{l_p^2}) \tag{19}$$

where, the entropy is Bekenstein-Hawking with corrections. As the area of the horizon increases, the corrections decrease, and the leading term indeed is a log area correction term. However, the complete correction, is decreasing *exponentially in area.*
(ii)The second situation arises, when the entropy is obtained, from a 'generalised coherent state' which is a superposition of all the orthogonal graph Hilbert, coherent states. This superposition allows for transition from one graph-Hilbert space to another. The entropy in this case is

$$S_{BH} = \ln\left(\frac{1}{\sqrt{5}}\left(\frac{3+\sqrt{5}}{2}\right)^{2A_H/l_p^2} - \left(\frac{3-\sqrt{5}}{2}\right)^{2A_H/l_p^2}\right) \tag{20}$$

For large areas the leading term is indeed Bekenstein-Hawking, and the corrections are all exponentially decreasing in area.

$$S_{BH} = 2\frac{A_H}{l_p^2}\ln\left(\frac{3+\sqrt{5}}{2}\right) + (6.854)^{-2A_H/l_p^2} + . \tag{21}$$

Thus, even semiclassically, the different ways of counting give different corrections,though the leading term is universally acknowledged to be Bekenstein-Hawking term. In particular for this derivation, the entropy is different, as per the 'coherent state' used to describe the same classical space-time.
Thus the correction terms and the proportionality constant (which is fixed to $1/4l_p^2$ using the Immirzi parameter, and can be done in this formalism also) in the entropy counting remain ambiguous. A experimental verification of the correction term will determine the specific 'generalised coherent state' the system is in.

5. Quantum Fluctuations

In this section we examine the consequences of quantum fluctuations on the density matrix. We then interpret the results in terms of information emerging from behind the horizon. The density matrix as was defined in [7] is obtained from the wavefunction which is written in the tensor product of Hilbert spaces of edges outside the horizon, at the horizon and within the horizon. The exact form of that is written in the area operator eigenstate basis $|jmn>$ for each edge.

$$|\Psi>=\sum \psi_{\{j_o\},\{j_H\},\{j_I\}}|\{j_o\}>|\{j_H\}>|\{j_I\}> \quad (22)$$

where $\psi_{\{j_o\},\{j_H\},\{j_I\}}$ are the correlated coefficients in this basis (This is just a re-writing of (15) in the area eigenstate basis).

The resultant density matrix from that is obtained as

$$\rho = |\Psi>< \Psi| \quad (23)$$

and the reduced density matrix is obtained by tracing over the internal $|\{j_I\}>$, which in the limit $t \to 0$ has diagonal components dominating. However taking the next order corrections, one finds off-diagonal terms easily obtained from $\sum_{\{j_I\}} \psi^*_{\{j_o'\}\{j_H'\}\{j_I\}}\psi_{\{j_o\}\{j_H'\}\{j_I\}}$. These off-diagonal terms are responsible for 'decoherence' in ordinary quantum systems. These terms are also the 'evolution' terms as they give a non-trivial commutator of the density matrix with the Hamiltonian. This is under investigation, and it is interesting to find why including the next order fluctuation implies emergence of information from behind the horizon. At this order the apparent horizon equation gets corrected [11].

$$<\psi|\hat{H}|\psi>= t\, F(h(v_O, v_I), P(v_O, v_I)) + O(t^2) \quad (24)$$

where F is a function of the classical holonomy and momenta of edges linked at both the vertices v_O (outside the horizon) and v_I (vertices inside the horizon). Thus, information about the phase space variables will start to emerge from behind the horizon. These fluctuations can be included in the classical Hamiltonian, and the density matrix evolved in time. Such time evolutions usually lead to a thermalisation of the density matrix, and hence we are discovering the semi-classical origin of Hawking radiation.

References

[1] J. Bardeen, B. Carter S. W. Hawking *Comm. Math. Phys.* 31, 161 (1973).

[2] B. Dittrich and T. Thiemann *Testing the Master Constraint Program for Loop Quantum Gravity: I-IV* arXiv: gr-qc/0411138-42

[3] B. Hall, *J. Func. Anal.* 122 103 (1994).

[4] T. Thiemann, *Class. Quan. Grav.* 18 3293 (2001).
H. Sahlmann, T. Thiemann, O. *Winkler Nucl. Phys. B* 606 401 (2001).

[5] T. Thiemann, B. Bahr *Gauge invariant coherent states for loop quantum gravity II: Non-abelian gauge gauge groups* arXiv: gr-qc/0709.4636.

[6] A. Dasgupta, *JCAP* 0308 004 (2004).

[7] A. Dasgupta, *Class. Quant. Grav.* 23 635 (2006).

[8] C. Rovelli, L. *Smolin Nucl. Phys. B* 442 4070 (1990).

[9] A. Dasgupta, H. Thomas, *Generalised Coherent States and Combinatorics of Horizon Entropy* arXiv: gr-qc/0602006.

[10] A. Ashtekar, J. Baez, K. Krasnov, A. Corichi, *Phys. Rev. Lett.* 80 904 (1998). R. K. Kaul, P. Majumdar, *Phys. Rev. Lett.* **84** 5255 (2000), S. Das, P. Majumdar, R. Bhaduri, *Class. Quant. Grav.* **19** 2355 (2002).

[11] A. Dasgupta, *Semiclassical Horizons* to be published in CJP arXiv: 0711.0714 [gr-qc]

In: Classical and Quantum Gravity Research
Editors: M.N. Christiansen et al, pp. 319-370
ISBN 978-1-60456-366-5

Chapter 8

Black Hole Evaporation as a Nonequilibrium Process

Hiromi Saida*
Department of Physics, Daido Institute of Technology,
Minami-ku, Nagoya 457-8530, Japan

Abstract

According to the black hole thermodynamics, a black hole itself is regarded as a self-gravitating system being in thermal equilibrium state of Hawking temperature. If the outside environment around black hole is cooler than the Hawking temperature, then the black hole loses its mass energy. This is the black hole evaporation. When a black hole evaporates, there arises a net energy flow from black hole into its outside environment due to the Hawking radiation and the energy accretion onto black hole. The existence of energy flow means that the black hole evaporation is a nonequilibrium process: Look at each moment during the evaporation process of a black hole whose horizon scale is larger than the Planck scale. Then it is recognized that, although the black hole itself is regarded as in an (quasi-)equilibrium state, the outside environment is in a nonequilibrium state whose nonequilibrium nature arises by the net energy flow. Therefore, to study the detail of evaporation process, nonequilibrium effects of the net energy flow in outside environment should be taken into account. The nonequilibrium nature of black hole evaporation is a challenging topic which includes not only black hole physics but also nonequilibrium physics.

In this chapter we simplify the situation so that the Hawking radiation consists of non-self-interacting massless matter fields and also the energy accretion onto the black hole consists of the same fields. Then the nonequilibrium nature of black hole evaporation is described by a nonequilibrium state of that field. Hence we formulate nonequilibrium thermodynamics of non-self-interacting massless fields. Then, by applying it to black hole evaporation, followings are shown: (1) Nonequilibrium effects of the energy flow tends to accelerate the black hole evaporation, and, consequently, a specific nonequilibrium phenomenon of semi-classical black hole evaporation is suggested. Furthermore a suggestion about the end state of quantum size black hole evaporation is proposed in the context of information loss paradox. (2) Negative heat capacity of

*E-mail address: saida@daido-it.ac.jp

black hole is the physical essence of the generalized second law of black hole thermodynamics, and self-entropy production inside the matter around black hole is not necessary to ensure the generalized second law. Furthermore a lower bound for total entropy at the end of black hole evaporation is given.

1. Introduction

Black hole evaporation is one of interesting phenomena in black hole physics [1]. A direct treatment of time evolution of the evaporation process suffers from mathematical and conceptual difficulties; the mathematical one will be seen in the dynamical Einstein equation in which the source of gravity may be a quantum expectation value of stress-energy tensor of Hawking radiation, and the conceptual one will be seen in the definition of dynamical black hole horizon. Therefore an approach based on the black hole thermodynamics is useful [2, 3].

Exactly speaking, dynamical evolution of any system is a nonequilibrium process. If and only if thermodynamic state of the system under consideration passes near equilibrium states during its evolution, its dynamics can be treated by an approximate method, the so-called *quasi-static process*. In this approximation, it is assumed that the thermodynamic state of the system evolves on a path lying in the state space which consists of only thermal equilibrium states, and the time evolution is described by a succession of different equilibrium states. However once the system comes far from equilibrium, the quasi-static approximation breaks down. In that case a nonequilibrium thermodynamic approach is necessary. For dissipative systems, the heat flow inside the system can quantify the degree of nonequilibrium nature [4, 5, 6, 7].

For the black hole evaporation, when its horizon scale is larger than Planck size, it is relevant to describe the black hole itself by equilibrium solutions of Einstein equation, Schwarzschild, Reissner-Nortström and Kerr black holes, because the evaporation proceeds extremely slowly and those equilibrium solutions are stable under gravitational perturbations [8]. The slow evolution is understandable by the Hawking temperature [1] which is regarded as an equilibrium temperature of black hole,

$$T_g := \frac{m_{pl}^2}{8\pi M}, \tag{1}$$

where M is the black hole mass, m_{pl} is the Planck mass and the units $c = \hbar = k_B = 1$ and $G = 1/m_{pl}^2$ are used. Obviously a classical size black hole ($M \gg m_{pl}$) has a very low temperature. This means a very weak energy emission rate by the Hawking radiation which is proportional to $(2GM)^2 T_g^4$ due to the Stefan-Boltzmann law. Therefore the quasi-static approximation works well for the black hole itself during its evaporation process. However the outside environment around black hole may not be described by the quasi-static approximation because of the energy flow due to the Hawking radiation. The Hawking radiation causes an energy flow in the outside environment, and that energy flow drives the outside environment out of equilibrium. As indicated by eq.(1), the black hole temperature and the energy emission rate by black hole increase as M decreases along the evaporation. The stronger the energy emission, the more distant from equilibrium the outside environment. Therefore the nonequilibrium nature of the outside environment becomes stronger as the black hole evaporation proceeds. At the same time, the quasi-static approximation is applicable to the black hole itself since equilibrium black hole solutions are stable under gravitational perturbation. Hence, in studying detail of evaporation process, while the black hole itself is described by quasi-static approximation, but the nonequilibrium effects of the energy flow in the outside environment should be taken into account.

In the above paragraph, the energy accretion onto black hole is ignored. However if the temperature of outside environment is non-zero and lower enough than the black hole temperature, then the black hole evaporates under the effect of energy exchange due to the Hawking radiation and the energy accretion. In this case the same consideration explained above holds and we recognize the importance of the net energy flow from black hole to outside environment. Dynamical behaviors of black hole evaporation will be described well by taking nonequilibrium nature of the net energy flow into account.

In section 2., we introduce a simple model of black hole evaporation to examine the net energy flow in the outside environment, where the matter fields of Hawking radiation and energy accretion are represented by non-self-interacting massless fields for simplicity. Section 3. is devoted to construction of nonequilibrium thermodynamics of that field. Then it is applied to the black hole evaporation. Section 4. reveals that the nonequilibrium effect tends to accelerate the evaporation process and, consequently, gives a suggestion about the end state of quantum size black hole evaporation in the context of the information loss paradox. Section 5. reveals that the generalized second law is guaranteed not by self-interactions of matter fields around black hole which cause self-production of entropy inside the matters, but by the self-gravitational effect of black hole appearing as its negative heat capacity in eq.(5). Readers can read sections 4. and 5. separately, and may skip over section 4. to see discussions on generalized second law in section 5.. Finally section 6. concludes this chapter with comments for future direction of this study.

Throughout this chapter except for eq.(1), Planck units $c = \hbar = G = k_B = 1$ are used. Then the Stefan-Boltzmann constant becomes $\sigma := \pi^2/60$ which is appropriate for photon gas. When one consider non-self-interacting massless matter fields, as indicated in section 3., it is necessary to replace σ by its generalization,

$$\sigma' := \frac{N\,\pi^2}{120} = \frac{N}{2}\,\sigma\,, \tag{2}$$

where $N := n_b + (7/8)\,n_f$. Here n_b is the number of inner states of massless bosonic fields and n_f is that of massless fermionic fields. ($n_b = 2$ for photons.) Furthermore, at least when the black hole temperature is lower than 1 TeV (upper limit by present accelerator experiments), it is appropriate to estimate the order of N by the standard particles (inner states of quarks, leptons and gauge particles of four fundamental interactions),

$$N \simeq 100\,. \tag{3}$$

This denotes $\sigma' \simeq 10$. Throughout this chapter, we simply assume that N is independent of black hole temperature and this estimate (3) holds always for semi-classical black hole evaporation ($T_g < 10^{16}$ TeV).

2. Thermodynamic Model of Black Hole Evaporation

According to the black hole thermodynamics [1, 2, 3], a stationary black hole is regarded as an object in thermal equilibrium, a black body. For simplicity, let us consider a Schwarzschild black hole. Its equations of states as a black body are

$$E_g = \frac{1}{8\pi T_g} = \frac{R_g}{2} \quad , \quad S_g = \frac{1}{16\pi T_g^2} = \pi R_g^2 \, , \tag{4}$$

where E_g, R_g, T_g, S_g correspond respectively to mass energy, horizon radius, Hawking temperature and Bekenstein-Hawking entropy. Obviously the radius R_g decreases when this body loses its energy E_g. The black hole evaporation is represented by the energy loss of this black body.

The heat capacity C_g of this body is negative,

$$C_g := \frac{dE_g}{dT_g} = -\frac{1}{8\pi T_g^2} = -2\pi R_g^2 < 0 \, . \tag{5}$$

The negative heat capacity is a peculiar property of self-gravitating systems [9]. Therefore the energy E_g includes self-gravitational effects of a black hole on its own thermodynamic state. Furthermore it has already been revealed that, using the Euclidean path-integral method for a black hole spacetime and matter fields on it, an equilibrium entropy of whole gravitational field on a black hole spacetime is given by the Bekenstein-Hawking entropy [10]. This means S_g in eq.(4) is the equilibrium entropy of whole gravitational field on black hole spacetime, and the gravitational entropy vanishes if there is no black hole horizon. Hence we find that energetic and entropic properties of a black hole are encoded in the equations of states (4). Hereafter we call this black body *the black hole*.

As mentioned section 1., the nonequilibrium nature of black hole evaporation arises in the matter fields around black hole due to the net energy flow by Hawking radiation and energy accretion onto the black hole. When we consider arbitrary dissipative matter fields as the Hawking radiation and energy accretion, we immediately face a very difficult problem how to construct a nonequilibrium thermodynamics for arbitrary dissipative matters. This is one of the most difficult subjects in physics [4, 5, 6, 7]. To avoid such a difficult problem and for simplicity, let us consider non-self-interacting massless fields to represent the Hawking radiation and energy accretion. For example, photon, graviton, neutrino (if it is massless) and free Klein-Gordon field ($\Box\Phi = 0$) are candidates of such matter fields, and they possess the generalized Stefan-Boltzmann constant σ' given in eq.(2). Hereafter we call these fields *the radiation fields*.

As mentioned above, nonequilibrium phenomenon is one of the most difficult subjects in physics. It is impossible at present to treat the nonequilibrium nature of black hole evaporation in a full general relativistic framework. Hence we resort to a simplified model to examine the nonequilibrium effects of net energy flow in the outside environment around black hole [11]:

Nonequilibrium Evaporation (NE) model: Put a spherical black body of temperature T_g in a heat bath of temperature $T_h (< T_g)$, where the equations of states of the spherical black body is given by eq.(4) and we call the black body *the black hole*. Let the

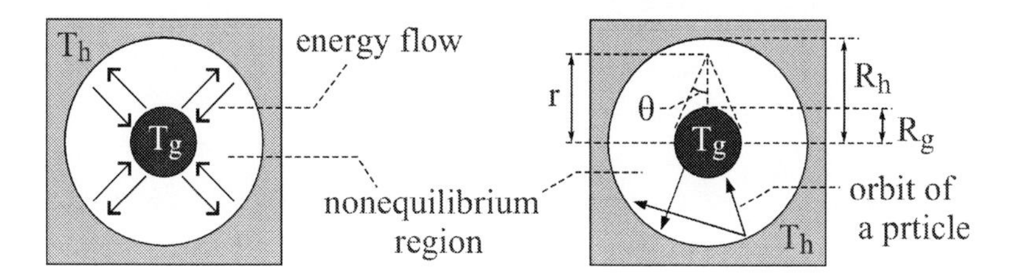

Figure 1. NE model. Left panel shows energy flow between black hole and heat bath. Right one shows some variables and particle orbits of radiation fields. The radiation fields sandwiched by black bodies are in two-temperature steady state.

> heat bath (the outer black body of temperature T_h) be made of ordinary materials of positive heat capacity. Then hollow a spherical region out of the heat bath around black hole as seen in fig.1. The hollow region is a shell-like region which is concentric with the black hole and separates the black hole from the heat bath. This region is filled with matter fields emitted by black hole and heat bath. Those matter fields are the non-self-interacting massless fields possessing generalized Stefan-Boltzmann constant σ' given in eq.(2), and we call these fields *the radiation fields*. This model consists of three parts, black hole, heat bath and radiation fields. Furthermore we consider the case that the whole system is isolated and the total energy of the three parts is conserved.

The isolated condition reflects the whole universe including an evaporating black hole. The temperature difference ($T_g > T_h$) causes a net energy flow from the black hole to the heat bath. This energy flow drives a relaxation process of the whole system because of the isolated condition. For ordinary systems of positive heat capacity, relaxation process reaches an equilibrium state at the end of the process. But the relaxation in NE model may not reach a total equilibrium state of the whole system, because the temperature difference $T_g - T_h$ may increase along the decrease of E_g due to the negative heat capacity C_g given in eq.(5). This causes the decrease of R_g due to the equations of states (4). Therefore the relaxation process in NE model corresponds to the black hole evaporation.

Here let us comment about terminology. There may be an objection that the term "relaxation" is not suitable for the case of increasing temperature difference. But in this chapter, please understand it means the time evolution arising in isolated inhomogeneous systems.

If a full general relativistic treatment is possible, the Hawking radiation experiences the curvature scattering to form a spacetime region filled with interacting matters and some fraction of Hawking radiation is radiated back to the black hole from that region. The heat bath in the NE model is understood as a simple representation of not only matters like accretion disk but also such region formed by curvature scattering.

Here, as an objection to the NE model, one may remember the so-called *Tolman factor* which appears in the "equilibrium" temperature of radiation fields around black hole: Exactly speaking, equilibrium of any matter fields around black hole is "local" equilibrium. It has already been known that, when the radiation fields around black hole are in local equilibrium with the black hole, the local equilibrium temperature $T_{eq}(r)$ of radiation fields (not of black hole) at a spacetime point of areal radius r from the center of black hole is $T_{eq}(r) = T_g/\sqrt{1 - R_g/r}$, where T_g is the Hawking temperature and R_g is the horizon radius. This $T_{eq}(r)$ is obtained in a full general relativistic framework of equilibrium thermodynamics, and the factor $1/\sqrt{1 - R_g/r}$ is the Tolman factor [12]. One may think that, because $T_{eq} \to \infty$ as $r \to R_g$, it is unreasonable to assign T_g to black hole as its temperature. But let us emphasize that $T_{eq}(r)$ is not black hole temperature but the equilibrium temperature of radiation fields. We can understand $T_{eq}(r)$ as follows: In order to retain equilibrium of radiation fields against external gravitational force by black hole, a temperature of radiation fields higher than the asymptotic value T_g is required, since the higher temperature denotes the higher pressure against external gravitational force. The Tolman factor describes the effect of external gravity on the equilibrium radiation fields, and becomes unity if the external gravity vanishes. The local equilibrium temperature of radiation fields may count an "intrinsic" temperature of black hole and an additional gravitational effect in Tolman factor. It may be reasonable to regard the asymptotic value T_g as an intrinsic black hole temperature. Hence we assign T_g to the black hole in NE model. But it is ture that the NE model is not a full general relativistic model, and ignoring, for example, gravitational redshift and curvature scattering on the "nonequilibrium" radiation fields propagating in the hollow region. Although the NE model may be too simple, let us try to investigate nonequilibrium effects of the net energy flow from black hole to its outside environment in the framework of NE model.

Furthermore, we put the quasi-equilibrium assumption to utilize the "equilibrium" equations of states (4) for black hole, and the fast propagation assumption to treat the nonequilibrium nature as simple as possible:

Quasi-equilibrium assumption: Time evolution in the NE model is not so fast that the evolutions of black hole and heat bath are approximated well by the quasi-static process individually, while the quasi-static approximation is not valid for the radiation fields due to the net energy flow from black hole to heat bath. Then, it is reasonable to use eq.(4) as the equations of states for black hole. Furthermore, since Schwarzschild black hole is not a quantum one, following relation is required,

$$R_g \gtrsim 1 \,. \tag{6}$$

Fast propagation assumption: The volume of hollow region is not so large that particles of radiation fields travel very quickly across the hollow region. Then the retarded effect on radiation fields during propagating in the hollow region is ignored.

There are two points which we should note here. The first point is about the thermodynamic states of black hole and heat bath under the quasi-equilibrium assumption. This denotes the temperatures T_g and T_h are given by equilibrium temperatures at each moment of evaporation process. Therefore we can regard T_g and T_h as constants within a time scale that one particle of radiation fields travels in the hollow region until absorbed by black hole or heat bath. This is consistent with the fast propagation assumption.

The second point is about the thermodynamic state of radiation fields. In the hollow region, the radiation fields of different temperatures T_g and T_h are simply superposed, since the radiation fields are of non-self-interacting (collisionless particles gas). This means the radiation fields are in a two-temperature nonequilibrium state. Furthermore, because T_g and T_h are constant while one particle of radiation fields travel across the hollow region, it is reasonable to consider that the radiation fields have a stationary energy flow from black hole to heat bath within that time scale. Hence, at each moment of time evolution of NE model, the thermodynamic state of radiation fields is well approximated to a macroscopically stationary nonequilibrium state, which we call *the steady state* hereafter. Consequently, time evolution of radiation fields is described by *the quasi-steady process* in which the thermodynamic state of radiation fields evolves on a path lying in the state space which consists of steady states, and the time evolution is described by a succession of different steady states. Therefore we need a thermodynamic formalism of two-temperature steady states for radiation fields. The steady state thermodynamics for radiation fields has already been formulated in [13], which is summarized in next section.

3. Steady State Thermodynamics for Radiation Fields

Before proceeding to the black hole evaporation, two-temperature steady state thermodynamics for radiation fields [13] is explained in this section. Subsection 3.1. introduces the minimum tools required to apply to the NE model. Readers may skip over subsection 3.2. to read remaining sections 4., 5. and 6.. Furthermore, since sections 4. and 5. are written separately, one can also skip over section 4. to see the generalized second law in section 5..

Subsection 3.2. exhibits a more detail of two-temperature steady state thermodynamics for radiation fields. Although a full understanding of it is not necessary for black hole evaporation, but it may be helpful to understand, for example, the free streaming in the universe like cosmic microwave background and/or the radiative energy transfer inside a star and among stellar objects. Keeping future expectation of such applications in mind, subsection 3.2. is placed here.

3.1. Minimum Tools for Applying to NE Model

To concentrate on investigating two-temperature steady states of radiation fields, we consider a model which can be realized in laboratory experiments. According to the NE model, we introduce the following model named SST after the Steady State Thermodynamics. Then, after constructing the steady state thermodynamics for radiation fields, we will modify the SST model to the NE model in sections 4. and 5..

SST model: Make a vacuum cavity in a large black body of temperature T_{out} and put an another smaller black body of temperature T_{in} ($\neq T_{out}$) in the cavity as seen in fig.2. For the case $T_{in} > T_{out}$, the radiation fields emitted by two black bodies causes a net energy flow from the inner black body to the outer one. When the outer black body is isolated from outside world and the heat capacities of two black bodies are positive definite, the whole system which consists of two black bodies and radiation fields between them relaxes to a total equilibrium state in which two black bodies and radiation fields have the same equilibrium temperature.

It should be emphasized that T_{in} and T_{out} are "equilibrium" temperatures of black bodies, respectively. Outer body is in an equilibrium state, and inner one is also. But their

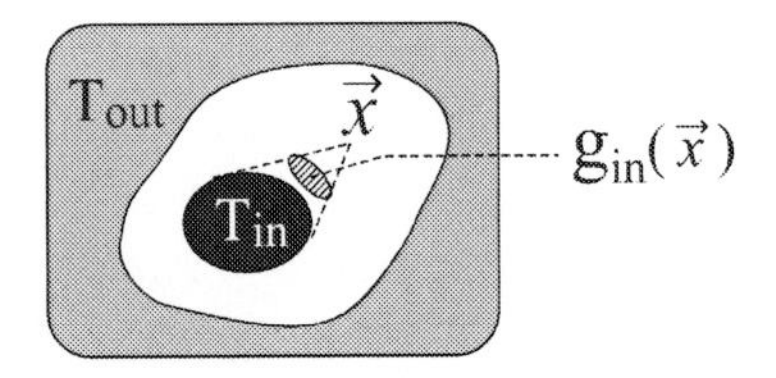

Figure 2. SST model. Radiation fields between outer and inner black bodies are in steady state.

equilibrium sates are different from each other since $T_{in} \neq T_{out}$. Then the radiation fields between them is in a nonequilibrium state. By keeping temperatures T_{in} and T_{out} constant, the nonequilibrium state of radiation fields becomes a macroscopically stationary nonequilibrium state, which we call a steady state.

The difference of SST model from the NE model is that heat capacities of two black bodies are positive definite and the system does not necessarily has symmetric geometry. Because of the positive heat capacity, the relaxation process in the SST model leads the whole system to a total equilibrium state, while the black hole evaporation increases the temperature difference between black hole and heat bath due to the negative heat capacity of black hole. But as for the NE model, the radiation fields in the SST model are also non-self-interacting massless fields. The SST model ignores gravitational interactions among two bodies and radiation fields. Furthermore we consider the case satisfying following two conditions according to the quasi-equilibrium and fast propagation assumptions; the first is that the evolution of each black body is of quasi-static during the relaxation process of the whole system, and the second condition is that the volume of cavity is so small that retarded effect on radiation fields is ignored. Then, as discussed section 2., the evolution of radiation fields is of quasi-steady. Due to the Stefan-Boltzmann law, each steady state composing the quasi-steady process of the radiation fields has a net energy flow

$$J_{sst} = \sigma \left(T_{in}^4 - T_{out}^4 \right) A_{in} \,, \tag{7}$$

where A_{in} is the surface area of inner black body. This J_{sst} equals the net energy exchanged par a unit time between the two black bodies via the radiation fields.

A consistent thermodynamic framework for the steady states of radiation fields has already been constructed [13]. The outline of the construction of steady state internal energy and steady state entropy is as follows.

For the first we consider a massless bosonic gas. The energy of the gas E_b is given by

$$E_b := \int \frac{dp^3}{(2\,\pi)^3} dx^3 \; g_{\vec{p},\vec{x}} \; \epsilon_{\vec{p},\vec{x}} \; d_{\vec{p},\vec{x}} \,, \tag{8}$$

where $\vec{p}$ is a momentum of particle, $\vec{x}$ is a spatial point, $\epsilon_{\vec{p},\vec{x}}$ is an energy of particle of $\vec{p}$ at $\vec{x}$, and $g_{\vec{p},\vec{x}}$ and $d_{\vec{p},\vec{x}}$ are respectively the number of states and the average number of particles at a point $(\vec{p}, \vec{x})$ in the phase space of the gas. For the entropy, we refer to the Landau-Lifshitz type definition of nonequilibrium entropy S_b for bosonic gas (see §55 in chapter 5 of [14]),

$$S_b := \int \frac{dp^3}{(2\,\pi)^3} dx^3\, g_{\vec{p},\vec{x}} \left[\, \left(1 + d_{\vec{p},\vec{x}} \right) \ln \left(1 + d_{\vec{p},\vec{x}} \right) - d_{\vec{p},\vec{x}} \ln d_{\vec{p},\vec{x}} \right] . \tag{9}$$

It has also been shown in §55 of [14] that the maximization of S_b for an isolated system ($\delta S_b = 0$) gives the equilibrium Bose distribution. This is frequently referred in many works on nonequilibrium systems as the *H-theorem*. However in §55 of [14], concrete forms of $g_{\vec{p},\vec{x}}$ and $d_{\vec{p},\vec{x}}$ are not specified, since an arbitrary system is considered.

In the SST model, the radiation fields are sandwiched by two black bodies, and its particle is massless and "collisionless". Then we can determine $\epsilon_{\vec{p},\vec{x}}$, $g_{\vec{p},\vec{x}}$ and $d_{\vec{p},\vec{x}}$ to be

$$\epsilon_{\vec{p},\vec{x}} = \omega \quad , \quad g_{\vec{p},\vec{x}} = n_b \quad , \quad d_{\vec{p},\vec{x}} = \frac{1}{\exp[\omega/T(\vec{p},\vec{x})] - 1} , \tag{10}$$

where the frequency of a particle $\omega = |\vec{p}|$, n_b is the number of inner states of bosonic gas which is assumed to be constant as mentioned in eq.(3), and $T(\vec{p},\vec{x})$ is given by

$$T(\vec{p},\vec{x}) := \begin{cases} T_{in} & \text{for } \vec{p} = \vec{p}_{in} \text{ at } \vec{x} \\ T_{out} & \text{for } \vec{p} = \vec{p}_{out} \text{ at } \vec{x} \end{cases} , \tag{11}$$

where $\vec{p}_{in}$ is momentum of a particle emitted by the inner black body and $\vec{p}_{out}$ emitted by the outer one. The $\vec{x}$-dependence of $T(\vec{p},\vec{x})$ arises, because the directions in which particles of $\vec{p}_{in}$ and $\vec{p}_{out}$ can come to a point $\vec{x}$ vary from point to point.

From the above, we obtain the steady state energy and entropy,

$$E_b = \int dx^3 e_b(\vec{x}) \quad , \quad e_b(\vec{x}) := 4\,\sigma_b \left(g_{in}(\vec{x})\, T_{in}^4 + g_{out}(\vec{x})\, T_{out}^4 \right) \tag{12a}$$

$$S_b = \int dx^3 s_b(\vec{x}) \quad , \quad s_b(\vec{x}) := \frac{16\,\sigma_b}{3} \left(g_{in}(\vec{x})\, T_{in}^3 + g_{out}(\vec{x})\, T_{out}^3 \right) , \tag{12b}$$

where integrals in eq.(17) are used, $\sigma_b := n_b \pi^2/120$, $g_{in}(\vec{x})$ is the solid angle (divided by 4π) covered by directions of $\vec{p}_{in}$ at $\vec{x}$ as shown in fig.2, and $g_{out}(\vec{x})$ is similarly defined with $\vec{p}_{out}$. By definition, we have $g_{in}(\vec{x}) + g_{out}(\vec{x}) \equiv 1$. On the other hand, if the radiation fields are in equilibrium of temperature T, the ordinary equilibrium thermodynamics determine equilibrium energy density e and entropy density s to be $e = 4\sigma_b T^4$ and $s = (16\sigma_b/3)\, T^3$. We find that the steady state energy and entropy are given by a simple linear combination of the values which are calculated as if the radiation fields are in equilibrium of temperature T_{in} or T_{out}. This is consistent with the collisionless nature of radiation fields.

Next we consider a massless fermionic gas, the formula of energy (8) holds for fermions as well. But the formula of entropy (9) is replaced by [14]

$$S_f := -\int \frac{dp^3}{(2\,\pi)^3} dx^3\, g_{\vec{p},\vec{x}} \left[\, \left(1 - d_{\vec{p},\vec{x}} \right) \ln \left(1 - d_{\vec{p},\vec{x}} \right) + d_{\vec{p},\vec{x}} \ln d_{\vec{p},\vec{x}} \right] . \tag{13}$$

Then, following the same procedure as for the massless bosonic gas with replacing the distribution function by $d_{\vec{p},\vec{x}} = \left[\exp[\omega/T(\vec{p},\vec{x})] + 1\right]^{-1}$, we can obtain the steady state energy and entropy of a massless fermionic gas,

$$E_f = \frac{7}{8} \frac{n_f}{n_b} E_b \quad , \quad S_f = \frac{7}{8} \frac{n_f}{n_b} S_b , \tag{14}$$

where integrals in eq.(17) are used, and n_f is the number of inner states of fermionic gas which is assumed to be constant as mentioned in eq.(3). Hence, when the black bodies in SST model emit n_f massless fermionic modes and n_b massless bosonic modes, the steady state energy E_{sst} and entropy S_{sst} of radiation fields are given by eqs.(12) with replacing σ_b by σ' given in eq.(2),

$$E_{sst} = \int dx^3 e_{sst}(\vec{x}) \quad , \quad e_{sst}(\vec{x}) := 4\,\sigma' \left(g_{in}(\vec{x})\, T_{in}^4 + g_{out}(\vec{x})\, T_{out}^4 \right) \tag{15a}$$

$$S_{sst} = \int dx^3 s_{sst}(\vec{x}) \quad , \quad s_{sst}(\vec{x}) := \frac{16\,\sigma'}{3} \left(g_{in}(\vec{x})\, T_{in}^3 + g_{out}(\vec{x})\, T_{out}^3 \right) , \tag{15b}$$

At equilibrium limit $T_{in} = T_{out}$, these densities e_{sst} and s_{sst} become equilibrium ones $e = 4\sigma' T^4$ and $s = (16\sigma/3)T^3$ respectively, where an identical equation $g_{in}(\vec{x}) + g_{out}(\vec{x}) \equiv 1$ is used.

By defining the other variables like free energy with somewhat careful discussions, it has already been checked that the 0th, 1st, 2nd and 3rd laws of ordinary equilibrium thermodynamics are extended to include two-temperature steady states of radiation fields [13]. This means that the steady state thermodynamics for radiation fields has already been constructed in a consistent way. Especially on the steady state entropy, the total entropy of the whole system which consists of two black bodies and radiation fields increases monotonously during the relaxation process of the whole system,

$$dS_{in} + dS_{out} + dS_{sst} \geq 0 \,, \tag{16}$$

where the equality holds for total equilibrium states, S_{in} and S_{out} are entropies of two black bodies which are given by ordinary equilibrium thermodynamics, and a simple sum of total entropy $S_{in} + S_{out} + S_{sst}$ is assumed for not only equilibrium states but also steady states. This indicates that the second law holds for steady states of radiation fields.

Here for the convenience to follow eqs.(12) and (14), we list useful formulae;

$$\begin{aligned}
&\int_0^\infty dx \frac{x^3}{e^x - 1} = \frac{\pi^4}{15} && \int_0^\infty dx \frac{x^3}{e^x + 1} = \frac{7\pi^4}{120} \\
&\int_0^\infty dx \frac{x^2}{1 - e^{-x}} \ln(1 - e^{-x}) = -\frac{\pi^4}{36} && \int_0^\infty dx \frac{x^2}{1 + e^{-x}} \ln(1 + e^{-x}) = \frac{11\pi^4}{180} - F \\
&\int_0^\infty dx \frac{x^2}{e^x - 1} \ln(e^x - 1) = \frac{11\pi^4}{180} && \int_0^\infty dx \frac{x^2}{e^x + 1} \ln(e^x + 1) = \frac{\pi^4}{60} + F
\end{aligned} \tag{17}$$

where $F = [(\ln 2)^2 - \pi^2]\,(\ln 2)^2/6 + 4\,\phi(4, 1/2) + (7 \ln 2/2)\,\zeta(3)$. Here $\zeta(z)$ is zeta function and $\phi(z, s)$ is modified zeta function (Apell's function).

3.2. A More Detail

This subsection exhibits a more detail of two-temperature steady state thermodynamics for radiation fields. A peculiar point of radiation fields is the collisionless nature of composite particles. Radiation fields are non-dissipative matter. A representative of them is a photon gas. Before a construction of steady state thermodynamics for radiation fields, there were some existing works on nonequilibrium radiation fields.

A traditional treatment of radiative energy transfer, for example that in a star [9], has been applied to a mixture of radiation fields (photon gas) with a matter like a dense gas or other continuum medium. In such a traditional treatment, successive absorption and emission of photons by components of medium matter makes it possible to consider radiation fields as if in local equilibrium states whose temperatures equal those of local equilibrium states of medium matter. However, when the radiation fields are in "vacuum" space, the idea of local equilibrium is never applicable to radiation fields because they are non-dissipative (see §63 in [14]).

Here let us look at ordinary dissipative systems very briefly. For ordinary dissipative systems, the so-called *extended irreversible thermodynamics* treats successfully nonequilibrium states [4]. However this is applicable not to any highly nonequilibrium state but to a state whose entropy flux is well approximated up to second order in the expansion by the heat flux of the nonequilibrium state. On the other hand, a steady state thermodynamics has already been suggested for dissipative systems like heat conduction, shear flow, electrical conduction and so on [5]. Also a heat flux appears as a consistent state variable in those steady state thermodynamics. However the range of its application is limited [6, 7]. Although present nonequilibrium thermodynamics for dissipative systems have some restriction on their applicability, the point is that a heat flux plays the role of consistent state variable which quantifies a degree of nonequilibrium nature of ordinary dissipative systems. When one deals with nonequilibrium dissipative systems, it is usual to place an interest on the heat flux.

Notable works on nonequilibrium radiation fields in "vacuum" are given by Essex [15]. Here it may be helpful to point out that a "heat" arises by dissipation. No heat flux exists in nonequilibrium radiation fields, but an energy flux in radiation fields may correspond to a heat flux in dissipative systems. It was natural that Essex considered an energy flux in nonequilibrium radiation fields (photon gas) in vacuum. Essex has shown that, contrary to the success in extended irreversible thermodynamics, the entropy flux for nonequilibrium radiation fields is NOT expressed by the expansion by energy flux of those radiation fields. Even if the same method of extended irreversible thermodynamics is applied to nonequilibrium radiation fields, the energy flux becomes inconsistent with the nonequilibrium free energy. This inconsistency will be looked over in eq.(24) later in this section. This means the energy flux does not work as a consistent state variable for nonequilibrium radiation fields in vacuum.

Apart from Essex's works, there were other works on nonequilibrium radiation fields (photon gas) in vacuum emitted by nonequilibrium ordinary matters [16]. They used the *information theory*. The basis of information theory is the assumption that nonequilibrium entropy is given by $\ln d_{ne}$, where d_{ne} is a nonequilibrium distribution function defined case by case according to the system under consideration (see for example [4] in which the information theory is also explained). Applying the information theory to the total system composed by nonequilibrium radiation fields and its source matter, complicated distribution functions for general nonequilibrium radiation fields and source matter have been suggested in [16]. However, after those works were reported, it is revealed in [7] that, at least for a matter whose components are colliding and interacting with each other, the distribution function for a steady state of that matter derived by the information theory does NOT qualitatively agree with that derived by a steady state Boltzmann equation. Furthermore it is also concluded in [7] that the nonequilibrium temperature of that matter determined by the information theory has no physical meaning. The information theory does not always work well. Therefore, because the suggested distribution function of nonequilibrium radiation fields depends on nonequilibrium temperature of source matter derived by the information theory, the reliability of the distribution function may not be given in [16]. There is no confirmed form of distribution function of nonequilibrium radiation fields emitted by nonequilibrium matter. Hence, to avoid the difficult problem on nonequilibrium state of source matter, we simply assume in the SST model that the source bodies are in equilibrium states.

From the above, we recognize the following three facts: (1) The traditional treatment of radiative transfer is applicable only to a mixture of radiation fields with a matter which is dense enough to ignore the vacuum region among components of the matter. (2) Energy flux is not a consistent state variable for nonequilibrium radiation fields in vacuum, and therefore a special approach different from that to dissipative systems is required to understand nonequilibrium radiation fields. (3) A consistent thermodynamic formulation for a system including nonequilibrium radiation fields in vacuum has not been accomplished, and therefore a consistent noncquilibrium order parameter for radiation fields has not been obtained so far.

At least for two-temperature steady states of radiation fields in vacuum, a thermodynamic formalism is accomplished and a consistent nonequilibrium order parameter for steady states is obtained in [13] in the framework of SST model. Exactly speaking, the radiation fields in SST model is in *local steady states*, because the distribution function in eq.(10) and its fermion version have $\vec{x}$-dependence. The radiation fields in a sufficiently small region are in a local steady state, but that local steady state may be different from a local steady state in the other small region. Therefore state variables for SST model should be defined as a function of spatial point $\vec{x}$. The extensive variable is to be understood as a density.

Let us exhibit consistent steady state variables for radiation fields. See [13] for detail discussions to justify the following definitions of state variables.

Steady state internal energy density $e_{sst}(\vec{x})$ and entropy density $s_{sst}(\vec{x})$: These are already defined in eq.(15).

Steady state Pressure tensor P^{ij}_{sst} (in 3-dim. space): One may naively expect that the pressure of "steady" state is a global quantity, since a pressure gradient in an ordinary dissipative system accelerates components of that system to cause a dynamical evolution. However this is not true of radiation fields whose particle is "collisionless". As seen below, P^{ij}_{sst} becomes a function of $\vec{x}$ because of $\vec{x}$-dependence of distribution function $d_{\vec{p},\vec{x}}$.

In general, pressure is defined by the momentum flux, the amount of momentum carried by composite particles par unit area and unit time. For equilibrium states, momentum flux is homogeneous and isotropic, then equilibrium pressure becomes a scalar quantity. However for nonequilibrium states, momentum flux is not homogeneous and/or isotropic, then the pressure should be defined as a tensor,

$$P^{ij}_{sst}(\vec{x}) := N \int dp^3 \, \frac{1}{p} \, p^i \, p^j \, d_{\vec{p},\vec{x}} \, , \tag{18}$$

where p and p^i are, respectively, spatial magnitude and components of momentum of a particle of radiation fields. Trace of P^{ij}_{sst} becomes

$$\frac{1}{3}\Sigma_i P^{ii}_{sst}(\vec{x}) = \frac{4\sigma'}{3}\left(g_{in}(\vec{x})\,T_{in}^4 + g_{out}(\vec{x})\,T_{out}^4\right) . \tag{19}$$

At equilibrium limit $T_{in} = T_{out}$, this trace becomes equilibrium pressure, $(4\sigma'/3)T^4$, where an identical equation $g_{in}(\vec{x}) + g_{out}(\vec{x}) \equiv 1$ is used.

Steady state free energy density f_{sst} : By a requirement that the differential of free energy by volume gives the minus of pressure as for ordinary equilibrium thermodynamics, we can obtain

$$f_{sst}(\vec{x}) := -\frac{1}{3}\Sigma_i P^{ii}_{sst} = -\frac{4\sigma'}{3}\left(g_{in}(\vec{x})\,T_{in}^4 + g_{out}(\vec{x})\,T_{out}^4\right) = -\frac{1}{3}e_{sst}(\vec{x}) . \tag{20}$$

At equilibrium limit $T_{in} = T_{out}$, this becomes equilibrium free energy, $-(4\sigma'/3)T^4$.

Steady state chemical potential : Chemical potential in general can be interpreted as a work needed to add one particle to the system under consideration. Because particles of radiation fields are collisionless, no work is needed to add a new one into radiation fields. This is the case for either equilibrium or steady states. Indeed, the chemical potential of radiation fields (photon gas) in equilibrium disappears. Therefore the steady state chemical potential of radiation fields disappears as well.

Steady state temperature $T_{sst}(\vec{x})$: By a requirement that the differential of f_{sst} by T_{sst} gives the minus of entropy density ($\partial f_{sst}/\partial T_{sst} = -s_{sst}$) as for equilibrium ordinary thermodynamics, we can obtain

$$T_{rad}(\vec{x}) := g_{in}(\vec{x})\,T_{in} + g_{out}(\vec{x})\,T_{out} . \tag{21}$$

At equilibrium limit $T_{in} = T_{out}$, this becomes equilibrium temperature.

Intensive steady state order parameter τ : Energy flux $\vec{j}(\vec{x})$ is defined by

$$\vec{j}(\vec{x}) := j\,\vec{n} , \tag{22}$$

where $\vec{n}$ is a unit vector in the direction of total momentum of particles at $\vec{x}$, and

$$j := N \int dp^3 \omega\, d_{\vec{p},\vec{x}} \cos\phi , \tag{23}$$

where $\omega = p$ is the energy (frequency) of a particle of radiation fields, and ϕ is the angle between $\vec{n}$ and $\vec{p}$. If $\vec{j}(\vec{x})$ is adopted as an intensive steady state order parameter, its conjugate state variable should also be defined well. Such a conjugate variable would be defined by the differential of free energy by $j(\vec{x})$. However it has shown in [13] such a conjugate variable vanishes,

$$\frac{\partial f_{sst}}{\partial j} \equiv 0 . \tag{24}$$

This means $\vec{j}(\vec{x})$ is not a consistent state variable, since its thermodynamic conjugate variable does not exist.
Hence, instead of energy flux, we adopt the temperature difference as an intensive steady state order parameter,

$$\tau := T_{in} - T_{out} \,. \tag{25}$$

This is obviously intensive variable and satisfies a natural requirement $\tau = 0$ at equilibrium limit $T_{in} = T_{out}$. This τ is consistent with the first law and the concavity of free energy as looked over in eqs.(28) and (29).

Extensive steady state order parameter density $\psi(\vec{x})$: After ordinary equilibrium thermodynamics, we define an extensive steady state order parameter as a thermodynamic conjugate variable to τ using the differential of free energy density. Hence we define as

$$\psi(\vec{x}) := -\frac{\partial f_{sst}}{\partial \tau} = \frac{16\sigma'}{3}\, g_{in}(\vec{x})\, g_{out}(\vec{x})\, \left(T_{in}^3 - T_{out}^3\right) \,. \tag{26}$$

This satisfies a natural requirement $\psi = 0$ at equilibrium limit $T_{in} = T_{out}$.
It may be useful to rewrite $\psi(\vec{x})$ as

$$\psi(\vec{x}) = g_{in}(\vec{x})\, g_{out}(\vec{x})\, \left(s_{eq}(T_{in}) - s_{eq}(T_{out})\right) \,, \tag{27}$$

where $s_{eq}(T) = (16\sigma'/3)T^3$ is the equilibrium entropy density of radiation fields of temperature T. It seems very natural and reasonable that a difference of entropies quantifies a degree of nonequilibrium nature of steady states.

From zeroth to third laws and concavity of free energy : Zeroth law, the existence of steady states, is the existence of systems which realize steady state radiation fields as shown in fig.2.
First law can be checked from the above definitions of state variables. We can obtain the following equation,

$$de_{sst}(\vec{x})\Big|_{\vec{x}=\text{fixed}} = T_{sst}(\vec{x})\, ds_{sst}(\vec{x}) + \tau\, d\psi(\vec{x})\Big|_{\vec{x}=\text{fixed}} \,, \tag{28}$$

where we fixed $\vec{x}$ since a local steady state is considered, and consequently a "work term" including pressure does not explicitly appear in this relation. A work term will appear if the above equation is integrated over the hollow region in SST model. Eq.(28) denotes the first law.
Second law is satisfied as mentioned in eq.(16).
Third law is satisfied by definition of $T_{sst}(\vec{x})$, if the third law of ordinary equilibrium thermodynamics holds for inner and outer black bodies.
Furthermore, as for ordinary equilibrium thermodynamics, we can find the free energy density is concave with intensive state variables,

$$\frac{\partial^2 f_{sst}}{\partial T_{sst}^2} \le 0 \quad , \quad \frac{\partial^2 f_{sst}}{\partial \tau^2} \le 0 \,. \tag{29}$$

From the above, we have obtained a consistent two-temperature steady state thermodynamics for radiation fields.

4. Black Hole Evaporation with Energy Accretion

Now we apply the steady state thermodynamics for radiation fields to the NE model. Contents of this section are based on [11]. This section is not necessary to read next section 5.. Readers interested in the generalized second law may skip over this section.

4.1. From SST to NE Model

The NE model is obtained from the SST model by setting the system spherically symmetric and assigning eq.(4) to the inner black body as its equations of states. Then the inner black body is regarded as a black hole, and the net energy flow from black hole to heat bath via radiation fields causes the black hole evaporation.

Before proceeding to the NE model, let us review here about existing works. In the framework of ordinary equilibrium thermodynamics, equilibrium states of black hole in a heat bath have already been investigated. It has already been revealed that an equilibrium state of the total system composed of a black hole and a heat bath is unstable for a sufficiently small black hole and stable for a sufficiently large black hole [17, 18]. If the instability occurs for a small black hole and the system starts to evolve towards the other stable state, there are two possibilities of its evolution: The first is that, due to the statistical (and/or quantum) fluctuation, the temperature of heat bath exceeds that of black hole and a net energy flow into black hole arises. Then the black hole swallows a part of heat bath and settles down to a stable equilibrium state of a larger black hole in heat bath. The second possible evolution is that, due to the statistical (and/or quantum) fluctuation, the temperature of heat bath becomes lower than that of black hole and a net energy flow from black hole arises. Then the black hole evaporates and settles down to some other stable state. However we do not know the detail of end state of the second possibility, since the final fate of black hole evaporation is an unresolved issue at present.

When one distinguishes the phase of the equilibrium system by a criterion whether a black hole can exist stably in an equilibrium with a heat bath or not, the phase transition of the system occurs in varying the black hole radius. This phenomenon is known as *the black hole phase transition* [17, 18]. So far cosmological constant is not considered. But the black hole phase transition has also been found for asymptotically anti-de Sitter black holes, which is known as *Hawking-Page phase transition* [19]. However, we do not consider cosmological constant throughout this chapter.

The black hole phase transition is one of interesting issues in black hole thermodynamics. However this section concentrates on a black hole evaporation in a heat bath after an instability of equilibrium occurs. We investigate a detail of black hole evaporation in the framework of NE model and try to extract some insight into the final fate of black hole evaporation. Readers interested also in the black hole phase transition may see [11] in which its equilibrium and nonequilibrium versions are also discussed.

4.2. Energy Transport in the NE Model

We discuss energetics of NE model. Total energy of the whole system is

$$E_{tot} := E_g + E_h + E_{rad} , \tag{30}$$

where E_g is the energy of black hole given in eq.(4), E_h is the energy of heat bath defined by ordinary thermodynamics, and E_{rad} is the steady state energy of radiation fields given in eq.(15a),

$$E_{rad} = 4\sigma' \left(G_g T_g^4 + G_h T_h^4 \right) , \tag{31}$$

where

$$G_g := \int_{V_{rad}} dx^3 \, g_g(\vec{x}) \quad , \quad G_h := \int_{V_{rad}} dx^3 \, g_h(\vec{x}) \, , \tag{32}$$

where $g_g(\vec{x})$ is the solid angle (divided by 4π) covered by directions of particles which are emitted by black hole and come to a point $\vec{x}$ (see figs.1 and 2), and $g_h(\vec{x})$ is defined similarly by particles emitted by heat bath. By definition $g_g(\vec{x}) + g_h(\vec{x}) \equiv 1$ holds, and consequently $V_{rad} := G_g + G_h$ gives the volume of hollow region. Furthermore, since the black hole is concentric with the hollow region, we obtain $g_g(\vec{x}) = (1 - \cos\theta)/2$ and $g_h(\vec{x}) = (1 + \cos\theta)/2$, where θ is the zenith angle which covers the black hole at a point of radial distance r (see right panel in fig.1). Then G_g and G_h are expressed as

$$G_g = \frac{2\pi}{3} \left[R_h^3 - R_g^3 - \left(R_h^2 - R_g^2 \right)^{3/2} \right] \tag{33a}$$

$$G_h = \frac{2\pi}{3} \left[R_h^3 - R_g^3 + \left(R_h^2 - R_g^2 \right)^{3/2} \right] , \tag{33b}$$

where R_h is the outermost radius of hollow region (see right panel in fig.1).

To understand the energy flow in NE model, we divide the whole system into two sub-systems X and Y as follows: Sub-system X is composed of the black hole and the "out-going" radiation fields emitted by black hole, and sub-system Y is composed of the heat bath and the "in-going" radiation fields emitted by heat bath (see left panel in fig.1). The sub-system X is a combined system of components of NE model which share the temperature T_g, and Y is that which share the temperature T_h. Then the total energy is expressed as

$$E_{tot} = E_X + E_Y , \tag{34}$$

where E_X and E_Y are respectively the energies of sub-systems X and Y,

$$E_X = E_g + E_{rad}^{(g)} \quad , \quad E_{rad}^{(g)} := 4\sigma' G_g T_g^4 \tag{35a}$$

$$E_Y = E_h + E_{rad}^{(h)} \quad , \quad E_{rad}^{(h)} := 4\sigma' G_h T_h^4 , \tag{35b}$$

where $E_{rad}^{(g)}$ and $E_{rad}^{(h)}$ are respectively the energies of out-going and in-going radiation fields. It is easily found that E_X has no T_h-dependence, while E_Y has T_g- and T_h-dependence. The energy flow in NE model can be understood as an energy transport between sub-systems X and Y. Because this energy transport is carried by the out-going and in-going radiation fields, the Stefan-Boltzmann law works well to give an explicit expression of energy transport,

$$\frac{dE_X}{dt} = -\sigma' \left(T_g^4 - T_h^4 \right) A_g \tag{36a}$$

$$\frac{dE_Y}{dt} = \sigma' \left(T_g^4 - T_h^4 \right) A_g , \tag{36b}$$

where $A_g = 4\pi R_g^2$ is the surface area of black hole, and t is a time coordinate which corresponds to a proper time of a rest observer at asymptotically flat region if we can extend the NE model to a full general relativistic model. Because some particles emitted by heat bath are not absorbed by the black hole but return to the heat bath (see right panel in fig.1), the effective surface area through which Y exchanges energy with X is equal to the surface area of black hole. Therefore A_g appears in eq.(36b). Furthermore eqs.(36) are formulated to be consistent with the isolated setting of the NE model, $E_{tot} \equiv constant$.

It is useful to rewrite the energy transport (36) to a more convenient form for later discussions. By eqs.(5), (33) and (35), we find eq.(36) becomes

$$C_X \frac{dT_g}{dt} = -J \quad , \quad C_X C_Y^{(h)} \frac{dT_h}{dt} = \left(C_X + C_Y^{(g)} \right) J \, , \tag{37}$$

where

$$J := \sigma' \left(T_g^4 - T_h^4 \right) A_g \, , \tag{38}$$

and

$$C_X := \frac{dE_X}{dT_g} = C_g + C_{rad}^{(g)} \tag{39a}$$

$$C_{rad}^{(g)} := \frac{dE_{rad}^{(g)}}{dT_g} = 16\,\sigma' G_g T_g^3 + \frac{\sigma'}{2\,\pi} \left(R_g - \sqrt{R_h^2 - R_g^2} \right) T_g \tag{39b}$$

$$C_Y^{(g)} := \frac{\partial E_Y}{\partial T_g} = \frac{\sigma'}{2\,\pi} \left(R_g + \sqrt{R_h^2 - R_g^2} \right) \frac{T_h^4}{T_g^3} \tag{39c}$$

$$C_Y^{(h)} := \frac{\partial E_Y}{\partial T_h} = C_h + 16\,\sigma' G_h T_h^3 \tag{39d}$$

$$C_h := \frac{dE_h}{dT_h} \, (> 0) \, , \tag{39e}$$

where $C_g = -2\pi R_g^2$ is given in eq.(5) and it is assumed for simplicity that $R_h \equiv constant$ and E_h depends on T_h but not on T_g. C_h is the heat capacity of heat bath, and we assume $C_h \equiv constant > 0$ for simplicity. $C_Y^{(g)}$ is the heat capacity of sub-system Y under the change of T_g with fixing T_h, and $C_Y^{(h)}$ is that under the change of T_h with fixing T_g. $C_{rad}^{(g)}$ is the heat capacity of out-going radiation fields, and C_X is the heat capacity of sub-system X. In analyzing the nonlinear differential equations (37), behaviors of various heat capacities (39) are used. Some useful properties of these heat capacities are explained in next subsection 4.3..

From the above, we find that an inequality $C_X + C_Y^{(g)} < 0$ has to hold in order to guarantee the validity of NE model. To understand this requirement, consider the case $T_g > T_h$ for the first. Due to the temperature difference, energy flows from black hole to heat bath via radiation fields, $dE_g < 0$ and $dE_h > 0$. Then $dT_g > 0$ and $dT_h > 0$ hold due to $C_g < 0$ and $C_h > 0$. Recall that the whole system is isolated, $E_{tot} \equiv constant$, which means $(C_X + C_Y^{(g)})\, dT_g + C_Y^{(h)}\, dT_h = 0$. Therefore, because of $C_Y^{(h)} > 0$ by definition, it is concluded that the inequality $C_X + C_Y^{(g)} < 0$ must hold. And an inequality $C_X < 0$ follows immediately due to $C_Y^{(g)} > 0$ by definition. The similar discussion holds for the case $T_g < T_h$, and gives the same inequality. Hence the following inequality must hold in the framework of NE model,

$$C_X + C_Y^{(g)} < 0 \quad (\Rightarrow C_X < 0)\,. \tag{40}$$

This inequality is the condition which guarantees the validity of NE model. A more detailed property of the combined heat capacity $C_X + C_Y^{(g)}$ is explained in next subsection 4.3.4., which shows that inequality (40) can hold for a sufficiently small T_h. Therefore we assume T_h is small enough so that the inequality (40) holds.

Concerning the validity of NE model, what the quasi-equilibrium assumption implies is important. This assumption requires the time evolution is not so fast. Therefore, when a black hole evaporates, the shrinkage speed of black hole surface is less than unity,

$$v := \left| \frac{dR_g}{dt} \right| < 1\,. \tag{41}$$

This inequality is also the condition which guarantees the validity of NE model. Hence, in the framework of NE model, our analysis should be restricted within the situations satisfying conditions (40) and (41).

It is helpful for later discussions to consider what a violation of validity conditions (40) and (41) denotes. Firstly consider if condition (40) is not satisfied. Then the system, especially the radiation fields, can never be described with steady state thermodynamics. The radiation fields are neither equilibrium nor steady (stationary nonequilibrium). This means that the radiation fields should be in a highly nonequilibrium dynamical state. Furthermore the quasi-equilibrium assumption is violated, because it is this assumption that lead us to utilize the steady state thermodynamics. Therefore highly nonequilibrium radiation fields make a black hole dynamical, and the black hole can not be treated by equilibrium solutions of Einstein equation. Next consider if condition (41) is not satisfied. Then the black hole evolves so fast that the quasi-equilibrium assumption is violated. The black hole can not be described by equilibrium solutions of Einstein equation, but described by some unknown dynamical solution. Therefore, because the source of radiation fields becomes dynamical, radiation fields evolve into a highly nonequilibrium dynamical state and the steady state thermodynamics is not applicable. Hence, when one of the conditions (40) or (41) is violated, the system evolves into a highly nonequilibrium dynamical state which can not be treated in the framework of NE model.

4.3. Properties of Some Heat Capacities

This subsection summarizes some properties of various heat capacities (39) which we will use in remaining subsections.

4.3.1. $C_X(R_g)$ as a Function of R_g

Here we show a behavior of heat capacity C_X of sub-system X as a function of black hole radius R_g. By eqs.(4) and (33), $C_X(R_g)$ is rewritten into the following form,

$$C_X(x) = -2\pi R_h^2\, x^2 + \frac{\sigma'}{8\pi^2}\, f(x)\,, \tag{42}$$

where $x := R_g/R_h$ and

$$f(x) := \frac{4}{3}\frac{1}{x^3}\left[1 - x^3 - \left(1 - x^2\right)^{3/2}\right] + \frac{1}{x}\left(x - \sqrt{1 - x^2}\right) . \tag{43}$$

By definition, $0 < R_g < R_h$, i.e., $0 < x < 1$, and we find

$$\begin{cases} C_X \to \infty & \text{as } x \to 0 \\ C_X = -2\pi R_h^2 + \dfrac{\sigma'}{8\pi^2} & \text{at } x = 1 \end{cases} . \tag{44}$$

By eq.(3) and eq.(6) which denotes $R_h > 1$, we find $C_X(x = 1) < 0$ holds for the NE model.

The differential of $C_X(x)$ is

$$\frac{dC_X(x)}{dx} = -4\pi R_h^2\, x + \frac{\sigma'}{8\pi^2}\frac{df(x)}{dx} , \tag{45}$$

where

$$\frac{df(x)}{dx} = \frac{4 - 3\,x^2 - 4\,\sqrt{1 - x^2}}{x^4\,\sqrt{1 - x^2}} \quad \Rightarrow \quad \begin{cases} \dfrac{df(x)}{dx} < 0 & \text{for } x < \dfrac{2\sqrt{2}}{3} \\ \dfrac{df(x)}{dx} > 0 & \text{for } x > \dfrac{2\sqrt{2}}{3} \end{cases} , \tag{46}$$

where $2\sqrt{2}/3 \simeq 0.943$. A schematic graph of $C_X(R_g)$ is shown in fig.3 (left panel), where $\tilde{R}_g$ is the solution of $C_X = 0$,

$$\tilde{R}_g := R_h\,\tilde{x} \quad , \quad \tilde{x} := \{x \,|\, C_X(x) = 0\} . \tag{47}$$

Because of $C_X(x = 1) < 0$, equation $C_X = 0$ has only one solution.

Finally, we estimate the value of $\tilde{x}$. Since $x < 1$ by definition, we apply the Taylor expansion, $(1 - x^2)^\alpha = 1 - \alpha\, x^2 + O(x^4)$, to eq.(42) and obtain

$$C_X(x) = -2\pi R_h^2 \left[x^2 - \varepsilon\,\frac{1}{x}\left(1 - \frac{x}{3} + O(x^2)\right)\right] , \tag{48}$$

where $\varepsilon = \sigma'/16\pi^3 R_h^2 = N/1920\pi R_h^2$. By eq.(3) and eq.(6) which denotes $R_h > 1$, we find $\varepsilon < 1$. Then $C_X(x)$ becomes

$$C_X(x) = -2\pi R_h^2\,\frac{1}{x}\left(x^3 - \varepsilon + \varepsilon\, O(x)\right) . \tag{49}$$

Hence, by taking leading terms of x and ε, we find an approximate expression for $\tilde{x}$ as

$$\tilde{x} \simeq \varepsilon^{1/3} . \tag{50}$$

This gives an approximate value of $\tilde{R}_g$,

$$\tilde{R}_g \simeq \left(\frac{N R_h}{1920\pi}\right)^{1/3} \simeq 0.055 \times (N R_h)^{1/3} . \tag{51}$$

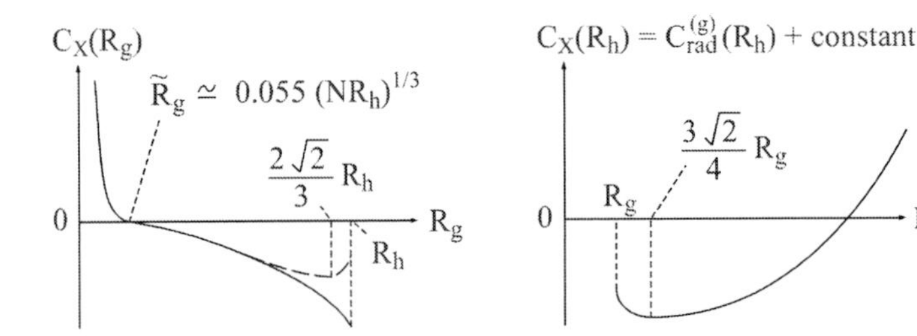

Figure 3. Left panel is for $C_X(R_g)$ as a function of R_g, where $2\sqrt{2}/3 \simeq 0.943$. Right is for $C_X(R_h)$ as a function of R_h, where $3\sqrt{2}/4 \simeq 1.06$. $C_X(R_g = R_h) < 0$ holds due to $\sigma' \simeq 10$ and $1 \lesssim R_g \leq R_h$.

4.3.2. $C_X(R_h)$ as a Function of R_h

Here we show a behavior of heat capacity C_X of sub-system X as a function of outermost radius R_h of the hollow region. We make use of the calculations done in previous subsection. According to eq.(42), $C_X(R_h)$ is expressed as

$$C_X(R_g) = -2\pi R_g^2 + C_{rad}^{(g)}(R_h) \quad , \quad C_{rad}^{(g)}(R_h) = \frac{\sigma'}{8\pi^2} f(x) \, , \tag{52}$$

where $x := R_g/R_h$, $0 < x < 1$ by definition, and $f(x)$ is given in eq.(43). This denotes $C_X(R_h)$ behaves as $C_{rad}^{(g)}(R_h) + constant$. The differential becomes

$$\frac{dC_X(R_h)}{dR_h} = -\frac{\sigma'}{8\pi} x^2 \frac{df(x)}{dx} \, . \tag{53}$$

Using eq.(46), we find

$$\begin{cases} \dfrac{dC_X}{dR_h} \to -\infty & \text{as } R_h \to R_g + 0 \; (x \to 1+0) \\ \dfrac{dC_X}{dR_h} \to \dfrac{\sigma'}{8\pi^2} & \text{as } R_h \to \infty \; (x \to 0) \end{cases} \, . \tag{54}$$

Hence referring to limit values (44), a schematic graph of $C_X(R_h)$ is obtained as shown in fig.3 (right panel). $C_X(R_h)$ is monotone increasing for $R_h > (3\sqrt{2}/4)R_g \simeq 1.06R_g$.

4.3.3. Proof of the Inequality $C_g/C_X > 1$

Here we prove inequality $C_g/C_X > 1$ under the condition $C_X < 0$ (see condition (40)). We make use of the calculations done in subsection 4.3.1.. By eq.(42) and definition of C_X given in eq.(39), $C_{rad}^{(g)}$ is expressed as $C_{rad}^{(g)}(x) = (\sigma'/8\pi^2)\, f(x)$, where $x := R_g/R_h$, $0 < x < 1$ by definition, and $f(x)$ is given in eq.(43). Then eq.(46) indicates $C_{rad}^{(g)}(x) \geq C_{rad}^{(g)}(2\sqrt{2}/3)$, and eq.(43) gives $f(2\sqrt{2}/3) \simeq 0.845 > 0$. Therefore we find $C_{rad}^{(g)} > 0$ for $0 < x < 1$, which is consistent with a naive expectation that an ordinary matter like radiation fields has a positive heat capacity.

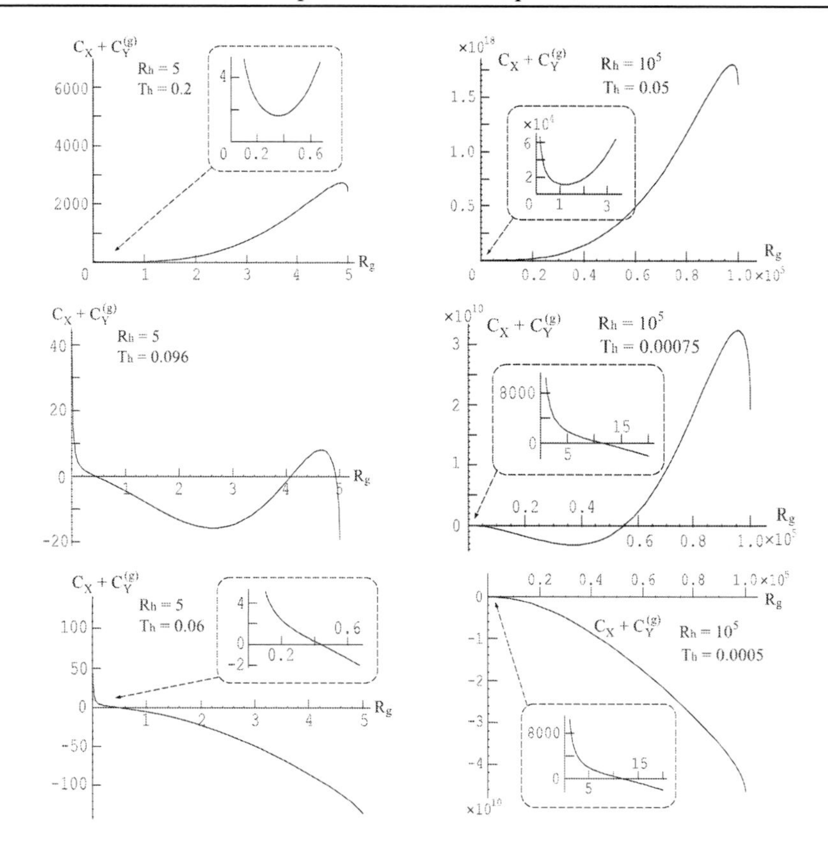

Figure 4. Numerical plots of $C_X + C_Y^{(g)}$ as a function of R_g with $N = 100$. Left three panels are for $R_h = 5$ and $T_h = 0.2, 0.096$ and 0.06 downwards. Right three panels are for $R_h = 10^5$ and $T_h = 0.05, 0.00075$ and 0.0005 downwards. As far as the author checked, the same behavior is observed for every value of $R_h > 1$. $C_X + C_Y^{(g)} < 0$ holds for a sufficiently small T_h.

On the other hand a required condition $C_X\ (= C_g + C_{rad}^{(g)}) < 0$ indicates $0 < C_{rad}^{(g)} < |C_g|$. This gives $|C_X| = |C_g| - C_{rad}^{(g)} < |C_g|$. Hence we find

$$\frac{C_g}{C_X} > 1\,. \tag{55}$$

4.3.4. $C_X + C_Y^{(g)}$ as a Function of R_g

Here we show a behavior of combined heat capacity $C_X + C_Y^{(g)}$ as a function of black hole radius R_g. We make use of the calculations done in subsection 4.3.1.. By eq.(42) and definitions of C_X and $C_Y^{(g)}$ given in eq.(39), $C_X + C_Y^{(g)}$ is expressed as

$$C_X(R_g) + C_Y^{(g)}(R_g) = -2\pi R_h^2\, x^2 + \frac{\sigma'}{8\pi^2}\left[f(x) + q\,x^3\left(x + \sqrt{1-x^2}\right)\right], \tag{56}$$

where $q := (4\pi R_h T_h)^4$, $x := R_g/R_h$, $0 < x < 1$ by definition, and $f(x)$ is given in eq.(43). Then, using limit values (44), we find

$$\begin{cases} C_X + C_Y^{(g)} \to \infty & \text{as } x \to 0 \\ C_X + C_Y^{(g)} = -2\pi R_h^2 + \dfrac{\sigma'}{8\pi^2} + \dfrac{\sigma'}{8\pi^2}\, q & \text{at } x = 1 \end{cases} . \tag{57}$$

The first two terms in $C_X + C_Y^{(g)}$ at $x = 1$ is negative in the framework of NE model as discussed in eq.(44). Therefore, if T_h is sufficiently small, then q can be small enough so that $C_X + C_Y^{(g)}$ at $x = 1$ becomes negative.

The differential of $C_X + C_Y^{(g)}$ is complicated and not suitable for analytical utility. Instead of analytical discussion, we show some numerical examples of $C_X + C_Y^{(g)}$ in fig.4. It is recognized with this figure that, for a sufficiently small T_h, the condition (40) is satisfied for a certain range of R_g. Although fig.4 shows only some examples, the same behavior is observed for every value of $R_h > 1$ as far as the author checked. Therefore the validity condition (40), $C_X + C_Y^{(g)} < 0$, holds for a sufficiently small T_h. During a semi-classical and quasi-equilibrium stage of black hole evaporation ($R_g(t) > 1$), it is reasonable to require $T_g < 1$ due to eq.(4), and also $T_h < T_g < 1$ for black hole evaporation. We assume throughout this chapter that T_h is small enough so that the equation $C_X(R_g) + C_Y^{(g)}(R_g) = 0$ has one, two or three solutions of R_g.

Finally we discuss what happens along a black hole evaporation for the case that the equation $C_X(R_g) + C_Y^{(g)}(R_g) = 0$ has two or three solutions. For the first consider the case of three solutions, and denotes these solutions by $R_{g0}^{(s)}$, $R_{g0}^{(m)}$ and $R_{g0}^{(l)}$ in increasing order, $R_{g0}^{(s)} < R_{g0}^{(m)} < R_{g0}^{(l)}$. If the evaporation starts with initial black hole radius $R_g(0)$ in the range $R_{g0}^{(l)} < R_g(0) < R_h$ (see for example left-center panel in fig.4), then the evaporation process is treated in the framework of NE model until R_g decreases to $R_{g0}^{(l)}$. Then, when R_g reaches $R_{g0}^{(l)}$, the NE model becomes inapplicable to the evaporation process because the validity condition (40) is violated in the range $R_g^{(m)} < R_g < R_{g0}^{(l)}$. However we can expect R_g decreases to $R_{g0}^{(m)}$ even if NE model is not applicable. Then the NE model becomes applicable again after R_g decreases less than $R_{g0}^{(m)}$. The NE model is applied to the evaporation process in the range $R_{g0}^{(s)} < R_g < R_{g0}^{(m)}$.

Next consider the case that the equation $C_X(R_g) + C_Y^{(g)}(R_g) = 0$ has two solutions. When the black hole evaporation starts with initial radius larger than the larger solution of $C_X(R_g) + C_Y^{(g)}(R_g) = 0$ (see for example right-center panel in fig.4), then the similar discussion given in previous paragraph holds. The NE model is not applicable to the evaporation process until R_g becomes smaller than the larger solution of $C_X(R_g) + C_Y^{(g)}(R_g) = 0$. However after R_g becomes less than the larger solution, the NE model becomes applicable until R_g decreases to the smaller solution of $C_X(R_g) + C_Y^{(g)}(R_g) = 0$.

From the above we find that, for both of the cases that equation $C_X(R_g) + C_Y^{(g)}(R_g) = 0$ has two and three solutions, it is the lowest solution of that equation at which the evaporation process goes out of the framework of NE model and proceeds to a highly nonequilibrium dynamical stage (see last paragraph in subsection 4.2.).

4.4. Nonequilibrium Effects of Energy Flow

4.4.1. General Aspect of the NE Model

We discuss the black hole evaporation in the framework of NE model. To analyze energy transport equations (37) from energetic viewpoint, we consider the energy emission rate J_{NE} by black hole,

$$J_{NE} := -\frac{dE_g}{dt} = \sigma' \frac{C_g}{C_X} \left(T_g^4 - T_h^4\right) A_g \,, \tag{58}$$

where eqs.(5) and (37) are used. The stronger J_{NE}, the more rapidly the mass energy of black hole E_g decreases along its evaporation process. The stronger emission rate J_{NE} denotes the acceleration of black hole evaporation.

As mentioned in eq.(39), we assume $R_h \equiv constant$ for simplicity. R_h is the parameter which controls the size of nonequilibrium region around black hole. To understand the nonequilibrium nature of black hole evaporation, it is useful to compare two situations which differ only by the value of R_h with sharing the same values of the other parameters of NE model, R_g, T_h, C_h and N. To do this comparison, we note the following three points; firstly $C_g < 0$ by definition, secondly $C_X < 0$ given in the condition (40), and finally that $|C_X|$ is monotone decreasing function of R_h for $R_h \geq \left(3\sqrt{2}/4\right) R_g \simeq 1.06 R_g$ under the condition $C_X < 0$ (see subsection 4.3.2.). The first and second points denote $C_g/C_X > 0$, then the third point concludes that C_g/C_X is monotone increasing function of R_h for $R_h \geq \left(3\sqrt{2}/4\right) R_g$ under the condition $C_X < 0$. Hence it is recognized that, for the case $R_h > \left(3\sqrt{2}/4\right) R_g$, the larger we set the nonequilibrium region, the stronger the emission rate J_{NE} and the faster the black hole evaporation proceeds. Numerical examples are shown later in subsection 4.4.4..

The above discussion is a comparison of NE model of a certain value of R_h with that of a different value of R_h. In the following subsections, we compare the NE model with the other models of black hole evaporation, the equilibrium model used in [17] and the black hole evaporation in an empty space (a situation without heat bath originally considered by Hawking in [1]).

4.4.2. Comparison with the Equilibrium Model Used in [17]

In the original work [17] suggesting the black hole phase transition, only the equilibrium of the system which consists of a black hole and a heat bath is considered. This equilibrium model is obtained from the NE model by setting $R_h = R_g$ (no hollow region) and $T_h = T_g$ (equilibrium). Obviously the equilibrium model does not include nonequilibrium nature of black hole evaporation, since the radiation fields disappear. Exactly speaking, the evaporation process is not described by the "equilibrium model". However even in the framework of equilibrium model, we can find the black hole evaporation occurs for a sufficiently small black holes as mentioned in subsection 4.1.. By extrapolating the equilibrium model to the evaporation process, we may set $T_h < T_g$ with keeping the condition $R_g = R_h$ (see left panel in fig.5). Then the energy emission rate J_{eq} by black hole in equilibrium model is given by setting $R_h = R_g$ in J_{NE},

$$J_{eq} := J_{NE}|_{R_h = R_g} = \sigma' \left(T_g^4 - T_h^4\right) A_g = J \,, \tag{59}$$

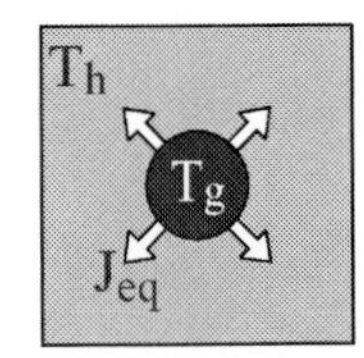

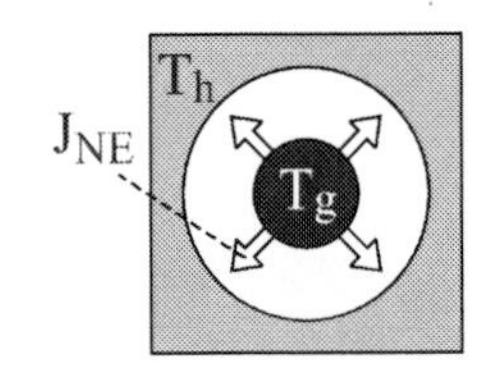

Figure 5. Energy emission rate by black hole $-dE_g/dt$. Left panel is J_{eq} for equilibrium model extrapolated to a black hole evaporation, $T_h < T_g$. Center is J_{NE} for NE model. Right is J_{empty} for a black hole evaporation in an empty space with ignoring grey body factor. White arrow in each panel expresses energy flow from black hole.

where J is given in eq.(38). We find $J_{NE} = (C_g/C_X)\, J_{eq}$. Here note that $C_g/C_X > 1$ is shown in subsection 4.3.3.. Therefore, when the values of R_g, T_h and N are the same for the NE and equilibrium models, then $J_{NE} > J_{eq}$ holds. This implies that the black hole evaporation in NE model proceeds faster than that in equilibrium model. We can recognize that the nonequilibrium effect of energy exchange between black hole and heat bath accelerates the black hole evaporation.

4.4.3. Comparison with the Black Hole Evaporation in an Empty Space with Ignoring Grey Body Factor

In the original work [1] of the Hawking radiation, Hawking considered mainly a simple situation as seen in right panel in fig. 5; a black hole in an empty space (a situation without heat bath) with ignoring curvature scattering of Hawking radiation. This describes a black hole evaporation in an empty space with ignoring the so-called *grey body factor*. There is no energy accretion onto black hole in this simple situation, and time evolution is given by the Stefan-Boltzmann law, $dE_g/dt = -\sigma' T_g^4 A_g$. Usually in most of the existing works on black hole physics, the time scale of black hole evaporation is estimated by assuming this simple situation.

It is interesting to compare the NE model with the black hole evaporation in an empty space with ignoring grey body factor. The energy emission rate J_{empty} by black hole in an empty space with ignoring grey body factor is given by the Stefan-Boltzmann law as follows,

$$J_{empty} := \sigma' \, T_g^4 \, A_g \,, \tag{60}$$

where it is assumed that matter fields of Hawking radiation is the non-self-interacting massless matter fields as for the NE model. Then we find

$$J_{NE} = \frac{C_g}{C_X} \left(1 - \frac{T_h^4}{T_g^4} \right) J_{empty} \,. \tag{61}$$

Recall that $T_g > T_h$ holds generally for a black hole evaporation, and $C_g/C_X > 1$ holds in the framework of NE model (see subsection 4.3.3.). Then the factor $(C_g/C_X)\left(1 - T_h^4/T_g^4\right)$ may be greater or less than unity. It is not definitely clear which of J_{NE} or J_{empty} is larger than the other.

One may naively expect that the incoming energy flow from heat bath to black hole in NE model never enhance the energy emission rate by black hole, and that the relation $J_{NE} > J_{empty}$ is impossible but $J_{NE} < J_{empty}$ must hold necessarily. It is true if the black hole heat capacity C_g is positive. However in the NE model, $C_g < 0$ as shown in eq.(5) and a naive sense based on ordinary systems of positive heat capacity is not always true. An inverse sense against the naive sense may be offered; the more amount of energy is extracted from black hole by heat bath, the more rapidly the black hole emits its mass energy. Furthermore the energy emitted by Hawking radiation is absorbed by the heat bath and affects the incoming energy flow from heat bath to black hole. The energetic interaction (energy exchange) between black hole and heat bath determines the energy emission rate J_{NE}. When the negative heat capacity and energetic interaction are taken into account, a naively unexpected relation $J_{NE} > J_{empty}$ is also possible as discussed in following paragraphs:

To analyze the energy emission rate by black hole J_{NE}, it is useful to recall the decomposition of the whole system of NE model into sub-systems X and Y, as considered in subsection 4.2.. On the other hand, the black hole evaporation in an empty space with ignoring grey body factor is regarded as the system obtained by removing the sub-system Y from the NE model. This means that, from energetic viewpoint, the black hole evaporation in an empty space with ignoring grey body factor can be thought of as a relaxation process of the "isolated sub-system X" keeping $E_X \equiv constant$. Therefore, for the black hole evaporation in an empty space with ignoring grey body factor, the energy emission rate by black hole J_{empty} is the energy transport just inside the sub-system X (from black hole to out-going radiation fields), and no energy flows out of X. However the energy transport (36) in NE model is the energy exchange between sub-systems X and Y. This indicates that, in the NE model, the energy E_X of X is extracted by Y due to the temperature difference $T_g > T_h$, and energy flows from X to Y. The energetic interaction (energy exchange) between X and Y makes the black hole evaporation in NE model quit different from the black hole evaporation in an empty space with ignoring grey body factor. This difference is recognized significantly by considering a limit of the energy transport (36) as follows: One may expect that the energy emission by black hole in an empty space with ignoring grey body factor, $dE_g/dt = -J_{empty}$, should be obtained from eq.(36) by the limit operations, $T_h \to 0$, $E_h \to 0$ (remove the sub-system Y) and $R_h \to \infty$ (infinitely large volume of out-going radiation fields). However these operations transform eq.(36) into the set of equations, $dE_g/dt = -J_{empty}$ and $0 \equiv J_{empty}$. This gives an unphysical result $E_g \equiv constant\,(= \infty)$ which contradicts the "evaporation", $dE_g/dt < 0$. The black hole evaporation in an empty space with ignoring grey body factor can not be described as some limit situation of the NE model.

In addition to the naive expectation $J_{NE} < J_{empty}$, the opposite relation $J_{NE} > J_{empty}$ may be expected due to the negative heat capacity of black hole (5) and energetic interaction as follows: In the NE model, because the energy extraction ($dE_X < 0$) occurs along the black hole evaporation ($dT_g > 0$), the heat capacity of X is always negative $C_X = dE_X/dT_g = C_g + C_{rad}^{(g)} < 0$, where $C_{rad}^{(g)} > 0$ as indicated in subsection 4.3.3.. Furthermore the larger the volume of hollow region, the larger the heat capacity $C_{rad}^{(g)}$ and the smaller the absolute value $|C_X|$ because of $C_g < 0$. This implies that the more thick the hollow region, the more accelerated the increase of T_g due to the relation $dT_g = |dE_X/C_X|$. Therefore, for sufficiently large R_h, the energy extraction from X by Y (the increase of T_g) can dominate over the in-coming energy flow onto black hole. This means the energy emission rate J_{NE} is enhanced by the energy extraction from X by Y, then $J_{NE} > J_{empty}$ is implied.

The above discussion can also be supported by the following rough analysis: When a black hole evaporates in NE model, the temperature difference $\delta T := T_g - T_h$ should grow infinitely, $\delta T \to \infty$, due to the negative heat capacity of black hole. Then, because of eq.(61) together with the facts $1 - (T_h/T_g)^4 \to 1$ (as $\delta T \to \infty$) and $C_g/C_X > 1$ (see subsection 4.3.3.), the larger the temperature difference δT, the larger the ratio J_{NE}/J_{empty}. Hence for the black hole evaporation in NE model, it is expected that the relation $J_{NE} > J_{empty}$ comes to be satisfied during the evaporation process even if the relation $J_{NE} < J_{empty}$ holds at initial time. Furthermore, if the relation $J_{NE} > J_{empty}$ holds for a sufficiently long time during the evaporation process, the evaporation time scale in NE model can be shorter than that in an empty space. Hence it is possible that the black hole evaporation in NE model proceeds faster than that in an empty space, where black holes of the same initial mass are considered in both cases. In next subsection, numerical examples support this discussion.

4.4.4. Numerical Example

We show numerical solutions $T_g(t)$ and $T_h(t)$ of energy transport equations (37). The initial conditions are

$$R_g(0) = 100 \quad , \quad T_h(0) = 0.0001 \,. \tag{62}$$

$R_g(0) = 100$ gives $T_g(0) \simeq 0.00079$. The other parameters are set

$$C_h = 1000 \quad , \quad N = 100 \,, \tag{63}$$

where see eq.(3) for N. Furthermore we have to specify the outermost radius R_h of hollow region. As mentioned in eq.(58), by comparison of a numerical solution of energy transport (37) of a certain value of R_h with that of a different value of R_h, we can observe the nonequilibrium effect of energy exchange between black hole and heat bath. The numerical results are shown in fig.6, and the value of R_h is attached in each panel. Time coordinate τ in this figure is a time normalized as

$$\tau := \frac{t}{t_{empty}} \,, \tag{64}$$

where t_{empty} is the evaporation time (life time) of a black hole in an empty space with ignoring grey body factor. t_{empty} is determined by the energy emission rate by black hole in an empty space,

$$\frac{dE_g}{dt} = -J_{empty} \,, \tag{65}$$

where no energy accretion exists due to the absence of heat bath and ignoring grey body factor, E_g corresponds to the mass energy of black hole and J_{empty} is given in eq.(60). This and eq.(4) give

$$R_g(t) = R_g(0) \left(1 - \frac{N\,t}{1280\,\pi\,R_g(0)^3}\right)^{1/3} \,, \tag{66}$$

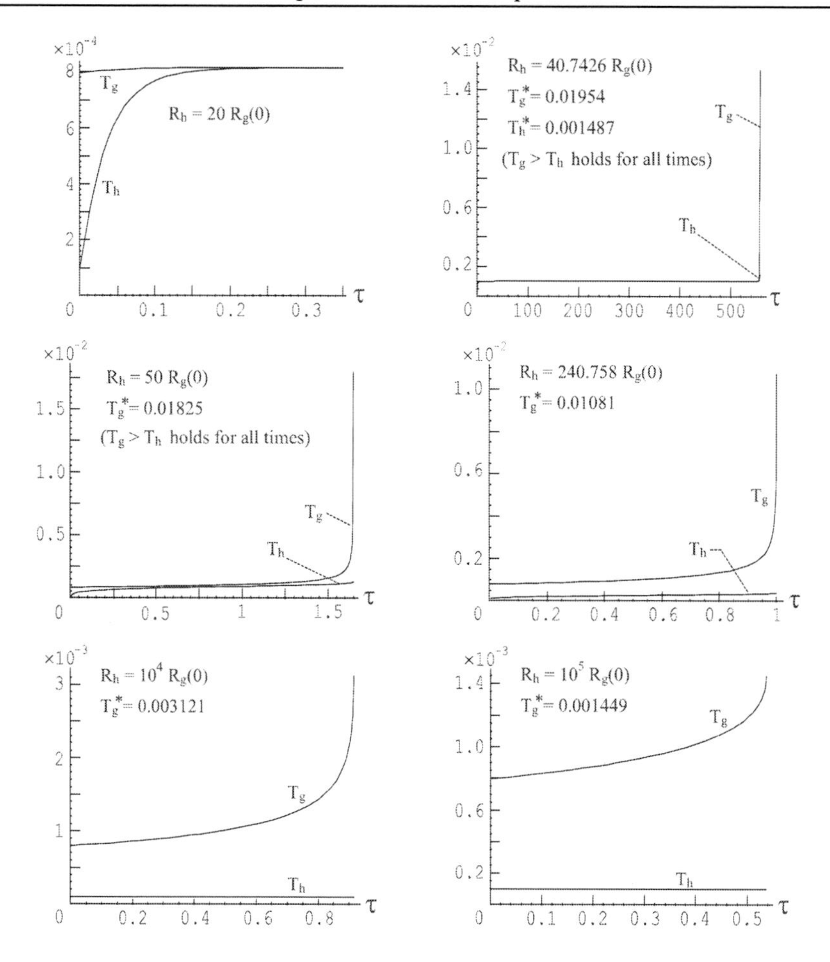

Figure 6. Numerical solutions of energy transport (37) for $C_h = 10^3$, $N = 100$ and the initial conditions, $R_g(0) = 100$ and $T_h(0) = 0.0001$. Horizontal line denotes the normalized time $\tau := t/t_{empty}$. The outermost radius of hollow region R_h determines the size of nonequilibrium region.

and we obtain

$$t_{empty} := \frac{1280\,\pi}{N} R_g(0)^3 \simeq 4.02124 \times 10^7 \,, \tag{67}$$

where conditions (62) and (63) are used in the second equality. This t_{empty} is usually adopted as the time scale of black hole evaporation in many existing works on black hole physics.

Furthermore we consider the other time t_{NE} at which one of the validity conditions of NE model (40) or (41) breaks down,

$$t_{NE} := \min \left[t_1, t_2 \,\middle|\, C_X(t_1) + C_Y^{(g)}(t_1) = 0 \,,\, v(t_2) = 1 \right] , \tag{68}$$

where $C_X + C_Y^{(g)}$ is regarded as a function of time t through $R_g(t)$, and $v(t) := |dR_g(t)/dt|$ is the shrinkage speed of black hole radius. Each panel in fig.6 shows time evolutions of

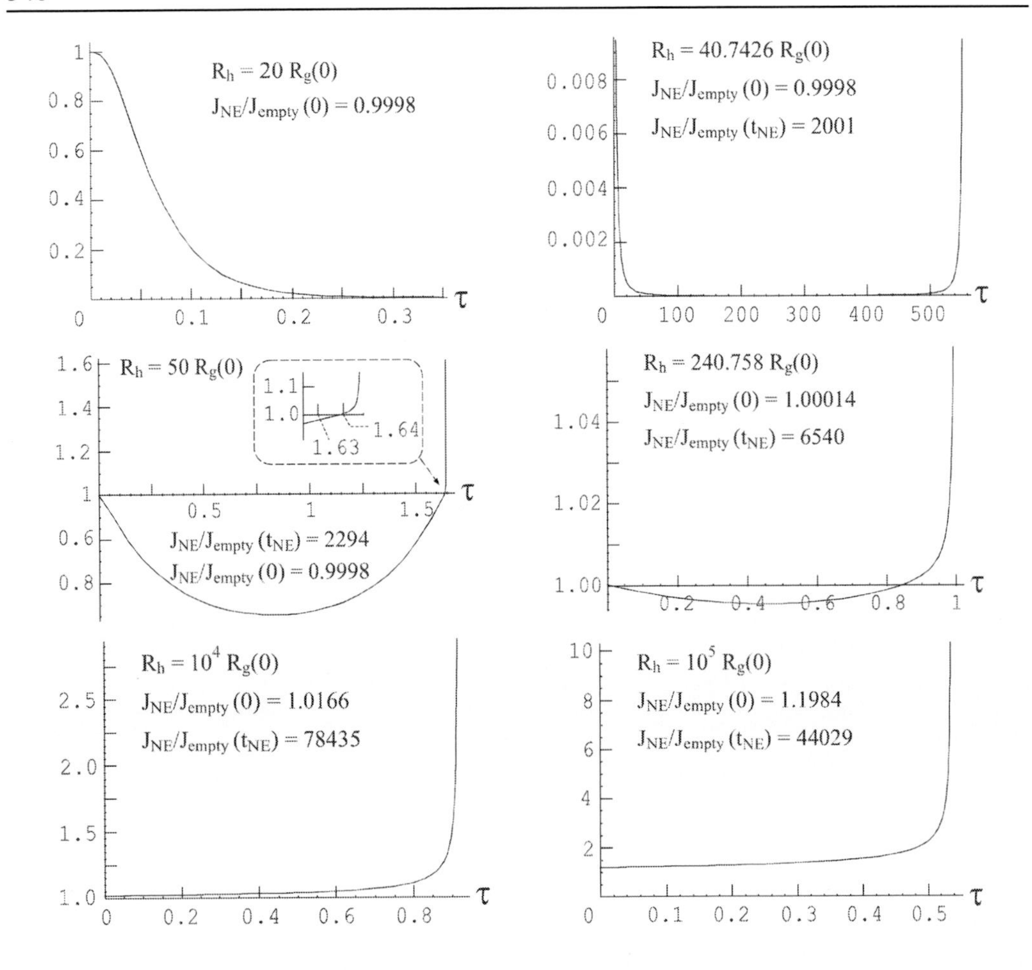

Figure 7. J_{NE}/J_{empty} for each case in fig.6. The graph in left-center panel is continuous and smooth.

$T_g(t)$ and $T_h(t)$ for $0 < t < t_{NE}$. The black hole temperature at t_{NE} is denoted by T_g^*,

$$T_g(t_{NE}) =: T_g^* = \frac{1}{4\pi R_g^*}, \tag{69}$$

which is also attached in each panel in fig.6. The time t_{NE} and the other quantities obtained from our numerical results are listed;

$R_h/R_g(0)$	40.7426	50	240.758	10^4	10^5
t_{NE}/t_{empty}	557.370	1.64705	1.00000	0.915560	0.536364
R_g^*	4.07277	4.36041	7.36293	25.4986	54.9354
$J_{NE}/J_{empty}(t_{NE})$	2001	2294	6540	78435	44029
R_h/t_{NE}	1.8×10^{-7}	7.5×10^{-5}	6.0×10^{-4}	2.7×10^{-2}	4.6×10^{-1}

(70)

For cases of $R_h/R_g(0) = 40.7462$, 50, 240.758 and 10^4, the numerical plots stopped by $v(t_{NE}) = 1$, but for a case of $R_h/R_g(0) = 10^5$, it stopped by $C_X(t_{NE}) + C_Y^{(g)}(t_{NE}) = 0$.

The second line in this list for t_{NE}/t_{empty} supports the discussion given in subsection 4.4.1. that the larger the radius R_h, the faster the black hole evaporation proceeds and the shorter the time t_{NE}. The third and fourth lines in list (70) for R_g^* and $J_{NE}/J_{empty}(t_{NE})$ give an important information in next subsection. The lowest line in list (70) for R_h/t_{NE} shows that our numerical results are consistent with the fast propagation assumption. To understand this, consider a typical time scale t_{rad} in which one particle of radiation fields travels in the hollow region from black hole to heat bath. This t_{rad} is given by $t_{rad} := R_h$, since the radiation fields are massless. If $t_{rad} < t_{NE}$, then the fast propagation assumption is reasonable. In fact, the ratio t_{rad}/t_{NE} $(= R_h/t_{NE})$ shown at the lowest line indicates the validity of fast propagation assumption.

Concerning a case $R_h = 20\,R_g(0)$, it is helpful to recognize that our conditions (62) and (63) give $2\,\pi \times 100^2 > 1000$ which denotes $|C_g| > C_h$. This $|C_g| > C_h$ is the condition for stable equilibrium of black hole and heat bath in the framework of the equilibrium model (see [17] or [11]). Therefore, if the nonequilibrium region is ignored and the equilibrium model used in [17] is considered with the same setting parameters of eqs.(62) and (63), then the black hole stabilizes with heat bath to settle down into a total equilibrium state $T_g - T_h \to 0$. Indeed, for the case $R_h = 20\,R_g(0)$, the hollow region is not large enough and the black hole stabilizes with heat bath. Hence the occurrence of accelerated increase of temperature T_g in fig.6 is obviously the nonequilibrium effect of energy exchange between black hole and heat bath.

Then as the outermost radius R_h of hollow region is set larger and the nonequilibrium region becomes larger, the black hole evaporation comes to be observed as an accelerated increase of temperature difference $T_g - T_h$ for the case $R_h = 40.7426\,R_g(0)$. t_{NE} is very longer than t_{empty} in this case. Furthermore as R_h is set larger, the time t_{NE} becomes shorter and we find $t_{NE} \simeq t_{empty}$ for $R_h \simeq 240.758\,R_g(0)$. At last t_{NE} becomes shorter to $t_{NE} \simeq 0.536364\,t_{empty}$ for $R_h = 10^5\,R_g(0)$. If we set $R_h \gtrsim 10^6$, the combined heat capacity becomes positive $C_X + C_Y^{(g)} > 0$ which violates the validity condition (40). A black hole with a large nonequilibrium region of $R_h \gtrsim 10^6$ can not be treated in the framework of NE model, and, as discussed in the last paragraph in subsection 4.2., a black hole evaporation for such case is a highly nonequilibrium dynamical process in which the black hole can not be treated with equilibrium black hole solutions of Einstein equation. The larger the nonequilibrium region, the faster the black hole evaporation process evolves into a highly nonequilibrium dynamical stage.

Fig.7 shows time evolutions of the ratio of energy emission rates J_{NE}/J_{empty} given by eq.(61). This figure indicates that, when a black hole evaporates, the relation $J_{NE} > J_{empty}$ comes to hold even if the converse relation $J_{NE} < J_{empty}$ holds at initial time. Therefore the discussion given in previous subsection 4.4.3. is supported. For the case $R_h = 20R_g(0)$ in which a black hole stabilizes with heat bath, the energy emission rate J_{NE} disappears, $J_{NE}/J_{empty} \to 0$, as easily expected by the behavior $T_g \to T_h \Rightarrow J_{NE} \to 0$.

4.5. Beyond the NE Model I: Abrupt Catastrophic Evaporation

When the nonequilibrium region around black hole is not so large, the evaporating black hole is well approximated to equilibrium solutions of Einstein equation and the NE model is applicable to such quasi-equilibrium evaporation stage. However the condition (40) or

(41) is violated at the time t_{NE} and the evaporation process becomes highly nonequilibrium dynamical stage (see last paragraph in subsection 4.2.). Therefore, since a relation $R_g^* > 1$ is expected from the list (70), we find the semi-classical evaporation stage ($R_g \gtrsim 1$) is divided into two stages, quasi-equilibrium one and highly nonequilibrium dynamical one. The former is described by the NE model, but the latter is beyond the range of NE model. This subsection extrapolates NE model to the latter stage and suggests a specific nonequilibrium phenomenon.

On the other hand, for the equilibrium model used in [17] and the black hole evaporation in an empty space with ignoring grey body factor, the highly nonequilibrium dynamical stage does not exist and the semi-classical stage is always described as the quasi-equilibrium stage. This is explained in next subsection 4.5.1.. Then the highly nonequilibrium dynamical stage and a suggestion about that stage in the framework of NE model are discussed in subsection 4.5.2..

4.5.1. On Semi-classical Evaporation Stage, Except for NE Model

This subsection shows that the highly nonequilibrium dynamical stage does not occur for the equilibrium model used in [17] and for the black hole evaporation in an empty space with ignoring grey body factor.

For the first consider the equilibrium model used in [17]. As explained in subsection 4.4.2., this model is obtained from NE model by setting $R_h = R_g$ and $T_g = T_h$. Exactly speaking, the evaporation process is not described by the "equilibrium model". However by extrapolating the equilibrium model to the evaporation process, we may set $T_h < T_g$ with keeping the condition $R_g = R_h$. Then the energy transport equations from black hole to heat bath are given by

$$\frac{dE_g}{dt} = -\sigma'\left(T_g^4 - T_h^4\right)A_g \quad , \quad \frac{dE_h}{dt} = \sigma'\left(T_g^4 - T_h^4\right)A_g \, , \tag{71}$$

where $A_g = 4\pi R_g^2$ is the surface area of black hole. The equation of dE_g/dt together with eq.(4) give

$$v := \left|\frac{dR_g}{dt}\right| = 2\left|\frac{dE_g}{dt}\right| = 2\sigma'\left(T_g^4 - T_h^4\right)A_g \, . \tag{72}$$

Since equations of states (4) are used, the quasi-equilibrium assumption is also necessary here and $v < 1$ is required. The inequality $v < 1$ corresponds to the validity condition (41) of NE model. Obviously $v = 1$ occurs for non-zero radius $R_g > 0$. On the other hand, from eq.(5), vanishing heat capacity of black hole $C_g = 0$ corresponds to zero radius $R_g = 0$. This means that the equilibrium model has no validity condition which corresponds to the condition (40) of NE model. Hence we can recognize that, if $v = 1$ occurs for a black hole of semi-classical size $R_g > 1$, it is concluded that the highly nonequilibrium dynamical stage occurs at semi-classical level. But if $v = 1$ does not occur for a semi-classical black hole, then the highly nonequilibrium dynamical stage does not occur at semi-classical level in the framework of equilibrium model.

The validity condition $v < 1$ is rewritten as

$$\sigma'\left(T_g^4 - T_h^4\right) < 2\pi T_g^2 \, , \tag{73}$$

and this gives

$$R_g^2 > \frac{1}{8\pi^2\beta}\left(1+\sqrt{1+\left(2T_h^2/\beta\right)^2}\right)^{-1}, \tag{74}$$

where $\beta := 2\pi/\sigma' = 240/\pi N$ and $R_g = 1/4\pi T_g$ is used. Here eq.(3) gives $\beta \simeq 1$. Recall that $T_g > T_h$ holds generally for any evaporation process and $T_g < 1$ holds due to the quasi-equilibrium assumption. Then we find $T_h < 1$ and $(2T_h^2/\beta)^2 < 1$. Therefore we can approximate inequality (74) to $R_g^2 > 1/(16\pi^2\beta) \simeq 10^{-2}$, where $\beta \simeq 1$ is used. This gives

$$R_g \gtrsim 0.1\,. \tag{75}$$

This denotes that the fast evaporation $v = 1$ occurs at $R_g \simeq 0.1$. Hence, as discussed in previous paragraph, the highly nonequilibrium dynamical stage does not occur in the framework of equilibrium model used in [17].

Next consider a black hole evaporation in an empty space with ignoring grey body factor. The Stefan-Boltzmann law gives eq.(65). This together with equations of states (4) give

$$v := \left|\frac{dR_g}{dt}\right| = 2\left|\frac{dE_g}{dt}\right| = 2\sigma' T_g^4 A_g\,. \tag{76}$$

Hence, following a similar discussion given in previous paragraph with setting $T_h = 0$, we require $v < 1$ and obtain $R_g^2 > 1/(16\pi^2\beta) \simeq 10^{-2}$, where $\beta \simeq 1$ is used. This gives

$$R_g \gtrsim 0.1\,. \tag{77}$$

This denotes that the fast evaporation $v = 1$ occurs at $R_g \simeq 0.1$, and that the highly nonequilibrium dynamical stage does not occur for a black hole evaporation in an empty space with ignoring grey body factor.

4.5.2. Abrupt Catastrophic Evaporation

Let us point out again a numerical evidence shown in list (70) that the black hole radius R_g^* at time t_{NE} is greater than unity $R_g^* > 1$. According to the discussion in last paragraph in subsection 4.2., we can expect a highly nonequilibrium dynamical stage of evaporation process will occur at semi-classical level $R_g^* > 1$ in the framework of NE model. After the time t_{NE}, the black hole and radiation fields should be described as highly nonequilibrium dynamical ones.

In the following discussion, we make two steps: Firstly, to confirm the numerical evidence, we show $R_g^* \gtrsim 1$ analytically. Secondly, a physical implication of $R_g^* \gtrsim 1$ is discussed and a specific nonequilibrium phenomenon is suggested.

For the first, we analyze the energy transport equations (37). Due to definition (68) of time t_{NE}, we consider two cases, $t_{NE} = t_1$ and $t_{NE} = t_2$, where t_1 and t_2 are given in definition (68). But before proceeding to the analysis of these cases, we should point out the following: As explained at the end of subsection 4.3.4., equation $C_X(R_g) + C_Y^{(g)}(R_g) = 0$ has one, two or three solutions of R_g for a sufficiently small T_h. However even if there are two or three solutions of R_g for our choice of T_h, it is the lowest solution at which a highly nonequilibrium dynamical evaporation stage starts towards a quantum evaporation stage. Hence the time t_1 in definition (68) is the lowest solution of equation $C_X(t) + C_Y^{(g)}(t) = 0$.

Here we estimate the order of R_g^*. Consider the case $t_{NE} = t_1$, where $R_g^* = R_g(t_1)$ and $C_X(R_g^*) + C_Y^{(g)}(R_g^*) = 0$. Because of $C_Y^{(g)} > 0$ by definition, $C_X(R_g^*) = -C_Y^{(g)}(R_g^*) < 0$ holds. Consequently, according to a behavior of $C_X(R_g)$ explained in subsection 4.3.1. (left panel in fig.3), we find $R_g^* > \tilde{R}_g \simeq 0.055 \times (N R_h)^{1/3}$. Hence together with eq.(3), we find $R_g^* \gtrsim 1$ for the situation $R_h \gtrsim 60$. And next consider the other case $t_{NE} = t_2$, where $|\dot{R}_g(t_2)| = 1$. Because of $t_2 < t_1$, we find $C_X(t_2) + C_Y^{(g)}(t_2) < 0$. Then, because it is assumed that T_h is small enough so that the validity condition (40) holds, we find by subsection 4.3.4. (fig.4) that $R_g^* = R_g(t_2) > R_{g0}$ holds, where R_{g0} is the lowest solution of $C_X(R_{g0}) + C_Y^{(g)}(R_{g0}) = 0$. Therefore, following the same discussion given for the case $t_{NE} = t_1$, we obtain $R_g^* > R_{g0} > \tilde{R}_g \simeq 1$ for the situation $R_h \gtrsim 60$. In summary, the black hole radius R_g^* at time t_{NE} is greater than unity $R_g^* \gtrsim 1$, when the black hole evaporates in the framework of NE model under the condition $R_h \gtrsim 60$. Here we have to note two points: First is that the condition $R_h \gtrsim 60$ is not a necessary condition but a sufficient condition for $R_g^* \gtrsim 1$, and there may remain a possibility that $R_g^* \gtrsim 1$ holds even if $R_h \lesssim 60$. Second point is that, if the nonequilibrium nature of black hole evaporation is not taken into account, the radius R_g^* can not be greater than unity but it becomes less than Planck length as seen in eqs.(75) and (77). The relation $R_g^* \gtrsim 1$ is a peculiar property of the NE model.

We proceed to the second part of this subsection, an implication of the above result, $R_g^* \gtrsim 1$. Recall a highly nonequilibrium dynamical stage of evaporation process occurs after the time t_{NE}. Because of $R_g^* \gtrsim 1$, a semi-classical (but not quasi-equilibrium) discussion is available for the highly nonequilibrium dynamical stage while the black hole radius shrinks from R_g^* to Planck length $l_{pl} := 1$. Then it is appropriate to consider that the mass energy of black hole evolves from E_g^* $(= R_g^*/2)$ to $E_p := l_{pl}/2$. Energy difference $\Delta E_g := E_g^* - E_p$ is emitted during highly nonequilibrium dynamical stage. Furthermore, for example, fig.7 and fourth line in list (70) of our numerical example imply a very strong luminosity J_{NE} of Hawking radiation in the NE model in comparison with the luminosity J_{empty} in the evaporation in an empty space with ignoring grey body factor. On the other hand J_{empty} is very strong as explained in §1 of [1]. Hence, J_{NE} may be a huge luminosity. The energy emission by a black hole in NE model may be understand as a strong "burst".

In addition to the luminosity of Hawking radiation, we consider the duration δt_{dyn} of the highly nonequilibrium dynamical stage. Since the shrinkage speed of black hole radius $v := |dR_g/dt|$ is approximately unity $v \sim 1$ during highly nonequilibrium dynamical stage (see condition (41)), the duration is estimated as $\delta t_{dyn} \sim R_g^*/v \sim R_g^*$, and the following relation is obtained,

$$\delta\tau := \frac{\delta t_{dyn}}{t_{empty}} \sim \frac{R_g^*}{t_{empty}} < \frac{R_g(0)}{t_{empty}} = \frac{N}{1280\pi R_g(0)^2} \sim \frac{1}{40 R_g(0)^2} \ll 0.025\,, \tag{78}$$

where eq.(3) is used, and, since the initial radius $R_g(0)$ should be large enough to consider a semi-classical evaporation stage, we introduced relations $R_g^* < R_g(0)$ and $1 \ll R_g(0)$. This denotes $\delta t_{dyn} \ll t_{empty}$. Furthermore, for example, we find the shortest $t_{NE} = O(0.1) \times t_{empty}$ from list (70). This together with (78) imply $\delta t_{dyn} \ll t_{NE}$. Hence it seems reasonable to consider that δt_{dyn} is very shorter than t_{NE}. (For example it seems that the tangent dT_g/dt seen in each panel in fig.6 becomes very large quickly as $t \to t_{NE}$, and T_g will reach Planck temperature quickly just after t_{NE}.) Hence it is suggested that the energy $\Delta E_g := E_g^* - E_p$ bursts out of black hole with a very strong luminosity within δt_{dyn} which is negligibly short in comparison with t_{NE}.

From the above, we suggest the following: When a black hole evaporates in the framework of NE model under the condition $R_h \gtrsim 60$, a quasi-equilibrium evaporation stage continues until t_{NE}. Then a highly nonequilibrium dynamical evaporation stage occurs at t_{NE}. In that stage, a semi-classical black hole of radius $R_g^* \gtrsim 1$ evaporates abruptly (within a negligibly short time scale δt_{dyn}) to become a quantum one. This abrupt evaporation in the highly nonequilibrium dynamical stage is accompanied by a burst of energy ΔE_g. We call this phenomenon *the abrupt catastrophic evaporation* at semi-classical level $R_g^* \gtrsim 1$, where "catastrophic" means the shrinkage speed of black hole radius is very high $v \sim 1$ and the energy ΔE_g bursts out of black hole with a huge luminosity $J_{NE} \gg J_{empty}$ within a negligibly short time scale $\delta t_{dyn} \ll t_{NE}$.

The above discussion is based on the NE model. As shown in previous subsection, for the equilibrium model used in [1] and the black hole evaporation in an empty space with ignoring grey body factor, the black hole radius becomes Planck size before the shrinkage speed of black hole radius reaches unity. The highly nonequilibrium dynamical stage and the abrupt catastrophic evaporation at semi-classical level do not occur in those models. Hence the abrupt catastrophic evaporation at semi-classical level seems to be a specific nonequilibrium phenomenon suggested by NE model.

Here we discuss about a black hole evaporation in an empty space in a full general relativistic framework. Note that, even if a black hole is in an empty space, there should exist an incoming energy flow onto the black hole due to the curvature scattering. When the curvature scattering is taken into account for the case of a black hole evaporation in an empty space, we can interpret the whole system as if a black hole is surrounded by some nonequilibrium matter fields which possess outgoing and incoming energy flows of Hawking radiation under the effects of curvature scattering. Furthermore, since the curvature scattering occurs whole over the spacetime, it is expected that the nonequilibrium region is so large that a condition corresponding to $R_h \gtrsim 60$ in NE model holds. Hence, if the NE model is extended to a full general relativistic model, we can expect that a black hole evaporation in an empty space can be treated in the framework of full general relativistic version of NE model (with removing the heat bath), and that the abrupt catastrophic evaporation at semi-classical level may occur as well since the nonequilibrium region is sufficiently large.

Finally we estimate a typical time scale of black hole evaporation with energy accretion. It is reasonable to consider the duration of quantum evaporation stage is about one Planck time. Then, the time scale of black hole evaporation t_{ev} is estimated as

$$t_{ev} \simeq t_{NE} + \delta t_{dyn} + 1 \sim t_{NE} \,. \tag{79}$$

The time t_{NE} gives a typical time scale of black hole evaporation with energy accretion.

4.6. Beyond the NE Model II: Final Fate of Quantum Black Hole Evaporation

So far we have considered semi-classical evaporation stages in the framework of NE model, and found it consists of two stages, quasi-equilibrium one and highly nonequilibrium dynamical one. This subsection discusses the quantum evaporation stage following the highly nonequilibrium dynamical stage.

Concerning quantum black hole evaporation, the so-called *information loss paradox* is an interesting and important issue [20]: If the black hole mass is radiated out completely by the Hawking radiation, then the thermal spectrum of Hawking radiation implies that only a matter field which is in a thermal equilibrium state may be left after the evaporation. This implies that the initial condition of black hole formation (gravitational collapse) is completely smeared out. For example, even if the initial state of collapsing matter is a pure quantum state, the final state after a black hole evaporation must be transformed to a thermal state. This contradicts the unitary invariance of quantum theory. This is the information loss paradox.

The study on the final fate of black hole evaporation is usually carried out in the context of quantum gravity theory. However, since no complete theory of quantum gravity has yet been constructed, it is meaningful to some extent to study the final fate of black hole evaporation with an appropriate model of black hole evaporation without referring to present incomplete quantum gravity theories. Therefore, we use the NE model and try to extract what we can suggest about the final fate of black hole evaporation.

In this subsection, we consider whether some remnant remains after the quantum evaporation or not. If a remnant remains, then it is implied that the information loss paradox does not exit, because the complete evaporation may not be true and the remnant may preserve the information about initial condition of black hole formation to guarantee the unitary evolution of the system. For the time being, we assume that a black hole evaporates out completely and only equilibrium radiation fields remain at the end state of quantum evaporation stage. If a contradiction results from this assumption, we may conclude that a remnant will remain after a black hole evaporation. This subsection aims to suggest a necessity of some remnant by the reductive absurdity.

As discussed at the end of previous subsection, it is expected that the abrupt catastrophic evaporation at semi-classical level occurs in a full general relativistic framework. Therefore we consider the case $R_h \gtrsim 60$ in the NE model which denotes the occurrence of abrupt catastrophic evaporation. Then, under the assumption of complete evaporation of black hole, we can draw a scenario in the framework of NE model as follows:

As a quasi-equilibrium stage of black hole evaporation proceeds until t_{NE}, the temperature T_g approaches a critical value $T_g^* = 1/4\pi R_g^*$, where $R_g^* := R_g(t_{NE})$. Then a highly nonequilibrium dynamical stage and a quantum evaporation stage follow successively. Under the assumption of complete evaporation, these successive stages are described as one phenomenon; a black hole emits completely its mass energy $E_g^*\ (= R_g^*/2)$ within a negligibly short time scale $\delta t_{dyn} + 1\ (\ll t_{NE})$, and equilibrium radiation fields remain. Hereafter in this subsection until a contradiction will be derived, the abrupt catastrophic evaporation under the assumption of complete evaporation means those successive stages, highly nonequilibrium dynamical one and quantum evaporation one. Then the abrupt catastrophic evaporation is described by a replacement of a black hole of radius R_g^* by equilibrium radiation fields of volume $V_g^* = (4\pi/3)\, R_g^{*3}$. In this replacement, thermodynamic states of heat bath and nonequilibrium radiation fields around the region of V_g^* are not changed (see the upper part in fig.8). Here we assume the energy conservation so that the energy of

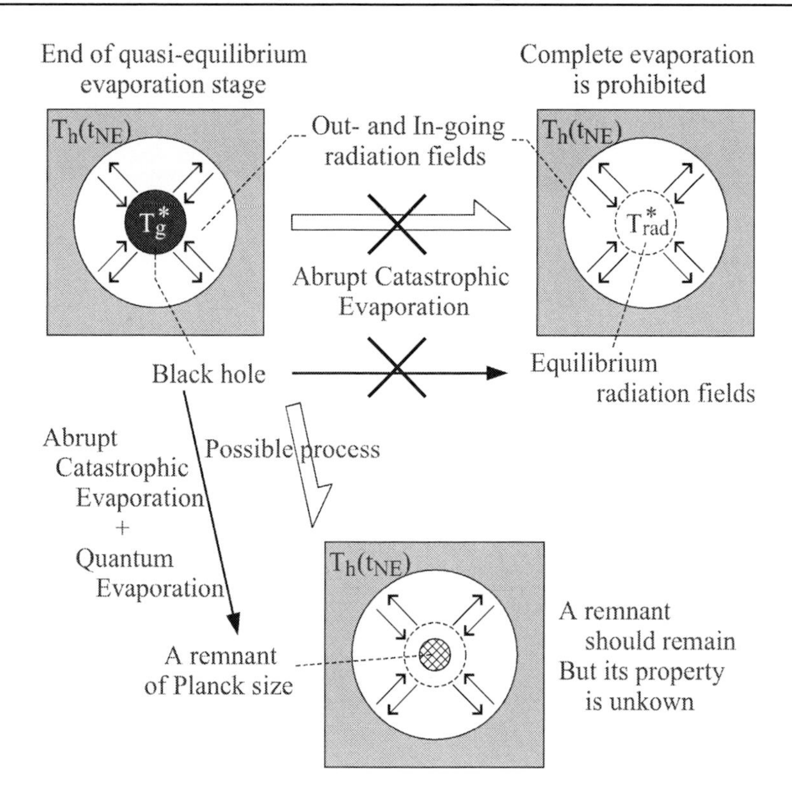

Figure 8. If a black hole evaporates out completely, when a black hole temperature T_g reaches T_g^*, the black hole is suddenly replaced by radiation fields of temperature T_{rad}^* which is determined by the energy conservation. However it is impossible from entropic viewpoint, and a remnant should remain. But its equations of states are unknown.

equilibrium radiation fields in volume V_g^* equals the mass energy E_g^* of black hole,

$$E_g^* = 4\,\sigma'\,T_{rad}^{*\,4}\,V_g^*\,, \tag{80}$$

where $4\sigma'\,T^4\,V$ is the equilibrium energy of radiation fields. This gives

$$T_{rad}^* = \left(\frac{3}{32\,\sigma'\,\pi\,R_g^{*\,2}}\right)^{1/4} . \tag{81}$$

When a sufficiently long time has passed after the abrupt catastrophic evaporation (under the assumption of complete evaporation), the whole system reaches an equilibrium state in which a heat bath and radiation fields in the hollow region have the same equilibrium temperature. However, without considering such totally equilibrium end state of the whole system, but with considering the states just before and just after the abrupt catastrophic evaporation, we can discuss whether a remnant remains or not after a quantum black hole evaporation.

It is reasonable to require the increase of total entropy along black hole evaporation, because of the isolated condition of the whole system in NE model. Therefore $S_{tot}^* < S_{tot}^{*\,\prime}$

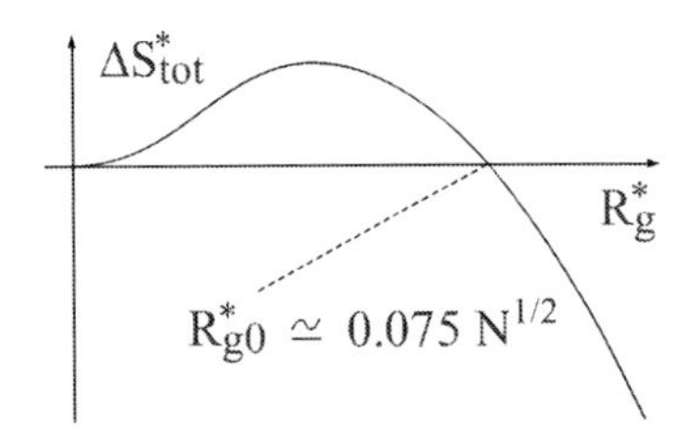

Figure 9. Schematic graph of ΔS^*_{tot}. A prohibition against complete evaporation of black hole is suggested.

holds, where S^*_{tot} is the total entropy of the whole system at time t_{NE} just before the abrupt catastrophic evaporation, and $S^{*\prime}_{tot}$ is the total entropy of the whole system just after the complete evaporation of black hole. Because the abrupt catastrophic evaporation is described by a simple replacement of a black hole by equilibrium radiation fields without changing thermodynamics states of heat bath and nonequilibrium radiation fields, the entropy difference is given by

$$\Delta S^*_{tot} := S^{*\prime}_{tot} - S^*_{tot} = S^*_{rad} - S^*_g \,, \tag{82}$$

where S^*_g is black hole entropy of temperature T^*_g and S^*_{rad} is equilibrium entropy of radiation fields of volume V^*_g and temperature T^*_{rad}. If the complete evaporation of black hole is true of the case, a relation $\Delta S^*_{tot} > 0$ must hold. However it is impossible as follows:

Using equations of states (4) and the equilibrium entropy of radiation fields $(16\sigma'/3)\, T^{*\,3}_{rad} V^*_g$ together with eq.(81), the entropy difference becomes

$$\Delta S^*_{tot} = \frac{2}{3}\left(\frac{32\,\pi\,\sigma'}{3}\right)^{1/4} R^{*\,3/2}_g - \pi\, R^{*\,2}_g \,. \tag{83}$$

A schematic graph of ΔS^*_{tot} is shown in fig.9, where $R^*_{g0} := (8/27\sqrt{5\pi})\sqrt{N} \simeq 0.075\sqrt{N}$ is given by $\Delta S^*_{tot} = 0$ and eq.(2). We find $\Delta S^*_{tot} > 0$ for $R^*_g < R^*_{g0}$ and $\Delta S^*_{tot} < 0$ for $R^*_g > R^*_{g0}$. Here recall that, because of $C^{(g)}_Y > 0$ by definition and a behavior of $C_X(R_g)$ explained in subsection 4.3.1. (left panel in fig.3), an inequality $R^*_g > \tilde{R}_g$ holds due to the validity condition (40), where $\tilde{R}_g$ is given by $C_X(R_g) = 0$. Furthermore, using eq.(51), we find $\tilde{R}_g/R^*_{g0} \simeq (27\sqrt{5\pi}/32(30\pi)^{1/3}) \times (R_h^2/N)^{1/6} \simeq 0.734 \times (R_h^2/N)^{1/6}$. Then, by eq.(3) and requirement $R_h \gtrsim 60$ of the occurrence of abrupt catastrophic evaporation, we obtain

$$\frac{R^*_g}{R^*_{g0}} > \frac{\tilde{R}_g}{R^*_{g0}} \simeq 1 \,. \tag{84}$$

This indicates $\Delta S^*_{tot} < 0$. Consequently, by the reductive absurdity as explained in fourth paragraph in this subsection, a complete evaporation of black hole is impossible (see fig.8).

From the above, we give a suggestion as follows: *a complete evaporation of black hole is prohibited and a remnant should remain after a black hole evaporation. The entropy of remnant should guarantee the increase of total entropy*. This implies disappearance of the information loss paradox.

Here we comment about the mass of remnant. One may think that the mass energy (or internal energy) of remnant may be extracted by some energetics. One may obtain a minimum entropy $S_{min}^{(rem)}$ of remnant as $S_{min}^{(rem)} = S_g^*$, and a mass energy of remnant may be obtained from $S_{min}^{(rem)}$. However, because equations of states of remnant is unknown, it is impossible to obtain an energy of remnant from $S_{min}^{(rem)}$. On the other hand, as discussed in subsection 4.5. (in eq.(78)), the mass of quantum black hole is expected to be about one Planck energy. Therefore, as long as the NE model is extrapolated over its validity, the mass of remnant seems to be of the order of one Planck energy.

In the rest of this subsection, we discuss more about the validity of discussion given in deriving inequality (84). There are three points we are going to discuss here. For the first, we discuss about the usage of NE model in this subsection. If we want to analyze details of the final stage of black hole evaporation process, then a quantum gravity is necessary, since a black hole becomes quantum, $R_g < 1$. On the other hand, because of the quasi-equilibrium assumption, the black hole radius has to be restricted as $(R_g(0) >) R_g^* \gtrsim 1$ in the framework of NE model. However this restriction $R_g^* \gtrsim 1$ does not mean an impossibility of NE model for considering the final fate of black hole evaporation. Although the NE model is based on a classical Schwarzschild black hole, once one assumes a complete evaporation, then it is true that radiation fields remain and their entropy can be counted after a quantum evaporation process ended. This means, even though we never refer to any detail of present incomplete quantum gravity theories, we can compare an entropy of black hole before the start of quantum evaporation process with an entropy of radiation fields after the end of quantum evaporation process. Hence our analysis done above seems to be reasonable to approach the final fate of black hole evaporation, and does not include the "uncertainty" of present incomplete quantum gravity theories. Our result seems more universal than the other discussions based on present incomplete quantum gravity theories.

For the second point, we turn our discussion to the starting assumption that a black hole evaporates out completely and equilibrium radiation fields remain. As seen above, eq.(83) has led a contradiction to deny the assumption and to result in necessity of a remnant. Here recall that eq.(83) is calculated with using the equilibrium entropy of radiation fields. Then, one may think it is more general to modify the starting assumption so that radiation fields after complete evaporation are not necessarily in an equilibrium. However, as explained in detail in [13], the nonequilibrium entropy of radiation fields should be smaller than the equilibrium one S_{rad}^*, since the equilibrium state is the maximum entropy state. Hence it is reasonable to expect that an inequality $\Delta S_{tot}^* < 0$ holds stronger for a modified starting assumption which requires nonequilibrium radiation fields after complete evaporation of black hole.

For the third point, we comment about black hole evaporation in an empty space (a situation without heat bath). As discussed at the end of subsection 4.5., the abrupt catastrophic evaporation at semi-classical level is expected to occur for a black hole evaporation in an empty space in a full general relativistic framework. Recall that calculations of entropy difference ΔS_{tot}^* in eq.(83) do not depend on the outside of black hole, but depend only on the black hole entropy S_g^* and the equilibrium entropy S_{rad}^* of radiation fields. This means, without respect to the degree of nonequilibrium nature of the environment around black hole, once the abrupt catastrophic evaporation of black hole occurs, the same discussion as given in this subsection seems applicable to a black hole evaporation in an empty space. Therefore, a full general relativistic treatment of black hole evaporation in an empty space may result in necessity of a remnant at the end state of black hole evaporation.

4.7. Summary of This Section

In this section, using the NE model, we found followings:

- The semi-classical evaporation stage consists of two stages, quasi-equilibrium one ($t < t_{NE}$) and highly nonequilibrium dynamical one ($t_{NE} < t$). The former is treated by the NE model, while the latter is not. If the steady state thermodynamics for radiation fields were not applied, the latter stage did not found. The existence of the latter stage is the very result of NE model.

- In the quasi-equilibrium evaporation stage, if the thickness of the hollow region is thin enough, then $J_{NE}/J_{empty} < 1$ holds during the black hole evaporation process and $t_{NE} > t_{empty}$ is obtained. However if the hollow region is thick enough, then $J_{NE}/J_{empty} > 1$ holds during the evaporation process and $t_{NE} < t_{empty}$ is obtained. The nonequilibrium effect of energy exchange between black hole and its outside environment tends to accelerate the evaporation process. Because of the negative heat capacity of black hole, an inverse sense against a naive sense based on positive heat capacity is offered; the more amount of energy is extracted from black hole by heat bath, the more rapidly the black hole emits its mass energy.

- Duration of the highly nonequilibrium dynamical stage δt_{dyn} is negligibly shorter than t_{NE}. Energy emission rate by black hole in NE model J_{NE} at t_{NE} is very stronger than that in an empty space J_{empty}. These imply a huge energy burst during highly nonequilibrium dynamical stage. Such huge burst is the very result of NE model, and we call it the abrupt catastrophic evaporation.

- Since the duration of quantum evaporation stage following the highly nonequilibrium dynamical stage will be about one Planck time, the time scale of black hole evaporation t_{ev} is estimated as $t_{ev} \simeq t_{NE} + \delta t_{dyn} + 1$. This gives $t_{ev} \sim t_{NE}$.

- By extrapolating the NE model to the highly nonequilibrium dynamical evaporation stage and quantum evaporation stage, a complete evaporation of black hole after the quantum evaporation stage is prohibited. This denotes a remnant of Planck size may remain at the end of quantum evaporation stage in order to guarantee the increase of total entropy along the whole process of evaporation. This implies a disappearance of the information loss paradox due to the nonequilibrium effect of energy exchange between black hole and its outside environment. This suggestion does not depend on present incomplete theories of quantum gravity.

5. Physical Essence of the Generalized Second Law

Now we apply the steady state thermodynamics for radiation fields to the NE model, and modify it to reveal the physical essence which guarantees the validity of generalized second law of black hole thermodynamics. Contents of this section are based on [21], and written without referring to previous section 4.. Readers interested in the generalized second law may skip over previous section.

5.1. GSL in the Context of Black Hole Evaporation

According to the black hole thermodynamics [1, 2, 3], a classical size black hole ($R_g > 1$) is regarded as an object in thermal equilibrium, whose equations of states are given in eq.(4). In a general relativistic framework, the statement of generalized second law in the context of black hole evaporation is as follows [3]:

Generalized second law (GSL): When a black hole evaporates, its mass energy E_g decreases and consequently the black hole entropy S_g decreases because of the equations of states (4). However the total entropy of the whole system which consists of the evaporating black hole, the Hawking radiation and any other matters around black hole, must increase.

In this statement we consider all matter fields on black hole spacetime plus the black hole itself. This denotes the whole system under consideration is isolated. Hence the GSL requires that the time evolution of that system is a relaxation process and the total entropy increases.

The original form of GSL was a conjecture that the horizon area (divided by 4) is regarded as a true black hole entropy [3]. However refer to the work [10] which were written after the suggestion of original GSL and revealed the horizon area (divided by 4) is the "equilibrium" entropy of whole gravitational field on black hole spacetime. Therefore, throughout this chapter, we consider the horizon area is a true equilibrium entropy of black hole and the GSL has already been proven [10, 22]. The equilibrium black hole entropy given by the horizon area is S_g in eq.(4), which is called the Bekenstein-Hawking entropy. This section aims not to prove GSL, but to reveal a physical essence which guarantees the validity of GSL.

When we consider GSL in the context of black hole evaporation, the nonequilibrium nature in the outside environment around black hole should be taken into account. This means that, while a classical size evaporating black hole is described by an equilibrium solution of Einstein equation under the quasi-equilibrium assumption, the outside environment should be treated as a nonequilibrium matter. Then we have to define the total entropy of the whole system, which consists of the evaporating black hole and the nonequilibrium matter fields around black hole including Hawking radiation. It has already been shown in [23] that, for the equilibrium state of a black hole with general "self-interacting" matter fields (a heat bath), the total "equilibrium" entropy is given by a simple sum of Bekenstein-Hawking entropy S_g and equilibrium entropy of those matter fields. This simple sum of equilibrium entropies is obtained in a general relativistic framework using the Euclidean path-integral method for a black hole spacetime and equilibrium matter fields on it. Extending those equilibrium results, we assume that the total entropy S_{tot} of the whole system which consists of evaporating black hole and nonequilibrium matter fields is also given by a simple sum,

$$S_{tot} := S_g + S_m \, , \tag{85}$$

where S_g is the black hole entropy given in eq.(4) and S_m is the nonequilibrium entropy of the general self-interacting matter fields including Hawking radiation. Here we consider a quasi-equilibrium regime of black hole evaporation process and use the equilibrium Bekenstein-Hawking entropy for black hole entropy S_g. The GSL requires the following inequality during the black hole evaporation process,

$$dS_{tot} > 0\,. \tag{86}$$

More discussions on the validity of additivity (85) and of the existence of a well-defined nonequilibrium entropy S_m for an arbitrary matter are given later in subsection 5.3.2..

The problem is how to deal with the nonequilibrium entropy S_m of matter fields. However, because a general definition of nonequilibrium entropy is not formulated, the existing proofs of GSL consider an equilibrium between a black hole and a heat bath surrounding it, or assume the existence of a well-defined nonequilibrium entropy of arbitrary matters [10, 22]. The equilibrium settings succeeded in proving the GSL, but unfortunately the physical essence of GSL remains veiled. To understand the veil over GSL, it is important to know explicitly the situation considered in the equilibrium settings; the existing proofs of GSL consider self-interacting matter fields as Hawking radiation and any other matters around black hole. Then, shift from equilibrium to black hole evaporation, we can recognize there are three physical origins of GSL:

Origin of GSL (a): Self-interactions of the matter fields of Hawking radiation and around black hole. These interactions are collision of composite particles and self-gravitation of matter fields. The self-interactions causes a self-relaxation of matter fields and produce their entropy. Self-interacting matter fields have a positive entropy self-production rate.

Origin of GSL (b): Gravitational interaction between black hole and matter fields. This interaction consist of curvature scattering, gravitational redshift and so on. The gravitational field around black hole works as if a virtual medium on which matter fields propagate, then the composite particles of matter fields interact with the virtual medium to result in a relaxation towards some equilibrium state. Therefore the gravitational interaction between black hole and matter fields produces matter entropy as well as the origin (a). A positive entropy production rate of matters results.

Origin of GSL (c): Increase of black hole temperature T_g along black hole evaporation due to the negative heat capacity $C_g < 0$ given in eq.(5), becomes one of origins of GSL as follows: When $dE_g < 0$ due to the evaporation, T_g increases since $dT_g = dE_g/C_g > 0$. Then, because the matter fields of Hawking radiation are ordinary matters of positive specific heat, we find that, the more evaporation process proceeds, the more matter entropy the evaporating black hole radiates out in Hawking radiation. This is not the entropy production inside the matter fields during propagating outside the evaporating black hole like origins (a) and (b), but the growth of the entropy emission rate by black hole along its evaporation. Because the negative heat capacity is a peculiar property of self-gravitating systems, this origin (c) is a self-gravitational effect of black hole on its own thermodynamic state.

In the existing proofs of GSL, all of these origins are included and it remains unclear which of these dominates over the others. If the GSL would be proven by considering a situation keeping one of them and discarding the others, then we can conclude that the one kept is the physical essence which guarantees the validity of GSL. This section aims to reveal the origin (c) is the physical essence of GSL.

To do so, we utilize the NE model. To pick up the origin (c), we remove the heat bath from NE model, and put the black hole in an empty and infinitely large flat spacetime. The black hole is bared in this situation. We consider this situation throughout the present section, and call it *the bare NE model*.

The bare NE model includes the self-gravitational effect of black hole on its own thermodynamic state through equations of states (4), but ignores the self-interactions (including self-gravitational interaction) of radiation fields and the gravitational interaction between black hole and radiation fields. This means that the bare NE model includes only the origin (c) and ignores the so-called *grey body factor*.

Here it should be pointed out that, while the quasi-equilibrium assumption is also valid for the bare NE model for classical size black hole evaporation, but the fast propagation assumption breaks down since the radiation fields spread out into an infinitely large space. Hence, we have to take the retarded effect on radiation fields into account. However because of the quasi-equilibrium assumption, we ignore the special relativistic Doppler effects due to the shrinkage of black hole surface.

Here one may show an example as an objection: When an inter-stellar gas collapses to form a star, the self-gravitational effect of that gas decreases its entropy. Then the origin (c), self-gravitational effects of black hole, may not be the essence of GSL. However as discussed later in subsection 5.3.3., this objection is not true of the black hole evaporation.

Because the bare NE model does not include the origins (a) and (b), the radiation fields have zero entropy production rate during propagating in an empty and infinitely large space. Therefore we find

$$dS_{tot} > dS_{NE} \,, \tag{87}$$

where S_{tot} is the total entropy with general self-interacting matter fields given in eq.(85), and S_{NE} is the total entropy of the bare NE model. Under the assumption of simple sum for nonequilibrium entropies considered in eq.(85), S_{NE} is decomposed as

$$S_{NE} := S_g + S_{rad} \,, \tag{88}$$

where S_{rad} is the nonequilibrium entropy of radiation fields propagating in an empty and infinitely large flat spacetime. Therefore, if an inequality $dS_{NE} > 0$ holds, then the GSL $dS_{tot} > 0$ follows and we can conclude the origin (c) is the physical essence of GSL. Next subsection shows $dS_{NE} > 0$ holds.

5.2. Time Evolution of Total Entropy S_{NE}

In this subsection, we calculate a time evolution of $S_{NE}(t)$ to show the inequality $dS_{NE} > 0$ along black hole evaporation. Here the time t corresponds to a proper time of a rest observer at asymptotically flat region if we can extend the bare NE model to a full general relativistic model. In the followings, we obtain explicit forms of $S_g(t)$ and $S_{rad}(t)$ as functions of t, then calculate the time evolution of $S_{NE}(t)$ under the assumption of simple sum given in eq.(88).

5.2.1. S_g as a Function of Time

Time evolution of black hole radius R_g is given by the Stefan-Boltzmann law,

$$\frac{dE_g}{dt} = -\sigma' T_g^4 A_g , \tag{89}$$

where σ' is the generalized Stefan-Boltzmann constant and $A_g := 4\pi R_g^2$ is the surface area of black hole. This together with eq.(4) gives

$$R_g(t) = R_0 \left(1 - \frac{N t}{1280\,\pi\, R_0^3}\right)^{1/3} , \tag{90}$$

where $R_0 := R_g(0)$ is the initial radius, and it is assumed that the emission of Hawking radiation starts at $t = 0$. This $R_g(t)$ leads the time evolution of thermodynamic states of black hole and radiation fields. Eq.(90) gives the evaporation time (life time) of black hole in the framework of the bare NE model,

$$t_{empty} := 1280\,\pi\, \frac{R_0^3}{N} . \tag{91}$$

Using eqs.(90) and (4), we obtain $S_g(t)$ as a function of time.

Here let us consider about the quasi-equilibrium assumption. To validate this assumption, a sufficiently slow evaporation is required. Then we discuss t_{spread} and v, where t_{spread} is a time scale of radiation fields to spread out into infinitely large space, and v is a shrinkage speed of black hole radius. For the first consider the time scale. t_{spread} is given by a particle of radiation fields traveling across the size of black hole,

$$t_{spread} := R_0 . \tag{92}$$

This t_{spread} gives a typical time scale of radiation fields to spread out into the empty and infinitely large space. If t_{empty} is longer than t_{spread}, we can consider the black hole evaporation proceeds slowly. Hence to validate the quasi-equilibrium assumption, we consider the case satisfying the following inequality,

$$\lambda := \frac{t_{empty}}{t_{spread}} = 1280\pi \frac{R_0^2}{N} > 1 . \tag{93}$$

For classical size initial condition $R_0 > 1$ together with eq.(3), this requirement (93) is relevant. For the second consider the shrinkage speed of black hole radius,

$$v := \left| \frac{dR_g(t)}{dt} \right| = \frac{R_0^3}{3\, t_{empty}} \frac{1}{R_g(t)^2} , \tag{94}$$

where eq.(90) is used. To validate the quasi-equilibrium assumption, v should be slow enough, $v < 1$. This gives

$$t < t_{empty} - \frac{\sqrt{N}}{48\sqrt{15\,\pi}} . \tag{95}$$

This requirement together with eq.(3) means that, because of $\sqrt{N/15\pi}/48 < 1$, the quasi-equilibrium assumption is valid at least until one Planck time before evaporation time t_{empty}.

5.2.2. S_{rad} as a Function of Time

We will obtain the nonequilibrium entropy S_{rad} as a function of t by applying the steady state thermodynamics for radiation fields. But before proceeding to that calculation, the retarded effect on radiation fields has to be introduced into the steady state thermodynamics, since the fast propagation assumption breaks down in the bare NE model. In this subsection, firstly we construct a modified distribution function for composite particles of radiation fields, then obtain S_{rad} by substituting that distribution function into formulae (9) and (13). Because of the quasi-equilibrium assumption, we ignore the special relativistic Doppler effects due to the shrinkage of black hole surface.

Hereafter r denotes the areal radius from the center of black hole. To take the retarded effect into account, consider a particle of radiation fields emitted by black hole at time $\tilde{t}$ and reaches a spatial point of r at time $t\,(> \tilde{t})$. The emission time $\tilde{t}$ depends on not only the coordinates (t, r) but also an angle θ between radial direction of r and momentum of the particle under consideration (see left panel in fig.10). This emission time $\tilde{t}(t, r, \theta)$ is obtained as a root of the equation,

$$R_g(\tilde{t})^2 = \left(t - \tilde{t}\right)^2 + r^2 - 2\left(t - \tilde{t}\right) r \cos\theta \,. \tag{96}$$

This is the equation of degree six about $\tilde{t}$, and an appropriate root as the emission time is the maximum root in range, $0 \leq \tilde{t} \leq t$. Although another root may exist in this range, however a non-maximum root corresponds to a particle emitted at point q in left panel in fig.10 which is obviously unphysical. The other four roots of eq.(96) may be of complex valued.

Furthermore the angle θ has an upper bound, $0 \leq \theta \leq \theta_m(t, r)$. To find an explicit expression for θ_m, look at a particle of radiation fields emitted by black hole at the initial time $t = 0$, which we call *an initial particle*. Obviously, we have $\theta_m = 0$ for $t < r - R_0$, since no particle has been reached a spatial point of radial distance r. The initial particles emitted in radial direction form the boundary of region filled with radiation fields. The initial particles emitted in off-radial directions propagate behind the initial particles emitted in radial directions. Therefore, the initial particles reach a point of radial distance $r > R_0$ in a time interval, $r - R_0 \leq t \leq \sqrt{r^2 - R_0^2}$. Within this time interval, the upper bound θ_m is given by initial particles, which is obtained with setting $\tilde{t} = 0$ in left panel in fig.10,

$$\cos\theta_m(t, r) = \frac{t^2 + r^2 - R_0^2}{2\,t\,r} \quad , \quad \text{for } r > R_0 \text{ and } r - R_0 \leq t \leq \sqrt{r^2 - R_0^2}\,. \tag{97}$$

For $t > \sqrt{r^2 - R_0^2}$, the upper bound θ_m is not given by any initial particle but by a particle emitted at point b at time t_m shown in right panel in fig.10. Then we find

$$\cos\theta_m(t, r) = \frac{t - t_m(t, r)}{r} \quad , \quad \text{for } r > R_0 \text{ and } \sqrt{r^2 - R_0^2} < t \,, \tag{98}$$

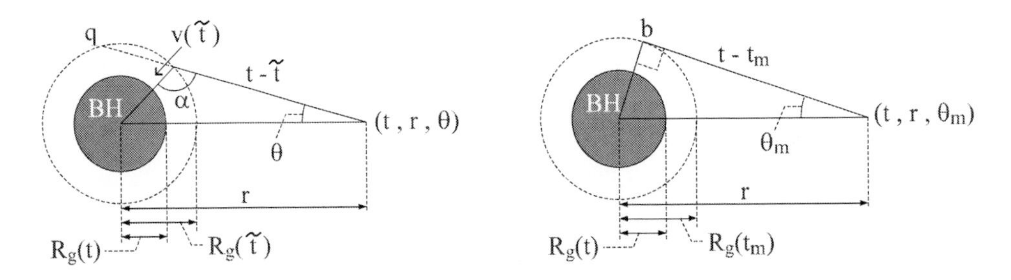

Figure 10. Left panel shows retarded effect on radiation fields. Right one shows the upper bound of θ for $r < R_0$, or $r < R_0$ and $t > \sqrt{r^2 - R_0^2}$.

where the time $t_m(t, r)$ is a real valued root of the equation,

$$r^2 = R_g(\, t_m \,)^2 + (\, t - t_m \,)^2 \,. \tag{99}$$

This is the equation of degree six about t_m. The appropriate root for t_m should be in the range, $0 \le t_m \le t$. This root may be degenerated, since the trajectory of this particle is tangent to the sphere of radius $R_g(t_m)$, and the other four roots may be of complex valued. Finally turn to a point of radial distance $r \le R_0$. It is obvious for this point that the upper bound θ_m is also given by formula (98),

$$\cos\theta_m(t, r) = \frac{t - t_m(t, r)}{r} \quad , \quad \text{for } r \le R_0 \,. \tag{100}$$

From the above, the distribution function of radiation fields is given as

$$d(t, r; \omega, \theta) = \begin{cases} \dfrac{1}{\exp\left[\omega/\tilde{T}\right] \pm 1} & , \quad \text{for } \theta \le \theta_m(t, r) \\ 0 & , \quad \text{for } \theta > \theta_m(t, r) \end{cases} \tag{101}$$

where $\tilde{T} = T_g(\, \tilde{t}(t, r, \theta)\,)$, $\omega = |\vec{p}|$, and the signatures "$-$" and "$+$" are respectively for bosons and fermions. Here we note, while (t, r)-dependence in $d(t, r; \omega, \theta)$ expresses a spacetime dependence, (ω, θ)-dependence expresses a dependence on momentum $\vec{p}$ of particles of radiation fields.

The nonequilibrium entropy of radiation fields S_{rad} is obtained by substituting eq.(101) into eqs.(9) and (13),

$$S_{rad}(t) = \int_{R_g(t)}^{t+R_0} dr\, 4\,\pi\, r^2\, s_{rad}(t, r) \,, \tag{102}$$

where

$$s_{rad}(t, r) := \frac{N}{2880\,\pi} \int_{y_m(t,r)}^{1} dy\, \frac{1}{R_g(\,\tilde{t}(t, r, y)\,)^3} \,, \tag{103}$$

where the integrals in eq.(17) are used, and we set $y := \cos\theta$, $y_m := \cos\theta_m$ and $g_{\vec{p},\vec{x}} = n_b$ and n_f for bosons and fermions respectively. Since the emission of radiation fields starts at $t = 0$, the radiation fields fill the space in range,

$$R_g(t) < r < t + R_0 \,. \tag{104}$$

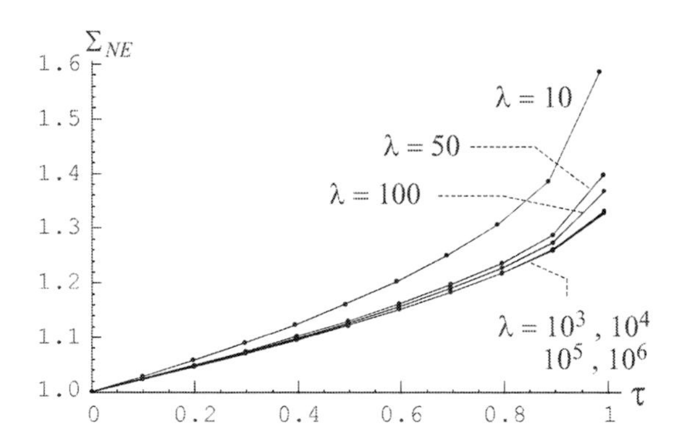

Figure 11. Time evolution of normalized total entropy $\Sigma_{NE}(\tau) := S_{NE}(t)/S_{NE}(0)$, where τ is a time normalized by evaporation time t_{empty}. The plotted curves are converging as λ increases.

5.2.3. Total Entropy

Now we obtain the total entropy S_{NE} of the bare NE model under the assumption of simple sum as in eq.(88),

$$S_{NE}(t) := S_g(t) + S_{rad}(t) , \tag{105}$$

where $S_g(t) = \pi\, R_g(t)^2$ is given by eq.(90), $S_{rad}(t)$ is by eq.(102) and time t corresponds to a proper time of a rest observer at asymptotically flat region if we can extend the bare NE model to a full general relativistic model. Analytic proof for $dS_{NE}/dt > 0$ is difficult. So we try to show it numerically. To do so, normalize S_{NE} as follows,

$$\tau := \frac{t}{t_{ev}} \quad , \quad \tilde{\tau} := \frac{\tilde{t}}{t_{ev}} \quad , \quad x := \frac{r}{R_0} \quad , \quad X_g(\tau) := \frac{R_g(t)}{R_0} = (\, 1 - \tau\,)^{1/3} \tag{106}$$

$$\Sigma_{NE}(\tau) := \frac{S_{NE}(t)}{S_{NE}(0)} \quad , \quad \sigma_{rad}(\tau, x) := \frac{s_{rad}(t, r)}{S_{NE}(0)} . \tag{107}$$

Then the normalized total entropy is

$$\Sigma_{NE}(\tau) = X_g(\tau)^2 + \frac{16}{9\,\lambda} \int_{X_g(\tau)}^{\lambda\tau+1} dx \int_{y_m(\tau,x)}^{1} dy\, \frac{x^2}{X_g(\,\tilde{\tau}(\tau, x, y)\,)^3} , \tag{108}$$

where λ is given in eq.(93). Because Σ_{NE} does not explicitly depend on R_0 (or N), then, once the value of λ is fixed, the choice of R_0 (or N) is arbitrary with adjusting the value of N (or R_0) appropriately to match with λ. We assume $\lambda > 1$ as mentioned in eq.(93).

Numerical results are shown in figs.11 and 12. Parameters R_0 and N are chosen so that inequality (93) holds and time interval $0 \le \tau \le 0.999$ is included in range (95). Those numerical calculations are carried out in that interval $0 \le \tau \le 0.999$ using *Mathematica* version 5.2.

Fig.11 shows time evolution of total entropy Σ_{NE} for $\lambda = 10, 50, 100, 10^3, 10^4, 10^5$ and 10^6. Here let us make a technical comment; since calculations by *Mathematica* took a very long time for double integral, fig.11 plots Σ_{NE} for $\tau = 0,\ 0.1,\ 0.2, \cdots, 0.9$, and 0.999

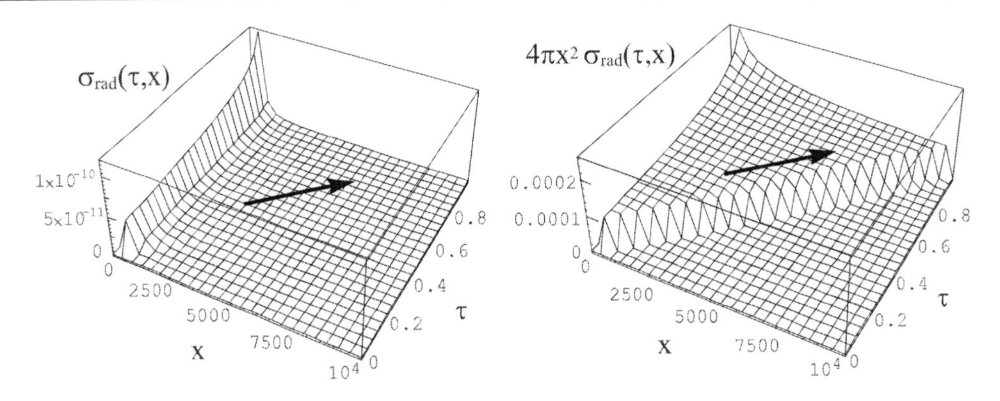

Figure 12. Spatial and temporal evolution of entropy density $\sigma_{rad}(\tau, x)$ and its spherical shell density $4\pi x^2\sigma_{rad}(\tau, x)$ for $\lambda = 10^4$, where $\sigma_{rad}(\tau, x) := s_{rad}(t, r)/S_{NE}(0)$ is a normalized spatial density of radiation fields, $x := r/R_0$ is a normalized spatial distance and $\tau := t/t_{empty}$ is a normalized time. The arrows denote direction of null world lines (geodesics).

for each λ. However any singular behavior (oscillation, divergence and so on) of Σ_{NE} is unexpected. Therefore, for the purpose of finding monotone increasing nature of Σ_{NE}, it is enough to show plots of some representative points like fig.11.

The plotted curves in fig.11 are converging as λ increases. Those plots for $\lambda \geq 10^3$ are almost coincident. What we can find with this figure is as follows:

- Because the plotted curves are converging, it is concluded that the total entropy Σ_{NE} is monotone increasing for $\lambda > 1$, and the GSL is well supported as explained in eqs.(87) and (88).
- Since the normalized black hole entropy $X_g(\tau)^2$ is obviously monotone decreasing, the entropy of radiation fields increases monotonously faster than the decrease of X_g^2.
- It is suggested rather well that $\Sigma_{NE}(\tau)$ gives almost the same curves in Σ-τ graph for all values of sufficiently large $\lambda \gtrsim 10^3$, and the final value is approximately given by $\Sigma_{NE}(0.999) \simeq 1.33$.

Left panel in fig.12 shows the normalized spatial entropy density for radiation fields $\sigma_{rad}(\tau, x)$, and right panel shows its normalized spherical shell density $4\pi x^2\sigma_{rad}(\tau, x)$. Fig.12 is made with $\lambda = 10^4$, however the same behavior is obtained for all values of sufficiently large $\lambda \gtrsim 10^3$ as far as the author checked. The arrows in fig.12 denote a direction of null world lines of particles of radiation fields. What we can find with this figure is as follows:

- Tracking the graphs in fig.12 along a null word line, we recognize the following: Although the spatial entropy density of radiation fields decreases (left panel), the spherical shell density on each spherical shell remains constant while it spreads out into an empty and infinitely large flat spacetime (right panel). This is consistent with the absence of self-relaxation of radiation fields and gravitational interaction between black hole and radiation fields.

- The entropy of radiation fields at black hole surface (right panel), which equals the amount of matter entropy emitted by black hole at each moment, increases monotonously as the black hole evaporation proceeds. This means an accelerated entropy emission by black hole.

Hence we find from figs.11 and 12 that the entropy emission rate by black hole increases faster than the decrease of black hole entropy. This indicates the physical essence of GSL is neither the origin (a) nor (b), but the origin (c) which is the self-gravitational effect of black hole causing a negative heat capacity and an increase of black hole temperature along its evaporation.

5.3. Summary and Supplementary Discussions of This Section

5.3.1. Summary

Considering the GSL in the context of black hole evaporation, we recognize there are three physical origins of entropy increase; (a) self-interactions of matter fields around black hole, (b) gravitational interaction between the matter and black hole, and (c) self-gravitational effect of black hole. The origins (a) and (b) give positive entropy production rates inside the matter fields. The origin (c) appears as an increase of black hole temperature due to the negative heat capacity and gives an increasing entropy emission rate by black hole. Then we consider the bare NE model constructed by removing heat bath from the NE model. The bare NE model describes a black hole evaporation including only the origin (c) and discarding the others (a) and (b). Applying the steady state thermodynamics to radiation fields in the bare NE model, we can calculate explicitly a time evolution of total entropy S_{NE}, and find S_{NE} increases monotonously. This denotes the GSL holds as explained in eqs.(87) and (88). Hence we conclude as follows:

Physical essence of GSL: The self-gravitational effect of black hole which appears as an increasing entropy emission rate by black hole guarantees a validity of GSL. The entropy emission rate by black hole increases faster than the decrease of black hole entropy. The entropy production inside the Hawking radiation and the other matter fields around black hole is not necessary for a validity of GSL.

The increase of entropy emission rate is shown by an accelerated increase of entropy density s_{rad} at black hole surface shown in fig.12 and causes the accelerated increase of total entropy S_{NE} shown in fig.11.

Furthermore we find another interesting result from fig.11: An inequality $\lambda > 10^3$ together with eq.(3) corresponds to black holes of initial radius $R_0 \gtrsim 5$. Then fig.11 shows that the normalized total entropy $\Sigma_{NE}(\tau)$ for black holes of initial radius $R_0 \gtrsim 5$ evolves in almost the same increasing fashion and reaches almost the same final value $\simeq 1.33$. On the other hand inequality (87) denotes S_{NE} is the lowest estimate for total entropy of a full general relativistic black hole evaporation including self-interaction of matter fields around black hole and gravitational interaction between black hole and the matters. Hence we find a result as follows:

Lower bound for growth of the total entropy: For black hole evaporation in a full general relativistic framework including self-interaction of matter fields and gravitational interaction between black hole and the matters, the final value of total entropy S_{tot} (not its lowest estimate S_{NE}) should be larger than $1.33 \times S_g(0)$.

Exactly speaking our analysis based on the bare NE model covers only a semi-classical evaporation stage until one Planck time before the time t_{empty} as discussed in eq.(95). However it is reasonable to expect that the total entropy after the end of quantum evaporation stage is greater than the total entropy just at the onset of quantum evaporation stage, since a well-defined entropy has to be of non-decreasing. Therefore the above result on the lower bound of total entropy should be true of the end state of quantum evaporation stage.

5.3.2. Supplementary Discussion 1

We discuss about two key assumptions remained to be proven, the existence of a well-defined nonequilibrium entropy S_m of an arbitrary self-interacting matter fields and the additivity of equilibrium and nonequilibrium entropies (85). These two assumptions seem to be reasonable as follows:

On the existence of a well-defined nonequilibrium entropy S_m, we refer to the present status of study on laboratory systems which have self-interactions but not self- and external-gravitational interactions. For example, consider a laboratory system is in a nonequilibrium state which is far from an equilibrium but whose heat flux is not extremely strong. Then the extended irreversible thermodynamics [4] gives a well-defined nonequilibrium entropy flux up to the second order in expansion by the heat flux. On the other hand, the evaporation time t_{empty} of a black hole has a very long time scale as explained in eq.(93). The black hole evaporation proceeds so slowly that the nonequilibrium state of self-interacting matter fields around black hole is not extremely far from an equilibrium. Therefore we can expect that the expansion of their thermodynamic quantities by the heat flux up to the second order is a good approximation. Hence it is reasonable to assume the existence of a well-defined nonequilibrium entropy of an arbitrary self-interacting matter fields according to extended irreversible thermodynamics.

Next consider about the additivity (85). As mentioned in the previous paragraph, the nonequilibrium entropy has already been defined well up to the second order in expansion by heat flux [4]. Furthermore that nonequilibrium entropy satisfies the additivity as in ordinary equilibrium thermodynamics. Therefore we may expect that, when an equilibrium entropy satisfies the additivity even under the effect of gravity, the nonequilibrium entropy in the framework of extended irreversible thermodynamics also satisfies the additivity. Hence, because of the additivity of equilibrium entropies of black hole and matter fields [23], it seems reasonable to assume the additivity (85).

5.3.3. Supplementary Discussion 2

Finally we try to answer an objection: When an inter-stellar gas collapses to form a star, the self-gravitational effect of that gas decreases its entropy. Then the black hole evaporation under the self-gravitational effect of black hole may not result in an increase of total entropy.

When an interstellar gas collapses to form a star, it is commonly believed that there arises the increase of net entropy of total system which consists of the collapsing gas and the radiated matters from the gas. However the self-gravitational effect of the collapsing gas causes the decrease of entropy of the collapsing gas. It is briefly explained as follows: Since the pressure of that gas at its surface is zero, the loss of energy of collapsing gas par a unit time ΔE due to energy emission is actually the loss of heat due to the first law of thermodynamics,

$$\Delta E \sim T\,\Delta S \tag{109}$$

where T is the temperature of collapsing gas, and ΔS is the "loss" of entropy of collapsing gas par a unit time. The energy loss ΔE is the minus of the luminosity L of collapsing gas, $L = -\Delta E$. Then, with the assumption of local mechanical and thermal equilibrium of collapsing gas at each moment of its collapse, relation (109) is rewritten more exactly to the following form [9],

$$L = -\frac{d[U+\Omega]}{dt} \sim -T\Delta S\,, \tag{110}$$

where U is the total internal energy of collapsing gas, and Ω is the total self-gravitational potential given by

$$\Omega = \int_0^M dm\left(-\frac{G\,m}{r(m)}\right)\,, \tag{111}$$

where G is Newton's constant, M is mass of collapsing gas, and the radial distance from the center of gas $r(m)$ is expressed as a function of mass m inside a sphere of radius r.

From the above, we recognize that the radius $r(m)$ becomes smaller as the gas collapses, then the self-gravitational potential Ω decreases to result in the radiation of energy $L > 0$ and the loss of entropy $\Delta S < 0$. This is a similar phenomenon to the evaporation of black hole itself with decreasing its entropy $dS_g < 0$. On the other hand, if we consider not a collapsing but an expanding self-gravitating gas, the radius $r(m)$ increases to result in the increase of entropy $\Delta S > 0$. Here we point out that the Hawking radiation in black hole evaporation corresponds not to a collapsing gas but to an expanding gas. If we consider the radiation fields of Hawking radiation including their self-gravitational effect, the entropy of radiation will increase during spreading out into an infinitely large space. This is just the origin (b) of GSL. Hence it seems that the entropy of self-gravitating matter fields of Hawking radiation is larger than that of a non-self-gravitating Hawking field. This implies $S_m > S_{rad}$, and the objection mentioned above is not true of the black hole evaporation process.

Finally let us recall a statement given in the second paragraph; it is commonly "believed" that there arises the increase of net entropy of total system which consists of the collapsing gas and the radiated matters from the gas. One of the reasons why it is not proven but "believed" is that there has not been nonequilibrium thermodynamics to treat the net entropy. Although we considered the black hole evaporation in this section, a similar method based on the steady state thermodynamics will be applicable to a star formation process including radiations from collapsing gas. Then "believed" will become "proven".

6. Conclusion

Exactly speaking the black hole evaporation is a nonequilibrium process. We used the NE model as a simplified thermodynamic model of black hole evaporation, and applied the steady state thermodynamics for radiation fields [13]. It is this steady state thermodynamics that enables us to treat the nonequilibrium nature of black hole evaporation. In the framework of NE model, we can find a detailed picture of evaporation process with energy accretion onto black hole [11], which is summarized in subsection 4.7.. Also, by modifying the NE model to describe the evaporation process in an empty space ignoring grey body factor for the Hawking radiation, we can find the essence of the generalized second law of black hole thermodynamics in the context of black hole evaporation [21], which is summarized in subsection 5.3..

One of the results in section 4. is that the larger the nonequilibrium region around black hole, the more accelerated the black hole evaporation with interacting outside environment. On the other hand, section 5. considered the bare NE model which is obtained by removing the heat bath from the NE model. The bare NE model is the black hole evaporation in an empty space with ignoring grey body factor, which is considered in subsection 4.4. to compare with the NE model. Then one may think that the bare NE model is an extreme case of NE model which has the largest nonequilibrium region around black hole, and the evaporation time of the bare NE model would be zero. However, as discussed in subsection 4.4.3., the bare NE model is not described as an extreme situation of NE model. The bare NE model is quite different from the NE model, since no energy accretion is in bare NE model while it is in NE model. Hence the bare NE model is not regarded as the largest nonequilibrium situation of NE model.

As mentioned section 2., the NE model is not a full general relativistic model, and ignoring gravitational redshift and curvature scattering on radiation fields propagating in hollow region. Towards a general relativistic NE model, we have to extend the steady state thermodynamics to its general relativistic version. If we will construct a general relativistic steady state thermodynamics for radiation fields, the gravitational effects on radiation fields will be included in the NE model. Even when the extended steady state thermodynamics for radiation fields are applied, the semi-classical evaporation stage is divided into two stages as shown in section 4., quasi-equilibrium one and highly nonequilibrium dynamical one. If we concentrate on the quasi-equilibrium evaporation stage, then the evaporating black hole can be expressed approximately by equilibrium solutions of Einstein equation. In that case, we can avoid the mathematical and conceptual difficulties for describing evaporating black holes (see the beginning of section 1.). However, in considering the highly nonequilibrium dynamical evaporation stage in a general relativistic setting, we will face those difficulties.

All of the results in this chapter are obtained by NE model and based on the steady state thermodynamics for radiation fields. Nonequilibrium thermodynamic approach may be a powerful tool for investigating the nonlinear and dynamical phenomena. While we considered the radiation fields which is of non-self-interacting, however nonequilibrium thermodynamics for ordinary dissipative systems has already been established in, for example, the extended irreversible thermodynamics [4]. It is applicable not to any highly nonequilibrium state but to state whose entropy flux is well approximated up to second order in the expansion by the heat flux of a nonequilibrium state under consideration. It is interesting to consider self-interacting matter field for the Hawking radiation and energy accretion, and apply the extended irreversible thermodynamics to those fields. Then we may find a variety of black hole evaporation phenomena. Furthermore apart from black hole evaporation, since accretion disks around black hole seem to consist of dissipative matters in realistic settings, the extended irreversible thermodynamics may give a unique new approach to investigate black hole astrophysics.

Apart from black hole physics, the steady state thermodynamics for radiation fields may be helpful to understand, for example, the free streaming in the universe like cosmic microwave background and/or the radiative energy transfer inside a star and among stellar objects. Also an example of possible application to a star formation process is explained at the end of subsection 5.3.3.. Keeping future expectation of such applications in mind, subsection 3.2. is devoted to exhibit a detail of the steady state thermodynamics for radiation fields.

References

[1] S.W.Hawking *Commun.Math.Phys.* 1975, 43, 199–220

[2] S.W.Hawking *Phys.Rev.Lett.* 1971, 26, 1344–1346
J.M.Bardeen, B.Carter and S.W.Hawking Commun.Math.Phys. 1973, 31, 161–170
P.C.W.Davies *Proc.R.Soc.Lond.* 1977, A353, 499–521
B.T.Sullivan and W.Israel *Phys.Lett.* 1980, A79, 371–372
W.Israel *Phys.Rev.Lett.* 1986, 57, 397–399
V.Iyer and R.M.Wald *Phys.Rev.* 1994, D50, 846–864

[3] J.D.Bekenstein *Phys. Rev.* 1973, D7, 2333–2346
J.D.Bekenstein *Phys. Rev.* 1974, D9, 3292–3300

[4] D.Jou, J.Casas-Vázquez and G.Lebon *Extended Irreversible Thermodynamics*; Springer, 1993

[5] Y.Oono and M.Paniconi *Prog.Theor.Phys.* 1998, suppl.130, 29–44
T.Hatano and S.Sasa *Phys.Rev.Lett.* 2001, 86, 3463–3466
S.Sasa and H.Tasaki (2001). Steady state thermodynamics for heat conduction. cond-mat/0108365
K.Hayashi and S.Sasa *Phys.Rev.* 2003, E68, 035104(R)
K.Hayashi and S.Sasa *Phys.Rev.* 2004, E69, 066119
S.Sasa and H. Tasaki (2004). Steady state thermodynamics. cond-mat/0411052

[6] K.Heyon-Deuk and H.Hayakawa *J.Phys.Soc.Jpn.* 2003, 72, 1904–1916
H.Hayakawa, T.H.Nishino and K.Heyon-Deuk (2004). Kinetic theory of dilute gases under nonequilibrium conditions. cond-mat/0412011

[7] H.-D.Kim and H.Hayakawa *J.Phys.Soc.Jpn.* 2003, 72, 2473–2476

[8] S.Chandrasekhar The Mathematical Theory of Black Holes; *International series of monographs on physics* **69**; Oxford Univ. Press, 1983

[9] J.Binney and S.Tremaine *Galactic Dynamics*; Princeton Univ. Press, 1987
R.Kippenhahn and A.Weigert *Stellar Structure and Evolution*; Springer, 1994

[10] G.W.Gibbons and S.W.Hawking *Phys.Rev.* 1977, D15, 2752–2756

[11] H.Saida *Class.Quant.Grav.* 2007, 24, 691–722

[12] R.C.Tolman *Relativity, Thermodynamics and Cosmology*; Dover, 1987 (its original edition; Oxford Univ. Press, 1934)

[13] H.Saida *Physica* 2005, A356, 481–508

[14] L.D.Landau and E.M.Lifshitz *Statistical Physics. Prat I*; Pergamon, 1980

[15] C.Essex *Planet.Space.Sci.* 1984, 32, 1035–1043
C.Essex *Adv.Thermodynamics* 1990, 3, 435–447

[16] R.Dominguez-Cascante and J.Faraudo *Phys.Rev.* 1996, E54, 6933–6935
J.Fort, D.Jou and J.E.Llebot *Physica* 1999, A269, 439–454
J.Fort *Phys.Rev.* 1999, E59, 3710–3713

[17] S.W.Hawking *Phys.Rev.* 1976, D13, 191–197

[18] J.W.York,Jr. *Phys.Rev.* 1086, D33, 2092–2099
H.W.Barden, B.F. Whiting and J.W.York,Jr. *Phys.Rev.* 1987, D36, 3614–3625
B.F.Whiting and J.W.York,Jr. *Phys.Rev.Lett.* **61** (1988) 1336–1339

[19] S.W.Hawking and D.N.Page *Commun.Math.Phys.* 1983 87, 577–588
S.Surya, K.Schleich and D.M.Witt *Phys.Rev.Lett.* 2001, 86, 5231–5234

[20] S.W.Hawking *Phys.Rev.* 1976, D 14, 2460–2473
U.H.Danielsson and M.Schiffer *Phys.Rev.* 1993, D 48, 4779–4784
C.R.Stephens, G.'tHooft and B.F.Whiting *Class.Quantum Grav.* 1994, 11, 621–647
A.Strominger (1995). *Les Houches Lectures on Black Holes*. hep-th/9501071
T.P.Shigh and C.Vaz *Int.J.Mod.Phys.* 2004, D13, 2369–2373

[21] H.Saida *Class.Quant.Grav.* 2006, 23, 6227–6243

[22] W.G.Unruh and R.M.Wald *Phys.Rev.* 1982, D25, 942–958
W.G.Unruh and R.M.Wald *Phys.Rev.* 1983, D27, 2271–2276
V.P.Frolov and D.N.Page *Phys.Rev.Lett.* 1993, 71, 3902–3905
E.E.Flanagan, D.Marolf and R.M.Wald *Phys.Rev.* 2000, D62, 084035

[23] E.A.Martinez and J.W.York,Jr. *Phys.Rev.* 1989, D40, 2124–2127

In: Classical and Quantum Gravity Research
Editors: M.N. Christiansen et al, pp. 371-426
ISBN 978-1-60456-366-5

Chapter 9

DEVELOPMENTS IN BLACK HOLE RESEARCH: CLASSICAL, SEMI-CLASSICAL, AND QUANTUM

A. DeBenedictis[*]
Pacific Institute for the Mathematical Sciences,
Simon Fraser University Site
and
Department of Physics, Simon Fraser University
Burnaby, British Columbia, V5A 1S6, Canada

Abstract

The possible existence of black holes has fascinated scientists at least since Michell and Laplace's proposal that a gravitating object could exist from which light could not escape. In the 20th century, in light of the general theory of relativity, it became apparent that, were such objects to exist, their structure would be far richer than originally imagined. Today, astronomical observations strongly suggest that either black holes, or objects with similar properties, not only exist but may well be abundant in our universe. In light of this, black hole research is now not only motivated by the fascinating theoretical properties such objects must possess but also as an attempt to better understand the universe around us. We review here some selected developments in black hole research, from a review of its early history to current topics in black hole physics research. Black holes have been studied at all levels; classically, semi-classically, and more recently, as an arena to test predictions of candidate theories of quantum gravity. We will review here progress and current research at all these levels as well as discuss some proposed alternatives to black holes.

1. Introduction

This paper presents some developments in black hole research from its very early history to modern day. Any manuscript undertaking such a task is bound to be incomplete, the subject matter being enormous. What is intended here is a coherent, reasonably self-contained (and relatively brief) article capturing essential features in black hole history and research at the

[*]E-mail address: adebened@sfu.ca

classical, semi-classical, and quantum level. The goal is to give the interested researcher or student an overview of some of the research that has been done and is currently being pursued in all these areas within a single manuscript. The style is more of a survey than an in-depth study and it is hoped the interested reader will find the references useful for further information. Given limited space and time, there are regrettably many, sometimes glaring, omissions and entire fascinating areas of research had to be left out. It was therefore decided that the bulk of the effort go into reviewing a few selected topics in four dimensional black holes within the context of the original general relativity theory of Einstein and Hilbert and their natural extensions into the quantum realm. A sincere apology goes to the authors of works not included here or accidentally missed. Some of the topics are chosen due to their lasting impact in the field as can be seen, for example, by the number of papers appearing on the arXiv related to these topics, and are not necessarily new. It is hoped that this type of review will give researchers in other areas of gravity a brief overview of the phenomena that these recurring topics comprise.

Black hole research has turned from an obscure, almost ignored area of research to one of the most studied segments in gravitational field theory. Today it is common to see more than a few black hole papers appear on the pre-print archive on a daily basis. It seems that the black hole still has many interesting surprises, from the purely classical, to the purely quantum, and everything in between.

The presentation here is done in a somewhat historical perspective. However, the bulk of the results focus on more recent developments as there are a number of excellent books and reviews from a purely historical point of view ([1], [2], [3] and references therein).

In section 2 we give a brief history of black hole research, which dates at least back to the 1780s. In section 3 we present the classical black holes and the fascinating research that accompanies them to this day. In section 4 we look at semi-classical research which also includes a section on Hawking's amazing result of black hole radiation. In section 5 we study results from quantum gravity, primarily the loop approach, which is not to imply that other approaches are not fruitful. In starting to write that section of the manuscript it seemed that with several differing theories, either no justice could be paid to any of them, or else the section had to focus on the research occurring in just one of them. It seems that loop quantum gravity is closest to the spirit of the rest of the paper and has also produced a number of interesting results.

Of course, black holes would not be nearly as interesting if not for the fact that there is now reason to believe that they may well exist (perhaps even in abundance) in the universe. We therefore focus on current astrophysical black hole research in section 6 (much of the discussion in this section also applies to the possible detection of *primordial* black holes [4] - [6], although many would have evaporated by the present era.) This is perhaps the fastest changing area in black hole research and it seems that the activity in this field will not be dying down any time soon. Finally, in section 7 we also discuss some alternative theories of collapse which avoid the formation of a singularity or even the black hole altogether.

2. Black Holes: A Short History

The idea that a gravitating object could exist from which not even light could escape seems to date at least back to the work of the Reverend John Michell in 1783 [7]. At this time it was already known that light traveled at a finite speed from Roemer's studies of Jupiter's moon Io [8]. If light behaved like a particle, with finite speed, why then could it not be affected by gravity like other objects?

In 1796 Pierre Laplace, apparently unaware of Michell's work postulated exactly the same thing; that a "dark star" could exist from which no light would escape [9]. Both Michell and Laplace calculated that an amount of mass M must be present within a radius $R = \frac{2GM}{c^2}$ in order for light not to escape from the object. The circumference corresponding to this radius was called the *critical circumference*. As is well known, this value is in (surprising) agreement with the value given by general relativity theory.

Although it was initially believed that these dark stars could be populous in the universe, they were later considered to be at best an academic curiosity, as the size or density such a body would possess was considered unphysical. Michell originally calculated that an object with a similar average density to that of the sun, would need to be approximately 500 times larger in order to stop light. Put another way, an object with the same mass as the sun would need to possess a diameter of a mere 20 kilometers. As well, studies in the 1800's indicated that light was a wave possessing no mass and therefore it was believed that it would not be influenced by gravitational effects.

The situation changed in 1915 when Einstein and Hilbert formulated the now famous field equations of general relativity [10], [11], which in the notation of this manuscript are presented as [1]:

$$R_{\mu\nu} - \frac{1}{2} R\, g_{\mu\nu} = \frac{8\pi G}{c^4} T_{\mu\nu} \,. \tag{1}$$

It was not long after the formulation of the field equations that astrophysicist Karl Schwarzschild came up with an exact solution which described the gravitational exterior of a perfectly spherical star [12]. The Schwarzschild solution is probably the most famous non-trivial solution of the field equations and it admits the following well-known line element in the spherical coordinate chart:

$$ds^2 = -\left(1 - \frac{2GM}{c^2 r}\right) dt^2 + \frac{dr^2}{\left(1 - \frac{2GM}{c^2 r}\right)} + r^2\, d\theta^2 + r^2 \sin^2\theta\, d\phi^2 \,. \tag{2}$$

Two things are immediately apparent in (2): (i) A singularity is present when $r = 0$. This singularity was believed to be of exactly the same nature as the corresponding one in the Newtonian theory and was of little concern. (ii) There exists a singularity in the metric when $r = \frac{2GM}{c^2}$, exactly the same radius value for which an object in Newton's theory yields an escape velocity equal to c. However, even though it was well known that light *was* affected by gravity in this new theory, it was still believed that stars whose radius was less than $\frac{2GM}{c^2}$ were to be considered as pathological and not existing in nature.

This period of complacency did not last particularly long. Many scientists at the time began to worry that something seemingly unnatural appeared in the gravitational field theory and appealed to Schwarzschild's constant density interior solution [13]. For a constant density sphere of density ρ_0 and radius a, Einstein's equations along with the conservation

[1] In subsequent sections we shall be utilizing geometrized units where $G = c \equiv 1$.

law $T^{\mu\nu}{}_{;\nu} = 0$ yield the following for the (isotropic) pressure:

$$p = \rho_0 \left[\frac{\sqrt{1 - 2Mr^2/a^3} - \sqrt{1 - 2M/a}}{3\sqrt{1 - 2M/a} - \sqrt{1 - 2Mr^2/a^3}} \right], \tag{3}$$

with $r < a$. In this expression it was noted that as a approaches $\frac{9}{4}M$, the central pressure becomes infinite. Thus, it was argued, systems with smaller radius-to-mass ratios could not exist in nature. This argument, however, relied on the use of an unphysical matter model and perhaps could not be trusted.

Einstein himself was uncomfortable with the fact that the solution to his field equations admitted such bizarre structures [1]. He attempted to disprove the existence of such compact objects by studying the circular orbits of massive particles in the Schwarzschild space-time. As he made the orbits smaller and smaller, the particle velocities increased until they finally reached the speed of light when located at a radius of 1.5 times the critical radius. The conclusion was that a spherical object could not exist whose radius was less than this one since no particle could exceed the speed of light [14]. This study, though perfectly correct, neglected radial motions of the particles which are inevitably present for all particles with geodesic orbital radii of less than 1.5 times the critical radius.

In the same year as Einstein's paper was published (1939), Robert Oppenheimer and Hartland Snyder performed an extremely difficult and pioneering calculation. They studied the gravitational collapse of a spherically symmetric isotropic dust cloud within full general relativity. Their seminal results were published in an article entitled *"On Continued Gravitational Contraction"* [15]. In this study they demonstrated how the dust cloud could collapse and how, when viewed from an external vantage point, the collapse would slow down and asymptotically halt as the critical radius was reached. They also showed that, to an observer co-moving with the dust, the collapse takes place in finite time. Although dust is by no means a realistic matter field, Oppenheimer and Snyder's calculations provided the most complete argument of gravitational condensation at the time.

One of the continuing opponents to black hole formation was J. A. Wheeler. Wheeler and many other scientists at the time were quite certain that some physical process must intervene during the collapse, and that the scenario played out by Oppenheimer and Snyder's idealized calculation would not occur. One proposal was that the nucleons present in a realistic structure would, under extreme conditions, radiate away [1]. Interestingly, this idea turns out to be partially correct in the paradigm of black hole evaporation.

By the 1960s computers were at the stage where a more realistic matter model could be utilized in gravitational collapse calculations. Such a calculation was carried out by Colgate and his collaborators at Livermore laboratory [16]. These calculations, although still perfectly spherically symmetric, took into account now well known nuclear physics processes inside the matter. These calculations indicated that for a star of mass greater than approximately two solar masses, collapse into a black hole was inevitable. Similar calculations were carried out in the Soviet Union by Zel'dovich and collaborators yielding similar results [1]. Such simulations along with the discovery of new coordinate systems to describe black holes [17] [18] aided in easing the scientific community's skepticism about black holes. In fact, it is well known that Wheeler became a big believer and is the originator of the term "black hole".

With the possible formation of black holes now accepted by much of the scientific community, black hole research saw the birth of the now famous Hawking-Penrose singularity theorems [19], the no-hair theorem [20], and of course, the Kerr solution to the field equations with all its interesting properties and peculiarities, and the laws of black hole mechanics.

The laws of black hole mechanics arose from the analysis due to various talented scientists and were put in their final form by J. Bardeen, B. Carter and S. Hawking [21]. In a nutshell, they can be stated as follows:

0. The surface gravity is constant on any surface corresponding to a black hole event horizon.

1. If an amount of material of mass δM, angular momentum δJ and charge δQ accretes into a black hole, the area of the event horizon responds according to

$$\frac{\kappa}{8\pi G}\delta A = \Omega\,\delta J + \Phi\,\delta Q - \delta\,M\,, \tag{4}$$

   where κ is the surface gravity of the horizon, Φ the electrostatic potential at the horizon and Ω the angular velocity of the horizon.

2. The area, A, of the event horizon cannot decrease.

3. It is impossible to reduce κ to zero by a finite sequence of operations.

Although some of these laws have caveats, it was not lost on the scientists of the day their amazing resemblance to the laws of thermodynamics. Specifically, the first law of black hole mechanics and the first law of thermodynamics would be analogous if one associated entropy with the area and temperature with the surface gravity. It was J. Bekenstein who took this analogy most seriously and today black hole entropy is commonly associated with his name. (For an interesting survey of black hole thermodynamics see [22], [23] and references therein.)

Some research at this time started to focus on *quantum* properties of black holes. A theory of quantum gravity is notoriously difficult to come by, although today there are some serious candidate theories. In the late 70s and 80s, in the absence of a full theory of quantum gravity, scientists started to study black holes from a *semi-classical* perspective. That is, the geometry of space-time was treated classically, but the matter fields propagating on the space-time were treated quantum mechanically. The fields were usually "test fields" in much the same way as test particles are used. They are quantized on the background space-time using techniques similar to many of those employed in quantum field theory in Minkowski space-time. There are, however, some issues, ambiguities and subtleties in curved space-time that are not present in the corresponding theory in flat space-time. Some of this will be discussed in later sections. Excellent expositions of this subject may be found in [22] and [24].

One very important result that has emerged from semi-classical studies is that black holes do indeed, as Wheeler suspected, evaporate. This black hole evaporation was first suggested for rotating black holes by Zel'dovich and later, in 1974, Hawking quantitatively discovered that *all* black holes must radiate [25]. This was confirmed and extended by

D. Page and W. Unruh [26] [27]. To this thermal radiation one could associate a temperature, which for the Schwarzschild hole is given by

$$T_{Sbh} = \frac{\hbar c^3}{8\pi GMk}, \tag{5}$$

as well as an entropy,

$$S_{Sbh} \approx \frac{kc^3}{4\pi G\hbar} A, \tag{6}$$

with A being the area of the black hole's horizon. The temperature is very small, approximately 10^{-7} K for a black hole of the order of a solar mass. The entropy, on the other hand, is very large, being of the order of 10^{54} for a solar mass black hole (in J$\cdot$K^{-1}). As we will see, theories of quantum gravity may explain the origin of this large entropy from the gravitational degrees of freedom associated with the horizon.

Since these results, the arena of semi-classical black hole research has been a very fruitful one. However, a more fundamental problem remained; a full theory of quantum gravity was still (and in many ways still is) elusive.

3. Classical Black Hole Research

3.1. Review of Important Solutions

Here we briefly review some of the important classical black hole solutions and their properties. These solutions are amongst the most studied metrics in general relativity theory. A detailed exposition on the mathematical aspects of classical black holes may be found in [28] and [29].

3.1.1. The Kottler-Reissner-Nordström Black Hole

Perhaps the most famous solution to the gravitational field equations is the Schwarzschild metric. With charge and cosmological constant it yields the line element:

$$ds^2 = -\left(1 - \frac{2M}{r} + \frac{Q^2}{r^2} - \frac{\Lambda}{3}r^2\right) dt^2 + \frac{dr^2}{1 - \frac{2M}{r} + \frac{Q^2}{r^2} - \frac{\Lambda}{3}r^2} + r^2\, d\theta^2 + r^2 \sin^2\theta\, d\varphi^2 . \tag{7}$$

Horizons exist where $g_{tt} = 0$.

It is actually not difficult to derive this solution. Consider a general, static, spherically symmetric line element:

$$ds^2 = -e^{\gamma(r)}\, dt^2 + e^{\alpha(r)}\, dr^2 + r^2\, d\theta^2 + r^2 \sin^2(\theta)\, d\varphi^2 . \tag{8}$$

Expression (8) yields the following, from the field equations (1):

$$G^t_{\ t} = \frac{e^{-\alpha(r)}}{r^2}\left(1 - r\alpha(r)_{,r}\right) - \frac{1}{r^2} = 8\pi T^t_{\ t} \tag{9a}$$

$$G^r_{\ r} = \frac{e^{-\alpha(r)}}{r^2}\left(1 + r\gamma(r)_{,r}\right) - \frac{1}{r^2} = 8\pi T^r_{\ r} \tag{9b}$$

$$G^\theta_{\ \theta} \equiv G^\varphi_{\ \varphi} = \frac{e^{-\alpha(r)}}{2}\Big(\gamma(r)_{,r,r} + \frac{1}{2}\left(\gamma(r)_{,r}\right)^2 + \frac{1}{r}\left(\gamma(r) - \alpha(r)\right)_{,r} - \frac{1}{2}\alpha(r)_{,r}\gamma(r)_{,r}\Big) = 8\pi T^\theta_{\ \theta} = 8\pi T^\varphi_{\ \varphi}, \tag{9c}$$

For simplicity the cosmological term and the charge term are set to zero, however, it is straight-forward to implement these, by including them as part of $T^\mu_{\ \nu}$.

Equation (9a) may be utilized to give the following:

$$e^{-\alpha(r)} = \frac{8\pi}{r}\int (r')^2 \left(T^t_t(r')\right) dr' + 1 =: 1 - \frac{2m(r)}{r}. \tag{10}$$

Since the system of equations is under-determined, two functions may be prescribed. Since the Schwarzschild solution, if considered in the domain $0 \le r < \infty$, corresponds to the gravitational field of a point mass of mass M, we can postulate a stress-energy tensor for a point mass with the following $T^0_{\ 0}$ component:

$$T^t_t(r) = -\frac{M}{4\pi r^2}\,\delta(r)\,. \tag{11}$$

The r dependence is motivated by dimensionality arguments and neglecting the factor of 4π would simply correspond to a rescaling of the mass. There is some arbitrariness on what the other function to be prescribed can be. However, to satisfy junction conditions implied by the above equations supplemented with the conservation law, $T^r_{\ r}$ should be continuous, and therefore set to zero. The remaining unknowns may be solved for by straight-forward manipulation of the field equations and conservation law:

$$e^{\gamma(r)} = e^{-\alpha(r)}e^{h_0}, \tag{12a}$$

$$T^\theta_{\ \theta} \equiv T^\varphi_{\ \varphi} := \frac{M\,\delta(r)}{16\pi r}\gamma_{,r}\,. \tag{12b}$$

The constant h_0 can be absorbed into the definition of the time coordinate.

For the case of a charged black hole ($Q \neq 0$, $\Lambda = 0$) there are two horizons, one at $r = r_+ = M + \sqrt{M^2 - Q^2}$ and another, inner, horizon at $r = r_- = M - \sqrt{M^2 - Q^2}$. In the extremal case, $Q = M$ and the horizons are coincident.

In closing this sub-section we quote the form of the line element in several other well known coordinate systems which historically have shed light on the causal structure (now with $Q = 0 = \Lambda$). In Painlevé-Güllstrand coordinates, the Schwarzschild metric is cast in the form:

$$ds^2 = -d\tilde{t}^2 + \left(dr + \sqrt{\frac{2M}{r}}\,d\tilde{t}^2\right)^2 + r^2\,d\theta + r^2\sin^2\theta\,d\varphi^2\,. \tag{13}$$

These coordinates are regular at the horizon (although g_{00} still vanishes there) and readily display the no-escape property of this surface. Considering radial light-rays ($ds = 0,\ d\theta = d\varphi = 0$), the equation of motion yields:

$$\frac{dr}{d\tilde{t}} = \pm 1 - \sqrt{\frac{2M}{r}}\,.$$

Note that for $r < 2M$ the quantity $\frac{dr}{d\tilde{t}}$ is negative for both solutions, indicating that both ingoing and "outgoing" null rays approach $r = 0$.

The ingoing Eddington-Finkelstein coordinates are also particularly useful in elucidating the "no escape" property of the event horizon. In these coordinates, the Schwarzschild metric yields:

$$ds^2 = -\left(1 - \frac{2M}{r}\right) dv^2 + 2\,dv\,dr + r^2\,d\theta^2 + r^2 \sin^2\theta\,d\varphi^2\,. \tag{14}$$

The nature of these coordinates, along with the causal structure is demonstrated in figure 1. Of course, there also exists the outgoing Eddington-Finkelstein coordinates:

$$ds^2 = -\left(1 - \frac{2M}{r}\right) du^2 - 2\,du\,dr + r^2\,d\theta^2 + r^2 \sin^2\theta\,d\varphi^2\,, \tag{15}$$

which are also illustrated in figure 1.

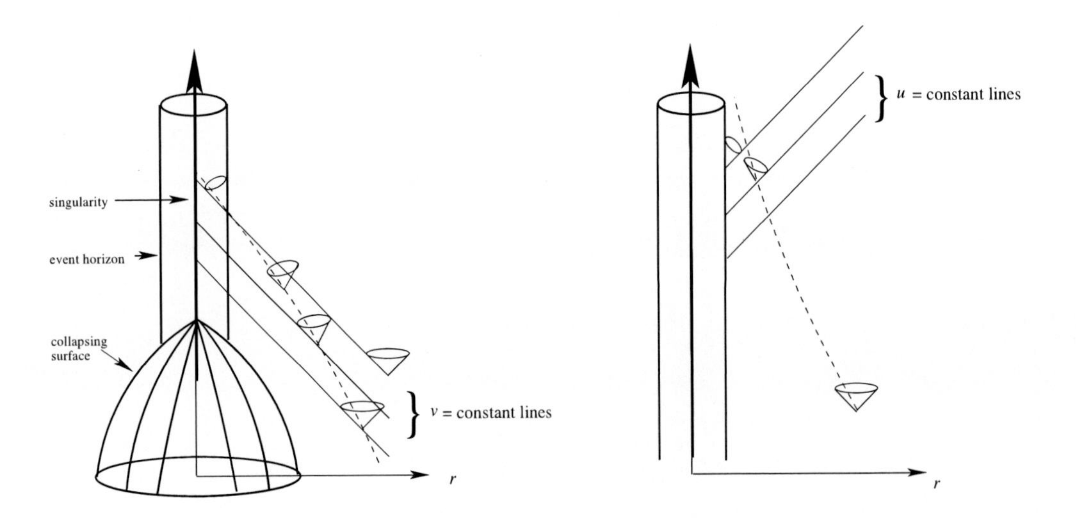

Figure 1. The ingoing Eddington-Finkelstein coordinates (left) and the outgoing Eddington-Finkelstein coordinates (right). The dashed line represents an in-falling time-like particle.

Finally we present the line element in Kruskal-Szekeres-Synge coordinates, which eliminate the coordinate pathology at $r = 2M$ altogether:

$$ds^2 = -32\frac{M^3}{r}e^{-r/2M}\,d\tilde{u}d\tilde{v} + r^2\,d\theta^2 + r^2 \sin^2\theta\,d\varphi^2\,, \tag{16}$$

where r now represents the solution to:

$$\tilde{u}\tilde{v} = \left(1 - \frac{r}{2M}\right) e^{r/2M} .$$

The Kruskal diagram (also known as the maximally extended Schwarzschild space-time) is illustrated in figure 2. The two coordinate patches corresponding to $r > 2M$ and $r < 2M$ in metric (7) (recall $Q = 0$, $\Lambda = 0$ in this discussion) are capable of describing regions I and II respectively of the maximally extended Schwarzschild space-time.

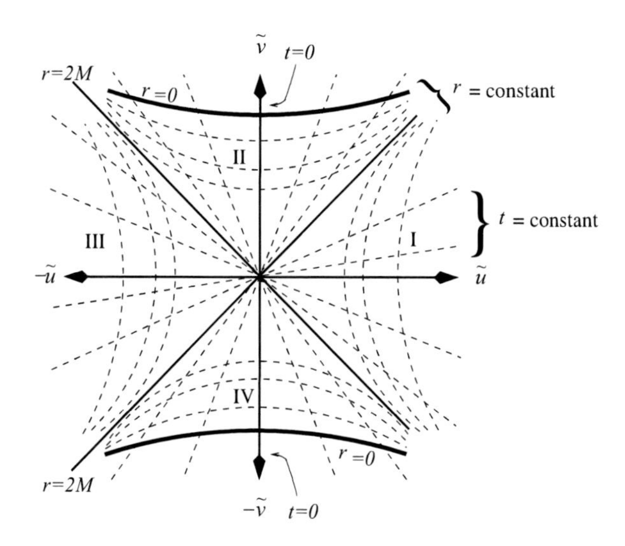

Figure 2. The maximally extended Schwarzschild space-time.

Further discussion of horizon-regular coordinate systems may be found in [30].

3.1.2. The Kerr-Newman-de Sitter Black Hole

This space-time describes a charged rotating ring or, in its asymptotic regime, the space-time outside a rotating charged star. It is the rotational analogue of the Kottler-Reissner-Nordström black hole. Unlike the previous case, here there is no Birkhoff's theorem [31] guaranteeing uniqueness of this solution as the exterior of a realistic rotating star.

The solution, without charge or cosmological constant, was discovered by Roy Kerr [32] in 1963, although several others had attempted to find such a solution before him. Kerr's original metric yielded the line element:

$$\begin{aligned} ds^2 = &- \left(1 - \frac{2Mr}{\rho^2}\right) \left(du + a\sin^2\theta\, d\varphi\right)^2 \\ &+ 2\left(du + a^2\sin^2\theta\, d\varphi\right)\left(dr + a\sin^2\theta\, d\varphi\right) \\ &+ \rho^2\left(d\theta^2 + \sin^2\theta\, d\varphi^2\right), \end{aligned} \tag{17}$$

where

$$\rho^2 := r^2 + a^2\cos^2\theta\,, \tag{18}$$

with a the angular momentum per unit mass and M the black hole mass.

With the addition of charge (Q), and in the presence of a cosmological constant (Λ), the line element is expressible in the following formidable form, utilizing the Boyer-Lindquist coordinates:

$$\begin{aligned} ds^2 &= -\frac{1}{\rho^2\Sigma^2}\left(\Delta_r - \Delta_\theta a^2 \sin^2\theta\right) dt^2 + \frac{\rho^2}{\Delta_r}\, dr^2 + \frac{\rho^2}{\Delta_\theta}\, d\theta^2 \\ &+\frac{1}{\rho^2\Sigma^2}\left[\Delta_\theta\left(r^2+a^2\right)^2 - \Delta_r a^2 \sin^2\theta\right]\sin^2\theta\, d\varphi^2 \\ &-\frac{2a}{\rho^2\Sigma^2}\left[\Delta_\theta(r^2+a^2) - \Delta_r\right]\sin^2\theta\, dt\, d\varphi\,, \end{aligned} \tag{19}$$

with

$$\Delta_r = \left(r^2+a^2\right)\left(1-\frac{\Lambda}{3}r^2\right) - 2Mr + Q^2\,, \tag{20a}$$

$$\Delta_\theta = 1 + \frac{1}{3}\Lambda a^2\cos^2\theta, \qquad \Sigma = 1 + \frac{1}{3}\Lambda a^2\,. \tag{20b}$$

This metric is quite complicated to work with and we will therefore focus attention on the case $Q = 0$ and $\Lambda = 0$. It will also be assumed that $a < M$.

In summary, the $Q = \Lambda = 0$ metric possesses the following properties: In the limit $a \to 0$ this solution goes over to the Schwarzschild solution. The $M = 0$ limit yields flat Minkowski space-time in oblate spheroidal coordinates. As well, a true singularity exists at $\rho = 0$ as can be seen from the computation of the Kretschmann scalar

$$R_{\alpha\beta\gamma\delta}R^{\alpha\beta\gamma\delta} = \frac{1}{\rho^{12}} 48M(r^2 - a^2\cos^2\theta)\left[\rho^4 - 16r^2a^2\cos^2\theta\right]\,. \tag{21}$$

The singularity is located at $r = 0$, $\theta = \pi/2$, which in a Cartesian-type coordinate system corresponds to $x^2+y^2 = a^2$, $z = 0$, indicating a ring-like structure to the singularity. Other domains of interest are:

i) $r = r_\pm := M \pm \sqrt{M^2 - a^2}$; these are the inner and outer event horizons.

ii) The coordinates t and φ are not orthogonal. This implies that a geodesic observer "moving" in the time direction must necessarily move in the φ direction. That is, the observer is dragged around the black hole in the direction of rotation. This is the famous Lense-Thirring effect [33].

iii) Related to the previous item, $r = r_{\mathrm{E}\pm} := M \pm \sqrt{M^2 - a^2\cos^2\theta}$ are inner and outer ergosurfaces. A time-like observer cannot resist the dragging effects while in this region. The light cone structure is studied in, for example, [19], [28] and [34].

Figure 3 illustrates the relative locations of these domains, in a Cartesian-like set of coordinates.

There are relatively few treatises on the Kerr geometry, mainly due to the technical difficulties involved in working with the metric. The interested reader is referred to [35], [36].

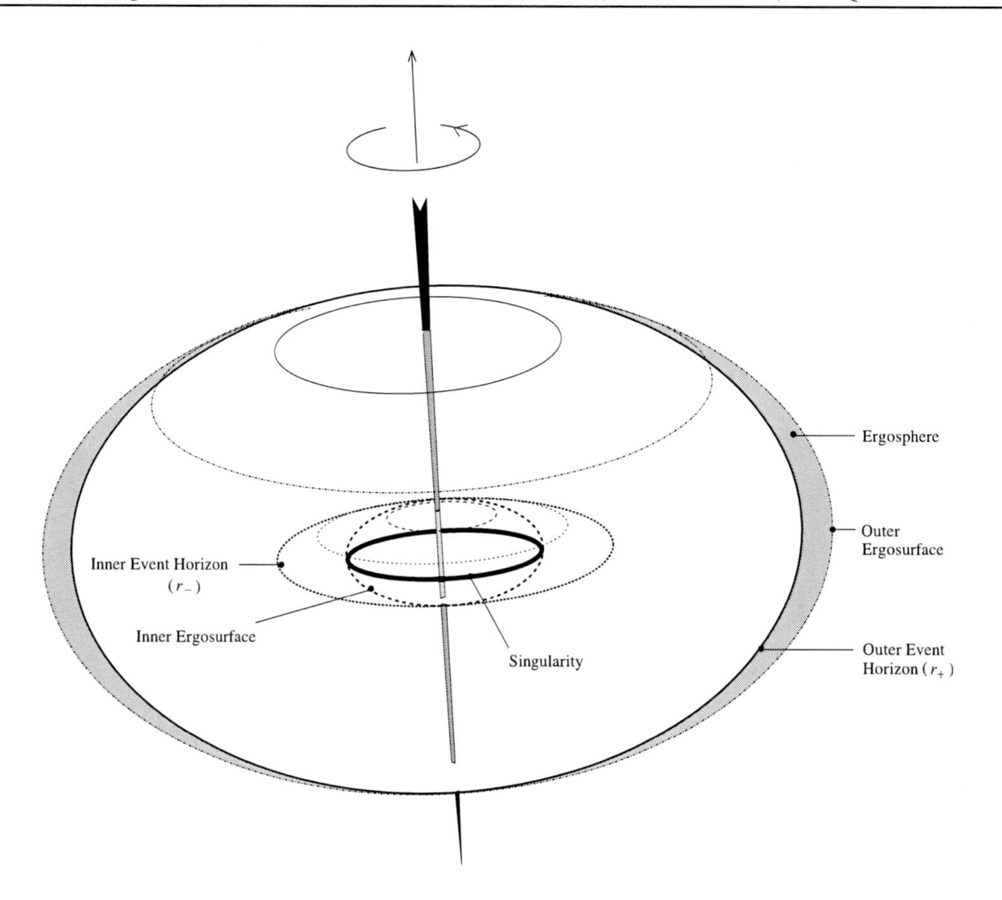

Figure 3. Various regions of a Kerr black hole in a Cartesian-type coordinate system.

3.1.3. Exotic Solutions

Here we briefly describe some solutions which are "exotic" in some sense. By exotic we mean solutions that are unlikely to exist in nature but are still of great interest on a theoretical basis. If one allows for a *negative* cosmological constant, then solutions exist in general relativity which can describe black holes with *planar*, *cylindrical*, or *higher genus* topology. These are sometimes known as topological black holes. A sufficiently general metric to describe such solutions (here with vanishing charge and angular momentum) is given by:

$$ds^2 = -\left(\alpha^2 r^2 - b - \frac{2M}{r}\right) dt^2 + \frac{d\rho^2}{\left(\alpha^2 r^2 - b - \frac{2M}{r}\right)} + r^2 \left(d\theta^2 + d\sinh^2(\sqrt{b}\,\theta)\, d\varphi^2\right) , \tag{22}$$

with $0 < \varphi \leq 2\pi$. Here, α is related to the cosmological constant via $\alpha^2 = -\Lambda/3$, M is the mass parameter, and d and b are constants that determine the topology of $t, r =$constant surfaces. The cases are as follows:

i) $b = -1$, $d = -1$: In this case constant $(t,\, r)$ surfaces are spheres (the Kottler solution).

ii) $b = 0$, $\lim\limits_{b\to 0} d = \frac{1}{b}$: In this case constant $(t,\, \rho)$ surfaces are tori.

iii) $b = 1$, $d = 1$: In this case constant (t, r) surfaces are surfaces of constant negative curvature of genus $g > 1$, depending on the identifications chosen.
An event horizon exists when $\left(\alpha^2 r^2 - b - \frac{2M}{r}\right) = 0$. Such topologies were studied in detail in [37]. The formation of such black holes from gravitational collapse was studied in [38].

The metric (22) does not uniquely describe such topological black holes. Lemos and Zanchin, for example, have studied a class of black holes which can be cast in the following form [39] [40]:

$$ds^2 = -\left(\alpha^2 r^2 - \frac{B_o M}{\alpha r}\right) dv^2 + 2\, dv\, dr + r^2 \left(d\theta^2 + d\varphi^2\right) . \tag{23}$$

Here, the values of B_0 and the identifications make up either toroidal, cylindrical or planar topologies. Specifically:
i) $0 \leq \theta < 2\pi, 0 \leq \varphi < 2\pi, B_0 = \frac{2\alpha}{\pi}$ yields the flat torus model.
ii) $-\infty < \theta < \infty$, $0 \leq \varphi < 2\pi$, $B_0 = 4$ yields the cylinder, with M the mass per unit length and the linear axis coordinate, z, is related to θ via $\theta = \alpha z$.
iii) $-\infty < \theta < \infty$, $-\infty < \varphi < \infty$, $B_0 = \frac{2}{\alpha}$ yields the planar case, with M the mass per unit area.
This metric differs slightly from the metric (22). Gravitational collapse forming such black holes was studied in [40]. It can be checked that in all the above cases, the Kretschmann scalar blows up at $r = 0$. Another construction of topological black holes may be found in [41]. Research involving the black holes described in this section will be presented below.

3.2. Developments in Classical Black Hole Research

3.2.1. Quasi-normal Mode Analysis

Broadly defined, black hole quasi-normal modes arise from perturbations in some black hole space-time. However, the modes are affected by the emission of gravitational radiation, which generally has a damping effect on the modes, and therefore these modified modes are named quasi-normal [2]. The oscillations can in theory be reconstructed via the analysis of their corresponding gravitational wave emissions. Reversing the argument, the vibrational modes of the black hole are closely linked to the corresponding emitted gravitational wave pattern and they are therefore important in light of gravitational wave astronomy. (In the case of pure gravitational perturbations, the oscillations are by definition the gravitational wave patterns.) It is believed that the gravitational waves due to oscillations produced during black hole formation may be strong enough to detect with current or near future gravitational wave detectors. The wave signature will be unique, the frequencies tending not to depend strongly on the perturbing process and which would yield a direct measurement of a black hole's existence and give information on its mass and angular momentum. Two or more modes may give useful information such as helping discern if general relativity is valid. In particular it may prove or disprove the no-hair theorem of general relativity [42]. An excellent exposé on quasi-normal modes may be found in the thesis by Cardoso

[2]Here we ignore subtle, but important, mathematical questions regarding the completeness of quasi-normal modes.

[43] as well as the works [44] and [45]. Also, a good reference on non-spherical metric perturbations of the Schwarzschild black-hole spacetime may be found in [46].

Perturbations of black holes were originally studied by Regge and Wheeler [47]. For the Schwarzschild black hole one may, for example, perturb the metric in a way appropriate for "axial" perturbations:

$$ds^2 = -\left(1-\frac{2M}{r}\right)dt^2 + \frac{dr^2}{1-\frac{2M}{r}} + r^2\,d\theta^2 + r^2\sin^2\theta\,[d\varphi - \omega\,dt - q_2\,dr - q_3\,d\theta]^2 . \tag{24}$$

We write

$$\omega(r,\,\theta,\,t) = \tilde{\omega}(r,\,\theta)e^{i\sigma t}, \tag{25}$$

and similarly for q_2 and q_3.

For perturbations of this form, it can be shown that the system governing the perturbations can be reduced to a single second-order differential equation, which can be solved by separating the variables r and θ (for example, see [28]). Briefly, the perturbed field equations give a relation between ω and q_2 and q_3, allowing the elimination of ω. The quantity $Q := (1-2M/r)r^2\sin^3\theta\,(\partial_\theta q_2 - \partial_r q_3)$ is written as $Q = R(r)\Theta(\theta)$ and the resulting radial equation is

$$r^2\left(1-\frac{2M}{r}\right)\frac{d}{dr}\left[\frac{1-\frac{2M}{r}}{r^2}\frac{dR}{dr}\right] - \mu_l^2\frac{\left(1-\frac{2M}{r}\right)R}{r^2} + \sigma^2 R = 0, \tag{26}$$

where μ_l^2 is the eigenvalue of the angular equation and may take on the values $\mu_l^2 := (l+2)(l-1)$ for $l = 2,\ 3,\ \ldots$.

We can make a change of coordinates,

$$r_* = r + 2M\ln\left(\frac{r}{2M}-1\right),$$

so that (26) reduces to a Schrödinger-type equation:

$$\left[\frac{d^2}{dr_*^2} - V_l(r)\right]Z_l = -\sigma_l^2 Z_l \tag{27}$$

where

$$Z_l(r) = \frac{R_l(r_*)}{r},$$

and

$$V_l(r) = \left(1-\frac{2M}{r}\right)\left[\frac{l(l+1)}{r^2} - \frac{6M}{r^3}\right]. \tag{28}$$

The equation (27) is often referred to the Regge-Wheeler equation with (28) the Regge-Wheeler potential. The above expression is valid for scalar perturbations, the coefficient of the last term differing for vector and tensor perturbations (see [48] for details). Later, Zerilli derived a similar equation for polar perturbations with a more complicated potential [49]:

$$V_l(r)_{pol} = \frac{2(r-2M)}{r^4(nr+3M)}\left[n^2(n+1)r^3 + 3Mn^2r^2 + 9M^2nr + 9M^3\right], \tag{29}$$

with $n := \frac{1}{2}(l+2)(l-1)$.

One can compute frequencies for functions which possess the asymptotic form [50]

$$Z_l(r) \rightarrow \begin{cases} e^{i\sigma r_*} & \text{for} \quad r_* \rightarrow \infty \, , \\ e^{-i\sigma r_*} & \text{for} \quad r_* \rightarrow -\infty \, . \end{cases} \tag{30}$$

The frequencies are complex and will be denoted as $\sigma = \sigma_1 + i\sigma_2$. For every value of l there exist a tower of modes, denoted here by the integer m. Some of the lowest frequencies for the Schwarzschild black hole are summarized in table 1 which hold for both the axial and polar perturbations.

Table 1. A summary of the first four quasi-normal mode frequencies in units of $M\sigma$ for $l = 2$, 3 and 4. The conversion to Hz is given by multiplying the given numbers by $2\pi\frac{M_\odot}{M}5142$Hz. This data is from [48] and [51].

	$l=2$		$l=3$		$l=4$	
	$M\sigma_1$	$M\sigma_2$	$M\sigma_1$	$M\sigma_2$	$M\sigma_1$	$M\sigma_2$
$m=0$	0.3737	0.0890	0.5994	0.0927	0.8092	0.0942
$m=1$	0.3467	0.2739	0.5826	0.2813	0.7966	0.2844
$m=2$	0.3011	0.4783	0.5516	0.4791	0.7727	0.4799
$m=3$	0.2515	0.7051	0.5120	0.6903	0.7398	0.6839

One item of particular interest in modern quasi-normal mode research is that of radiative tails. The tail refers to the non-trivial fall-off properties (in time) of the perturbation. This phenomenon is sometimes known as black hole ringing, due to the oscillatory behavior of the tail. An example of such a tail is given in figure 4, for an infalling particle perturbing a Schwarzschild black hole. The vertical axis represents the gravitational wave amplitude. The dashed line represents an analytical fit using a linear combination of the first two $l = 2$ modes.

The Kerr black hole is much more difficult to analyze. Some work has been performed in [52] - [56]. Studies find that as the angular momentum increases, σ_1 is bounded, whereas σ_2 is not. However, as a approaches the value of M, σ_2 tends to zero and the oscillations would therefore continue without damping [52]. These modes may not be realized though as some studies indicate that the amplitudes of these modes also tend to zero [57].

Another interesting phenomenon associated with the Kerr black hole is that of super-radiance. For massless vector and tensor perturbations due to scattering off of the Kerr potential, the reflection coefficient can exceed unity if the incoming wave possesses a frequency below a certain critical value. This is believed to be due to an interplay between particle creation in black holes and the Penrose energy extraction process [58] [59].

The understanding of quasi-normal modes from black holes is becoming a very important issue in black hole physics due to the possibility of detection with gravitational wave detectors. There are several astrophysical processes that can give rise to quasi-normal modes. For example, an infalling particle could provide such a perturbation. The perturbation due to infalling extended bodies has also been studied [60]. In this case, the effect

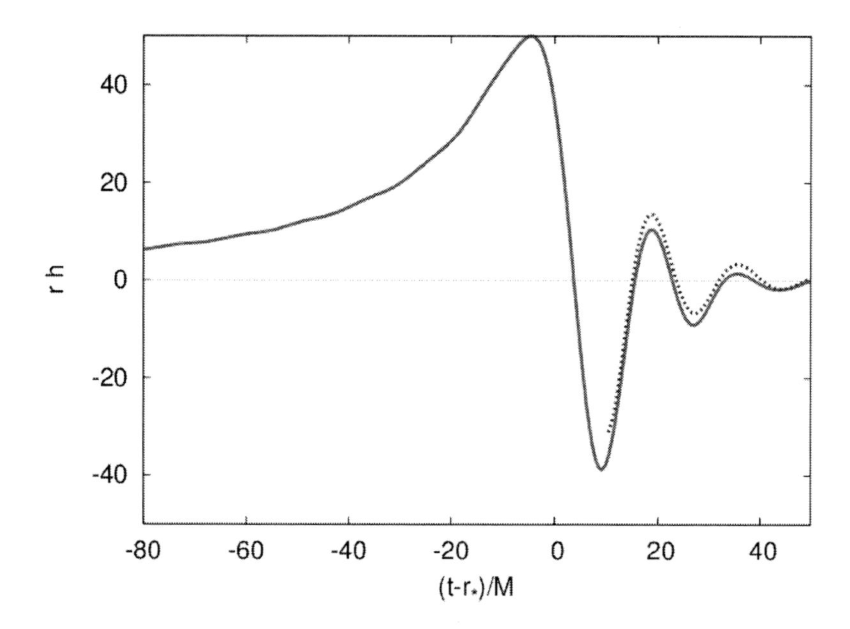

Figure 4. A computation of the gravitational wave amplitude of a Schwarzschild black hole perturbed by an infalling particle (solid line) along with an analytic fit (dashed line) utilizing the first two modes corresponding to $l = 2$. Figure courtesy of V. Ferrari and L. Gualtieri, Roma. (From [51].)

is smaller than in the point particle case due to interference effects. In the case of fluid material orbiting the black hole, modes are only significantly excited when $r < 4M$, and is therefore important in the case of unstable orbits. More realistic processes involve large scale computations and make up an important area of study in modern black hole research. The collapse of a neutron star core has been studied [61] as have the modes caused by thick accretion disks [62] and [63]. We will discuss the case of black hole mergers separately below.

3.2.2. Critical Behavior

From the point of view of classical black hole physics there are two possible outcomes that result from the evolution of regular Cauchy data: Either a black hole forms or it does not. In this paradigm, an interesting question to ask is what happens in the regime that straddles this bifurcation? This question was originally studied by Choptuik utilizing a spherically symmetric massless scalar field minimally coupled to gravity [64]. This choice of matter field is convenient as one does not need to worry about possible formation of field condensates (stars) in this model. Complete field dispersion or black hole formation are the only possible outcomes. Samples of the two outcomes are displayed in figure 5.

The problem was tackled as follows: The initial data contained one tunable parameter, usually denoted as p. By changing this parameter the numerical evolution would either

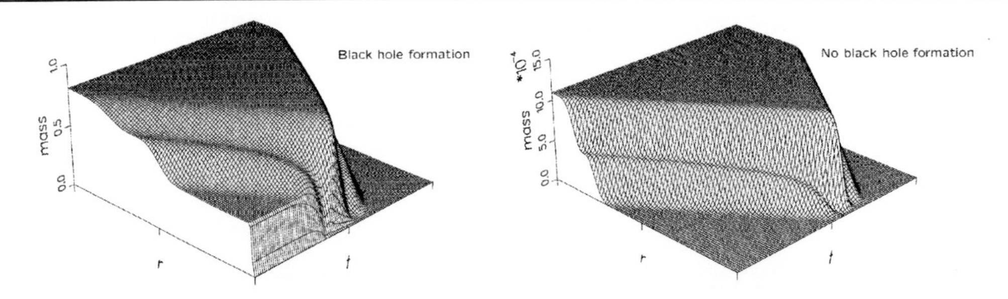

Figure 5. Evolution of scalar field data undergoing gravitational collapse. In the left diagram a black hole forms at late time whereas in the right diagram the field disperses. Figures courtesy of M. W. Choptuik, Yukawa International Seminar.

form a black hole or not. Ideally, the problem is set up so that there is a simple relationship between the size of this parameter's numerical value and the "strength" of the gravitational interaction as there is no way a priori to know if the evolution will result in an event horizon. Therefore many evolutions need to be run, in each trial changing the value of the parameter, before the critical point is reached.

For the minimally coupled massless scalar, ϕ, the field equations yield:

$$R_{\mu\nu} - \frac{1}{2} R\, g_{\mu\nu} = 8\pi \left(\phi_{,\mu}\phi_{,\nu} - \frac{1}{2}\phi_{,\alpha}\phi^{;\alpha}\, g_{\mu\nu} \right) , \tag{31}$$

whereas the conservation law gives rise to the wave equation:

$$\phi_{,\alpha}^{\;;\alpha} = 0 \,. \tag{32}$$

With a metric admitting the line element as in (8), except that now $e^{\gamma(r)} \to e^{\gamma(r,t)} =: b^2$ and $e^{\alpha(r)} \to e^{\alpha(r,t)} =: a^2$, the field equations go over to:

$$\frac{a_{,r}}{a} + \frac{a^2 - 1}{2r} - 2\pi r \left[\frac{a^2}{b^2} (\phi_{,t})^2 + \phi_{,r} \right] = 0 \,, \tag{33a}$$

$$\frac{b_{,r}}{b} - \frac{a_{,r}}{a} - \frac{a^2 - 1}{r} = 0 \,, \tag{33b}$$

$$\frac{a_{,t}}{b} - 4\pi r \phi_{,r} \frac{a}{b} = 0 \,. \tag{33c}$$

The wave equation, after some mild manipulation yields the condition:

$$\left(\frac{a}{b}\phi_{,t} \right)_{,t} = \frac{1}{r^2} \left(r^2 \frac{b}{a} \phi_{,r} \right)_{,r} \,. \tag{34}$$

Dirichlet and Neumann boundary conditions were used for the metric and the scalar field. At $r = 0$, the condition of space-time regularity was imposed. As well, the initial ($t = 0$) scalar field profile is also fully specified along with an initial velocity.

Choptuik found some unexpected and very interesting behavior when studying the above system of equations. In the cases that were barely super-critical (i.e. cases where the parameter was tuned to values just above the black hole formation limit), Choptuik found an unexpected relationship. He found that the masses of the "mini" black holes formed followed a scaling law of the form

$$M = C(p - p_c)^{\eta} . \tag{35}$$

Here, p_c is the critical value of the adjustable parameter. That is the value (within numerical precision, which was 10^{-15} for the original study) p attains when it is exactly at the bifurcation point. C is a constant that depends on the particular initial data profile chosen and, perhaps most interestingly, the exponent η possesses a value that is *universal* for all initial scalar field data[3]. For the minimally coupled, massless scalar field, Choptuik found that $\eta \approx 0.374$. No matter what the initial data profile, the masses of small black holes followed the law (35) with this numerical value of the exponent.

Since Choptuik's original work, various authors have extended the domain of study to other coordinate systems [65], different matter models ([66] - [75]) and different numbers of dimensions ([73], [76] - [79]). Also, work in non-spherical symmetry has been done as well as other couplings and some analytic analysis of the problem ([80] - [87]). Some of the possibilities and results are summarized in table 2. Further, the collapse can be classified as type-I, where there is a minimum finite black hole size, or as type-II, where it is possible to form arbitrarily small mass black holes. The phenomenon discussed here applies to the type-II case.

Table 2. A summary of some fields studied and their critical exponents.

Field	η
Massless real scalar field	0.37
Massless complex scalar field	0.39
Massless complex scalar field (angular momentum)	0.11
Massive real scalar field (small mass)	0.37
Radiation	0.36
SU(2)	0.20
Gravitational waves	0.36

The behavior discussed in this section is reminiscent of certain phase transitions in statistical mechanics. One example is the liquid-gas phase transition. Near the critical temperature, T_c, a substance on its boiling curve will possess a discontinuity in the density of the two phases of the form

$$\rho_l - \rho_g \propto (T_c - T)^{\eta} , \tag{36}$$

where ρ_g and ρ_l are the densities in the gas and liquid phases respectively. Ferromagnetic materials obey a similar law near the Curie temperature, T_c,

$$m \propto (T_c - T)^{\eta} , \tag{37}$$

[3]This exponent is often labeled as γ in the literature. However, in this manuscript γ plays multiple roles and therefore η is used to avoid confusion.

where m represents the magnitude of the magnetization vector.

There is another interesting phenomenon associated with the threshold of black hole formation, namely that of self-similarity of the space-time. As this phenomenon is mainly of interest on the side of the transition where black holes do not occur (i.e. slightly sub-critical) it is beyond the scope of this review and is omitted due to article size considerations. As well, we have only scratched the surface in citing the large number of works in this field. The interested reader is referred to the thorough reviews in [88] and [89] and references therein.

In closing this section we quote the words of C. Gundlach [88] on this topic "Critical phenomena are arguably the most important contribution from numerical relativity to new knowledge in general relativity to date."

3.2.3. Black Hole Mergers

The two-body problem in general relativity is extremely difficult to analyze, mainly due to the non-linearity of the field equations. Any, even somewhat realistic, system needs to be evolved numerically in the strong-field regime and even with modern computational technology it is still a taxing problem. Black holes, being "simple" objects, and having the possibility of producing a strong enough gravitational wave signal to detect are natural objects to consider in the two body problem. This system provides an excellent arena to study the strong-field effects in G. R. without the complications introduced by material (i.e. non-gravitational) effects. Since the two black hole system is unstable, it is expected that at late times the solution will approach that of a Kerr black hole. The case of a black hole-neutron star merger in full general relativity has been recently analyzed in [90].

Binary black hole systems were first numerically modeled in 1964 by Hahn and Lindquist [91] on a 51×151 mesh utilizing axial symmetry in a head-on collision scenario. Within 4 hours the evolution had proceeded 50 time steps in their simulation. A decade later Smarr [92] and Eppley [93] also constructed simulations of head-on collisions in the hope of studying the gravitational wave emissions. With the announcement in 1990 of the LIGO gravitational wave observatories to be constructed, gravitational wave simulations and the two body problem were taken up in force. In such studies the numerical methods almost invariably involve some form of $3+1$ split or null variants of it.

The black hole merger is usually separated into 3 regimes: The inspiral stage, the merger stage and, finally, the ringdown stage (see [94] for full details, which includes an earlier fourth stage, the Newtonian phase).

In the inspiral stage, gravitational wave emission has the greatest effects on the dynamics. It is expected that in the case of small to medium mass ratios, large eccentricities in the orbit will decay, yielding a roughly circular orbit by the end of this phase [95], [96]. Extreme mass ratios are expected to possess high eccentricities [97], [98]. This phase is often well modeled by post-Newtonian and higher-order methods. If the mass ratio is extreme, the smaller partner may be viewed as a test particle in the background space-time of the larger partner. Methods exist to calculate gravitational wave emission in this "test-particle" case (see, for example [99] and references therein).

An interesting point is that the orbit of the small companion in an extreme mass-ratio scenario will not generally lie in a plane, due to frame-dragging effects. Therefore, the geometry of the surrounding space-time can be well probed by the small companion and this information would be transmitted by the gravitational wave emission which is believed to lie within the future LISA detector's bandwidth.

The merger stage is extremely complicated and requires full numerical investigations. This is a very short lasting phase, perhaps lasting two cycles of the emitted gravitational waves. The luminosity of gravitational waves emitted at this stage may approach the order of 10^{52} J/s and the frequency of the gravitational wave approaches that of the dominating quasi-normal mode of the resulting black hole. An enormous amount of energy is liberated in this phase which totals in the neighborhood of three percent of the rest mass energy of the system [94].

Finally, the ringdown phase refers to the settling down of the single black hole formed to a Kerr black hole. Quasi-normal mode analysis is quite important in this stage as one has a perturbed black hole space-time radiating away energy via gravitational wave emission. The ringdown is dominated by the following frequencies [42], [94], [100]:

$$\frac{\sigma_1}{2\pi} \approx (32\,\text{kHz})\frac{M_\odot}{M}\left[1 - 0.63\left(1 - \frac{J}{M^2}\right)^{0.3}\right], \tag{38a}$$

$$\sigma_2^{-1} \approx (20\mu\text{s})\frac{M}{M_\odot}\frac{1}{\left(1-\frac{J}{M^2}\right)^{0.45}\left[1 - 0.63\left(1 - \frac{J}{M^2}\right)^{0.3}\right]}, \tag{38b}$$

with J the angular momentum of the final resulting black hole. The equivalent of one or two percent of the rest mass energy is radiated during ringdown. At late time the emission is dominated by the radiative power-law tail.

As discussed above, much work in this area is numerical, due to the complications presented by the Einstein field equations. However, there exist constraints in the system of equations which must hold throughout any evolution scheme for it to be valid and this complicates the numerical evolution. Mathematically, these constraints arise as follows: Consider a metric with elements $g_{\mu\nu}$ and define a unit normal, n_μ , to a class of time slices. We also define $\tilde{g}_{\mu\nu} := g_{\mu\nu} - n_\mu n_\nu$. On a space-like hypersurface the following (Gauss-Codazzi) relations hold:

$$\tilde{R}^{\mu}{}_{\nu\rho\sigma} = \tilde{g}^{\mu}{}_{\lambda} R^{\lambda}{}_{\alpha\beta\gamma}\, \tilde{g}^{\alpha}{}_{\nu}\, \tilde{g}^{\beta}{}_{\rho}\, \tilde{g}^{\gamma}{}_{\sigma} - K^{\mu}{}_{\rho}K_{\nu\sigma} + K^{\mu}{}_{\sigma}K_{\nu\rho}\,, \tag{39a}$$

$$D_\mu K_{\nu\sigma} - D_\nu K_{\mu\sigma} = -\,n_\lambda R^{\lambda}{}_{\alpha\beta\gamma}\, \tilde{g}^{\alpha}{}_{\sigma}\, \tilde{g}^{\beta}{}_{\mu}\, \tilde{g}^{\gamma}{}_{\nu}\,. \tag{39b}$$

Here $\tilde{R}^{\mu}{}_{\nu\rho\sigma}$ is the curvature tensor constructed with $\tilde{g}$, D is built with the connection of $\tilde{g}$, and $K_{\mu\nu}$ is the extrinsic curvature of the hyper-surface. It should be noted that these relations hold regardless of the field equations. By pulling back (39a) and (39b) to the hyper-surface, and utilizing the field equations, one obtains the constraints for the Cauchy problem in general relativity. Therefore, the constraint equations enforce solutions to be the allowable data sub-sets permitted within general relativity theory.

It is a major focus of modern numerical research to find a scheme which ensures that the constraints are not seriously violated in the subsequent evolution. As the continuum ADM form of the field equations are weakly hyperbolic, this is no trivial task and generally some clever method needs to be devised such as consistent modifications to the equations of motion. An evolution program often "crashes" or becomes "unstable" when constraint violating modes grow to a size that is considered unacceptably larger than discretization/truncation errors. A breakthrough in the long-evolution stability problem occurred recently in 2005-06 with the advent of two methods that yield particularly stable evolutions (at least relatively). These are the generalized harmonic coordinates with constraint damping (GHCCD) method [101] [102] and an improved variant of the Baumgarte-Shapiro-Shibata-Nakamura method with moving punctures (BSSN) [103], [104], [105], [106]. Briefly, in the GHCCD method one adds to the field equations a function of the constraints which is designed to minimize or dampen the constraint violation. A recent proposal for such a counter-term in the case of the harmonic coordinates was devised in [107]. In the BSSN scenario one defines a conformal metric,

$$\bar{g}_{ij} := e^{-\Phi}\tilde{g}_{ij}, \quad e^{\Phi} = \tilde{g}^{1/3}, \tag{40}$$

with $\tilde{g}$ the spatial three-metric in the ADM decomposition, as well as a conformal trace-free extrinsic curvature quantity

$$\bar{A}_{ij} := e^{-\Phi}\left[K_{ij} - \frac{1}{3}\tilde{g}_{ij}K\right], \tag{41}$$

with K the trace of the extrinsic curvature. The conformal connection is given by

$$\bar{\Gamma}^{i} := \bar{g}^{jk}\bar{\Gamma}^{i}{}_{jk}\,. \tag{42}$$

In the BSSN scheme Φ, $\bar{A}_{ij}$, $\bar{\Gamma}^{i}$ and the lapse and shift are considered the basic variables of the evolution. The reason this scheme is utilized is that the long ranged degrees of freedom can easily be isolated from the non-radiative ones. Another reason is that the constraints can easily be substituted into some of the evolution equations thus implementing some of the constraints at the dynamic level. Also, with appropriate implementation of gauge, the evolution equations are hyperbolic, which is related to the fact that the connection (42) is treated as an independent quantity. The "moving punctures" refer to the fact that the black hole singularities are represented by punctures which move within the grid although there is no evolution at the puncture itself.

An excellent review of the progess in black hole mergers may be found in Pretorius' review [94]. We summarize here some recent results.

For the scenario involving two equal mass black holes with minimal spin and eccentricities the amount of energy released in the last stages before a Kerr black hole remnant is approximately 3.5% of the system's total energy. The resulting Kerr black hole possesses a spin parameter of $a \approx 0.69$ ([94], [105], [106], [108] and references therein). After the "collision" the gravitational waveform is dominated by the fundamental of the quadrupole moment of the quasi-normal mode of the final black hole. Also, when the flux is near its maximum, and subsequently, the waveform may be closely represented as a sum of quasi-normal modes, which is surprising as this is expected to be a highly non-linear regime.

If one removes the restriction of equal mass, but maintains minimal eccentricity and spin there is a decrease in the total energy emitted as well as the final spin of the resulting black hole. Also, although the quadrupole is still dominant, higher modes become non-negligible due to the reduction in the symmetry of the problem. This symmetry reduction is also responsible for an asymmetric gravitational radiation beaming. This delivers a recoil to the produced black hole in the orbital plane as there is net momentum carried away by the radiation. The maximum velocity imparted due to this effect seems to be approximately 175 Km/s when the mass ratio is 3:1 [94].

For the case of equal mass, nominal eccentricity but non-minimal spin, a new degree of freedom is introduced in this case as the individual black hole spins can be aligned in various directions compared to the orbital angular momentum. If the net spin angular momentum has a component parallel to the orbital angular momentum then more energy will be emitted than in the corresponding zero-spin case. If there is a component anti-parallel, less energy is emitted. Some particular studies indicate that a pair of holes, each with $a \approx 0.76$, radiated approximately 7% of their rest mass energy when the spins were aligned with the orbital angular momentum. In the case where the spins were anti-aligned, only approximately 2% of the rest mass energy was radiated. The final black hole spins were approximately 0.89 and 0.44 respectively [109], [110].

The next case presented is for two equal mass black holes with minimal spin but large eccentricity. This is the case studied in the first complete merger simulation by Pretorius in [101]. In this study two localized scalar field profiles were initially employed which collapsed to form black holes. The outcome of the collapse essentially yields a two black hole vacuum. The two field profiles were given equal magnitude but opposite direction boosts, with zero boost corresponding to a head-on collision. The final result, merger or separation, depends on the value of the single parameter, k, measuring the strength of the boost. An interesting result was found. Near the threshold value of the boost parameter, for a given class of initial profiles, the number of orbits, n scale as

$$e^n \propto |k - k_c|^{-\beta} , \tag{43}$$

where k is the value of the boost parameter and k_c is the threshold value of this parameter. The exponent β was found to possess a value of approximately 0.34. As it was noted in [94], this behavior is similar to that of test particles in equatorial orbit around a Kerr black hole. Those test particles which are near the capture threshold approach unstable circular orbits of the Kerr space-time and possess a scaling behavior similar to (43).

Recently, there have been studies providing simple formulas for the calculation of the final spin in a binary merger [111], [112].

4. Semi-Classical Black Hole Research

4.1. Review of Semi-classical Theory

Before discussing semi-classical relativity, we shall need a result from flat space-time for future use. Let us begin by studying the real massless scalar field in Minkowski space-time. Such a field obeys the wave equation:

$$\phi_{,\mu}^{\ ;\mu} = 0. \tag{44}$$

The field can be decomposed into Fourier components of positive and negative frequency

$$\phi = \sum_{n=0}^{\infty} \left[a_n f_n(\mathbf{x}) e^{-i\omega_n t} + a_n^{\dagger} f_n^*(\mathbf{x}) e^{i\omega_n t} \right] . \tag{45}$$

As is well known, the number of particle excitations associated with a particular state may be deduced from the number operator:

$$\hat{N} = \sum_n a_n^\dagger a_n . \tag{46}$$

Under proper Lorentz transformations we have (dropping subscripts on ω)

$$\sum_{n=0}^{\infty} \left[a_n f_n(\mathbf{x}) e^{-i\omega t} + a_n^\dagger f_n^*(\mathbf{x}) e^{i\omega t} \right] = \sum_{m=0}^{\infty} \left[a'_m f'_m(\mathbf{x}') e^{-i\omega' t'} + a_m^{\dagger\prime} f_m^{*\prime}(\mathbf{x}') e^{i\omega' t'} \right] , \tag{47}$$

where the prime denotes quantities calculated in the boosted frame. The relation between the primed and unprimed modes is

$$\begin{aligned} f'_m &= \sum_n \left[\alpha_{mn} f_n + \beta_{mn} f_n^* \right] , \\ f_n &= \sum_m \left[\alpha_{nm}^* f'_m - \beta_{nm} f_m^{\prime *} \right] , \end{aligned} \tag{48}$$

where α and β are the Bogoliubov coefficients of the transformation.

The unprimed and primed vacua are defined via:

$$a_n \left|0\right\rangle = 0, \;\; a'_m \left|0'\right\rangle = 0 . \tag{49}$$

In particular, the unprimed and primed creation and annihilation operators are related as

$$a_n = \sum_m \left[\alpha_{nm} a'_m + \beta_{nm}^* a_m^{\prime\dagger} \right] \tag{50}$$

under Lorentz transformations, as can be seen by inserting (48) into (47). Note that if the β coefficients are not zero, $\left|0'\right\rangle \neq \left|0\right\rangle$ and the two vacua will *not* be the same. That is, one observer's zero particle state will not be a zero particle state for a Lorentz transformed observer. Another way to view this is that the number operator in the primed frame does not agree with the number operator in the unprimed frame. In Minkowski space-time, proper Lorentz transformations respect the condition $\beta_{nm} = 0$ and therefore uniformly boosted observers will agree on particle content. This is not true for the case of observers which are accelerating, even in flat space-time. This leads to an interesting effect known as Unruh radiation. Discussion of this is beyond the scope of this manuscript.

The semi-classical approach is based on the premise that the gravitational field remains classical, but the matter content is quantized. The most straight-forward way to incorporate this into Einstein's theory is to write Einstein's equations as

$$R_{\mu\nu} - \frac{1}{2} R\, g_{\mu\nu} = 8\pi \left\langle \psi \right| T_{\mu\nu}(\phi_i) \left| \psi \right\rangle =: 8\pi \left\langle T_{\mu\nu} \right\rangle . \tag{51}$$

That is, the expectation value of a stress-energy tensor operator that depends on quantized fields, ϕ_i, is utilized as a source term in the gravitational field equations. Of course, one may still add a purely classical piece to the right-hand-side of the equations, $T_{\mu\nu}$, so that

$\langle\psi|\, T_{\mu\nu}(\phi_i)\, |\psi\rangle$ yields quantum corrections to the classical theory. The simplicity of equation (51) is deceptive. The matter fields themselves are, of course, quantized on the curved space-time, generally leading to a complicated dependence on $g_{\mu\nu}$ on the right-hand-side. There is also the issue of what state the fields should be in. As well, the quantity $\langle T_{\mu\nu}\rangle$ will diverge and therefore a regularization of the effective action leading to $\langle T_{\mu\nu}\rangle$ needs to be performed.

There are several states of particular importance in semi-classical black hole physics. These include, but are certainly not limited to, the Hartle-Hawking vacuum, the Unruh vacuum, and the Boulware vacuum. The Hartle-Hawking vacuum corresponds to a black hole in thermal equilibrium with a bath of thermal radiation, the Unruh vacuum corresponds to a state with a particle flux at future infinity, and the Boulware vacuum to a state where particles do not traverse to infinity, as measured from near the black hole.

In the case of a scalar field, the regularization leads to a renormalized value of the cosmological constant and a renormalized value of the Newtonian gravitational constant. As well, the divergence in the effective action also leads to a term possessing fourth-order derivatives of the metric and terms quadratic in the curvature as the divergences possess the form:

$$\langle T_{\mu\nu}\rangle_{div} \propto \frac{1}{\sigma}\left[A g_{\mu\nu} + B G_{\mu\nu}\right] + \left({}_{(1)}C\, {}_{(1)}H_{\mu\nu} + {}_{(2)}C\, {}_{(2)}H_{\mu\nu}\right) \ln\sigma\,. \tag{52}$$

Here ${}_{(1)}H_{\mu\nu}$ and ${}_{(2)}H_{\mu\nu}$ are given by:

$${}_{(1)}H_{\mu\nu} := 2R_{;\mu;\nu} - 2g_{\mu\nu}R^{;\rho}_{\ ;\rho} + \frac{1}{2}g_{\mu\nu}R^2 - 2RR_{\mu\nu}\,, \tag{53a}$$

$${}_{(2)}H_{\mu\nu} := 2R_{\mu\ \ ;\nu;\alpha}^{\ \alpha} - R_{\mu\nu\ \ ;\rho}^{\ \ ;\rho} - \frac{1}{2}g_{\mu\nu}R^{;\rho}_{\ ;\rho} - 2R_{\mu}^{\ \alpha}R_{\alpha\nu} + \frac{1}{2}g_{\mu\nu}R^{\alpha\beta}R_{\alpha\beta}\,, \tag{53b}$$

and σ is Synge's world function [113], which is equal to one-half the square of the geodesic distance between two nearby points. The limit $\sigma \to 0$ needs to be taken and the logarithmic divergence may be removed by renormalization of the constants ${}_{(1)}C$ and ${}_{(2)}C$. Although these terms are due to the metric dependence of $\langle T_{\mu\nu}\rangle$ itself, they are often written on the left-hand-side of the field equations:

$$G_{\mu\nu(ren)} + \Lambda_{(ren)}g_{\mu\nu} + {}_{(1)}C_{(ren)}\, {}_{(1)}H_{\mu\nu} + {}_{(2)}C_{(ren)}\, {}_{(2)}H_{\mu\nu} = 8\pi\left[T_{\mu\nu} + \langle T_{\mu\nu}\rangle_{(ren)}\right]\,, \tag{54}$$

where the subscript (ren) denotes renormalized values. Equations (54) are sometimes known as the semi-classical Einstein field equations.

An approximation often employed in semi-classical research is a perturbative approach. That is, the metric is written as:

$$g_{\mu\nu} = {}_{(0)}g_{\mu\nu} + \epsilon h_{\mu\nu}\,, \tag{55}$$

where ${}_{(0)}g_{\mu\nu}$ is a classical "background" metric and the small constant parameter ϵ vanishes in the limit $\hbar \to 0$ so that $h_{\mu\nu}$ represents first-order quantum corrections to the classical metric. One customarily limits calculations to order linear in ϵ. The Einstein equations then yield an Einstein tensor of the form:

$$G_{\mu\nu} = {}_{(0)}G_{\mu\nu} + \epsilon\, {}_{(\hbar)}G_{\mu\nu}\,, \tag{56}$$

with ${}_{(\hbar)}G_{\mu\nu}$ being due to the quantum part of the metric and ${}_{(0)}G_{\mu\nu}$ satisfies the usual Einstein equation for the classical stress-energy tensor. One then constructs a stress-energy tensor on the classical background by some means, denoted here as $\epsilon\,{}_{(0)}\langle T_{\mu\nu}\rangle$, and attempts to solve ${}_{(\hbar)}G_{\mu\nu} = 8\pi\,{}_{(0)}\langle T_{\mu\nu}\rangle$ for $h_{\mu\nu}$. This is the *back-reaction problem*.

4.2. Developments in Semi-classical Black Hole Research

Perhaps one of the most amazing predictions to come out of semi-classical theory is the evaporation of black holes. The calculation was first carried out in Hawking's paper *"Particle creation by black holes"* in 1975 [25]. Although this is now an old result, given its importance it is appropriate to review it in this section.

We will study this effect in the Schwarzschild black hole (2) in the forms given by (14) and (15). We will also be considering Hawking's original argument where he considered a collapsing star with asymptotically flat regions. The Penrose diagram for the collapsing star is shown in figure 6. Any direct interaction of the quantum particles with the material making up the stellar body will be ignored since the gravitational field of the star is the main ingredient for this effect and not the star itself.

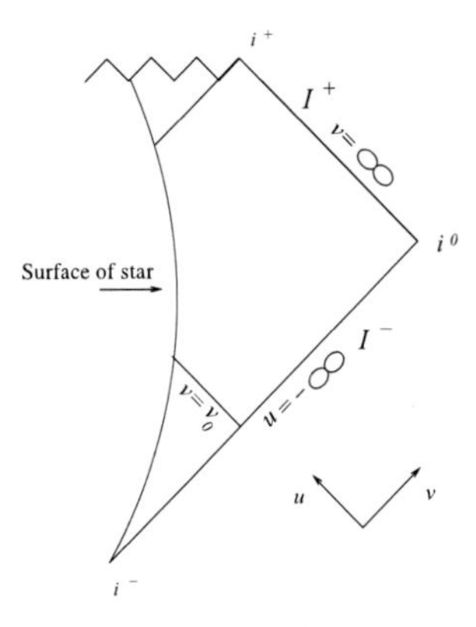

Figure 6. Penrose diagram depicting the surface of a collapsing star.

For simplicity a null scalar field will be used to illustrate this effect. At $\mathcal{I}^-$, there are no particles present and the state will be denoted as $|0_{in}\rangle$. It will be found that, at $\mathcal{I}^+$, this state corresponds to a state with a *non-zero* particle content. This is due to the presence of the non-trivial gravitational field.

Since $\mathcal{I}^-$ corresponds to approximately Minkowski space-time, we can define modes there with frequency ω' and therefore the solutions to (44) in Schwarzschild space-time may be written as

$$\varphi'_{\omega',v} = f'_{\omega'}(\mathbf{x})e^{-i\omega' v}\,, \tag{57}$$

i.e. the standard particle states of flat space-time. The field ϕ is written as before, with appropriate creation and annihilation operators satisfying $a_\omega\,|0_{in}\rangle = 0$:

$$\phi = \int \left[a'_{\omega'} f'_{\omega'}(\mathbf{x})e^{-i\omega' v} + a'^{\dagger}_{\omega'} f'^{*}_{\omega'}(\mathbf{x})e^{i\omega' v}\right] d\omega'\,. \tag{58}$$

(Note that in this argument we only have inward propagating waves so only inward components are necessary here.)

At $\mathcal{I}^+$ the space-time is also Minkowski. We can define modes there as well, which in terms of the null coordinate u have the form:

$$\varphi_{\omega,u} = f(\mathbf{x})_\omega e^{-i\omega u}\,, \tag{59}$$

which again correspond to the usual flat space-time modes with positive frequency. These are outgoing modes.

Now, we can express φ in terms of φ' utilizing a similar transformation as in (48):

$$\varphi = \int \left[\alpha^*_{\omega'\omega}\varphi' - \beta_{\omega'\omega}\varphi'^*\right] dv\,. \tag{60}$$

As in the flat space-time case, should the second Bogoliubov coefficient not vanish, there will be a disagreement in particle content between an "in" observer and an "out" observer.

Let us consider the transformation between the in and out observers. For a single outgoing mode we have:

$$e^{-i\omega u} = e^{-i\omega(v-2r^*)} = e^{-i\omega\left[v-2r-4M\ln\left(\frac{r}{2M}-1\right)\right]}. \tag{61}$$

Consider now tracing back the path of a particle (outgoing) as it skims the horizon, near $r = 2M$. If the star is collapsing, the level surface representing the stellar boundary (assumed to be near $r = 2M$) changes with time as the particle traverses through the star. The equation describing this surface is given by $r \approx 2M + \kappa_0(v - v_0) + \mathcal{O}(v - v_0)^2$. Adopting the eikonal approximation, and assuming most of the change in the eikonal takes place near this surface we get

$$\omega u \approx \omega\left[v - 4M + 2\kappa_0(v - v_0) + ... - 4M\ln\left(\frac{\kappa_0(v_0 - v)}{2M} + ...\right)\right], \tag{62}$$

with $v < v_0$.

Concentrating on the dominant part of the eikonal, we can see that the ingoing eikonal (which we denote as Ω) corresponding to the outward eikonal (62) is given by:

$$\Omega(v) \approx -4M\omega\ln\left((v_0 - v)\right) + \text{constant}\,, \tag{63}$$

the constant shift being irrelevant. Comparing this expression with (60) it may be seen that, as a Fourier transform, $\varphi_{\omega,u}$ has both $e^{-i\omega' v}$ and $e^{+i\omega' v}$ components and therefore the corresponding $\beta_{\omega'\omega}$ *does not vanish* in the transformation. There are therefore out particles even in the absence of in particles. The conformal diagram corresponding to the collapse into, and subsequent evaporation of, a black hole is shown in figure 7. A much more detailed calculation, which includes a derivation of the actual particle spectrum, may be found in [114].

This evaporation issue is related to what is sometimes known as the information loss problem where "information" falling into a purely classical black hole is presumably lost due to the fact that a black hole is described solely by its mass, charge and angular momentum and therefore the evolution is not unitary. It is thought that if the resulting black hole

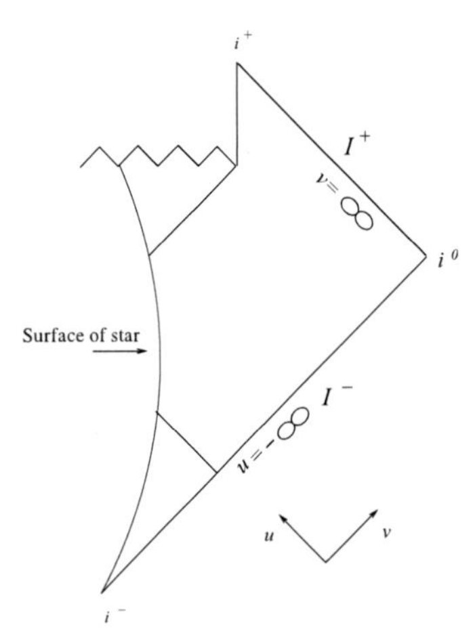

Figure 7. Penrose diagram of stellar collapse taking into account subsequent black hole evaporation.

radiation spectrum is not truly thermal then the information may re-emerge from the black hole in this way. Later it will be shown that a full theory of quantum gravity may resolve this in another way.

Since the Hawking result, much work has been done in semi-classical gravity, major advancements having been made in the 1980s. Much of the early works concentrated on methods to obtain sensible, convergent quantities within semi-classical theories. The interested reader is referred to [24] and references therein. In the following we limit our survey to some issues which directly involve the back-reaction problem as the literature and topics in the field of semi-classical gravity are almost as vast as in the classical theory making anything resembling a complete coverage nearly impossible.

More recently, semi-classical black hole research has focused on modeling the perturbations on the classical background geometry due to quantum fields and their fluctuations. Study of the latter effect makes up the arena of stochastic gravity. These problems require computation of the renormalized stress-energy tensor on the background, which is then to be used as a source for the metric perturbations.

In the black hole context, Hiscock, Larson and Anderson have calculated the back-reaction effects of scalar, spinor and vector fields inside a Schwarzschild black hole's event horizon [115]. To construct $\langle T_{\mu\nu}\rangle$[4] they utilized various approximation schemes developed in the literature [116] - [120] and the DeWitt-Schwinger expansion [121]. They studied the quantum field's back reaction on the anisotropy of the interior as well as the first-order correction to the Kretschmann scalar. In summary they found the following: Spinors and minimal and conformally coupled scalars tended to decrease the anisotropy as the singularity was approached whereas vector fields tended to increase the anisotropy. Regarding the effect on the Kretschmann scalar, it was found that the minimally coupled massive scalar

[4]From now on in this section we drop the (0) subscript in front of $\langle T_{\mu\nu}\rangle$ and it is understood that $\langle T_{\mu\nu}\rangle$ refers to the first-order quantum correction to the stress-energy tensor and is constructed with the zeroth-order (classical) background metric.

field and spinor fields tend to slow down the rate of increase of curvature as one approaches the singularity whereas other couplings and fields tended to increase the rate of curvature growth.

A similar problem was studied for the case of cylindrical black holes (or black strings) in [122] utilizing the stress-tensor approximation in [116] for quantum scalar fields. It was found there that the conformally coupled scalar also tended to increase the growth of curvature near the horizon. In this case, utilizing the cylindrical version of the metric (23) as a background space-time, with cylindrical (rather than one null coordinate) it was found that

$$\delta K \approx \left[\frac{3}{10}\frac{\alpha^4}{\pi} - 2\frac{\alpha^5}{\pi}\left(\frac{2}{M}\right)^{1/3}\left(T - \frac{(4M)^{1/3}}{\alpha}\right)\right], \tag{64}$$

where δK is the perturbation on the background Kretschmann scalar. Here T is the interior time coordinate (corresponding to the radial coordinate outside the black string) and the horizon is located at $T = \frac{(4M)^{1/3}}{\alpha}$.

Of course, the above results are only valid insofar as the perturbation is valid and the results cannot be extrapolated right down to the singularity.

In spherical black holes an enormous amount of work has been done in calculating field expectation values and stress-energy tensors of various fields and couplings. There has also been much work in trying to produce the $A/4$ entropy of black holes from semi-classical theory. The amount of work in this fascinating area is too large to even begin reviewing here. We refer the interested reader to [159] and references therein along with the books [24] [22].

For cylindrical black holes the amount of work is much more modest. Very briefly, in the context of the cylindrical black holes, Piedra and de Oca have studied the quantization of massive scalar and spinor fields over static black string backgrounds [123] [124]. They have calulated $\langle T_{\mu\nu}\rangle$ up to second order in the inverse mass value. Dias and Lemos have studied magnetic strings in anti-de Sitter general relativity [125] and the scalar expectation value, $\langle\phi^2\rangle$, has been computed in [126].

Interestingly, critical behavor has also been studied in the context of semi-classical gravity. For technical reasons, much of this work has been done in two dimensional tensor-dilaton gravity as $\langle T_{\mu\nu}\rangle$ may be determined from the trace anomaly along with the mild assumption of conservation. Ayal and Piran, for example, obtained a critical scaling exponent of $\eta \approx 0.409$ [127]. A slightly different model, utilizing a conformally coupled scalar, was analyzed in [128] and a critical exponent of 0.5 was found. In [129] yet another model was employed which allows one to turn the quantum effects on and off. A critical scaling exponent of $\eta \approx 0.53$ was found in this study. In [130] the authors calculated the quantum stress tensor on a classical background spacetime with perfect fluid source. The quantum effects then were treated as perturbations of the classical fluid gravitating system. They found that a mass gap exists when $\eta \geq 0.5$ so that there is a minimum size to the black holes formed. The case $\eta < 0.5$ could not be studied as the semi-classical approximation breaks down in that regime.

In stochastic gravity one takes the level of quantum approximation one step further, considering the effects of the field fluctuations. Given limited space we cannot cover stochastic gravity here. The interested reader is referred to the reviews [131] - [134] and references therein.

5. Quantum Black Hole Research

5.1. Quantum Gravity

Attempts to quantize gravity go almost as far back as the dawn of quantum mechanics. One of the earliest arguments for the quantization of gravity, in fact almost as old as general relativity itself, is that if $T_{\mu\nu}$ is inherently quantum, then so it should be with the gravitational field which it produces [135]. However it can be argued that this is not necessarily the case [136]. Although it is now generally believed that the gravitational field must be quantized in some way, there is still some debate on this necessity.

In the early days of quantum theory, the first person to realize that there would be serious problems applying those techniques to gravity seems to have been Matevi Bronstein. Bronstein had the insight to deduce that quantum theory could not be applied in any obvious way to a theory that was background independent. It was possible, according to Bronstein, that the ordinary notions of space and time would have to be abandoned [137]. Other pioneers in the field included P. Bergmann and P. Dirac.

After quantum field theoretic techniques were sufficiently developed, an obvious approach to quantizing gravity was implemented as a simple background expansion:

$$g_{\mu\nu} = \eta_{\mu\nu} + \epsilon h_{\mu\nu} + \mathcal{O}(\epsilon^2)\,. \tag{65}$$

As is now well known, treating the perturbations as fields on the background metric ($\eta_{\mu\nu}$) yields a non-renormalizable quantum field theory with divergences commencing at the one-loop level for gravity with matter couplings.

In the early 60s, Feynman, working at tree-level, computed transition amplitudes and demonstrated that reasonable results are obtained. This gave hope for this line of quantum gravity research. However, he noted that at loop level problems began to arise which required the introduction of Faddeev-Popov ghosts by DeWitt [139]- [142].

In the 70s it became generally accepted that gravity coupled to matter will be non-renormalizable. It was, however, found that one could add a spin $\frac{3}{2}$ particle to general relativity which yields a theory finite at two-loops. Thus began the field of study known as supergravity [138].

Finally, it was shown explicitly in 1986, by Goroff and Sagnoti that, at two-loop order, finite S matrix elements could be attained if the gravity action contained a counter-term of the form[5] [143]:

$$\mathcal{L}_{ct} = \frac{1}{(d-4)}\frac{209}{2880(16\pi)^2}\sqrt{-g}R^{\alpha\beta}{}_{\gamma\delta}\,R^{\gamma\delta}{}_{\mu\nu}\,R^{\mu\nu}{}_{\alpha\beta}\,, \tag{66}$$

with d the effective regularized dimension of the space-time, thus explicitly showing the divergent properties at two-loop level.

[5]At one loop order the divergence possesses terms proportional to R^2 and $R_{\mu\nu}R^{\mu\nu}$ and is therefore finite in the absence of matter.

On the canonical side, DeWitt in 1967 publishes what was originally thought of as the "Einstein-Schrödinger equation" also known as the Wheeler-DeWitt equation [144]. Some argued at the time that the problem of quantizing the gravitational field had been solved.

This Hamiltonian approach begins with the familiar ADM decomposition of space-time, as illustrated in figure 8. In the figure q_{ij} is the metric of the three-surface Σ and N and N^j are the usual lapse function, and shift vector associated with the ADM decomposition.

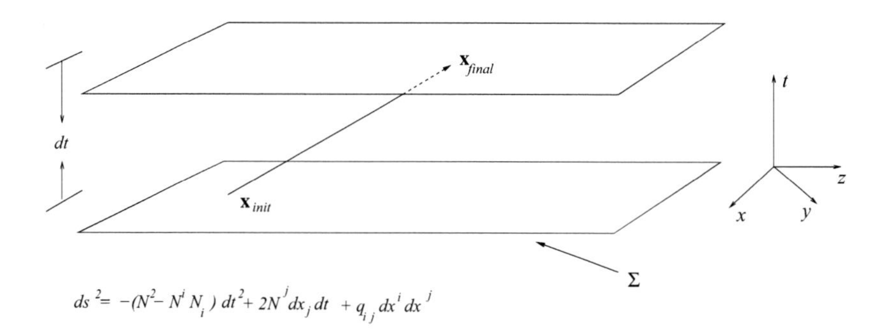

Figure 8. ADM decomposition of space-time into "space" and "time".

With this decomposition, the gravitational action may be written as:

$$
\begin{aligned}
I &= \frac{1}{16\pi}\int dt \int_{\Sigma} dx^3 \left[\Pi^{ab}\dot{q}_{ab}\right. \\
&\left. +2N_b\nabla_a^{(3)}\left(q^{-1/2}\Pi^{ab}\right) + Nq^{1/2}\left(R^{(3)} - q^{-1}\Pi_{cd}\Pi^{cd} + \frac{1}{2}q^{-1}\Pi^2\right)\right], \quad (67)
\end{aligned}
$$

with $\Pi := \Pi^{ab}q_{ab}$, $\Pi^{ab} := q^{-1/2}\left[K^{ab} - K\,q^{ab}\right]$, $K := K^{ab}q_{ab}$. K_{ab} is the extrinsic curvature tensor and over-dots denote differentiation with respect to the slicing time, t. In this scheme, the field variable is q^{ab} and its conjugate momentum is Π^{ab}. The equations of motion go over to

$$
\begin{aligned}
2\nabla_i^{(3)}\left[q^{-1/2}\Pi^{ij}\right] &= 0 =: -V^j\,, \\
q^{1/2}\left[R^{(3)} - q^{-1}\Pi_{ab}\Pi^{ab} + \frac{1}{2}q^{-1}\Pi^2\right] &= 0 =: -S\,.
\end{aligned}
$$

The action leads to a Hamiltonian density:

$$
\mathcal{H}_G = \frac{1}{16\pi}\Pi^{ab}\dot{q}_{ab} - \mathcal{L}_G = N_bV^b + N\,S\,,
$$

and the following symplectic structure:

$$
\begin{aligned}
\left\{\Pi^{ab}(\mathbf{x}),\, q_{cd}(\mathbf{y})\right\} &= 16\pi\delta^a_{(c}\delta^b{}_{d)}\delta(\mathbf{x},\,\mathbf{y}), \\
\left\{\Pi^{ab}(\mathbf{x}),\, \Pi^{cd}(\mathbf{y})\right\} &= 0 = \left\{q_{ab}(\mathbf{x}),\, q_{cd}(\mathbf{y})\right\}\,,
\end{aligned}
$$

which after quantization leads to the famous Wheeler-DeWitt equation:

$$\left[\left[-q_{ab}q_{cd} + \frac{1}{2}q_{ac}q_{bd}\right]\frac{\delta}{\delta q_{ac}}\frac{\delta}{\delta q_{bd}} + q^{1/2}R^{(3)}\right]\Psi(q) = 0. \tag{68}$$

The Wheeler-DeWitt formulation suffers from some problems. The configuration field (3-metric) does not appear as a gauge field. As well, there are inconsistencies with certain transition probabilities in the path-integral version.

There are several other candidate theories of quantum gravity. These include the sum-over-Euclidean geometries developed by Hawking [145] and its Lorentzian counterpart, the causal set approach of Sorkin [146] [147], dynamical triangulations [148], and other theories, including loop quantum gravity. Interestingly, the causal set approach predicts the existence of a small positive cosmological constant of the order of that required to provide the observed acceleration of the universe.

5.1.1. The Loop Quantum Gravity Program in Brief

Loop quantum gravity provides a promising quantization scheme for general relativity. There is a Hamiltonian approach and a covariant approach, yielding a spin-foam model, so named due to the resemblance of the Feynman diagram analogs to a foam of bubbles. We will concentrate here on the Hamiltonian approach, which is perhaps a bit more perspicuous. For a nice review of the spin-foam approach, the reader is referred to [149].

It was noted by Ashtekar [150] that general relativity can be very neatly reformulated in terms of a densitized triad, E^b_j instead of the metric:

$$q\, q^{ab} = E^a_i E^b_j \delta^{ij}$$

with

$$E^a_i := \frac{1}{2}\epsilon^{abc}\epsilon_{ijk}e^j_{\ b}e^k_{\ c}, \quad \text{where} \quad q_{ab} = e^i_{\ a}e^j_{\ b}\delta_{ij}.$$

This is sometimes known as the phase-space representation and the indices i, j etc. denote the orthonormal components. In terms of the new variables, the ADM action (67) may be written as

$$I = \frac{1}{8\pi}\int dt \int_\Sigma \left[E^a_i \dot{K}^i_{\ a} - N_b V^b - NS - N^i\epsilon_{ijk}E^{aj}K^k_{\ a}\right] d^3x\,, \tag{69}$$

with $K^i_{\ a} := \frac{1}{\sqrt{\det(E)}}K_{ab}E^b_{\ j}\delta^{ij}$. In this scheme, the canonically conjugate variables are E^a_i and $K^i_{\ a}$.

The symmetry group in Σ is $SO(3)$. We can write $K^i_{\ a}$ in terms of the fiducial $so(3)$ connection[6] on Σ, $\Gamma^i_{\ a}$:

$$\gamma K^i_{\ a} = A^i_{\ a} - \Gamma^i_{\ a}\,,$$

[6]The fiducial connection is that yielded by the solution of Cartan's structural equation, $\partial_{[a}e^i_{\ b]} + \epsilon^i_{\ jk}\Gamma^j_{\ [a}e^k_{\ b]} = 0$, where $e^i_{\ a}$ is the standard (non-densitized) triad, $q_{ab} = e^i_{\ a}e^j_{\ b}\delta_{ij}$.

where γ is known as the *Immirzi parameter*. The "modified" connection $A^i_{\ a}$ can be defined by this equation. In terms of this new connection the action may be written as

$$
\begin{aligned}
I &= \frac{1}{8\pi}\int dt \int_\Sigma \left[E^a_i \dot{A}^i_{\ a} - N_b V_b - NS - N^i G_i\right] d^3x\,, \qquad (70)\\
G_i &:= \partial_a E^a_i + \epsilon_{ij}{}^k A^j_{\ a} E^a_k\,,
\end{aligned}
$$

with symplectic structure:

$$
\begin{aligned}
\{E^a_j(\mathbf{x}),\, A^i_{\ b}(\mathbf{y})\} &= 8\pi\gamma\delta^a_b\delta^i_{\ j}\,\delta(\mathbf{x},\mathbf{y})\,,\\
\left\{E^a_i(\mathbf{x}),\, E^b_j(\mathbf{y})\right\} &= \{A^j_{\ a}(\mathbf{x}),\, A^i_{\ b}(\mathbf{y})\} = 0\,.
\end{aligned}
$$

The above formulae do not distinguish between $SO(3)$ and $SU(2)$ and both these groups possess the same algebra so it is customary to work in $SU(2)$, the indices i, j now coupling quantities to the $su(2)$ algebra. The quantization of this system yields the canonical version of loop quantum gravity. E^a_i and $A^i_{\ a}$ are the *Ashtekar - Barbero variables*. The quantum versions of the equations of motion yield the *quantum Einstein equations*:

$$\hat{G}_i\,|\Psi\rangle = \widehat{D_a E^a_i}\,|\Psi\rangle = 0\,, \qquad (71a)$$

$$\hat{V}_a\,|\Psi\rangle = \left[\widehat{E^a_i F^i_{\ ab}} - (1-\gamma^2)\widehat{K^i_{\ b} G_i}\right]|\Psi\rangle = 0\,, \qquad (71b)$$

$$\hat{S}\,|\Psi\rangle = \left[\frac{1}{\sqrt{\det(E)}}\widehat{E^a_i E^b_j \epsilon^{ij}{}_k F^k_{\ ab}} - 2(1+\gamma^2)\widehat{K^i_{\ [a} K^j_{\ b]}}\right]|\Psi\rangle = 0\,, \qquad (71c)$$

with

$$
\begin{aligned}
F^i_{\ ab} &:= \partial_a A^i_{\ b} - \partial_b A^i_{\ a} + \epsilon^i{}_{jk} A^j_{\ a} A^k_{\ b}\,,\\
D_a v_i &:= \partial_a v_i - \epsilon_{ijk} A^j_a v^k\,.
\end{aligned}
$$

Note that now there is a constraint equation for G_i (the Gauss constraint). This reflects the fact the triad possesses a rotational freedom; one can choose different frames locally by rotating the triad. This redundancy is eliminated in the new Gauss constraint. To the above system one could add matter couplings by supplementing the action with a matter term and quantizing appropriately. The problems associated with (68) are not present in this representation.

Before continuing, we shall make a few comments about this quantization scheme:

1. The scheme is background independent and respects diffeomorphism invariance. The choice of time slicing is arbitrary and does not affect the physics.

2. A superpartner can be accommodated and therefore supersymmetry can be incorporated. This has been done [151].

3. Instead of the Einstein-Hilbert action, one can accommodate geometric actions made up of arbitrary curvature invariants. The scheme is generally similar to that outlined above.

4. Higher dimensions can be accommodated.

It should be noted that 2, 3, and 4 are *not required* but simply can be accommodated.

What is of interest is the holonomy of A as it is transported around what are known as *spin-networks*, (first introduced by Penrose [152] and utilized early in loop quantum gravity by Jacobson and Smolin [153]) and the state vectors $|\Psi\rangle$ which are functions of this holonomy. The concept of time evolution is now encoded in terms of how the interrelationship of the network, which describes space, evolves. The details are beyond the scope of this manuscript but the interested reader may find them in [154], [155], [156]. What is of particular interest in the context of black hole research is that this theory predicts that on the small scale, space is *discrete*![7] Classically, the area may be constructed out of the triad via

$$A(\mathcal{S}) = \int_{\mathcal{S}} \sqrt{n^a E^i_a n_b E^b_i}\, d^2 s \,, \tag{72}$$

with the n vectors denoting normals to the 2-surface $\mathcal{S}$. To go over to the quantum theory one replaces the classical triad with the corresponding quantum operator. The picture that arises is that each fiber of the spin-network that pierces a surface $\mathcal{S}$ endows it with a certain amount of area and geometry, via the introduction of an angular defect (see figure 9).

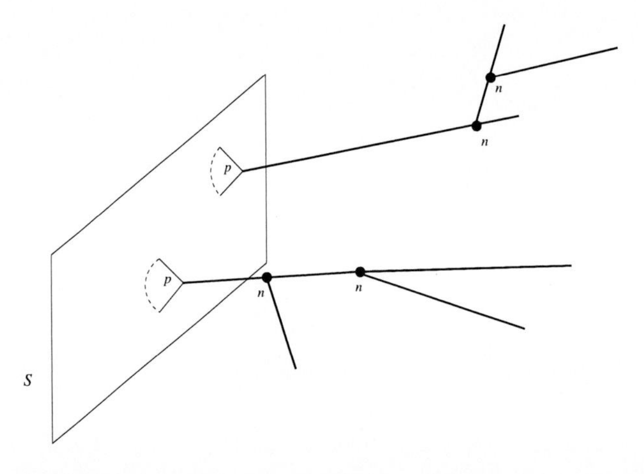

Figure 9. A gravitational spin-network endowing a surface S with area and geometry. The spin-network punctures the surface $\mathcal{S}$ at p. The surface is flat everywhere except at the punctures. There, the spin-network can be pictured as "tugging" on the surface, endowing it with geometry and introducing a local angular defect on the surface. The n's are the nodes, which are associated with volumes.

The triad possesses the following spectrum when acting on the state functions of loop quantum gravity:

$$\hat{E}_i(S_I)\hat{E}^i(S_I)\,\Phi^j\left(h_e[A]\right) = (8\pi l_p^2\gamma)^2\left[j_p(j_p+1)\right]\Phi^j\left(h_e[A]\right) \,,$$

where $\Phi^j\left(h_e[A]\right)$ are state functions which depend on the holonomy of A, $h_e[A]$, along an edge e of the spin-network. The j_p are half-integers and l_p is the Planck length. Therefore,

[7] It should be emphasized that this is a *prediction* of the theory and is not put in "by hand".

for the area operator[8] :

$$\widehat{Area_S}\,|\Psi\rangle = 8\pi l_p^2 \gamma \sum_p \sqrt{j_p(j_p+1)}\,|\Psi\rangle \ . \tag{73}$$

Notice that the eigenvalues of area are *discrete*!

One can also construct the classical volume utilizing the triad:

$$V = \int_M \sqrt{\left| \frac{1}{3!} \epsilon_{abc} E_i^a E_j^b E_k^c \epsilon^{ijk} \right|}\, d^3x \ . \tag{74}$$

This can be replaced by its quantum analog and volume eigenvalues can be calculated. The results are somewhat complicated and we omit them here. However, the volume is also discrete.

We will see below that these two operators are of extreme importance in loop quantum gravity black hole research.

5.2. Developments in Loop Quantum Black Hole Research

There are a number of results regarding black holes in loop quantum gravity. We shall concentrate here on what are arguably the two most significant results; namely the source of black hole entropy and the resolution of the singularity problem. These are of importance because it has long been believed that any viable theory of quantum gravity should explain where the enormous entropy of a black hole comes from and it should also eliminate singularities present in the classical theory.

5.2.1. Black Hole Entropy

The subject of black hole entropy has been one of intense interest ever since Bekenstein's calculations [157]. Many methods have since been utilized to calculate the entropy (see [158], [159] and references therein for excellent reviews of the subject). One belief is that the source of this entropy is strictly gravitational in origin. That is, one should be able to define microstates in a full quantum theory of gravity which, when counted, yields the correct entropy law. This has been done within the framework of loop quantum gravity.

The basic idea is as follows: The gravitational spin-network pierces the surface corresponding to the horizon of the black hole. As described above, this endows the surface with area and geometry (see figure 10). The entropy is given by the logarithm of the number of loop quantum gravity states that give the surface a fixed area, a_0. This counting is non-trivial as for a black hole of reasonable size there could be an enormous number of punctures, with various values of j_p. The total area is given by summing up all the contributions from all of the punctures. The total area is therefore given by[9]

$$a_0 = 8\pi l_p^2 \gamma \sum_p \sqrt{j_p(j_p+1)}\,, \tag{75}$$

[8] We are making an assumption here regarding how the spin-network pierces the surface S. The general case yields eigenvalues which are slightly more complicated than (73).

[9] We are making a similar assumption here as in the previous footnote.

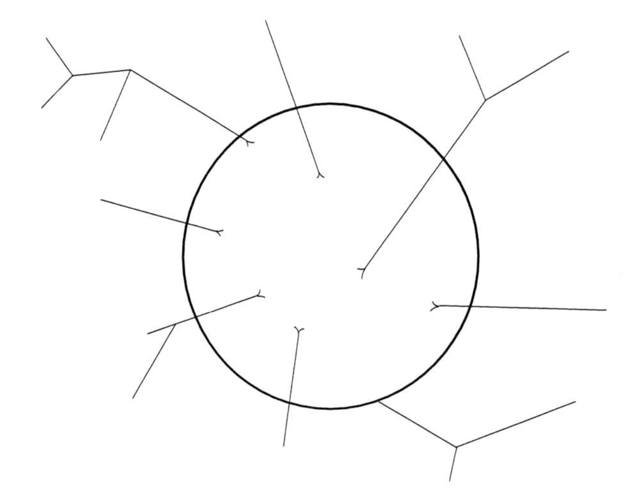

Figure 10. A gravitational spin-network giving a spherical black hole horizon its geometry and area.

which is obviously the sum of eigenvalues associated with the edges puncturing the surface. Notice that this formulation yields the entropy as a function of γ. The quantity γ therefore has to be set by other means or else it can be set by demanding that the leading order term in the entropy calculation agrees with the Hawking-Bekenstein result of $a_0/4$. Several studies have set the Immirzi parameter in this way [160] - [163].

There is another subtlety which complicates the expression which is topological in origin. To illustrate this we begin by noting that the action giving rise to the surface states (and thus the surface Hilbert space) is a Chern-Simons action:

$$I_{|\partial M} = -\frac{i\,a_0}{32\pi\gamma(g-1)} \int_{\partial M} \mathrm{Tr}\left[A \wedge dA + \frac{2}{3} A \wedge A \wedge A\right] , \tag{76}$$

where the trace of the $SU(2)$ connection, ${A^i}_a$ is taken over the $su(2)$ indices. The form of this surface term was first calculated in the pioneering work of [164]. Later this was generalized to arbitrary genus g surfaces, as above [165]. This term arises from the fact that at the inner boundary (the isolated horizon [166]), the triad and the connection cannot be fixed independently and are actually related (hence only the connection appears in (76)). The isolated horizon boundary conditions reduce the degrees of freedom and the above action can be written in terms of a $U(1)$ connection [164]:

$$I_{|\partial M} = \frac{i\,a_0}{16\pi^2(1-g)} \int_{\partial M} W \wedge dW , \tag{77}$$

where W represents $U(1)$ connections on the boundary surface, which are restricted by the value of the bulk $SU(2)$ connection penetrating the surface at that particular point on the horizon. Finally one can construct the symplectic structure on the boundary as was done in [164]:

$$\Omega_{|\partial M\, grav}(..,..) = k \oint_{\partial M} \delta W \wedge \delta' W , \tag{78}$$

with $k := \frac{a_0}{4\pi(g-1)\gamma}$ (an integer) known as the Chern-level and with δW and $\delta' W$ tangent vectors in the space of $U(1)$ connections defined on the horizon. Note that we now have a topological $U(1)$ theory on the boundary. The number degrees of freedom are related to the number of topologically independent closed paths one can construct on the punctured surface.

In the case of a surface with spherical topology S^2, one can place punctures on the sphere and then define a closed path around each puncture. This would seemingly yield a $2N$ dimensional phase-space as each closed path (cycle) also has a conjugate open path associated with it (chain) (roughly speaking each cycle represents a configuration variable that must have a conjugate momentum, represented by chains). However, these are not all independent degrees of freedom. On a sphere, note that going around the N-th puncture is the same as going around all the other punctures but in the opposite direction. If the cycles are denoted as η_i, this may be expressed as:

$$\eta_1 \cdot \eta_2 \cdot \ldots \cdot \eta_{N-1} = \eta_N^{-1} , \tag{79}$$

Therefore, the topology reduces the number of independent degrees of freedom.

When one quantizes the system, quantum states ψ_m are obtained for integers $m = (m_1, .., m_{N-1})$ with $m_i \in \{1, .., k\}$ [167]. Note that in this sense, the integers m_i play a role similar to the magnetic quantum number in ordinary quantum mechanics. The condition (79) gives rise to a constraint:

$$m_1 + \ldots + m_{N-1} = -m_N . \tag{80}$$

This restriction is the quantum analogue of the Gauss-Bonnet theorem for a sphere. Note that one now has N generators and one constraint. Thus, for a spherical horizon, states can be labeled with $m = (m_1, .., m_N)$ subject to constraint (80). In other words, *not all* states that yield classical area a_0 are allowable. Only the ones meeting the condition (80) are to be counted. The details of the counting may be found in [168] and references therein. Only the results will be cited here.

A very careful numerical counting of acceptable states for a spherical horizon was performed by Corichi et al in [162]. Those authors performed the counting with the projection constraint (80) and without considering it. They found the following: *Without* considering the projection constraint they found that the entropy obeys the $a_0/4$ law, provided the Immirzi parameter is set equal to $\gamma \approx 0.274$. This is the same value found in [161] and [168]. *With* the projection constraint it was found that the number of acceptable states is reduced in such a way that does not involve γ. The result is

$$S = \frac{a_0}{4} - \frac{1}{2}\ln(a_0) . \tag{81}$$

For the case of a genus g surface, the situation is slightly more subtle as the paths around the punctures can also be related to paths around the genus holes of the surface. This yields a quantum Gauss-Bonnet theorem of the following type [165]:

$$\eta_{g+1} \cdot \eta_{g+2} \cdot \ldots \cdot \eta_{g+N} = \eta_1 \gamma_1 \eta_1^{-1} \gamma_1^{-1} \cdot \ldots \cdot \eta_g \gamma_g \eta_g^{-1} \gamma_g^{-1} , \tag{82}$$

where η_1 through η_g denote the paths associated with the genus holes and η_{g+1} to η_{g+N} denote with the paths associated with the spin-network punctures. Utilizing this topological condition, the entropy of a $g > 1$ horizon is given by [165]

$$S = \frac{a_0}{4} + (g-1)\left[\ln(a_0) - \ln(4\pi\gamma(g-1))\right] , \tag{83}$$

provided that the Immirzi parameter is set to the same value as in the spherical case. Therefore, the same value of the Immirzi parameter yields the first-order $a_0/4$ term for all cases whereas the sub-leading term depends on topolgy and is independent of γ. This behavior is consistent with other, non-quantum gravity approaches to calculating black hole entropy of $g > 0$ horizons [169] [170] [171] [172].

An ambiguity exists for $g = 1$ due to the decoupling of the triad and the connection at the horizon in this case. However, one may analytically extend (83) to $g = 1$ yielding an $a_0/4$ entropy *without* logarithmic correction. This result is consistent with studies of $g = 1$ horizon entropy utilizing non loop quantum gravity techniques [170], [172]. The $g > 1$ result however, is qualitatively different from the $g = 0$ case, and therefore cannot be extended to reliably encompass the $g = 0$ horizons. This is due to the non-trivial interplay between the spin-network punctures and the genus holes of the surface (note that the coefficient of the logarithmic correction in (83) differs from (81) by a factor of 2 for $g = 0$).

5.2.2. Removal of the Classical Singularity

Another problem that a quantum theory of gravitation is expected to resolve is that of the singularities that exist in the classical theory. This is also related to the problem of information loss associated with black holes. This singularity issue has been studied in the framework of loop quantum gravity in the case of mini-superspace models. That is, models where the full system is first reduced to a mechanical system which consists of only the relevant degrees of freedom. One then quantizes this reduced system. This is done due to the technical difficulty involved when trying to work with the full theory. There are currently several black hole studies available within the symmetry reduced models [173] - [175] and here we shall be outlining the approach in [175]. A related method was studied in [176] where it was shown that a Nariai universe replaces the classical singularity.

The basic ideas are as follows: Construct an evolution equation utilizing the Hamiltonian constraint and check if the evolution remains finite at the point corresponding to the classical singularity. Also, one may compute operators and their expectation values which encode the information about curvature and which classically diverge at the classical singularity. If they remain finite in the quantization the singularity is avoided in the quantum theory.

We will focus on the most studied black hole in quantum gravity, the Schwarzschild black hole ($\Lambda = 0 = Q$), whose line element for $r < 2M$ can be written as:

$$ds^2 = -\frac{dT}{\frac{2M}{T} - 1} + \left(\frac{2M}{T} - 1\right) dR^2 + T^2\, d\theta^2 + T^2 \sin^2\theta\, d\varphi^2 , \tag{84}$$

where T is the interior time coordinate (corresponding to r in (7)) and R is an interior spatial coordinate (corresponding to t in (7)).

Recall that the conjugate variables in loop quantum gravity are the densitized triad and the modified $SU(2)$ connection. A pair is is constructed which respects the symmetry of the space, $\mathbb{R} \times SO(3)$ [175], [177]:

$$A^{i}_{\ a}\tau_i\, dx^a = c\tau_3\, dR + (a\tau_1 + b\tau_2)d\theta + (a\tau_2 - b\tau_1)\sin\theta\, d\varphi + \tau_3 \cos\theta\, d\varphi\,, \tag{85a}$$

$$E^a_i \tau^i \partial_a = p_c\tau_3 \sin\theta \partial_R + (p_a\tau_1 + p_b\tau_2)\sin\theta\, \partial_\theta + (p_a\tau_2 - p_b\tau_1)\partial_\varphi\,, \tag{85b}$$

where the τ_i denote the standard $su(2)$ basis. The quantities $a,\ b,\ c$ and $p_a,\ p_b,\ p_c$ are to be determined and act as conjugate "position-momentum" pairs. The classical analog of the Gauss constraint can be satisfied but not in a unique way. Any pair that satisfy:

$$ap_b - bp_a = 0\,, \tag{86}$$

will satisfy the Gauss constraint [175]. Therefore, it is useful to set $a = p_a = 0$ in the sequel. There is still some residual gauge freedom but we shall not discuss it here.

The co-triad can also be constructed as:

$$\omega_a^{\ i}\tau_i dx^a = \frac{\mathrm{sgn}\, p_c\, |p_b|}{\sqrt{|p_c|}}\tau_3\, dR + \sqrt{|p_c|}\frac{p_b}{|p_b|}\tau_2\, d\theta - \sqrt{|p_c|}\frac{p_b}{|p_b|}\tau_1 \sin\theta\, d\varphi\,. \tag{87}$$

By comparison of (85b) (or (87)) with (84) one can see that the following identification may be made:

$$p_b = \sqrt{T\,(2M - T)}\,, \quad p_c = \pm T^2\,. \tag{88}$$

Therefore, the degeneracy in (87) at $p_b = 0 \neq p_c$ corresponds to the classical horizon whereas $p_c = 0$ corresponds to the classical singularity.

Next a basis is defined in the Hilbert space, these are denoted as $\frac{1}{\sqrt{2}}\left[|\mu,\,\tau\rangle + |-\mu,\,\tau\rangle\right]$ and are made up of eigenstates of the operators corresponding to p_b and p_c:

$$\hat{p}_b\,|\mu,\,\tau\rangle = \frac{1}{2}\gamma\mu\,|\mu,\,\tau\rangle\,, \quad \hat{p}_c\,|\mu,\,\tau\rangle = \frac{1}{2}\gamma\tau\,|\mu,\,\tau\rangle\,. \tag{89}$$

The volume operator is also needed:

$$V = \int d^3x\, \sqrt{|\mathrm{det}E|} = 4\pi\sqrt{|p_c|}|p_b| \to \hat{V} = 4\pi\sqrt{|\hat{p}_c|}|\hat{p}_b|\,, \tag{90}$$

which is diagonal in the $|\mu,\,\tau\rangle$ basis and possesses eigenvalues:

$$V_{\mu\tau} = 2\pi\gamma^{3/2}|\mu|\sqrt{|\tau|}\,. \tag{91}$$

As well, the co-triad operator can be created. In the notation of [175], $\omega_c := \frac{\mathrm{sgn}\, p_c\, |p_b|}{\sqrt{|p_c|}}$ and $\omega_b := \mathrm{sgn} p_b \sqrt{|p_c|}$. It is noted that the co-triad can be written in terms of the holonomy and volume:

$$\omega_c = (2\pi\gamma)^{-1}\,\mathrm{Tr}\left(\tau_3 h_R \left\{h_R^{-1},\, V\right\}\right)\,,$$

where h_R corresponds to the holonomy along an interval in the R-direction. The operator version is given by:

$$\hat{\omega}_c\,|\mu,\,\tau\rangle = -i(2\pi\gamma)^{-1}\mathrm{Tr}\left(\tau_3\hat{h}_R\left\{\hat{h}_R^{-1},\,\hat{V}\right\}\right)|\mu,\,\tau\rangle = \frac{\sqrt{\gamma}}{2}|\mu|\left(\sqrt{|\tau+1|} - \sqrt{|\tau-1|}\right)|\mu,\,\tau\rangle\,. \tag{92}$$

Similarly, for ω_b one can construct [175]

$$\hat{\omega_b}\,|\mu,\,\tau\rangle = \sqrt{\gamma}\,\mathrm{sgn}(\mu)\sqrt{|\tau|}\,|\mu,\,\tau\rangle \tag{93}$$

As is usual, a general state can be expanded in terms of the above eigenstates: $|\psi\rangle = \sum\limits_{\mu\tau} c_{\mu\tau}\,|\mu,\,\tau\rangle$.

Next the Hamiltonian constraint is constructed. One may pursue this in two ways. One way is to write the extrinsic curvature connection, K in (71c), in terms of the modified connection A. The second way is to regard K as the connection to be used. However, in this second case, the holonomies are to be constructed as functions of K, not A. It was shown in [175] that utilizing the second method results in a Hamiltonian constraint of the form:

$$C_{Ham} = \frac{1}{\gamma^2}\frac{1}{\sqrt{\det(E)}}E^{ai}E^{bj}\epsilon_{ijk}\left[\gamma^2\Omega^k_{ab} - {}^oF^k{}_{ab}\right]. \tag{94}$$

Here ${}^oF := dK + [K,\,K]$ and $\Omega := d\Gamma = -\sin\theta\,\tau_3 d\theta\wedge d\varphi$ (which can be calculated utilizing the triad associated with the standard two-sphere metric). Written in terms of the co-triad one has:

$$\frac{1}{\sqrt{\det(E)}}\epsilon_{ijk}\tau^i E^{aj}E^{bk} = -\left(4\pi\gamma\mathcal{L}_{(k)}\right)^{-1}\epsilon^{abc}\,{}^o\omega^k_c\,h^\delta_k\left\{h^{\delta-1}_k,\,V\right\}. \tag{95}$$

Here, h_k corresponds to the holonomy along an edge in the k-direction ($k = R,\,\theta,\,\phi$) (the index k is summed over). $\mathcal{L}_{(R)} = \mathcal{L}_{(\theta)} = \mathcal{L}_{(\varphi)} = \delta$, where δ is the length of the curve in the coordinate directions over which the holonomy is measured (superscripts o denote that the quantity is in the K connection representation). As well, the last part of the Hamiltonian constraint can be written in terms of the co-triad:

$${}^oF^i{}_{ab}\tau_i \approx \frac{{}^o\omega^i_a\,{}^o\omega^j_b}{\mathcal{A}_{(ij)}}\left(h^A_{(ij)} - 1\right) + \mathcal{O}(\delta), \tag{96}$$

where $\mathcal{A}_{R\theta} = \mathcal{A}_{R\varphi} = \mathcal{A}_{\theta\varphi} = \delta^2$ and $h_{(ij)} := h_i h_j h_i^{-1} h_j^{-1}$.

At this stage, all quantities are constructed out of holonomies (in ${}^o\omega$) and tetrads (in ${}^o\omega$ and V) and therefore we can go to the quantum picture by simply replacing these with their operator analogues. In order to make the constraint Hermitian, the *gravitational constraint* is defined as $\hat{C}^\delta_{\text{grav}} := \frac{1}{2}\left[\hat{C}^\delta + \hat{C}^{\delta\,\dagger}\right]$. Without details, we quote the main results of this construction: The Hamiltonian constraint operator, acting on the states yields a *difference* equation for $c_{\mu\tau}$:

$$\begin{aligned}\hat{C}^\delta_{\text{grav}}\,c_{\mu,\,\tau} =& 2\delta\left(\sqrt{|\tau+2\delta|}+\sqrt{|\tau|}\right)\left[c_{\mu\,+\,2\delta,\,\tau\,+\,2\delta} - c_{\mu\,-\,2\delta,\,\tau\,+\,2\delta}\right]\\ &+\left(\sqrt{|\tau+\delta|}-\sqrt{|\tau-\delta|}\right)\left[(\mu+2\delta)c_{\mu\,+\,4\delta,\,\tau} - (1+2\gamma^2\delta^2)\mu\,c_{\mu,\,\tau}\right.\\ &\left.+\,(\mu-2\delta)c_{\mu-4\delta,\tau}\right] + 2\delta\left(\sqrt{|\tau-2\delta|}+\sqrt{|\tau|}\right)\left[c_{\mu\,-\,2\delta,\,\tau\,-\,2\delta}\right.\\ &\left.-c_{\mu\,+\,2\delta,\,\tau\,-\,2\delta}\right] = 0\,.\end{aligned} \tag{97}$$

To analyze the behavior at the singularity, one starts at some positive value of τ and utilizes (97) to evolve the $c_{\mu,\tau}$ to smaller values of τ. The singularity resolution issue is insensitive to the choice of initial conditions (provided, of course, that they are not pathalogical).

It turns out that the coefficients are always regular throughout the evolution for all values of $\tau > 0$ as well as $\tau \leq 0$. From (88) and (89), $\tau = 0$ corresponds to the classical singularity. Therefore, the evolution of the $c_{\mu,\tau}$ coefficients is regular and may proceed *beyond* the classical singularity. In this theory of quantum geometry then, the quantum analogue of $T = 0$ is no longer a boundary of the space-time. In essence there is a smooth "bounce" which evolves to another large region of space-time. Expectation values of curvature encoding quantities are also finite. Interestingly, (though perhaps not surprising) mini-superspace reduced LQG cosmological models possess similar behavior where the big-bang is replaced by a smoothly evolving quantum bounce. (See [178] and references therein.)

6. Black Holes in Astrophysics

Although there exists no unquestionable proof that black holes exist in nature, there is mounting evidence suggesting that black holes are present in our universe. The evidence must be, by definition, indirect. We overview here some of the observational evidence for the existence of black holes as well as how researchers utilize observations to deduce the properties of these fascinating objects.

Perhaps the first evidence that some extreme objects exist in the universe was the observation of X-ray sources outside the solar system and of quasars in the 1960s [179]. Quasars are objects which possess luminosities on the order of 10^{14} that of the sun. Matter accreting into a black hole could most easily explain such emissions of electromagnetic radiation. It is now also believed that black holes are related in some way to the observed gamma ray bursts. Since it is expected that almost all black holes have some amount of rotation, the Kerr solution provides a viable background in which to study these emissions. Astrophysical theory is suggestive that there is a limit on the angular momentum parameter of $-0.998M \leq a \leq 0.998M$ [180] with high rotations amost certainly containing an event horizon and therefore unlikely to be an alternative to a black hole without event horizon [181].

One of the earliest (and brightest) X-ray sources detected was Cygnus X-1 [182], [183], which was noted to vary with time. A large number of X-ray sources have since been discovered, many of them associated with optically faint, distant stars. In such cases, it was not possible that the star itself could be emitting the X-rays. Instead, an argument was put forward that the X-rays originated from the accretion of the star's outer material onto a yet unseen companion object, likely a neutron star [184] [185]. Further, it was postulated that the slightest amount of angular momentum in the material (likely inevitable) would preclude anything resembling radial in-fall and the material would, rather, be forced into a disk around the compact companion (see figure 11). Viscous drag forces would then heat up the material to high enough temperatures to emit the radiation observed. The Uhuru satellite confirmed that the stars indeed must be orbiting some companion object, which must be very compact. If the mass of the companion is above the neutron star limit (approximately 1.5 - 4 solar masses) then the likely alternative is a black hole.

The evidence for a binary system comes from the periodicity of the visible star's spectrum. One then needs to determine if the properties of the unseen companion allow it to be a neutron star or some more compact object. Consider a binary system as shown in figure 12. As an approximation, Kepler's law can be utilized:

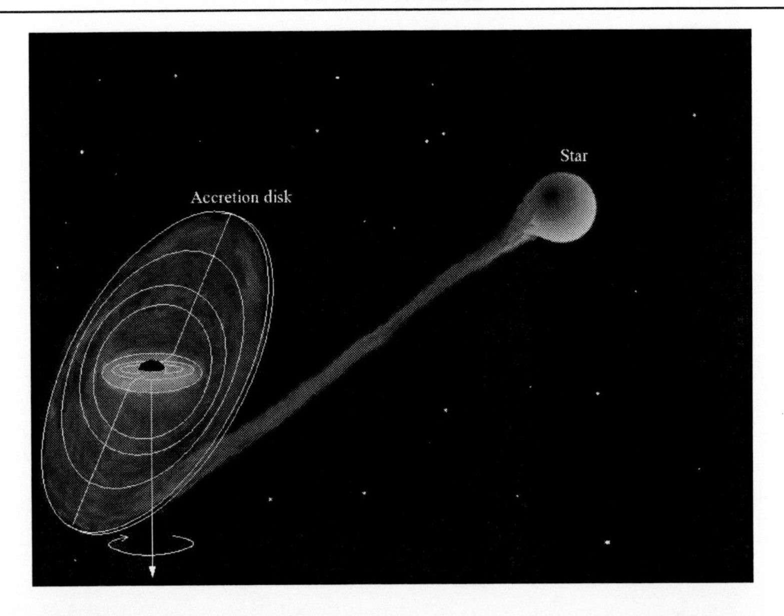

Figure 11. A rotating black hole accreting matter from a nearby star. Although the outskirts of the accretion disk is tilted with respect to the orbital plane, the inner regions are forced into alignment with the orbital plane of the black hole.

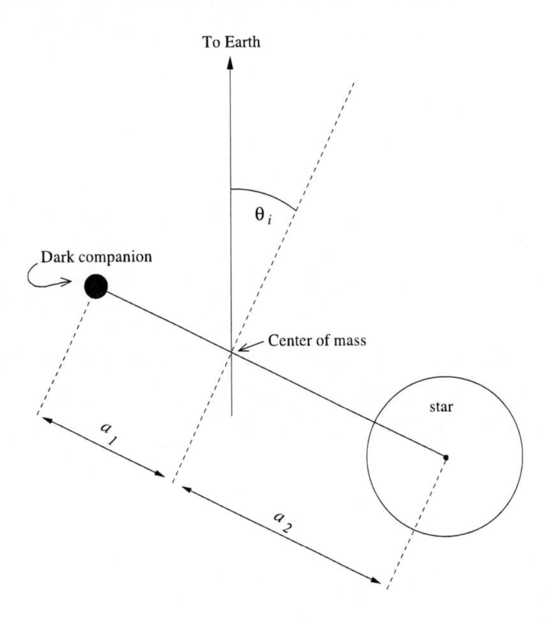

Figure 12. An example of a binary system.

$$T^2 = \frac{M_1 4\pi^2 a_2^3}{(M_1 + M_2)^2}, \tag{98}$$

where T is the orbital period and M_1 and M_2 represent the mass of the dark companion and observable star respectively. If the period, M_2, and a_2 are provided, the mass of the

the companion may be determined. In practice it is usually difficult to determine a_2 as it requires knowledge of the inclination angle, θ_i.

For the case of Cygnus X-1, the optical member is a hot OB star (HDE 226868) which typically have very large masses ($\gtrsim$ 20 solar masses). With best estimates for the parameters, the conclusion is that the dark companion in this binary system possesses a mass of the order 10 solar masses, much greater than the neutron star limit. It is therefore likely that the companion is a black hole. One of the sources of uncertainty is the value of the mass of HDE 226868.

A class of X-ray objects are known as the X-ray transients. These objects emit X-rays periodically, followed by long periods of no emission. This allows one to study the optical companion in detail during the X-ray-quiet periods without the noise from the X-rays interfering with the observations. This is useful since detailed studies of the optical member of the binary will yield tighter constraints on the mass of the star and therefore on the mass of the companion object. In such a system, if the compact companion possessed a solid surface, it should be possible to see a characteristic emission of energy as the accreting material is brought to rest on the surface [186], [187], [188]. Observations of various sources seem to show no indication of such emission, and therefore the presence of an event horizon, as opposed to an object with a solid surface, is favored.

As mentioned earlier, the discovery of quasars in the 1960s has led to speculation that some compact object must be responsible for the emission of so much energy. Over the years the evidence has become compelling that the gravitational sources are likely supermassive black holes at the center of galaxies. These objects are now generally referred to as active galactic nuclei (AGN). These sources typically emit $10^{12} - 10^{14}$ solar luminosities and have length scales on the order of less than a light-year and in many cases the scale may be measured in light-hours. These scales are based on the fact that appreciable changes in a system can not occur on time-scales shorter than it takes light to cross the system.

The physics involved in AGNs is similar to the compact binary described above. Namely, nearby matter is accreted into a disk around the black hole and X-rays are emitted via a friction mechanism. A natural question arises in the case of AGNs: Could the gravitational effects required to produce such X-ray emissions be due to the large number of stars and galactic matter near the core of the galaxy instead of a black hole? The constraints on the size, the lack of periodicities in the signals and the stability of the signals seem to favor a single central object (with masses of the order of 10^{10} solar masses!) instead of a widespread, non-uniform source [189]. Also, there are "jets" of material present in may AGNs which remain aligned for time periods on the order of 10^6 light-years (see figure 13). This indicates that a preferred axis must have been present in the system for at least that long, making a gravitationally bound compound object unlikely.

Modeling these jets is an extremely difficult task involving general relativistic magnetohydrodynamics. Large-scale computing must be employed in order to produce reliable results from the models. The jets can arise from a complex interplay between gas evaporating off of the accretion disk and magnetic fields present, known as the Blandford-Znajek process [190].

One of the major sources of the X-rays is the iron $K\alpha$ line and there are several effects on this line. One effect is not strictly speaking gravitational in origin. It is the special relativistic doppler shift, where the line is blueshifted on one side of the accretion disk and

Figure 13. Jets of ionized gas being ejected from the galactic core of M87. Permission to use image, made available through NASA and STScI, kindly provided by Tod R. Lauer (National Optical Astronomy Observatory), Sandra M. Faber (UC Observatories/ Lick Observatory).

redshifted on the other side. This will yield a doppler broadened line. Another effect is also present in the absence of gravity, this is the special relativistic beaming effect whereby the radiation intensity is amplified in the direction of particle motion compared to the intensity in the rest-frame. This effect is also differential in that the effect enhances the intensity on the approaching side of the disk. Another effect is strictly gravitational in origin. This is the gravitational redshift and time dilation. Unlike the doppler shift, this shift does not depend on what side of the disk (approaching or receding side) the material is residing. Also, the gravitational time dilation has the effect of reducing the overall flux since the emitter is "slowing down" compared to an observer at infinity. These effects act to skew the line profile, as is illustrated in 14. The difference between the X-ray spectrum of matter accreting into a spinning versus non-spinning black hole is also displayed in this figure. This difference due to the spin arises from the fact that stable circular orbits in Schwarzschild geometry do not exist below $r = 6M$, whereas for a rotating black hole this value is much smaller. Therefore, the gravitational redshift and time dilation can be much more pronounced in the case of a Kerr black hole as it is generally expected that the bulk of X-ray emission occurs in orbits at or above the stable orbit limit. With this assumption, data from the XMM-Newton satellite, analyzing the X-ray iron line of Seyfert galaxy MCG-6-30-15 constrains the Kerr spin parameter to be $|a| > 0.93$ [191]. The rotational dragging of a Kerr black hole also has the effect of forcing the portion of the accretion disk that is close to the black hole to orbit in the equatorial plane (see figure 11).

Interestingly, there is strong evidence that there is a super-massive black hole at the center of our own galaxy (Sgr A*). The "close" proximity of this black hole allows astronomers to directly measure its influence on its stellar neighbors [192] - [195]. The optical range of frequencies cannot penetrate the galactic center so studies of the galactic black hole are usu-

Figure 14. An example of an X-ray spectrum of iron atoms in the vicinity of a non-spinning (left) and spinning (right) black hole. A spinning black hole drags the particles around it, via the Lense-Thirring effect, this allows for particles to orbit nearer to the black hole. Kind permission to display this figure, from the Harvard Chandra website, was granted by NASA/CXC/SAO.

ally performed utilizing the radio or the infra-red region. Knowing the orbital parameters between the black hole and its nearby stars, Kepler's law allows the determination of the approximate mass of the black hole. A mass of approximately 3.5×10^6 solar masses is calculated for the mass of the galactic black hole [196], [197]. Observational data relating to these orbits, along with Keplerian fits, are displayed in figure 15.

Other possible methods to detect black holes include: Gamma ray bursts, gravitational lensing (of background objects as well as of the orbital features of the accompanying binary star) [198], [199], and hopefully in the near future, gravitational wave signatures. Other proposals may be found in [200], [201] and [202].

We have only scratched the surface here regarding the observations and theoretical techniques used to study black holes in astrophysical contexts. There exist a number of excellent books and reviews on the subject. The interested reader is referred to the (much more thorough) review articles [203], [204] and [205] and the large number of references therein.

7. Alternatives to Black Holes

In this final section we will mention a few proposed alternatives to black holes along with possible measurements that may be performed in order to distinguish these objects from black holes. Some of these objects are black holes in the strict sense of the word. That is, they may contain a horizon but there is no singularity hiding behind it. It would be difficult, but not impossible, to distinguish some of these models from black holes [206]. We list some of the alternatives here with a brief description.

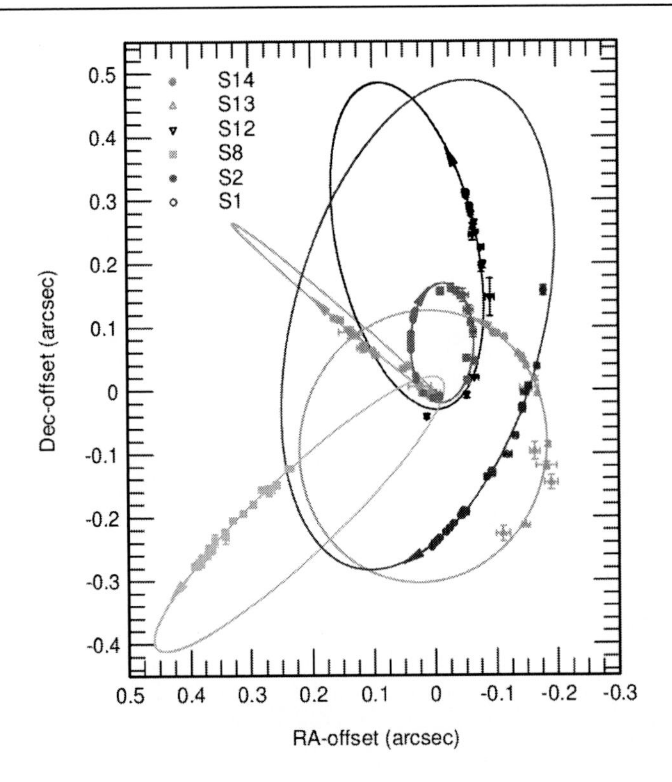

Figure 15. Observation of the orbits of stars near the galactic center. This figure displays data points and best fit Keplerian orbits. Figure reproduced from [196] with kind permission from F. Eisenhauer and The Astrophysical Journal.

Neutron stars with non-standard equation of state: Perhaps the greatest Achille's heel to the arguments in favor of black holes as the likely candidate in the binary systems discussed above is the uncertainty in the equation of state. This is the main reason for the large uncertainty in the neutron star limit. The regime of neutron star density is above what can reliably be studied in a lab and therefore any properties at these densities are not well constrained. Since pressure is also a source of gravity in general relativity, modification of the equation of state could increase the maximum mass that neutron stars may possess and therefore some large mass objects thought to be black holes could turn out to be neutron stars. A general form of the equation of state was studied in detail by Rhoades and Ruffini [207]. They made mild assumptions such as the speed of sound being bounded $0 \leq c_s \leq 1$, and that at lower densities (below some value ρ_0), it should produce equations of state thought to be well understood. In their study, they found a neutron star limit of approximately 3.2 solar masses. Adding rotation to the picture yields [208]:

$$M_{\text{max}} \approx 8.4 \left(\frac{\rho_0}{10^{14}}\right)^{-1/2} M_{\odot}\,. \tag{99}$$

It is therefore not inconceivable that the neutron star limit could be as high as 8-10 solar masses or higher.

A possible scenario is the "Q-star", which allows for the possibility of nucleon confinement under extreme conditions. In these theories, it is expected that under certain conditions, the equation of state differs strongly from standard ones, even at for relatively low densities (low values of ρ_0). Therefore, the assumptions used to derive the expression above are no longer valid. Such stars may possess masses on the order of 100 solar masses yet possess radii which are approximately 1.4 times the corresponding Schwarzschild radius [209].

However, as mentioned previously, the absence of flare-ups due to material being brought to rest on a hard surface is in favor of a black hole instead of a neutron star or Q-star. As well, no reasonable value of ρ_0 would allow a neutron star scenario for the super-massive galactic black holes thought to be responsible for AGNs.

Repulsive interiors: This is not an alternative to a black hole as much as it is a possible alternative to the standard picture of an event horizon shielding a singularity. These scenarios basically stem from the fact that there is no reason to believe that the Schwarzschild solution is valid down to $r = 0$. In the T-domain (the $r < 2M$ domain of the Schwarzschild solution) it is possible, for example, to patch the Schwarzschild solution to a deSitter metric via a shell located at some space-like surface. The idea is to preserve the properties of the event horizon, which seems to fit observational data, but modify the interior. This idea seems to date back to Sakharov and Gliner who considered the possibility that, under extreme conditions, matter would possess an equation of state of the form $\rho = -p$ [210], [211]. Explicit constructions of this model were performed in [212] and alternates of this model were also considered in [213]- [215]. A conformal diagram of Schwarzschild space-time with a deSitter interior is shown in figure 16.

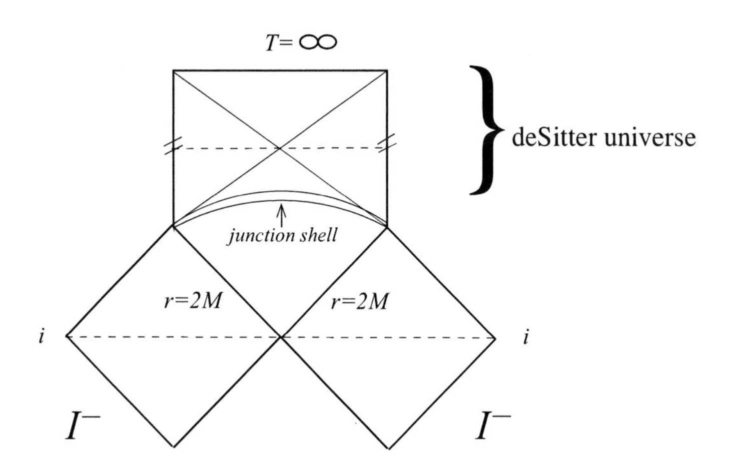

Figure 16. A conformal diagram illustrating the space-time that results from patching the T-domain of the Schwarzschild space-time to a deSitter universe at a junction shell. Presumably a phase-transition of the collapsing matter occurs at the shell yielding the deSitter interior. T denotes the interior time coordinate whereas r denotes the exterior radial coordinate.

Gravastars: A recent extension of the above idea is the gravitational vacuum star, or gravastar. The gravastar idea originated with P. Mazur and E. Mottola as an alternative to a black hole and possesses *no* event horizon [216] - [218]. In the gravastar picture, quantum vacuum fluctuations are expected to play a non-trivial role in the collapse dynamics. A phase transition is believed to occur yielding a repulsive de Sitter core which aids in balancing the collapsing object and thus preventing horizon (and singularity) formation. This transition is expected to occur very close to the limit $2m(r)/r = 1$ so that, to an outside observer, it would be very difficult to distinguish the gravastar from a true black hole. The idea of a phase transition of the vacuum from a $\Lambda \approx 0$ state to a non-negligible Λ state is motivated from the behavior of Bose-Einstein condensates. The final gravastar configuration would also possess much less entropy than a black hole of similar size and therefore the problem of where the enormous black hole entropy comes from is alleviated.

The original Mazur - Mottola model consisted of a deSitter interior separated from a Schwarzschild exterior via a finite shell with an equation of state satisfying $\rho = +p$ (with thin shells on either side for patching purposes). It was later shown that, were a transition between a deSitter center and Schwarzschild exterior to be smooth and yield physically reasonable outer layers, anisotropic pressures must be present within the structure [219]. Models with continuous pressures satisfying various equations of state were explicitly constructed in [220]. Examples of the pressure and density profiles for a sample gravastar (originally displayed in [220]) are displayed in figure 17. Lobo and Arellano have studied several variants of gravastars or gravastar-like objects in [221] [222].

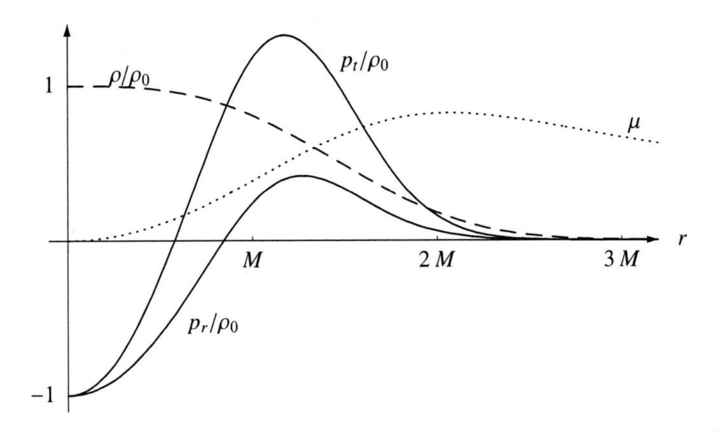

Figure 17. The gravastar with energy density profile $\rho = \rho_0 \exp[-(r/r_0)^n]$ and anisotropy $\tilde{\Delta} = \alpha^2 \ (\rho/\rho_0) \ \mu/12$. Displayed are: radial (lower solid line) and transversal (upper solid line) pressures, energy density (dashed line) and the compactness (dotted line). In this example the parameters are $n = 3$, total mass of configuration $M = 1$ and maximal compactness within the gravastar is $\mu_{\max} = 0.80$. Notation is as follows: $\tilde{\Delta} := \frac{p_t - p_r}{\rho_0} = \frac{\alpha^2}{12}\frac{2m(r)}{r}\frac{\rho}{\rho_0}$, with α and ρ_0 constants. μ is the "compactness" function $2m(r)/r$. The transverse pressure, radial pressure and energy density are denoted as p_t, p_r and ρ respectively.

Note that by definition, the strong energy condition cannot be satisfied in any model with a deSitter region. A thorough discussion of gravastar energy conditions may be found in [223].

In the case of anisotropic models, one way to distinguish their presence from either a black hole or a neutron star is via their surface redshift. It is known (see [224] and references therein) that stability in anisotropic spheres allows for a higher maximum redshift than for a stable perfect fluid sphere of similar mass, due to increased allowable compactness. Also, Chirenti and Rezzolla have discussed how to distinguish a gravastar from a black hole via quasi-normal mode analysis [225].

The list presented above is not exhaustive but should cover some of the most popular alternatives to black hoes. Another interesting alternative for example, put forward by Robertson and Leiter, is the magnetic eternally collapsing object (or MECO) (see references [226], [227], [228] and references therein.)

Acknowledgments

I am grateful to S. Kloster and J. Brännlund for discussions and help with improving the manuscript. I thank M. Bojowald for kindly clarifying an issue in the quantum singularity resolution. I would also like to thank the various authors, journals and institutions which have granted permission to use their figures in this paper (acknowledged individually in the figure captions).

References

[1] Thorne, K. S. *Black Holes and Time Warps - Einstein's Outrageous Legacy* (W. W. Norton and Co., New York, 1994).

[2] Rovelli, C., Presented at the *9th Marcel Grossmann Meeting*, (Roma, Italy, 2000).

[3] Ashtekar, A., *Curr. Sci.* 89, 2064 (2005).

[4] Khlopov, M. Y. and Polnarev, A. G., *Phys. Lett.* **B97**, 383 (1980).

[5] Polnarev, A. G. and Khlopov, M. Y., *Astron. Zh.* (1981) 58, 706 (1981). English translation: *Sov. Astron.* 25, 406 (1981).

[6] Polnarev, A. G. and Khlopov, M. Y., *Astron. Zh.* (1982), V. 59, 639 (1982). English translation: *Sov. Astron.* (1982) 26, 391 (1982).

[7] Michell, J., *Phil. Trans. Roy. Soc. London* 74, 35 (1784).

[8] Roemer, O., *Phil. Trans. Roy. Soc.* 136, 887 (1677).

[9] LaPlace, P. S., *Exposition du Systèm du Monde* (Paris, 1796). English translation: (W. Flint, London, 1809).

[10] Einstein, A., *Sitz. Deutsch. Akad. Wis. Berlin* 1915, 844 (1915).

[11] Hilbert, D., *Nachr. von der König. Ges. der Wiss. Göttingen*, 395 (1916).

[12] Schwarzschild, K., *Sitz. Deutsch. Akad. Wis. Berlin* 1916, 189 (1916). English translation by S. Antoci and A. Loinger: *arXiv:physics/9905030v1*.

[13] Schwarzschild, K., *Sitz. Deutsch. Akad. Wis. Berlin* 1916, 424 (1916).

[14] Einstein, A., *Ann. Math.* 40, 922 (1939).

[15] Oppenheimer, J. R. and Snyder, H., *Phys. Rev.* 56, 455 (1939).

[16] May, M. M. and White, R. H., *Phys. Rev.* 141, 1232 (1966).

[17] Finkelstein, D., *Phys. Rev.* 110, 965 (1958).

[18] Kruskal, M. D., *Phys. Rev.* 119, 1743 (1960).

[19] Hawking, S. W. and Ellis, G. F. R., *The Large Scale Structure of Space-Time*, (Cambridge University Press, Cambridge, 1973).

[20] Price, R. H., *Phys. Rev.* D5, 2419 (1972).

[21] Bardeen, J. M., Carter, B. and Hawking, S. W., *Commun. Math. Phys.* 31, 161 (1973).

[22] Wald, R. M., *Quantum Field Theory in Curved Spacetime and Black Hole Thermodynamics*, (University of Chicago Press, Chicago, 1994).

[23] Padmanabhan, T., *Phys. Repts.* 406, 49 (2005).

[24] Birrell, N. D. and Davies, P. C. W., *Quantum Fields in Curved Space*, (Cambridge University Press, Cambridge, 1982).

[25] Hawking, S. W., *Comm. Math. Phys.* 43, 199 (1975).

[26] Page, D., *Phys. Rev.* D13, 198 (1976).

[27] Unruh, W. G., *Phys. Rev.* D14, 870 (1976).

[28] Chandrasekhar, S., *The Mathematical Theory of Black Holes*, (Oxford University Press, Oxford, 1992).

[29] Frolov, V. P. and Novikov I. D., *Black Hole Physics: Basic Concepts and New Developments*, (Kluwer Academic Publishers, Dordrecht, 1998).

[30] Martel, K. and Poisson, E., *Am. J. Phys.* 69, 476 (2001).

[31] Birkhoff, G., *Relativity and Modern Physics*, (Harvard University Press, Cambridge, 1923).

[32] Kerr, R., *Phys. Rev. Let.* 11, 237 (1963).

[33] Lense, J. and Thirring, H., *Physik. Zeit.* 19, 156 (1918).

[34] Schee, J., Zdeněk, S. and Juráň, J., *Proc. RAGtime 6/7: Workshop on Black Holes and Neutron Stars*, (Opava, Czech Republic, 2005).

[35] O'Neill, B., *Geometry of Kerr Black Holes*, (A. K. Peters Ltd., Wellesley, 1995).

[36] Scott, S., Visser, M. and Wiltshire, D., (Eds.) *The Kerr Space-time* (in progress) and *arXiv:0706.0622v1* [gr-qc].

[37] Balazs, N. L. and Voros A., *Phys. Rep.* 143, 109 (1986).

[38] Smith, W. L. and Mann R. B., *Phys. Rev.* **D56**, 4942 (1997).

[39] Lemos, J. P. S. and Zanchin V. T., *Phys. Rev.* **D54**, 3840 (1996).

[40] Lemos, J. P. S., *Phys. Rev.* **D57**, 4600 (1998).

[41] Aminneborg, S., Bengtsson, I., Sören Holst and Peldán, P., *Class. Quant. Grav.* 13, 2707 (1996).

[42] Berti, E., Cardoso, V. and Will, C. M., *Phys. Rev.* D73, 064030 (2006).

[43] Cardoso, V., *arXiv:gr-qc/0404093v1*.

[44] Padmanabhan, T.,*Class. Quant. Grav.* **21**, L1 (2004).

[45] Roy Choudhury, T. and Padmanabhan, T., *Phys. Rev.* **D69**, 064033 (2004).

[46] Nagar, A. and Rezzolla, L., *Class. Quant. Grav.* **22**, R167 (2005).

[47] Regge, T. and Wheerler, J. A., *Phys. Rev.* 108, 1063 (1957).

[48] Kokkotas, K. D. and Schmidt, B., *Liv. Rev. Rel.* 2, 2 (1999).

[49] Zerilli, J. F., *Phys. Rev.* D2, 2141 (1970).

[50] Vishveshwara, C. V., *Phys. Rev.* D1, 2870 (1970).

[51] Ferrari, V. and Gualtieri, L., *arXiv:0709.0657v1* [gr-qc]. To appear in *Gen. Rel. Grav.*

[52] Detweiller, S. L., *Astroph. J.* 225, 687 (1980).

[53] Leaver, N. M., *Proc. Roy. Soc. Lond.* A402, 285 (1985).

[54] Seidel, E. and Iyer, S., *Phys. Rev.* 41, 374 (1990).

[55] Kokkotas, K. D., *Class. Quant. Grav.* 8, 2217 (1991).

[56] Onozawa, H., *Phys. Rev.* D55, 3593 (1997).

[57] Ferrari, V. and Mashoon, B., *Phys. Rev. Lett.* 52, 1361 (1984).

[58] Starobinski, A. A. and Churilov, S. M., *Sov. J. E. P. T.* 38, 1 (1973).

[59] Press, W. H. and Teukolsky, S., *Astroph. J.* 185, 649 (1973).

[60] Berti, E., Cardoso, V. and Will, C. M., *A. I. P. Conf. Proc.* 848, 687 (2006).

[61] Baiotti, L., Hawke, I., Rezzolla L. and Schnetter, E., *Phys. Rev. Lett.* 94, 131101 (2005).

[62] Zanotti, O., Font. J. A., Rezzolla, L. and Montero, P. J., *M. N. R. A. S.* 356, 1371 (2005) and
Nagar, A., Zanotti, O., Font, J. A. and Rezzolla, L., *Phys. Rev.* D75, 044016 (2007).

[63] Ferrari, V., Gualtieri, L. and Rezzolla L., *Phys. Rev.* D73, 124028 (2006).

[64] Choptuik, M. W., *Phys. Rev. Let.* 70, 9 (1993).

[65] Garfinkle, D., *Phys. Rev.* D51 5558, (1995).

[66] Abrahams, A. M. and Evans, C. R., *Phys. Rev. Let.* 70, 2980 (1993).

[67] Evans, C. R. and Coleman, J. S., *Phys. Rev. Lett.* 72, 1782 (1994).

[68] Hirschmann, E. W. and Eardley, D. M., *Phys. Rev.* D51, 4198 (1995).

[69] Hod, S. and Piran, T., *Phys. Rev.* D55, 3485 (1997).

[70] Choptuik, M. W., Chmaj, T. and Bizon, P., *Phys. Rev. Lett.* 77, 424 (1996).

[71] Gundlach, C., *Phys. Rev.* D55, 6002 (1997).

[72] Koike, T., Hara, T. and Adachi, S., *Phys. Rev.* D59, 104008 (1999).

[73] Villas da Rocha, J. and Wang, A., *arXiv:gr-qc/9910109v2*.

[74] Alcubierre, M. et al., *Phys. Rev.* D61, 041501, (2000).

[75] Husa, S., Lehner, C., Purrer, M., Thornburg, J. and Aichelburg, A., *Phys. Rev.* D62, 104007 (2000).

[76] Maki, T. and Shiraishi, K., *Class. Quant. Grav.* 12, 159 (1995).

[77] Muniain, J. and Piriz, D., *Phys. Rev.* D53, 816 (1996).

[78] Soda, J. et al., *Phys. Lett.* B387, 271 (1996).

[79] Garfinkle, D., Cutler, C. and Duncan, C. G., *Phys. Rev.* D60, 104007 (1999).

[80] Evans, C. R., *Phys. Rev.* D49, 3998 (1994).

[81] Brady, P. R., *Class. Quant. Grav.* 11, 1255 (1994).

[82] Chiba, T. and Soda, J., *Prog. Theor. Phys.* 96, 567 (1996).

[83] Oliviera, H. P. and Cheb-Terrab, E. S., *Class. Quant. Grav.* 13, 425 (1996).

[84] Price, R. H. and Pullin, J., *Phys. Rev.* D54, 3792 (1996).

[85] Wang, A., *arXiv:gr-qc/9901044v1*.

[86] Choptuik, M. W., Hirschmann, E. W., Liebling, S. L. and Pretorius, F., *Phys. Rev. Lett.* 93, 131101 (2004).

[87] Harada, T. and Mahajan, A., *arXiv:0707.3000v2*. To appear in *Gen. Rel. Grav.*

[88] Gundlach, C., *Liv. Rev. Rel.* 2, 1 (1999).

[89] Gundlach, C., *Phys. Rept.* 376, 339 (2003).

[90] Shibata, M. and Taniguchi, K., *arXiv:0711.1410v1* [gr-qc].

[91] Hahn, S. G. and Lindquist, R. W., *Ann. Phys.* 29, 304 (1964).

[92] Smarr, L., *Ph.D. thesis*, (U. Texas, Austin, 1975).

[93] Eppley, K. R., *Ph. D. thesis*, (Princeton U., Princeton, 1977).

[94] Pretorius, F., *arXiv:0710.1338v1* [gr-qc].

[95] Peters, P. C. and Mathews, J., *Phys. Rev.* 131, 435 (1963).

[96] Peters, P. C., *Phys. Rev.* 136, B1224 (1964).

[97] Gair, J. R. et al, *Class. Quant. Grav.* 21, S1595 (2004).

[98] Amaro-Seoane, P. et al, *arXiv.org:astro-ph/0703495*.

[99] Gair, J. R., Kennefick, D. J. and Larson, S. L., *Phys. Rev.* D72, 084009 (2005).

[100] Echeverria, F., *Phys. Rev.* D40, 3194 (1997).

[101] Pretorius, F., *Phys. Rev. Lett.* 95, 121101 (2005).

[102] Pretorius, F., *Class. Quant. Grav.* 23, S529 (2006).

[103] Shibata, M. and Nakamura, T., *Phys. Rev.* D52, 5428 (1995).

[104] Baumgarte, T. W. and Shapiro, S. L., *Phys. Rev.* D59, 024007 (1999).

[105] Campanelli, M., Lousto, C. O., Maronetti, P. and Zlochower, Y., *Phys. Rev. Lett.* 96, 111101 (2006).

[106] Baker, J. G., Centrella, J., Choi, D., Koppitz, M. and van Meter, J., *Phys. Rev. Lett.* 96, 111102 (2006).

[107] Gundlach, C., Martin-Garcia, J. M., Calabrese, G. and Hinder, I., *Class. Quant. Grav.* 22, 3767 (2005).

[108] Baker, J. G., Centrella, J., Choi, D. I., Koppitz, M. and van Meter, J., *Phys. Rev.* D73, 104002 (2006).

[109] Campanelli, C., Lousto, C. O. and Zlochower, Y., *Phys. Rev.* D74, 041501 (2006).

[110] Pollney, D. et al, *arXiv:0707.2559* [gr-qc].

[111] Rezzolla, L. et al., *arXiv:0708.3999v2* [gr-qc].

[112] Rezzolla, L. et al., *Astroph. J.* **674**, L29 (2008).

[113] Synge, J. L., *Relativity: The General Theory*, (North-Holland Publishing, Amsterdam, 1960).

[114] Traschen, J., in *Mathematical Methods of Physics, proceedings of the 1999 Londrina Winter School*, (World Scientific, Singapore, 2000).

[115] Hiscock, W. A., Larson, S. L. and Anderson, P. R., *Phys. Rev.* D56, 3571 (1997).

[116] Page, D. N., *Phys. Rev.* D25, 1499 (1982).

[117] Brown, M. R. and Ottewill, A. C., *Phys. Rev.* D31, 2514 (1985).

[118] Brown, M. R., Ottewill, A. C. and Page, D. N., *Phys. Rev.* D33, 2840 (1986).

[119] Frolov, V. P. and Zel'nikov, A.I., *Phys. Rev.* D35, 3031 (1987).

[120] Anderson, P. R., Hiscock, W. A. and Samuel, D. A., *Phys. Rev. Lett.* 70, 1739 (1993).

[121] DeWitt, B. S., *Dynamical Theory of Groups and Fields*, (Gordon and Breach, New York, 1965).

[122] DeBenedictis, A., *Class. Quant. Grav.* 16, 1955 (1999).

[123] Piedra, O. P. F. and de Oca, A. C. M., *Phys. Rev.* D75, e107501 (2007)

[124] Piedra, O. P. F. and de Oca, A. C. M., *arXiv:0707.0708v2* [gr-qc].

[125] Dias, O. J. C. and Lemos, J. P. S., *Class. Quant. Grav.* 19, 2265 (2002).

[126] DeBenedictis, A., *Gen. Rel. Grav.* 31 1549, (1999).

[127] Ayal, S. and Piran, T., *Phys. Rev.* D56, 4768 (1997).

[128] Strominger, A. and Thorlacius, L., *Phys. Rev. Lett.* 72, 1584 (1994).

[129] Peleg, Y., Bose, S. and Parker, L., *Phys. Rev.* D55, 4525 (1997).

[130] Brady, P. R., Ottewill, A. C., *Phys. Rev.* D58, 024006 (1998).

[131] Hu, B. L., *Int. J. Theor. Phys.* 38, 2987 (1999).

[132] Hu, B. L. and Verdaguer, E., *Class. Quant. Grav.* 20, R1 (2003).

[133] Hu, B. L. and Verdaguer, E., *Liv. Rev. Rel.* 7, 3 (2004).

[134] Ford, L. H. and Wu, C. H., *arXiv:0710.3787v1* [gr-qc].

[135] Einstein, A., *Preuss. Akad. der Wiss. Sitz.* 688 (1916).

[136] Rosenfeld, L., *Nucl. Phys.* 40, 353 (1963).

[137] Bronstein, M. P., *Physik. Zeit. Sow.* 9, 140 (1936).

[138] Ferrara, S., van Nieuwenhuizen, P. and Freedman, D. Z., *Phys. Rev.* D13, 3214 (1976).

[139] Feynman, R., *Act. Phys. Polon.* XXIV, 697 (1963).

[140] DeWitt, B. S., *Dynamical Theory of Groups and Fields*, (Wiley Publishers, New York, 1965).

[141] Mandlestan, S., *Phys. Rev.* 175, 1604 (1968).

[142] Faddeev, L. D. and Popov, V. N., *Phys. Lett.* B25, 30 (1967).

[143] Goroff, M. H., and Sagnotti, A., *Nucl. Phys.* B266, 709 (1986).

[144] DeWitt, B. S., *Phys. Rev.* **160**, 1113 (1967).

[145] Gibbons G. W. and Hawking S. W. (eds.), *Euclidean Quantum Gravity*, (World Scientific, Singapore, 1993).

[146] Bombelli, L., Lee, J., Meyer, D. and Sorkin, R. D., *Phys. Rev. Lett.* 59, 521 (1997).

[147] Henson, J., *arXiv:gr-qc/0601121v2*.

[148] Ambjorn, J., Jurkiewicz, J. and Loll, R., *Phys. Rev.* D72, 064014 (2005).

[149] Baez, J. C., *Lect. Notes Phys.* 543, 25 (2000).

[150] Ashtekar, A., *Phys. Rev. Lett.* 57, 2244 (1986).

[151] Ling, Y. and Smolin, L., *Phys. Rev.* D61, 044008 (2000).

[152] Penrose, R., in *Quantum theory and beyond*, (Cambridge University Press, Cambridge, 1971).

[153] Jacobson, T. and Smolin, L., *Nucl. Phys.* B299, 295 (1988).

[154] Rovelli, C., *Quantum Gravity*, (Cambridge University Press, Cambridge, 2004).

[155] Rovelli, C., *Liv. Rev. Rel.* 1, 1 (1998).

[156] Perez, A., Lectures presented at the *2nd International Conference of Fundamental Interactions*, (Pedra Azul, Brazil, 2004).

[157] Bekenstein, J. D, *Phys. Rev.* **D7**, 2333 (1973).

[158] Majumdar, P., *arXiv:gr-qc/9807045*.

[159] Wald, R. M., *Liv. Rev. Rel.* **4**, 1 (2001).

[160] Meissner, K. A., *Class. Quant. Grav.* **21**, 5245 (2004).

[161] Ghosh, A. and Mitra, P., *Phys. Let.* **B616**, 114 (2005).

[162] Corichi, A., Díaz-Polo J. and Fernández-Borja, E., *Class. Quant. Grav.* **24**, 243 (2007).

[163] Jacobson, T., *arXiv:0707.4026v1* [gr-qc].

[164] Ashtekar, A., Corichi, A. and Krasnov, K., *Adv. Theor. Math. Phys.* **3**, 419 (2000).

[165] Kloster, S., Brannlund, J., and DeBenedictis, A., *Class. Quant. Grav.* **25**, 065008 (2008).

[166] Ashtekar, A. and Krishnan, B., *Living Rev. Rel.* 7, 1 (2004).

[167] Ashtekar A., Baez, J. C. and Krasnov K., *Adv. Theor. Math. Phys.* **4**, 1 (2000).

[168] Corichi, A., Díaz-Polo J. and Fernández-Borja, E., *Proceedings of the NEB XII International Conference* (Napflio, Greece, 2006).

[169] Govindarajan, T. R., Kaul R. K. and Suneeta V., *Class. Quant. Grav.* **18**, 2877 (2001).

[170] Vanzo, L., *Phys. Rev.* **D56**, 6475 (1997).

[171] Mann, R. B. and Solodukhin, S. N., *Nucl. Phys.* **B523**, 293 (1998).

[172] Liko, T., *Phys. Rev.* D77, 064004 (2008).

[173] Modesto, L., *Phys. Rev.* D70, 124009 (2004).

[174] Modesto, L., *Class. Quant. Grav.* 23, 5587 (2006).

[175] Ashtekar, A. and Bojowald, M., *Class. Quant. Grav.* 23, 391 (2006).

[176] Böhmer, C. G. and Vandersloot, K., *Phys. Rev.* D76, 104030 (2007).

[177] Witten, E., *Phys. Rev. Lett.* 38, 121 (1976).

[178] Bojowald, M., *Liv. Rev. Rel.* 8, 11 (2005).

[179] Greenstein, J. and Schmidt, M., *Astroph. J.* 140, 1 (1964).

[180] Thorne, K. S., *Astroph. J.* 191, 507 (1974).

[181] Cardoso, V., Pani, P., Cadoni, M. and Cavaglia, M., *arXiv.org:0709.0532 [gr-qc]*.

[182] Giacconi, R. H., Gursky, H., Paolini, F. R. and Rossi, B. B., *Phys. Rev. Lett.* 9, 439 (1962).

[183] Bolton, C. T., *Nature* 235, 271 (1971).

[184] Shklovskii, I. S., *Astron. Zuhr.* 44, 930 (1967).

[185] Prendergast, K. H. and Burbidge, G. R., *Astroph. J. Lett.* 151, L83 (1968).

[186] Narayan, R. and Ti, I., *Astroph. J.* 428, L13 (1994).

[187] Abramowicz, M. A., Chen, X., Kato, S., Lasota, J. and Regev, O., *Astroph. J.* 438, L37 (1995).

[188] Menou, K., Quataert, E. and Narayan, R., in *Black Holes, Gravitational Radiation, and the Universe* (Kluwer, New York, 1999).

[189] Rees, M. J., *Ann. Rev. Astro. Astroph.* 22 471 (1984).

[190] Blandford, R. and Znajek, R., *M. N. R. A. S.* 179, 433 (1977).

[191] Reynolds, C. S., Brenneman, L. W. and Garofalo, D., *Astrophys. Space Sci.* 300, 71 (2005).

[192] Eckart, A., Genzel, R., Hofmann, R., ven der Werf, P. P. and Drapatz, S., *Nature* 355, 526 (1992).

[193] Eckart, A. and Genzel, R., *Nature* 383, 415 (1996).

[194] Ghez, A. M., Klein, B. L., Morris, M. and Becklin, E. E., *Astroph. J.* 509, 678 (1998).

[195] Ghez, A. M., Salim, S., Hornstein, S. D., Tanner, A., Lu, J. R., Morris, M., Becklin, E. E. and Duchene, G., *Astoph. J* 620, 744 (2005).

[196] Eisenhauer, F., et al., *Astroph. J.* 628, 246 (2005).

[197] Paumard, T., et al., *Am. Nat.* 326, 568 (2005).

[198] Bennet, D. P., et al., *Astroph. J.* 579, 639 (2002).

[199] Mao, S., et al., *M. N. R. A. S.* 329, 349 (2002).

[200] Salpeter, E. E., *Astroph. J.* 140, 796 (1964).

[201] Zel'dovich, Y. B. and Novikov, I. D., *Usp. Fizich. Nauk* 84, 877 (1964). English translation: *Sov. Phys.-Usp.* 7, 763 (1965).

[202] Zel'dovich, Y. B. and Novikov, I. D., *Usp. Fizich. Nauk* 86, 447 (1965). English translation: *Sov. Phys.-Usp.* 8, 522 (1966).

[203] Celotti, A., Miller, J. C. and Sciama, D. W., *Class. Quant. Grav.* 16, A3 (1999).

[204] Falcke, H. and Hehl, F. W., (Eds.) *The Galactic Black Hole - Lectures on General Relativity and Astrophysics* (I.O.P. publishing, Bristol, 2003).

[205] Müller, A., *Proc. Sci.* 017 (2007).

[206] Broderick, A. E. and Narayan, R., *Class. Quant. Grav.* **24**, 659 (2007),

[207] Rhoades, C. E. and Ruffini, R., *Phys. Rev. Lett.* 32, 324 (1974).

[208] Friedman, J. L. and Ipser, J. R., *Astroph. J.* 314, 594 (1987).

[209] Miller, J. C., Shahbaz, T. and Nolan, L. A., *M. N. R. A. S.* 294, L25 (1998).

[210] Sakharov, A. D., *Sov. Phys. JETP.* 22, 241 (1965).

[211] Gliner, E. B., *Sov. Phys. JETP.* 22, 378 (1966).

[212] Frolov, V. P., Markov, M. A. and Mukhanov, V. F., *Phys. Rev.* D41, 383 (1990).

[213] Dymnikova, I. G., *Gen. Rel. Grav.* 24, 235 (1992).

[214] Dymnikova, I. G. and Soltysek, B., *Gen. Rel. Grav.* 30, 1775 (1998).

[215] Dymnikova, I. G., *Phys. Lett.* B472, 33 (2000).

[216] Mazur P. O. and Mottola, E., *arXiv.org:gr-qc/0109035*.

[217] Mazur P. O. and Mottola, E., *Proceedings of the Sixth Workshop on Quantum Field Theory Under the Influence of External Conditions* (Rinton Press, Princeton, 2003).

[218] Mazur P. O. and Mottola, E., *Proc. Nat. Acad. Sci.* **111**, 9545 (2004).

[219] Cattoen, C., Faber, T. and Visser, M., *Class. Quantum Grav.* **22**, 4189 (2005).

[220] DeBenedictis, A., Horvat, D., Ilijić, S., Kloster, S. and Viswanathan, K. S., *Class. Quant. Grav.* 23, 2303 (2006).

[221] Lobo, F. S. N. and Arellano, A. V. B., *Class. Quant. Grav.* 24, 1069 (2007).

[222] Lobo, F. S. N., *Proc. 11th Marcel Grossmann Meeting*, (Berlin, Germany, 2006).

[223] Horvat, D. and Ilijić, S., *Class. Quant. Grav.* **24**, 5637 (2007).

[224] Boehmer, C. G. and Harko, T., *Class. Quant. Grav.* 23, 6479 (2006).

[225] Chirenti, C. B. M. H. and Rezzolla, L., *Class. Quant. Grav.* 24, 4191 (2007).

[226] Robertson, S. L. and Leiter, D. J., *arXiv.org:astro-ph/0208333*.

[227] Robertson, S. L. and Leiter, D. J., *arXiv.org:astro-ph/0307438*.

[228] Robertson, S. L. and Leiter, D. J., in *New Developments in Black Hole Research* (Nova Science Publishers, New York, 2005).

In: Classical and Quantum Gravity Research
Editors: M.N. Christiansen et al, pp. 427-464
ISBN 978-1-60456-366-5

Chapter 10

THE DYNAMICS OF ANISOTROPIC UNIVERSES

Sigbjørn Hervik*
Dalhousie University, Dept. of Mathematics and Statistics,
Halifax, NS, Canada B3H 3J5
and
University of Stavanger, Dept. of Mathematics and Natural Sciences,
N-4036 Stavanger, Norway

Abstract

The aim of the study of anisotropic universe models is to explain the apparent isotropy the Universe currently has. Here we will review what we know about the evolution of the anisotropic Bianchi universes. In this regard the investigation of ever-expanding universes of Bianchi type with a tilted perfect fluid as a source have recently been completed using the dynamical systems approach. The dynamical systems approach has proven to be extremely useful for this analysis and, therefore, some detail will be given in determining the evolution equations for the various Bianchi models. We will also review some of the different behaviours found for these models and we will discuss what we have learnt regarding the evolution of anisotropies from this analysis. Furthermore, we will discuss the issue of isotropy in detail and point out which of these models isotropise in the future (and in what sense). From a more mathematical point of view, the analysis has also shown a wealth of different phenomena, like centre manifolds, attracting closed curves and even attracting tori. We will also discuss some aspects of alternative theories of gravity and point out certain different behaviours that might appear for these theories.

1. Introduction

The standard isotropic and spatially homogeneous Friedmann-Robertson-Walker (FRW) models assume a high degree of symmetry. These models are therefore quite special and restrictive; for example, if a FRW model contains matter (or fluid) it is necessary for this matter to be isotropic. If one wants to describe the evolution of fluids with peculiar velocities, for example, one is inevitably led to the study anisotropic universes. Furthermore,

*E-mail address: herviks@mathstat.dal.ca

anisotropic fluids like that of magnetic and/or electric fields, are incompatible with the isotropic FRW geometry; hence, the influence of such fluids on the evolution of the Universe can only be studied within the anisotropic models of our Universe.

The systematic study of the so-called anisotropic Bianchi universes really picked up speed with the corner-stone work of Ellis and MacCallum [1]. Before this work there had been a few authors studying individual Bianchi models, like Taub [2]; however, Ellis and MacCallum's work was the first that approached the Bianchi models in a unified way using the orthonormal frame method.

The Bianchi models assume that the Universe can be foliated into space-like hypersurfaces parameterised with a time-variable t: Σ_t. It is common to choose this time-variable to be the proper time of the geodesics being orthogonal to the hypersurfaces Σ_t. The metric of such a cosmology can then be written:

$$\mathrm{d}s^2 = -\mathrm{d}t^2 + g_{ij}(t)\boldsymbol{\omega}^i\boldsymbol{\omega}^j, \tag{1}$$

where $\boldsymbol{\omega}^i$ are one-forms on Σ_t. The simplest choice of Σ_t arises when we assume both spatial homogeneity and isotropy (in accordance with the cosmological principles) in which case we get the FRW models:

$$\mathrm{d}s^2_{\text{FRW}} = -\mathrm{d}t^2 + a^2(t)\left[\frac{\mathrm{d}r^2}{1-kr^2} + r^2(\mathrm{d}\theta^2 + \sin^2\theta \mathrm{d}\phi^2)\right], \tag{2}$$

where $k < 0$, $k = 0$, $k > 0$ correspond to the open, flat and closed FRW model, respectively. Although these models have been relatively successful in understanding many features of our universe, they cannot explain the high-degree of isotropy our universe has. In order to explain this isotropy, one must assume that the models do not necessarily possess this property.

A natural choice would be to assume that the model is spatially homogeneous, but *not necessarily isotropic*. These models would describe, in general, anisotropic universes and it is these models what will be the focus of this review.

A spatially homogeneous (SH) model can be divided into 3 categories according to the dimension of the isotropy group of the spatial hypersurfaces, Σ_t [3,4]:

1. FRW models: These are isotopic.

2. Locally Rotationally Symmetric (LRS) models: These are, in general, anisotropic but have a rotational symmetry.

3. Bianchi models: These are spatially homogeneous but have, in general, no additional symmetries.

It turns out that the all models in category 1. and 2. can be considered special cases of models in category 3., except for the Kantowski-Sachs (KS) universe model [5]:

$$\mathrm{d}s^2_{\text{KS}} = -\mathrm{d}t^2 + a^2(t)\mathrm{d}x^2 + b^2(t)(\mathrm{d}\theta^2 + \sin^2\theta \mathrm{d}\phi^2). \tag{3}$$

In this review, we will be concerned with the remaining SH models, namely the Bianchi models. These models are anisotropic, but are, by construction, spatially homogeneous.

2. The Effect of Vorticity and Tilt

Before we embark on the formal analysis of the full Einstein equations governing the tilted Bianchi universes it is helpful to use some elementary physical arguments to determine when we might expect critical changes in the behaviour of velocities and vorticities to occur as the equation of state varies [6, 7]. Using a fairly simple argument we can give a rough estimate of the stability properties to be expected of non-comoving perfect fluids [8]. A perfect fluid with equation of state, $p = (\gamma - 1)\rho$, and constant γ, will evolve according to $\rho \propto a^{-3\gamma} \propto \theta^2$, where $a(t)$ is the mean scale factor and $\theta = 3\dot{a}/a$ is the volume expansion rate. Consider an expanding universe that is close to isotropy: an eddy of fluid, with mass m, will conserve angular momentum, $I\omega$, if

$$ma^2\omega = \text{constant}.$$

For tilted fluids with velocity $v = a\omega$ and mass $m \propto \rho a^3$, this implies

$$\rho a^5\omega = \rho a^4 v = \text{constant}.$$

Thus for perfect fluids

$$v \propto a^{3\gamma-4}, \quad \omega \propto a^{3\gamma-5}.$$

The ratio of the distorting *rotational energy density* ω^2 to the isotropic density evolves as

$$\frac{\omega^2}{\theta^2} \propto \frac{\omega^2}{\rho} \propto a^{9\gamma-10}.$$

In relativity we also can consider the ratio of the time-like vorticity component to the expansion scalar whose square can be related to the entity:

$$\frac{v^2\omega^2}{\theta^2} \propto \frac{v^2\omega^2}{\rho} \propto a^{3(5\gamma-6)}.$$

We see that we expect particular values of γ to mark thresholds in the evolution of models containing non-comoving velocities and vorticity. In particular a threshold exists when $\gamma = 4/3$: rotational velocities grow with $a(t)$ for $\gamma > 4/3$ [6]. The influence of vorticity on the expansion dynamics grows when $\gamma > 10/9$. The significance of the value $\gamma = 10/9$ can be seen, for example, in the exceptional model, Bianchi type $\text{VI}^*_{-1/9}$ [9], or the type VII_0 considered in section 6.. The value $\gamma = 6/5$ occur in the analysis of type VI_0 models [10], but also as a threshold value of the behaviour of monotone functions (see, e.g., the monotone function Z_2 in [11]).

In the analysis of Bianchi type universes we find that these critical values of γ recur, along with the value $\gamma = 2/3$, which marks a transition between universes that isotropise at late times (for $\gamma < 2/3$) and those that do not (for $\gamma > 2/3$). Of course, the detailed behaviour is more complicated than this discussion implies because the Bianchi types need not be close to isotropy and there are geometrical constraints on how the different components of the 3-velocity are able to evolve if energy and momentum are to be conserved.

3. The Bianchi Models

Let us first review how the Bianchi models are constructed.

The Bianchi models are spatially homogeneous so there exist 3 space-like Killing vectors $\boldsymbol{\xi}_i$. Recall that a Killing vector, $\boldsymbol{\xi}$, obeys the Killing equation [4, 12]:

$$\xi_{\mu;\nu} + \xi_{\nu;\mu} = 0.$$

For the Bianchi models these Killing vectors form a 3-dimensional Lie algebra, meaning that there exist constants, $C^i_{\ jk}$, such that

$$[\boldsymbol{\xi}_j, \boldsymbol{\xi}_k] = C^i_{\ jk}\boldsymbol{\xi}_i. \tag{4}$$

Furthermore,

$$\begin{aligned} C^i_{\ jk} + C^i_{\ kj} &= 0, \quad \text{(antisymmetry)}, & (5)\\ C^i_{\ jk}C^j_{\ lm} + C^i_{\ jl}C^j_{\ mk} + C^i_{\ jm}C^j_{\ kl} &= 0. \quad \text{(Jacobi identity)}. & (6) \end{aligned}$$

The Killing vectors are the infinitesimal generators of a Lie group, G, and assuming that hypersurfaces Σ_t are connected and simply connected, we can identify $\Sigma_t = G$ (or the universal cover of G). The Bianchi classification is a classification of Lie algebras, so, given a set of Killing vectors, $\{\boldsymbol{\xi}_i\}$, we would like to define a metric being invariant under the (left-)action of G. The symmetry group would then be G acting from the left on itself; i.e., for $g \in G, p \in \Sigma_t = G$: $p \mapsto g \cdot p$ (see, e.g. [13]).

A left-invariant metric can be defined as follows. We first define a set of left-invariant vectors $\{\mathbf{e}_j\}$ by Lie transport:

$$\pounds_{\boldsymbol{\xi}_i}\mathbf{e}_j = [\boldsymbol{\xi}_i, \mathbf{e}_j] = 0.$$

The corresponding one-forms, defined by $\boldsymbol{\omega}^i(\mathbf{e}_j) = \delta^i_{\ j}$, are consequently also left-invariant:

$$\pounds_{\boldsymbol{\xi}_i}\boldsymbol{\omega}_j = 0.$$

The forms $\boldsymbol{\omega}^i$ are therefore called left-invariant one-forms. The corresponding cosmological model would now be:

$$\mathrm{d}s^2 = -\mathrm{d}t^2 + g_{ij}(t)\boldsymbol{\omega}^i\boldsymbol{\omega}^j. \tag{7}$$

In the *metric approach* we choose all the time-dependence to be in the 'matrix' $g_{ij}(t)$. In the *orthonormal frame formalism*, on the other hand, we choose the matrix $g_{ij}(t)$ to time-independent and equal to δ_{ij}. All the time dependence will now be in the left-invariant frame $\{\boldsymbol{\omega}^i\}$. However, these left-invariant one-forms will obey

$$\mathrm{d}\boldsymbol{\omega}^i = -\frac{1}{2}c^i_{\ jk}(t)\boldsymbol{\omega}^j \wedge \boldsymbol{\omega}^k, \tag{8}$$

where $c^i_{\ jk}(t)$ fulfil the same anti-symmetry, and Jacobi identity as $C^i_{\ jk}$. Indeed, the Lie algebras given by $c^i_{\ jk}(t)$ and $C^i_{\ jk}$ are, as can be shown, isomorphic as Lie algebras.

Example: Let us consider an example. Assume we have the Killing vectors:

$$\boldsymbol{\xi}_1 = \frac{\partial}{\partial x}, \quad \boldsymbol{\xi}_2 = \frac{\partial}{\partial y}, \quad \boldsymbol{\xi}_3 = x\frac{\partial}{\partial x} + y\frac{\partial}{\partial y} + z\frac{\partial}{\partial z}.$$

These Killing vector correspond to the Bianchi type V model. A set of left-invariant vectors can now be found, by solving $[\boldsymbol{\xi}_i, \mathbf{e}_j] = 0$. These can be taken to be

$$\mathbf{e}_1 = z\frac{\partial}{\partial x}, \quad \mathbf{e}_2 = z\frac{\partial}{\partial y}, \quad \mathbf{e}_3 = z\frac{\partial}{\partial z}.$$

Hence, the left-invariant one-forms are

$$\boldsymbol{\omega}^1 = \frac{\mathrm{d}x}{z}, \quad \boldsymbol{\omega}^2 = \frac{\mathrm{d}y}{z}, \quad \boldsymbol{\omega}^3 = \frac{\mathrm{d}z}{z}.$$

A Bianchi type V model is now given by eq.(7) where the $\boldsymbol{\omega}^i$s are given above.

3.1. The Equations of Motion in the Orthonormal Frame Approach

For the Bianchi models Einstein's field equations,

$$R_{\mu\nu} - \frac{1}{2}Rg_{\mu\nu} + \Lambda g_{\mu\nu} = \kappa T_{\mu\nu}, \tag{9}$$

simplify radically and turn into a set of ordinary differential equations. The orthonormal frame approach was considered with a non-tilted perfect fluid in a pioneering work by Ellis and MacCallum [1]. Later, King and Ellis generalised the method to a tilted fluid [14]. General references to the orthonormal frame approach are, for example, the books [3, 15], however, these do not consider tilted fluids in any detail.

Let us consider models with an energy-momentum tensor given by

$$T_{\mu\nu} = (\rho + p)u_\mu u_\nu + pg_{\mu\nu} + 2q_{(\mu}u_{\nu)} + \pi_{\mu\nu}. \tag{10}$$

Here, ρ is the energy density of the fluid; p the isotropic stress; u_μ is a time-like four-velocity (for example, for an observer); q_μ is the energy flux; and $\pi_{\mu\nu}$ is the anisotropic stress tensor. The anisotropic stress tensor is symmetric and trace-free:

$$\pi_{\mu\nu} = \pi_{\nu\mu}, \qquad \pi^\mu_{\ \mu} = 0.$$

In addition, we have

$$\pi_{\mu\nu}u^\nu = q_\mu u^\mu = 0.$$

For the Bianchi models we assume the spacetime is of the form

$$M = \mathbb{R} \times \Sigma_t, \tag{11}$$

and it is convenient to choose u^μ to be orthogonal to the spatial surfaces of transitivity, Σ_t. Furthermore, we choose an orthonormal frame $\{\mathbf{u}, \mathbf{e}_a\}$, where $\mathbf{e}_a$ span the tangent spaces of the spatial surfaces of transitivity [1]. For the Bianchi models, these will obey

$$[\mathbf{e}_a, \mathbf{e}_b] = c^d_{\ ab}\mathbf{e}_d, \tag{12}$$

[1]In the following, Greek indices will run over the full spacetime manifold, while Latin run over the spatial surfaces.

Table 1. The Bianchi types in terms of the algebraic properties of the structure coefficients. Here, n_1, n_2 and n_3 are the eigenvalues of n_{ab} and $a^2 = a_b a^b$. Class A is defined by $a = 0$, while class B are those with $a \neq 0$.

Class	Type	a	n_1	n_2	n_3
A	I	0	0	0	0
	II	0	0	0	+
	VI_0	0	0	−	+
	VII_0	0	0	+	+
	VIII	0	−	+	+
	IX	0	+	+	+
B	V	+	0	0	0
	IV	+	0	0	+
	VI_h	+	0	−	+
	VII_h	+	0	+	+

where $c^d{}_{ab}$ are the structure constants of the Bianchi model under consideration. These may only depend on time.

The structure constants for the 3-dimensional Lie algebras can be expressed in terms of a symmetric 2-tensor, n_{ab}, and a vector a_d:

$$c^d{}_{ab} = \varepsilon_{abc} n^{cd} + \delta^d_b a_a - \delta^d_a a_b. \tag{13}$$

This is called the *Behr decomposition*. The Jacobi identity implies that the vector a_b lies in the kernel of the symmetric tensor n^{ab}:

$$n^{ab} a_b = 0. \tag{14}$$

The Bianchi types are enumerated I-IX and can be given in terms of the algebraic types of n^{ab} and a_b, see table 1. In this table the type III model is considered as the case VI_{-1}. The parameter h is given by

$$a^2 = h n_2 n_3.$$

Moreover, the Jacobi identity implies the following time evolution of the structure constants (dot means derivative with respect to the vector $\mathbf{u}$; i.e., $\dot{X} = \mathbf{u}(X) \equiv \frac{\partial X}{\partial t}$):

$$\dot{a}_b + \frac{1}{3}\theta a_b + \sigma_{bc} a^c + \varepsilon_{bcd} a^c \Omega^d = 0, \tag{15}$$

$$\dot{n}_{ab} + \frac{1}{3}\theta n_{ab} + 2n^k{}_{(a}\varepsilon_{b)kl}\Omega^l - 2n_{k(a}\sigma^k_{b)} = 0, \tag{16}$$

where Ω^a is the local angular velocity of a Fermi-propagated axis with respect to the triad $\mathbf{e}_a$. It is purely spatial and is defined by

$$\Omega^\mu = \frac{1}{2}\varepsilon^{\mu\alpha\beta\gamma} u_\alpha \mathbf{e}_\beta \cdot \dot{\mathbf{e}}_\gamma.$$

Note that eqs.(15) and (16) are purely geometrical and will therefore be theory independent.

We define the *extrinsic curvature*, or, equivalently, the *volume-expansion tensor* of the unit normal $\mathbf{u}$, by

$$u_{\mu;\nu} = \theta_{\mu\nu}. \tag{17}$$

This tensor is symmetric and purely spatial; i.e. $\theta_{\mu\nu} = \theta_{\nu\mu}$ and $\theta_{\mu\nu}u^{\mu} = 0$. We split the expansion tensor into a trace part and a trace-free part

$$\theta_{ab} = \frac{1}{3}\theta h_{ab} + \sigma_{ab}, \tag{18}$$

where h_{ab} is the metric on the surfaces Σ_t, $\theta = \theta^{\mu}_{\ \mu}$ is the *volume-expansion scalar*, and σ_{ab} is the *shear tensor*. The volume-expansion scalar is related to the Hubble scalar via $\theta = 3H$. Therefore, since the Hubble scalar is somewhat more used, we shall use the Hubble scalar instead the volume-expansion scalar.

Einstein's equations can now be expressed in terms of the above variables. Including a cosmological constant, Λ, they are as follows:
The R_{0a}-equations

$$3a^b\sigma_{ba} - \varepsilon_{abc}n^{cd}\sigma^b_{\ d} = \kappa q_a; \tag{19}$$

the shear propagation equations

$$\dot{\sigma}_{ab} + 3H\sigma_{ab} - 2\sigma^d_{\ (a}\varepsilon_{b)cd}\Omega^c + {}^{(3)}R_{ab} - \frac{1}{3}h_{ab}{}^{(3)}R = \kappa\pi_{ab}; \tag{20}$$

Raychaudhuri's equation

$$\dot{H} + H^2 + \frac{1}{3}\sigma_{ab}\sigma^{ab} + \frac{\kappa}{6}(\rho + 3p) - \frac{\Lambda}{3} = 0; \tag{21}$$

and the generalised Friedmann equation

$$3H^2 = \frac{1}{2}\sigma_{ab}\sigma^{ab} - \frac{1}{2}{}^{(3)}R + \kappa\rho + \Lambda, \tag{22}$$

where the three-curvature is ($n = n^a_{\ a}$)

$$\begin{aligned} {}^{(3)}R_{ab} &= -2\varepsilon^{cd}_{\ \ (a}n_{b)c}a_d + 2n_{ad}n^d_{\ b} - nn_{ab} - h_{ab}\left(2a^2 + n_{cd}n^{cd} - \frac{1}{2}n^2\right), \\ {}^{(3)}R &= {}^{(3)}R^a_{\ a} = -\left(6a^2 + n_{cd}n^{cd} - \frac{1}{2}n^2\right). \end{aligned} \tag{23}$$

Note that eqs.(19) and (23) do not contain any time-derivatives and for this reason these are often called the linear (or momentum) constraint, and Friedmann (or Hamiltonian) constraint, respectively.

For the Bianchi type I-VIII models it can be shown that the scalar 3-curvature ${}^{(3)}R$ is non-positive; i.e., ${}^{(3)}R \leq 0$ [13]. For the type IX model, on the other hand, the scalar 3-curvature can be either positive, zero, or negative. This makes the type IX model a bit trickier to analyse as this model can undergo a turning point in the evolution, and recollapse. This has consequences for the dynamical systems approach, as we will see later.

An important special case of the above is when there exists a time-like velocity $\hat{\mathbf{u}}$ such that the energy-momentum tensor simplifies to

$$T_{\mu\nu} = (\hat{\rho} + \hat{p})\hat{u}_\mu \hat{u}_\nu + \hat{p} g_{\mu\nu}. \tag{24}$$

This is the energy-momentum tensor of a perfect fluid where $\hat{\mathbf{u}}$ is the four-velocity of the fluid. This fluid is isotropic with respect to $\hat{\mathbf{u}}$. If the fluid is *co-moving* with the spatial surfaces of transitivity, Σ_t, then $\hat{\mathbf{u}} = \mathbf{u}$ and we say that the fluid is *non-tilted*. The more general case where $\hat{\mathbf{u}} \neq \mathbf{u}$, the fluid has a peculiar velocity with respect to the co-moving observers, and in this case we say that the fluid is *tilted* [14]. Note that $q_a = \pi_{ab} = 0$ for the non-tilted case.

In the tilted case we therefore have two naturally defined time-like vector fields and either of these can, in principle, be chosen to be the four-velocities of the fundamental observers. The fluid four-velocity, $\hat{\mathbf{u}}$, corresponds to observers who are associated with the fluid; i.e., a fluid point-of-view. Observers following the geometric congruence, $\mathbf{u}$, are co-moving with the spatially homogeneous hypersurfaces. Because these congruences are not identical, the different observers may observe and interpret the Universe quite differently. Here, we will adopt the standard choice of letting the fundamental observers follow the normal congruence $\mathbf{u}$ (for more on this issue, see e.g., [16–18]).

In the tilted case we can introduce the *tilt velocity*, v^a, so that, in component form,

$$\hat{\mathbf{u}} = \frac{1}{\sqrt{1-V^2}}(1, v^a), \quad V^2 \equiv v^a v_a,$$

with respect to the normal congruence. The velocity v^a can therefore be interpreted as the 3-velocity of the fluid with respect to the observers with four-velocity $\mathbf{u} = (1, 0)$ (hence, comoving with Σ_t). We will also assume that the perfect fluid obeys a barotropic equation of state:

$$\hat{p} = (\gamma - 1)\hat{\rho},$$

where the equation of state parameter γ is constant. With these assumptions we get the following relations:

$$\hat{\rho} = \frac{(1-V^2)}{G_+}\rho, \tag{25}$$

$$p = \frac{(\gamma-1) + \frac{1}{3}(3-2\gamma)V^2}{G_+}\rho, \tag{26}$$

$$q_a = \frac{\gamma \rho v_a}{G_+}, \tag{27}$$

$$\pi_{ab} = \frac{\gamma\rho}{G_+}\left(v_a v_b - \frac{1}{3}V^2\delta_{ab}\right). \tag{28}$$

Here, we have defined $G_+ = 1 + (\gamma - 1)V^2$. The equations for the fluid can now be determined from the energy-momentum conservation equations $T^\mu_{\nu;\mu} = 0$.

3.2. The FRW Models as Special Bianchi Models

As we pointed out earlier, the FRW models can be considered a special case of the Bianchi models. The FRW models have a 6-dimensional symmetry group and, mathematically,

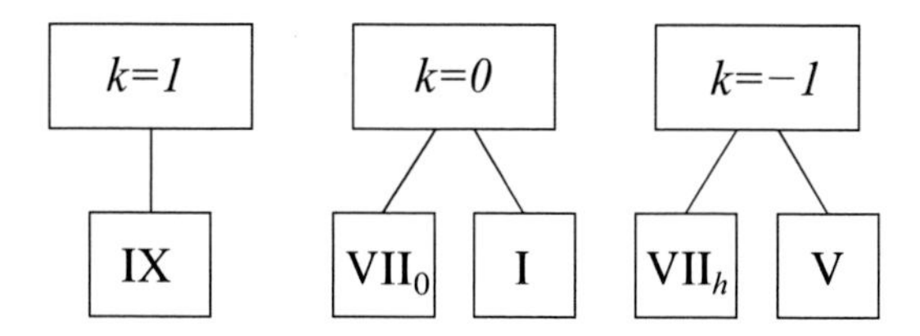

Figure 1. The Bianchi generalisations of the closed ($k = 1$), flat ($k = 0$) and open ($k = -1$) FRW model.

the question whether which FRW model is a special case of a Bianchi model is which 3-dimensional subgroup acts transitively on the FRW spatial geometry. Now, there may be several subgroups acting transitively so there might be several Bianchi model generalising a particular FRW model. Indeed, this is the case with both the flat and open model; the flat model allows for 2 simply transitive models (type I and VII$_0$), while the open model allows for an infinite number of groups (type V and VII$_h$).

For illustration, consider the open FRW model. This spatial geometry is the hyperbolic space, which, in Poincaré coordinates, can be written as

$$\mathrm{d}s^2 = \frac{1}{z^2}\left(\mathrm{d}x^2 + \mathrm{d}y^2 + \mathrm{d}z^2\right). \tag{29}$$

This space possesses 6 Killing vectors of which 4 are:

$$\boldsymbol{\xi}_1 = \frac{\partial}{\partial x}, \quad \boldsymbol{\xi}_2 = \frac{\partial}{\partial y}, \quad \boldsymbol{\xi}_3 = x\frac{\partial}{\partial x} + y\frac{\partial}{\partial y} + z\frac{\partial}{\partial z}, \quad \boldsymbol{\xi}_4 = y\frac{\partial}{\partial x} - x\frac{\partial}{\partial y}. \tag{30}$$

It can be easily shown that the following sets are acting transitively:

$$\begin{aligned} \mathrm{V}: &\quad \{\boldsymbol{\xi}_1, \boldsymbol{\xi}_2, \boldsymbol{\xi}_3\}, \\ \mathrm{VII}_h: &\quad \left\{\boldsymbol{\xi}_1, \boldsymbol{\xi}_2, \boldsymbol{\xi}_3 + \tilde{h}\boldsymbol{\xi}_4\right\}, \end{aligned}$$

where $\tilde{h}$ is any real number. The corresponding Bianchi types are indicated. This implies that the open model allows for generalisations of Bianchi type V and VII$_h$.

The Figure 1 indicates which of the Bianchi types that allow FRW as special cases. As pointed out, types V and VII$_h$ generalise the open FRW model. Moreover, types I and VII$_0$ generalise the flat, while type IX generalises the closed FRW model. These models therefore have a special significance regarding the FRW models. It should be emphasised, however, that all the remaining models could come *arbitrary close* to the flat FRW model.

In terms of the geometric frame variables, a^c, and n_{ab}, the isotropic cases have:

$$a^c = (a, 0, 0), \quad n_{ab} = \mathrm{diag}(0, n, n), \tag{31}$$

or

$$a^c = 0, \quad n_{ab} = \mathrm{diag}(n, n, n). \tag{32}$$

In addition, the universe must be non-shearing; i.e., $\sigma_{ab} = 0$. In these cases the equations of motion reduce to the standard Friedmann equations.

4. The Dynamical Systems Approach

We will now consider a very fruitful approach to the behaviour of cosmological models in general and Bianchi models in particular. The method we will use is that of *dynamical systems* [3, 15, 19]. A dynamical system is a set of autonomous differential equations [20]:

$$\mathsf{X}' = \mathsf{F}(\mathsf{X}),$$

where $\mathsf{X}(\tau) \in \mathbb{R}^n$ is called the *state vector*. There can also be a set of constraints:

$$\mathcal{C}_i(\mathsf{X}) = 0,$$

where the constraints are invariant subspaces; i.e., they are preserved by the equations of motion.

It is desirable to choose the state vector to be a set of dimensionless variables. As long as $H \neq 0$, we can accomplish this by dividing with appropriate powers of the Hubble scalar, H. In the following, we will also assume that we are considering the Bianchi models of type I-VII$_h$ (i.e., excluding VIII and IX). The case of type VIII a similar analysis can be done but some amendments are needed [11]. The investigation of type IX requires a different analysis altogether as these models may recollapse and, hence, H may be zero.

In the following we will use a dimensionless time, τ, defined as:

$$\frac{\mathrm{d}\tau}{\mathrm{d}t} = H.$$

This dynamical time is related to the *length scale*, ℓ, as follows: $\ell = \ell_0 e^{\tau}$. In an ever-expanding universe, the dynamical time is therefore ever-increasing: $\tau \to \infty$. Furthermore, it is useful to introduce the *deceleration parameter*, q, defined as $q \equiv -\dot{H}/H^2 - 1$, so that,

$$H' = -(1+q)H,$$

where prime denotes derivative with respect to τ; i.e., $' \equiv \mathrm{d}/\mathrm{d}\tau$.

Let us choose the spatial frame such that $a^c = a\delta^1_c$. Furthermore, we choose the angular velocity Ω^a such that

$$\Omega^2 = -\sigma_{13}, \quad \Omega^3 = \sigma_{12},$$

while Ω^1 is still left undetermined. This component represents a remaining frame degree of freedom, or "gauge", as we will see later.

We introduce the following expansion-normalised variables:

$$(\Sigma_{ab}, A, N_{ab}, R_1, \Omega, v_a, \Omega_\Lambda) = \left(\frac{\sigma_{ab}}{\sqrt{3}H}, \frac{a}{H}, \frac{n_{ab}}{H}, \frac{\Omega^1}{H}, \frac{\kappa\rho}{3H^2}, v_a, \frac{\Lambda}{3H^2}\right). \tag{33}$$

Furthermore, we will parameterise the expansion-normalised shear, Σ_{ab}, and curvature, N_{ab}, as follows:

$$\begin{aligned}
\Sigma_{ab} &= \begin{bmatrix} -2\Sigma_+ & \sqrt{3}\Sigma_{12} & \sqrt{3}\Sigma_{13} \\ \sqrt{3}\Sigma_{12} & \Sigma_+ + \sqrt{3}\Sigma_- & \sqrt{3}\Sigma_{23} \\ \sqrt{3}\Sigma_{13} & \sqrt{3}\Sigma_{23} & \Sigma_+ - \sqrt{3}\Sigma_- \end{bmatrix}, \\
N_{ab} &= \sqrt{3}\begin{bmatrix} 0 & 0 & 0 \\ 0 & \bar{N} + N_- & N_{23} \\ 0 & N_{23} & \bar{N} - N_- \end{bmatrix}.
\end{aligned} \tag{34}$$

To facilitate the remaining gauge freedom we will utilise complex variables [21] (which will be written in bold typeface) :

$$\begin{aligned} \mathbf{N}_\times &= N_- + iN_{23}, \quad \mathbf{\Sigma}_\times = \Sigma_- + i\Sigma_{23} \\ \mathbf{\Sigma}_1 &= \Sigma_{12} + i\Sigma_{13}, \quad \mathbf{v} = v_2 + iv_3. \end{aligned} \tag{35}$$

A gauge transformation is then given by

$$(\mathbf{N}_\times, \mathbf{\Sigma}_\times, \mathbf{\Sigma}_1, \mathbf{v}) \mapsto \left(e^{2i\phi}\mathbf{N}_\times, e^{2i\phi}\mathbf{\Sigma}_\times, e^{i\phi}\mathbf{\Sigma}_1, e^{i\phi}\mathbf{v}\right). \tag{36}$$

Here, the phase ϕ can be a function of time and corresponds to a rotation of frame. From the above we see that the variables $\mathbf{N}$ and $\mathbf{\Sigma}_\times$ are spin-2 objects and $\mathbf{\Sigma}_1$ and $\mathbf{v}$ are spin-1 objects under the frame rotations. The complex conjugates, which we will denote with an asterisk, transform similarly:

$$\left(\mathbf{N}^*_\times, \mathbf{\Sigma}^*_\times, \mathbf{\Sigma}^*_1, \mathbf{v}^*\right) \mapsto \left(e^{-2i\phi}\mathbf{N}^*_\times, e^{-2i\phi}\mathbf{\Sigma}^*_\times, e^{-i\phi}\mathbf{\Sigma}^*_1, e^{-i\phi}\mathbf{v}^*\right). \tag{37}$$

In the following, we will explicitly leave the gauge function ϕ in the equations of motion and we will assume that ϕ is given with respect to a frame for which $\phi' = 0$ implies $R_1 = 0$. Thus we will replace the function R_1 with the function ϕ. Explicitly, we have set $R_1 = -\phi'$. At this stage it is important to note that all the *physical variables* have to be independent of the gauge. Hence, objects like $\mathbf{N}_\times$ and $\mathbf{v}$ are themselves not physical variables; only scalars constructed from these are gauge independent. Scalars are therefore objects like $\mathbf{\Sigma}^*_\times \mathbf{N}_\times$, $\mathbf{N}^*_\times \mathbf{v}^2$, etc. [21].

For the expansion-normalised variables the equations of motion for types I-VII $_h$ are:

$$\Sigma'_+ = (q-2)\Sigma_+ + 3|\mathbf{\Sigma}_1|^2 - 2|\mathbf{N}_\times|^2 + \frac{\gamma\Omega}{2G_+}\left(-2v_1^2 + |\mathbf{v}|^2\right) \tag{38}$$

$$\mathbf{\Sigma}'_\times = (q-2+2i\phi')\mathbf{\Sigma}_\times + \sqrt{3}\mathbf{\Sigma}_1^2 - 2\mathbf{N}_\times(iA + \sqrt{3}\bar{N}) + \frac{\sqrt{3}\gamma\Omega}{2G_+}\mathbf{v}^2 \tag{39}$$

$$\mathbf{\Sigma}'_1 = \left(q - 2 - 3\Sigma_+ + i\phi'\right)\mathbf{\Sigma}_1 - \sqrt{3}\mathbf{\Sigma}_\times\mathbf{\Sigma}^*_1 + \frac{\sqrt{3}\gamma\Omega v_1}{G_+}\mathbf{v} \tag{40}$$

$$\mathbf{N}'_\times = \left(q + 2\Sigma_+ + 2i\phi'\right)\mathbf{N}_\times + 2\sqrt{3}\mathbf{\Sigma}_\times\bar{N} \tag{41}$$

$$\bar{N}' = (q + 2\Sigma_+)\bar{N} + 2\sqrt{3}\mathrm{Re}\left(\mathbf{\Sigma}^*_\times\mathbf{N}_\times\right) \tag{42}$$

$$A' = (q + 2\Sigma_+)A \tag{43}$$

$$\Omega'_\Lambda = 2(q+1)\Omega_\Lambda \tag{44}$$

The equations for the fluid are

$$\Omega' = \frac{\Omega}{G_+}\left\{2q - (3\gamma - 2) + 2\gamma A v_1 + \left[2q(\gamma-1) - (2-\gamma) - \gamma S\right]V^2\right\} \tag{45}$$

$$v'_1 = (T + 2\Sigma_+)v_1 - 2\sqrt{3}\mathrm{Re}\left(\mathbf{\Sigma}_1\mathbf{v}^*\right) - A|\mathbf{v}|^2 + \sqrt{3}\mathrm{Im}(\mathbf{N}^*_\times\mathbf{v}^2) \tag{46}$$

$$\mathbf{v}' = \left(T - \Sigma_+ + i\phi' + Av_1 - i\sqrt{3}\bar{N}v_1\right)\mathbf{v} - \sqrt{3}\left(\mathbf{\Sigma}_\times + i\mathbf{N}_\times v_1\right)\mathbf{v}^* \tag{47}$$

$$V' = \frac{V(1-V^2)}{1-(\gamma-1)V^2}\left[(3\gamma-4) - 2(\gamma-1)Av_1 - S\right] \tag{48}$$

where

$$\begin{aligned} q &= 2\Sigma^2 + \frac{1}{2}\frac{(3\gamma-2)+(2-\gamma)V^2}{1+(\gamma-1)V^2}\Omega - \Omega_\Lambda \\ \Sigma^2 &= \Sigma_+^2 + |\mathbf{\Sigma}_\times|^2 + |\mathbf{\Sigma}_1|^2 \\ \mathcal{S} &= \Sigma_{ab}c^a c^b, \quad c^a c_a = 1, \quad v^a = V c^a, \\ V^2 &= v_1^2 + |\mathbf{v}|^2, \\ G_+ &= 1+(\gamma-1)V^2, \\ T &= \frac{[(3\gamma-4)-2(\gamma-1)Av_1](1-V^2)+(2-\gamma)V^2\mathcal{S}}{1-(\gamma-1)V^2}. \end{aligned} \tag{49}$$

These variables are subject to the constraints

$$1 = \Sigma^2 + A^2 + |\mathbf{N}_\times|^2 + \Omega + \Omega_\Lambda \tag{50}$$

$$0 = 2\Sigma_+ A + 2\mathrm{Im}(\mathbf{\Sigma}_\times^* \mathbf{N}_\times) + \frac{\gamma\Omega v_1}{G_+} \tag{51}$$

$$0 = \mathbf{\Sigma}_1(i\bar{N} - \sqrt{3}A) + i\mathbf{\Sigma}_1^* \mathbf{N}_\times + \frac{\gamma\Omega\mathbf{v}}{G_+} \tag{52}$$

$$0 = A^2 + 3h\left(|\mathbf{N}_\times|^2 - \bar{N}^2\right) \tag{53}$$

The parameter γ will be assumed to be in the interval $\gamma \in (0,2)$.

4.1. The State Space

The state vector can be considered as $\mathsf{X} = (\Sigma_+, \mathbf{\Sigma}_\times, \mathbf{\Sigma}_1, \mathbf{N}_\times, \bar{N}, A, \Omega_\Lambda, v_1, \mathbf{v})$ modulo the constraints eqs. (51), (52) and (53). The latter is the group constraint, which we assume determines the parameter h and consequently is not a free variable. The Bianchi identities ensure that the constraints (50)-(52) are first integrals; thus satisfying the constraints on a initial hypersurface is sufficient to ensure that they are satisfied at all times.

The expansion-normalised energy density, Ω, is determined from the Hamiltonian constraint, eq.(50), and can thus be eliminated. The tuple X is 13-dimensional, but the constraints reduce the number of independent components to 9. In the above equations the function ϕ carries the choice of gauge and does not have an evolution equation. Specifying this function determines the gauge completely. We can use this function to eliminate one degree of freedom. This reduces the dimension of the state space to 8 (for a given h); i.e., the physical state space is 8-dimensional. In the case where the cosmological constant is absent, i.e., $\Omega_\Lambda = 0$, the dimension of the state space is reduced to 7.

The various Bianchi models correspond to the following invariant sets:

1. $T(\mathcal{A})$ The full state space of tilted solvable Bianchi models.
2. $T(VI_h)$ Type VI$_h$: $|\mathbf{N}_\times|^2 - \bar{N}^2 > 0$.
3. $T(VII_h)$ Type VII$_h$: $|\mathbf{N}_\times|^2 - \bar{N}^2 < 0$.
4. $T(VI_0)$ Type VI$_0$: $|\mathbf{N}_\times|^2 - \bar{N}^2 > 0, A = 0$.

5. $T(VII_0)$ Type VII$_0$: $|\mathbf{N}_\times|^2 - \bar{N}^2 < 0, A = 0$.

6. $T(V)$ Type V: $|\mathbf{N}_\times| = \bar{N} = 0, A \neq 0$.

7. $T(IV)$ Type IV: $|\mathbf{N}_\times|^2 - \bar{N}^2 = 0, A \neq 0$.

8. $T(II)$ Type II: $|\mathbf{N}_\times|^2 - \bar{N}^2 = 0, A = 0$.

9. $B(I)$ Type I: $|\mathbf{N}_\times| = \bar{N} = A = V = 0$.

We will assume that the energy-densities are non-negative which means $\Omega, \Omega_\Lambda \geq 0$. Note that the Hamiltonian constraint, ensures that

$$\Sigma_+^2 + |\mathbf{\Sigma}_\times|^2 + |\mathbf{\Sigma}_1|^2 + |\mathbf{N}_\times|^2 + A^2 + \Omega_\Lambda \leq 1. \tag{54}$$

In addition we will require that the tilt velocities are not superluminal; i.e.

$$v_1^2 + |\mathbf{v}|^2 \leq 1. \tag{55}$$

The variable $\bar{N}$ may or may not be bounded. However, from eq.(53) we see that $\bar{N}$ can only be unbounded for $T(VII_0)$. Hence, for all models except type VII$_0$, the state space is compact.

5. The Evolution of Anisotropic Universes

With the papers [9,22,23] the study of ever-expanding non-tilted ($V = 0$) Bianchi models was completed (for the other models, see [3,24,25]). There were several interesting features with these models, for example, the chaotic initial regime present in the Bianchi models VI$^*_{-1/9}$, VIII and IX, and self-similarity breaking at late times for the models VII$_0$ and VIII [22,23]. Other papers that have considered vacuum, or non-tilted Bianchi models are, for example, [26,27].

The evolution of *tilted* Bianchi models (with $\Omega_\Lambda = 0$) have been extensively studied in a sequence of papers [10,11,21,28–37]. Indeed, the investigation of the late-time behaviour of all ever-expanding tilted Bianchi models have just recently been completed. The evolution of such universes have shown a rich variety of new behaviours. Here in this section, and the following ones, we will focus on a few of them and try to give a flavour of the dynamics of tilted Bianchi models.

5.1. The Cosmic No-hair Theorem

The equations of motion describe a Bianchi model with a tilted fluid and a cosmological constant. In the presence of a cosmological constant it is expected that the cosmological constant will eventually dominate the evolution. Indeed, this statement is manifest in the so-called *cosmic no-hair* theorem. This theorem was originally stated by Wald [38]; however, he did not include tilted fluids in his theorem (in this review we have only considered types I-VII$_h$, for proofs for type VIII, see [11]).

In the presence of a $\Lambda > 0$ we have:

Cosmic no-hair theorem I: *Consider a (tilted) Bianchi model of type I-VIII with* $\Lambda > 0$*. Then,*

$$\Omega_\Lambda \to 1, \quad (\Sigma_{ab}, A, N_{ab}, \Omega) \to (0, 0, 0, 0).$$

To prove this theorem we note that, using the Hamiltonian constraint,

$$(q+1) = 3\Sigma^2 + |\mathbf{N}_\times|^2 + A^2 + \frac{\gamma(3+V^2)}{2G_+}\Omega \geq 0,$$

which is zero if and only if $\Sigma = \mathbf{N}_\times = A = \Omega = 0$. The equation $\Omega'_\Lambda = 2(q+1)\Omega_\Lambda$ implies therefore that Ω_Λ is ever-increasing. However, since $\Omega_\Lambda \leq 1$, we must have $\Omega_\Lambda \to 1$ and $q \to -1$. To show that $\bar{N} \to 0$ also, we note that

$$\left(\bar{N}^2 - |\mathbf{N}_\times|^2\right)' = 2(q + 2\Sigma_+)\left(\bar{N}^2 - |\mathbf{N}_\times|^2\right),$$

which implies that at sufficiently late times, $\left(\bar{N}^2 - |\mathbf{N}_\times|^2\right) \to 0$ and the theorem follows.

This theorem is extremely powerful and states that the cosmological constant Λ will eventually dominate the evolution. At late times the universe will asymptotically approach a de Sitter universe implying the universe experiences accelerated (in fact, exponential) expansion.

Regarding the tilt velocity the situation is a little bit more subtle. At sufficiently late times we can approximate the tilt equation:

$$V' \approx \frac{V(1-V^2)}{1-(\gamma-1)V^2}(3\gamma - 4). \tag{56}$$

Hence, we expect a change of behaviour at $\gamma = 4/3$ (radiation). Indeed, as can be shown (except for type I which does not allow for tilt),

$$V \to \begin{cases} 0, & 0 < \gamma < 4/3, \\ V_0, & \gamma = 4/3, \\ 1, & 4/3 < \gamma < 2. \end{cases} \tag{57}$$

This implies, in contrary to common belief, that fluids stiffer that radiation will not isotropise in the presence of a positive Λ! Note that not even radiation-type of fluids will in general isotropise. Hence, in the presence of cosmological constant, the geometry will isotropise, but the fluid may, or may not, depending on the equation of state parameter.

Consider now the case $\Lambda = 0$ but where the equation of state parameter obeys $0 < \gamma < 2/3$. Also in this case the universe will isotropise:

Cosmic no-hair theorem II: *Consider a (tilted) Bianchi model of type I-VIII with* $\Lambda = 0$ *and* $0 < \gamma < 2/3$*. Then,*

$$\Omega \to 1, \quad (\Sigma_{ab}, A, N_{ab}, V) \to (0, 0, 0, 0).$$

This theorem is a little bit trickier to prove. The idea of the proof is to show first that $V \to 0$ which means that the universe becomes asymptotically non-tilted. This again implies $\Omega \to 1$ at sufficiently late times. The details of the proof will be omitted here but the proof in its entirety can be found in [11,21].

This theorem also implies that the universe experiences accelerated expansion. However, in this case only as a power-law. We can see this by noting that the deceleration parameter approaches $q \to (3\gamma-2)/2$; and hence, the Hubble scalar is at late times $H \approx 2/(3\gamma t)$. This gives the scale factor: $a(t) \propto t^{2/3\gamma}$.

Let us also mention the relation between the Bianchi IX and these no-hair theorems. Unlike the Bianchi types I-VIII, the Bianchi type IX model allows for positive scalar 3-curvature. Due to this fact, these models may recollapse [39–41], even in the presence of a cosmological constant, $\Lambda > 0$. Thus the no-hair theorems, as stated above, are not valid. However, for $\Lambda > 0$, or $0 < \gamma < 2/3$, there still exist ever-expanding Bianchi type IX universes, and for such ever-expanding universes a similar isotropisation mechanism will occur (for inhomogeneous case, see [42]) .

5.2. Self-similarity

In dynamical systems theory *equilibrium points* are of special importance. Equilibrium points are fixed points, X_0, such that, given a dynamical system $\mathsf{X}' = \mathsf{F}(\mathsf{X})$, we have

$$\mathsf{F}(\mathsf{X}_0) = 0.$$

In particular, these are therefore exact solutions to the equations of motion. It is very useful to study such equilibrium points because they may act as potential attractors/sources for the system of equations. They may also represent intermediate behaviour. Consider an equilibrium point X_0 in state space. We can now expand the equations of motion with respect to this point:

$$\mathsf{X}' = \mathsf{F}(\mathsf{X}_0) + \mathsf{L}\,(\mathsf{X}-\mathsf{X}_0) + \mathcal{O}\left(|\mathsf{X}-\mathsf{X}_0|^2\right),$$

where $\mathsf{L} \equiv \frac{\partial \mathsf{F}}{\partial \mathsf{X}}\Big|_{\mathsf{X}_0}$ is the linearised matrix with components $L^i_{\ j} = \partial F^i/\partial X^j\big|_{\mathsf{X}_0}$. Hence, since we are considering an equilibrium point, we get to lowest order:

$$(\mathsf{X}-\mathsf{X}_0)' \approx \mathsf{L}(\mathsf{X}-\mathsf{X}_0).$$

If all the eigenvalues of L are non-zero, the *local* stability of X_0 is determined by the eigenvalues of L. If there are zero-eigenvalues, we have to go to higher orders to determine the local stability. These zero-eigenvalues give rise to so-called centre manifolds [43].

Let us consider the meaning of these equilibrium points for the corresponding spacetime. Consider therefore an equilibrium point X_0. In our case, a component of X is a certain expansion-normalised variable, $X^i = x^i/H^n$, say. This means that the following rescaling leaves the component invariant:

$$(H, x^i) \mapsto (e^{\lambda}H,\ e^{n\lambda}x^i).$$

For the equilibrium points such a rescaling is generated by time evolution:

$$H' = -(1+q)H \quad \Rightarrow H = H_0 e^{-(1+q)\tau},$$

since q is a constant. The equilibrium points are thus solutions for which

$$(H, x^i) = (e^{-(1+q)\tau} H_0,\ e^{-n(1+q)\tau} x^i_0).$$

Therefore, *equilibrium points correspond to spacetimes having a scaling symmetry which is generated by time-translation.*

This scaling symmetry manifests itself in term of a time-like homothety; i.e., a vector $\boldsymbol{\xi}_H$ such that

$$\pounds_{\boldsymbol{\xi}_H} \mathbf{g} = \kappa^2 \mathbf{g},$$

where κ is a constant and $\mathbf{g}$ is the metric. For the spatially homogeneous models we also have a set of Killing vectors $\boldsymbol{\xi}_i$ which together with the homothety forms the *similarity algebra*. In our case this similarity algebra will act transitively on the spacetime and we say that the space time is *self-similar*.

In cosmology there have been a long outstanding question, or rather, conjecture, called the self-similarity hypothesis: *at sufficiently late times the universe will approach a self-similar spacetime* [44, 45]. We know that this is not true in general (as we will see later) but it is still an interesting question which models, and for what initial conditions, this hypothesis is valid.

For our dynamical system we can restate the hypothesis in terms of equilibrium points: For which models and initial conditions do the solutions approach an equilibrium point?

For many of the Bianchi models this hypothesis is true. For example, the tilted models (with or without Λ) of type I-VII$_{h\neq 0}$ are asymptotically self-similar. However, the models VII$_0$ and VIII in the absence of Λ are not. As pointed out earlier, the state space for these models are non-compact which may allow the solutions (in state space) to fly off to infinity. This divergence is essentially what causes the self-similarity breaking at late times; the solutions are forever moving and do not settle down at any equilibrium point. The Bianchi type VII$_0$ model will be considered in some detail later where this fact will be investigated.

In our case, a word of caution is in order. We still have a free "gauge function" ϕ left in our equations of motion. Since neither the physics nor the geometry are dependent on this function we should really define equilibrium points as the constancy of scalars rather than components. More on this issue, see [21].

5.3. Isotropisation

Let us now discuss the issue of isotropisation. First, we have to define what we mean by isotropisation and what characterises a universe that isotropises.

Let us consider a FRW universe which is exactly isotropic with respect to the co-moving observers. A FRW universe is characterised by $\Sigma_{ab} = 0$, and the curvature variables obeying eq.(31) or eq.(32). Alternatively, we can use the the Weyl tensor for an invariant characterisation:

$$\text{FRW: } C_{\mu\nu\alpha\beta} = 0, \quad \Rightarrow \nabla^{(k)} C_{\mu\nu\alpha\beta} = 0.$$

Here, $\nabla^{(k)}$ means the kth covariant derivative. For an anisotropic universe we need to consider if one, or several, of these conditions are satisfied in the far future.

There are several definitions of isotropisation. It is clear that in order to define this concept we need to define a set of observers. Isotropy is usually referred to the isotropy group being $SO(3)$. The problem with this is that this is clearly dependent on the set of observers. Furthermore, in a general spacetime there is no preferred set of observers. For a spatially homogeneous universe, which we will consider here, there are two preferred sets of observers: the hypersurface orthogonal observers (the geometrically defined congruence); and the fluid observers (the matter congruence). It is common to choose the geometric congruence, u^μ, so we will do the same here.

The work of Collins and Hawking [46] was a pioneering work which tried to address the issue of isotropisation. This was a study of the stability of isotropic ever-expanding universes against spatially homogeneous perturbations of Bianchi type VII$_h$ and was not restricted to perfect fluids and comoving fluids. In the case of VII$_h$ perturbations to open FRW universes, isotropy was shown to be unstable. However, this well known result requires careful interpretation. The definition of stability used was asymptotic stability in the sense of Lyapunov; that is, as $t \to \infty$ we need to have $\Sigma \to 0$. However, in general it is found that the VII$_h$ perturbations of the open FRW universe approach the Lukash plane-wave spacetimes (see later) which have $\Sigma = \mathrm{constant}$. Thus although isotropy is not asymptotically stable it is stable in the sense that the deviations from isotropy ($\Sigma = 0$) are always bounded [47,48].

Geometrically, we shall distinguish between different kinds of isotropisation:

- Shear and curvature isotropisation (SI): $\Sigma_{ab} \to 0$ and (A_c, N_{ab}) approach either (31) or (32).
- Weyl isotropisation: $C_{\mu\nu\alpha\beta}/H^2 \to 0$.
- Weyl^k isotropisation: $\nabla^{(k)}C_{\mu\nu\alpha\beta}/H^{2+k} \to 0$.
- Weyl$^\infty$ isotropisation: $\nabla^{(k)}C_{\mu\nu\alpha\beta}/H^{2+k} \to 0$ for any k.

We can also consider the isotropisation of the matter[2]

- Weak matter-isotropisation (WMI): $\pi_{ab}/H^2 \to 0$, $q_a/H^2 \to 0$.
- Strong matter-isotropisation (SMI): $\pi_{ab}/T_{00} \to 0$, $q_a/T_{00} \to 0$.

When all of these are satisfied we will call it:

- Perfect (or smooth) Isotropisation (I): Shear and curvature, Weyl$^\infty$, weak and strong matter-isotropisation.

Now, it can also be advantageous to consider when the evolution comes *arbitrary* close to isotropy, as pointed out above. An archetypical example of this is the instability found by Collins and Hawking for the Bianchi type VII$_h$ models. In the dynamical systems approach this 'instability' is due to a line of equilibrium points. So, for example, a perturbation away from the Milne universe will generally evolve towards a point on this line. However, due to the fact that the Milne universe is a point on this line we can end up *arbitrary close* to the Milne universe and, in this sense, arbitrary close to isotropy. If this is the case, we will use the symbol ϵ to indicate that although Σ_{ab}, say, fails to approach zero, we can get Σ_{ab} arbitrary close to zero by a sufficiently small perturbation.

[2]Here, we will consider the cosmological constant as part of the matter; i.e., we collect Λ and matter terms into one energy-momentum tensor. However, we could, in principle, also consider the isotropisation of all matter terms separately.

Table 2. Isotropisation at late-times with respect to an arbitrary small perturbation. The $\star$ indicates the details are complicated and depends on the parameters γ and h. A 'No' indicates an instability and a departure from isotropy. Recall that for $\Lambda > 0$ the cosmological constant is considered a part of the energy-momentum tensor.

Type		SI	Weyl	Weyl$^\infty$	WMI	SMI	I
I-IX	$\Lambda > 0$	Yes	Yes	Yes	Yes	Yes	Yes
I-IX	$0 < \gamma < 2/3$	Yes	Yes	Yes	Yes	Yes	Yes
I	$2/3 < \gamma < 2$	Yes	Yes	Yes	Yes	Yes	Yes
II	$2/3 < \gamma < 10/7$	No	No	No	Yes	Yes	No
	$10/7 < \gamma < 2$	No	No	No	No	No	No
III	$2/3 < \gamma < 1$	No	No	No	Yes	Yes	No
	$\gamma = 1$	No	Yes	Yes	Yes	Yes	No
	$1 < \gamma < 2$	No	Yes	Yes	Yes	No	No
IV	$2/3 < \gamma < 4/3$	ϵ	ϵ	ϵ	Yes	Yes	ϵ
	$4/3 < \gamma < 2$	ϵ	ϵ	ϵ	Yes	No	No
V	$2/3 < \gamma < 4/3$	Yes	Yes	Yes	Yes	Yes	Yes
	$4/3 < \gamma < 2$	Yes	Yes	Yes	Yes	No	No
VI_0	$2/3 < \gamma < 6/5$	No	No	No	Yes	Yes	No
	$6/5 \leq \gamma < 2$	No	No	No	No	No	No
$\text{VI}_{h\neq-1}$	$2/3 < \gamma < 2$	No	No	No	$\star$	$\star$	No
VII_0	$2/3 < \gamma < 1$	Yes	Yes	No	Yes	Yes	No
	$1 \leq \gamma < 4/3$	Yes	No	No	Yes	Yes	No
	$4/3 \leq \gamma < 2$	No	No	No	No	No	No
VII_h	$2/3 < \gamma < 4/3$	ϵ	ϵ	ϵ	Yes	Yes	ϵ
	$4/3 < \gamma < 2$	ϵ	ϵ	ϵ	Yes	No	No
VIII	$2/3 < \gamma < 4/5$	No	Yes	No	Yes	Yes	No
	$4/5 \leq \gamma < 1$	No	No	No	Yes	Yes	No
	$\gamma = 1$	No	No	No	Yes	Yes	No
	$1 < \gamma < 2$	No	No	No	Yes	No	No

There might be various applications where the different versions of isotropy might be useful. For example, if the measurements of the observers can only capture the C^2 structure of the universe, then the Weyl-isotropisation is of essence. In this case the higher-derivative structure, which is captured in Weyl^k- and Weyl$^\infty$-isotropisation, would be beyond the scope of the measurement. Furthermore, regarding the matter, if we have WMI but not SMI (for example, for the type V model with matter $4/3 < \gamma < 2$) the geometry will isotropise; hence, any measurements of the geometry only, will indicate an isotropic universe. Only when we are considering the matter will be able to note that there is a preferred direction (in particular, for the type V model with $4/3 < \gamma < 2$, the matter tends to vacuum but the matter will flow in a particular direction). Only measurements of the fluid will capture this anisotropy of the fluid.

With regards to isotropisation, it is also of interest when this may happen to the past. In general, the models studied here would be dominated by the shear as we go to the past, so the issue of past-isotropisation is related to the issue of initial conditions. This has been studied by several authors [49–51], especially in the context of Penrose's Weyl curvature hypothesis [52–54].

6. The Type VII_0 Model

One of the most interesting models is the Bianchi type VII_0 model. This model is the most general Bianchi model having the flat FRW model as a special case. In addition, this model shows some very interesting behaviour. As pointed out earlier the state space for this model is non-compact which allows for a self-similarity breaking at late times [22, 35] (the type VIII model exhibits a similar behaviour [11,23])

In the following we will assume that the cosmological constant it absent: $\Omega_\Lambda = 0$. The case of a non-vanishing cosmological constant is covered by the no-hair theorem. Furthermore, the type VII_0 model is given by $A = h = 0$, and $\bar{N}^2 - |\mathbf{N}_\times|^2 > 0$. Let us also in the following choose the gauge where $\phi' = 0$.

One of the most important features with this model (for $\Omega_\Lambda = 0$) is the behaviour of $\bar{N}$. As pointed out, this variable has no upper bound and can therefore become arbitrary large. Indeed, it is conjectured that, for non-inflationary perfect fluids ($0 < \gamma < 2/3$, $\Omega_\Lambda = 0$) [35]:

$$\lim_{\tau\to\infty} |\bar{N}| = \infty. \tag{58}$$

This behaviour of the curvature variable $\bar{N}$ causes two phenomena: a self-similarity breaking, and a rapidly increasing oscillation at late times.

This oscillation manifests itself in a oscillation of the variables $\mathbf{\Sigma}_\times$ and $\mathbf{N}_\times$, and also the tilt $\mathbf{v}$. Introducing $M = 1/\bar{N}$, so that $M \to 0$, this oscillation is more manifest if we introduce the following variables:

$$\mathbf{\Sigma}_\times + i\mathbf{N}_\times \equiv e^{2i\psi}\mathbf{X}, \quad \mathbf{\Sigma}_\times - i\mathbf{N}_\times \equiv e^{-2i\psi}\mathbf{Y}, \quad \mathbf{v} \equiv e^{-i\theta}v_2, \tag{59}$$

where

$$\psi' = \frac{\sqrt{3}}{M}, \tag{60}$$

$$\theta' = \frac{\sqrt{3}}{M}\left\{v_1 + \frac{1}{2}M\mathrm{Im}\left[(1+v_1)\mathbf{X}e^{2i(\theta+\psi)} + (1-v_1)\mathbf{Y}e^{2i(\theta-\psi)}\right]\right\}. \tag{61}$$

The angular variables ψ and θ are introduced to take care of the rapid oscillation as $M \to 0$. We note that the variables ψ and θ are not in synchronization since $v_1 < 1$. Hence, in general, we expect two different oscillations with different frequencies. The equation of motions for these variables are given in [35]. In this case we can explicitly see how the oscillatory terms enter into the equations of motion. Moreover, we note that both of these rapid oscillations are observable; e.g., by considering the scalars:

$$\begin{aligned} S_1 &= (\mathbf{\Sigma}_\times + i\mathbf{N}_\times)(\mathbf{\Sigma}^*_\times + i\mathbf{N}^*_\times) = e^{4i\psi}\mathbf{X}\mathbf{Y}^*, \\ S_2 &= (\mathbf{\Sigma}^2_\times + \mathbf{N}^2_\times)(\mathbf{v}^*)^4 = e^{4i\theta}\mathbf{X}\mathbf{Y}v_2^4. \end{aligned}$$

Now this increasingly rapid oscillation enters the equations of motion in such a way that they will effectively 'cancel' at sufficiently late times. In this way, we can rewrite the system as a reduced system of equations whose solutions, $\mathsf{X}_{\text{eff}}(\tau)$, are approximate solutions to the full system:

$$\mathsf{X}(\tau) = \mathsf{X}_{\text{eff}}(\tau) + \mathcal{O}(M).$$

Since $M \to 0$ these effective solutions become increasingly accurate and thus are good approximations for the real system (the details are somewhat hairy but the interested reader can consult [35]).

6.1. The Type VII$_0$ Reduced System

By defining

$$\begin{aligned} \sigma_1 &\equiv |\mathbf{N}_\times|^2 + |\mathbf{\Sigma}_\times|^2 = \frac{1}{2}\left(|\mathbf{X}|^2 + |\mathbf{Y}|^2\right), \qquad (62)\\ \sigma_2 &\equiv 2\mathrm{Im}\left(\mathbf{\Sigma}_\times \mathbf{N}_\times^*\right) = -\frac{1}{2}\left(|\mathbf{X}|^2 - |\mathbf{Y}|^2\right), \qquad (63)\end{aligned}$$

the system effectively reduces to the following system at sufficiently late times:

$$\begin{aligned} \Sigma_+' &= (Q-2)\Sigma_+ - \sigma_1 + \frac{\gamma\Omega}{2G_+}\left(-2v_1^2 + v_2^2\right) \qquad (64)\\ \sigma_1' &= 2(Q + \Sigma_+ - 1)\sigma_1 \qquad (65)\\ M' &= -\left(Q + 2\Sigma_+\right)M \qquad (66)\\ \Omega' &= \frac{\Omega}{G_+}\left\{2Q - (3\gamma-2) + \left[2Q(\gamma-1) - (2-\gamma) - \gamma\mathcal{S}\right]V^2\right\} \qquad (67)\\ v_1' &= \left(T + 2\Sigma_+\right)v_1 \qquad (68)\\ v_2' &= (T - \Sigma_+)v_2 \qquad (69)\end{aligned}$$

where

$$\begin{aligned} Q &= 2\Sigma_+^2 + \sigma_1 + \frac{1}{2}\frac{(3\gamma-2) + (2-\gamma)V^2}{1 + (\gamma-1)V^2}\Omega, \qquad (70)\\ V^2\mathcal{S} &= \left(-2v_1^2 + v_2^2\right)\Sigma_+. \qquad (71)\end{aligned}$$

These variables are subject to the constraint

$$1 = \Sigma_+^2 + \sigma_1 + \Omega. \qquad (72)$$

Furthermore, σ_2 is determined from

$$\sigma_2 = -\frac{\gamma\Omega v_1}{G_+},$$

which gives the bound

$$\sigma_1 \geq \frac{\gamma\Omega|v_1|}{G_+}. \qquad (73)$$

Using the constraint (72) we can solve for σ_1 or Ω.

This system of equations can now be analysed in a standard manner using dynamical systems theory.

6.2. Late-Time Behaviour

Let us consider the case $2/3 < \gamma < 4/3$ for illustration. In this case the solutions will have the following late-time behaviour:

$$\lim_{\tau \to \infty} (\Sigma_+, V, \sigma_1) = (0, 0, 0),$$

which means the universe experiences shear-isotropisation (SI). Furthermore, since the fluid becomes non-tilted at late times, the matter isotropises both weakly and strongly (WMI and SMI).

By assuming an ansatz with coefficients and exponents to be determined, we obtain the decay rates when $\frac{2}{3} < \gamma < \frac{4}{3}$ as follows (hatted variables are constants):

$$\Sigma_+ \approx -\frac{2\hat{\sigma}_1}{3\gamma - 2} e^{-(4-3\gamma)\tau} + \hat{\Sigma}_+ e^{-\frac{3}{2}(2-\gamma)\tau}, \tag{74}$$

$$\sigma_1 \approx \hat{\sigma}_1 e^{-(4-3\gamma)\tau}, \tag{75}$$

$$v_1 \approx \hat{v}_1 e^{-(4-3\gamma)\tau}, \tag{76}$$

$$v_2 \approx \hat{v}_2 e^{-(4-3\gamma)\tau}, \tag{77}$$

$$M \approx \hat{M} e^{-\frac{1}{2}(3\gamma-2)\tau}. \tag{78}$$

The angular variables are given asymptotically by:

$$\psi \approx \hat{\psi} + \frac{2\sqrt{3}}{(3\gamma - 2)\hat{M}} e^{\frac{1}{2}(3\gamma-2)\tau}, \tag{79}$$

$$\theta \approx \begin{cases} \hat{\theta}, & 2/3 < \gamma < 10/9 \\ \hat{\theta} + \frac{\sqrt{3}\hat{v}_1}{\hat{M}}\tau, & \gamma = 10/9 \\ \hat{\theta} + \frac{2\sqrt{3}\hat{v}_1}{(9\gamma-10)\hat{M}} e^{\frac{1}{2}(9\gamma-10)\tau}, & 10/9 < \gamma < 4/3. \end{cases} \tag{80}$$

The meaning of the bifurcation value $\gamma = 10/9$ is more clear if we calculate the fluid vorticity; in particular, to leading order we have:

$$W^a W_a \propto e^{(9\gamma-10)\tau}, \qquad W^0 W_0 \propto e^{3(5\gamma-6)\tau}. \tag{81}$$

This means that, for $\gamma > 10/9$, the universe does not isotropise in terms of the vorticity. We note that the bifurcation values for the vorticity and tilt at $\gamma = 10/9$ and $\gamma = 4/3$, respectively, coincide with with the values found in section 2. and [55]. Note also that the bifurcation value $\gamma = 6/5$ agrees with the one found in section 2..

The Hubble-normalised Weyl scalars $\mathcal{W}_1$ and $\mathcal{W}_2$, defined by

$$\mathcal{W}_1 = \frac{C_{abcd}C^{abcd}}{48H^4}, \quad \mathcal{W}_2 = \frac{C_{abcd}{}^*C^{abcd}}{48H^4}, \tag{82}$$

evolve as

$$\begin{aligned} \mathcal{W}_1 + i\mathcal{W}_2 &\approx -\frac{12}{M^2} S_1 \\ \Rightarrow \quad (\mathcal{W}_1, \mathcal{W}_2) &\approx -\frac{12}{\hat{M}^2}\sqrt{\hat{\sigma}_1^2 - \gamma^2 \hat{v}_1^2}\, e^{6(\gamma-1)\tau} (\cos 2\psi_2, \sin 2\psi_2), \end{aligned} \tag{83}$$

where $0 \leq \hat{v}_1 \leq \frac{\hat{\sigma}_1}{\gamma}$, $\hat{M}$ is the constant from the variable M, and ψ_2 is defined by $S_1 = |S_1|e^{2i\psi_2}$. The angular variable ψ_2 is related asymptotically to the variable ψ via $\psi_2 \approx 2\psi + \psi_0$, where ψ_0 is a constant.

We note that the first term in (74) is dominant, and the second term is included to display the constant $\hat{\Sigma}_+$. The Weyl scalars diverge for $1 \leq \gamma < \frac{4}{3}$ (hence, no Weyl-isotropisation). In the case $\gamma = 1$, the magnitude of the Weyl scalars are asymptotically constant.

For radiation ($\gamma = 4/3$) or stiffer fluids ($4/3 < \gamma < 2$) the fluid will be asymptotically tilted. The corresponding decay rates for these cases can also be calculated in an analogous manner (see [35]).

7. The Great Plane Waves

Let us also consider a class of spacetimes which are the late-time attractors for the type VII$_h$ model (which is the most general open model). For non-inflationary fluids the late-time attractors are the so-called *vacuum plane wave* solutions (also known as Lukash plane waves).

These plane-wave solutions are given by

$$ds^2 = e^{2T}(-dT^2 + dx^2) + e^{2s(x+T)}\left[e^{\beta}(\cos v dy - \sin v dz)^2 + e^{-\beta}(\sin v dy + \cos v dz)^2\right],$$

where $v = b(x+T)$, and b, β and s are constants satisfying $b^2\sinh^2\beta = s(1-s)$, $0 < s < 1$. The Killing vectors ∂_y and ∂_z span the wave-fronts. We also note that the case $\beta = 0$, $s = 1$ is the isotropic limit for which the metric reduces to the Milne universe.

In terms of the expansion-normalised variables these solutions are given by (where the gauge $\phi' = \sqrt{3}\lambda\Sigma_-$ is used, so that $\mathbf{N}_\times = iN_{23}$):

$$\begin{aligned} & q = 2\chi,\ \Sigma_+ = -\chi,\ A = 1-\chi,\ \Sigma_-^2 = N_{23}^2 = \chi(1-\chi), \\ & \bar{N} = \lambda N_{23}, \Sigma_{23} = \Sigma_{12} = \Sigma_{13} = N_- = \Omega = 0, \\ & \text{where } 0 < \chi < 1,\ \lambda^2 > 1,\ (1-\chi) = 3h(\lambda^2-1)\chi. \end{aligned} \tag{84}$$

Note that these solutions correspond to vacuum spacetimes. Furthermore, the spacetimes allow for a homothetic vector field, and hence, these are self-similar (in fact, they are space-time homogeneous).

Regarding the fluid, we see that $\Omega = 0$ which implies that the tilt-equations decouple and turns into three equations in (v_1, v_2, v_3), in a plane-wave background. This background can be parameterised in terms of 3 parameters, γ, h and λ. Dynamically, as the universe approach a plane-wave solution, the geometric variables Σ_{ab}, N, etc. 'freeze in' to particular values Σ^*_{ab}, N^* and $\Omega \to 0$. As this happens the tilt equations decouple and turns effectively into a 3-dimensional dynamical system with 3 parameters. This reduced 3-dimensional system turns out to have a very sensitive dependence on the parameters and exhibits rich behaviour.

In order to shed some light on this interesting behaviour we utilise the identity

$$(v_2^2 + v_3^2)^2 = (2v_2v_3)^2 + (v_2^2 - v_3^2)^2,$$

and define (x, ρ, θ) by

$$x = v_1, \quad \rho \equiv v_2^2 + v_3^2, \quad 2v_2v_3 = \rho\sin\theta, \quad (v_2^2 - v_3^2) = \rho\cos\theta. \tag{85}$$

Then we have for the reduced system

$$\begin{aligned} x' &= (T + 2\Sigma_+^*)x - (A^* + \sqrt{3}N^*\cos\theta)\rho, \\ \rho' &= 2\left(T - \Sigma_+^* + A^*x - \alpha\cos\theta\right)\rho, \\ \theta' &= 2\alpha(\lambda^* + \sin\theta), \end{aligned} \tag{86}$$

where $\alpha = \sqrt{3}(1-x)N^*$ and, by use of the discrete symmetry $(v_2, v_3) \mapsto (-v_2, -v_3)$, we can assume that $0 \leq \theta < 2\pi$. Furthermore, these variables are bounded by

$$0 \leq \rho, \quad V^2 = x^2 + \rho \leq 1. \tag{87}$$

An asterisk has been added to the variables to emphasise that these should be thought of as the limit values for the full system.

These variables are essentially doubly-covering cylindrical coordinates. The angle θ corresponds to the winding angle around the x-axis. We note that since $|\lambda^*| > 1$, the angle θ is monotone which means that the solutions wind around the x-axis indefinitely. This is an interesting observation and makes us wonder if there might be closed periodic orbits for this dynamical system.

Indeed, for various values of the parameters, this system does possess closed period orbits. Furthermore, in a tiny region of parameter space, evidence for attracting tori has been found. In the latter case, the late-time asymptote will be a curve winding on a torus.

The detailed analysis of these closed curves will be omitted here (we refer the interested reader to [29]); however, we will present some results and numerics to illustrate the various possibilities.

For a periodic orbit, c, it is useful to introduce the average of a variable B:

$$\langle B\rangle \equiv \frac{1}{T_n}\oint_c B\mathrm{d}\tau, \quad T_n \equiv \oint_c \mathrm{d}\tau. \tag{88}$$

We can use this to say something about the stability of a closed periodic orbit. For example, consider the evolution equation for Ω, which we write as $\Omega' = \Omega\lambda_\Omega$. Assume that $c(\tau)$ is a closed periodic orbit with period T_n. Then for every τ_0 and $\epsilon > 0$, there exists a solution curve, $\tilde{c}(\tau) = c(\tau) + \delta(\tau)$, for which

$$|\delta(\tau)| < \epsilon, \quad \tau_0 < \tau < \tau_0 + T_n. \tag{89}$$

(This follows from Proposition 4.2, page 104, in [3].) Hence, the curve $\tilde{c}(\tau)$ can be used to approximate the closed curve $c(\tau)$. Using $\tilde{c}(\tau)$ we get

$$\ln\left[\frac{\Omega(\tau_0 + T_n)}{\Omega(\tau_0)}\right] = \int_{\tau_0}^{\tau_0+T_n}\frac{\Omega'}{\Omega}d\tau = \int_{\tau_0}^{\tau_0+T_n}\lambda_\Omega d\tau = T_n\langle\lambda_\Omega\rangle + \mathcal{O}(\epsilon). \tag{90}$$

Since ϵ can be arbitrarily small, we can approximate

$$\Omega(\tau_0 + T_n) \approx \Omega(\tau_0)\exp(T_n\langle\lambda_\Omega\rangle). \tag{91}$$

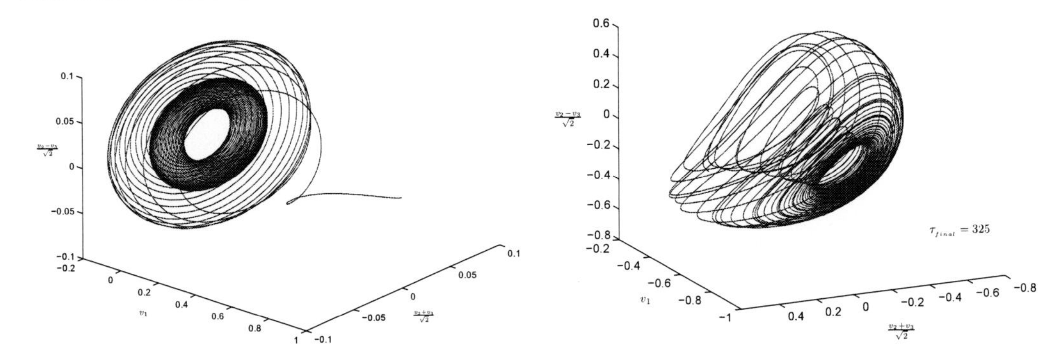

Figure 2. Shown is a closed orbit acting as the attractor (left), and a torus acting as the attractor (right). On the right we can see how the solution curve winds itself around on a torus.

Hence, if $\langle\lambda_\Omega\rangle < 0$, then $\Omega(\tau_0 + T_n) < \Omega(\tau_0)$ for a sufficiently small perturbation; i.e., the closed curve is *stable* with respect to Ω. If, on the other hand, $\langle\lambda_\Omega\rangle > 0$ then the closed curve is *unstable* with respect to a perturbation of Ω. This result is the corresponding local stability criterion for periodic orbits.

It is now possible to show (see [29] for a proof):

Theorem: *Assume that there is a closed properly periodic orbit, $c(\tau)$, for the dynamical system (86). Then either*

$$\mathcal{C}: \quad \langle x\rangle = -\frac{(3\gamma-4)-\Sigma_+^*}{(1+\Sigma_+^*)(3-2\gamma)}, \quad \langle\lambda_\Omega\rangle = -\frac{(1-2\Sigma_+^*)(5\gamma-6)}{3-2\gamma}, \tag{92}$$

or $V = 1$ and

$$\mathcal{E}: \quad \langle x\rangle = \frac{1+\Sigma_+^*}{3\Sigma_+^*}, \quad \langle\lambda_\Omega\rangle = \frac{(1-2\Sigma_+^*)(1+2\Sigma_+^*)}{\Sigma_+^*}. \tag{93}$$

This theorem states exactly when a closed periodic orbit, if it exists, is stable with respect to the variable Ω. Equation (91) implies that small deviations from $\Omega = 0$ will effectively decay exponentially if $\langle\lambda_\Omega\rangle < 0$.

These are necessary criteria for closed periodic orbits. However, in some cases we can even prove existence of such orbits.

Theorem (existence of closed curves): *For every $\Sigma_+^* \in (-1, -1/4)$, $(\lambda^*)^2 > 1$, and $\gamma \in (0,2)$ there exists one, and only one up to winding number, closed periodic orbit, $\widetilde{\mathcal{E}}(VII_h)$, with $V = 1$ for the dynamical system (86).*

The proof of this theorem uses the Poincaré-Bendixson theorem (existence) and the previous theorem (uniqueness) [29].

We will not go into the details of the existence of closed curves here. However, we will just point out that such curves exist for a wide range of parameters. In Fig.2 we have shown one case where a closed orbit is acting as the attractor, and a case there a torus is acting as the attractor. This indicates that that there is a rich, and extremely sensitive, behaviour as we change the parameters of this 3-dimensional system. For details of this see [29].

8. Anisotropic Universes in Alternative Theories of Gravity

In the previous sections we have reviewed some of the behaviour seen for anisotropic universes in standard Einstein gravity. During the last few years, the interest for alternative theories of gravity has increased drastically. In these theories the equations of motion will be altered and hence, the behaviour of anisotropic universes may also change. Let therefore also, in brief, discuss how anisotropic universe may appear in alternative theories of gravity. In particular, we going to point out two alternative theories of gravity which have attracted much attention lately and point out some changes to the behaviour of anisotropic universes for these models.

8.1. Brane-Worlds

In brane-world cosmology, matter fields and gauge interactions are confined to a four-dimensional brane moving in a higher-dimensional "bulk" spacetime. This paradigm is motivated by string and M-theory; in particular, generalized Randall-Sundrum-type models [56] are relatively simple phenomenological five-dimensional (5D) models which capture some of the essential features of the dimensional reduction of Ho řava-Witten theory [57]. In a recent analysis of the asymptotic dynamical evolution of perfect fluid brane-world cosmological models close to the initial singularity, it was found that for an appropriate range of the equation of state parameter an isotropic singularity is a past-attractor [58]. It was subsequently argued that the initial cosmological singularity is isotropic in brane-world cosmological models.

The 5D field equations are

$$^{(5)}G_{ab} = -\Lambda_5 \, ^{(5)}g_{ab} + \kappa_5^2 \, ^{(5)}T_{ab} \,, \quad \Lambda_5 = -\frac{6}{\ell^2} \,. \tag{94}$$

Here, $a, b = 0, ..., 4$, Λ_5 and g_{ab} is the 5D cosmological constant and metric, respectively. The projected field equations on the brane are [59]:

$$G_{\mu\nu} = -\Lambda g_{\mu\nu} + \kappa^2 T_{\mu\nu} + 6\frac{\kappa^2}{\lambda} S_{\mu\nu} - \mathcal{E}_{\mu\nu} \,, \tag{95}$$

where μ, ν are brane indices (i.e. $\mu, \nu = 0, .., 3$) $\Lambda = \Lambda_5/2 + \lambda^2\kappa_5^4/12$ and $\kappa^2 = \lambda\kappa_5^4/6$. The term $S_{\mu\nu}$ is quadratic in $T_{\mu\nu}$ and dominates at high energies ($T_{00} = \rho > \lambda$), and the five-dimensional Weyl tensor is felt on the brane via its projection, $\mathcal{E}_{ab} = C_{acbd}n^c n^d$, where n^a is the unit normal vector to the brane. There may also be terms that arise from from 5D sources in the bulk other than the vacuum energy Λ_5, such as a bulk dilaton field. In general, in the 4-dimensional picture the conservation equations do not determine all of the independent components of $\mathcal{E}_{\mu\nu}$ on the brane (and a complete higher-dimensional analysis, including the dynamics in the bulk, is necessary) [60].

In these models the gravitational field can also propagate in the extra dimensions (i.e., in the 'bulk'). For example, there might occur thermal radiation of bulk gravitons [61]. In particular, at sufficiently high energies particle interactions can produce 5D gravitons which are emitted into the bulk. Conversely, in models with a bulk black hole, there may be gravitational waves hitting the brane. At sufficiently large distances from the black hole these gravitational waves may be approximated as of type N. Alternatively, if the brane has low energy initially, energy can be transferred onto the brane by bulk particles such as gravitons; an equilibrium is expected to set in once the brane energy density reaches a limiting value. (For an alternative approach see, e.g., [62].) Here we shall study the consequences of assuming that in some appropriate regimes the bulk gravitational waves can be approximated by plane waves [63].

8.1.1. An Example of a Type N Bulk with a Plane Wave Brane

First, let us consider an example of an exact 5D solution of type N [64] with a negative cosmological constant. This particular example will give us some hints what we can expect from the brane-world analysis.

Consider the Siklos metric [65] (where we have dropped an overall scaling)

$$\mathrm{d}s^2 = \frac{1}{z^2}\left[2\mathrm{d}v\mathrm{d}u + H(u,x,y,z)\mathrm{d}u^2 + \mathrm{d}x^2 + \mathrm{d}y^2 + \mathrm{d}z^2\right], \quad z > 0. \tag{96}$$

This metric is an Einstein space (i.e., $R_{ab} = \Lambda g_{ab}$) if the function H solves the equation:

$$\left(\frac{\partial^2}{\partial x^2} + \frac{\partial^2}{\partial y^2} + \frac{\partial^2}{\partial z^2} - \frac{3}{z}\frac{\partial}{\partial z}\right) H(u,x,y,z) = 0. \tag{97}$$

These solutions describe gravitational waves propagating in a negatively curved Einstein space. Hence, these metrics generalise the AdS_5 spaces to solutions of the Einstein equation with a negative cosmological constant containing gravitational waves. These are exactly the type of models we want to investigate for the bulk.

To see how these can generate anisotropic stresses in the bulk, we can calculate the Weyl tensor of the above solutions. In general the Weyl tensor has the following non-vanishing components (in a coordinate basis):

$$\begin{aligned} C_{uiuj} &= -\frac{1}{2z^2}\frac{\partial^2}{\partial x^i \partial x^j}H, \\ C_{uiui} &= -\frac{1}{6z^2}\left(2\frac{\partial^2}{\partial x^{i2}} - \frac{\partial^2}{\partial x^{j2}} - \frac{\partial^2}{\partial x^{k2}}\right)H, \end{aligned} \tag{98}$$

where (i,j,k) is a permutation of (x,y,z). Since $\mathcal{E}_{ab} = C_{acbd}n^c n^d$, the Weyl tensor can induce anisotropic stresses on the brane.

Let us, for the sake of illustration, consider a simple exact 5-dimensional solution of a brane of this type. We consider a case in which a brane is embedded in the 5-dimensional solution eq.(96). The brane is located at $z = z_0$ with normal unit vector $\mathbf{n} = z\partial/\partial z$. Note that if

$$\left.\frac{\partial H}{\partial z}\right|_{z=z_0} = 0, \tag{99}$$

we obtain for the extrinsic curvature $K_{\mu\nu} \propto g_{\mu\nu}$, where $g_{\mu\nu}$ is the induced metric on the brane. Hence, we see from the junction conditions (see, e.g., [12]) that there are some vacuum solutions[3] by suitably tuning the brane tension λ in the usual way. The induced metric on the brane can be written

$$ds^2_{\text{brane}} = 2\mathrm{d}v\mathrm{d}u + \mathcal{H}(u,x,y)\mathrm{d}u^2 + \mathrm{d}x^2 + \mathrm{d}y^2, \tag{100}$$

and hence, is a plane wave. This form of the plane waves are called the Brinkmann form and they include the plane-waves studied earlier. Using eq.(97) and the junction conditions, the function $\mathcal{H}(u,x,y) \equiv H(u,x,y,z_0)$ obeys

$$\left(\frac{\partial^2}{\partial x^2} + \frac{\partial^2}{\partial y^2}\right)\mathcal{H}(u,x,y) = -\left.\frac{\partial^2}{\partial z^2}H(u,x,y,z)\right|_{z=z_0}. \tag{101}$$

These brane solutions are therefore not the usual vacuum solutions; they are in general sourced by the projected Weyl tensor of the bulk (see e.g. [4]). The gravitational waves in the bulk propagate along the brane and induce, via the projected Weyl tensor, a non-local anisotropic stress. This stress mimics the effect of an electromagnetic wave travelling on the brane at the speed of light.

We note that *any* plane wave brane can be embedded (locally) in a Siklos spacetime. Given $\mathcal{H}(u,x,y)$, a solution can be found by choosing

$$H(u,x,y,z) = \mathsf{D}(z,\nabla_2^2)\mathcal{H}(u,x,y), \quad \nabla_2^2 \equiv \frac{\partial^2}{\partial x^2} + \frac{\partial^2}{\partial y^2}, \tag{102}$$

where $\mathsf{D}(z,\nabla_2^2)$ is the operator

$$\mathsf{D}(z,\nabla_2^2) = \sum_{i=0}^{\infty} F_i(z)\left(\nabla_2^2\right)^i, \tag{103}$$

and $F_i(z)$ are defined iteratively:

$$F_0(z) = 1, \quad F_{i+1}(z) = -\int_{z_0}^{z} z^3\left(\int_{z_0}^{z}\frac{1}{z^3}F_i(z)\mathrm{d}z\right)\mathrm{d}z. \tag{104}$$

The brane is located at z_0 and H is chosen such that eq.(99) is fulfilled. In the special case where $\mathcal{H}(u,x,y)$ is a polynomial in x and y, the above sum will terminate and H will only contain a finite number of terms.

Homogeneous plane-wave branes: By choosing the function $\mathcal{H}(u,x,y)$ appropriately, we note that the above example can include branes of Bianchi type III, IV, V, VI$_h$ and VII$_h$. The following example includes these. Let us assume that the brane is a space-time homogeneous plane wave. All such can be written

$$\mathcal{H}(u,x,y) = A(u)(x^2+y^2) + B(u)(x^2-y^2) + 2C(u)xy,$$

where $A(u)$, $B(u)$ and $C(u)$ are some functions of u.

[3]Also note that a propagating electromagnetic wave on the brane can support a non-zero $\left.\frac{\partial H}{\partial z}\right|_{z=z_0}$.

We also have:

$$F_0(z) = 1, \quad F_1(z) = -\frac{1}{8z_0^2}(z^2 - z_0^2)^2,$$

and hence, using eq.(102)

$$H(u,x,y,z) = A(u)(x^2+y^2) + B(u)(x^2-y^2) + 2C(u)xy - \frac{A(u)}{2z_0^2}(z^2 - z_0^2)^2. \quad (105)$$

These metrics therefore represent homogeneous plane-wave branes in a Siklos bulk.

8.1.2. A General Type N Bulk

Let us now be more systematic and assume that the 5D bulk is algebraically special and of type N. This puts a constraint on the 5D Weyl tensor which makes it possible to deduce the form of the non-local stresses from a brane point of view. For a 5D type N spacetime there exists a frame ℓ_a, $\tilde{n}_a$, $m^i_{\ a}$, $i = 2,3,4$ such that [64]:

$$C_{abcd} = 4C_{1i1j}\ell_{\{a}m^i_{\ b}\ell_c m^j_{\ d\}}. \quad (106)$$

The frame ℓ_a, $\tilde{n}_a$, $m^i_{\ a}$ (the frame vector $\tilde{n}^a$ is not to be confused with the brane normal vector n^a) is defined via the only non-zero contractions:

$$\ell_a\tilde{n}^a = \ell^a\tilde{n}_a = 1, \quad m^i_{\ a}m^{ja} = \delta^{ij}.$$

The "electric part" of the Weyl tensor, $\mathcal{E}_{ab} = C_{acbd}n^c n^d$, where n^c is the normal vector on the brane, can for the type N bulk be written as

$$\mathcal{E}_{ab} = C_{1i1j}\left[\ell_a(m^i_{\ c}n^c) - m^i_{\ a}(\ell_c n^c)\right]\left[\ell_b(m^j_{\ c}n^c) - m^j_{\ b}(\ell_c n^c)\right]. \quad (107)$$

Furthermore, one can easily check that $\mathcal{E}_{ab}n^b = \mathcal{E}^a_{\ a} = 0$. Note that for a type N bulk we also have

$$\mathcal{E}_{ab}\ell^b = 0. \quad (108)$$

This can be rewritten using the projection operator on the brane, $\tilde{g}^a_{\ b} = g^a_{\ b} - n^a n_b$,

$$\mathcal{E}_{ac}\tilde{g}^c_{\ b}\ell^b = \mathcal{E}_{ab}\hat{\ell}^b = 0, \quad \hat{\ell}^b \equiv \tilde{g}^b_{\ c}\ell^c. \quad (109)$$

Hence, the vector $\hat{\ell}_b$ is the projection of the null vector ℓ_b onto the brane. By contracting this vector with itself, we get

$$\hat{\ell}^b\hat{\ell}_b = -(\ell^a n_a)^2. \quad (110)$$

The following analysis splits into two cases, according to whether $\ell^a n_a$ equals zero or not:

1. $\ell^a n_a = 0$: $\hat{\ell}_a = \ell_a$ and null.

2. $\ell^a n_a \neq 0$: $\hat{\ell}_a$ time-like.

These cases have to be treated separately and have different interpretations on the brane. From a 5D point of view they are of the same type, but since the 5D spacetime is anisotropic the orientation of the brane with respect to ℓ_a is of significance. More precisely, considering plane-wave spacetimes, the case $\ell^a n_a = 0$ corresponds to when the wave propagates parallel to the brane, and in the case $\ell^a n_a \neq 0$ the wave hits the brane.

The case $\ell^a n_a = 0$: Let us first investigate the consequence of $\hat{\ell}_\mu$ being a null vector. We note that $\mathcal{E}_{\mu\nu}$, the four-dimensional projected Weyl tensor on the brane, can be written in this case

$$\mathcal{E}_{\mu\nu} = -\left(\frac{6}{\lambda\kappa^2}\right)\epsilon\hat{\ell}_\mu\hat{\ell}_\nu, \tag{111}$$

where ϵ is some appropriate function. Hence, this is formally equivalent to the energy-momentum tensor of a *null* fluid. Equivalently we can consider it as the energy-momentum tensor of an extreme tilted perfect fluid. Using a covariant decomposition of $\mathcal{E}_{\mu\nu}$ with respect to a preferred time-like vector u_μ being orthogonal to some 3-surface with metric $h_{\mu\nu}$, we have

$$\mathcal{E}_{\mu\nu} = -\left(\frac{6}{\lambda\kappa^2}\right)\left[\mathcal{U}\left(u_\mu u_\nu + \frac{1}{3}h_{\mu\nu}\right) + \mathcal{P}_{\mu\nu} + 2\mathcal{Q}_{(\mu}u_{\nu)}\right]. \tag{112}$$

The non-local energy terms are thus given by [60]:

$$\begin{aligned} \mathcal{U} &= \epsilon(\hat{\ell}_\nu u^\nu)^2, \\ \mathcal{Q}_\mu &= \epsilon(\hat{\ell}_\nu u^\nu)\hat{\ell}_\mu, \\ \mathcal{P}_{\mu\nu} &= \epsilon\hat{\ell}_{\langle\mu}\hat{\ell}_{\nu\rangle}. \end{aligned} \tag{113}$$

Here, angled brackets $\langle\cdots\rangle$ denotes the projected, symmetric and tracefree part with respect to the metric $h_{\mu\nu}$ of the spatial 3-surfaces. The equations on the brane now close and the dynamical behaviour can be analysed (see [63]). Note that

$$\mathcal{U}\mathcal{P}_{\mu\nu} = \mathcal{Q}_\mu\mathcal{Q}_\nu - \frac{1}{3}g_{\mu\nu}\mathcal{Q}_\lambda\mathcal{Q}^\lambda, \tag{114}$$

so that in this case $\mathcal{E}_{\mu\nu}$ is determined completely by $\mathcal{U}$ and $\mathcal{Q}_\mu$.

The case $\ell^a n_a \neq 0$: In this case the vector ℓ^a has a component orthogonal to the brane. This implies that the vector $\hat{\ell}^a$ is time-like. This vector lives on the brane, and hence we can set u^μ parallel to $\hat{\ell}^\mu$. In this frame, the requirement $\hat{\ell}^\mu\mathcal{E}_{\mu\nu} = 0$ implies that we have $\mathcal{U} = 0 = \mathcal{Q}_\mu$ and hence, we can write

$$\mathcal{E}_{\mu\nu} = -\left(\frac{6}{\lambda\kappa^2}\right)\mathcal{P}_{\mu\nu}. \tag{115}$$

This is formally equivalent to a fluid which possesses anisotropic stresses with no energy density or energy-flux. From a brane point of view it appears as if these stresses are super-luminal; however, as can be seen from from a 5D point of view, this is just an artefact of living on a brane in a higher-dimensional spacetime. The stresses *do* have a gravitational origin, namely from gravitational waves in the 5D bulk.

8.2. Quadratic Theories of Gravity

In this section we will briefly discuss the dynamical evolution of anisotropic universes in quadratic theories of gravity. Past studies of these generalisations of general relativity (GR) have focussed on the isotropic Friedmann metrics where it is sufficient to consider only the effects of the R^2 contribution to the field equations [66]. However, the R^2 contribution has fairly predictable cosmological consequences because the resulting quadratic vacuum theory is conformally equivalent to GR with a scalar field with a single asymmetric minimum. This type of solution has been well studied in connection with inflation [67]. The addition of the $R_{\mu\nu}R^{\mu\nu}$ Ricci term to the Lagrangian in the anisotropic case creates a far richer diversity of cosmological behaviours that are far harder to summarise in terms of modifications of the general-relativistic situation. The effective stresses that are contributed to the field equations by the Ricci terms can mimic a wide range of fluids which violate the strong and weak energy conditions. This allows completely different behaviour to occur than is familiar from GR and the quadratic theory with R^2 contributions. In particular, the cosmic no-hair theorems (see section 5.1.) no longer hold: vacuum universes with positive cosmological constant do not necessarily approach de Sitter but can inflate anisotropically [68]. Moreover, the addition of quadratic Ricci terms to the Lagrangian can create cosmological models which have no counterpart in the GR limit.

To this end we consider the gravitational action

$$S_G = \frac{1}{2\kappa}\int_M \mathrm{d}^4x\sqrt{|g|}\left(R+\alpha R^2+\beta R_{\mu\nu}R^{\mu\nu}-2\Lambda\right). \tag{116}$$

Variation of this action leads to the following generalised Einstein equations (see, e.g., [69]):

$$G_{\mu\nu}+\Phi_{\mu\nu}+\Lambda g_{\mu\nu}=\kappa T_{\mu\nu}, \tag{117}$$

where $T_{\mu\nu}$ is the energy-momentum tensor of the ordinary matter sources, which we here will assume to be zero, for simplicity, and where

$$G_{\mu\nu} \equiv R_{\mu\nu}-\frac{1}{2}Rg_{\mu\nu}, \tag{118}$$

$$\begin{aligned}\Phi_{\mu\nu} \equiv\ & 2\alpha R\left(R_{\mu\nu}-\frac{1}{4}Rg_{\mu\nu}\right)+(2\alpha+\beta)\left(g_{\mu\nu}\Box-\nabla_\mu\nabla_\nu\right)R \\ & +\beta\Box\left(R_{\mu\nu}-\frac{1}{2}Rg_{\mu\nu}\right)+2\beta\left(R_{\mu\sigma\nu\rho}-\frac{1}{4}g_{\mu\nu}R_{\sigma\rho}\right)R^{\sigma\rho},\end{aligned} \tag{119}$$

with $\Box \equiv \nabla^\mu\nabla_\mu$. The tensor $\Phi_{\mu\nu}$ incorporates the deviation from regular Einstein gravity introduced by the quadratic terms in the action, and we see that $\alpha=\beta=0$ implies $\Phi_{\mu\nu}=0$, although the converse need not be true.

8.2.1. Exact Anisotropic Solutions

We now present two classes of exact vacuum anisotropic and spatially homogeneous universes of Bianchi types II and VI_h with $\Lambda>0$. These are exact solutions of the eqns. (117) with $(\alpha,\beta)\neq(0,0)$.

Bianchi type II solutions:

$$\mathrm{d}s^2_{\rm II} = -\mathrm{d}t^2 + e^{2bt}\left[\mathrm{d}x + \frac{a}{2}(z\mathrm{d}y - y\mathrm{d}z)\right]^2 + e^{bt}(\mathrm{d}y^2 + \mathrm{d}z^2), \tag{120}$$

where

$$a^2 = \frac{11 + 8\Lambda(11\alpha + 3\beta)}{30\beta}, \quad b^2 = \frac{8\Lambda(\alpha + 3\beta) + 1}{30\beta}. \tag{121}$$

These solutions are spacetime homogeneous with a 5-dimensional isotropy group. They have a one-parameter family of 4-dimensional Lie groups, as well as an isolated one (with Lie algebras $A^q_{4,11}$ and $A^1_{4,9}$, respectively, in Patera et al's scheme [70]) acting transitively on the spacetime. An interesting feature of this family of solutions is that there is a lower bound on the cosmological constant, given by $\Lambda_{\rm min} = -1/[8(\alpha+3\beta)] = -a^2/8$ for which the spacetime is static. For $\Lambda > \Lambda_{\rm min}$ the spacetime is inflating and shearing. The inflation does not result in approach to isotropy or to asymptotic evolution close to the de Sitter metric. Interestingly, even in the case of a vanishing Λ the universe inflates exponentially but anisotropically. We also note from the solutions that the essential term in the action causing this solution to exist is the $\beta R_{\mu\nu}R^{\mu\nu}$-term and the distinctive behaviour occurs when $\alpha = 0$. The solutions have no well defined $\beta \to 0$ limit, and do not have a general relativistic counterpart. They are non-perturbative. Similar solutions exist also in higher dimensions. Their existence seem to be related to so-called Ricci nilsolitons [71, 72].

Bianchi type VI$_h$ solutions:

$$\mathrm{d}s^2_{{\rm VI}_h} = -\mathrm{d}t^2 + \mathrm{d}x^2 + e^{2(rt+ax)}\left[e^{-2(st+a\tilde{h}x)}\mathrm{d}y^2 + e^{+2(st+a\tilde{h}x)}\mathrm{d}z^2\right], \tag{122}$$

where

$$\begin{aligned} r^2 &= \frac{8\beta s^2 + (3+\tilde{h}^2)(1+8\Lambda\alpha) + 8\Lambda\beta(1+\tilde{h}^2)}{8\beta\tilde{h}^2}, \\ a^2 &= \frac{8\beta s^2 + 8\Lambda(3\alpha+\beta) + 3}{8\beta\tilde{h}^2}. \end{aligned} \tag{123}$$

and r, s, a, and $\tilde{h}$ are all constants. These are also homogeneous universes with a 4-dimensional group acting transitively on the spacetime. Both the mean Hubble expansion rate and the shear are constant. Again, we see that the solution inflates anisotropically and is supported by the existence of $\beta \neq 0$. It exists when $\alpha = 0$ and $\Lambda = 0$ but not in the limit $\beta \to 0$.

8.2.2. Avoidance of the No-hair Theorem

The no-hair theorem in section 5.1. states that for Bianchi types I-VIII the presence of a positive cosmological constant drives the late-time evolution towards the de Sitter spacetime. It requires the matter sources (other than Λ) to obey the strong-energy condition. It has been shown that if this condition is relaxed then the cosmic no-hair theorem cannot be proved and counter-examples exist [73]. In [74], the cosmic no-hair conjecture was discussed for Bianchi cosmologies with an axion field with a Lorentz Chern-Simons term. Interestingly, exact Bianchi type II solutions, similar to the ones found here, were found which avoided the cosmic no-hair theorem. However, unlike for our solutions, these violations were driven by an axion field whose energy-momentum tensor violated the strong and dominant energy condition. The no-hair theorem for spatially homogeneous solutions of Einstein gravity also requires the spatial 3-curvature to be non-positive. This condition ensures that universes do not recollapse before the Λ term dominates the dynamics but it also excludes examples like that of the Kantowski-Sachs $S^2 \times S^1$ universe, eq.(3), which has an exact solution with $\Lambda > 0$ which inflates in some directions but is static in others. These solutions, found by Weber [75], were used by Linde and Zelnikov [76] to model a higher-dimensional universe in which different numbers of dimensions inflate in different patches of the universe. However, it was subsequently shown that this behaviour, like the Weber solution, is unstable [77, 78]. We note that the above solutions to gravity theories with $\beta \neq 0$ possess anisotropic inflationary behaviour without requiring that the spatial curvature is positive and are distinct from the Kantowski-Sachs phenomenon.

The Bianchi type solutions given above inflate in the presence of a positive cosmological constant Λ. However, they are neither de Sitter, nor asymptotically de Sitter; nor do they have initial singularities. Let us examine how these models evade the conclusions of the cosmic no-hair theorem. Specifically, consider the type II solution, eq.(120). We introduce an orthonormal frame as before and consider the expansion scalar, θ and the shear, σ_{ab}, in the standard way. For the type II metric, we find (in the orthonormal frame)

$$\theta = 2b, \quad \sigma_{ab} = \frac{1}{6}\text{diag}(2b, -b, -b).$$

The expansion-normalised shear is now:

$$\Sigma_{ab} = \text{diag}\left(\frac{1}{2}, -\frac{1}{4}, -\frac{1}{4}\right).$$

Interestingly, the expansion-normalised shear components are constants (and independent of the parameters α, β, and Λ) and this shows that these solutions violate the cosmological no-hair theorem (which requires $\sigma_{\mu\nu}/\theta \to 0$ as $t \to \infty$). To understand how this solution avoids the no-hair theorem of section 5.1., rewrite eq.(117) as follows:

$$G_{\mu\nu} = \widetilde{T}_{\mu\nu}, \quad \widetilde{T}_{\mu\nu} \equiv -\Lambda g_{\mu\nu} - \Phi_{\mu\nu} + \kappa T_{\mu\nu}.$$

In this form the higher-order curvature terms can be interpreted as matter terms contributing a fictitious energy-momentum tensor $\widetilde{T}_{\mu\nu}$. For the Bianchi II solution we find

$$\begin{aligned} \widetilde{T}_{\mu\nu} &= \tfrac{1}{4}\text{diag}\left(5b^2 - a^2, -3b^2 + 3a^2, -7b^2 - a^2, -7b^2 - a^2\right) \\ &= \text{diag}(\widetilde{\rho}, \widetilde{p}_1, \widetilde{p}_2, \widetilde{p}_3). \end{aligned} \tag{124}$$

where $\widetilde{\rho}$ and $\widetilde{p}_i$ are the energy density and the principal pressures, respectively. The no hair theorems require the dominant energy condition (DEC) and the strong energy condition (SEC) to hold. However, since $\tilde{\rho} + \tilde{p}_1 + \tilde{p}_2 + \tilde{p}_3 = -3b^2 < 0$ the SEC is always violated when $b \neq 0$. The DEC is violated when $\tilde{\rho} < 0$ and the weak energy condition (WEC) is also violated because $\tilde{\rho} + \tilde{p}_2 = \tilde{\rho} + \tilde{p}_3 = -(a^2 + b^2)/2 < 0$.These violations also ensure that the singularity theorems will not hold for these universes and they have no initial or final singularities.

Are these solutions stable? In [79] the dynamical systems approach was used to study stability of certain solutions in these kind of theories. Compared to regular Einstein gravity, the complexity of the phase space increases dramatically due to the higher-derivative terms. However, in spite of this complication the type II solution above was shown to possess, at least, one unstable eigenvalue. Therefore, these solutions do not serve as general attractors in the full quadratic theory. Nonetheless, these solutions are good examples of surprising behaviours that may arise in such higher-curvature theories.

9. Conclusion and Outlook

Here we have given a review of some of the recent developments within the study of spatially homogeneous, but anisotropic, universes. We have focussed on the late-time behaviour and exact solutions. It was the aim of this review to give some flavour of the rich behaviour of the various models and to show some of the techniques which have enabled us to unveil some of these interesting and fascinating behaviours.

One of the models not considered here is the Bianchi type IX model. This model shows a different behaviour as it may recollapse [39–41]. This causes the expansion-normalised variables to be ill-defined near the turning point of the expansion; hence, a different set of variables are needed to explore this model in detail.

One thing we have not considered in detail is the early time behaviour which is believed to exhibit chaos [80–85]. Although it is believed that the initial singularity consists of oscillatory behaviour, the details of these oscillations are not completely known. A next step of this program would therefore be to try to understand this initial behaviour better.

Another next step might be to consider the evolution of inhomogeneous universes. For example, we could consider the so-called G_2 [86–88], or even G_0 [83,84] models which are inhomogeneous models possessing 2 abelian, or none Killing vectors, respectively. Only some stability results for some special solutions are known for these models, among them, the plane-wave solutions of Bianchi type VII$_h$ [89,90]. Recently, these models have also shown some interesting behaviour near the initial singularity, like the presence of spikes [88,91].

Acknowledgements

I would like to thank John D. Barrow, Alan A. Coley, Robert J. van den Hoogen and Woei Chet Lim whose various collaborations this review is partly based on.

References

[1] G.F.R. Ellis and M.A.H. MacCallum, *Comm. Math. Phys.* 12 (1969) 108

[2] A.H. Taub, *Ann. Math.* 53 (1951) 472

[3] C.G. Hewitt and J. Wainwright in *Dynamical Systems in Cosmology*, eds: J. Wainwright and G.F.R. Ellis, Cambridge University Press (1997)

[4] H. Stephani, D. Kramer, M. MacCallum, C. Hoenselaers and E. Herlt, *Exact Solutions of Einstein's Field Equations, 2nd Ed.* Cambridge University Press (2003)

[5] R. Kantowski and R.K. Sachs, *J. Math. Phys.* 7 (1966) 443

[6] J.D. Barrow, *MNRAS* 179 (1977) 47P

[7] J.D. Barrow and S. Hervik, *Class. Quantum Grav.* 20 (2003) 2841

[8] J.D. Barrow and D.H. Sonoda, *Phys. Reports* 139 (1986) 1

[9] C. G. Hewitt, J. T. Horwood, J. Wainwright, *Class.Quant.Grav.* 20 (2003) 1743-1756

[10] S. Hervik, *Class. Quantum Grav.* 21 (2004) 2301

[11] S. Hervik and W.C. Lim, *Class. Quantum Grav.* 23 (2006) 3017

[12] Ø. Grøn and S. Hervik, *Einstein's general theory of relativity: with modern applications in Cosmology*, Springer, NY, 2007.

[13] J. Milnor, *Adv. Math.* 21 (1976) 293

[14] A.R. King and G.F.R. Ellis, *Commun. Math. Phys.* 31 (1973) 209

[15] A.A. Coley, *Dynamical Systems and Cosmology*, Kluwer, Academic Publishers (2003)

[16] G.F.R. Ellis and A.R. King, *Comm. Math. Phys.* 38 (1974) 119.

[17] C.B. Collins and G.F.R. Ellis, *Phys. Rep.* 56 (1979) 65

[18] A.A. Coley, S. Hervik and W.C. Lim, *Phys. Lett. B* 638 (2006) 310-313; A.A. Coley, S. Hervik and W.C. Lim, *Class. Quant. Grav.* 23 (2006) 3573-3591; A.A. Coley, S. Hervik and W.C. Lim, *Int. J. Mod. Phys. D* 15 (2006); W.C. Lim, A.A. Coley and S. Hervik, , *Class. Quant. Grav.* 24 (2007) 595-604

[19] O.I. Bogoyavlenskii *Methods in the Qualitative Theory of Dynamical Systems in Astrophysics and Gas Dynamics*, Springer-Verlag (1985).

[20] D.K. Arrowsmith and C.M. Place, *An introduction to Dynamical Systems*, Cambridge University Press (1990).

[21] A.A. Coley and S. Hervik, *Class. Quantum Grav.* 22 (2005) 579

[22] J. Wainwright, M.J. Hancock and C. Uggla, *Class. Quant. Grav.* 16 (1999) 2577; U.S. Nilsson, M.J. Hancock, J. Wainwright, *Class. Quant. Grav.* 17 (2000) 3119

[23] J.T. Horwood, M.J. Hancock, D. The, J. Wainwright, *Class. Quant. Grav.* 20 (2003) 1757

[24] C. G. Hewitt and J. Wainwright, *Class. Quant. Grav.* 10, 99 (1993).

[25] J. Wainwright and L. Hsu, *Class. Quant. Grav.* 6, 1409 (1989).

[26] J.D. Barrow and Y. Gaspar, *Class. Quant. Grav.* 18 (2001) 1809

[27] H. Ringström, *Class. Quant. Grav.* 17 (2000) 713; H. Ringström, *Class. Quant. Grav.* 18 (2001) 3791; H. Ringström, *Class. Quant. Grav.* 20 (2003) 1943

[28] C.G. Hewitt, R. Bridson, J. Wainwright, *Gen. Rel. Grav.* 33 (2001) 65

[29] S. Hervik, R.J. van den Hoogen and A.A. Coley, *Class. Quantum Grav.* 22 (2005) 607

[30] I.S. Shikin, *Sov. Phys. JETP* 41 (1976) 794

[31] C.B. Collins, *Comm. Math. Phys.* 39 (1974) 131

[32] C.G. Hewitt and J. Wainwright, *Phys. Rev.* D46 (1992) 4242

[33] D. Harnett, *Tilted Bianchi type V cosmologies with vorticity*, Master's thesis, University of Waterloo, Canada (1996)

[34] A.A. Coley and S. Hervik, *Class. Quantum Grav.* 21 (2004) 4193

[35] S. Hervik, R.J. van den Hoogen, W.C. Lim and A.A. Coley, *Class. Quantum Grav.* 23 (2006) 845

[36] S. Hervik, R.J. van den Hoogen, W.C. Lim and A.A. Coley, *Class. Quantum Grav.* 24 (2007) 3859

[37] S. Hervik, R.J. van den Hoogen, W.C. Lim and A.A. Coley, *Class. Quantum Grav.*, 25 (2008) 015002

[38] R. Wald, *Phys. Rev.* D28 (1983) 2113.

[39] J.D. Barrow, G.J. Galloway and F.J. Tipler *MNRAS* 223 (1986) 835

[40] J.D. Barrow and F.J. Tipler *MNRAS* 216 (1985) 395

[41] X. Lin and R. Wald, *Phys. Rev.* D40 (1989) 3280; X. Lin and R. Wald, *Phys. Rev.* D41 (1990) 2444

[42] W. C. Lim, H. van Elst, C. Uggla and J. Wainwright, *Phys. Rev.* D 69, 103507 (2004)

[43] J. Carr, *Applications of Center manifold theory*, Springer Verlag (1981)

[44] B.J. Carr and A.A. Coley, *Class. Quant. Grav.* 16 (1999) R31

[45] B.J. Carr and A.A. Coley, *Gen. Rel. Grav.* 37 (2005) 2165

[46] C.B. Collins and S.W. Hawking, *Astrophys. J.* 180 (1973) 317

[47] J.D. Barrow, *Quart. Jl. Roy. astr. Soc.* 23 (1982) 344

[48] J.D. Barrow and F.J. Tipler, *The Anthropic Cosmological Principle*, Oxford University Press (1986)

[49] J. Wainwright and P. J. Anderson, *Gen. Rel. Grav.* 16, 609 (1984).

[50] S. W. Goode and J. Wainwright, *Class. Quant. Grav.* 2, 99 (1985).

[51] S. W. Goode, A. A. Coley and J. Wainwright, *Class. Quant. Grav.* 9, 445 (1992).

[52] R. Penrose, in *Proc. First Marcel Grossmann Meet. Gen. Rel. (ICTP Trieste)*, ed. R. Ruffini, North-Holland (1977); R. Penrose, in *General Relativity, an Einstein centenary survey*, eds. S.W. Hawking and W. Israel, Cambridge Univ. Press (1979); R. Penrose, *J. Stat. Phys,*, 77 (1994) 217

[53] Ø. Grøn and S. Hervik, *Int. J. Theo. Phys. Group Th. Non-L. Opt.* 10 (2003) 29, arXiv:gr-qc/0205026.

[54] J.D. Barrow and S. Hervik, *Class. Quantum Grav.* 19 (2002) 5173

[55] J.D. Barrow and F.J. Tipler, *Nature* 276 (1978) 453

[56] L. Randall and R. Sundrum, *Phys. Rev. Lett.* 83, 3370 (1999); ibid 83, 4690 (1999).

[57] P. Horava and E. Witten, *Nucl. Phys.* B460, 506 (1996).

[58] A. A. Coley, *Class. Quant. Grav.* 19, L45 (2002); A. A. Coley, Y. He and W. C. Lim, *Class. Quant. Grav.* 21, 1311 (2004)

[59] T. Shiromizu, K. Maeda and M. Sasaki, *Phys.Rev.* D62 029012 (2000)

[60] R. Maartens, *Living Rel. Rev.* 7 (2004) 7

[61] D. Langlois, arXiv:astro-ph/0403579 (2004)

[62] E. Kiritsis, G. Kofinas, N. Tetradis, T.N. Tomaras and V. Zarikas, *JHEP* 0302, 035 (2003)

[63] A. Coley and S. Hervik *Class. Quantum Grav.* 21 (2004) 5759

[64] A. Coley, R. Milson, V. Pravda, A. Pravdova, *Class. Quant. Grav.* 21 (2004) L35; A. Coley *et al.*, *Phys. Rev.* D 67, 104020 (2002)

[65] S.T.C. Siklos, Lobatchevski plane gravitational waves, in *Galaxies, axisymmetric systems and relativity* ed. M.A.H. MacCallum, Cambridge University Press, 1985.

[66] J.D. Barrow and A.C. Ottewill, *J. Phys. A* 16, 2757 (1983).

[67] A. Guth, *Phys. Rev. D* 23, 347 (1981); A.D. Linde, *Phys.Lett.* B129, 177 (1983); V. Müller, H.-J. Schmidt and A.A. Starobinskii, *Phys. Lett.* B202, 198 (1988); A. Berkin and K.I. Maeda, *Phys. Rev. D* 44, 1691 (1991); S. Gottlöber, V. Müller and A.A. Starobinskii, *Phys. Rev. D* 43, 2510 (1991); A.A. Starobinskii and H.-J. Schmidt, *Class. Quantum Grav.* 4, 695 (1987); H.-J. Schmidt, *Class. Quantum Grav.* 5, 233 (1988).

[68] J.D. Barrow and S. Hervik, *Phys. Rev.* D 73, 023007 (2006)

[69] S. Deser and B. Tekin, *Phys. Rev.* D67, 084009 (2003).

[70] J. Patera, R.T. Sharp, P. Winternitz and H. Zassenhaus, *J. Math. Phys.* 17, 986 (1976).

[71] J. Lauret, *Math. Ann.* 319, 715 (2001); J. Lauret, *Quart. J. Math.* 52, 463 (2001); J. Lauret, *Math. Z.* 241, 83 (2002); J. Lauret, *Diff. Geom. Appl.* 18, 177 (2003).

[72] S. Hervik, talk held at the CMS Summer 2004 meeting, Halifax, NS, Canada.

[73] J.D. Barrow, *Phys. Lett.* B187, 12 (1987); J.D. Barrow, *Phys. Lett.* B 180, 335 (1987); J.D. Barrow, *Nucl. Phys.* B 310, 743 (1988); J.D. Barrow, *Phys. Lett.* B 183, 285 (1987).

[74] N. Kaloper, *Phys. Rev. D* 44, 2380 (1991).

[75] E. Weber, *J. Math. Phys.* 25, 3279 (1984).

[76] A.D. Linde and M.I. Zelnikov, *Phys. Lett.* B 215, 59 (1988).

[77] J.D. Barrow and J. Yearsley, *Class. Quantum Grav.* 13, 2693 (1996).

[78] H.H. Soleng, *Class. Quantum Grav.* 6, 1387 (1989).

[79] J.D. Barrow and S. Hervik, *Phys. Rev.* D74 (2006) 124017

[80] J.D. Barrow, *Phys. Rev. Lett.* 46 (1981) 963; D.F. Chernoff and J.D. Barrow, *Phys. Rev. Lett.* 50 (1983) 134;

[81] R.T. Jantzen, pages 61-147 in *Proc. Int. Sch. Phys "E. Fermi" Course LXXXVI*, eds. Ruffini and Melchiorri, North Holland, 1987, arXiv:gr-qc/0102035

[82] D. Hobill, A. Burd. and A. Coley, eds, *Deterministic Chaos in General Relativity*, New York:Plenum Press (1994).

[83] C. Uggla, H. van Elst, J. Wainwright and G. F. R. Ellis, *Phys. Rev.* D 68, 103502 (2003)

[84] L. Andersson, H. van Elst, W. C. Lim and C. Uggla, *Phys. Rev. Lett.* 94, 051101 (2005)

[85] D. Garfinkle, *Phys. Rev. Lett.* 93 (2004) 161101; D. Garfinkle, *Int. J. Mod. Phys.* D 13, 2261 (2004)

[86] C. G. Hewitt and J. Wainwright, *Class. Quant. Grav.* 7, 2295 (1990)

[87] H. van Elst, C. Uggla and J. Wainwright, *Class. Quant. Grav.* 19, 51 (2002)

[88] W. C. Lim, *The dynamics of inhomogeneous cosmologies*, PhD thesis, University of Waterloo, 2004, arXiv:gr-qc/0410126.

[89] S. Hervik and A.A. Coley, *Class. Quantum Grav.* 22 (2005) 3391

[90] J. Barrow and C. Tsagas, *Class.Quant.Grav.* 22 (2005) 825-840

[91] W. C. Lim, arXiv:0710.0628 [gr-qc].

In: Classical and Quantum Gravity Research
Editors: M.N. Christiansen et al, pp. 465-509
ISBN 978-1-60456-366-5

Chapter 11

COFRAME GEOMETRY AND GRAVITY

Yakov Itin*
Institute of Mathematics, Hebrew University of Jerusalem
and Jerusalem College of Technology,
Jerusalem 91904, Israel

Abstract

The possible extensions of GR for description of fermions on a curved space, for supergravity and for loop quantum gravity require a richer set of 16 independent variables. These variables can be assembled in a coframe field, i.e., a local set of four linearly independent 1-forms. In this chapter we study the gravity field models based on a coframe variable alone. We give a short review of the coframe gravity. This model has the viable Schwarzschild solutions even being an alternative to the standard GR. Moreover, the coframe model treating of the gravity energy may be preferable to the ordinary GR where the gravity energy cannot be defined at all. A principal problem that the coframe gravity does not have any connection to a specific geometry even being constructed from the geometrical meaningful objects. A geometrization of the coframe gravity is an aim of this chapter. We construct a complete class of the coframe connections which are linear in the first order derivatives of the coframe field on an n dimensional manifolds with and without a metric. The subclasses of the torsion-free, metric-compatible and flat connections are derived. We also study the behavior of the geometrical structures under local transformations of the coframe. The remarkable fact is an existence of a subclass of connections which are invariant when the infinitesimal transformations satisfy the Maxwell-like system of equations. In the framework of the coframe geometry construction, we propose a geometrical action for the coframe gravity. It is similar to the Einstein-Hilbert action of GR, but the scalar curvature is constructed from the general coframe connection. We show that this geometric Lagrangian is equivalent to the coframe Lagrangian up to a total derivative term. Moreover there is a family of coframe connections in which Lagrangian does not include the higher order terms at all. In this case, the equivalence is complete.

*E-mail address: itin@math.huji.ac.il

1. Introduction. Why Do We Have to Go beyond Riemannian Geometry?

General relativity (GR) is, probably, the best of the known theories of gravity. From mathematical and aesthetic points of view, it can be used as a standard of what a physical theory has to be. Up to this day, the Einstein theory is in a very good agreement with the observation data. Probably the main idea of Einstein's GR is that the physical properties of the gravitational field are in one-to-one correspondence with the geometry of the base manifold. The standard GR is based on a Riemannian geometry with a unique metric tensor and a unique Levi-Civita connection constructed from this tensor. Hence, the gravity field equations of GR predicts a unique (up to diffeomorphism transformations) metric tensor and consequently a unique geometry. Therefore any physical field, except for gravity, can not have an intrinsic geometrical sense in the Riemannian geometry.

After the classical works of Weyl, Cartan and others, we know that the Riemannian construction is not a unique possible geometry. A most general geometric framework involves independent metric and independent connection. A gravity field model based on this general geometry (Metric-affine gravity) was studied intensively, see [1]— [17] and the references given therein. Probably a main problem of this construction is a huge number of geometrical fields which do not find their physical partner.

In this chapter we study a much more economical construction based on a unique geometrical object — coframe field. Absolute (teleparallel) frame/coframe variables (repèr, vierbein, ...) were introduced in physics by Einstein in 1928 with an aim of a unification of gravitational and electromagnetic fields (for classical references, see [18]). The physical models for gravity based on the coframe variable are well studied, see [19]— [42]. In some aspects such models are even preferable from the standard GR. In particular, they involve a meaningful definition of the gravitational energy, which is in a proper correspondence with the Noether procedure. Moreover some problems inside and beyond Einstein's gravity require a richer set of 16 independent variables of the coframe. In the following issues of gravity, the coframe is not only a useful tool but often it cannot even be replaced by the standard metric variable: (i) Hamiltonian formulation [43], [44]; (ii) positive energy proofs [45]; (iii) fermions on a curved manifold [46], [47]; (iv) supergravity [48]; (v) loop quantum gravity [49].

Unfortunately, in the coframe gravity models, the proper connection between physics and the underlying geometry is lost. In this chapter, we propose a way of geometrization of the coframe gravity. In particular, we study which geometric structure can be constructed from the vierbein (frame/coframe) variables and which gravity field models can be related to this geometry.

The organization of the chapter is as follows:

In the first section, we give a brief account of the gravity field model based on the coframe field instead of the pure metrical construction of GR. We discuss the following features: (i) The coframe gravity is described by a 3-parametric set of models; (ii) All the coframe models are derivable from a Yang-Mills-type Lagrangian; (iii) The coframe field equations are well defined for all values of the parameters. Only for the pure GR case, the system id degenerated to 10 equations for 16 variables; (iv) The energy-momentum tensor of the coframe field is well defined for all models except GR. In the latter case the tensor nature of the energy-momentum expression is lost; (v) There is a subset of viable fields with a unique spherical symmetric solution, which corresponds to Schwarzschild metric; (vi) The same subset is derived by the requirement of the free field limit approximation. All these positive properties make the coframe gravity a relevant subject of investigation.

In section 2, we construct a geometrical structure based on a coframe variable as unique building block. In an addition to the coframe volume element and metric, we present a most general coframe connection. The Levi-Civita and flat connections are special cases of it. The torsion and nonmetricity tensors of the general coframe connection are calculated. We identify the subclasses of symmetric (torsion-free) connections and of metric-compatible connections. The unique symmetric metric-compatible connection is of Levi-Civita. We study the transformations of the coframe field and identify a subclass of connections which are invariant under restricted coframe transformations. Quite remarkable that restriction conditions are approximated by a Maxwell-type system.

In section 3, we are looking for a geometric representation of the gravity coframe model. The main result is that the free-parametric gravity coframe Lagrangian can be replaced by a standard Einstein-Hilbert Lagrangian, when the curvature scalar is calculated on a general coframe connection. The standard GR Lagrangian contains a second order derivative term which appears in the form of the total derivative. This term does not influence the field equation, but it cannot be consistently removed. We show that there is a set of coframe connections in which the Einstein-Hilbert Lagrangian does not involve the second order derivative term at all.

In the last section, some proposals of possible developments of a geometrical coframe construction and its applications to gravity are presented.

2. Coframe Gravity

Let us give a brief account of gravity field models based on a coframe field. We refer to such models as *coframe gravity*. This is instead of the Einsteinian *metric gravity* based on a metric tensor field. We will use here mostly the notations accepted in [33].

2.1. Coframe Lagrangian

Consider a smooth, non-degenerated coframe field $\{\vartheta^\alpha,\ \alpha = 0,1,2,3\}$ defined on a $4D$ smooth differential manifold M. The 1-forms ϑ^α are declared to be pseudo-orthonormal. Thus a metric on M is defined by

$$g = \eta_{\alpha\beta}\vartheta^\alpha \otimes \vartheta^\beta\,, \qquad \eta_{\alpha\beta} = (-1,1,1,1)\,. \tag{1}$$

So, the coframe field ϑ^α is considered as a basic dynamical variable while the metric g is treated as a secondary structure.

The coframe field is defined only up to *global pseudo-rotations*, i.e. $SO(1,3)$ transformations. Consequently, the truly dynamical variable is an equivalence class of coframes $[\vartheta^\alpha]$, while the global pseudo-rotations produce an equivalence relation on this class. Hence, in addition to the invariance under the diffeomorphic transformations of the manifold M, the basic geometric structure has to be global (rigid) $SO(1,3)$ invariant.

Gravity is described by differential invariants of the coframe structure. There is an important distinction between the diffeomorphic invariants of the metric and of the coframe structures. Since the metric invariants of the first order are trivial, the metric structure admits diffeomorphic invariants only of the second order or greater. A unique invariant of the second order is the scalar curvature. This expression is well known to play the key role of an integrand in the Einstein-Hilbert action. The coframe structure admits diffeomorphic and rigid $SO(1,3)$ invariants even of the first order. A simple example is the expression $e_\alpha \rfloor d\vartheta^\alpha$, see Appendix for notations and basic definitions. The operators, which are diffeomorphic invariants and global covariants, can contribute to a general coframe field equation. A rich class of such equations is constructed in [27]. A requirement of derivability of the field equations from a Lagrangian strictly restricts the variety of possible options.

We restrict the consideration to odd, quadratic (in the first order derivatives of the coframe field ϑ^α), diffeomorphic, and global $SO(1,3)$ invariant Lagrangians. A general Lagrangian of such a type is represented by a linear combination of three 4-forms which are referred to as the Weitzenböck invariants. Consider the exterior differentials of the basis 1-forms $d\vartheta^\alpha$ and introduce the coefficients of their expansion in the basis of even 2-forms $\vartheta^{\alpha\beta}$

$$d\vartheta^\alpha = \vartheta^\alpha_{i,j} dx^i \wedge dx^j = \frac{1}{2} C^\alpha{}_{\beta\gamma} \vartheta^{\beta\gamma} . \tag{2}$$

We use here the abbreviation $\vartheta^{\alpha\beta\cdots} = \vartheta^\alpha \wedge \vartheta^\beta \wedge \cdots$. By definition, the coefficients $C^\alpha{}_{\beta\gamma}$ are antisymmetric, $C^\alpha{}_{\beta\gamma} = -C^\alpha{}_{\gamma\beta}$. Their explicit expression can be given by the differential form notations (see Appendix)

$$C^\alpha{}_{\beta\gamma} = e_\gamma \rfloor (e_\beta \rfloor d\vartheta^\alpha) . \tag{3}$$

The symmetric form of a general second order coframe Lagrangian is given by [25]

$$^{(\mathtt{cof})}L = \frac{1}{2\ell^2} \sum_{i=1}^{3} \rho_i \, {}^{(i)}L , \tag{4}$$

where ℓ denotes the Planck length constant, while ρ_i are dimensionless parameters. The partial Lagrangian expressions are

$$\begin{aligned}
{}^{(1)}L &= d\vartheta^\alpha \wedge * d\vartheta_\alpha = \frac{1}{2} C_{\alpha\beta\gamma} C^{\alpha\beta\gamma} * 1 , && (5)\\
{}^{(2)}L &= (d\vartheta_\alpha \wedge \vartheta^\alpha) \wedge * \left(d\vartheta_\beta \wedge \vartheta^\beta\right) = \frac{1}{2} C_{\alpha\beta\gamma} \left(C^{\alpha\beta\gamma} + C^{\beta\gamma\alpha} + C^{\gamma\alpha\beta}\right) * 1 , && (6)\\
{}^{(3)}L &= (d\vartheta_\alpha \wedge \vartheta^\beta) \wedge * (d\vartheta_\beta \wedge \vartheta^\alpha) = \frac{1}{2} \left(C_{\alpha\beta\gamma} C^{\alpha\beta\gamma} - 2C^\alpha{}_{\alpha\gamma} C_\beta{}^{\beta\gamma}\right) * 1 . && (7)
\end{aligned}$$

The 1-forms ϑ^α are assumed to carry the dimension of length, while the coefficients ρ_i are dimensionless. Hence the total Lagrangian $^{(\mathtt{cof})}L$ is dimensionless. In order to simplify the formulas below we will use the Lagrangian $L = \ell^{2} {}^{(\mathtt{cof})}L$ which dimension is length square. In other worlds the geometrized units system with $G = c = \hbar = 1$ is applied.

Every term of the Lagrangian (4) is independent of a specific choice of a coordinate system and invariant under a global (rigid) $SO(1,3)$ transformation of the coframe field. Thus, different choices of the free parameters ρ_i yield different rigid $SO(1,3)$ and diffeomorphic invariant classical field models. Some of them are known to be applicable for description of gravity.

Let us rewrite the coframe Lagrangian in a compact form

$$^{(\texttt{cof})}L = \frac{1}{4}C_{\alpha\beta\gamma}C_{\alpha'\beta'\gamma'}\lambda^{\alpha\beta\gamma\alpha'\beta'\gamma'} * 1\,, \tag{8}$$

where the constant symbols

$$\begin{aligned}\lambda^{\alpha\beta\gamma\alpha'\beta'\gamma'} &= (\rho_1+\rho_2+\rho_3)\eta^{\alpha\alpha'}\eta^{\beta\beta'}\eta^{\gamma\gamma'} + \rho_2(\eta^{\alpha\beta'}\eta^{\beta\gamma'}\eta^{\gamma\alpha'} + \eta^{\alpha\gamma'}\eta^{\beta\alpha'}\eta^{\gamma\beta'}) \\ &\quad -2\rho_3\eta^{\alpha\gamma}\eta^{\alpha'\gamma'}\eta^{\beta\beta'}\end{aligned} \tag{9}$$

are introduced. It can be checked, by straightforward calculation, that these λ-symbols are invariant under a transposition of the triplets of indices:

$$\lambda^{\alpha\beta\gamma\alpha'\beta'\gamma'} = \lambda^{\alpha'\beta'\gamma'\alpha\beta\gamma}\,. \tag{10}$$

We also introduce an abbreviated notation

$$F^{\alpha\beta\gamma} = \lambda^{\alpha\beta\gamma\alpha'\beta'\gamma'}C_{\alpha'\beta'\gamma'}\,. \tag{11}$$

The total Lagrangian (4) reads now as

$$^{(\texttt{cof})}L = \frac{1}{4}C_{\alpha\beta\gamma}F^{\alpha\beta\gamma} * 1\,. \tag{12}$$

This form of the Lagrangian will be used in sequel for the variation procedure. The Lagrangian (12) can also be rewritten in component free notations. Define one-indexed 2-forms: a *field strength form*

$$\mathcal{C}^\alpha := \frac{1}{2}C^{\alpha\beta\gamma}\vartheta_{\beta\gamma} = d\vartheta^\alpha\,, \tag{13}$$

and a *conjugate field strength form* $\mathcal{F}^\alpha := \frac{1}{2}F^{\alpha\beta\gamma}\vartheta_{\beta\gamma}$

$$\mathcal{F}^\alpha = (\rho_1+\rho_3)\mathcal{C}^\alpha + \rho_2 e^\alpha \rfloor(\vartheta^\mu\wedge\mathcal{C}_\mu) - \rho_3\vartheta^\alpha\wedge(e_\mu\rfloor\mathcal{C}^\mu)\,. \tag{14}$$

Another form of $\mathcal{F}^\alpha$ can be given via the irreducible (under the Lorentz group) decomposition of the 2-form $\mathcal{C}^\alpha$ (see [5], [4]). Write

$$\mathcal{C}^\alpha = {}^{(1)}\mathcal{C}^\alpha + {}^{(2)}\mathcal{C}^\alpha + {}^{(3)}\mathcal{C}^\alpha, \tag{15}$$

where

$$^{(2)}\mathcal{C}^\alpha = \frac{1}{3}\vartheta^\alpha\wedge(e_\mu\rfloor\mathcal{C}^\mu)\,, \qquad {}^{(3)}\mathcal{C}^\alpha = \frac{1}{3}e^\alpha\rfloor(\vartheta_\mu\wedge\mathcal{C}^\mu)\,, \tag{16}$$

while $^{(1)}\mathcal{C}^\alpha$ is the remaining part. Substitute (16) into (14) to obtain

$$\mathcal{F}^\alpha = (\rho_1+\rho_3)^{(1)}\mathcal{C}^\alpha + (\rho_1-2\rho_3)^{(2)}\mathcal{C}^\alpha + (\rho_1+3\rho_2+\rho_3)^{(3)}\mathcal{C}^\alpha. \tag{17}$$

The coefficients in (17) coincide with those calculated in [25].
The 2-forms $\mathcal{C}^\alpha$ and $\mathcal{F}^\alpha$ do not depend on a choice of a coordinate system. They change as vectors by global $SO(1,3)$ transformations of the coframe. Using (13) the coframe Lagrangian can be rewritten as

$$^{(\texttt{cof})}L = \frac{1}{2}\mathcal{C}_\alpha \wedge *\mathcal{F}^\alpha\,. \tag{18}$$

Observe that the Lagrangian (18) is of the same form as the standard electromagnetic Lagrangian $^{(\texttt{cof})}L = \frac{1}{2}F \wedge *F$. Observe, however, that the coframe Lagrangian involves the vector valued 2-forms of the field strength, while the electromagnetic Lagrangian is constructed of the the scalar valued 2-forms.

2.2. Variation of the Lagrangian

The Lagrangian (18) depends on the coframe field ϑ^a and on its first order derivatives only. Thus the first order variation formalism guarantee the corresponding Euler-Lagrange equation to be at most of the second order. Consider the variation of the coframe Lagrangian (12) with respect to small independent variations of the 1-forms ϑ^α. The λ-symbols (9) are constants and obey the symmetry property (10). Thus

$$C_{\alpha\beta\gamma}\delta F^{\alpha\beta\gamma} = C_{\alpha\beta\gamma}\lambda^{\alpha\beta\gamma\alpha'\beta'\gamma'}\delta C_{\alpha'\beta'\gamma'} = \delta C_{\alpha\beta\gamma}F^{\alpha\beta\gamma}\,. \tag{19}$$

Consequently the variation of the Lagrangian (12) takes the form

$$\delta L = \frac{1}{2}\delta C_{\alpha\beta\gamma}F^{\alpha\beta\gamma} * 1 - L * \delta(*1)\,. \tag{20}$$

The variation of the volume element is

$$\delta(*1) = -\delta(\vartheta^{0123}) = -\delta\vartheta^0 \wedge \vartheta^{123} - \cdots = -\delta\vartheta^0 \wedge *\vartheta^0 - \cdots = \delta\vartheta^\alpha \wedge *\vartheta_\alpha\,.$$

Thus the second term of (20) is given by

$$L * \delta(*1) = (\delta\vartheta^\alpha \wedge *\vartheta_\alpha) * L = -\delta\vartheta^\alpha \wedge (e_\alpha \rfloor L)\,. \tag{21}$$

As for the variation of the C-coefficients, we calculate them by equating the variations of the two sides of the equation (2)

$$\delta d\vartheta_\alpha = \frac{1}{2}\delta C_{\alpha\mu\nu}\vartheta^{\mu\nu} + C_{\alpha\mu\nu}\delta\vartheta^\mu \wedge \vartheta^\nu\,. \tag{22}$$

Use the formulas (A.12) and (A.13) to derive

$$\begin{aligned}
\delta d\vartheta_\alpha \wedge *\vartheta_{\beta\gamma} &= \frac{1}{2}\delta C_{\alpha\mu\nu}\vartheta^{\mu\nu} \wedge *\vartheta_{\beta\gamma} + C_{\alpha\mu\nu}\delta\vartheta^\mu \wedge \vartheta^\nu \wedge *\vartheta_{\beta\gamma} \\
&= -\frac{1}{2}\delta C_{\alpha\mu\nu}\vartheta^\mu \wedge *(e^\nu \rfloor \vartheta_{\beta\gamma}) - C_{\alpha\mu\nu}\delta\vartheta^\mu \wedge *(e^\nu \rfloor \vartheta_{\beta\gamma}) \\
&= \delta C_{\alpha\beta\gamma} * 1 - 2\delta\vartheta^\mu \wedge C_{\alpha\mu[\beta} * \vartheta_{\gamma]}\,.
\end{aligned}$$

Therefore

$$\delta C_{\alpha\beta\gamma} * 1 = \delta(d\vartheta_\alpha) \wedge *\vartheta_{\beta\gamma} + 2\delta\vartheta^\mu \wedge C_{\alpha\mu[\beta} * \vartheta_{\gamma]}\,. \tag{23}$$

After substitution of (21–23) into (20) the variation of the Lagrangian takes the form

$$\delta L = \frac{1}{2} F^{\alpha\beta\gamma}\Big(\delta(d\vartheta_\alpha)\wedge *\vartheta_{\beta\gamma} + 2\delta\vartheta^\mu \wedge C_{\alpha\mu[\beta} * \vartheta_{\gamma]}\Big) + \delta\vartheta^\mu \wedge (e_\mu \rfloor L)\,.$$

Extract the total derivatives to obtain

$$\delta L = \frac{1}{2}\delta\vartheta_\mu \wedge \Big(d(*F^{\mu\beta\gamma}\vartheta_{\beta\gamma}) + 2F^{\alpha\beta\gamma}C_{\alpha\mu[\beta} * \vartheta_{\gamma]} + 2e_\mu \rfloor L\Big)$$

$$+\frac{1}{2}d\Big(\delta\vartheta_\alpha \wedge *F^{\alpha\beta\gamma}\vartheta_{\beta\gamma}\Big)\,. \tag{24}$$

The variation relation (24) plays a basic role in derivation of the field equation and of the conserved current. We rewrite it in a compact form by using the 2-forms (13) and (14). The terms of the form $F \cdot C$ can be rewritten as

$$F^{\alpha\beta\gamma}C_{\alpha\mu[\beta} * \vartheta_{\gamma]} = (F^{\alpha\beta\gamma} - F^{\alpha\beta\gamma})C_{\alpha\mu[\beta} * \vartheta_{\gamma]} = C_{\alpha\mu\beta} * (e^\beta \rfloor \mathcal{F}^\alpha) = -(e_\mu \rfloor \mathcal{C}_\alpha) \wedge *\mathcal{F}^\alpha\,.$$

Hence, (24) takes the form

$$\delta L \;=\; \delta\vartheta^\mu \wedge \Big(d(*\mathcal{F}_\mu) - (e_\mu \rfloor \mathcal{C}_\alpha)\wedge *\mathcal{F}^\alpha + e_\mu \rfloor L\Big) + d(\delta\vartheta^\mu \wedge \mathcal{F}_\mu)\,. \tag{25}$$

Collect now the quadratic terms into a differential 3-form

$$\mathcal{T}_\mu := (e_\mu \rfloor \mathcal{C}_\alpha)\wedge *\mathcal{F}^\alpha - e_\mu \rfloor L\,. \tag{26}$$

Consequently, the variational relation (24) results in a compact form

$$\delta L = \delta\vartheta^\mu \wedge \Big(d * \mathcal{F}_\mu - \mathcal{T}_\mu\Big) + d\Big(\delta\vartheta^\mu \wedge \mathcal{F}_\mu\Big)\,. \tag{27}$$

2.3. The Coframe Field Equations

We are ready now to write down the field equations. Consider independent free variations of a coframe field vanishing at infinity (or at the boundary of the manifold ∂M). The variational relation (27) yields *the coframe field equation*

$$d * \mathcal{F}^\mu = \mathcal{T}^\mu\,. \tag{28}$$

Observe that the structure of coframe field equation is formally similar to the structure of the standard electromagnetic field equation $d * F = J$. Namely, in both equations, the left hand sides are the exterior derivative of the dual field strength while the right hand sides are odd 3-forms. Thus the 3-forms $\mathcal{T}^\mu$ serves as a source for the field strength $\mathcal{F}^\mu$, as well as the 3-form of electromagnetic current J is a source for the electromagnetic field F. There are, however, some important distinctions: (i) The coframe field current $\mathcal{T}_\mu$ is a vector-valued 3-form while the electromagnetic current J is a scalar-valued 3-form. (ii) The field equation (28) is nonlinear. (iii) The electromagnetic current J depends on an exterior matter field, while the coframe current $\mathcal{T}^\mu$ is interior (depends on the coframe itself).

The exterior derivation of the both sides of field equation (28) yields the conservation law

$$dT^\mu = 0. \tag{29}$$

Note, that this equation obeys all the symmetries of the coframe Lagrangian. It is diffeomorphism invariant and global $SO(1,3)$ covariant. Thus we obtain a conserved total 3-form (26) which is constructed from the first order derivatives of the field variables (coframe). It is local and covariant. The 3-form $\mathcal{T}_\mu$ is our candidate for the coframe energy-momentum current.

2.4. Conserved Current and Noether Charge

The current $\mathcal{T}_\mu$ is obtained directly, i.e., by separation of the terms in the field equation. In order to identify the proper nature of this conserved 3-form we have to answer the question: *What symmetry this conserved current can be associated with?*

Return to the variational relation (27). On shell, for the fields satisfying the field equations (28), it takes the form

$$\delta L = d(\delta\vartheta^\alpha \wedge *\mathcal{F}_\alpha)\,. \tag{30}$$

Consider the variations of the coframe field produced by the Lie derivative taken relative to a smooth vector field X, i.e.,

$$\delta\vartheta^\alpha = \mathcal{L}_X\vartheta^\alpha = d(X\rfloor\vartheta^\alpha) + X\rfloor d\vartheta^\alpha\,. \tag{31}$$

The Lagrangian (12) is a diffeomorphic invariant, hence its variation is produced by the Lie derivative taken relative to the same vector field X, i.e.,

$$\delta L = \mathcal{L}_X L = d(X\rfloor L)\,. \tag{32}$$

Thus the relation (30) takes a form of a conservation law $d\Theta(X)$ for the Noether 3-form

$$\Theta(X) = \Big(d(X\rfloor\vartheta^\alpha) + X\rfloor\mathcal{C}^\alpha\Big) \wedge *\mathcal{F}_\alpha - X\rfloor L\,. \tag{33}$$

This quantity includes the derivatives of an arbitrary vector field X. Such a non-algebraic dependence of the conserved current is an obstacle for definition of an energy-momentum tensor. This problem is solved merely by using the canonical form of the current. Let us take $X = e_\alpha$. The first term of (33) vanishes identically. Thus

$$\Theta(e_\mu) = (e_\mu\rfloor\mathcal{C}^\alpha) \wedge *\mathcal{F}_\alpha - e_\mu\rfloor L\,. \tag{34}$$

Observe that the right hand side of the equation (34) is exactly the same expression as the source term of the field equation (28):

$$\Theta(e_\mu) = \mathcal{T}_\mu\,. \tag{35}$$

Thus the conserved current $\mathcal{T}_\mu$ defined in (26) is associated with the diffeomorphism invariance of the Lagrangian. Consequently the vector-valued 3-form (26) represents the *energy-momentum current of the coframe field.*

Let us look for an additional information incorporated in the conserved current (33). Extract the total derivative to obtain

$$\Theta(X) = d\Big((X\rfloor\vartheta^\alpha) * \mathcal{F}_\alpha\Big) - (X\rfloor\vartheta^\alpha)(d * \mathcal{F}_\alpha - \mathcal{T}_\alpha)\,. \tag{36}$$

Thus, up to the field equation (28), the current $\mathcal{T}(X)$ represents a total derivative of a certain 2-form $\Theta(X) = dQ(X)$. This result is a special case of a general proposition due to Wald [54] for a diffeomorphic invariant Lagrangians. The 2-form

$$Q(X) = (X\rfloor\vartheta^\alpha) * \mathcal{F}_\alpha\,. \tag{37}$$

can be referred to as the *Noether charge for the coframe field*. Consider $X = e_\alpha$ and denote $Q_\alpha := Q(e_\alpha)$. From (37) we obtain that this canonical Noether charge of the coframe field coincides with the dual of the conjugate strength

$$Q_\alpha = Q(e_\alpha) = *\mathcal{F}_\alpha\,. \tag{38}$$

In this way, the 2-form $\mathcal{F}_\alpha$, which was used above only as a technical device for expressing the equations in a compact form, obtained now a meaningful description. Note, that the Noether charge plays an important role in Wald's treatment of the black hole entropy [54].

2.5. Energy-Momentum Tensor

In this section we construct an expression for the energy-momentum tensor for the coframe field. Let us first introduce the notion of the energy-momentum tensor via the differential-form formalism. We are looking for a second rank tensor field of a type $(0,2)$. Such a tensor can always be treated as a bilinear map $T: \mathcal{X}(M)\times\mathcal{X}(M)\to\mathcal{F}(M)$, where $\mathcal{F}(M)$ is the algebra of C^∞-functions on M while $\mathcal{X}(M)$ is the $\mathcal{F}(M)$-module of vector fields on M. The unique way to construct a scalar from a 3-form and a vector is is to take the Hodge dual of the 3-form and to contract the result by the vector. Consequently, we define the energy-momentum tensor as

$$T(X,Y) := Y\rfloor * \mathcal{T}(X)\,. \tag{39}$$

Observe that this quantity is a tensor if and only if the 3-form current $\mathcal{T}$ depends linearly (algebraic) on the vector field X. Certainly, $T(X,Y)$ is not symmetric in general. The antisymmetric part of the energy-momentum tensor is known from the Poincaré gauge theory [1] to represent the spinorial current of the field. The canonical form of the energy-momentum $T_{\alpha\beta} := T(e_\alpha, e_\beta)$ tensor is

$$T_{\alpha\beta} = e_\beta\rfloor * \mathcal{T}_\alpha\,. \tag{40}$$

Another useful form of this tensor can be obtained from (40) by applying the rule (A.13)

$$T_{\alpha\beta} = *(\mathcal{T}_\alpha\wedge\vartheta_\beta)\,. \tag{41}$$

The familiar procedure of raising the indices by the Lorentz metric $\eta^{\alpha\beta}$ produces two tensors of a type $(1,1)$

$$T_\alpha{}^\beta = *(\mathcal{T}_\alpha\wedge\vartheta^\beta), \quad \text{and} \quad T^\alpha{}_\beta = *(\mathcal{T}^\alpha\wedge\vartheta_\beta)\,, \tag{42}$$

which are different, in general. By applying the rule (A.13) the first relation of (42) is converted into

$$\mathcal{T}_\alpha = T_\alpha{}^\beta * \vartheta_\beta\,. \tag{43}$$

Thus, the components of the energy-momentum tensor are regarded as the coefficients of the current $\mathcal{T}_\alpha$ in the dual basis $*\vartheta^\alpha$ of the vector space Ω^3 of odd 3-forms.

In order to show that (43) conforms with the intuitive notion of the energy-momentum tensor let us restrict to a flat manifold and represent the 3-form conservation law as a tensorial expression. Take a closed coframe $d\vartheta^\alpha = 0$, thus $d * \vartheta_\beta = 0$. From (43) we derive

$$d\mathcal{T}_\alpha = dT_\alpha{}^\beta \wedge *\vartheta_\beta = -T_\alpha{}^\beta{}_{,\beta} * 1\,.$$

Hence, in this approximation, the differential-form conservation law $d\mathcal{T}_\alpha = 0$ is equivalent to the tensorial conservation law $T_\alpha{}^\beta{}_{,\beta} = 0$.

Apply now the definition (40) to the conserved current (26) for the coframe field. The energy-momentum tensor $T_{\mu\nu} = e_\nu\rfloor * \mathcal{T}_\mu$ is derived in the form

$$T_{\mu\nu} = e_\nu\rfloor * \Big((e_\mu\rfloor\mathcal{C}_\alpha) \wedge *\mathcal{F}^\alpha - \frac{1}{2}e_\mu\rfloor(\mathcal{C}_\alpha \wedge *\mathcal{F}^\alpha)\Big)\,. \tag{44}$$

Using (A.13) we rewrite the first term in (44) as

$$e_\nu\rfloor * \Big((e_\mu\rfloor\mathcal{C}_\alpha) \wedge *\mathcal{F}^\alpha\Big) = - * \Big((e_\mu\rfloor\mathcal{C}_\alpha) \wedge *(e_\nu\rfloor\mathcal{F}^\alpha)\Big)\,.$$

As for the second term in (44) it takes the form

$$-\frac{1}{2}e_\nu\rfloor * \Big(e_\mu\rfloor(\mathcal{C}_\alpha \wedge *\mathcal{F}^\alpha)\Big) = \frac{1}{2}\eta_{\mu\nu} * (\mathcal{C}_\alpha \wedge *\mathcal{F}^\alpha)\,.$$

Consequently the energy-momentum tensor for the coframe field is

$$T_{\mu\nu} = - * \Big((e_\mu\rfloor\mathcal{C}_\alpha) \wedge *(e_\nu\rfloor\mathcal{F}^\alpha)\Big) + \frac{1}{2}\eta_{\mu\nu} * (\mathcal{C}_\alpha \wedge *\mathcal{F}^\alpha)\,. \tag{45}$$

Observe that this expression is formally similar to the known expression for the energy-momentum tensor of the Maxwell electromagnetic field:

$${}^{(\mathtt{em})}T_{\mu\nu} = - * \Big((e_\mu\rfloor F) \wedge *(e_\nu\rfloor F)\Big) + \frac{1}{2}\eta_{\mu\nu} * (F \wedge *F)\,. \tag{46}$$

The form (46) is no more than an expression of the electromagnetic energy-momentum tensor in arbitrary frame. In a specific coordinate chart $\{x^\mu\}$ it is enough to take the coordinate basis vectors $e_a = \partial_\alpha$ and consider $T_{\alpha\beta} := {}^{(e)}T(\partial_\alpha, \partial_\beta)$ to obtain the familiar expression

$${}^{(\mathtt{em})}T_{\alpha\beta} = -F_{\alpha\mu}F_\beta{}^\mu + \frac{1}{4}\eta_{\alpha\beta}F_{\mu\nu}F^{\mu\nu}\,. \tag{47}$$

The electromagnetic energy-momentum tensor is obviously traceless. The same property holds also for the coframe field tensor. In fact, the coframe energy-momentum tensor defined by (45) is traceless for all models described by the Lagrangian (4), i.e., for all values of the parameters ρ_i. Indeed, compute the trace $T^\mu{}_\mu = T_{\mu\nu}\eta^{\mu\nu}$ of (45):

$$\begin{aligned}{}^{(\mathtt{cof})}T^\mu{}_\mu &= - * \Big((e_\mu\rfloor\mathcal{C}_\alpha) \wedge *(e^\mu\rfloor\mathcal{F}^\alpha)\Big) + 2 * (\mathcal{C}_\alpha \wedge *\mathcal{F}^\alpha)\\ &= - * \Big(\vartheta^\mu \wedge (e_\mu\rfloor\mathcal{C}_\alpha) \wedge *\mathcal{F}^\alpha\Big) + 2 * (\mathcal{C}_\alpha \wedge *\mathcal{F}^\alpha) = 0\,.\end{aligned}$$

It is well known that the traceless of the energy-momentum tensor is associated with the scale invariance of the Lagrangian. The rigid (λ is a constant) scale transformation $x^i \to \lambda x^i$, is considered acting on a matter field as $\phi \to \lambda^d \phi$, where d is the dimension of the field. The transformation does not act, however, on the components of the metric tensor and on the frame (coframe) components. It is convenient to shift the change on the metric and on the frame (coframe) components, i.e., to consider

$$g_{ij} \to \lambda^2 g_{ij}\,, \qquad \vartheta^\alpha{}_i \to \lambda \vartheta^\alpha{}_i\,, \qquad \text{and} e_\alpha{}^i \to \lambda^{-1} \vartheta_\alpha{}^i \tag{48}$$

with no change of coordinates. In the coordinate free formalism the difference between two approaches is neglected and the transformation is

$$g \to \lambda^2 g\,, \qquad \vartheta^\alpha \to \lambda \vartheta^\alpha\,, \qquad \text{and} \qquad e_\alpha \to \lambda^{-1} e_\alpha\,. \tag{49}$$

The transformation law of the coframe Lagrangian is simple to obtain from the component-wise form (8). Under the transformation (49) the volume element changes as $*1 \to \lambda^4 * 1$. As for the C-coefficients, they transform due to (3) as $C^a{}_{bc} \to \lambda^{-1} C^a{}_{bc}$. Consequently, by (5), the transformation law of the Lagrangian 4-form is $L \to \lambda^2 L$, which is the same as for the Hilbert-Einstein Lagrangian $L_{HE} = R\sqrt{-g}d^4x \to \lambda^2 L_{HE}$. After rescaling the Planck length the scale invariance is reinstated. Hence, for the pure coframe field model the energy-momentum tensor have to be traceless in accordance with the proposition above.

2.6. The Field Equation for a General System

The coframe field equation have been derived for a pure coframe field. Consider now a general minimally coupled system of a coframe field ϑ^α and a matter field ψ. The matter field can be a differential form of an arbitrary degree and can carry arbitrary number of exterior and interior indices. Take the total Lagrangian of the system to be of the form ($\ell =$ Planck length)

$$L = \frac{1}{\ell^2} {}^{(\texttt{cof})}L(\vartheta^\alpha, d\vartheta^\alpha) + {}^{(\texttt{mat})}L(\vartheta^\alpha, \psi, d\psi)\,, \tag{50}$$

where the coframe Lagrangian ${}^{(\texttt{cof})}L$, defined by (4), is of dimension length square. The matter Lagrangian ${}^{(\texttt{mat})}L$ is dimensionless.

The minimal coupling means here the absence of coframe derivatives in the matter Lagrangian. Take the variation of (50) relative to the coframe field ϑ^α to obtain

$$\delta L = \frac{1}{\ell^2} \delta\vartheta^\alpha \wedge \left(d * \mathcal{F}_\alpha - {}^{(\texttt{cof})}\mathcal{T}_\alpha - \ell^{2\,(\texttt{mat})}\mathcal{T}_\alpha \right), \tag{51}$$

where the 3-form of coframe current is defined by (28). The 3-form of matter current is defined via the variation derivative of the matter Lagrangian taken relative to the coframe field ϑ^α:

$$^{(\texttt{mat})}\mathcal{T}_\alpha := -\frac{\delta}{\delta\vartheta^\alpha} {}^{(\texttt{mat})}L\,. \tag{52}$$

Introduce the total current of the system ${}^{(\texttt{tot})}\mathcal{T}_\alpha = {}^{(\texttt{cof})}\mathcal{T}_\alpha + \ell^{2\,(\texttt{mat})}\mathcal{T}_\alpha$, which is of dimension length (mass). Consequently, the field equation for the general system (50) takes the form

$$d * \mathcal{F}_\alpha = {}^{(\texttt{tot})}\mathcal{T}_\alpha\,. \tag{53}$$

Using the energy-momentum tensor (43) this equation can be rewritten in a tensorial form

$$e_\beta \rfloor * d * \mathcal{F}_\alpha = {}^{(\mathtt{tot})}T_{\alpha\beta}\,, \tag{54}$$

or equivalently

$$\vartheta_\beta \wedge d * \mathcal{F}_\alpha = {}^{(\mathtt{tot})}T_{\alpha\beta} * 1\,. \tag{55}$$

The conservation law for the total current $d\mathcal{T}_\alpha = 0$ is a straightforward consequence of the field equation (53). The form (53) of the field equation looks like the Maxwell field equation for the electromagnetic field $d * F = J$. Observe, however, an important difference. The source term in the right hand side of the electromagnetic field equation depends only on external fields. In the absence of the external sources $J = 0$, the electromagnetic strength $*F$ is a closed form. As a consequence, its cohomology class interpreted as a charge of the source. The electromagnetic field itself is uncharged.

As for the coframe field strength $\mathcal{F}^\alpha$ its source depends on the coframe and of its first order derivatives. Consequently, the 2-form $*\mathcal{F}^\alpha$ is not closed even in absence of the external sources. Hence the gravitational field is massive (charged) itself.

On the other hand the tensorial form (54) of the coframe field equation is similar to the Einstein field equation for the metric tensor $G_{\alpha\beta} = 8\pi{}^{(\mathtt{mat})}T_{\alpha\beta}$. Indeed, the left hand side in both equations are pure geometric quantities. Again, the source terms in the field equations are different. The source of the Einstein gravity is the energy-momentum tensor only of the matter fields. The conservation of this tensor is a consequence of the field equation. Thus even if some meaningful conserved energy-momentum current for the metric field existed it would have been conserved regardless of the matter field current. Consequently, any redistribution of the energy-momentum current between the matter and gravitational fields is forbidden in the framework of the traditional Einstein gravity.

As for the coframe field equation, the total energy-momentum current plays a role as the source of the field. Consequently the coframe field is completely "self-interacted" - the energy-momentum current of the coframe field produces an additional field. The conserved current of the coframe-matter system is the total energy-momentum current, not only the matter current. Thus in the framework of general coframe construction the redistribution of the energy-momentum current between the matter field and the coframe field is possible, in principle.

2.7. Spherically Symmetric Solution

Let us look for a static spherically symmetric solution to the field equation (28). We will use the isotropic coordinates $\{x^{\hat{i}}\,,\, \hat{i} = 1, 2, 3\}$ with the isotropic radius ρ. Denote

$$s = \rho^2 = \delta_{\hat{i}\hat{j}} x^{\hat{i}} x^{\hat{j}} = x^2 + y^2 + z^2\,. \tag{56}$$

Recall that we identify the gravity variable with the coframe field defined up to an infinitesimal Lorentz transformation. It is equivalent to the metric field. So it is enough to look for a coframe solution of a "diagonal" form [29]

$$\vartheta^0 = f(s)\, dx^0\,, \qquad \vartheta^{\hat{i}} = g(s)\, dx^{\hat{i}}\,. \tag{57}$$

Although this ansatz is not the most general one, it is enough because (57) corresponds to a most general static spherical symmetric metric

$$ds^2 = e^{2f(s)}dt^2 - e^{2g(s)}(dx^2 + dy^2 + dz^2)\,. \tag{58}$$

Substitution of (57) into the field equation (28) we obtain an over-determined system of three second order ODE for two independent variables $f(s)$ and $g(s)$

$$\begin{cases} \rho_1\Big(2f''s + 3f' + 2f'g's - 2(g')^2s + (f')^2s\Big) + 2\rho_3\Big(2g''s + 3g' + (g')^2s\Big) = 0 \\ \rho_1\Big(2g'' + 2f'g' - 2(f')^2 - 2(g')^2\Big) + 2\rho_3\Big(f'' + g'' + (f')^2 - 2f'g' - (g')^2\Big) = 0 \\ \rho_1\Big(4g''s + 4g' + 4f'g's - 2(f')^2s\Big) + 2\rho_3\Big(2f''s + 2f' + 2g''s + 2g' + 2(f')^2s\Big) = 0. \end{cases} \tag{59}$$

This system has a solution with the Newtonian behavior on infinity $f \sim 1 - C/\rho$ only if the parameter ρ_1 is equal to zero. In this case, the system (59) has a unique solution

$$f = \ln\frac{1 - \frac{1}{c\rho}}{1 + \frac{1}{c\rho}}, \qquad g = 2\ln\left(1 + \frac{1}{c\rho}\right). \tag{60}$$

By taking the parameter of integration to be inversely proportional to the mass of the central body $c = \frac{2}{m}$ we obtain the coframe field in the form

$$\vartheta^0 = \frac{1 - \frac{m}{2\rho}}{1 + \frac{m}{2\rho}}dt, \qquad \vartheta^i = \left(1 + \frac{m}{2\rho}\right)^2 dx^i, \qquad i = 1, 2, 3\,. \tag{61}$$

This coframe field yields the Schwarzschild metric in isotropic coordinates

$$ds^2 = \left(\frac{1 - \frac{m}{2\rho}}{1 + \frac{m}{2\rho}}\right)^2 dt^2 - \left(1 + \frac{m}{2\rho}\right)^4 (dx^2 + dy^2 + dz^2)\,. \tag{62}$$

Note that the values of the parameters ρ_2, ρ_3 are not determined via the "diagonal" ansatz. Thus the Schwarzschild metric is a solution for a family of the coframe field equations which defined by the parameters:

$$\rho_1 = 0\,, \quad \rho_2, \rho_3 - \text{arbitrary}. \tag{63}$$

The ordinary GR is extracted from this family by requiring of the *local* $SO(1, 3)$ invariance, which is realized by an additional restriction of the parameters:

$$\rho_1 = 0\,, \quad 2\rho_2 + \rho_3 = 0\,. \tag{64}$$

2.8. Weak Field Approximation

Linear approximation of coframe models was usually applied to study the deviation from the standard GR, and for comparison with the observation data, see [20], [21], [23]. We will use this approach to study the meaning of the condition $\rho_1 = 0$, see [39]. Recall that this condition guarantees the existence of viable solutions.

To study the approximate solutions to (37), we start with a trivial exact solution, a *holonomic coframe*, for which

$$d\vartheta^a = 0\,. \tag{65}$$

Consequently, $\mathcal{F}^a = \mathcal{C}^a = 0$, so both sides of Eq. (37) vanish. By Poincaré's lemma, the solution of (65) can be locally expressed as $\vartheta^a = d\tilde{x}^a(x)$, where $\tilde{x}^a(x)$ is a set of four smooth functions defined in a some neighborhood U of a point $x \in \mathcal{M}$. The functions $\tilde{x}^a(x)$, being treated as the components of a coordinate map $\tilde{x}^a : U \to \mathbb{R}^4$, generate a local coordinate system on U. The metric tensor reduces, in this coordinate chart, to the flat Minkowski metric $g = \eta_{ab} d\tilde{x}^a \otimes d\tilde{x}^b$. Thus the holonomic coframe plays, in the coframe background, the same role as the Mankowski metric in the (pseudo-)Riemannian geometry. Moreover, a manifold endowed with a (pseudo-)orthonormal holonomic coframe is flat. The weak perturbations of the basic solution $\vartheta^a = dx^a$ are

$$\vartheta^a = dx^a + h^a = (\delta^a_b + h^a{}_b)\, dx^b\,. \tag{66}$$

The indices in $h^a{}_b$ can be lowered and raised by the Mankowski metric

$$h_{ab} := \eta_{am} h^m{}_b\,, \qquad h^{ab} := \eta^{bm} h^a{}_m\,. \tag{67}$$

The first operation is exact (covariant to all orders of approximations), while the second is covariant only to the first order, when $g^{ab} \approx \eta^{ab}$. The symmetric and the antisymmetric combinations of the perturbations

$$\theta_{ab} := h_{(ab)} = \frac{1}{2}(h_{ab} + h_{ba}), \qquad \text{and} \qquad w_{ab} := h_{[ab]} = \frac{1}{2}(h_{ab} - h_{ba})\,. \tag{68}$$

as well as the trace $\theta := h^m{}_m = \theta^m{}_m$ are covariant to the first order. The components of the metric tensor, in the linear approximation, involve only the symmetric combination of the coframe perturbations

$$g_{ab} = \eta_{ab} + 2\theta_{ab}\,. \tag{69}$$

When the decomposition

$$h_{ab} = \theta_{ab} + w_{ab} \tag{70}$$

is applied, the field strength is splitted to a sum of two independent strengths — one defined by the symmetric field θ_{ab} and the second one defined by the antisymmetric field w_{ab}

$$\mathcal{F}_a(\theta_{mn}, w_{mn}) = {}^{(\mathtt{sym})}\mathcal{F}_a(\theta_{mn}) + {}^{(\mathtt{ant})}\mathcal{F}_a(w_{mn})\,, \tag{71}$$

where

$${}^{(\mathtt{sym})}\mathcal{F}_a = -\Big[(\rho_1+\rho_3)\theta_{a[b,c]} + \rho_3 \eta_{a[b}\theta_{c]m}{}^{,m} - \rho_3\eta_{a[b}\theta_{,c]}\Big]\,\vartheta^b \wedge \vartheta^c\,, \tag{72}$$

and

$${}^{(\mathtt{ant})}\mathcal{F}_a = -\Big[(\rho_1+\rho_3)w_{a[b,c]} + 3\rho_2 w_{[ab,c]} - \rho_3\eta_{a[b}w_{c]m}{}^{,m}\Big]\,\vartheta^b \wedge \vartheta^c\,. \tag{73}$$

Hence, for arbitrary values of the parameters ρ_i, the field strengths of the fields θ_{ab} and w_{ab} are independent.

The linearized field equation takes the form

$$\begin{aligned}(\rho_1+\rho_3)\big(\Box\,\theta_{ab} - \theta_{am,b}{}^{,m}\big) + \rho_3\big(-\eta_{ab}\Box\,\theta - \theta_{mb}{}^{,m}{}_{,a} + \theta_{,a,b} + \eta_{ab}\theta_{mn}{}^{,m,n}\big) +\\ (\rho_1+2\rho_2+\rho_3)\big(\Box\,w_{ab} - w_{am,b}{}^{,m}\big) + (2\rho_2+\rho_3)w_{bm,a}{}^{,m} = 0\,.\end{aligned} \tag{74}$$

Proposition 1: *For the case $\rho_1 = 0$, the linearized coframe field equation (74), splits, in arbitrary coordinates, into two independent systems*

$$^{(\mathtt{sym})}\mathcal{E}_{(ab)}(\theta_{mn}) = \Box\,\theta_{ab} = 0\,, \qquad \textit{and} \qquad ^{(\mathtt{ant})}\mathcal{E}_{[ab]}(w_{mn}) = \Box\,w_{ab} = 0\,.$$

If $\rho_1 \neq 0$, Eq.(74) does not split in any coordinate system.

Consequently, for $\rho_1 = 0$ and for generic values of the parameters ρ_2, ρ_3, the field equation of the coframe field is split into two independent field equations for two independent field variables. This splitting emerges also for the Lagrangian and the energy-momentum current.

Proposition 2: *For $\rho_1 = 0$, the Lagrangian of the coframe field is reduced, up to a total derivative term, to the sum of two independent Lagrangians*

$$\mathcal{L}(\theta_{ab}, w_{ab}) = {}^{(\mathtt{sym})}\mathcal{L}(\theta_{ab}) + {}^{(\mathtt{ant})}\mathcal{L}(w_{ab})\,. \tag{75}$$

Moreover, the coframe energy-momentum current is reduced, on shell, in the first order approximation, as

$$\mathcal{T}_a(\theta_{mn}, w_{mn}) = {}^{(\mathtt{sym})}\mathcal{T}_a(\theta_{mn}) + {}^{(\mathtt{ant})}\mathcal{T}_a(w_{mn})\,, \tag{76}$$

up to a total derivative.

The result of our analysis is as following: In the linear approximation the field variable is split into a sum of two independent fields. These fields do not interact only in the case $\rho_1 = 0$. Remarkable that this condition coincides with the viable condition (63), which is necessary for Schwarzschild metric.

3. Coframe Geometry

The coframe gravity represented above is not related to a certain specific geometric structure. In this section we are looking for a geometry that can be constructed from the coframe field. It is well known that, on a Riemannian manifold there exists a unique linear connection of Levi-Civita [50]. Already this statement indicates that when we want to deal with some other connection, for instance with the flat one, we have to use some other non-Riemannian geometric structure. In this section, we define a geometry based on a coframe field. It is instead used of the standard Riemannian geometry based on a metric tensor field.

3.1. Coframe Manifold. Definitions and Notations

Our construction will repeat the main properties of the Riemannian structure. Let us start with the basic definitions.

Differential manifold. Let M be a smooth $n+1$ dimensional differentiable manifold, which is locally (in an open set $U \subset M$) parametrized by a coordinate chart $\{x^i; i = 0, 1, \ldots n\}$. The set of $n+1$ differentials dx^i provides a coordinate basis for the module of the differential forms on U. Similarly, the set of $n+1$ vector fields $\partial_i = \partial/\partial x^i$ forms the coordinate basis for the module of the vector fields on U. Arbitrary smooth transformations of the coordinates $x^i \to y^i(x^j)$ are admissible. Under these transformations, the elements of the coordinate bases transform by the tensorial law

$$dx^i \to dy^i = \frac{\partial y^i}{\partial x^j}\, dx^j\,, \qquad \frac{\partial}{\partial x^i} \to \frac{\partial}{\partial y^i} = \frac{\partial x^j}{\partial y^i}\frac{\partial}{\partial x^j}\,. \tag{77}$$

The Jacobian matrix $\partial y^i/\partial x^j$ is assumed to be smooth and invertible. The coordinate bases dx^i and $\partial_i = \partial/\partial x^i$ are referred to as *holonomic bases*. They satisfy the relations $d(dx^i) = 0$ and $[\partial_i, \partial_j] = 0$.

For a compact representation of geometric quantities, it is useful to have an alternative description via *nonholonomic bases*. Denote by θ^a a generic nonholonomic basis of the module of the 1-forms on U. Its dual f_a is a basis of of the module of the vector fields on U. In general, $d\theta^a \neq 0$ and $[f_a, f_b] \neq 0$. Relative to the coordinate bases, the elements of the nonholonomic bases are locally expressed as

$$\theta^a = \theta^a{}_i\, dx^i\,, \qquad f_a = f_a{}^i\, \partial_i\,. \tag{78}$$

Here the matrices $\theta^a{}_i$ and $f_a{}^i$ are the inverse to each-other, i.e.,

$$\theta^a{}_i\, f_a{}^j = \delta^j_i\,, \qquad \theta^a{}_i\, f_b{}^i = \delta^a_b\,. \tag{79}$$

Arbitrary smooth pointwise transformations of the nonholonomic bases

$$\theta^a \to A^a{}_b(x)\theta^b\,, \qquad f_a \to A_a{}^b(x) f_b\,. \tag{80}$$

are admissible. Here $A_a{}^b$ denotes, as usual, the matrix inverse to $A^a{}_b$.

Although the basis indices change in the same range $a, b, \cdots = \{0, 1, \ldots, n\}$ they are distinguished from the coordinate indices $i, j, \cdots$. In particular, the contraction of the indices in the quantities $\theta^a{}_i$ or $f_a{}^i$ is forbidden since the result of such an action is not a scalar. The base transformations (80) are similar to the coordinate transformations (77). Note that the basis θ^a can be changed to an arbitrary other basis, for instance to the coordinate one. Indeed, the formulas (78) can be treated as certain transformations of the bases. Consequently, θ^a cannot be given any intrinsic geometrical sense. In particular, it cannot be used as a model of a physical field.

Coframe field. Let the manifold M be endowed with a smooth nondegenerate coframe field ϑ^α. It comes together with its dual — the frame field e_α. In an arbitrary chart of local coordinates $\{x^i\}$, these fields are expressed as

$$\vartheta^\alpha = \vartheta^\alpha{}_i dx^i\,, \qquad e_\alpha = e_\alpha{}^i \partial_i\,, \tag{81}$$

i.e., by two nondegenerate matrices $\vartheta^\alpha{}_i$ and $e_\alpha{}^i$ which are the inverse to each-other. In other words, we are considering a set of n^2 independent smooth functions on M. Also the coframe indices change in the same range $\alpha, \beta, \ldots = 0, \ldots, n$ as the coordinate indices $i, j, \ldots$ and the basis indices $a, b, \ldots$. They all however have to be strictly distinguished. In particular, the indices in $\vartheta^\alpha{}_i$ or $e_\alpha{}^i$ cannot be contracted.

Coframe transformation. For most physical models based on the coframe field, this field is defined only up to global transformations. It is natural to consider a wider class of coframe fields related by local pointwise transformations

$$\vartheta^\alpha \to L^\alpha{}_\beta(x)\vartheta^\beta\,, \qquad e_\alpha \to L_\alpha{}^\beta(x)e_\beta\,. \tag{82}$$

Here $L^\alpha{}_\beta(x)$ and $L_\alpha{}^\beta(x)$ are inverse to each-other at arbitrary point x. Denote the group of matrices $L^\alpha{}_\beta(x)$ by G. Note two specially important cases: (i) G is a group of global transformations with a constant matrix $L^\alpha{}_\beta$; (ii) G is a group of arbitrary local transformations such that the entries of $L^\alpha{}_\beta$ are arbitrary functions of a point. In the latter case, the difference between the coframe field ϑ^α and the reference basis θ^a is completely removed and the coframe structure is trivialized.

Consequently we involve an additional element of the coframe structure — *the coframe transformations group*

$$G = \Big\{L^\alpha{}_\beta(x) \in GL(n+1,\mathbb{R});\ \text{for every}\, x \in M\Big\}. \tag{83}$$

On this stage, we only require the matrices $L^\alpha{}_\beta(x)$ to be invertible at an arbitrary point $x \in M$. The successive specializations of the coframe transformation matrix will be involved in sequel.

Coframe field volume element. We assume the coframe field to be non-degenerate at an arbitrary point $x \in M$. Consequently, a special $n+1$-form, *the coframe field volume element*, is defined and nonzero. Define

$$\text{vol}(\vartheta^\alpha) = \frac{1}{n!}\,\varepsilon_{\alpha_o\cdots\alpha_n}\vartheta^{\alpha_o}\wedge\cdots\wedge\vartheta^{\alpha_n}\,, \tag{84}$$

where $\varepsilon_{\alpha_o\cdots\alpha_n}$ is the Levi-Civita permutation symbol normalized by $\varepsilon_{01\cdots n} = 1$. Treating the coframe volume element as one of the basic elements of the coframe geometric structure, we apply the following invariance condition.

> *Volume element invariance postulate:* Volume element $\text{vol}(\vartheta^\alpha)$ is assumed to be invariant under pointwise transformations of the coframe field
>
> $$\text{vol}\left(\vartheta^\alpha\right) = \text{vol}\left(L^\alpha{}_\beta\vartheta^\beta\right)\,. \tag{85}$$
>
> This condition is satisfied by matrices with the unit determinant. Consequently, the coframe transformation group (83) is restricted to
>
> $$G = \Big\{L^\alpha{}_\beta(x) \in SL(n+1,\mathbb{R});\ \text{for every}\, x \in M\Big\}. \tag{86}$$

Metric tensor. For a meaningful physical field model, it is necessary to have a metric structure on M. Moreover, the metric tensor has to be of the Lorentzian signature. In a coordinate basis and in an arbitrary reference basis, a generic metric tensor is written correspondingly as

$$g = g_{ij}dx^i \otimes dx^j\,, \qquad g = g_{ab}\theta^a \otimes \theta^b\,, \tag{87}$$

where the components g_{ij} and g_{ab} are smooth functions of a point $x \in M$.

On a coframe manifold, a metric tensor is not an independent quantity. Instead, we are looking for a metric explicitly constructed from a given coframe field, $g = g(\vartheta^\alpha)$. We assume the metric tensor to be quadratic in the coframe field components and independent of its derivatives. Moreover, it should be of the Lorentzian type, i.e., should be reducible at a point to the Lorentzian metric $\eta_{\alpha\beta} = \text{diag}(-1, 1, \cdots, 1)$. These requirements are justified by an almost flat approximation: for an almost holonomic coframe, ${\vartheta^\alpha}_i \approx \delta^\alpha_i$, we have to reach the flat Lorentzian metric. With these restrictions, we come to a definition of the *coframe field metric tensor*

$$g = \eta_{\alpha\beta}\vartheta^\alpha \otimes \vartheta^\beta , \qquad g_{ij} = \eta_{\alpha\beta}{\vartheta^\alpha}_i{\vartheta^\beta}_j . \tag{88}$$

Note that the equations (88) often appear as a definition of a (non unique) orthonormal basis of reference for a given metric. Another interpretation treats (88) as an expression of a given metric in a special orthonormal basis of reference, as in (87). In our approach, (88) has a principally different meaning. It is a definition of the metric tensor field via the coframe field. Certainly the form of the metric $\eta_{\alpha\beta}$ in the tangential vector space T_xM is an additional axiom of our construction. With an aim to define an invariant coframe geometric structure we require:

Metric tensor invariance postulate: Metric tensor is assumed to be invariant under pointwise transformations of the coframe field, i.e.,

$$g\left(\vartheta^\alpha\right) = g\left({L^\alpha}_\beta \vartheta^\beta\right) . \tag{89}$$

This condition is satisfied by pseudo-orthonormal matrices,

$$\eta_{\mu\nu}{L^\mu}_\alpha L^\nu_\beta = \eta_{\alpha\beta}. \tag{90}$$

Consequently, the invariance of the coframe metric restricts the coframe transformation group to

$$G = \left\{{L^\alpha}_\beta(x) \in O(1, n, \mathbb{R}); \text{ for every } x \in M\right\} . \tag{91}$$

In order to have simultaneously a metric and a volume element structure both constructed from the coframe field, we have to assume a successive restriction of the coframe transformation group:

$$G = \left\{{L^\alpha}_\beta(x) \in SO(1, n, \mathbb{R}); \text{ for every } x \in M\right\} . \tag{92}$$

Topological restrictions. A global smooth coframe field may be defined only on a parallelizable manifold, i.e., on a topological manifold of a zero second Whitney class. This topological restriction is equivalent to existence of a spinorial structure on M. In this chapter, we restrict ourselves to a local consideration, thus the global definiteness problems will be neglected. Moreover, we assume the coframe field to be smooth and nonsingular only in a "weak" sense. Namely, the components ${\vartheta^\alpha}_i$ and ${e_\alpha}^i$ are required to be differentiable and linearly independent at almost all points of M, i.e., except in a zero measure set. So, in general, the coframe field can degenerate at singular points, on singular lines (strings), or even on singular submanifolds (p-branes). This assumption leaves room for the standard singular solutions of the physics field equations such as the Coulomb field, the Schwarzschild metric, the Kerr metric etc.

3.2. Coframe Connections

From the geometrical point of view, a differential manifold endowed with a coframe field is a rather poor structure. In particular, we can not determine if two vectors attached at distance points are parallel to each-other or not. In order to have a meaningful geometry and, consequently, a meaningful geometrical field model for gravity, we have to consider a reacher structure. In this section we define a coframe manifold with a linear coframe connection. The connection 1-form $\mathit{\Gamma}_a{}^b$ will not be an independent variable, as in the Cartan geometry or in MAG [5]. Alternatively in our construction the connection will be explicitly constructed from the coframe field and its first order derivatives. Thus we are dealing with a category of *coframe manifolds with a linear coframe connection:*

$$\Big\{M\,,\vartheta^\alpha,G\,,\Gamma_a{}^b(\vartheta^\alpha)\Big\}\,. \tag{93}$$

We start with a coframe manifold without an addition metric structure. Metric contributions to the connection will be considered in sequel.

Affine connection. Recall the main properties of a generic linear affine connection on an $(n+1)$ dimensional differential manifold. Relative to a local coordinate chart x^i, a connection is represented by a set of $(n+1)^3$ independent functions $\Gamma^k{}_{ij}(x)$ — *the coefficients of the connection.* The only condition these functions have to satisfy is to transform, under a change of coordinates $x^i \mapsto y^i(x)$, by an inhomogeneous linear rule:

$$\Gamma^i{}_{jk} \mapsto \left(\Gamma^l{}_{mn}\frac{\partial y^m}{\partial x^j}\frac{\partial y^n}{\partial x^k}+\frac{\partial^2 y^l}{\partial x^j\partial x^k}\right)\frac{\partial x^i}{\partial y^l}\,. \tag{94}$$

When an arbitrary reference basis $\{\theta^a\,,f_b\}$ is involved, the coefficients of the connection are arranged in a $GL(n,\mathbb{R})$-valued *connection 1-form*, which is defined as [50]

$$\mathit{\Gamma}_a{}^b = f_a{}^k\left(\theta^b{}_i\Gamma^i{}_{jk}-\theta^b{}_{k,j}\right)dx^j\,. \tag{95}$$

In a holonomic coordinate basis, we can simply use the identities $\theta^a{}_i=\delta^a_i$ and $f_a{}^i=\delta^i_a$. Consequently, in a coordinate basis, the derivative term is canceled out and (95) reads

$$\mathit{\Gamma}_j{}^i = \Gamma^i{}_{jk}\,dx^k\,. \tag{96}$$

Due to (94), this quantity transforms under the coordinate transformations as

$$\mathit{\Gamma}_j{}^i \to \left[\Gamma_m{}^l\frac{\partial y^m}{\partial x^j}+d\left(\frac{\partial y^l}{\partial x^j}\right)\right]\frac{\partial x^i}{\partial y^l}\,. \tag{97}$$

Alternatively, the connection 1-form (95) is invariant under smooth transformations of coordinates. The inhomogeneous linear behavior is shifted here to the transformations of $\varGamma_a{}^b$ under a linear local map of the reference basis (θ^a, f_a) given in (82):

$$\varGamma_a{}^b \mapsto \left(\Gamma_c{}^d A_a{}^c + dA_a{}^d\right) A^b{}_d \,. \tag{98}$$

On a manifold with a given coframe field ϑ^α, the connection 1-form (95), can also be referred to this field. We denote this quantity by $\varGamma_\alpha{}^\beta$. It is defined similarly to (95):

$$\varGamma_\alpha{}^\beta = \left(\vartheta^\beta{}_i \Gamma^i{}_{jk} - \vartheta^\beta{}_{k,j}\right) e_\alpha{}^k \, dx^j \,. \tag{99}$$

This quantity can be treated as an expression of a generic connection (95) in a special basis. Note an essential difference between two very similar equations (95) and (99). In (95), we must be able to apply arbitrary pointwise linear transformations of the basis. The coefficients of the connection $\Gamma^i{}_{jk}$ are independent on the basis (θ^a, f_a) used in (95). On the other hand, in (99), we permit only the transformations of the coframe field ϑ^α that are restricted by some invariance requirements. Moreover, we will require the connection $\Gamma^i{}_{jk}$ to be constructed explicitly from the derivatives of the coframe field itself.

Linear coframe connections. We restrict ourselves to the quasi-linear $\Gamma^i{}_{jk}(\vartheta^\alpha)$, i.e., we consider a connection constructed as a linear combination of the first order derivatives of the coframe field. The coefficients in this linear expression may depend on the frame/coframe components. In other words, we are looking for a coframe analog of an ordinary Levi-Civita connection.

Let us assist ourselves with a similar construction from the Riemannian geometry. So let us look now for a most general connection that can be constructed from the metric tensor components. Consider a general linear combination of the first order derivatives of the metric tensor:

$$g^{mi}(\alpha_1 g_{mj,k} + \alpha_2 g_{mk,j} + \alpha_3 g_{jk,m}) \,. \tag{100}$$

Although this expression has the same index content as $\Gamma^i{}_{jk}$, it is a connection only for some special values of the parameters $\alpha_1, \alpha_2 . \alpha_3$. Indeed, any two connections differ by a tensor. Thus an arbitrary connection can be expressed as a certain special connection plus a tensor

$$\Gamma^i{}_{jk} = \overset{*}{\Gamma}{}^i{}_{jk} + K^i{}_{jk} \,. \tag{101}$$

Use for $\overset{*}{\Gamma}{}^i{}_{jk}$ the Levi-Civita connection

$$\overset{*}{\Gamma}{}^i{}_{jk} = \frac{1}{2} g^{im}(g_{mj,k} + g_{mk,j} - g_{jk,m}) \,. \tag{102}$$

However in Riemannian geometry, a tensor constructed from the first order derivatives of the metric does not exist. Therefore $K^i{}_{jk} = 0$, thus the Levi-Civita connection is a unique connection that can be constructed from the first order derivatives of the metric tensor. It is evidently symmetric and metric compatible.

In an analogy to this construction, we will look for a most general coframe connection of the form

$$\Gamma^i{}_{jk}(\vartheta^\alpha) = \overset{\circ}{\Gamma}{}^i{}_{jk}(\vartheta^\alpha) + K^i{}_{jk}(\vartheta^\alpha) \,. \tag{103}$$

Here $\overset{\circ}{\Gamma}{}^i{}_{jk}$ is a certain special connection, while $K^i{}_{jk}$ is a tensor. To start with, we need a certain analog of the Levi-Civita connection, i.e., a special connection constructed from the coframe field.

The flat Weitzenböck connection. On a bare differentiable manifold M, without any additional structure, the notion of parallelism of two vectors attached to distance points depends on a curve joint of the points. Oppositely, on a coframe manifold $\{M\,,\vartheta^\alpha\}$, a certain type of the parallelism of distance vectors may be defined in an absolute (curve independent) sense [52]. Namely, two vectors $u(x_1)$ and $v(x_2)$ may be declared parallel to each other, if, being referred to the local elements of the coframe field $u(x_1) = u_\alpha(x_1)\vartheta^\alpha(x_1)$ and $v(x_2) = v_\alpha(x_2)\vartheta^\alpha(x_2)$, they have the proportional components $u_\alpha(x_1) = Cv_\alpha(x_2)$. This definition is independent on the coordinates used on the manifold and on the nonholonomic frame of reference. It depends on the coframe field. Since, by local transformations, the coframes at distance points change differently, only rigid linear coframe transformations preserve such type of a parallelism.

This geometric picture may be reformulated in terms of a special connection. The elements of the coframe field attached to distinct points have to be assumed parallel to each other. It means that a special connection $\overset{\circ}{\Gamma}{}^i{}_{jk}$ exists such that the corresponding covariant derivative of the coframe field components is zero:

$$\vartheta^\alpha{}_{j;k} = \vartheta^\alpha{}_{j,k} - \overset{\circ}{\Gamma}{}^i{}_{jk}\vartheta^\alpha{}_i = 0\,. \tag{104}$$

Multiplying by $e_\alpha{}^i$, we have an explicit expression

$$\overset{\circ}{\Gamma}{}^i{}_{jk} = e_\alpha{}^i\vartheta^\alpha{}_{k,j}\,. \tag{105}$$

Under a smooth transformation of coordinates, this expression is transformed in accordance with the inhomogeneous linear rule (94). Consequently, (105) indeed gives the coefficients of a special connection which is referred to as the *Weitzenböck flat connection*. This connection is unique for a class of coframes related by rigid linear transformations.

In an arbitrary nonholonomic reference basis (θ^a, f_a), we have correspondingly a unique Weitzenböck's connection 1-form which is constructed by (95) from (105)

$$\Gamma_a{}^b = f_a{}^k\left(\theta^b{}_i\,\overset{\circ}{\Gamma}{}^i{}_{jk} - \theta^b{}_{k,j}\right)dx^j\,. \tag{106}$$

Substituting the coframe field ϑ^α instead of the nonholonomic basis θ^a we have

$$\overset{\circ}{\Gamma}{}_\alpha{}^\beta = \left(-\vartheta^\beta{}_{i,j} + \vartheta^\beta{}_k e_\alpha{}^k\vartheta^\alpha{}_{i,j}\right)e_\alpha{}^i\,dx^j = 0\,. \tag{107}$$

Thus the Weitzenböck connection 1-form is zero, when it is referred to the coframe field $(\vartheta^\alpha, e_\alpha)$ itself. Certainly, this property is only a basis related fact. It yields, however, vanishing of the curvature of the Weitzenböck connection, which is a basis independent property.

General coframe connections. Recall that we are looking for a general coframe connection constructed from the first order derivatives of the coframe field components. In the Riemannian geometry, the analogous construction yields a unique connection of Levi-Civita. In the coframe geometry, however, the situation is different.

Proposition 3: *The general linear connection constructed from the first order derivatives of the coframe field is given by a 3-parametric family:*

$$\Gamma^i{}_{jk} = \overset{\circ}{\Gamma}{}^i{}_{jk} + \alpha_1 C^i{}_{jk} + \alpha_2 C_j \delta^i_k + \alpha_3 C_k \delta^i_j \,. \tag{108}$$

Proof: The difference of two connections is a tensor of a type $(1,2)$, so an arbitrary connection can be expressed as the Weitzenböck connection plus a tensor

$$\Gamma^i{}_{jk} = \overset{\circ}{\Gamma}{}^i{}_{jk} + K^i{}_{jk} \,. \tag{109}$$

Since $\overset{\circ}{\Gamma}{}^i{}_{jk}$ is already a linear combination of the first order derivatives, the additional tensor also has to be of the same form. Observe that $K^i{}_{jk}$ involves only coordinate indices, while the partial derivatives $\vartheta^\alpha{}_{j,i}$ have a coframe index α. This coframe index has to be suppressed. Hence the first order derivatives of the coframe components may appear in $K^i{}_{jk}$ only by the expressions $e_\alpha{}^i \vartheta^\alpha{}_{j,k}$. Notice that this quantity coincides with the coefficients of Weitzenböck's connection (105), which is not a tensor. Since the matrix of the frame field components $e_\alpha{}^j$ is the inverse of $\vartheta^\alpha{}_i$, the derivatives of the frame field $e_\alpha{}^j{}_{,k}$ are linear combinations of $\vartheta^\alpha{}_{j,k}$. Thus we do not need to involve additional derivatives of the frame field into $K^i{}_{jk}$. Consequently, the components of the tensor $K^i{}_{jk}$ have to be linear in $\overset{\circ}{\Gamma}{}^i{}_{jk}$. Write a general expression of such a type:

$$K^i{}_{jk} = \frac{1}{2} \chi_{jkl}{}^{imn} \overset{\circ}{\Gamma}{}^l{}_{mn} \,. \tag{110}$$

Since, under a transformation of coordinates, the connection $\overset{\circ}{\Gamma}{}^l{}_{mn}$ changes by inhomogeneous rule, it can appear in the tensor $K^i{}_{jk}$ only in the antisymmetric combination. Thus the most general expression for this tensor is

$$K^i{}_{jk} = \frac{1}{2} \chi_{jkl}{}^{imn} \overset{\circ}{\Gamma}{}^l{}_{[mn]} = \frac{1}{2} \chi_{jkl}{}^{imn} C^l{}_{mn} \,. \tag{111}$$

Hence, the symmetry relation $\chi_{jkl}{}^{imn} = \chi_{jkl}{}^{i[mn]}$ holds. The coefficients $\chi_{jkl}{}^{imn}$ have to be constructed from the components of the absolute basis $\vartheta^\alpha{}_m$ and $e_\alpha{}^m$. Again, since $\chi_{jkl}{}^{imn}$ involves only coordinate indices, it has to be constructed from the traced products of the frame and the coframe components. However all such products are equal to the Kronecker symbol. Thus $\chi_{jkl}{}^{imn}$ has to be a tensor expressed only by the Kronecker symbols. Consequently, the general expression for $\chi_{jkl}{}^{imn}$ can be written as

$$\chi_{jkl}{}^{imn} = \alpha_1 \delta^{[m}_j \delta^{n]}_k \delta^i_l + \alpha_2 \delta^{[m}_l \delta^{n]}_k \delta^i_j + \alpha_3 \delta^{[m}_l \delta^{n]}_j \delta^i_k \,. \tag{112}$$

Substituting into (111) we have

$$K^i{}_{jk} = \alpha_1 C^i{}_{jk} + \alpha_2 C_j \delta^i_k + \alpha_3 C_k \delta^i_j \,. \tag{113}$$

Consequently (108) is proved. ■

By (95), the connection 1-form corresponded to the coefficients (108), being referred to as a nonholonomic basis, takes the form

$$\Gamma_a{}^b = f_a{}^k \left(-\theta^b{}_{k,m} + \theta^b{}_l \overset{\circ}{\Gamma}{}^l{}_{mk} + K^l{}_{mk}\theta^b{}_l\right) dx^m \,. \tag{114}$$

When this quantity is referred to the coframe field itself, the first two terms are canceled. In this special basis, the expression is simplified to

$$\Gamma_\alpha{}^\beta = K^i{}_{jk} e_\alpha{}^k \vartheta^\beta{}_i dx^j \,, \tag{115}$$

where $K^i{}_{jk}$ is given in (113). Since the 1-form (115) depends only on antisymmetric combinations of the first order derivatives, it can be expressed by the exterior derivative of the coframe:

$$\Gamma_\alpha{}^\beta = \left(\alpha_1 C^\beta{}_{\gamma\alpha} + \alpha_2 C_\gamma \delta^\beta_\alpha + \alpha_3 C_\alpha \delta^\beta_\gamma\right) \vartheta^\gamma \,. \tag{116}$$

Also a components free expression is available

$$\Gamma_\alpha{}^\beta = -\frac{1}{2}\left[\alpha_1 e_\alpha \rfloor d\vartheta^\beta + \alpha_2 (e_\alpha \rfloor \mathcal{A})\vartheta^\beta + \alpha_3 \delta^\beta_\alpha \mathcal{A}\right] \,. \tag{117}$$

Metric-coframe connection. Consider a manifold endowed with the coframe metric tensor (88). Again, we are looking for a most general coframe connection that can be constructed from the first order derivatives of the coframe field. We will refer to it as the *metric-coframe connection*. Thus we are dealing with a category of *coframe manifolds with a coframe metric and a linear coframe connection:*

$$\left\{M\,, \vartheta^\alpha, G\,, g(\vartheta^a), \Gamma_a{}^b(\vartheta^\alpha)\right\} . \tag{118}$$

Now the connection expression will involve some additional terms which depend on the metric tensor (88). To describe all possible combinations of the metric tensor components and frame/coframe components it is useful to pull down all the indices. Define:

$$\Gamma_{ijk} = g_{im}\Gamma^m{}_{jk} \,, \qquad C_{ijk} = g_{im} C^m{}_{jk} \,. \tag{119}$$

Proposition 4: *The most general metric-coframe connection constructed from the first order derivatives of the coframe field is represented by a 6-parametric family:*

$$\Gamma_{ijk} = \overset{\circ}{\Gamma}_{ijk} + \alpha_1 C_{ijk} + \alpha_2 g_{ik} C_j + \alpha_3 g_{ij} C_k + \beta_1 g_{jk} C_i + \beta_2 C_{jki} + \beta_3 C_{kij} \,. \tag{120}$$

Proof: Similarly to the case of a pure coframe connection, a metric-coframe connection can be represented as the Weitzenböck connection plus an arbitrary tensor. So we can write

$$\Gamma_{ijk} = \overset{\circ}{\Gamma}_{ijk} + K_{ijk} \,. \tag{121}$$

The tensor K_{ijk} has to be proportional to the derivatives of the coframe field $\vartheta^\alpha{}_{i,j}$. Repeating the consideration given above we come to the same conclusion: the first order derivatives of the coframe field can appear in the tensor K_{ijk} only via the antisymmetric combination of the flat connection $\overset{\circ}{\Gamma}{}_{l[mn]} = C_{lmn}$. Consequently we have a relation

$$K_{ijk} = \frac{1}{2}\chi_{ijk}{}^{lmn}C_{lmn}\,. \qquad (122)$$

The tensor $\chi_{ijk}{}^{lmn}$ may involve now the components of the metric tensor in addition to the Kronecker symbol. Using the symmetry relation $\chi_{ijk}{}^{lmn} = \chi_{ijk}{}^{l[mn]}$ we construct a most general expression of such a type

$$\begin{aligned}\chi_{ijk}{}^{lmn} &= \alpha_1\delta^l_i\delta^{[m}_j\delta^{n]}_k + \beta_2\delta^l_j\delta^{[m}_k\delta^{n]}_i + \beta_3\delta^l_k\delta^{[m}_i\delta^{n]}_j + \\ &\quad \alpha_2 g_{ik}g^{l[m}\delta^{n]}_j + \alpha_3 g_{ij}g^{l[m}\delta^{n]}_k + \beta_1 g_{jk}g^{l[m}\delta^{n]}_i\,. \end{aligned} \qquad (123)$$

Consequently, the additional tensor takes the required form

$$K_{ijk} = \alpha_1 C_{ijk} + \alpha_2 g_{ik}C_j + \alpha_3 g_{ij}C_k + \beta_1 g_{jk}C_i + \beta_2 C_{jki} + \beta_3 C_{kij}\,. \qquad (124)$$

■

The expression (120) can be rewritten in a

$$\Gamma^i{}_{jk} = \overset{\circ}{\Gamma}{}^i{}_{jk} + \alpha_1 C^i{}_{jk} + \alpha_2\delta^i_k C_j + \alpha_3\delta^i_j C_k + \beta_1 g^{il}g_{jk}C_l + \beta_2 g^{il}C_{jkl} + \beta_3 g^{il}C_{klj}\,. \quad (125)$$

In fact, this expression is a proper form of the coefficients of the coframe connection. Here we can identify two groups of terms: (i) The terms with the coefficient α_i that do not depend on the metric; (ii) The terms with the coefficient β_i that can be constructed only by use of the metric tensor.

With respect to a nonholonomic basis (f_a, θ^a), the coefficients of a connection (125) correspond to a connection 1-form (95)

$$\Gamma_a{}^b = \overset{\circ}{\Gamma}{}_a{}^b + K^i{}_{jk}f_a{}^k\theta^b{}_i dx^j\,. \qquad (126)$$

When (126) is referred to the coframe field itself, it is simplified to

$$\Gamma_\alpha{}^\beta = K^i{}_{jk}e_\alpha{}^k\vartheta^\beta{}_i dx^j\,. \qquad (127)$$

This expression depends only on the antisymmetric combinations of the first order derivatives of the coframe components. So it can be expressed by the exterior derivative of the coframe. We have

$$\Gamma_\alpha{}^\beta = \left(\alpha_1 C^\beta{}_{\gamma\alpha} + \alpha_2 C_\gamma\delta^\beta_\alpha + \alpha_3 C_\alpha\delta^\beta_\gamma + \beta_1 C^\beta\eta_{\alpha\gamma} + \beta_2 C_{\gamma\alpha\nu}\eta^{\beta\nu} + \beta_3 C_{\alpha\nu\gamma}\eta^{\beta\nu}\right)\vartheta^\gamma\,, \qquad (128)$$

or, equivalently,

$$\begin{aligned}\Gamma_\beta{}^\alpha &= -\frac{1}{2}\Big[\alpha_1 e_\beta\rfloor d\vartheta^\alpha + \alpha_2\vartheta^\alpha(e_\beta\rfloor\mathcal{A}) + \alpha_3\delta^\alpha_\beta\mathcal{A} + \beta_1(e^\alpha\rfloor\mathcal{A})\vartheta_\beta + \\ &\quad \beta_2 e^\alpha\rfloor(e_\beta\rfloor d\vartheta_\mu)\vartheta^\mu + \beta_3 e^\alpha\rfloor d\vartheta_\beta\Big]\,. \end{aligned} \qquad (129)$$

3.3. Torsion of the Coframe Connection

Torsion tensor and torsion 2-form. Definitions. Consider a connection 1-form $\Gamma_b{}^a$ referred to an arbitrary basis (θ^a, f_a). For a tensor valued p-form of a representation type $\rho\left(A_a{}^b\right)$, the *covariant exterior derivative* operator $D : \Omega^p(\mathcal{M}) \to \Omega^{p+1}(\mathcal{M})$ is defined as [24], [5]

$$D = d + \Gamma_b{}^a \rho\left(A_a{}^b\right) \wedge . \tag{130}$$

In particular, the covariant exterior derivative of a scalar-valued form ϕ is $D\phi = d\phi$. For a vector-valued form ϕ^a, it is given by $D\phi^a = d\phi^a + \Gamma_b{}^a \wedge \phi^b$, etc.

For a connection 1-form $\Gamma_a{}^b$ written with respect to a nonholonomic basis, the *torsion 2-form* $\mathcal{T}^a$ is defined as

$$\mathcal{T}^a = D\theta^a = d\theta^a + \Gamma_b{}^a \wedge \theta^b . \tag{131}$$

On a D dimensional manifold, this covector valued 2-form has $D(D^2 - D)/2$ independent components. Substituting (95) into (131), we observe that the coframe derivative term $d\vartheta^a$ cancels out. Hence,

$$\mathcal{T}^a = \Gamma^i{}_{jk}\theta^a{}_i dx^j \wedge dx^k = \Gamma^i{}_{[jk]}\theta^a{}_i dx^j \wedge dx^k . \tag{132}$$

In a coordinate coframe, this expression is simplified to

$$\mathcal{T}^i = \Gamma^i{}_{[jk]} dx^j \wedge dx^k . \tag{133}$$

Consequently, the torsion 2-form $\mathcal{T}^a$ is completely determined by an antisymmetric combination of the coefficients of the connection. Observe that such combination is a tensor. Thus, the torsion 2-form is completely equivalent to a $(1, 2)$-rank *torsion tensor* which is defined as

$$T^i{}_{jk} = 2\Gamma^i{}_{[jk]} . \tag{134}$$

In holonomic and nonholonomic bases, the torsion 2-form is expressed respectively as

$$\mathcal{T}^i = \frac{1}{2} T^i{}_{jk} dx^j \wedge dx^k , \qquad \mathcal{T}^a = \frac{1}{2} T^i{}_{jk}\theta^a{}_i dx^j \wedge dx^k . \tag{135}$$

It is useful to define also a quantity

$$\mathcal{T}^\alpha = \frac{1}{2} T^i{}_{jk}\vartheta^\alpha{}_i dx^j \wedge dx^k . \tag{136}$$

Observe that this set of 2-forms cannot be regarded as a vector-valued form since the transformations of the coframe field ϑ^α are restricted. However, the proper vector valued torsion 2-forms (135) are related to the quantity (136) by the following simple equations

$$\mathcal{T}^i = e_\alpha{}^i \mathcal{T}^\alpha , \qquad \mathcal{T}^a = \theta^a{}_i e_\alpha{}^i \mathcal{T}^\alpha . \tag{137}$$

With respect to the coframe field, the torsion 2-form of the Weitzenböck connection (136) reads

$$\overset{\circ}{\mathcal{T}}{}^\alpha = d\vartheta^\alpha . \tag{138}$$

Torsion of the metric-coframe connection. For the metric-coframe connection (120), the covariant components $T_{ijk} = 2g_{im}\Gamma^m{}_{[jk]}$ of the torsion tensor take the form

$$T_{ijk} = 2(1+\alpha_1)C_{ijk} + (\alpha_2 - \alpha_3)(g_{ik}C_j - g_{ij}C_k) + (\beta_2 + \beta_3)(C_{jki} + C_{kij})\,. \quad (139)$$

The corresponded torsion 2-form is expressed in the coordinate basis as

$$T^i = \Big[(1+\alpha_1)C^i{}_{jk} + (\alpha_2 - \alpha_3)C_j\delta^i_k + (\beta_2+\beta_3)g^{im}C_{jkm}\Big]\,dx^j \wedge dx^k\,. \quad (140)$$

In terms of the differential forms $\mathcal{A}$ and $\mathcal{B}$ (see Appendix) we derive

$$\mathcal{T}^\alpha = (1+\alpha_1)\,d\vartheta^a - \frac{1}{2}\,(\alpha_2-\alpha_3)\vartheta^\alpha \wedge \mathcal{A} - \frac{1}{2}\,(\beta_2+\beta_3)\Big(d\vartheta^a - e^\alpha\rfloor\mathcal{B}\Big)\,. \quad (141)$$

Irreducible decomposition of the torsion. On a manifold of a dimension $D \geq 3$ endowed with a metric tensor, the torsion 2-form admits an irreducible decomposition into three independent pieces [5]

$$\mathcal{T}^a = {}^{(1)}\mathcal{T}^a + {}^{(2)}\mathcal{T}^a + {}^{(3)}\mathcal{T}^a\,. \quad (142)$$

Here *the trator* and *the axitor* parts [5] are defined correspondingly as

$${}^{(2)}\mathcal{T}^a = \frac{1}{n-1}\,\theta^a \wedge (f_b\rfloor\mathcal{T}^b)\,, \qquad {}^{(3)}\mathcal{T}^a = \frac{1}{3}f^a\rfloor(\theta^b \wedge \mathcal{T}_\beta)\,. \quad (143)$$

The remainder ${}^{(1)}\mathcal{T}^a$ is referred to as a *tensor part*. The irreducible decomposition means that the different pieces transform independently by the same tensorial rule as the total quantity. Particularly, we can check straightforwardly that for every part of the torsion tensor

$${}^{(p)}\mathcal{T}^a = {}^{(p)}\mathcal{T}^\alpha\theta^a{}_i e_\alpha{}^i\,, \qquad p = 1,2,3. \quad (144)$$

So it is enough to provide the calculations of the irreducible pieces with respect to the coframe field itself. We have the second piece of the torsion as

$${}^{(2)}\mathcal{T}^\alpha = \frac{1}{n-1}\,\vartheta^\alpha \wedge (e_\beta\rfloor\mathcal{T}^\beta) = \frac{\tau_2}{2(n-1)}\,\vartheta^\alpha \wedge \mathcal{A}\,, \quad (145)$$

where

$$\tau_2 = 2(1+\alpha_1) - (\beta_2+\beta_3) - (\alpha_2-\alpha_3)(n-1)\,. \quad (146)$$

The third piece of torsion is given by

$${}^{(3)}\mathcal{T}^\alpha = \frac{1}{3}e^\alpha\rfloor(\vartheta^\beta \wedge \mathcal{T}_\beta) = \frac{\tau_3}{3}\,e^\alpha\rfloor\mathcal{B}\,, \quad (147)$$

where

$$\tau_3 = (1+\alpha_1) + (\beta_2+\beta_3)\,. \quad (148)$$

The first part takes the form

$${}^{(1)}\mathcal{T}^\alpha \;=\; \mathcal{T}^\alpha - {}^{(2)}\mathcal{T}^\alpha - {}^{(3)}\mathcal{T}^\alpha = \tau_1\left(d\vartheta^a - \frac{1}{n-1}\,\vartheta^\alpha \wedge \mathcal{A} - \frac{1}{3}\,e^\alpha\rfloor\mathcal{B}\right)\,, \quad (149)$$

where

$$\tau_1 = (1+\alpha_1) - \frac{1}{2}(\beta_2+\beta_3)\,. \qquad (150)$$

Torsion-free metric-coframe connection. Let us look for which values of the parameters the torsion of the metric-coframe connection is identically zero. The corresponded connection is called *the symmetric or torsion-free connection*. It is clear from (139) that the metric-coframe connection is symmetric if

$$\alpha_1 = -1\,, \qquad \alpha_2 = \alpha_3\,, \qquad \beta_2 = -\beta_3\,. \qquad (151)$$

The necessity of this condition can be derived from the irreducible decomposition. Indeed, since the three pieces of the torsion are mutually independent, they have to vanish simultaneously. Hence we have a condition $\tau_1 = \tau_2 = \tau_3 = 0$ which is equivalent to (151). Note that this requirement is necessary only for a manifold of the dimension $D \geq 3$. On a two-dimensional manifold, the metric-coframe connection is symmetric under a weaker condition

$$2(1+\alpha_1) + (\alpha_2 - \alpha_3) - (\beta_2+\beta_3) = 0\,. \qquad (152)$$

On a curve, every connection is unique and symmetric.

Thus on a manifold of the dimension $D \geq 3$ there exists a 3-parametric family of the symmetric (torsion-free) connections:

$$\Gamma^i{}_{jk} = \overset{\circ}{\Gamma}{}^i{}_{jk} - C^i{}_{jk} + \alpha_2\left(\delta^i_k C_j + \delta^i_j C_k\right) + \beta_1 g_{jk} g^{im} C_m + \beta_2 g^{im}\left(C_{jkm} - C_{kmj}\right) \qquad (153)$$

3.4. Nonmetricity of the Metric-Coframe Connection

Nonmetricity tensor and nonmetricity 2-form. Definition. When Cartan's manifold is endowed with a metric tensor, the connection generates an additional tensor field called *the nonmetricity tensor*. It is expressed as a covariant derivative of the metric tensor components. For a metric given in a local system of coordinates as $g = g_{ij}dx^i \otimes dx^j$, the nonmetricity tensor is defined as

$$Q_{kij} = -\nabla_k g_{ij} = -g_{ij,k} + \Gamma^m{}_{ik} g_{mj} + \Gamma^m{}_{jk} g_{im}\,, \qquad (154)$$

or,

$$Q_{kij} = -g_{ij,k} + \Gamma_{jik} + \Gamma_{ijk}\,. \qquad (155)$$

Evidently, this tensor is symmetric in the last pair of indices $Q_{kij} = Q_{kji}$. Hence, on a D dimensional manifold, the nonmetricity tensor has $D(D^2+D)/2$ independent components.

For the exterior form representation, it is useful to define *the nonmetricity 1-form*. In a coordinate basis, it is given by

$$Q_{ij} = Q_{kij}dx^k = -dg_{ij} + \Gamma_{ij} + \Gamma_{ji}\,. \qquad (156)$$

In an arbitrary reference basis $(f_a\,, \theta^a)$, the metric tensor is expressed as $g = g_{ab}\theta^a \otimes \theta^b$. Correspondingly, the nonmetricity 1-form reads

$$Q_{ab} = -dg_{ab} + \Gamma_{ab} + \Gamma_{ba}\,. \qquad (157)$$

With respect to the coframe field ϑ^α, the components of the metric are constants $\eta_{\alpha\beta}$, thus the nonmetricity is merely the symmetric combination of the connection 1-form components

$$Q_{\alpha\beta} = \Gamma_{\alpha\beta} + \Gamma_{\beta\alpha} \,. \tag{158}$$

Note, that this expression is not a usual tensorial quantity. In fact, it is an expression of a tensor-valued 1-form of nonmetricity with respect to a special class of bases. Its relation to a proper tensorial valued 1-form (156) is, however, very simple. By a substitution of (99) into (158) we have

$$Q_{ij} = Q_{\alpha\beta}\vartheta^\alpha{}_i\vartheta^\beta{}_j \,. \tag{159}$$

The following generalization of the Levi-Civita theorem from the Riemannian geometry provides a decomposition of an arbitrary affine connection [53]. Its simple proof is instructive for our construction.

Proposition 5: *Let a metric g on a manifold M be fixed and two tensors T_{ijk} and Q_{ijk} with the symmetries*

$$T_{ijk} = -T_{ikj} \,, \qquad Q_{kij} = Q_{kji} \,. \tag{160}$$

be given. A unique connection Γ_{ijk} exists on M such that T_{ijk} is its torsion and Q_{ijk} is its nonmetricity. Explicitly,

$$\Gamma_{ijk} = \overset{*}{\Gamma}_{ijk} - \frac{1}{2}\big(Q_{ijk} - Q_{jki} - Q_{kij}\big) + \frac{1}{2}\big(T_{ijk} + T_{jki} - T_{kij}\big) \,, \tag{161}$$

where

$$\overset{*}{\Gamma}_{ijk} = \frac{1}{2}\big(g_{ij,k} + g_{ik,j} - g_{jk,i}\big) \tag{162}$$

are the components of the Levi-Civita connection.

Proof: On a D-dimensional manifold definitions of the torsion and the nonmetricity tensors

$$T_{ijk} = 2\Gamma_{i[jk]} \,, \qquad Q_{kij} = -g_{ij,k} + \Gamma_{ijk} + \Gamma_{jik} \tag{163}$$

can be viewed as a linear system of D^3 linear equations for D^3 independent variables Γ_{ijk}

$$\Gamma_{i[jk]} = \frac{1}{2}T_{ijk} \,, \qquad \Gamma_{(ij)k} = \frac{1}{2}(Q_{kij} + g_{ij,k}) \,. \tag{164}$$

For $T_{ijk} = Q_{kij} = 0$, the system has a unique solution — the Levi-Civita connection $\overset{*}{\Gamma}_{ijk}$. Thus the determinant of the matrix of the system (164) is nonsingular. Consequently also for arbitrary tensors T_{ijk} and Q_{kij}, the system has a unique solution. In order to check the specific form of the solution (161), it is enough to substitute the definitions (163). ■

Nonmetricity of the metric-coframe connection. We calculate now the nonmetricity tensor of the metric-coframe connection (120)

$$\begin{aligned} Q_{kij} = {} & \Big(-g_{ij,k} + \overset{\circ}{\Gamma}_{ijk} + \overset{\circ}{\Gamma}_{jik}\Big) + \\ & (\alpha_1 - \beta_2)(C_{ijk} - C_{jki}) + (\alpha_2 + \beta_1)(g_{ik}C_j + g_{jk}C_i) + 2\alpha_3 g_{ij}C_k \,. \end{aligned} \tag{165}$$

The first parenthesis represents the nonmetricity tensor of the Weitzenböck connection. This expression vanishes identically, i.e., the Weitzenböck connection is metric-compatible. Indeed, we have

$$g_{ij,k} = \eta_{\alpha\beta}(\vartheta^\alpha{}_{i,k}\vartheta^\beta_j + \vartheta^\alpha{}_i\vartheta^\beta_{j,k}) = \overset{\circ}{\Gamma}{}_{ijk} + \overset{\circ}{\Gamma}{}_{jik}\,. \tag{166}$$

Consequently, (165) is simplified to

$$Q_{kij} = (\alpha_1 - \beta_2)(C_{ijk} + C_{jki}) + (\alpha_2 + \beta_1)(g_{ik}C_j + g_{jk}C_i) + 2\alpha_3 g_{ij}C_k\,. \tag{167}$$

Relative to the coframe field, we have, using (128,158), the 1-form of nonmetricity

$$\begin{aligned} Q_{\alpha\beta} \;&=\; -\frac{1}{4}\,(\alpha_1-\beta_2)\Big(e_\alpha\rfloor d\vartheta_\beta + e_\beta\rfloor d\vartheta_\alpha\Big) + \frac{1}{2}\,\alpha_3\eta_{\alpha\beta}A + \\ &\quad \frac{1}{4}\,(\alpha_2+\beta_1)\Big[(e_\alpha\rfloor A)\vartheta_\beta + (e_\beta\rfloor A)\vartheta_\alpha\Big]\,. \end{aligned} \tag{168}$$

Irreducible decomposition of the nonmetricity. We are looking now for an irreducible decomposition of the nonmetricity 1-form Q_{ab} under the pseudo-orthogonal group. Since Q_{ab} is a tensor-valued 1-form it can be calculated in an arbitrary basis. Certainly, the basis of the coframe field is the best for these purposes. We have only to remember that for a transformation to an arbitrary basis we have to simply multiply the corresponding quantity $Q_{\alpha\beta}$ by the matrix of the transformation. We cannot, however, transform the coframe basis to an arbitrary basis. This is because the coframe field is a fixed building block of our construction.

The irreducible decomposition of the nonmetricity 1-form under the pseudo-orthogonal group $SO(1,n)$ is constructed by the in correspondence to the Young diagrams. For actual calculations we use the algorithm given in [5]. The resulting decomposition is given as a sum of four independent pieces

$$Q_{\alpha\beta} = {}^{(1)}Q_{\alpha\beta} + {}^{(2)}Q_{\alpha\beta} + {}^{(3)}Q_{\alpha\beta} + {}^{(4)}Q_{\alpha\beta}\,. \tag{169}$$

For the nonmetricity 1-form (158), the irreducible parts are

$${}^{(1)}Q_{\alpha\beta} = \mu_1\Big[(n-1)e_{(\alpha}\rfloor d\vartheta_{\beta)} + (e_{(\alpha}\rfloor A)\vartheta_{\beta)} - 4\eta_{\alpha\beta}A\Big]\,, \tag{170}$$

$${}^{(2)}Q_{\alpha\beta} = \mu_2\Big[(n-1)e_{(\alpha}\rfloor d\vartheta_{\beta)} + (e_{(\alpha}\rfloor A)\vartheta_{\beta)} + 2\eta_{\alpha\beta}A\Big]\,, \tag{171}$$

$${}^{(3)}Q_{\alpha\beta} = \mu_3\Big[(e_{(\alpha}\rfloor A)\vartheta_{\beta)} + \frac{2}{n}\,\eta_{\alpha\beta}A\Big]\,, \tag{172}$$

$${}^{(4)}Q_{\alpha\beta} = \mu_4\left[\frac{1}{n}\,\eta_{\alpha\beta}A\right]\,. \tag{173}$$

The coefficients of these quantities depend on the parameters of the general connection as

$$\mu_1 \;=\; -\frac{1}{6(n-1)}\,(\alpha_1-\beta_2)\,, \qquad \mu_2 = \frac{1}{2}\,\mu_1\,, \tag{174}$$

$$\mu_3 \;=\; \frac{1}{4}\left[\frac{1}{n-1}\,(\alpha_1-\beta_2) + (\alpha_2+\beta_1)\right]\,, \tag{175}$$

$$\mu_4 \;=\; \frac{1}{2}\Big[-(\alpha_1-\beta_2) + n\alpha_3 + (\alpha_2+\beta_1)\Big]\,. \tag{176}$$

Metric compatible metric-coframe connection. Let us look for which values of the coefficients the connection is *metric-compatible*, i.e., has an identically zero non-metricity tensor. Recall that both quantities, the metric tensor and the connection, are constructed from the same building block — the coframe field ϑ^α. It is clear from (168) that the metric-coframe connection is metric-compatible if

$$\alpha_1 = \beta_2\,, \qquad \alpha_2 = -\beta_1\,, \qquad \alpha_3 = 0\,. \tag{177}$$

The necessity of this condition can be derived from the irreducible decomposition of the nonmetricity tensor. Four irreducible pieces of the non-metricity tensor are mutually independent, so they have to vanish simultaneously. Hence we have a condition $\mu_1 = \mu_2 = \mu_3 = \mu_4 = 0$ which turns out to be equivalent to (177). Note that this requirement is necessary only for a manifold of the dimension $D \geq 3$, where the irreducible decomposition (169) is valid. On a two-dimensional manifold, the metric-coframe connection is metric-compatible if and only if

$$\alpha_1 - \beta_2 = \alpha_2 + \beta_1 = \alpha_3\,. \tag{178}$$

On a one-dimensional manifold, every connection is metric-compatible.

Metric compatible and torsion-free metric-coframe connection. Let us look now for a general coframe connection of a zero torsion and zero non-metricity, i.e., for a symmetric metric compatible connection constructed from the coframe field. The system of conditions (151) and (177) has a unique solution

$$\alpha_1 = \beta_2 = -\beta_3 = -1\,, \qquad \beta_1 = \alpha_2 = \alpha_3 = 0\,. \tag{179}$$

Consequently, a metric-compatible symmetric connection is unique. This is in a correspondence to the original Levi-Civita theorem, and the unique connection is of Levi-Civita. Moreover, substituting (179) into (120) we can express now the standard Levi-Civita connection $\overset{*}{\Gamma}{}^i{}_{jk}$ via the flat connection of Weitzenböck $\overset{\circ}{\Gamma}{}^i{}_{jk}$ —

$$\overset{*}{\Gamma} ijk \;=\; \overset{\circ}{\Gamma}{}_{i(jk)} + C_{kij} - C_{jki}\,. \tag{180}$$

In the basis constructed from the coframe field itself, the nonmetricity 1-form for the Levi-Civita connection reads

$$\overset{*}{\Gamma}{}_{\alpha\beta} \;=\; e_\alpha \rfloor d\vartheta^\beta - e_\beta \rfloor d\vartheta^\alpha - \frac{1}{2}\, e_\alpha \rfloor e_\beta \rfloor B\,. \tag{181}$$

It is in a correspondence with a formula given in [5].

3.5. Gauge Transformations

Local transformations of the coframe field. The geometrical structure considered above is well defined for a fixed coframe field e_α. Moreover, it is invariant under rigid coframe transformations. The gauge paradigm suggests now to look for a localization of such transformations:

$$\vartheta^\alpha \mapsto L^\alpha{}_\beta\, \vartheta^\beta\,, \qquad e_\alpha \mapsto L_\alpha{}^\beta\, e_\beta\,, \tag{182}$$

or, in the components,

$$\vartheta^\alpha{}_i \mapsto L^\alpha{}_\beta \vartheta^\beta{}_i \,, \qquad e_\alpha{}^i \mapsto L_\alpha{}^\beta \, e_\beta{}^i \,. \tag{183}$$

Here the matrix $L^\alpha{}_\beta$ and its inverse $L_\alpha{}^\beta$ are functions of a point $x \in M$. We require the volume element (84) and the metric tensor (88) both to be invariant under the pointwise transformations (182). Consequently, $L^\alpha{}_\beta$ is assumed to be a pseudo-orthonormal matrix whose tensors are smooth functions of a point. We will also use an infinitesimal version of the transformation (183) with $L^\alpha{}_\beta = \delta^\alpha_\beta + X^\alpha{}_\beta$. In the components, it takes the form

$$\vartheta^\alpha{}_i \mapsto \vartheta^\alpha{}_i + X^\alpha{}_\beta \vartheta^\beta{}_i \,, \qquad e_\alpha{}^i \mapsto e_\alpha{}^i - X^\beta{}_\alpha \, e_\beta{}^i \,. \tag{184}$$

As the elements of the algebra $so(1,n)$, the matrix $X_{\alpha\beta} = \eta_{\alpha\mu} X^\mu{}_\beta$ is antisymmetric. We define a corresponded antisymmetric tensor

$$F_{ij} = \vartheta^\alpha{}_i \vartheta^\beta{}_j X_{\alpha\beta} \,. \tag{185}$$

Connection invariance postulate. Recall that we are looking for a most general geometric structure that can be explicitly constructed from the coframe field. Moreover, we are interested not in a one fixed coframe field, but rather in a family of fields related by the left action of the elements of some continuous group G.

In a general setting, the different geometrical structures such as the volume element, the metric tensor, and the field of affine connections, are completely independent. We have already postulated the invariance of the volume element and of the metric tensor under the coframe transformations. It is natural to involve now an additional invariance requirement concerning the affine connection.

> *Connection invariance postulate:* Affine coframe connection is assumed to be invariant under pointwise transformations of the coframe field
>
> $$\Gamma^i{}_{jk}\left(\vartheta^\alpha\right) = \Gamma^i{}_{jk}\left(L^\alpha{}_\beta \vartheta^\beta\right) . \tag{186}$$

Since the coframe connection is constructed from the first order derivatives of the coframe field, (186) is a first order PDE for the elements of the group G and for the components of the coframe field.

Weitzenböck connection transformation. Since the Weitzenböck connection is a basis tool of our construction, it is useful to calculate the change of this quantity under the coframe transformations (182). We have

$$\Delta \overset{\circ}{\Gamma}{}^i{}_{jk} = e_\alpha{}^i \vartheta^\beta{}_k Y^\alpha{}_{\beta j} \,, \qquad \text{where} \qquad Y^\alpha{}_{\beta j} = L^\alpha{}_\gamma L^\gamma{}_{\beta,j} \,. \tag{187}$$

All matrices involved here are nonsingular, consequently the Weitzenböck connection is preserved only under the rigid transformations of the coframe field with $L^\gamma{}_{\beta,j} = 0$.

Let us rewrite (187) in alternative forms. Since the metric tensor is invariant under the transformations (182) we have

$$\Delta \overset{\circ}{\Gamma}{}_{ijk} = \Delta \left(g_{im} \overset{\circ}{\Gamma}{}^m{}_{jk} \right) = g_{im} \Delta \overset{\circ}{\Gamma}{}^m{}_{jk} \,. \tag{188}$$

Consequently

$$\Delta \overset{\circ}{\Gamma}_{ijk} = \vartheta^{\alpha}{}_{i}\vartheta^{\beta}{}_{k}Y_{\alpha\beta j}\,, \qquad \text{where} \qquad Y_{\alpha\beta j} = \eta_{\alpha\mu}Y^{\mu}{}_{\beta j}\,. \tag{189}$$

In the infinitesimal approximation, (187) takes the form

$$\Delta \overset{\circ}{\Gamma}{}^{i}{}_{jk} = e_{\alpha}{}^{i}\vartheta^{\beta}{}_{k}X^{\alpha}{}_{\beta,j}\,. \tag{190}$$

while (188) with $X_{\alpha\beta} = \eta_{\alpha\mu}X^{\mu}{}_{\beta}$ reads

$$\Delta \overset{\circ}{\Gamma}_{ijk} = \vartheta^{\alpha}{}_{i}\vartheta^{\beta}{}_{k}X_{\alpha\beta,j}\,. \tag{191}$$

Note that since $X_{\alpha\beta}$ is antisymmetric, we have in this approximation

$$\Delta \overset{\circ}{\Gamma}_{ijk} = -\Delta \overset{\circ}{\Gamma}_{kji}\,. \tag{192}$$

We will also consider an additional physical meaningful approximation when the derivatives of the coframe is considered to be small relative to the derivatives of the transformation matrix. In this case, (190) and (191) read

$$\Delta \overset{\circ}{\Gamma}{}^{i}{}_{jk} = F^{i}{}_{k,j}\,, \qquad \text{where} \qquad F^{i}{}_{k} = e_{\alpha}{}^{i}\vartheta^{\beta}{}_{k}X^{\alpha}{}_{\beta}\,, \tag{193}$$

and

$$\Delta \overset{\circ}{\Gamma}_{ijk} = F_{ik,j}\,, \qquad \text{where} \qquad F_{ij} = \vartheta^{\alpha}{}_{i}\vartheta^{\beta}{}_{j}X_{\alpha\beta}\,. \tag{194}$$

Transformations preserved the geometric structure. Since the coframe field appears in the coframe geometrical structure only implicitly, (182) is a type of a gauge transformation. Invariance of the metric tensor and of the volume element restricts $L^{\alpha}{}_{\beta}$ to a pseudo-orthonormal matrix $G = SO(1,n)$. Let us ask now, under what conditions the general coframe connection (120) is invariant under the coframe transformations (182). First we rewrite (120) via the Levi-Civita connection. Using (179) we have

$$\overset{\circ}{\Gamma}_{ijk} = \overset{*}{\Gamma}_{ijk} + C_{ijk} - C_{kij} + C_{jki}\,. \tag{195}$$

Thus (120) takes the form

$$\Gamma_{ijk} = \overset{*}{\Gamma}_{ijk} + (\alpha_1+1)C_{ijk} + \alpha_2 g_{ik}C_j + \alpha_3 g_{ij}C_k + \beta_1 g_{jk}C_i + (\beta_2+1)C_{jki} + (\beta_3-1)C_{kij}\,. \tag{196}$$

Since the Levi-Civita connection $\overset{*}{\Gamma}_{ijk}$ is invariant under the transformations (182), the equation $\Delta\Gamma_{ijk} = 0$ takes the form

$$(\alpha_1+1)\Delta C_{ijk} + \alpha_2 g_{ik}\Delta C_j + \alpha_3 g_{ij}\Delta C_k + \beta_1 g_{jk}\Delta C_i + (\beta_2+1)\Delta C_{jki} + (\beta_3-1)\Delta C_{kij} = 0\,. \tag{197}$$

Hence in order to have an invariant coframe connection, we have to look for possible solutions of equation (197).

Trivial solutions of the invariance equation. Consider first two trivial solutions of (197) which turn out to be non-dynamical.

(i) *Arbitrary transformations — Levi-Civita connection.*
The equation (197) is evidently satisfied when all the numerical coefficients mutually equal zero. It is easy to check that these six relations are equivalent to (179). Thus the corresponded connection is of Levi-Civita. In this case, the elements of the matrix $L^\alpha{}_\beta$ are arbitrary functions of a point. Thus we come to a trivial fact that the Levi-Civita connection is a unique coframe connection which is invariant under arbitrary local $SO(1,n)$ transformations of the coframe field.

(ii) *Rigid transformations.*
Another trivial solution of the system (197) emerges when we require $\Delta C_{ijk} = 0$. All permutations and traces of this tensor are also equal to zero so (197) is trivially valid. Due to (189), it means that the matrix of transformations is independent on a point. In this case, an arbitrary coframe connection, in particular the Weitzenböck connection, remains unchanged. Thus we come to another trivial fact that the coframe connection is invariant under rigid transformations of the coframe field.

Dynamical solution. We will look now for nontrivial solutions of the system (197). Three traces of this system yield the equations of the type $\lambda\Delta C_i = 0$, where λ is a linear combination of the coefficients α_i, β_i. Thus we have to apply the first condition

$$\Delta C_i = 0\,. \tag{198}$$

The system (197) remains now in the form

$$(\alpha_1+1)\Delta C_{ijk} + (\beta_2+1)\Delta C_{jki} + (\beta_3-1)\Delta C_{kij} = 0\,. \tag{199}$$

Applying the complete antisymetrization in three indices we derive the second equation

$$\Delta C_{[ijk]} = 0\,. \tag{200}$$

The equation (199) remains now in the form

$$(\beta_2-\alpha_1)\Delta C_{jki} + (\beta_3-\alpha_1-2)\Delta C_{jki} = 0\,. \tag{201}$$

We have to restrict now the coefficients, otherwise we obtain $\Delta C_{ijk} = 0$, i.e., only the rigid transformations. Consequently we require

$$\beta_2 = \alpha_1\,, \qquad \beta_3 = \alpha_1 + 2\,. \tag{202}$$

Thus we have proved

Proposition 6: *The coframe connection*

$$\Gamma_{ijk} = \overset{*}{\Gamma}_{ijk} + (\alpha_1+1)C_{[ijk]} + \alpha_2 g_{ik}C_j + \alpha_3 g_{ij}C_k + \beta_1 g_{jk}C_i\,. \tag{203}$$

is invariant under the coframe transformations satisfied the equations

$$\Delta C_i = 0\,. \qquad \Delta C_{[ijk]} = 0\,. \tag{204}$$

Observe that this family includes the Levi-Civita connection, which is invariant under arbitrary transformations of the coframe field. The torsion tensor of the connection (203) is expressed as

$$T_{ijk} = (\alpha_1 + 1)C_{[ijk]} + (\alpha_2 - \alpha_3)(g_{ik}C_j - g_{ij}C_k)\,. \tag{205}$$

Thus a torsion-free subfamily of (203) is given by

$$\Gamma_{ijk} = \overset{*}{\Gamma}_{i(jk)} + \alpha_2(g_{ik}C_j + g_{ij}C_k) + \beta_1 g_{jk}C_i\,. \tag{206}$$

The nonmetricity tensor of the connection (203) reads

$$Q_{kij} = (\alpha_2 + \beta_1)(g_{ik}C_j + g_{jk}C_i) + 2\alpha_3 g_{jj}C_k\,. \tag{207}$$

Thus a metric compatible subfamily of (203) is given by

$$\Gamma_{ijk} = \overset{*}{\Gamma}_{i(jk)} + (\alpha_1 + 1)C_{[ijk]} + \alpha_2(g_{ik}C_j - g_{jk}C_i)\,. \tag{208}$$

From (205) and (207) we derive an interesting conclusions:

$$\Delta Q_{kij} = 0 \qquad \Longleftrightarrow \qquad \Delta C_i = 0\,. \tag{209}$$

and, together with this relation,

$$\Delta T_{ijk} = 0 \qquad \Longleftrightarrow \qquad \Delta C_{[ijk]} = 0\,. \tag{210}$$

Thus the relations (204) obtain a geometric meaning, they correspond to invariance of the torsion and nonmetricity tensors under coframe transformations.

3.6. Maxwell-Type System

Let us examine now what physical meaning can be given to the invariance conditions [41]

$$\Delta C_{[ijk]} = 0\,, \qquad \Delta C_i = 0\,. \tag{211}$$

Denote $K_{ijk} = \Delta C_{ijk}$. Thus (211) takes the form

$$K_{[ijk]} = 0\,, \qquad K^m{}_{im} = 0\,. \tag{212}$$

The tensor K_{ijk} depends on the derivatives of the Lorentz parameters $X_{\alpha\beta}$ and on the components of the coframe field

$$K_{ijk} = \frac{1}{2}\vartheta^\alpha{}_k\Big(X_{\alpha\beta,j}\vartheta^\beta{}_i - X_{\alpha\beta,i}\vartheta^\beta{}_j\Big)\,. \tag{213}$$

Thus, in fact, we have in (212), two first order partial differential equations for the entries of an antisymmetric matrix $X_{\alpha\beta}$. Let us construct from this matrix an antisymmetric tensor F_{ij}

$$F_{ij} = X_{\mu\nu}\vartheta^\mu{}_i\vartheta^\nu{}_j\,, \qquad X_{\mu\nu} = F_{ij}e_\mu{}^i e_\nu{}^j\,. \tag{214}$$

Substituting into (213), we derive

$$K_{ijk} = F_{k[i,j]} - \frac{1}{2}X_{\alpha\beta}\Big[(\vartheta^\alpha{}_k\vartheta^\beta{}_i)_{,j} - (\vartheta^\alpha{}_k\vartheta^\beta{}_j)_{,i}\Big]$$
$$= F_{k[i,j]} - F_{km}C^m{}_{ij} - \frac{1}{2}\left(F_{mi}\,\overset{\circ}{\Gamma}{}^m{}_{kj} - F_{mj}\,\overset{\circ}{\Gamma}{}^m{}_{ki}\right). \tag{215}$$

Consequently, the first equation from (212) takes the form

$$F_{[ij,k]} = \frac{2}{3}(C^m{}_{ij}F_{km} + C^m{}_{jk}F_{im} + C^m{}_{ki}F_{jm})\,, \tag{216}$$

while the second equation from (212) is rewritten as

$$F^i{}_{j,i} = -2F^i{}_mC^m{}_{ij} + F_{kj}g^{ki}{}_{,i} + F_{mj}g^{ki}\,\overset{\circ}{\Gamma}{}^m{}_{ki} - F_{mi}g^{ki}\,\overset{\circ}{\Gamma}{}^m{}_{kj}\,. \tag{217}$$

Observe first a significant approximation to (216—217). If the right hand sides in both equations are neglected, the equations take the form of the ordinary Maxwell equations for the electromagnetic field in vacuum —

$$F_{[ij,k]} = 0\,, \qquad F^i{}_{j,i} = 0\,. \tag{218}$$

In the coframe models, the gravity is modeled by a variable coframe field, i.e., by nonzero values of the quantities $\overset{\circ}{\Gamma}{}_{ij}{}^k$. Consequently, the right hand sides of (216—217) can be viewed as curved space additions, i.e., as the gravitational corrections to the electromagnetic field equations. In the flat spacetime, when a suitable coordinate system is chosen, these corrections are identically equal to zero. Consequently, in the flat spacetime, the invariance conditions (212) take the form of the vacuum Maxwell system.

On a curved manifold, the standard Maxwell equations are formulated in a covariant form. Let us show that our system (216—217) is already covariant. We rewrite (215) as

$$K_{ijk} = \frac{1}{2}(F_{ki,j} - F_{km}\,\overset{\circ}{\Gamma}{}^m{}_{ij} - F_{mi}\,\overset{\circ}{\Gamma}{}^m{}_{kj}) - \frac{1}{2}(\,i \longleftrightarrow j\,)\,. \tag{219}$$

Consequently,

$$K_{ijk} = F_{k[i;j]}\,, \tag{220}$$

where the covariant derivative (denoted by the semicolon) is taken relative to the Weitzenböck connection. Consequently, the system (216—217) takes the covariant form

$$F_{[ij;k]} = 0\,, \qquad F^i{}_{j;i} = 0\,. \tag{221}$$

These equations are literally the same as the electromagnetic sector field equations of the Maxwell-Einstein system. The crucial difference is encoded in the type of the covariant derivative. In the Maxwell-Einstein system, the covariant derivative is taken relative to the Levi-Civita connection, while, in our case, the corresponding connection is of Weitzenböck. Observe that, due to our approach, the Weitzenböck connection is rather natural in (221). Indeed, since the electromagnetic-type field describes the local change of the coframe field, it should itself be referred only to the global changes of the coframe. As we have shown, such global transformations correspond precisely to the teleparallel geometry with the Weitzenböck connections.

4. Geometrized Coframe Field Model

4.1. Generalized Einstein-Hilbert Lagrangian

One of the most important features of the Einstein gravity theory is its pure geometrical content. The basic field variable of this theory is the metric tensor field g_{ij}. The action integral is given by the Einstein-Hilbert Lagrangian

$$^{(\mathtt{GR})}\mathcal{A} = \int_M R\left(\overset{*}{\Gamma}{}^i_{jk}(g), g\right) * 1\,, \tag{222}$$

where R is the curvature scalar constructed from the metric tensor and its partial derivatives while $*1$ is the invariant volume element constructed from the metric tensor. When we restrict to the quasilinear second order field equations the Lagrangian (222) is a unique possibility.

The coframe field model also constructed from the geometrical field variable — coframe. Its Lagrangian however is taken as an arbitrary linear combination of the global $SO(1,3)$ invariants. The geometrical sense of this expression is not clear. Although the coframe Lagrangian can be written in terms of the torsion of the flat connection it does not mean that it corresponds to the Weitzenböck geometry with a flat curvature and a non-zero connection. Indeed also the standard Einstein-Hilbert Lagrangian (222) can be rewritten in such a form. Moreover, as we have seen in the previous section, there is a wide class of connections all constructed from Weitzenböck connection and its torsion. In particular, using the coframe Lagrangian in the form (4) we cannot answer the question: *What special geometry corresponds to the set of viable coframe models?*

Our proposal is to consider for the coframe Lagrangian an expression similar to (222)

$$^{(\mathtt{cof})}\mathcal{A} = \int_M R\Big(\Gamma^i_{jk}(\vartheta^\alpha), g(\vartheta^\alpha)\Big) * 1\,, \tag{223}$$

which is constructed from the general free parametric coframe connection. Also the invariant volume element $*1$ is constructed here from the coframe field. Since the Levi-Civita connection is included as a special case of general coframe connection we have in (223) a generalization of the standard GR.

4.2. Curvature of the Coframe Connection

Riemannian curvature 2-form. We start with the definitions of the Riemannian curvature machinery. Although it is a classical subject of differential geometry [50], in the case of a general connection of non-zero torsion and nonmetricity, slightly different notations are in use. Moreover, in this case, it is useful to apply the formalism of differential forms. We accept the agreements used in metric-affine gravity [5].

Let a connection 1-form $\mathit{\Gamma}_a{}^b$ referred to a general nonholonomic basis (θ^a, f_a) be given. The *curvature 2-form* is defined as

$$\mathcal{R}_a{}^b = d\mathit{\Gamma}_a{}^b - \mathit{\Gamma}_a{}^c \wedge \mathit{\Gamma}_c{}^b\,. \tag{224}$$

It satisfies two fundamental identities:

The first Bianchy identity involves the first order derivatives of the connection

$$DT^a - \mathcal{R}_b{}^a \wedge \theta^b = 0\,, \qquad \text{or} \qquad dT^a + \Gamma_b{}^a \wedge T^b - \mathcal{R}_b{}^a \wedge \theta^b = 0\,. \tag{225}$$

The second Bianchy identity involves the second order derivatives of the connection

$$D\mathcal{R}_b{}^a = 0\,, \qquad \text{or} \qquad d\mathcal{R}_a{}^b + \Gamma_a{}^c \wedge \mathcal{R}_c{}^b - \Gamma_c{}^b \wedge \mathcal{R}_a{}^c = 0\,. \tag{226}$$

It is useful to consider the Riemannian curvature of the coframe connection to be referred to a basis composed from the elements of the coframe field itself. The corresponded quantity

$$\mathcal{R}_\alpha{}^\beta = d\Gamma_\alpha{}^\beta - \Gamma_\alpha{}^\gamma \wedge \Gamma_\gamma{}^\beta\,. \tag{227}$$

is related to the generic basis expression by the standard tensorial rule with the matrices of transformation $\vartheta^\alpha{}_i f_a{}^i$

$$\mathcal{R}_a{}^b = \mathcal{R}_\alpha{}^\beta(\vartheta^\alpha{}_i f_a{}^i)(e_\beta{}^j \theta^b{}_j)\,. \tag{228}$$

From (227), we see that the Riemannian curvature of the Weitzenböck connection is zero being referred to a basis of the coframe field. Due to (227), it is zero in an arbitrary basis.

Being referred to a coordinate basis, the Riemannian curvature 2-form reads

$$\begin{aligned} \mathcal{R}_i{}^j &= d\Gamma_i{}^j - \Gamma_i^k \wedge \Gamma_k{}^j & (229)\\ &= d\Gamma^j{}_{in} \wedge dx^n - \Gamma^k{}_{im}\Gamma^j{}_{kn} dx^m \wedge dx^n & (230)\\ &= \left(\Gamma^j{}_{in,m} - \Gamma^k{}_{im}\Gamma^j{}_{kn}\right) dx^m \wedge dx^n\,. & (231) \end{aligned}$$

The components of the Riemannian curvature 2-form

$$\mathcal{R}_i{}^j = \frac{1}{2} R^j{}_{imn} dx^m \wedge dx^n \tag{232}$$

are arranged in the familiar expression of the *Riemannian curvature tensor*

$$R^j{}_{imn} = \Gamma^j{}_{in,m} - \Gamma^j{}_{im,n} + \Gamma^k{}_{in}\Gamma^j{}_{km} - \Gamma^k{}_{im}\Gamma^j{}_{kn}\,. \tag{233}$$

Curvature scalar density. Curvature scalar plays an important role in physical applications. In fact, it is used as an integrand in action of geometrical field models — Hilbert-Einstein Lagrangian density

$$\mathcal{L} = R\,\text{vol} = R * 1\,, \tag{234}$$

where star denotes the Hodge dual. In term of the curvature 2-form, this expression is rewritten as

$$\mathcal{L} = \mathcal{R}_{ij} \wedge * (dx^i \wedge dx^j) = \mathcal{R}_{\alpha\beta} \wedge * \vartheta^{\alpha\beta}\,. \tag{235}$$

where the abbreviation $\vartheta^{\alpha\beta} = \vartheta^\alpha \wedge \vartheta^\beta$ is used. Extracting in (235) the total derivative term we obtain

$$\begin{aligned} \mathcal{L} &= (d\Gamma_{\alpha\beta} - \Gamma_\alpha{}^\gamma \wedge \Gamma_{\gamma\beta}) \wedge * \vartheta^{\alpha\beta} \\ &= d\left(\Gamma_{\alpha\beta} \wedge * \vartheta^{\alpha\beta}\right) + \Gamma_{\alpha\beta} \wedge d * \vartheta^{\alpha\beta} - \Gamma_\alpha{}^\gamma \wedge \Gamma_{\gamma\beta} \wedge * \vartheta^{\alpha\beta}\,. \qquad (236) \end{aligned}$$

For actual calculation of this quantity, it is useful to express the connection 1-form in the basis of the coframe field. We denote

$$\Gamma_{\alpha\beta}=K_{\alpha\gamma\beta}\vartheta^\gamma\,. \tag{237}$$

Substituting it in the total derivative term of (236) we have

$$\begin{aligned} d\left(\Gamma_{\alpha\beta}\wedge *\vartheta^{\alpha\beta}\right) &= d\left(K_{\alpha\gamma\beta}\vartheta^\gamma\wedge *\vartheta^{\alpha\beta}\right)=(-1)^n d\left[K_{\alpha\gamma\beta}*\left(e^\gamma\rfloor\vartheta^{\alpha\beta}\right)\right] \\ &= (-1)^n d\left[\left(K^\alpha{}_{\alpha\beta}-K_{\beta\alpha}{}^\alpha\right)*\vartheta^\beta\right]. \end{aligned} \tag{238}$$

The second term of (236) reads

$$\Gamma_{\alpha\beta}\wedge d*\vartheta^{\alpha\beta}=K_{\alpha\gamma\beta}\vartheta^\gamma\wedge d*\vartheta^{\alpha\beta}=K_{\alpha\gamma\beta}\left[d\vartheta^\gamma\wedge *\vartheta^{\alpha\beta}-d\left(\vartheta^\gamma\wedge *\vartheta^{\alpha\beta}\right)\right]. \tag{239}$$

Calculate:

$$d\vartheta^\gamma\wedge *\vartheta^{\alpha\beta}=\frac{1}{2}C^\gamma{}_{\mu\nu}\vartheta^{\mu\nu}\wedge *\vartheta^{\alpha\beta}=(-1)^{n+1}C^{\gamma\alpha\beta}*1\,. \tag{240}$$

and

$$d\left(\vartheta^\gamma\wedge *\vartheta^{\alpha\beta}\right)=(-1)^n d*\left(\eta^{\alpha\gamma}\vartheta^\beta-\eta^{\beta\gamma}\vartheta^\alpha\right)=(-1)^n\left(\eta^{\beta\gamma}C^\alpha-\eta^{\alpha\gamma}C^\beta\right)*1\,. \tag{241}$$

Consequently the second term of (236) takes the form

$$\Gamma_{\alpha\beta}\wedge d*\left(\vartheta^\alpha\wedge\vartheta^\beta\right) = (-1)^n\left[K_{\alpha\gamma\beta}C^{\gamma\alpha\beta}-\left(K^\alpha{}_{\alpha\beta}-K_{\beta\alpha}{}^\alpha\right)C^\beta\right]*1 \tag{242}$$

The third term of (236) reads

$$\begin{aligned} \Gamma_\alpha{}^\gamma\wedge\Gamma_{\gamma\beta}\wedge *\left(\vartheta^\alpha\wedge\vartheta^\beta\right) &= K_{\alpha\mu}{}^\gamma K_{\gamma\nu\beta}\vartheta^{\mu\nu}\wedge *\vartheta^{\alpha\beta} \\ &= (-1)^n\left(K^{\alpha\beta\gamma}K_{\gamma\alpha\beta}-K^\alpha{}_{\alpha\gamma}K^{\gamma\beta}{}_\beta\right)*1\,. \end{aligned} \tag{243}$$

Consequently the Lagrangian density takes the form

$$\begin{aligned} \mathcal{L}(-1)^n &= d\left[\left(K^\alpha{}_{\alpha\beta}-K_{\beta\alpha}{}^\alpha\right)*\vartheta^\beta\right]+\left[K_{\alpha\gamma\beta}C^{\gamma\alpha\beta}-\left(K^\alpha{}_{\alpha\beta}-K_{\beta\alpha}{}^\alpha\right)C^\beta\right]*1 \\ &\quad -\left(K^{\alpha\beta\gamma}K_{\gamma\alpha\beta}-K^\alpha{}_{\alpha\gamma}K^{\gamma\beta}{}_\beta\right)*1\,. \end{aligned} \tag{244}$$

Due to (128), the tensor $K_{\alpha\gamma\beta}$ is of the form

$$K_{\alpha\gamma\beta}=\alpha_1C_{\beta\gamma\alpha}+\alpha_2C_\gamma\eta_{\alpha\beta}+\alpha_3C_\alpha\eta_{\beta\gamma}+\beta_1C_\beta\eta_{\alpha\gamma}+\beta_2C_{\gamma\alpha\beta}+\beta_3C_{\alpha\beta\gamma}\,, \tag{245}$$

Substituting this expression in (244) we obtain a total derivative term plus a sum of terms which are quadratic in $C_{\alpha\beta\gamma}$. Since (4) is the most general expression quadratic in $C_{\alpha\beta\gamma}$, the following statement is clear.

Proposition 7 *The Hilbert-Einstein Lagrangian of the general metric-coframe connection (128) is equivalent up to a total derivative term to the general coframe Lagrangian*

$$R(\Gamma_{\alpha\beta})*1=\zeta_0 d(C_\alpha*\vartheta^\alpha)+\left(\zeta_1C_{\alpha\beta\gamma}C^{\alpha\beta\gamma}+\zeta_2C_{\alpha\beta\gamma}C^{\beta\gamma\alpha}+\zeta_3C_\alpha C^\alpha\right)*1\,, \tag{246}$$

where the parameters ζ_i are expressed by second order polynomials of the coefficients α_i,β_i.

The actual expressions for the coefficients ζ_i are rather involved. We discuss the parameter ζ_0 in sequel.

4.3. Einstein-Hilbert Lagrangian without Second Order Derivatives

It is well known that in GR the Einstein-Hilbert Lagrangian involves the second order derivatives of the metric tensor. These terms join in a total derivative term which is not relevant for the field equation. Although, the total derivative terms cannot consistently be dropped out. In particular, the quantization procedure requires an addition of a boundary term in order to compensate the total derivative [56], [57]. Let us calculate the total derivative term in our model. With (245) we have

$$K^{\alpha}{}_{\alpha\beta} = \eta^{\alpha\gamma} K_{\alpha\gamma\beta} = [\alpha_2 + \alpha_3 + (n+1)\beta_1 + \beta_2 - \beta_3]\, C_\beta\,, \tag{247}$$

and

$$K_{\beta\alpha}{}^{\alpha} = \eta^{\beta\gamma} K_{\alpha\gamma\beta} = [\alpha_1 + \alpha_2 + (n+1)\beta_1 + \beta_2 - \beta_3]\, C_\beta\,. \tag{248}$$

Thus

$$d\left[(K^{\alpha}{}_{\alpha\beta} - K_{\beta\alpha}{}^{\alpha}) * \vartheta^\beta\right] = -\left[\alpha_1 + n(\alpha_3 - \beta_1) + 2\beta_2 - \beta_3\right] d\left(C_\beta * \vartheta^\beta\right). \tag{249}$$

Consequently, the coefficient ζ_0 in (246) takes the form

$$\zeta_0 = \alpha_1 + n(\alpha_3 - \beta_1) + 2\beta_2 - \beta_3\,. \tag{250}$$

For the Weitzenböck connection, this coefficient is zero together with all other terms of the Lagrangian. For the Levi-Civita connection, $\zeta_0 = -2$ on a manifold of an arbitrary dimension.

We can identify now a family of coframe connections without a total derivative term at all. It is enough to require

$$\alpha_1 + n(\alpha_3 - \beta_1) + 2\beta_2 - \beta_3 = 0\,. \tag{251}$$

The corresponding connection is given by

$$\begin{aligned} \Gamma_{ijk} &= \overset{\circ}{\Gamma}_{ijk} + \alpha_1(C_{ijk} + C_{kij}) + \alpha_2 g_{ik} C_j + \alpha_3(g_{ij}C_k + nC_{kij}) + \\ &\quad \beta_1(g_{jk}C_i - nC_{kij}) + \beta_2(C_{jki} + 2C_{kij})\,. \end{aligned} \tag{252}$$

This family includes the metric-compatible connections

$$\Gamma_{ijk} = \overset{\circ}{\Gamma}_{ijk} + \alpha_1(C_{ijk} + C_{jki} + 3C_{kij}) + \alpha_2(g_{ik}C_j - g_{jk}C_i + nC_{kij})\,, \tag{253}$$

and the symmetric (torsion-free) connections

$$\Gamma_{ijk} = \overset{\circ}{\Gamma}_{i(jk)} + \alpha_2\,(g_{ik}C_j + g_{ij}C_k) + \beta_1 g_{jk} C_i + [1 - n(\alpha_2 - \beta_1)]\,(C_{jki} + C_{kji})\,. \tag{254}$$

Also the gauge invariant connections (203) can be found into the family (252).

Consequently we identified a remarkable property of the coframe geometry. There is a family of coframe connections in which standard Einstein-Hilbert Lagrangian does not involve second order derivatives terms at all. It means that there is a family of coframe models with a geometrical Lagrangian which is completely equivalent to the Yang-Mills Lagrangians of particle physics.

5. Conclusion

GR is a well-posed classical field theory for 10 independent variables — the components of the metric tensor. Although, this theory is completely satisfactory in the pure gravity sector, its possible extensions to other physics phenomena is rather problematic. In particular, the description of fermions on a curved space and the supergravity constructions require a richer set of 16 independent variables. These variables can be assembled in a coframe field, i.e., a local set of four linearly independent 1-forms. Moreover, in supergravity, it is necessary to involve a special flat connection constructed from the derivatives of the coframe field. These facts justify the study of the field models based on a coframe variable alone.

The classical field construction of the coframe gravity is based on a Yang-Mills-type Lagrangian which is a linear combination of quadratic terms with dimensionless coefficients. Such a model turns out to be satisfactory in the gravity sector and has the viable Schwarzschild solutions even being alternative to the standard GR. Moreover, the coframe model treating of the gravity energy makes it even more preferable than the ordinary GR where the gravity energy cannot be defined at all. A principle problem is that the coframe gravity construction does not have any connection to a specific geometry even being constructed from the geometrical meaningful objects. A geometrization of the coframe gravity is an aim of this chapter.

We construct a general family of coframe connections which involves as the special cases the Levi-Civita connection of GR and the flat Weitzenböck connection. Every specific connection generates a geometry of a specific type. We identify the subclasses of metric-compatible and torsion-free connections. Moreover we study the local linear transformations of the coframe fields and identify a class of connections which are invariant under restricted coframe transformations. Quite remarkable that the restriction conditions are necessary approximated by a Maxwell-type system of equations.

On a basis of the coframe geometry, we propose a geometric action for the coframe gravity. It has the same form as the Einstein-Hilbert action of GR, but the scalar curvature is constructed from the general coframe connection. We show that this geometric Lagrangian is equivalent to the coframe Lagrangian up to a total derivative term. Moreover there is a family of coframe connections which Lagrangian does not include the higher order terms. In this case, the equivalence is complete.

However, the Hilbert-Einstein-type action itself is not enough to predict a unique coframe connection. Indeed, the coframe connection has six free parameters, while the action involves only four of their combinations. Moreover, one combination represents a total derivative term in Lagrangian which does not influence the field equations. So the gravity action itself is not defined uniquely the geometry on the base manifold. It should not be, however, a problem. Indeed, the gravitational field is not a unique physical field. Moreover, gravity does not even exist without matter fields as its origin. An action for an arbitrary (non-scalar) field necessarily involves the connection. So the problem can be formulated as the following: To find out which matter field has to be added to the coframe Lagrangian in order to predict uniquely the type of the coframe connection and consequently the geometry of the underlying manifold. This problem can serve as a basis for future investigation.

Acknowledgements

I thank Shmuel Kaniel, Yakov Bekenstein, Yaakov Friedman (Jerusalem), Friedrich W. Hehl (Cologne and Missouri-Columbia), Yuri N. Obukhov (Moscow and Cologne), and Roman Jackiw (MIT) for fruitful discussions of the coframe gravity.

6. Appendix — Differential Form Notations

We collect here some algebraic rules which are useful for calculations with the differential forms. Recall that we are working on an $n+1$ dimensional manifold.

1. Interior product

In a basis of 1-forms ϑ^α, a p-form Ψ is expressed as

$$\Psi = \frac{1}{p!}\Psi_{\alpha_1 \cdots \alpha_p}\vartheta^{\alpha_1} \wedge \cdots \wedge \vartheta^{\alpha_p} . \tag{A.1}$$

Interior product couples the basis vectors and basis 1-forms as

$$e_\alpha \rfloor \vartheta^\beta = \delta^b_\alpha . \tag{A.2}$$

By bilinearity and the Leibniz-type rule,

$$e_\alpha \rfloor (w_1 \wedge w_2) = (e_\alpha \rfloor w_1) \wedge w_2 + (-1)^{\deg w_1} w_1 \wedge (e_\alpha \rfloor w_2) , \tag{A.3}$$

the definition of the interior product is extended to forms of arbitrary degree. Mixed applications of the exterior and interior products to a p-form w satisfy the relations

$$\vartheta^\alpha \wedge (e_a \rfloor w) = pw , \tag{A.4}$$

and

$$e_a \rfloor (\vartheta^\alpha \wedge w) = (n-p)w . \tag{A.5}$$

2. Hodge star operator

The Hodge star operator maps p-forms into $(n+1-p)$-forms. In a pseudo-orthonormal basis ϑ^α, the metric tensor is represented by the constant components $\eta_{\alpha\beta} = \text{diag}(-1, 1, \cdots, 1)$. In this case, the Hodge star operator is defined as

$$*\Psi = \frac{1}{p!(n+1-p)!}\Psi_{\alpha_0 \cdots \alpha_p}\eta^{\alpha_0\beta_0} \cdots \eta^{\alpha_p\beta_p}\varepsilon_{\beta_0 \cdots \beta_n}\vartheta^{\beta_{p+1} \wedge \cdots \wedge \vartheta^{\beta_n}} , \tag{A.6}$$

where the permutation symbol is normalized as

$$\varepsilon_{0 \cdots n} = 1 , \qquad \varepsilon^{0 \cdots n} = -1 . \tag{A.7}$$

For the basis forms themselves, this formula can be rewritten as

$$*(\vartheta_{\alpha_0} \wedge \cdots \wedge \vartheta_{\alpha_p}) = \frac{1}{(n+1-p)!}\varepsilon_{\alpha_0 \cdots \alpha_p \beta_1 \cdots \beta_{n-p}}\vartheta^{\beta_1} \wedge \cdots \wedge \vartheta^{\beta_{n-p}} . \tag{A.8}$$

In particular,

$$*(\vartheta_{\alpha_0}\wedge\cdots\wedge\vartheta_{\alpha_n})=\varepsilon_{\alpha_0\cdots\alpha_n}\,,\qquad *1=\frac{1}{n!}\varepsilon_{\alpha_0\cdots\alpha_n}\vartheta^{\alpha_1}\wedge\cdots\wedge\vartheta^{\alpha_n}\,. \tag{A.9}$$

When the Hodge map defined by a Lorentzian-type metric $\eta_{\alpha\beta}$ it acts on a p-form w

$$**\,w=(-1)^{p(n+1-p)+1}w=(-1)^{pn+1}w\,. \tag{A.10}$$

For the forms w_1, w_2 of the same degree,

$$w_1\wedge *w_2=w_2\wedge *w_1\,. \tag{A.11}$$

With the Hodge map, the wedge product can be transformed into the interior product and vice versa by the relations

$$*(w\wedge\vartheta_\alpha)=e_\alpha\rfloor *w\,, \tag{A.12}$$

and

$$\vartheta_\alpha\wedge *w=(-1)^{n(n-p)}*(e_\alpha\rfloor w)\,. \tag{A.13}$$

3. Exterior derivative and coderivative of the coframe field
We express the exterior derivative of the coframe field as

$$d\vartheta^\alpha=\frac{1}{2}C^\alpha{}_{\beta\gamma}\vartheta^\beta\wedge\vartheta^\gamma\qquad C_\alpha=C^\mu{}_{\mu\alpha}\,. \tag{A.14}$$

The divergence of the coframe 1-form is

$$d*\vartheta_\alpha=-C_\alpha *1\,. \tag{A.15}$$

Indeed, using (A.8) we calculate

$$\begin{aligned} d*\vartheta_\alpha &= \frac{1}{n!}\varepsilon_{\alpha\beta_1\cdots\beta_n}d(\vartheta^{\beta_1}\wedge\cdots\wedge\vartheta^{\beta_n})\\ &= \frac{1}{2(n-1)!}\varepsilon_{\alpha\beta_1\cdots\beta_n}C^{\beta_1}{}_{\mu\nu}\vartheta^\mu\wedge\vartheta^\nu\wedge\vartheta^{\beta_2}\wedge\cdots\wedge\vartheta^{\beta_n}\,. \end{aligned} \tag{A.16}$$

Using (A.9) and (A.10) we have

$$\vartheta^\mu\wedge\vartheta^\nu\wedge\vartheta^{\beta_2}\wedge\cdots\wedge\vartheta^{\beta_n}=-\varepsilon^{\mu\nu\beta_2\cdots\beta_n}*1\,. \tag{A.17}$$

Consequently,

$$\begin{aligned} d*\vartheta_\alpha &= -\frac{1}{2(n-2)!}\varepsilon_{\alpha\beta_1\cdots\beta_{n-1}}\varepsilon^{\mu\nu\beta_2\cdots\beta_{n-1}}C^{\beta_1}{}_{\mu\nu}*1=\\ &= \frac{1}{2}(\delta^\mu_\alpha\delta^\nu_{\beta_1}-\delta^\nu_\alpha\delta^\mu_{\beta_1})C^{\beta_1}{}_{\mu\nu}*1=C^\mu{}_{\alpha\mu}*1=-C_\alpha*1\,. \end{aligned} \tag{A.18}$$

In a coordinate basis we consider the tensors

$$C^i{}_{jk}=\frac{1}{2}\left(\overset{\circ}{\Gamma}{}^i{}_{jk}-\overset{\circ}{\Gamma}{}^i{}_{kj}\right)\,,\qquad C_i=C^m{}_{mi}\,. \tag{A.19}$$

It is easy to check the relations

$$C^i{}_{jk} = C^\alpha{}_{\beta\gamma} e_\alpha{}^i \vartheta^\beta{}_j \vartheta^\gamma{}_k \,, \qquad C_i = C_\alpha \vartheta^\alpha{}_i \,. \tag{A.20}$$

Define a non-indexed (scalar-valued) 1-form

$$\mathcal{A} = e_\mu \rfloor d\vartheta^\mu = 2\vartheta^\mu{}_{[i,j]} e_\mu{}^i \, dx^j = 2C_i dx^i = 3C_\alpha \vartheta^\alpha \,. \tag{A.21}$$

On a manifold with a metric $g = \eta_{\mu\nu}\vartheta^\mu \otimes \vartheta^\nu$ (Section 3), we define, in addition, a scalar-valued 3-form

$$\begin{aligned} \mathcal{B} &= \eta_{\mu\nu} \, d\vartheta^\mu \wedge \vartheta^\nu = -\eta_{\mu\nu} \vartheta^\mu{}_{i,j} \vartheta^\nu{}_k \, dx^i \wedge dx^j \wedge dx^k \\ &= C_{ijk} dx^i \wedge dx^j \wedge dx^k = C_{\alpha\beta\gamma} \vartheta^\alpha \wedge \vartheta^\beta \wedge \vartheta^\gamma \,. \end{aligned} \tag{A.22}$$

The operations of symmetrization and antisymmetrization of tensors are used here in the normalized form:

$$(a_1 \cdots a_p) = \frac{1}{p!} \mathrm{Sym}(a_1 \cdots a_p) \,, \qquad [a_1 \cdots a_p] = \frac{1}{p!} \mathrm{Ant}(a_1 \cdots a_p) \,. \tag{A.23}$$

References

[1] F. W. Hehl, P. Von Der Heyde, G. D. Kerlick and J. M. Nester, *Rev. Mod. Phys.* 48, 393 (1976).

[2] S. Hojman, M. Rosenbaum, M. P. Ryan and L. C. Shepley, *Phys. Rev. D* 17, 3141 (1978).

[3] W. Kopczynski, *Acta Phys. Polon. B* 10 (1979) 365.

[4] J. D. McCrea, *Class. Quant. Grav.* 9, 553 (1992).

[5] F. W. Hehl, J. D. McCrea, E. W. Mielke and Y. Neeman, *Phys. Rept.* 258, 1 (1995);

[6] F. Gronwald and F. W. Hehl, arXiv:gr-qc/9602013.

[7] G. Giachetta and G. Sardanashvily, *Class. Quant. Grav.* 13, L67 (1996)

[8] Yu. N. Obukhov, E. J. Vlachynsky, W. Esser and F. W. Hehl, *Phys. Rev. D* 56, 7769 (1997).

[9] J. Socorro, C. Lammerzahl, A. Macias and E. W. Mielke, *Phys. Lett. A* 244, 317 (1998)

[10] A. Garcia, F. W. Hehl, C. Laemmerzahl, A. Macias and J. Socorro, *Class. Quant. Grav.* 15, 1793 (1998)

[11] F. Gronwald, *Int. J. Mod. Phys. D* 6, 263 (1997)

[12] F. W. Hehl and A. Macias, *Int. J. Mod. Phys. D* 8, 399 (1999)

[13] D. Puetzfeld, *Class. Quant. Grav.* 19, 3263 (2002)

[14] M. Godina, P. Matteucci and J. A. Vickers, *J. Geom. Phys.* 39, 265 (2001)

[15] D. Vassiliev, *Annalen Phys.* 14, 231 (2005)

[16] V. Pasic and D. Vassiliev, *Class. Quant. Grav.* 22, 3961 (2005)

[17] Y. N. Obukhov, *Phys. Rev. D* 73, 024025 (2006)

[18] H. F. M. Goenner, *Living Rev. Rel.* 7, 2 (2004).

[19] K. Hayashi and T. Shirafuji, *Phys. Rev. D* 19, 3524 (1979)

[20] E. Sezgin and P. van Nieuwenhuizen, *Phys. Rev. D* 21, 3269 (1980).

[21] J. Nitsch and F. W. Hehl, *Phys. Lett. B* 90, 98 (1980);

[22] F. Mueller-Hoissen and J. Nitsch, *Phys. Rev. D* 28, 718 (1983);

[23] R. Kuhfuss and J. Nitsch, *Gen. Rel. Grav.* 18, 1207 (1986).

[24] E. W. Mielke, *Annals Phys.* 219, 78 (1992);

[25] U. Muench, F. Gronwald and F. W. Hehl, *Gen. Rel. Grav.* 30, 933 (1998);

[26] R. S. Tung and J. M. Nester, *Phys. Rev. D* 60, 021501 (1999);

[27] Y. Itin and S. Kaniel, *J. Math. Phys.* 41, 6318 (2000)

[28] M. Blagojevic and M. Vasilic, *Class. Quant. Grav.* 17, 3785 (2000);

[29] Y. Itin, *Int. J. Mod. Phys. D* 10, 547 (2001)

[30] M. Blagojevic and I. A. Nikolic, *Phys. Rev. D* 62, 024021 (2000)

[31] I. L. Shapiro, *Phys. Rept.* 357, 113 (2001)

[32] R. T. Hammond, *Rept. Prog. Phys.* 65, 599 (2002).

[33] Y. Itin, *Class. Quant. Grav.* 19, 173 (2002);

[34] Y. Itin, *Gen. Rel. Grav.* 34, 1819 (2002);

[35] Y. N. Obukhov and J. G. Pereira, *Phys. Rev. D* 67, 044016 (2003)

[36] Y. Itin, *J. Phys. A* 36, 8867 (2003)

[37] M. Leclerc, *Phys. Rev. D* 71, 027503 (2005)

[38] Y. N. Obukhov, G. F. Rubilar and J. G. Pereira, *Phys. Rev. D* 74, 104007 (2006)

[39] Y. Itin, *J. Math. Phys.* 46 12501 (2005).

[40] F. B. Estabrook, *Class. Quant. Grav.* 23, 2841 (2006)

[41] Y. Itin, *Class. Quant. Grav.* 23 (2006) 3361.

[42] G. G. L. Nashed and T. Shirafuji, *Int. J. Mod. Phys. D* 16, 65 (2007)

[43] A. Ashtekar, *Phys. Rev. D* 36, 1587 (1987).

[44] S. Deser and C. J. Isham, *Phys. Rev. D* 14, 2505 (1976).

[45] J. M. Nester and R. S. Tung, *Phys. Rev. D* 49, 3958 (1994)

[46] S. Deser and P. van Nieuwenhuizen, *Phys. Rev. D* 10, 411 (1974).

[47] P. G. Bergmann, V. de Sabbata, G. T. Gillies and P. I. Pronin, *International School of Cosmology and Gravitation: 15th Course: Spin in Gravity: Is it Possible to Give an Experimental Basis to Torsion?*, Erice, Italy, 13-20 May 1997.

[48] P. Van Nieuwenhuizen, *Phys. Rept.* 68, 189 (1981).

[49] A. Perez and C. Rovelli, *Phys. Rev. D* 73, 044013 (2006)

[50] S. Kobayashi and K. Nomizu, *Foundations of Differential Geometry*, vol. 1 and 2, Interscience Tracts in Pure and Applied Mathematics, Interscience Publ., New-York, 1969.

[51] S. Kobayashi, *Transformation Groups in Differential Geometry*, Springer-Verlag, 1972.

[52] T.Y. Thomas: *The differential invariants on generalized spaces*, Cambridge, The University Press, 1934.

[53] J.A. Schouten, *Ricci-Calculus, An Introduction to Tensor Analysis and its Geometrical Applications* (2nd ed., Springer-Verlag, New York, 1954).

[54] V. Iyer, R.M. Wald *Phys.Rev.* , **D50**, (1994), 846-864.

[55] F.W. Hehl and Yu.N. Obukhov, *Lecture Notes in Physics* Vol. 562 (Springer: Berlin, 2001) pp. 479-504.

[56] J. W. . York, *Phys. Rev. Lett.* 28, 1082 (1972).

[57] G. W. Gibbons and S. W. Hawking, *Phys. Rev. D* 15, 2752 (1977).

Reviewer: Yuri Obukhov, email: yo@thp.uni-koeln.de
Institute for Theoretical Physics, University of Cologne 50923 Koln, Germany
Dept. of Theoretical Physics, Moscow State University 117234 Moscow, Russia

In: Classical and Quantum Gravity Research
Editors: M.N. Christiansen et al, pp. 511-532
ISBN 978-1-60456-366-5

Chapter 12

DOES RELATIONALISM ALONE CONTROL GEOMETRODYNAMICS WITH SOURCES?

Edward Anderson
Peterhouse, Cambridge CB2 1RD and DAMTP,
Centre for Mathematical Sciences, Wilberforce Road,
Cambridge CB3 OWA

Abstract

This paper concerns relational first principles from which the Dirac procedure exhaustively picks out the geometrodynamics corresponding to general relativity as one of a handful of consistent theories. This was accompanied by a number of results and conjectures about matter theories and general features of physics – such as gauge theory, the universal light cone principle of special relativity and the equivalence principle – being likewise picked out. I have previously shown that many of these matter results and conjectures are contingent on further unrelational simplicity assumptions. In this paper, I point out 1) that the exhaustive procedure in these cases with matter fields is slower than it was previously held to be. 2) While the example of equivalence principle violating matter theory that I previously showed how to accommodate on relational premises has a number of pathological features, in this paper I point out that there is another closely related equivalence principle violating theory that also follows from those premises and is less pathological. This example being known as an 'Einstein–aether theory', it also serves for 3) illustrating limitations on the conjectured emergence of the universal light cone special relativity principle.

PACS 04.20.Fy, 04.20.Cv.

1. Introduction

1.1. Relationalism

The relational perspective of Barbour [1, 2, 3, 4, 5] implements ideas of Leibniz [6] and Mach [7] (see also [8]) to modern physics. In this approach, one starts with a configuration space Q of (models of) whole-universe systems. One then adopts two relational postulates.

Configurational relationalism: that certain transformations acting on Q are physically meaningless. One way [9][1] of implementing this is to use arbitrary-G-frame-corrected quantities rather than bare Q configurations, where G is the group of physically meaningless motions. For, despite this augmenting Q to $\mathsf{Q} \times G$, variation with respect to each adjoined independent auxiliary G-variable produces a constraint which removes one G variable and one redundancy among the Q variables, so that one ends up on the quotient space Q/G (the desired reduced configuration space). This is widely a necessity in theoretical physics through working on the various reduced spaces directly often being technically unmanageable, for instance in particle physics theories with its internal gauge group G or in the split spacetime approach to general relativity with its spatial diffeomorphisms.

Temporal relationalism: that there is no meaningful primary notion of time for the universe as a whole. One implementation of temporal relationalism is through using manifestly reparametrization invariant actions that do not rely on any extraneous time-related variables either.

For $\mathsf{Q} = \{$n particle postions$\}$ and G the Euclidean group of translations and rotations, the relational postulates form plausible alternative foundations for a portion of Newtonian mechanics [11, 12, 13] (and admit also a scale-free counterpart for G the Similarity group of translations, rotations and dilatations [10, 14, 12]). The main idea in this paper concerns that (spatially compact without boundary) general relativity can be derived from these postulates in the case in which G is the group of 3-diffeomorphisms. (This derivation also assumes a set of mathematical simplicity postulates and observational checks [4, 15] described in Sec 1.3). This answers a question of Wheeler: *"if one did not know the Einstein–Hamilton–Jacobi equation, how might one hope to derive it straight off from plausible first principles without ever going through the formulation of the Einstein field equations themselves?"* ([16], p 273) (Hojman, Kuchař and Teitelboim [17] had previously provided a distinct answer in which spacetime structure was presupposed; the present answer presupposes less structure than that, being a 3-space rather than split spacetime approach). Finally relational particle models have a number of useful analogue features permitting them to serve as useful [18, 19, 13] toy model analogues for investigations of such as the problem of time in quantum gravity [18, 20].

1.2. General Relativity Admits a Relational Formulation

One should first demonstrate that general relativity can indeed be recast as a 3-space approach theory. The Einstein–Hilbert action for the spacetime formulation of general rela-

[1]Barbour's own way of conceptualing about configurational relationalism ('best matching'), see e.g. [1, 10] is that, given two configurations, one should be kept fixed and the other should be shuffled around until an identification is found that minimizes its incongruence with the first one. The arbitrary frame method described in the main text here permits the form of the shuffling correction to be derived. Both approaches can be carried out for multiplier or velocity of a cyclic coordinate interpretations of auxiliaries in simple cases (which include all of those covered in this paper).

tivity,[2]

$$\mathsf{I}^{\mathrm{EH}}_{\mathrm{GR}}[g_{AB}] = \int \mathrm{d}^4x\sqrt{|g|}\mathcal{R}\,, \tag{1}$$

when split with respect to a family of spatial hypersurfaces takes the conventional form [21, 22]

$$\mathsf{I}^{\mathrm{ADM}}_{\mathrm{GR}}[h_{ab}, \alpha, \beta_a, \dot{h}_{ab}] = \int \mathrm{d}\lambda \int \mathrm{d}^3x\sqrt{h}\alpha\left\{\frac{\mathsf{T}^{\mathrm{ADM}}_{\mathrm{GR}}}{4\alpha^2} + R\right\} \tag{2}$$

for

$$\mathsf{T}^{\mathrm{ADM}}_{\mathrm{GR}} = \frac{1}{\sqrt{h}}G^{abcd}\{\delta_\beta h_{ab}\}\delta_\beta h_{cd} \tag{3}$$

and

$$G^{abcd} = \sqrt{h}\{h^{ac}h^{bd} - h^{ab}h^{cd}\} \tag{4}$$

the inverse of the DeWitt supermetric [22].

A more useful prototype 3-space approach action [1] can be formed by Baierlein, Sharp and Wheeler's [23] procedure: solve the α-multiplier equation $R - \mathsf{T}^{\mathrm{ADM}}_{\mathrm{GR}}/4\alpha^2 = 0$ for α itself, $\alpha = \frac{1}{2}\sqrt{\mathsf{T}^{\mathrm{ADM}}_{\mathrm{GR}}/R}$, and then use this to algebraically eliminate α from the Arnowitt–Deser–Misner Lagrangian. Thus one obtains

$$\mathsf{I}^{\mathrm{BSW}}_{\mathrm{GR}}[h_{ab}, \beta_a, \dot{h}_{ab}] = \int \mathrm{d}\lambda \int \mathrm{d}^3x\sqrt{h}\sqrt{R\mathsf{T}^{\mathrm{ADM}}_{\mathrm{GR}}}\,. \tag{5}$$

This is not quite reparametrization invariant because the shift is considered to be a coordinate for the purposes of variation. However, the Arnowitt–Deser–Misner split can be replaced by a split in terms of an instant variable I (such that $\alpha = \dot{\mathrm{I}}$) and a grid variable F_a (such that $\beta_a = \dot{\mathrm{F}}_a$, which is an example of a frame variable) at the pre-variational level if one takes into careful account that the auxiliary variables should be varied with free end spatial hypersurfaces [24].[3] Then one has an action

$$\mathsf{I}^{\mathrm{A}}_{\mathrm{GR}}[h_{ab}, \dot{h}_{ab}, \dot{\mathrm{F}}_a, \dot{\mathrm{I}}] = \int \mathrm{d}\lambda \int \mathrm{d}^3x\sqrt{h}\dot{\mathrm{I}}\left\{\frac{\mathsf{T}^{\mathrm{A}}_{\mathrm{GR}}}{4\dot{\mathrm{I}}^2} + R\right\}\,. \tag{6}$$

for

$$\mathsf{T}^{\mathrm{A}}_{\mathrm{GR}} = \frac{1}{\sqrt{h}}G^{abcd}\{\delta_{\dot{\mathrm{F}}}\}h_{ab}\delta_{\dot{\mathrm{F}}}h_{cd}\,. \tag{7}$$

Then performing Routhian reduction on this to eliminate $\dot{\mathrm{I}}$ works out exactly the same as Baierlein–Sharp–Wheeler multiplier elimination, giving

$$\mathsf{I}^{\mathrm{A}'}_{\mathrm{GR}}[h_{ab}, \dot{h}_{ab}, \dot{\mathrm{F}}_a] = \int \mathrm{d}\lambda \int \mathrm{d}^3x\sqrt{h}\sqrt{R\mathsf{T}^{\mathrm{A}}_{\mathrm{GR}}}\,. \tag{8}$$

[2] Here, g_{AB} is the spacetime metric with determinant g and Ricci scalar $\mathcal{R}$. h_{ab} is the induced 3-metric on a positive-definite 3-surface Σ (interpreted, for the moment, as a spatial hypersurface within a spacetime), with determinant h, covariant derivative D_a and Ricci scalar R. α is the lapse and β_μ is the shift. $\delta_\beta = \dot{} - \pounds_\beta$ is the hypersurface derivative, where the dot is $\frac{\partial}{\partial\lambda}$ and $\pounds_\beta$ is the Lie derivative with respect to β_a.

[3] See [10, 25, 26, 27] for earlier and further discussion of these variational methods.

This may now be taken as a starting point as done in [4, 25, 9] that implements the relational principles, in which case I use the notation $\&_{\dot{\mathrm{F}}}$ for arbitrary G frame corrected derivative, here for G the 3-diffeomorphisms on Σ:

$$\mathsf{I}_{\mathrm{GR}}^{\mathrm{TSA}}[h_{ab}, \dot{h}_{ab}, \dot{\mathrm{F}}_a] = \int \mathrm{d}\lambda \int \mathrm{d}^3x\sqrt{h}\sqrt{R\mathsf{T}_{\mathrm{GR}}^{\mathrm{TSA}}[h_{ab}, \dot{h}_{ab}, \dot{\mathrm{F}}_a]}\ , \tag{9}$$

$$\mathsf{T}_{\mathrm{GR}}^{\mathrm{TSA}} = \{h^{ac}h^{bd} - h^{ab}h^{cd}\}\{\&_{\dot{\mathrm{F}}}h_{ab}\}\&_{\dot{\mathrm{F}}}h_{cd}\ , \tag{10}$$

rather than the hypersurface derivative notation $\delta_{\dot{\mathrm{F}}}$ that presupposes spacetime. Of course, in the present case, spacetime is nevertheless recovered.

1.3. The 'Relativity without Relativity' Result

Suppose next that one presupposes less structure: just 3-space notions rather than '3-space within spacetime' notions. Does general relativity then emerge? Does it emerge alone? One goes about investigating these questions using the *Dirac procedure* [28]. This involves taking the constraints that arise purely from the form of the Lagrangian without any variation (primary constraints) and those that have arisen so far by the variational process (secondary constraints), and demanding that these be propagated by the theory's evolution equations. This can lead to new constraints, in which case the Dirac procedure is applied again to these. Now, as each new constraint uses up some degrees of freedom (usually per space point in the present field-theoretic context) and the trial system has a finite amount of these, if the Dirac procedure runs through enough iterations, it uses up at least as many degrees of freedom as the trial theory had to start off with (see e.g. [29, 30]. In this case, the trial theory has been demonstrated to be undesirable in being inconsistent (less than no degrees of freedom left), trivial (no degrees of freedom left) or undersized (e.g. a few global degrees of freedom alone could survive due to the shapes of the restrictions caused by the constraints, see e.g. [9]). Then the only remaining way out is to restrict the trial theory by allowing some of the constraints to dictate how some of its hitherto free non-variational parameters should be fixed, and so one is exhaustively removing a number of the trial options. Thus the Dirac procedure lends itself to proofs by exhaustion.

By this method the 'relativity without relativity' result arises: if one does not presuppose general relativity but rather to start with a wide class of reparametrization-invariant actions built out of good 3-d space objects in accord with the relational principles [4, 31, 32, 9, 15], general relativity emerges. More concretely the input trial ansätze are

$$\mathsf{T}_{\mathrm{grav(trial)}} = \frac{1}{\sqrt{h}Y}G^{abcd}(W)\{\&_{\dot{\mathrm{F}}}h_{ab}\}\&_{\dot{\mathrm{F}}}h_{cd}\ , \tag{11}$$

for the gravitational kinetic term that is **homogeneous quadratic in the velocities**, where

$$G^{ijkl}(W) \equiv \sqrt{h}\{h^{ik}h^{jl} - Wh^{ij}h^{kl}\}\ ,\ W \neq \frac{1}{3}\ , \tag{12}$$

is the inverse of the most general (**invertible, ultralocal**) supermetric

$$G_{abcd}(X) = \frac{1}{\sqrt{h}}\left\{h_{ac}h_{bd} - \frac{X}{2}h_{ab}h_{cd}\right\}\ ,\ X = \frac{2W}{3W-1}\ , \tag{13}$$

and

$$\mathsf{V}_{\text{grav(trial)}} = A + BR \tag{14}$$

for the gravitational potential term. This is **second-order in spatial derivatives**. The **local square root** action is then

$$\mathsf{I}_{\text{grav(trial)}}[h_{ab}, \dot{h}_{ab}, \dot{\mathrm{F}}_i] = \int \mathrm{d}\lambda \int \mathrm{d}^3x\sqrt{h}\sqrt{\mathsf{V}_{\text{grav(trial)}}\mathsf{T}_{\text{grav(trial)}}} \,. \tag{15}$$

[I use bold font to denote what assumptions are being made; all of the assumptions in this Subsection are *mathematical simplicity postulates* rather than deep physical principles.]

Then, setting $\dot{\mathrm{M}}$ to be the emergent quantity $\frac{1}{2}\sqrt{\mathsf{T}^{\text{TSA}}_{\text{grav(trial)}}/\mathsf{V}_{\text{grav(trial)}}}$, the gravitational momenta are

$$\pi^{ab} \equiv \frac{\partial \mathsf{L}}{\partial \dot{h}_{ab}} = \frac{\sqrt{h}Y}{2\dot{\mathrm{M}}} G^{abcd}(W)\&_{\dot{\mathrm{F}}} h_{cd} \,, \tag{16}$$

which are related by a primary constraint

$$\mathcal{H}_{\text{grav(trial)}} \equiv YG_{abcd}(X)\pi^{ab}\pi^{cd} - \sqrt{h}\{A + BR\} = 0 \,. \tag{17}$$

Additionally, variation with respect to F_a leads to a secondary constraint that is the usual momentum constraint

$$\mathcal{H}_a = D_b{\pi_a}^b = 0 \tag{18}$$

thereby ensuring that the physical content of the theory is in the shape of the 3-geometry and not in the coordinate grid painted on it. The propagation of $\mathcal{H}_{\text{grav(trial)}}$ then gives [31, 9]

$$\dot{\mathcal{H}}_{\text{grav(trial)}} \approx \frac{2}{\dot{\mathrm{M}}}\{X - 1\}BYD_i\{\dot{\mathrm{M}}^2 D^i\pi\} \,, \tag{19}$$

[where $\approx$ is Dirac's notion of weak equality, i.e. equality up to (already-known) constraints].

From this, the main output is the 'relativity without relativity' result that the Hamiltonian constraint propagates if the coefficient in the supermetric takes the DeWitt value $X = 1 = W$. In this case, embeddability of the 3-space into spacetime is recovered. This in no way determines whether the emergent spacetime's signature is Lorentzian ($B = -1$) or Euclidean ($B = 1$): that is to be put in by hand.

1.4. General Relativity as Geometrodynamics Does not Arise Alone in the 3-space Approach

As it has 3 further factors [4, 31, 33, 15], (19) can vanish in 3 other ways.

1) $B = 0$ gives strong or 'Carrollian' gravity options regardless of whether $W = 1$ or not. The $W = 1$ case is the strong-coupled limit of general relativity [34], which is a regime in which distinct points are causally disconnected by their null cones being squeezed into lines. This is relevant as an approximation to general relativity near singularities. While, for $W \neq 1$, it is a similar limit of scalar–tensor theories [31]. In fact, all of these $B = 0$ options exist in two different forms: one without a momentum constraint which thus are temporally but not spatially relational *metrodynamics* and one with a momentum constraint

which are are other consistent theories of geometrodynamics different to that obtained from decomposing the spacetime formulation of general relativity.
2) $Y = 0$, gives 'Galilean' theories. Here, the null cones are squashed into planes and there is no gravitational kinetic term. Strictly speaking, for this option to arise, one should start with the Hamiltonian version of the 'Galilean' theory (as its degeneracy leads to there being no corresponding Lagrangian).
3) $\pi = 0$ or $\pi/\sqrt{h}$ = constant preferred slicing conditions make the fourth factor vanish. This gives rise to alternative theories of conformal gravity [25] and to a *derivation of general relativity, alongside the conformal method of treating its initial-value problem* [35], *from a relational perspective* [25, 26, 27]. These theories can be recast by restarting with an enlarged irrelevant group G that consists of the 3-diffeomorphisms together with some group of conformal transformations.

1.5. Inclusion of Matter in the 3-space Approach

The second theme of the 3-space approach papers concerns the inclusion of fundamental matter. This is important for the 3-space approach, both as a robustness test for the axiomatization and to establish whether special relativity and the equivalence principle are emergent or require presupposition in this approach.

The robustness test is passed: using first constructive techniques [4, 29] and then Kuchař's [36] split spacetime framework techniques ([32], see also Sec 3), all of minimally-coupled scalars, electromagnetism, Yang–Mills theory, Dirac theory, and all the associated gauge theories were found to be admitted by the 3-space approach. There were some claims as regards well-known matter field types and physical principles being picked out. For example, it was claimed that
1) That electromagnetism and Yang–Mills theory are uniquely picked out [4, 29].
2) That minimally coupled scalars and 1-forms share null cones among themeselves (which is evidence toward the emergence of the special relativity principle). This is through each being forced to share the gravitational null cone.
3) That the equivalence principle is emergent rather than assumed.

These were based on the simplicity assumptions of **matter kinetic terms homogeneous quadratic in their velocities, no metric-matter kinetic cross terms, no matter dependence in the kinetic metric**.

However, the split spacetime framework and allied techniques proved powerful enough to include massive (and other) vector fields [9, 33, 15] if these simplicity postulates are weakened in various ways, showing that the latter claims are partly based on mere simplicities that have nothing to do with relationalism. Hence 1) is false. Furthermore, this paper demonstrates that 2) and 3) are false. Essentially the split spacetime framework suggests further terms for the kinetic and potential ansatze with the inclusion of which further consistent theories can be cast in 3-space approach form. Overall, the relational postulates do not pick out the fields of nature, they include a wider range of fields.

A further new point I make in the present paper is that even within the simplicity postulates assumed, the claims were based on calculations that have two further tacit simplicity assumptions, without which the exhaustion rate would be slower than it was held to be.
1) **Linear combination constraint preclusion.** In the original calculations, constraints

arising as linear combinations of terms with different a priori free parameters were not considered to be a possibility. However, there is no good theoretical reason to preclude such constraints from arising.
2) **Second class constraint preclusion.** The original calculations' counting implicitly assumed that all constraints arising were first-class as regards how many degrees of freedom they used up.[4]
While preclusion 2) is a brief and mathematically well defined simplicity postulate, it is highly restrictive, e.g. it does not cover the usual presentation of the phenomenologically useful massive vector field. This sort of restriction makes it very desirable from a theoretical perspective to uplift this simplicity. One way to proceed as regards 2) (which is simple and rigorous, although it is clearly not the most efficient) is to only assume that each constraint uses up at least one degree of freedom.

1) and 2) are clearly then capable of increasing the number of iterations required before a theory is shown to be inconsistent. In particular, for a set of interacting vector fields, weakening 1) costs one the capacity to produce internal index valued constraints at each step, meaning that one can no longer can one work for 'arbitrary' gauge group.[5] All that is known now then is that for *fairly small* gauge groups the calculation excludes alternatives, the calculation remaining unfinished for larger gauge groups. Thankfully, the gauge groups that have been found to explain experimental particle physics are not too large... (One can thus work furthermore case-by-case for larger groups required for more speculative theories of particle physics such as grand unified theories, while one should also not rule out that some new efficiency trick could be found so as to recover the result for an 'arbitrary group'). One could likewise work case-by-case so as to safeguard other previous claims such as those about higher potential derivatives in vacuo in [4].

1.6. Outline of the Rest of This Paper

The constructive workings of [4, 29], all of which assume homogeneous quadratic kinetic terms with no metric–matter cross terms or matter field dependence in the kinetic metric, suffice as an arena in which to investigate the local emergence of special relativity (Sec 2), at least for simple 3-space approach theories. In Sec 3, I recollect (and add to) arguments against the assertion (p 3217 of [4])that in the 3-space approach *"self-consistency requires that any 3-vector field must satisfy ... the equivalence principle"*. These arguments involve casting scalar(–vector)–tensor theories into 3-space approach form to act as counterexamples. I add further to these arguments in Sec 4 by constructing a unit vector tensor theory in 3-space approach form that is free of some pathologies common to vector–tensor theories and is both equivalence principle violating and special relativity violating in the sense that it has more than one distinct finite fundamental propagation speed. Hence relationalism alone

[4]A constraint is *first-class* if its Poisson brackets with all the other constraints close, and *second-class* otherwise. First-class constraints use up two degrees of freedom each while second-class ones use up just one. However one does not know before the Dirac process terminates whether a constraint is first or second class – do its Poisson brackets with as yet unfound constraints from further along the Dirac progress close? Thus one cannot argue for emergent constraints to use up two degrees of freedom each (at least until the Dirac process has terminated and one has evaluated all those Poisson brackets).

[5]'Arbitrary' here is subject to the (usual) requirement of being a direct sum of compact simple and U(1) Lie subalgebras so that the Gell-Mann–Glashow theorem applies [37].

does not locally imply the special relativity principle.

2. The Position Hitherto about the Emergence of Special Relativity

On p4 of [28], Dirac explains that he uses actions so that relativity and gauge symmetry can be straightforwardly incorporated from the start. This is done by constructing one's action out of quantities that are Poincaré invariant for special relativity, diffeomophism–invariant for general relativity, U(1) gauge invariant for electromagnetic theory, and so on. The 3-space approach is in a sense is a reverse of this: neither spacetime structure nor its locally special relativistic element are presupposed and it is shown that most alternatives to this are inconsistent. The way in which the early 3-space approach papers [4, 29] include a range of standard bosonic matter fields minimally coupled to general relativity is a sufficient arena to investigate whether and how special relativity locally emerges in the 3-space approach. These papers make the homogeneously-quadratic kinetic ansatz $\mathsf{T} = \mathsf{T}_{\Psi} + \mathsf{T}_{\text{grav(trial)}}$, where the matter fields Ψ_{Δ} have kinetic term

$$\mathsf{T}_{\Psi} = G^{\Gamma\Delta}(h_{ij})\{\&_{\dot{\text{F}}}\Psi_{\Gamma}\}\&_{\dot{\text{F}}}\Psi_{\Delta} \, , \tag{20}$$

potential term denoted by U_{Ψ} and momenta denoted by Π^{Δ}.

Then the implementation of temporal relationalism by reparametrization invariance leads to a Hamiltonian-type constraint

$$\mathcal{H}_{\text{grav}-\Psi(\text{trial})} \equiv \sqrt{h}\{A + BR + \mathsf{U}_{\Psi}\} - YG_{abcd}(X)\pi^{ab}\pi^{cd} + \frac{G_{\Gamma\Delta}\Pi^{\Gamma}\Pi^{\Delta}}{\sqrt{h}} = 0 \, . \tag{21}$$

Applying Dirac's procedure and assuming that U_{Ψ} at worst depends on connections (rather than their derivatives, which is true for the range of fields in question), the propagation of $\mathcal{H}_{\text{grav},\Psi(\text{trial})}$ gives

$$\dot{\mathcal{H}}_{\text{grav}-\Psi(\text{trial})} \approx \frac{2}{\dot{\mathrm{M}}} D^a \left\{ \dot{\mathrm{M}}^2 \left\{ Y \left\{ B \left\{ D^b \pi_{ab} + \{X - 1\} D_a \pi \right\} + \left\{ \pi_{ij} - \frac{X}{2}\pi h_{ij} \right\} \left\{ \frac{\partial \mathsf{U}_{\Psi}}{\partial {\Gamma^c}_{ia}} h^{cj} - \frac{1}{2}\frac{\partial \mathsf{U}_{\Psi}}{\partial {\Gamma^c}_{ij}} h^{ac} \right\} \right\} + G_{\Gamma\Delta}\Pi^{\Gamma}\frac{\partial \mathsf{U}_{\Psi}}{\partial(\partial_a \Psi_{\Delta})} \right\} \right\} \, , \tag{22}$$

which is just an extension of (19) to include some matter fields. The terms in (22) are then required to vanish for consistency. This can occur according to various options, each of which imposes restrictions on $\mathcal{H}_{\text{grav}-\Psi(\text{trial})}$. Furthermore, these options turn out to be very much connected to those encountered in the usual development of special relativity.

There is now a three-pronged fork in the choice of a universal transformation law in setting up special relativity. Two prongs are the Galilean or Lorentzian fork that Einstein faced (infinite or finite universal maximum propagation speed c). The third prong is the Carrollian option $c = 0$. This last option occurs above through setting $B = 0$. The vanishing of the other factors is attained by 1) declaring that U_{Ψ} cannot contain connections. 2) It is then 'natural' for U_{Ψ} not to depend on $\partial_a \Psi_{\Delta}$ either (ultralocality in Ψ_{Δ}), whereby the last term is removed. Of course, we have good reasons to believe nature does not have

$c = 0$, but what this option does lead to is alternative dynamical theories of geometry to the usual general relativistic geometrodynamics. Some are spatially relational and some are not. This is an interesting fact from a broader perspective: it issues a challenge to why the 3-space approach insists on geometrodynamical theories since metrodynamical theories are also possible. But what happens in the general relativity option is that the momentum constraint is an integrability condition [38, 31] so one is stuck with geometrodynamics whether one likes it or not.

One could also enforce consistency above by the 'Galilean' strategy of choosing $Y = 0$. This removes all but the last term. It would seem natural to take this in combination with $\Pi^\Delta = 0$, whereupon the fields are not dynamical. Moreover this does not completely trivialize the matter fields since they would then obey analogues of Poisson's law, or Ampère's, and these are capable of governing a wide variety of complicated patterns. Thus one arrives at an entirely nondynamical 'Galilean' world. In vacuo, this possibility cannot be obtained from a Baierlein–Sharp–Wheeler-type Lagrangian (the kinetic factor is badly behaved) but the Hamiltonian description of the theory is unproblematic. Of course, the Hamiltonian constraint is now no longer quadratic in the momenta:

$$\mathcal{H}_{\text{grav}-\Psi(\text{trial})}(Y = 0) = A + BR + \mathsf{U}_\Psi = 0\,. \tag{23}$$

This option is not of interest if the objective is to find *dynamical* theories. Nevertheless, this option is a logical possibility, and serves to highlight how close parallels to the options encountered in the development of special relativity arise within the 3-space approach.

There is also a combined locally Lorentzian physics and spacetime structure strategy as follows. The signature is to be set by hand (one could just as well have any other non-degenerate signature for the argument below). Take (22) and introduce the concept of a gravity–matter momentum constraint $\mathcal{H}^a_{\text{grav}-\Psi(\text{trial})}$ by using $0 = -\frac{1}{2}\mathcal{H}^a_\Psi + \frac{1}{2}\mathcal{H}^a_\Psi$ and refactoring:

$$\begin{aligned}\dot{\mathcal{H}}_{\text{grav}-\Psi(\text{trial})} \approx \frac{2D^a}{\dot{\mathrm{M}}}\Bigg\{\dot{\mathrm{M}}^2\Bigg\{Y\Bigg\{B\Bigg\{\underline{\Bigg\{D^b\pi_{ab} - \frac{1}{2}\left\lfloor \Pi^\Delta\frac{\delta\pounds_{\dot{\mathrm{F}}}\Psi_\Delta}{\delta\dot{\mathrm{F}}^a}\right\rfloor\Bigg\}} + \underline{\frac{1}{2}\left\lfloor \Pi^\Delta\frac{\delta\pounds_{\dot{\mathrm{F}}}\Psi_\Delta}{\delta\dot{\mathrm{F}}^a}\right\rfloor}\Bigg\}\Bigg\} \\ \underline{+G_{\Gamma\Delta}\Pi^\Delta\frac{\partial\mathsf{U}_\Psi}{\partial(\partial_a\Psi_\Delta)}} + \underline{YB\{X-1\}D_a\pi} \\ \underline{+Y\left\{\pi_{ij} - \frac{X}{2}\pi h_{ij}\right\}\left\{\frac{\partial\mathsf{U}_\Psi}{\partial\Gamma^c{}_{ia}}h^{cj} - \frac{1}{2}\frac{\partial\mathsf{U}_\Psi}{\partial\Gamma^c{}_{ij}}h^{ac}\right\}}\Bigg\}\Bigg\}\,,\end{aligned} \tag{24}$$

so that the first two underlined terms are then proportional to $\mathcal{H}^a_{\text{grav}-\Psi(\text{trial})}$.[6] In the 'orthodox general covariance option', the third and fourth underlined terms cancel, amounting to the enforcement of a universal null cone. This requires supplementing by some means of discarding the fifth underlined term. Here, one can furthermore *choose* the orthodox option $X = 1$: the recovery of embeddability into spacetime corresponding to general relativity (the 'relativity without relativity' result), or, *choose* the alternative preferred-slicing worlds of $D_a\pi = 0$ which are governed by conformal mathematics. As both of these options

[6] $\frac{\delta A}{\delta B}$ denotes the functional derivative, and the special brackets $\lfloor\ \rfloor$ delineate the factors over which the implied integration by parts is applicable.

are valid, the recovery of locally-Lorentzian physics does not occur solely in generally-covariant theories. The connection terms (sixth underlined term) must also be discarded, but the Dirac procedure does this automatically for the given ansätze.

Thus in the 3-space approach, locally-Lorentzian general relativistic spacetime arises as one option; other permitted options include Carrollian worlds, Galilean worlds and locally-Lorentzian preferred slicing worlds. These alternatives all lack some of the features of generally relativistic spacetime. [15] went on to talk about hybrids of the above The ultralocal and nondynamical strategies for dealing with the last term in (22) are available in *all* the above options. So as things stand, one does derive that gravitation enforces a *unique finite* propagation speed, but the possibility of coexisting with fields with infinite and zero propagation speeds is not precluded by consistency, although it does read to undersized solution spaces.[7] And of making the fourth underlined term vanish algebraically along the lines of parallel E and B in Poynting vector. But none of these situations ruin the emergence of the special relativity principle in the sense that: any adjoined zero-momentum Galilean fields cannot propagate so that it does not matter that their propagation speed is in principle infinite, while adjoined Carrollian fields are precluded from propagating information away from any point by their ultralocal nature, and the parallel E and B field situation is also well-known to preclude the associated propagation (mutual orthogonality in the E and B fields causing each other to continue to oscillate).

However, we shall see in Sec 4 that the above fork breaks down for more complicated matter.

3. The Position Hitherto on the Equivalence Principle in the 3-space Approach

3.1. Equivalence Principle Violations at the Level of the Action

While this study started with partial evidence for the equivalence principle being emergent in the 3-space approach [4, 9], it then suffered the setback of counterexamples as more complete potential ansatze were devised. The counterexamples to date have, however, suffered from certain limitations. In this paper I extend the counterexamples to cases for which these limitations do not occur. I should first describe some symptoms at the level of action principles of whether a theory obeys or violates the equivalence principle. Coordinates can be provided at each particular point p such that the metric connection vanishes at p, so there is no obstruction in passage to the local special relativity form for curved spacetime matter field equations[8] that contain no worse than metric connection. However, the curvature tensor is an obstruction to such a passage if the field equations contain derivatives of the metric connection. Thus theories in which the matter terms contribute additional such terms are

[7]That such a dilemma exists was simply overlooked in [4, 29] papers since it was claimed that these ultralocal and nondynamical strategies only lead to trivial theories, in the latter case by counting arguments. Unfortunately, inspection of this triviality reveals it to mean 'less complicated than in conventional Lorentzian theories' rather than 'devoid of mathematical solutions'. In particular, the counting argument is insufficient in not taking into account the geometry of the restrictions on the solution space.

[8]N.B. that the gravitational field equations are given a special separate status in the equivalence principle ('all the laws of physics bar gravity') and thus do not interfere with the logic of this.

equivalence principle violating. One way in which derivatives of the metric connection in the field equations can arise from actions is through there already being such derivatives in the action e.g. in curvature–matter coupling terms. A second way is from integration by parts during the variational working causing mere metric connections in the action to end up as derivatives of metric connections in the field equations.

3.2. The Split Spacetime Framework

Rather than the previous sections' exhaustive Dirac procedure, this section requires the split spacetime framework, which does presuppose the general relativistic notion of spacetime. The point of this is that there is then a systematic treatment of Kuchař [36] by which the spacetime formulation of specific consistent matter theories can be recast in split spacetime framework form. It was using the split spacetime framework [32, 9] that many matter theories were found to admit formulations that conform to the 3-space approach's relational principles (Sec 1.5). I next provide (as a new result) the variant of the split spacetime framework that is in terms of instant-grid variables for the case relevant here: a 1-form matter field.

One is presupposing that one has a hypersurface Σ within a spacetime M. n_A is the normal to Σ and e^a_A the projector onto this hypersurface. Then it is meaningful to decompose each matter field into perpendicular and tangential parts with respect to Σ. In the case of the 1-form,

$$A_A = n_A A_\perp + e^a_A A_a \,. \tag{25}$$

Hypersurfaces can be re-gridded and deformed. Re-gridding kinematics involves Lie derivatives with respect to $\dot{\mathrm{F}}_a$; these appear as corrections to the velocities so that these feature in the action as 'hypersurface derivatives' rather than as 'bare velocities'. As regards deformations, the arbitrary deformation of a hypersurface near a point p splits into a *translation part* such that

$$\dot{\mathrm{I}}(p) \neq 0\,, \quad \{\dot{\mathrm{I}}_{,a}\}(p) = 0 \tag{26}$$

and a *tilt* part such that

$$\dot{\mathrm{I}}(p) = 0\,, \quad \{\dot{\mathrm{I}}_{,a}\}(p) \neq 0\,. \tag{27}$$

The translation piece further splits into a translation on a background spacetime piece and a *derivative coupling* piece which alters the nature of the background spacetime. Furthermore, the re-gridding, tilt and derivative-coupling kinematics pieces that arise within this spacetime-presupposing framework are *universal*: they depend solely on the rank of the tensor matter field rather than on any details of that particular field. It then so happens that two of these bear a tight relationship with what is needed to implement the configurational relationalism and temporal relationalism postulates.

1) Use of the arbitrary 3-diffeomorphism frame is none other than re-gridding.

2) The absence of tilt terms is a guarantee of an algebraic Routhian reduction procedure. Thereby an extraneous time variable free action can be obtained, at least in principle.[9]

[9]Were tilt terms not removable, as these contain spatial derivatives of $\dot{\mathrm{I}}$, they would compromise the algebraicity of the elimination of $\dot{\mathrm{I}}$ from the cyclic equation that arises from variation with respect to I. Of course, the algebraic equation might have no roots or only physically unacceptable (e.g. non-real) roots, in which cases the theory should be discarded. It might also not be explicitly soluble – that is what I mean by 'in principle'.

Tilts can however be removed from at least some actions by such as integration by parts or redefining field variables.

On the other hand, the derivative coupling universal feature is related to the equivalence principle, in that theories in which derivative coupling features in the split action are equivalence principle violating. Absense of derivative coupling is termed the **geometrodynamical equivalence principle** in [17]. It corresponds to the no metric–matter cross–term and no matter field dependence in the kinetic metric. Thus these particular mathematical simplicity postulates additionally have physical significance. [But demanding that these hold amounts to imposing aspects of the equivalence principle by hand, so that one can no longer claim that the equivalence principle is emergent in the 3-space approach.]

This paper requires the split spacetime form of the derivatives of a 1-form, which is also a good illustration of re-gridding, tilt and derivative coupling terms (the last of these are those terms that involve the extrinsic curvature

$$K_{ab} = -\frac{1}{2\dot{\mathrm{I}}}\delta_{\dot{\mathrm{F}}}h_{ab} \tag{28}$$

of the hypersurface).

$$\nabla_b A_\perp = D_b A_\perp - K_{bc}A^c\,, \tag{29}$$

$$\dot{\mathrm{I}}\nabla_\perp A_a = -\delta_{\dot{\mathrm{F}}}A_a - \dot{\mathrm{I}}K_{ab}A^b - A_\perp\partial_a\dot{\mathrm{I}}\,, \tag{30}$$

$$\nabla_b A_a = D_b A_a - A_\perp K_{ab}\,, \tag{31}$$

$$\dot{\mathrm{I}}\nabla_\perp A_\perp = -\delta_{\dot{\mathrm{F}}}A_\perp - A^a\partial_a\dot{\mathrm{I}}\,, \tag{32}$$

Intuitively, these relations come about because spacetime derivatives are not equal to spatial derivativess as the former have extra connection components, which this scheme interprets geometrically from the perspective of the hypersurface.

3.3. Scalar–Tensor Theories

The 3-space approach counterexample to date of the first type is Brans–Dicke theory [39] (see also [40] for a canonical treatment). While this was included in [4] by casting the theory in the Einstein frame[10] However, this transformation does away with the equivalence principle violation, so it is more instructive for the present context to work in Brans–Dicke theory's usual Jordan frame. For this, the spacetime action is

$$\mathsf{I}_{\mathrm{BD}}[g_{AB}, \chi] = \int \mathrm{d}^4x\sqrt{|g|}\mathrm{e}^{-\chi/2}\{\mathcal{R} - \omega\partial_A\chi\partial^A\chi\}\,. \tag{33}$$

The subsequent split spacetime action has a kinetic term proportional to

$$\left\{h^{ac}h^{bd} - \frac{X-2}{3X-4}h^{ab}h^{cd}\right\}\delta_{\dot{\mathrm{F}}}h_{ab}\delta_{\dot{\mathrm{F}}}h_{cd}$$

$$+\frac{4}{3X-4}h^{ab}\delta_{\dot{\mathrm{F}}}h_{ab}\delta_{\dot{\mathrm{F}}}\chi + \frac{3X-2}{(3X-4)(X-1)}\delta_{\dot{\mathrm{F}}}\chi\delta_{\dot{\mathrm{F}}}\chi \tag{34}$$

[10] Under this field redefinition, it is then a scalar field minimally coupled to gravity, which is clearly included among the 3-space approach castable cases listed in Sec 1.5.

for

$$X = \frac{2\{1+\omega\}}{2\omega+3} \,. \tag{35}$$

Thus it is equivalence principle violating as it contains metric–matter kinetic cross terms. Nevertheless it can be cast into 3-space approach form [33] (but was missed in [4] through the ansatze there not including metric–matter kinetic cross-terms).

However, this example suffers the observational weakness that its parameter ω is fixed, expected on grounds of theoretical naturality to be of order unity and yet is bounded by the Cassini data to be above 20000 [41]. This weakness can be removed by showing that the more general scalar–tensor theory with spacetime action[11]

$$\mathsf{I}_{\mathrm{STT}}[g_{AB},\chi] = \int \mathrm{d}^4x\sqrt{|g|}e^{-\chi/2}\{\mathcal{R} - \omega(\chi)\partial_A\chi\partial^A\chi + \mathsf{U}(\chi)\} \tag{36}$$

which one can likewise cast as a 3-space approach theory by performing the split with respect to a family of spatial hypersurfaces using instant–grid variables and then eliminating $\dot{\mathrm{I}}$ and writing $\&_{\dot{\mathrm{F}}}$ for $\delta_{\dot{\mathrm{F}}}$ in the usual fashion. That this can be cast in 3-space approach form is clear because adding a potential and replacing ω with $\omega(\chi)$ do not affect the split of the spacetime tensorial objects in the action or the form that the Routhian reduction that eliminates $\dot{I}$ are to take. While, this no longer suffers from the observational weakness because now ω varies and there is evidence that it tends dynamically to a large value in the late universe (toward the general relativity value of $+\infty$) [43].

3.4. Vector–Tensor Theories

An example of 3-space approach theory [15] in which the second kind of equivalence principle violation occurs can be found among the the vector–tensor theories considered in e.g. [44]. This class of theories has the spacetime form:

$$\mathsf{I}_{\mathrm{VTT}}[g_{AB},A_A] = \int \mathrm{d}\lambda \int \mathrm{d}^3x\alpha \left\{\mathcal{R} + \nu\left\{\nabla_A A^A \nabla_B A^B + m^2A^2\right\}\right\} \tag{37}$$

Now the split form of action (37) is (by the above derivative formulae and then using the field redefinition

$$\dot{\mathrm{I}}A^a = \dot{v} \tag{38}$$

to remove 'tilts' and also setting $A_\perp$ to be some ϕ):

$$\mathsf{I}^{\mathrm{ADM}}_{\mathrm{VTT}}[h_{ab},\dot{h}_{ab},v_i,\dot{v}_i,\phi,\dot{\phi},\dot{\mathrm{F}}_i,\dot{\mathrm{I}}] = \int\int \mathrm{d}\lambda\dot{\mathrm{I}}\sqrt{h}\mathrm{d}^3x \left\{\frac{\mathsf{T}^{\mathrm{A}}_{\mathrm{GR}}[h_{ab},\dot{h}_{ab},\dot{\mathrm{F}}_i]}{4\dot{\mathrm{I}}^2} + \right.$$

$$\left.\frac{\nu}{\dot{\mathrm{I}}^2}\left\{\left\{D_a\{\dot{v}^a\} + \frac{\phi}{2}h^{ij}\delta_\beta h_{ij} + \delta_{\dot{\mathrm{F}}}\phi\right\}^2 + m^2\dot{v}^2\right\} + R - \nu m^2\phi^2\right\} . \tag{39}$$

[11] This is not the most general scalar-tensor theory (see e.g. [42] and references therein). E.g. one could replace $e^{-\chi/2}$ by an arbitrary function of χ, or furthermore extend the theory to have more than 1 scalar. However, the example in this paper is general enough to illustrate the point in question.

Then a Routhian reduction of the same form as that mentioned in Sec 1.2 is possible, giving

$$\mathsf{I}^{\mathrm{A}'}_{\mathrm{VTT}}[h_{ab}, \dot{h}_{ab}, v_i, \dot{v}_i, \phi, \dot{\phi}, \dot{\mathrm{F}}_i] =$$

$$\int\int \mathrm{d}\lambda \mathrm{d}^3x\sqrt{h}\sqrt{\{R - \nu m^2\phi^2\}\left\{\mathsf{T}^{\mathrm{A}'}_{\mathrm{GR}} + 4\nu\left\{\left\{D_a\{\dot{v}^a\} + \frac{\phi}{2}h^{ij}\delta_{\dot{\mathrm{F}}}h_{ij} + \delta_{\dot{\mathrm{F}}}\phi\right\}^2 + m^2\dot{v}^2\right\}\right\}}\,. \tag{40}$$

[The equations encoded by this action happen to be weakly unaffected by whether $\dot{v}^a$ is replaced by $\delta_{\dot{\mathrm{F}}}v^a$.]

Thus if one starts with 3-space approach principles, and using the arbitrary 3-diffeomorphism frame symbol $\&_{\dot{\mathrm{F}}}$ in place of the hypersurface derivative symbol $\delta_{\dot{\mathrm{F}}}$, one obtains the 3-space approach action

$$\mathsf{I}^{\mathrm{TSA}}_{\mathrm{VTT}}[h_{ab}, \dot{h}_{ab}, v_i, \dot{v}_i, \phi, \dot{\phi}, \dot{\mathrm{F}}_i] = \int\int \mathrm{d}\lambda \mathrm{d}^3x\sqrt{h}\sqrt{\{R - \nu m^2\phi^2\}\left\{\mathsf{T}_{\mathrm{GR}}[h_{ab}, \dot{h}_{ab}, \dot{\mathrm{F}}_i]\right.}$$

$$\overline{\left.+4\nu\left\{\left\{D_a\{\&_{\dot{\mathrm{F}}}v^a) + \frac{\phi}{2}h^{ij}\&_{\dot{\mathrm{F}}}h_{ij} + \&_{\dot{\mathrm{F}}}\phi\right\}^2 + m^2\{\&_{\dot{\mathrm{F}}}v\}^2\right\}\right\}}\,. \tag{41}$$

Thus one has a consistent (by reverse of above working and the original spacetime formulation being consistent) and nontrivial equivalence principle violating theory for geometry, a scalar and a 1-form. It should be noted that [4] missed this not on relational grounds but on simplicity grounds: the theory has a kinetic term that is not ultralocal, has metric–matter cross-terms and field dependence.

This was missed in [4] through it having metric–matter kinetic cross-terms, matter field dependence in the kinetic metric and a mixture of 1-form and scalar modes from the 3-space perspective.

Many theories of this type have a number of undesirable features, such as classical and quantum instabilities [44, 45], non-positivness of total energy [47] and formation of shocks beyond which the evolution cannot be extended [45]. [15] speculated that some axiom that avoids such pathologies could be used to bring down this class of counterexample.

4. A New Example of 3-space Approach Theory That Is All of Equivalence Principle Violating, Special Relativity Violating and Less Pathological

4.1. Unit Vector–Tensor Theories (Einstein–Aether Theories)

[46, 47, 48] consider a general Einstein–Aether action of the form

$$\mathsf{I}_{\mathrm{EAT}}[g_{AB}, u_A] = \int \mathrm{d}^4x\sqrt{-g}\{\mathcal{R} + E_1\{\nabla_A u_B\}\nabla^A u^B + E_2\{\nabla_A u^A\}^2$$

$$+E_3\{\nabla_A u^B\}\nabla_B u^A + E_4 u^A u^B\{\nabla_A u_C\}\nabla_B u^C + \lambda\{u_A u^A - 1\}\}\,. \tag{42}$$

As compared to the general theories considered by Isenberg and Nester, this permits 1 further derivative term (though I do not make use of it in my specific examples), and furthermore interprets what was the mass now as a Lagrange multiplier and adds the multiplier again as an extra potential piece. [The Lagrange multiplier is there to implement the unit-field constraint.] At least some of these unit-field theories are less pathological [47].

These theories are in general equivalence principle violators, the exception being if all of

$$E_1 + E_3 = 0 \text{ (Maxwellian combination)}, \tag{43}$$

$$E_2 = 0 = E_4 \tag{44}$$

hold. These theories are also in general special relativity violating, for they contain [46, 47] spin-2 fields propagating at squared speeds

$$c_2 = \frac{1}{1 - \{E_1 + E_3\}}, \tag{45}$$

spin-1 fields propagating at speed

$$c_1 = \frac{E_1 - E_1^2/2 + E_3^2/2}{\{E_1 + E_4\}\{1 - \{E_1 + E_3\}} \tag{46}$$

and spin-0 fields propagating at speed

$$c_0 = \frac{\{E_1 + E_2 + E_3\}\{2 - \{E_1 + E_4\}\}}{\{E_1 + E_4\}\{1 - \{E_1 + E_3\}\}\{2 + E_1 + E_3 + 3E_2\}}. \tag{47}$$

These are fairly extensively finite and with at least one distinct from the speed of light $c = 1$ in these units. This is the case unless all of

$$E_1 + E_3 = 0, \tag{48}$$

$$E_4 = 0, \tag{49}$$

and

$$E_1^{-1} - E_2^{-1} = 2 \tag{50}$$

hold.

These squared speeds are also capable of going negative, corresponding to undesirable exponential-type instabilities. Positive linearized energy density requires

$$\{2E_1 - E_1^2 + E_3^2\}/\{1 - E_1 - E_3\} > 0 \tag{51}$$

(vector mode contribution) and

$$\{E_1 + E_4\}\{2 - E_1 - E_4\} > 0 \tag{52}$$

(trace mode contribution). One would also like the kinetic energy contributions to have the usual sign for matter kinetic terms.

4.2. Einstein–Aether Theories That Are Castable in 3-space Approach Form

To build a suitable 3-space approach example that is equivalence principle violating, special relativity violating in the sense of having 2 different finite fundamental propagation speeds and not subject to the above three pathologies, proceed as follows.

Consider first the theory with E_2 alone nonzero. Compared to the previous section's theory, the only difference is to the potential (which is trivial to split spacetime framework decompose), so the previous section's working will straightforwardly extend to the Einstein–aether theory case that is analogous to the above specially-chosen case. But for E_2 theory $c_2 = 1(= c)$ and the other two are not finite, so this does not constitute a special relativity violation of the type I am seeking.

But consider then furthermore including a Maxwell-type combination ($E_1 = -E_3 \neq 0$) in the action; as this has a very simple split spacetime framework decomposition, this addition does not ruin the algebraicity of the Routhian reduction. So the theory I choose to work with is identified in the spacetime picture as

$$\mathsf{I}_{\mathrm{EAT}}[g_{AB}, u_A] = \int \mathrm{d}^4x\sqrt{-g}\,\{\mathcal{R} + E_1\{\nabla_A u_B\}\nabla^A u^B + E_2\{\nabla_A u^A\}^2$$
$$-E_1\{\nabla_A u^B\}\nabla_B u^A + \lambda\{u_A u^A - 1\}\}\ . \tag{53}$$

split spacetime framework splitting this, adhering to the redefinition (38), using symmetry-antisymmetry cancellations on the new quadratic tilt terms and integration by parts on the new linear tilt terms, one indeed passes to a homogeneous quadratic action to which the usual Routhian reduction move can be carried out. The resulting action may, moreover be interpreted as (rewriting $\delta_{\dot{\mathrm{F}}}$ as $\&_{\dot{\mathrm{F}}}$ and adopting this action as one's new starting-point) a 3-space approach theory that follows from the configurational and temporal relationalism principles:

$$\mathsf{I}^{\mathrm{TSA}}_{\mathrm{EAT}}[h_{ab}, \dot{h}_{ab}, \dot{\mathrm{F}}_a, \dot{v}_a] = \int \mathrm{d}\lambda \int \mathrm{d}^3x\sqrt{h}\sqrt{\mathsf{TU}} \tag{54}$$

for

$$\mathsf{U} = R - \mu\{\phi^2 + 1\} \tag{55}$$

and

$$\mathsf{T} = \mathsf{T}^{\mathrm{BFO-A}}_{\mathrm{GR}} + \mathsf{T}^{\mathrm{BFO-A}}_{\mathrm{V}}(\nu \rightarrow E_2, m^2 \rightarrow \mu) + \mathsf{T}^{\mathrm{BFO-A'}}_{\mathrm{V}}\ , \tag{56}$$

where

$$\mathsf{T}^{\mathrm{BFO-A'}}_{\mathrm{V}} =$$
$$E_1\left\{\{h^{ac}h^{bd} - h^{ad}h^{bc}\}\partial_a\{\&_{\dot{\mathrm{F}}}v_b\}\partial_c\&_{\dot{\mathrm{F}}}v_d\right.$$
$$\left.+2D^b\left\{\&_{\dot{\mathrm{F}}}v^a\{\partial_b\&_{\dot{\mathrm{F}}}v_a - \partial_a\&_{\dot{\mathrm{F}}}v_b\}\right\} + \{\&_{\dot{\mathrm{F}}}\phi\}^2\right\}\ . \tag{57}$$

Compared to the original example I gave, this is more general in having E_2 and less general in being a unit vector field.

Now, indeed, there is in general more than 1 fundamental propagation speed, as

$$c_2 = c_1 = 1\ (\ = c)\ ,\ c_0 = \frac{E_2\{2 - E_1\}}{E_1\{2 + 3E_2\}}\ . \tag{58}$$

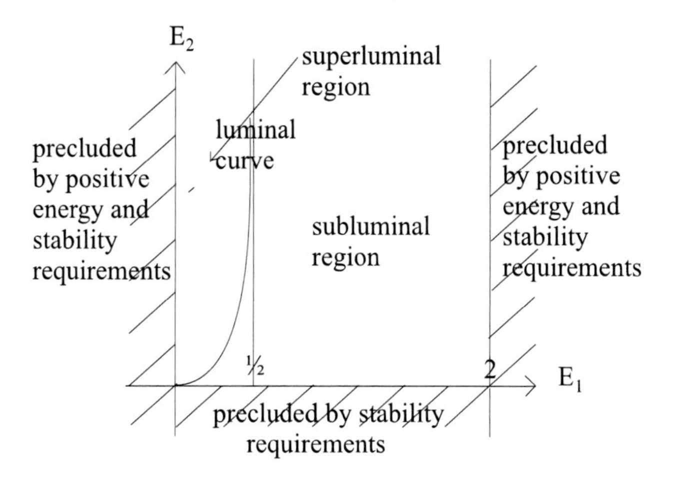

Figure 1.

I.e. this example contains 1) a non-generic case

$$E_1^{-1} - E_2^{-1} = 2 \tag{59}$$

which is *not* a counterexample to violation of the special relativity lightcone (this is a subcase of the above non-special relativity violating example). 2) The general case

$$E_1^{-1} - E_2^{-1} \neq 2 \tag{60}$$

for which

$$c_0 \neq 1 = c_1 = c_2 = c\,. \tag{61}$$

So there is a 1-parameter family (bar a single parameter value) of 3-space approach complying special relativity violating theories: there are scalar modes whose propagation speed in vacuo is different from the speed of light, so these have a null cone structure that is *not* shared with the other fields in this theory.

A fair portion of the above example's parameter space is able to comply with positive linearized energy: that for which

$$0 < E_1 < 2\,. \tag{62}$$

For $E_2 > 0$ or $< -2/3$, this remaining region complies with the stability criteria c_0^2, c_1^2, $c_2^2 > 0$ since these reduce to the trivial $1 > 0$ (twice) and

$$\frac{E_2\{2 - E_1\}}{E_1\{2 + 3E_2\}} > 0\,, \tag{63}$$

however the latter subregion should be discarded to ensure that the kinetic term is of the right characteristic sign for a matter contribution, thus leaving one with the 'region of non-pathology'

$$0 < E_1 < 2\,,\ E_2 > 0 \tag{64}$$

for the theory's coupling constants.

This region is split into two pieces by the curve of special relativity-compliance [i.e. of universal luminal fundamental speed (50)]; the subregion above this curve has the scalar modes propagate superluminally and the subregion below this curve has them propagate subluminally.

5. Conclusion

The 3-space approach is based on temporal and configurational relational principles. General relativity in geometrodynamical form can be derived as one consistent alternative that follows from these premises when applied to a theory for which the 3-metrics on a fixed spatial topology are redundant dynamical objects under the associated 3-diffeomorphisms. A sufficient set of fundamental matter fields to describe nature can be adjoined to this scheme. It was furthermore claimed that

1) working with matter fields alongside spatial 3-metrics *picks out* electromagnetism (and Yang–Mills theory) coupled to general relativity as the only consistent theories of one (and K interacting) 1-forms.

2) The equivalence principle is emergent.

3) The universal null cone of special relativity is locally recovered.

These were always subject to simplicity assumptions as well as the relational postulates. Claim 1) should be weakened, at least on basis of current workings. This is not only because lifting unrelational simplicities that were identified as such at the time of doing the calculation has been shown to destroy the result, but also because of two further tacit simplicities assumed in the proof, one of which is unmotivated and the other of which is unduly restrictive from a theoretical perspective. Without these the exhaustion goes more slowly and one then has to work case by case rather than once and for all with an arbitrary gauge group.

Also, examples including some new to this paper show that it is necessary for the relational postulates to be supplemented by non-relational simplicity assumptions in order for 2) and 3) to hold. As these necessary simplicity assumptions include what Hojman, Kuchař and Teitelboim identify as the geometrodynamical equivalence principle, the 3-space approach's claim of *deriving* the equivalence principle loses its credibility. Furthermore, from the split spacetime framework perspective, the geometrodynamical equivalence principle and statements equivalent to the relational postulates come as a neat package involving the three types of universal kinematics: hypersurface derivatives, the absense of tilts and the absense of derivative couplings, while taking the 3-space approach and the geometrodyamical equivalence principle as one's principles is more heterogeneous. [12] That said, the equivalence principle is separate from other postulates in Einstein-type spacetime approach, so one is doing no worse than what one does in taking the geometrodynamical equivalence principle alongside the relational postulates to be the heart of the axiomatization of general relativity. That reflects the primality of the equivalence principle as regards axiomatizations of general relativity – so far as the author (or Brown [49]) are aware, no derivations of the equivalence principle from more basic postulates are known (which is what merited my concentrated effort to bring down [4]'s conjecture otherwise).

'Simple' (in the sense of Sec 1.5) matter fields coupled to dynamical 3-metrics builds in the equivalence principle; one then encounters (an extension of) the fork Einstein encountered in setting up special relativity as the "roots of" an explicit equation arising from

[12] The 3-space approach assumes less structure than (split) spacetime approaches. That makes it 'more interesting' but also harder to work with as there being less structure makes proving thoerems harder in the 3-space approach than e.g. in Hojman–Kuchar–Teitelboim's approach that presupposes spacetime. E.g., their use of induction proofs specifically rely on additional spacetime structure.

the Hamiltonian-type constraint by the Dirac procedure. These correspond to Lorentzian relativity (single finite physical propagation speed), Galilean relativity (infinite propagation speed), and Carrollian relativity (zero propagation speed). But if the associated simplicities are dropped, this article's example shows that equivalence principle violation is possible including in otherwise relatively non-pathological situations, and more than 1 finite fundamental propagation speed can occur – consistency other than by the above fork becomes allowed.

A new issue to investigate – I dare not call it a conjecture – is whether each of the local recovery of special relativity and of gauge theory can be shown to follow from the equivalence principle free of the 3-space approach or even geometrodynamical formalism. Here, 'shown to follow' might mean that they are among the natural structures to emerge, and perhaps a further axiom or a collection of observations or demands from local quantum field theory could remove some (or all) of the structures that co-emerge with them. This does not look to be restricted to a specific geometrodynamical formulation or even to geometrodynamics, but would rather be a stronger formalism-independent result of general relativity.

Acknowledgments

I thank Julian Barbour and Harvey Brown for previous discussions and Peterhouse Cambridge for funding me in 2007-2008.

References

[1] J.B. Barbour and B. Bertotti, *Proc. Roy. Soc. Lond.* A382 295 (1982).

J.B. Barbour, in *Quantum Concepts in Space and Time* ed. R. Penrose and C.J. Isham (Oxford University Press, Oxford 1986).

[2] *Mach's principle: From Newton's Bucket to Quantum Gravity* ed. J.B. Barbour and H. Pfister (Birkhäuser, Boston 1995).

[3] J.B. Barbour, *Class. Quantum Grav.* 11 2853 (1994).

[4] J.B. Barbour, B.Z. Foster and N. Ó Murchadha, *Class. Quantum Grav.* 19 3217 (2002), gr-qc/0012089.

[5] J.B. Barbour, *The End of Time* (Oxford University Press, Oxford 1999).

[6] See e.g. *The Leibniz–Clark Correspondence*, ed. H.G. Alexander (Manchester 1956).

[7] E. Mach, *Die Mechanik in ihrer Entwickelung, Historisch-kritisch dargestellt* (J.A. Barth, Leipzig 1883). The Enlish translation is *The Science of Mechanics: A Critical and Historical Account of its Development* (Open Court, La Salle, Ill. 1960).

[8] J.B. Barbour, *Absolute or Relative Motion? Vol 1: The Discovery of Dynamics* (Cambridge University Press, Cambridge, 1989);

J.B. Barbour, forthcoming book.

[9] E. Anderson, in *General Relativity Research Trends, Horizons in World Physics* 249 ed. A. Reimer (Nova, New York 2005), gr-qc/0405022.

[10] J.B. Barbour, *Class. Quantum Grav.* **20**, 1543 (2003), gr-qc/0211021.

[11] J.B. Barbour and L. Smolin, unpublished, dating from 1989;

D. Lynden-Bell, in [2];

L.Á Gergely, *Class. Quantum Grav.* 17 1949 (2000), gr-qc/0003064;

L.Á Gergely and M. McKain, *Class. Quantum Grav.* 17 1963 (2000), gr-qc/0003065;

C. Kiefer, *Quantum Gravity* (Clarendon, Oxford 2004).

[12] E. Anderson, *AIP Conf. Proc.* 861 285 (2006), gr-qc/0509054;

E. Anderson, *Class. Quant. Grav.* 23 2491 (2006), gr-qc/0511069;

E. Anderson, *Class. Quantum Grav.* 24 5317 (2007), gr-qc/0702083;

E. Anderson, *"Quantum Mechanics in Triangleland"*, forthcoming.

[13] E. Anderson, *Class. Quant. Grav.* 23 2469 (2006), gr-qc/0511068;

E. Anderson, *"Dynamics without Dynamics: Relative Angular Momentum Exchange in Timeless Classical and Quantum Triangleland"*, forthcoming.

[14] J.B. Barbour, in *Decoherence and Entropy in Complex Systems (Proceedings of the Conference DICE, Piombino 2002* ed. H. -T. Elze (Springer Lecture Notes in Physics 2003), arXiv:gr-qc/0309089.

[15] E. Anderson, *Studies in History and Philosophy of Modern Physics*, 38 15 (2007), gr-qc/0511070.

[16] J.A. Wheeler, in *Battelle Rencontres: 1967 Lectures in Mathematics and Physics* ed. C. DeWitt and J.A. Wheeler (Benjamin, New York 1968).

[17] S.A. Hojman, K.V. Kuchař and C. Teitelboim, *Annals of Physics* 96 88 (1976).

[18] K.V. Kuchař, in *Proceedings of the 4th Canadian Conference on General Relativity and Relativistic Astrophysics* ed. G. Kunstatter, D. Vincent and J. Williams (World Scientific, Singapore 1992).

[19] E. Anderson, *Class. Quantum Grav.* 24 2935 (2007), gr-qc/0611007;

E. Anderson, *Class. Quantum Grav.* 24 2971 (2007), gr-qc/0611008.

E. Anderson, arXiv:0706.3934;

E. Anderson, arXiv:0709.1892;

E. Anderson, *2 papers on Semiclassical Quantum Gravity*, forthcoming;

E. Anderson, *"What is the Distance between two Shapes?"*, forthcoming.

[20] C.J. Isham, in *Integrable Systems, Quantum Groups and Quantum Field Theories* ed. L.A. Ibort and M.A. Rodríguez (Kluwer, Dordrecht 1993), gr-qc/9210011.

[21] R. Arnowitt, S. Deser and C.W. Misner, in *Gravitation: an Introduction to Current Research* ed L. Witten (Wiley, New York 1962).

[22] B.S. DeWitt, *Phys. Rev.* 160 1113 (1967).

[23] R.F. Baierlein, D. Sharp and J.A. Wheeler, *Phys. Rev.* 126 1864 (1962).

[24] E. Anderson, "New Interpretation of variational principles for gauge Theories. I. Cyclic coordinate alternative to ADM spilt, arXiv:0711.0288.

[25] E. Anderson, J.B. Barbour, B.Z. Foster and N. Ó Murchadha, *Class. Quantum Grav.* 20 157 (2003), gr-qc/0211022.

[26] E. Anderson, J.B. Barbour, B.Z. Foster, B. Kelleher and N. Ó Murchadha, *Class. Quantum Grav.* 22 1795 (2005), gr-qc/0407104.

[27] E. Anderson, *"New Interpretation of Variational Principles for Gauge Theories. II. General Auxiliary Coordinate in Conformogeometrodynamics."*, forthcoming.

[28] P.A.M. Dirac, *Lectures on Quantum Mechanics* (Yeshiva University, New York 1964).

[29] E. Anderson and J.B. Barbour, *Class. Quantum Grav.* 19 3249 (2002), gr-qc/0201092.

[30] T. Thiemann, *Living Rev. Rel.* (2001), gr-qc/0110034.

[31] E. Anderson, *Gen. Rel. Grav.* 36 255, gr-qc/0205118.

[32] E. Anderson, *Phys. Rev.* D68 104001 (2003), gr-qc/0302035.

[33] E. Anderson, *"Geometrodynamics: Spacetime or Space?"* (Ph.D. Thesis, University of London 2004), gr-qc/0409123.

[34] C.J. Isham, *Proc. R. Soc. Lond. A* 351 209 (1976);

M. Henneaux, *Bull. Soc. Math. Belg.* 31 47 (1979);

C. Teitelboim, *Phys. Rev.* D25 3159 (1982);

M. Pilati, in *Quantum Structure of Space and Time* ed. M. Duff and C.J. Isham (Cambridge University Press, Cambridge 1982);

M. Pilati, *Phys. Rev.* D26 2645 (1982);

M. Pilati, *Phys. Rev.* D28 729 (1983);

G. Francisco and M. Pilati, *Phys. Rev.* D31 241 (1985).

[35] J.W. York, *Phys. Rev.* Lett. 28 1082 (1972);

J.W. York, J. *Math. Phys.* 14 456 (1973).

[36] K.V. Kuchař, J. *Math. Phys.* 17 777 (1976);

K.V. Kuchař, J. *Math. Phys.* 17 792 (1976);

K.V. Kuchař, J. *Math. Phys.* 17 801 (1976);

K.V. Kuchař, J. *Math. Phys.* 18 1589 (1977).

[37] See e.g. S. Weinberg, *the Quantum Theory of Fields, Vol II* (Cambridge University Press, Cambridge 1995).

[38] V. Moncrief and C. Teitelboim, *Phys. Rev.* D6 966 (1972);
N. Ó Murchadha, *Int. J. Mod. Phys.* A 20 2717 (2002).

[39] C. Brans and R. Dicke, *Phys. Rev.* 124 925 (1961).

[40] C. Kiefer and E.A. Martinez, *Class. Quantum Grav.* 10 2511 (1993), gr-qc/9306029.

[41] This follows from the data in B. Bertotti, L. Iess and P. Tortora, *Nature* 425 374 (2003) applied to the theoretical calculations of e.g. [42].

[42] C.M. Will, *Living Rev. Rel.* 4 4 (2001), gr-qc/0103036.

[43] T. Damour and K. Nordvedt *Phys. Rev.* Lett. 70 2217 (1993); *Phys. Rev.* D48 3437 (1993).

[44] J. Isenberg and J. Nester, *Ann. Phys.* (New York) 108 368 (1977).

[45] M.A. Clayton, gr-qc/0104103.

[46] T. Jacobson, D. Mattingly, *Phys. Rev.* D70 024003 (2004), gr-qc/0402005.

[47] C. Eling, T. Jacobson, D. Mattingly, to appear in the Deserfest proceedings, gr-qc/0410001.

[48] C. Eling, T. Jacobson and M.C. Miller, arXiv:0705.1565;
D. Garfinkle, C. Eling and T. Jacobson, gr-qc/0703093;
C. Eling and T. Jacobson, *Phys. Rev.* D74 084027 (2006), gr-qc/0608052;
C. Eling and T. Jacobson, *Class. Quant. Grav.* 23 5643 (2006), gr-qc/0604088;
C. Eling and T. Jacobson, *Class. Quant. Grav.* 23 5625 (2006), gr-qc/0603058;
C. Eling, R. Guedens and T. Jacobson, *Phys. Rev.* Lett. 96 121301 (2006), gr-qc/0602001;
B.Z. Foster and T. Jacobson, *Phys. Rev.* D73 (2006) 064015, gr-qc/0509083.

[49] H. Brown, private communication to E.A. in 2004.

INDEX

#

A

B

C

D

E

F

G

H

I

J

K

L

Q

R

S

T

U

V

W

X

Y